W9-CBV-102

MIND AS MACHINE

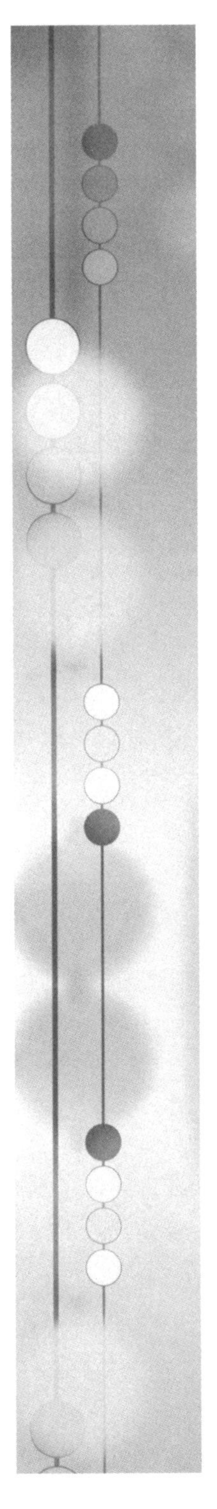

MIND AS MACHINE

A History of Cognitive Science

MARGARET A. BODEN

VOLUME 1

CLARENDON PRESS · OXFORD

OXFORD
UNIVERSITY PRESS

Great Clarendon Street, Oxford OX2 6DP

Oxford University Press is a department of the University of Oxford.
It furthers the University's objective of excellence in research, scholarship,
and education by publishing worldwide in

Oxford New York

Auckland Cape Town Dar es Salaam Hong Kong Karachi
Kuala Lumpur Madrid Melbourne Mexico City Nairobi
New Delhi Shanghai Taipei Toronto

With offices in

Argentina Austria Brazil Chile Czech Republic France Greece
Guatemala Hungary Italy Japan Poland Portugal Singapore
South Korea Switzerland Thailand Turkey Ukraine Vietnam

Oxford is a registered trade mark of Oxford University Press
in the UK and in certain other countries

Published in the United States
by Oxford University Press Inc., New York

First published 2006

British Library Cataloguing in Publication Data
Data available

Library of Congress Cataloging in Publication Data
Boden, Margaret A.
Mind as machine : a history of cognitive science / Margaret A. Boden.
p. cm.
Includes bibliographical references and indexes.
ISBN-13: 978–0–19–924144–6 (alk. paper)
ISBN-10: 0–19–924144–9 (alk. paper)
1. Cognitive science—History. I. Title.
BF311.B576 2006
153.09—dc22

2006011795

Typeset by Laserwords Private Limited, Chennai, India
Printed in Great Britain
on acid-free paper by
Biddles Ltd., King's Lynn, Norfolk

ISBN 978–0–19–929237–0 (Volume 1)
ISBN 978–0–19–924144–6 (Set)

10 9 8 7 6 5 4 3

For Ruskin and Claire,
Jehane and Alex,
and Byron, Oscar, and Lukas . . .

—and in memory of Drew Gartland-Jones (1964–2004)

Tell me where is fancy bred?
Or in the heart or in the head?
How begot, how nourished?

(William Shakespeare,
The Merchant of Venice, III. ii)

What a piece of work is a man! how noble in reason! how infinite in faculty! in form and moving, how express and admirable! in action how like an angel! in apprehension how like a god! the beauty of the world! the paragon of animals!

(William Shakespeare, *Hamlet*, II. ii)

Even Clerk Maxwell, who wanted nothing more than to know the relation between thoughts and the molecular motions of the brain, cut short his query with the memorable phrase, "but does not the way to it lie through the very den of the metaphysician, strewn with the bones of former explorers and abhorred by every man of science?" Let us peacefully answer the first half of this question "Yes," the second half "No," and then proceed serenely.

Our adventure is actually a great heresy. We are about to conceive of the knower as a computing machine.

(Warren McCulloch 1948: 143)

ACKNOWLEDGEMENTS

A number of friends and colleagues have made constructive comments on various parts of the manuscript, and I'd like to thank them here. I'm grateful to them all—none of whom, of course, is responsible for any errors that remain.

Some have read one or more entire chapters, and I'm especially obliged to them: Jane Addams Allen, Jack Copeland, Paul Harris, Inman Harvey, Gerry Martin, Pasha Parpia, Geoff Sampson, Elke Schreckenberg, Aaron Sloman, and Chris Thornton.

Many others have checked individual sections or part-sections, or sent me their published and/or unpublished papers, or provided other helpful information. The book has been hugely improved by their input.

Accordingly, I thank Igor Alexander, John Andreae, Michael Arbib, Alan Baddeley, Simon Baker, Paul Ballonoff, Nicole Barenbaum, Eileen Barker, Jerome Barkow, Horace Barlow, Daniel Bobrow, Robert Boyd, Pascal Boyer, Jerome Bruner, Alan Bundy, David Burraston, Ronald Chrisley, Patricia Churchland, Dave Cliff, Rob Clowes, Harold Cohen, Kate Cornwall-Jones, Kerstin Dautenhahn, Paul Davies, Dan Dennett, Zoltan Dienes, Ezequiel di Paolo, Jim Doran, Stuart Dreyfus, Shimon Edelman, Ernest Edmonds, Roy Ellen, Jeffrey Elman, Edward Feigenbaum, Uta Frith, Drew Gartland-Jones, Gerald Gazdar, Zoubin Ghahramani, Gerd Gigerenzer, Peter Gray, Richard Gregory, Richard Grimsdale, Stephen Grossberg, Derek Guthrie, Timothy Hall, Patrick Hayes, Douglas Hofstadter, Owen Holland, Keith Holyoak, Eric Horvitz, Buz Hunt, Phil Husbands, Edwin Hutchins, Steve Isard, Lewis Johnson, Phil Johnson-Laird, Bill Keller, David Kirsh, Janet Kolodner, Bob Kowalski, David Kronenfeld, Ben Kuipers, Mike Land, Pat Langley, Ronald Lemmen, Nigel Llewellyn, Aaron Lynch, John McCarthy, Thorne McCarty, Drew McDermott, Alan Mackworth, Matt Mason, John Maynard Smith, Wolfe Mays, Stephen Medcalf, Ryszard Michalski, Marvin Minsky, Steve Mithen, Gunalan Nadarajan, Mike O'Shea, Andras Pellionisz, David Perkins, Tony Prescott, Victor Raskin, Jasia Reichardt, Graham Richards, Peter Richerson, Edwina Rissland, Edmund Robinson, Yvonne Rogers, Oliver Selfridge, Marek Sergot, Anil Seth, Mike Sharples, Bradd Shore, Aaron Sloman, Karen Sparck Jones, Bob Stone, Doron Swade, Henry Thompson, Larry Trask, Brian Vickers, Des Watson, Barbara Webb, Michael Wheeler, Blay Whitby, Yorick Wilks, Peter Williams, Terry Winograd, David Young, Lofti Zadeh, and John Ziman.

In addition, I have profited immeasurably from the stimulating intellectual environment of the interdisciplinary Centre for Research in Cognitive Science (formerly the School of Cognitive and Computing Sciences) at the University of Sussex. Indeed, many of the names just listed are/were among my colleagues there.

Celia McInnes has patiently put my bibliography items into the required format, printed out countless draft chapters, and for the ten years that I have been working on this project has helped to keep me as near to sanity as I ever manage. Peter Momtchiloff and Laurien Berkeley of OUP, and Jacqueline Korn of David Higham Associates, have been very helpful too. Peter, in particular, has offered a rare degree of intellectual understanding, common sense, and (not least) patience.

I'm grateful to the various publishers who've given permission to reproduce diagrams originally published by them. And I've occasionally cannibalized some of my own writings, paraphrasing or reusing paragraphs here and there. Most of these brief passages are taken from my books *Artificial Intelligence and Natural Man* (1977/1987) and *Computer Models of Mind* (1988), or from my papers 'Horses of a Different Color?' (1991), 'Is Metabolism Necessary?' (1999), and 'Autopoiesis and Life' (2000).

Finally, I thank my children and close friends for their love and support, and for not being bored stiff when I talked about how the writing was going.

M.A.B.

CONTENTS

ANALYTICAL TABLE OF CONTENTS

VOLUME I

VOLUME II

PREFACE

Cardinal Mazarin's librarian had a low opinion of history books. In what time he could spare from his master's collection of 40,000 volumes—opened to the public in 1644, Thursdays only! (J. A. Clarke 1970)—Gabriel Naudé wrote some brief tracts himself. One was the first-ever book on library science (Naudé 1627/1644). Another, less well-known today, was a cry of outrage against historians (Naudé 1625/1657).

His specific accusation was that they'd maligned the growing band of automata makers as dangerous dabblers in illicit magic, instead of recognizing them as brave pioneers of the mathematical arts. His general complaint was that the authors of history books are "a sort of people seldome or never representing things truly and naturally, but shadowing them and making them according as they would have them appear" (1625/1657: 9).

That was a spot too fierce. ("Seldome"? "Never"??) After all, Naudé believed that his own history of automata making was close to the truth. But his basic point was correct. Every history is a narrative told for particular purposes, from a particular background, and with a particular point of view.

Someone who knows what those are is in a better position to understand the story being told. This preface, then, says what this history aims to do, and outlines the background and viewpoint from which it was written.

i. The Book

This is a historical essay, not an encyclopedia: it expresses one person's view of cognitive science as a whole. It's driven by my conviction that cognitive science today—and, for that matter, tomorrow—can't be properly understood without a historical perspective. In that sense, then, my account describes the field as it is *now*. It does this in a second sense too, for it features various examples of state-of-the-art research, all placed in their historical context.

Another way of describing it is to say that it shows how cognitive scientists have tried to answer myriad puzzling questions about minds and mental capacities. These questions are very familiar, for one doesn't need a professional licence to raise them. One just has to be intellectually alive. So although this story will be most easily read by cognitive scientists, I hope it will also interest others.

These puzzles are listed at the opening of Chapter 1. They aren't all about 'cognition', or knowledge. Some concern free will, for instance. What is it? Do we have it, or do we merely appear to have it? Under hypnosis, do we lose it? Does any other species have it? If not, why not? What is it about dogs' or crickets' minds, or brains, which denies them freedom? Above all, how is human free choice possible? *What type of system*, whether on Earth or Mars, is capable of freewill?

My account is focused on ideas, not anecdotes: it's not about who said what to whom over the coffee cups. Nevertheless, the occasional coffee cup does feature. Sometimes, a pithy personal reminiscence can speak volumes about what was going on at a certain time, and how different groups were reacting to it.

Nor does it explore sociopolitical influences at any length, although some are briefly mentioned—for instance, the seventeenth-century respect due to the word of a 'gentleman', the twentieth-century role of military funding, and the post-1960 counter-culture. In addition, I've said a little about how various aspects of cognitive science reached—or didn't reach—the general public, and how it was received by them. What's printed in the newspapers, accurate or (more usually) not, has influenced the field indirectly in a number of ways—and it has influenced our culture, too.

Mainly, however, I've tried to show how the central ideas arose—and how they came together. To grasp what cognitive science is trying to do, one needs to understand how the multidisciplinary warp and weft were interwoven in the one interdisciplinary field.

My text, too, holds together much as a woven fabric does. It's best read entire, as an integrated whole—not dipped into, as though it were a work of reference. Indeed, I can't resist quoting the King of Hearts' advice to the White Rabbit: "Begin at the beginning, and go on till you come to the end: then stop."

I realize, however, that many readers won't want to do that—though I hope they'll read the whole of Chapter 1 before starting on any of the others. Moreover, even reading a single chapter from beginning to end will typically leave lots of loose ends still hanging. Most of the important topics can't be properly understood without consulting *several* chapters. Freewill, for example, is addressed in more than one place (7.i.g–h, 14.x.b, and 15.vii). Similarly, nativism—alias the nature/nurture debate—is discussed in the context of:

* psychology: Chapters 5.ii.c and 7.vi;
* anthropology: 8.ii.c–d and iv–vi;
* linguistics: 9.ii–iv and vii.c–d;
* connectionism: 12.viii.c–e and x.d–e;
* neuroscience: 14.ix.c–d;
* and philosophy: 2.vi.a and 16.iv.c.

So besides the Subject Index, I've provided many explicit cross-references, to encourage readers to follow a single topic from one disciplinary chapter to another. Peppering the text with pointers saying "see Chapter *x*" isn't elegant, I'm afraid. But I hope it's useful, as the best I could do to emulate links in hypertext. (The King of Hearts, of course, hadn't heard of that.)

These pointers are intended as advice about what to look at next. They're helpful not least because I may have chosen to discuss a certain topic in a chapter *other than* the one in which you might expect to find it. (The theory of concepts as 'prototypes', for example, is discussed in the anthropology chapter, not the psychology one.) My placements have been decided partly in order to emphasize the myriad interdisciplinary links. So no chapter that's dedicated to one discipline avoids mention of several others.

History, it has been said, is "just one damn thing after another". Were that true, this account would be hardly worth the writing. In fact, any history is a constructed narrative, with a plot—or, at least, a reasonably coherent theme.

The plot can always be disputed (hence some of Naudé's scorn), and in any case usually wasn't obvious to the dramatis personae concerned. Several examples of work experienced at the time as thrilling new beginnings are described here, and with hindsight it's clear that some of them actually *were*. But I'll also describe examples where it looked as though the end had already come—or anyway, where it wasn't known whether/when there'd be a revival. As for future episodes of the story, no one can know now just what they'll be. I'll indicate some hunches (17.ii–iii), but with fingers firmly crossed.

In the case of cognitive science, *theme* is as problematic as *plot*. The field covers so many different topics that a single theme may not be immediately obvious. At a cursory glance, it can seem to be a hotch-potch of disparate items, more properly ascribed to quite distinct disciplines. Indeed, some people prefer to speak of "the cognitive *sciences*", accordingly (see 1.ii.a).

The key approaches are psychology, neuroscience, linguistics, philosophy, anthropology, AI (artificial intelligence), and A-Life (artificial life)—to each of which I've devoted at least one chapter. Control engineering is relevant too, for it provides one of the two theoretical 'footpaths' across the many disciplinary meadows of cognitive science (see Chapters 1.ii.a, 4.v–ix, 10.i.g, 12.vii, 14.viii–ix, and 15.viii.c.).

Ignorance of the field's history reinforces this ragbag impression. So does a specialist fascination with particular details. But my aim, here, is to see the wood as well as the trees. I want to help readers understand what cognitive science as a whole is trying to do, and what hope there is of its actually doing it.

Each discipline, in its own way, discusses the mind—asking what it is, what it does, how it works, how it evolved, and how it's even possible. Or, if you prefer to put it this way, each discipline asks about *mental processes* and/or about how the *mind/brain* works. (That doesn't prevent them asking also whether the emphasis on the mind/brain is too great: some say we should consider the mind—or rather, the person—as *embodied*, too. And some add that we should focus on *minds*, not on *mind*: that is, we should remember the essentially social dimension of humanity.)

Moreover, each discipline, in so far as it's relevant to cognitive science, focuses on computational and/or informational answers—whether to recommend them or to criticize them (see Chapter 1.ii).

These questions, and these answers, unify the field. In my view, the best way to think about it is as *the study of mind as machine*. As explained in Chapter 1.ii.a, however, more than one type of machine is relevant here. In a nutshell: some for digital computing, some for cybernetic self-organization or dynamical control. Much of the theoretical—and historical—interest in the field lies in the tension that follows from that fact.

In short, I've tried to give a coherent overview, showing how the several disciplines together address questions that most thinking people ask themselves, at some time in their lives.

Many trees would need to be felled for a fully detailed history of cognitive science, for every discipline would require at least one large volume. The prospect is daunting, the forests are already too empty, and life is too short. This account has a more modest aim: despite its length, it's a thumbnail sketch rather than a comprehensive record.

That means that decisions have to be made about what to mention and what to omit. So my story is unavoidably selective, not only in deciding what research to include but also in deciding which particular aspects of it to highlight.

Some of my selections may surprise you. On the one hand, you may find topics that you hadn't expected. For instance, the psychological themes include emotion, personality, social communication, and the brain's control of movement. (In other words, cognitive science *isn't* just the science of cognition: see Chapter 1.ii.) Other perhaps unexpected themes include evolutionary robotics, the mating calls of crickets, and the development of shape in embryos. However, all those topics are relevant if one wants to understand the nature of mind.

Moreover, quite a few of the people I discuss aren't in the mainstream. Some have been unjustly forgotten, while others hold views that are (currently) distinctly off-message.

Indeed, some aren't even in a sidestream, since they deny the possibility of *any* scientific explanation of mind. And some, such as Johann von Goethe, are highly unfashionable to boot. Other authors recounting the history of the field might not mention any of them. Nevertheless, I try to show that they're all relevant, in one way or another. Sometimes, admittedly, it's largely a question of *Know your enemy*! (see the 'aperitif' to Chapter 16). But even one's intellectual enemies usually have things of value to say.

On the other hand, I deliberately ignore some themes and names which you might have expected to encounter. In discussing linguistics, for example, I say almost nothing about phonetics, or about automatic speech processing. These aren't irrelevant, and they figure prominently in more specialist volumes. But the general points I want to stress can be better made by addressing other aspects of language.

Similarly, in my account of cybernetics only a few people feature strongly: Norbert Wiener, John von Neumann, Warren McCulloch, Gordon Pask, W. Grey Walter, W. Ross Ashby, and Kenneth Craik. Others (such as Gregory Bateson and Stafford Beer) are only briefly visible, but might have been featured at greater length. And some bit players don't appear in my pages at all. In a comprehensive volume devoted solely to cybernetics, one could try to mention all of them (see Heims 1991). In a history spanning cognitive science as a whole, one can't.

That space constraint applies in all areas, of course—so please forgive me if I haven't mentioned Squoggins! Indeed, please forgive me if I haven't mentioned someone *much* more famous than Squoggins: the characters in my narrative are numerous enough as it is.

Even those who do appear could have been discussed more fully, so as to do justice to the rich network of formative influences behind any individual's ideas. With respect to the origins of A-Life, for example, I mention the coffee-house conversations of von Neumann and Stanislaw Ulam (Chapter 15.v.a). But just how much credit should be given to Ulam? To answer that question—which I don't try to do—would require many more pages, including a discussion about how sowing an intellectual seed should be weighed against nurturing the developing plant. In short, to detail *every* researcher of any historical importance would be impossible.

Still less could one specify all *current* work. For such details, there are the numerous specialist textbooks—and, better still, the professional journals and conference proceedings. However, I've mentioned a range of up-to-date examples, in order to indicate how much—or, in some cases, how little—has changed since the early days.

Sir Herbert Read once said that whereas the art historian deals with the dead, the art critic deals with the living, an even more risky thing to do. Although I've written this book primarily as a historian (which of course involves a critical dimension), I've dipped my toes into the riskier waters of contemporary criticism too. That's implicit in my choices of what recent work to mention, and what to ignore. And in the final chapter, I've said which instances of current work I regard as especially promising. However, those choices are made from what's already a highly selective sample: contemporary cognitive science contains many more strands than I've had space to indicate.

So the bad news is that some things which merit discussion don't get discussed. The good news is that if you find the recent examples I've selected intriguing, you can be sure that there are more. Tomorrow, of course, there will be more still.

ii. The Background

One of the founders of cognitive science expressed Naudé's insight in less disgusted terms. As Jerome Bruner put it:

> The Past (with a capital letter) is a construction: how one constructs it depends on your perspective toward the past, and equally on the kind of future you are trying to legitimize. (Bruner 1997: 279).

The future I'm trying to legitimize here is one in which interdisciplinarity is valued and alternative theoretical approaches respected—and, so far as possible, integrated.

As for my perspective on the past, this springs from my own experience of the field over the past fifty years. Indeed, it's even longer than that if one includes my reasons for being drawn to it in the first place. For I was already puzzling over some of its central questions in my early-teenage years.

(I was born in 1936. I mention that, and give other researchers' years of birth whenever I could discover them, less to record the appearance of particular individuals on this planet than to indicate the passage of intellectual *generations*.)

I first encountered cognitive science in 1957, at the University of Cambridge. I'd just completed the degree in medical sciences there, during which time I'd been especially interested in neurophysiology and embryology.

The medical course was almost uniformly fascinating (although the biochemistry was fairly low on my list of priorities). I remember being intrigued by Lord Adrian's work on spinal reflexes and action potentials, and spellbound by Andrew Huxley's hot-off-the-press lecture on muscle contraction—which had earned him a standing ovation from the usually blasé medical students (2.viii.e). Likewise, I'd been amazed by Alan Turing's paper on morphogenesis, and entranced by D'Arcy Thompson's writings on mathematical biology (15.iii–iv).

I now had one year to spare before going—or so I thought—to St Thomas's Hospital in London. There, I would do my clinical training, as a prelude to a career in psychiatry.

My College expected me to spend the year specializing in neurophysiology, which indeed I found absorbing. And Cambridge was a superb place to do it. Besides the awe-inspiring Adrian–Huxley tradition, exciting new work was being done by Horace

Barlow: 14.iii.b. (He was one of my physiology demonstrators: many's the time he helped me to coax a frog's leg to move in a physiology practical.)

But that would have meant doing lengthy experiments on cats, and the comatose rabbits pinned out in my pharmacology practicals had been troubling enough. The neurophysiological experiments that could then be done were fairly broad-brush, since single-cell research had only just begun (2.viii.f). But I don't know that a unit-recording approach would have made much difference to the way I felt. (For a description of what this involves today, see J. A. Anderson *et al.* 1990*a*: 215.) My qualms were largely irrational, of course: not only would I not have felt quite the same about rats, but the cats would be anaesthetized, or decorticate, or even decerebrate. Nevertheless, I hesitated.

As for psychology as an alternative, I'd originally planned to do this in my third year—but the course at Cambridge had turned out to be too rat-oriented, and too optics-based, for my taste. I'd already gate-crashed all the psychopathology lectures, and for six weeks worked as a resident nursing assistant at Fulbourn mental hospital nearby. But mental illness, the psychological topic which interested me most, figured hardly at all in the curriculum. Perhaps that was because precious little could be done to help. (Psychotropic drugs were still a rarity: Largactil, aka chlorpromazine, was being given to schizophrenics on the ward I nursed on at Fulbourn, but that was because the hospital's director was exceptionally forward-looking.)

Moreover, I now knew something I hadn't realized until after my arrival at Cambridge, namely that universities offered degrees in philosophy. This was a revelation. (Without it, I'd probably have ignored my qualms and turned to the cats.)

I'd discovered philosophy while I was still at high school, and found it deeply engaging. I remember reading Bertrand Russell with excitement, cross-legged on the floor in the second-hand bookshops on London's Charing Cross Road. I also remember plaguing several of my schoolteachers with questions that were philosophical in intent. But I had no idea that one could study philosophy at university.

Now, some five years later, I'd discovered that it was an option available in the final year at Cambridge, after completing the exams in medical sciences. I hadn't lost my love for philosophy, and this seemed to be my one and only chance to do something about it.

So I decided, against all (and I do mean *all*) advice, to spend my interim year studying what was then called Moral Sciences—a label that elicited relentless teasing from my fellow medics. I planned to concentrate as far as possible on the philosophy of mind and of science. And despite opposition from an unimaginative Director of Studies, I insisted on being taught by Margaret Masterman—who was neither a Fellow of Newnham nor a University faculty member, and who was far too original and eccentric to be popular with the College authorities.

I found my philosophical studies so exciting that the 'one' year turned into two. Meanwhile, my medical contemporaries and I received our degrees in 1958 from Lord Adrian himself, who was Vice-Chancellor at the time. (We each knelt down with our two hands between his, transfixed—in my case, anyway—by the University's huge golden seal-ring on one of his long, slender fingers.)

During those two years, and alongside some (very different!) supervisions with the logician Casimir Lewy, Masterman taught me weekly at the Cambridge Language

Research Unit, or CLRU. This had been founded in 1954—one year before I arrived in Cambridge (and two years before artificial intelligence was named).

The Unit wasn't an official part of the University but an independent, and distinctly maverick, research group directed by Masterman. Most of its funding came from military agencies in the USA (11.i.a). Its home was a small brick building tucked away on 'the other side' of the river. There were apple trees in the garden, and Buddhist gods carved on the big wooden doors. ("The place is full of gods," Masterman had said to me when I first phoned her to ask for directions. I couldn't imagine what she might mean.)

It was an exciting place, and not just because of the gods. Nor even because several members, seeking to combine science and religion, had founded the Epiphany Philosophers. This was a small community for worship and discussion, who met sometimes in a chapel hidden behind a wall upstairs and sometimes in a fenland mill. It was later widely taken as the inspiration for Iris Murdoch's novel *The Bell.* (Murdoch had studied philosophy in Cambridge in 1947–8.) The Epiphany Philosophers notwithstanding, what was most exciting about CLRU was its intellectual diversity and originality.

Masterman's research group in 1957 included a number of people specializing in the study of language:

* Karen Sparck Jones, now a distinguished researcher in information and language processing (Sparck Jones 1988);
* Richard Richens, a pioneer of machine translation who was by then a senior figure in the Commonwealth Abstracts Bureau (Richens 1958);
* Robin Mackinnon Wood;
* and Frederick Parker-Rhodes, who could read proficiently in twenty-three languages and who (like Masterman) saw metaphorical, not literal, language as primary (Parker-Rhodes 1978).
* Several members of CLRU were then working on automatic Chinese–English translation (Parker-Rhodes 1956; Masterman 1953), helped by Michael Halliday, who became involved with CLRU while Lecturer in Chinese at Cambridge.

Yorick Wilks and Martin Kay, now professors of artificial intelligence and computational linguistics at Sheffield and Stanford universities, joined them very soon after I left.

Another member of the language group at that time was Roger Needham. He was working at the still-new Computer Laboratory at Cambridge, where Maurice Wilkes had built the first relatively easy-to-use computer only a few years before, in 1948–9. Much later, he succeeded Wilkes as its Head, and recently directed Microsoft's UK research laboratory (sadly, he died in 2003). He and his wife, Sparck Jones, immediately aroused my admiration, for building their house with their own four hands. They were living on-site in a caravan surrounded by mud—hence their well-worn wellington boots—while also doing high-level intellectual work.

Among the others I encountered in the Unit was physicist Ted Bastin. He and Parker-Rhodes were developing a highly maverick account of quantum theory, with quanta as self-organizing entities. This is now (so I'm told: I can't make head or tail of quantum physics) a standard alternative view, with several web sites devoted to Parker-Rhodes.

In addition, the exceptionally original cybernetician Pask—today, the object of even more numerous web sites—was literally a back-room boy. He was usually hunched over his DIY computer, which he'd cobbled together out of biscuit tins and string.

Last but not least, philosophers Richard Braithwaite—Masterman's husband—and Dorothy Emmet were frequently around. I shared their interest in the philosophy of religion (a subject I later taught for many years) and, above all, in a scientifically grounded philosophy of mind.

Braithwaite—whom I saw more often—was a leading philosopher of science, and also held the Knightsbridge Chair of Moral Philosophy at Cambridge. Much concerned to integrate science with other areas of life, he'd recently recommended the theory of games as a tool for the moral philosopher (Braithwaite 1955). And he combined a broadly positivistic philosophy of science with Christian beliefs (Braithwaite 1971)—or rather, with a practical commitment to the moral principles illustrated by Christian stories. (Rumour has it that when called upon to recite the Creed at his public baptism service, his full-voiced "I believe ..." was preceded *sotto voce* by "I will behave in all ways as if . . .".)

Emmet, who'd very recently (1950–3) given the Stanton Lectures in the Philosophy of Religion at Cambridge, held the Chair of Philosophy at the University of Manchester. She knew of the growing excitement about the potential of computing, for the prototype of the world's first stored-program electronic computer had been operational in Manchester since 1948. Indeed, Turing—who wrote some of the first programs for the full version of the Manchester machine—had worked there also. As early as 1949, Emmet's philosophy seminar had discussed 'The Mind and the Computing Machine', with Turing present as one of the discussants (16.ii.a).

That's not to say that she was a devoted Turing fan. True, her department had developed an electrical machine for teaching symbolic logic, already in use for some years. Designed by Wolfe Mays and Dietrich Prinz (who was closely involved in the design of the Manchester computer), it had been exhibited at the annual British philosophy conference (Mays and Prinz 1950; Mays *et al.* 1951). But Mays' device wasn't based on Turing's ideas. Rather, it was inspired by the keyboard-and-rods Logical Piano, originated in 1869 by Stanley Jevons to illustrate the formal principles of validity—see 2.ix.a. (Jevons had been Professor of Logic at Owens College, the forerunner of the University of Manchester.)

Nor did Emmet and her Manchester colleagues agree with Turing that there was no good reason to deny that some conceivable digital computer could *think*. In the departmental seminar he attended, she'd objected that a machine could not be conscious. Michael Polanyi had added that whereas a machine is fully specifiable, a mind is not. And Mays had argued trenchantly that computers are, as John Searle (1980) would later put it, all syntax and no semantics (Manchester Philosophy Seminar 1949).

In the Cambridge apple orchard, however, Turing's influence was strong (see Chapters 4.i and 16.ii). He'd died in 1954, only one year before my arrival in Cambridge. And he'd been close to Braithwaite. Soon after the publication of Turing's seminal paper in *Mind* (A. M. Turing 1950), Braithwaite had chaired a BBC radio debate, in which Turing participated, on the possibility of machine intelligence. (The transcript is in Copeland 1999: 445–76.) Some years before that, they'd been fellow Fellows at

King's College. Indeed, Braithwaite was one of the only two people to have requested an offprint of Turing's 'Computable Numbers' paper written in 1936 (Hodges 1983: 123–4). And, so he told me later, it was he who'd pointed out to Turing its relationship to Gödel's work (letter from R.B.B. to M.A.B., 21 Oct. 1982).

By the time of my becoming a once-to-thrice-weekly visitor to CLRU in 1957, Turing's vision was rarely discussed there in general terms. When it was, the emphasis was more on his technological predictions than on his philosophical views. Believing those predictions to be well grounded, the denizens of CLRU focused rather on the exciting challenges involved in bringing them to fruition.

In other words, the interdisciplinary community amidst the apple trees was making early attempts in the mechanization of thought. In particular, they were trying to identify, and formalize, some of the structural principles informing learning and language.

Pask, for example, was doing pioneering work on adaptive machines, using a wide variety of devices he'd built himself (Pask 1961). Some of his ideas may be viewed as early attempts in AI and A-Life, but he saw them as research in cybernetics (see Chapter 4.v.e). He was largely inspired by Ashby's self-equilibrating Homeostat of the 1940s (Ashby 1948). And he received strong encouragement from McCulloch, one of the founders of the cybernetics movement in the USA—whom I was to meet six years later (Pask 1961: 8).

Four years earlier, in 1953, Pask (with Mackinnon Wood) had constructed Musicolour, an array of lights that adapted to a musician's performance. It had toured various theatres, ending up in a Mecca dance hall. Being a devotee of Mecca dance halls at the time, I much regretted never having encountered it. (I didn't know that it had acquired a reputation for bursting into flames—Mallen 2005: 86.) And in 1958 he started building self-organizing chemical systems that "learned", "evolved", and grew their own "sensors" (sound detectors)—Pask (1961: 105–8; see 4.v.e and 15.vi.d).

But his main interest at that time was in adaptive teaching machines (Pask 1961, ch. 6). Rejecting the easy notion that one size fits all, he was trying to make his machines respond to individual differences between people's thought patterns, or cognitive styles (4.v.e). He'd been designing adaptive teaching machines since 1952, and his SAKI (Self-Adaptive Keyboard Instructor) of 1956, which taught people how to do key punching efficiently, was the first such system to go into commercial production (Pask 1958; 1961: 96 ff.).

Unfortunately, I saw only very little of Pask in his back room at CLRU. A few years after leaving Cambridge, however, I would visit his makeshift office–laboratory in Richmond, where he was exploring yet more ambitious automatic teaching aids (Pask 1975*a*).

Bastin, too, was interested in cybernetics. Much of his spare time went into building a self-equilibrating machine (Bastin 1960). This was inspired by Grey Walter's electromechanical "tortoises" (Grey Walter 1950*a*,*b*), which I'd seen exhibited at the Festival of Britain a few years earlier, in 1951 (4.viii.a). But it also involved ideas about hierarchy, which he was applying to quantum physics as well as to life (Bastin 1969).

The main efforts at CLRU, however, were in the study of language (9.x.a and d). Masterman's group was doing research on what's now called Natural Language

Processing, or NLP (Wilks, forthcoming). They ranged widely over topics later claimed for AI and cognitive science. These included machine translation, the representation of knowledge for information retrieval, and the nature and process of classification. Although their theory of classification was never described in print as computational "learning", it dealt with issues later so described by AI (10.iii.d and 13.iii.f).

Masterman was one of the first people in the world to attempt machine translation, and she made semantics, not syntax, the driving force. She was deeply influenced by certain aspects of Ludwig Wittgenstein's later philosophy of language. Despite her gender—Wittgenstein was notorious for his mysogyny—she'd been one of his favourite students, to whom he'd dictated the lectures later known as *The Blue Book* (Monk 1990: 336). Indeed, she described herself to me on our first meeting as "the only person in England who really understands Wittgenstein". (Modesty wasn't one of her virtues.)

Accordingly, she handled translation by way of a computational thesaurus (Masterman 1957, 1962). More subtle than word-for-word dictionary look-up, her approach enabled word ambiguities to be resolved by inspecting the penumbra of concepts associated with neighbouring words in the text. Or rather, it made this possible in principle. In practice, the method was far from infallible: she delighted in telling people that Virgil's sentence *agricola incurvo terram dimovit ararat* had come out as *ploughman crooked ground plough plough*. This couldn't have happened without the thesaurus, because only *ararat* has a root likely to be listed against *plough* in a dictionary.

The work was practical as well as theoretical, asking how concepts and their semantic interrelations could be implemented in computers—"could be", rather than "were": computing facilities in 1957 were primitive (see 3.v.b). Using CLRU's data, Needham did some classification experiments on the EDSAC-2 in the Computer Laboratory. But this machine (in use until 1964) was far too small to handle a comprehensive thesaurus like Roget's. Moreover, no machine-readable thesaurus existed.

Some genuinely computational, though very primitive, work was done at the CLRU in the 1950s, using a Hollerith punched-card sorter. It wasn't until 1964, five years after I left, that the Unit received its first electronic computer: an ICL 1202, with 200 registers on a drum (Sparck Jones, personal communication).

Because of these practical difficulties, the language team often had to do pseudo-computational tests. That is, they often worked 'mechanically' with paper lists, in the way required for the procedures using punched-card apparatus then being devised at the Unit (Masterman *et al.* 1957/1986: 2). (Perhaps the Buddhist gods were witnessing the first instantiation of Searle's Chinese Room?—16.v.c.)

Masterman was a stimulating, if often infuriating, presence. Her conversation and teaching were peppered with provocative, sometimes deeply insightful, remarks. She encouraged my interest in the philosophy of mind. At her urging, I sent an early essay on 'free will' (i.e. the nature of intentions) to Gilbert Ryle, who published it in *Mind* eighteen months later, in April 1959 (Boden 1959). And her computational thesaurus was highly intriguing: how *could* one get the farmer to plough his ground in English, as well as in Latin?

However, it seemed to me, as an occasional looker-on, to be a technological project, not a psychological one. It clearly rested on intuitions about how people understand language. But I never heard it described as an exercise in the psychology of language—still less, as part of a general project aiming to understand all mental processes in computational terms.

Nor did I have the wit to recognize that possibility for myself—although if I'd interacted more often with Pask, I probably would have done. Masterman's research emphasized (semantic) *structure* rather than *process*, and didn't immediately suggest a way of conceptualizing mental processes as such.

Although I felt that it must somehow be connected with the puzzle of how thought of any kind is possible in a basically material universe, I couldn't see how to generalize it to the mind as a whole. I found her work interesting. But—or so I thought at the time—it wasn't relevant to the issues that most concerned me, and which had fascinated me as a schoolgirl even before becoming a medical student.

These were the nature and evolution of mind, the mind–body problem in general, and free will and psychopathology in particular. I was intrigued by paranoia, multiple personality, automatisms, and hypnosis. And I was especially puzzled by psychosomatic phenomena, such as hysterical paralyses and anaesthesias.

In these cases, there's no bodily damage: under hypnosis, the 'paralysed' arm moves normally, and the 'anaesthetized' skin is sensitive. Still more puzzling, the bodily limits of the clinical syndrome are inconsistent with the gross neuromuscular anatomy, and seem to be determined instead by what the layman-patient *thinks of* as an 'arm' or a 'leg'. For example, the movements that the 'paralysed' patient is unable to make don't correspond to any specifiable set of spinal nerves. They can be described only by using the non-anatomist's concept of an arm, thought of as bounded by the line of a sleeveless shirt. In other words, the mind appears not only to be influencing the body, as in normal voluntary action, but even overcoming it. How could this be?

Machine translation didn't help me to answer such questions. The most promising avenues, I thought, lay elsewhere: in the philosophy of mind and psychology, and in psychiatric medicine.

My intention at that time was to become a psychiatrist. The foray into philosophy was merely temporary. But in May 1959, when I was revising for my Moral Sciences finals and looking forward to going on afterwards to St Thomas's Hospital, I was unexpectedly invited (at Braithwaite's suggestion) to apply for a philosophy lectureship at the University of Birmingham. This was "unexpected" in more senses than one. I'd never considered such a possibility for a moment. Nor was there much time to think about it, for the interviews were to take place only three days later.

Since I wasn't sure that I wanted the job, and didn't think I'd get it anyway, I was totally relaxed on the day. To my amazement, they offered it to me at interview. (It turned out that the little piece in *Mind* had helped.) But I asked for forty-eight hours to think it over: *medicine or philosophy?* was a difficult decision. Masterman's very strong support (she showed me her written reference, when she found that I was dithering) was one of many factors that influenced me to accept the offer.

(I also sought advice from my former pathology supervisor, today a distinguished Emeritus Professor of Pathology. He observed that having a wife with a medical degree, like his—whom I could see hanging out the washing in the garden with pegs between her teeth, while he smoked his pipe in his armchair—would always help a family to get a second mortgage if needed. This remark, in that all-too-familiar domestic context, was less persuasive than he'd intended.)

It was a strong department (Peter Geach was one of the luminaries), and I was very happy there. However, I soon got bored. For Birmingham's Chinese walls between disciplines impeded my interests in the philosophy of psychology and biology. I considered returning to medicine (St Thomas's said "Yes, come!"), but having forfeited my state studentship to earn my own living I could no longer afford to do so.

Instead, I followed the suggestion of my old Cambridge friend Charlie Gross (who a few years later would discover the 'monkey's hand' neurones in the monkey's brain: 14.iv.b). He said, "There's this man Bruner at Harvard, who's been doing some work I think you'd find interesting—and there are scholarships you can apply for to go to the States." So I applied for a Harkness Fellowship, which enabled me, in the autumn of 1962, to cross the pond to study cognitive and social psychology with Bruner. (When I first met him, he was chatting with George Miller. "Here's our double-first from Oxford," he said to him. "Cambridge!" I protested—and "Welcome to Yale!" came quick as a flash from Miller.)

By the time I left for the USA, I'd already decided to go to the just-initiated University of Sussex when I got back to England. This was because, most unusually, it was conceived from the start as an interdisciplinary institution. I was already committed to interdisciplinarity, of course. But, sailing happily through the storms on the magnificent *Queen Mary* (the roughest voyage for twelve years), I never imagined that my colleagues and I would eventually found Sussex's Cognitive Studies Programme (later the School and now the Centre for Cognitive Science), which in 1973 world-pioneered degrees integrating AI, philosophy, psychology, and linguistics. The idea couldn't even have occurred to me.

Barely a week after docking at Manhattan, however, it might have done. The conceptual leap from computation to psychology, and to the mind–body problem, happened (for me) a mere two days after arriving in the other Cambridge.

The occasion was my first sight of the remarkable book *Plans and the Structure of Behavior*, by Miller, Eugene Galanter, and Karl Pribram (1960). I picked it up while browsing in a second-hand bookshop on Massachusetts Avenue. Why I did so, I'll never know. It was a hideous object: bound in a roughly textured cloth, a dull rust in colour (my least-favourite hue), horribly coffee-stained, and defaced by heavy underlining on almost every page. But it changed my life.

Nor was I the only one, for it was highly influential (see Chapter 6.iv.c). I soon discovered that it was on Bruner's reading lists at Harvard's new Center for Cognitive Studies, founded only a few months before—and not just because Miller was the co-founder! It was recommended also for Phil Stone's seminar on 'Computer Simulation', for which I was to do my first programming. (We wrote our programs in Victor Yngve's early list processing language COMIT, using punched cards for MIT's pre-release prototype of the IBM-360—not officially announced until 1964.)

But all that was still to come. Already primed by Masterman and Pask, my thinking was instantly triggered by this coffee-stained volume. Leafing through it in the bookshop, it seemed to offer a way to tackle just those questions which had bothered me as a schoolgirl.

It was an intoxicating attempt to apply specific computational ideas—hierarchies of Test–Operate–Test–Exit procedures (TOTE-units)—to the whole of psychology. Unlike Masterman, it focused on process as well as structure. And it ranged from animal learning and instinct, through memory and language, to personality, psychopathology, and hypnosis. Self-confessedly vague and simplistic, and often careless to boot, it was nevertheless a work of vision.

Its computational ideas soon informed my own work. In 1963 I wrote a paper applying them to William McDougall's rich theory of the purposive structures underlying normal and abnormal personality (Boden 1965). And a few years later, I addressed one of my long-standing puzzles by outlining how a robot could have a paralysis conforming not to its actual wires-and-levers anatomy, but to its programmed 'concept' of what an arm is (Boden 1970). Its behaviour, I argued, would therefore be describable in intentional terms. That is, what it was 'doing', and how it might be 'cured', could be stated only by reference to the descriptions and instructions in its program.

In the interim, I'd returned to England (and moved to Sussex in 1965), and was writing my first book: *Purposive Explanation in Psychology* (1972). Begun as my Ph.D. thesis (the first purely theoretical thesis that the Harvard department had ever allowed), this took me eight years to finish. The delay was explained only partly by the amount of intellectual work involved: the publisher's airmailed advance copy reached me in hospital on the day after the birth of my second baby. (Both were deep purple on arrival.)

In that book, I developed a fundamentally physicalist but non-reductionist account of purpose, and other intentional concepts. That is, I offered an essentially functionalist philosophy of mind—though using my own terminology, not Hilary Putnam's (I came across his work later). I compared my account of mind, and of the mind–body relation, with a wide range of theories in psychology and philosophy. And I focused most closely on McDougall—not as an unquestioned guru, but as an intellectual sparring partner.

What had drawn me to McDougall was his deep insight into the complex structure of the human mind, and his many explicit arguments against psychological reductionism (2.x.b. and 5.ii.a). Most of these, but not all, I thought to be valid.

It turned out later that there were several personal links, as well. On researching McDougall's life history (1871–1938), I was intrigued to discover that, after completing his degree in medical sciences at Cambridge, he'd taken up the same clinical scholarship at St Thomas's Hospital which had been offered to me in 1959, and again in 1962. He, like me, had moved from medicine to psychology, with philosophy of mind constantly in the background. And he, too, had gone from one Cambridge to the other: he was a professor at Harvard for several years.

There was even a link that made me one of his intellectual grandchildren, by means of a sort of apostolic succession. For when I said to Bruner, in the spring of 1964, that instead of doing experiments on information density in disjunctive concepts (a mind-numbing topic which he, intending to be helpful, had suggested to me) I wanted to study McDougall's theory of purpose, he told me—after a gasp of amazement—that McDougall had been his teacher at Duke University.

Whereas I'd been recommended to go to Harvard Graduate School by Charlie Gross, Bruner—so he told me—had been specifically warned against it by McDougall. McDougall's broadly ranging psychology had been highly influential until it was suddenly eclipsed by behaviourism. He remained bitter about this for the rest of his life, and left the newly behaviourist Harvard in disgust to set up his own outfit at Duke.

When the young Bruner announced his intention of travelling north to Massachusetts, McDougall had gruffly warned him about the intellectual corruption (he never minced his words) that awaited him there.

By the time I arrived at Harvard, a quarter-century later, the behaviourist "corruption" was less strong. Or anyway, that was true in Bruner and Miller's Center for Cognitive Studies, if not in the Psychology Department as such—whose denizens included Burrhus Skinner and Richard Herrnstein. Even so, McDougall's name had vanished from the curriculum. (Hence Bruner's gasp of amazement.) It survived only as one of many items on a three-page mimeographed list of long-dead worthies, circulated as potential essay topics for Gordon Allport's seminar on the history of social psychology.

What alerted me to him initially was the title of his book *Body and Mind* (1911), which Allport had included on the list alongside the author's name. On consulting his work (long-unborrowed from the Harvard library), I found that McDougall's ideas had a welcome subtlety and depth, and a refreshing concern for real life, whether psychiatric syndromes or everyday pursuits. His psychology dealt not only with cognition, but with motivation and emotion too—and, significantly, with how these three types of mental function are closely integrated in individual personalities.

Besides those strong points, his writings abounded with philosophical as well as empirical questions. These attracted me, although they were anathema to most (Anglo-American) textbook writers of the 1950s and 1960s, who believed—wrongly—that psychology had finally 'escaped' from philosophy. In truth, they'd simply accepted the current philosophical fashion (operationalism, or logical positivism), without stopping to question it carefully.

That's not to say that I agreed with all his philosophical arguments, or that I accepted his robust defence of "animism". Far from it. McDougall had been combatively anti-mechanistic, even claiming that purposive behaviour requires a special form of energy (*horme*), intrinsically directed to instinctive goals. That, for me, was a step too far: purposive explanation is one thing, purposive energy quite another. But it was important to understand why, when considering intentional phenomena, he'd felt it necessary to say this.

As part of my critique (Boden 1972), I suggested—what probably made him turn in his grave—that his many insights about personality and psychopathology could be simulated in computational terms. If this was indeed possible, those theoretical insights could be saved, and even clarified, without positing any mysterious energy. And if it could be done for McDougall's avowedly anti-mechanistic psychology, it could in principle be done for any other. (Sigmund Freud's purposive theory would have been less suitable as an exemplar, for he believed that psychology does have a mechanistic base.)

In sketching specific ways of doing this, I had to extrapolate from the computer models that already existed. In 1963, when the book was begun, there were only a handful of candidates.

By the time it was finished, early in 1971, there were many more. These ranged from work in computer problem solving, through programs for vision and language, to models of analogy, learning, and various aspects of personality. For instance, preliminary reports on Terry Winograd's research, whose official publication in 1972 suddenly raised the visibility of computer modelling in the wider intellectual community (see 9.xi.b and 10.iv.a), had been made available informally in 1970.

These new examples, whether successful in their aims or not, were clearly relevant to my book's central claim: that purposive behaviour is intelligible in computational terms, and could in principle be simulated in computers. But although I was able to refer briefly to a few of them, they were in general too numerous—and many were too late—to be added to my already lengthy manuscript.

So I decided that, as soon as my first book was finished, I would write an extended footnote to it. This would detail what could—and, just as importantly, what could not—be done by computer modelling in the early 1970s, and what would be needed for the many remaining obstacles to be overcome.

The resulting footnote, *Artificial Intelligence and Natural Man* (Boden 1977), ran to 537 pages. (By this time, the term 'artificial intelligence' had largely replaced 'computer simulation'.) It devoted a chapter each to the philosophical, psychological, and social implications of AI. Indeed, in one sense they were what the book was really about. Most chapters, however, described the AI as such.

For the sake of readers knowing nothing about AI and highly suspicious of computers to boot, it contained not one line of code. It was also highly critical, in the sense that it identified countless mismatches between existing AI programs and real minds. Nevertheless, it was assigned—alongside Patrick Winston's very different *Artificial Intelligence* (1977)—as a compulsory text for AI courses at MIT and Yale. I was told it was the first time they'd assigned two books, rather than just one. It was also used as the basis of various psychology courses in the UK and USA, including the Open University's first Cognitive Psychology course. (Later, in 1993, I was delighted to be elected an early Fellow of the American Association for Artificial Intelligence, in part for having written it.)

That book *was* fully comprehensive. Had Squoggins been working at the time, he would very likely have been included. For it mentioned virtually every AI program of any interest, including many available only as privately circulated reports or working papers. And it gave closely detailed explanations and critiques of many of them. It ranged over diverse aspects of mind: from language and vision to neurosis and creativity—on which last I promised myself a whole volume, later (Boden 1990*a*, 2004). And it identified theoretical challenges many of which *still* remain to be met. (So the second edition, in 1987, was unchanged except for an added 'update' chapter.) In short, it provided a near-exhaustive description of the state of the art of AI at the time.

Such a project is no longer possible: even 1,080 A4-sized pages aren't enough for a fully comprehensive account (Russell and Norvig 2003). If it's not possible for AI, still less is it feasible for cognitive science as a whole. Too much water has flowed under the bridge. There are too many unsung Squogginses out there, and too many branching implications that could be explored. That's even more true if one takes a *historical* approach, for then the potential subjects multiply yet more relentlessly.

I hope these far-from-comprehensive pages will tell an illuminating story, nonetheless. As Naudé realized (fuelling his attack on historians in general), the facts one chooses to relate, and how one decides to link them, will depend on one's background and perspective. In this book, I've aimed to show how cognitive science has developed so as to help solve problems about mind, brain, and personality that have intrigued me ever since I was a girl.

M.A.B.

Brighton
January 2006

1

SETTING THE SCENE

Once upon a time there was a teddy bear called Twink—and with those few words, the scene is set. We know what we're talking about. Twink's story can begin . . .

This story can't begin so quickly, however. For we *don't* yet know what we're talking about. Some readers may know very little about cognitive science at this stage. Even more to the point, those who are already familiar with it think of it in varying ways. That was true right from the start, and it's even more true now. (So it's no accident that the summary chapter of a recent book is subtitled: 'It's Cognitive Science—But Not As We Know It'—M. W. Wheeler 2005: 283.)

One of the founders of the field, when asked to define it, confessed that "Trying to speak for cognitive science, as if cognitive scientists had but one mind and one voice, is a bum's game" (G. A. Miller 1978: 6). And twenty years afterwards, two long-time leaders edited a book called *What Is Cognitive Science?* (Lepore and Pylyshyn 1999). You'd think they'd know by now! But no: even in the textbooks, never mind coffee conversations and idle chat, definitions differ.

I shan't list them: the boredom barometer would shoot through the roof. However, the differences do make a difference. This will become clearer throughout the following pages, as we see how theory and practice have changed over the years (in some cases, coming full circle). Meanwhile, before starting the story, some scene setting may be helpful.

One way of saying what we're talking about is to give some examples of the wide-ranging questions studied by cognitive science. I'll do that in Section i. And I'll do it in everyday language: the technicalities can wait until later.

Another is to give a definition of the field, even if this can't be presented as *the* universally agreed definition. I'll do that in Section ii. This, I hope, will help to show why I've decided to tell the story in the way I do.

Finally, in Section iii, I'll identify a number of traps that lie in wait for anyone discussing the field's intellectual history.

1.i. Mind and its Place in Nature

A host of intriguing questions about mind and its place in nature occur to most thinking people. (The FAQs of the mind, Web-users might say.) As explained in the Preface, some have puzzled me for almost as long as I can remember—and I usually found that my friends were puzzled by them too. They centred on the nature of mind and the mind–body problem; the evolution of mind; freedom and purpose; and how various psychopathologies are possible.

Most of the topics studied in cognitive science fall under one of these broad categories. And those which don't, such as the nature of *computation*, are closely related to them.

a. Questions, questions . . .

We're intrigued by consciousness, for example. We know there are close correlations between brain events and conscious states—but why is that so? The answer seems to be that our brains generate our consciousness. But how do they do this, in practice? Even more puzzling, how *can* they do this, in principle?

Or maybe we only *think* we know this? Some people argue that it *doesn't even make sense* to suggest that there are correlations between conscious states and brain states. How could anyone with any common sense be led to make such a deeply counter-intuitive claim? Perhaps "common sense" itself is radically misguided here (and was radically different in other historical periods)?

What about dogs and horses: are they conscious? And snails, flies, newts . . . ? For that matter, what about newborn babies: are they conscious in *anything like* the sense in which adult humans are? What of machines? Could a machine be conscious—and if not, why not?

People often wonder whether a creature has to have a brain, or something very like one, to be intelligent. If so, why? Is a brain (as well as eyes) needed to *see*, for example? What do the visual brain cells do that the retinal cells don't? What about intelligent *action*? How, for instance, does the brain convert an Olympic diver's intention to dive into the finely modulated bodily movements that ensue? If we knew this, could we drop talk of intentions and refer only to brains instead?

Consider chimps, or cats: what can *their* brains do, and what can't they do? And what can they do without the mammalian (and avian) glory, the cerebral hemispheres? Given that *Homo sapiens* evolved from lower animals, what does this tell us about our mental powers? Can anything interesting be learnt about the human mind by studying distantly related species such as frogs, or insects?

As for machines, just how—if at all—must an artifice resemble a real brain if it's even to *seem* to support a mind? And *even if* studying insects can teach us something about ourselves, what about studying inanimate tin cans—like a Mars robot, or an automatic controller in a chemical factory? How could these things (*sic*) possibly be relevant?

What mental powers does a human brain provide, and how does it manage to do so? How is free will possible? And creativity? Are creative ideas unpredictable, and if so why? What are emotions—and do they conflict with rationality, or support it?

Are our abilities inborn, or determined by experience? And how does the brain get its detailed anatomical structure: from genetics or from the environment—or perhaps even from spontaneous self-organization? (Is that last suggestion mere hand-waving, more magic than science?)

Do we all share psychological properties that mould every human culture? Perhaps the same underlying sense of beauty: maybe in symmetry, or expanses of water? Or the same tendency towards religious belief? If so, is that because we've evolved that way? Or are evolutionary explanations of human psychology mere Just So stories, no more plausible than the delightful tale about The Cat Who Walked By Himself (Kipling 1902)? Superficially, at least, cultures are hugely diverse... but can they harbour *just any* conceivable idea?

In mental illnesses of various kinds, what's gone wrong: something in the brain, or something in the mind? What's the difference?

Sometimes, people say that only living things can have a mind. Is that true? If so, why? What is life, anyway? And how did it arise in the first place? Could a living thing be created by us?

Last, but by no means least, coffee-table chat abounds with puzzles about language. For instance, people wonder what *counts* as a language: why not birdsong? Can any non-human animals learn a language? If not, is that merely because we're better at learning, or because language is a human instinct? And what, exactly, does that mean? Is language needed for thought, or can some dumb animals think?

Can two different languages ever express exactly the same thought? Or is perfect translation impossible? Could a machine converse with us in English, or French—and would it understand us, even if it did? Imagine a machine that appeared to be solving problems and using language just like us: would that prove that it was truly intelligent?

None of these questions is new. (That's largely why listing them is a scene-setting equivalent of saying "Once upon a time, there was a teddy bear...".)

Some date back to Aristotle. Many, including those about language-using machines, were discussed in the 1630s by René Descartes. Others were considered by Immanuel Kant, Johann von Goethe, or Wilhelm von Humboldt in the late eighteenth century. The rest surfaced in the nineteenth, or very early twentieth, century (see Chapter 2).

Originally, then, most were discussed by philosophers. Some still are (the difference between *mind* and *brain*, for example). But even those need to be considered in light of the scientific data available.

Most of our Twink-questions were later developed—and some answered—by traditional scientific research in psychology, anthropology, neurophysiology, or biology. Since the 1940s, however, *every one* has been further sharpened by work in cognitive science.

b. How to find some answers

Cognitive science tries to answer these questions in two closely related ways. Both of them draw on machines. But the machines in question are very unlike what used to be thought of as a machine.

Forget steam-engines and telephones: these new machines can be hugely more complex even than an E-type Jag, or a jet plane. Indeed, the capacities of modern jets, from the much-lamented Concorde to stealth bombers, are largely due to their having these new machines inside them. It follows that to think of minds *as* machines, as cognitive scientists in general do, isn't so limiting—nor so absurd—as it may seem to someone who has only pre-1950 machines in mind.

Specifically, cognitive science uses abstract (logical/mathematical) concepts drawn from artificial intelligence (AI) and control theory, alias cybernetics (see Section ii.a, below).

* AI tries to make computers do the sorts of things that minds can do. These things range from interpreting language or camera input, through making medical diagnoses and constructing imaginary (virtual) worlds, to controlling the movements of a robot.
* Control theory studies the functioning of self-regulating systems. These systems include both automated chemical factories and living cells and organisms.

These concepts (of computation and control) sharpen psychological questions because they can express ideas about mental processes more clearly than verbal concepts can. Moreover, when implemented in computer models they can test the coherence and implications of those ideas more rigorously. Often, they show that a previously favoured theory has unsuspected gaps in it. Sometimes, they suggest how those gaps might be filled. They can show that a theory *might be*, *could be*, true—although to know whether it *is* true, we need psychological and/or neuroscientific evidence as well. Some important questions have been answered in this way which couldn't have been answered otherwise.

Consider language and machines, for instance. It's now clear that computers can (seem to) use natural language, up to a point. What's not yet clear is just where, in practice or principle, that point lies. How good can we expect future computer prose, or computer conversation, to be? And what problems will have to be overcome to get there? For that matter, what are the problems which have already been overcome, to get to where we are now? And are these problems linguistic, psychological, or philosophical—or perhaps a mixture of all three?

Thirty years ago, a medical friend told me he'd spent the afternoon visiting an immigrant family from India, whose 8-year-old son had been translating across three languages for his elders. "You'd never get a computer to do that!" he said.—Maybe, maybe not. But if not, why not? And if so, how?

Only five years after this pessimistic comment, the European Union's translation system achieved 78 per cent intelligibility for its 'raw' text, and 98 per cent for the tidied-up version. Unlike the little boy, this program could handle only two languages at a time. Ten years later, however, another one could switch between forty-two different language pairs. But the boy—by then, in his early twenties—still had the edge. He could translate remarks about anything, within reason, whereas these programs could deal only with relatively specialized topics.

What my friend didn't seem to realize was that if AI research could enable a computer to use even *one* language properly, translating it into another would be easy by comparison. Or rather, translating it helpfully, usefully, acceptably ... would be relatively easy. Translating it perfectly is another matter. But then, it's not clear that a human being, whether 8 years old or 80, could produce a *perfect* translation of anything interesting. Even *Please give me six cans of baked beans* will cause problems, if one of the languages codes the participants' social status by the particular word chosen for *Please*.

Nor did he stop to ask how the 8-year-old did it—still less, how his own children had learnt their mother tongue. He simply took it for granted that language learning happens. But how? After all, vocabulary isn't the only problem: there's grammar, too.

Different languages have different grammars. Or at least, they appear to. (The order of adjective and noun varies, for instance: think of *the red house* and *la maison rouge*.) But perhaps all languages share some underlying 'universal' grammar? If so, what is it? And how is it related to the syntactic rules that bedevil us when we encounter a new tongue? How did the family's young translator manage to cope with three distinct grammars, from different language groups? As for the rules of one's mother tongue, how are these learnt, and how are they represented in the mind/brain?

All these questions, and many more, have had to be faced by cognitive scientists working in psycholinguistics and/or natural language processing (NLP). And a great deal has been learnt in the process, even if many mysteries—and some bitter controversies—remain (see Chapters 9 and 12.vi.e and x.d–e).

"Not too fast!" you may say. "Computers can handle language to some extent, and even translate it usefully too. But do they *understand* the language they use?" (Don't let's stop, here, to ask what it is for *human beings* to understand the language they use—but see Chapters 7.ii.d, 12.x.g, and 16.)

You may even mention the Chinese Room, an intriguing idea that's hit the mass media worldwide (16.v.c). This example is intended to show that the answer to your question is "No". A monoglot English speaker could spend weeks following formal rules for shuffling slips of paper bearing *squiggles* and *squoggles*, without ever realizing that they are Chinese characters which, for readers of Chinese, can be used to deliver true answers to meaningful questions. The moral is supposed to be that AI programs are intrinsically meaningless, and that for understanding you need a brain.—And, it's often added, you need a brain for consciousness too: a robot, no matter how human-like, would be a non-conscious zombie.

An equally well-known argument claims that even if the language produced by a computer program, or a robot, were indistinguishable from that produced by a human being, that wouldn't prove that the thing was really intelligent. "Passing the Turing Test", as this is called, wouldn't guarantee intelligence, understanding, or consciousness. The Test-passer might simply be a zombie.

Both these arguments have been hotly debated within cognitive science—although each is much less important *for the practice of AI* than most people imagine (see 16.ii.c). There's still no unanimity on whether they're well founded. Indeed, some cognitive scientists hold that *there can be no such thing* as a zombie—not because the technology is too difficult (although in fact it may be), but because the very notion is incoherent. On this view, science-fiction novels and Hollywood scenarios about zombies are, literally, non-sense (14.xi and 16.iii–v).

Whether the technology really is too difficult is disputed also. The vast majority of cognitive scientists would say that it is, at least for the foreseeable future. But one leading research team, initially with a prominent philosopher on board, is betting that it isn't. They're hoping to build a (literally) conscious robot, with a mind like that of a young child (see Chapter 15.vii.a).

(*An aside*: That last sentence was true when I wrote it, in the mid-1990s. Now, in 2005, the project has ground to a halt. The roboticist team leader always had other fish to fry in his research time, and is now buckling under a heavy administrative load as well; as for the research students who were working on it, in snatched moments of their spare time, they've left to take up jobs elsewhere. However, the leader still believes that the project is feasible, and he might even revive it some day. Given that fact, the following paragraphs can stand, as though the plan were still being actively pursued.)

Even ignoring the issue of consciousness, they face hugely challenging problems. To build a robot that even *seems* to have the intelligence of a 5-year-old, they must provide all the relevant perceptual discriminations, motor skills, learning power, problem-solving ability, and language mastery.

For each of these, they must depend on work done by other cognitive scientists. For example, they need a computer vision system modelling the child's visual powers: so they need to know *what these are* and *how they work* (see Chapters 7.v, 14.iv and vi.b–d). They need a powerful theory of perceptuo-motor control, for generating appropriate movements of the eyes, head, and fingers (14.vii and x). They should also enable the robot to switch smoothly between stable gaits, such as crawling, standing, walking, and running (14.v and ix.b). (In fact, they've avoided those problems by giving their robot a pedestal in place of legs.) The system's capacity for learning must be built on the work that's been done in this area (see Chapters 10.iii.d, 12, 13.iii.d, and 14). As for enabling the robot to develop language, they must rely on research into some of the psycholinguistic questions outlined above (Chapters 7.ii and vi, 9, and 12.vi.e).

Strictly, they should also simulate the temper tantrums of 'the terrible twos', a stage of infancy that all parents will remember with a shudder. And, if only to preserve their own sanity, they should enable the robot to develop the greater self control—which is to say, the greater freedom (7.i.g)—of the 5-year-old. But the control of temper tantrums is even more difficult to model than stable walking or running is.

Indeed, you may think that the appropriate word here isn't "difficult", but "impossible". Certainly, many people believe that a computational psychology can't have anything to say about emotions, still less freedom.

Well, it can, and it does (see Chapter 7.i). This challenge was mounted over forty years ago, and was soon taken up by Herbert Simon, one of the high priests of computational psychology. At that time, too, a computer simulation of neurosis was developed in which different levels of 'anxiety' selected different defence mechanisms to repress the 'troubling' thought. In 1983 the authors of the Gifford Lectures on Natural Religion gave a computational analysis of personality and freedom (and religious belief) in which emotion figured prominently. Several philosophers have analysed human freedom in terms of a certain type of computational (cognitive and emotional) complexity. And a very recent program models the emotion-guided activity of a nursemaid caring for a dozen babies, each of whom has to be fed, watered, changed, cuddled, entertained, and prevented from falling into the river or crawling towards a busy road.

A nursemaid is free to choose what to do at every moment. But her choices are far from random. To the contrary, they're constrained by the goals she wants to achieve (which may conflict: she only has two hands); by the priorities she holds (feeding is necessary, lullabies aren't); by her deliberations about consequences (no-cuddles will produce an unhappy baby); by her judgements of urgency (even the hungriest baby can be temporarily ignored, if another is nearing the road); and by her emotional reactions (sometimes, she must rescue the baby *immediately*, without stopping to think). On some occasions, she 'has no choice': the danger *must* be averted, and it must be done *now*; and the baby *must* be fed, soon. But the sense in which she (sometimes) has no choice is fundamentally different from the sense in which a non-human animal, such as a cricket for example, (always) has no choice about what to do next (see Chapter 15.vii). She's free, it isn't. Moreover, her freedom doesn't depend on randomness, or on mysterious spiritual influences: to the contrary, it's an aspect of *how her mind works*.

The nursemaid research group has even analysed the computational structure of grief. The emotion of grief is more than mere feeling. It involves irrational behaviour driven by obsessional thoughts, continual distraction, depression, anger, and guilt—all of which gradually pass, over many months, as mourning does its work. (Just what "work" is that? These cognitive scientists suggest an answer: Chapter 7.i.f.)

Grief is possible only for humans, although dogs sometimes seem to *sorrow*. A cricket simply cannot grieve. It lacks the necessary mental architecture: the concepts, knowledge, motives, values, and social commitments required to generate—or to overcome—the deeply disturbing emotion of grief. And unlike a human baby, who can't grieve either (even though it can 'miss' an absent carer), it has no way of developing them. Its mind, if one wants to use that term at all here, is very simple. It can't even learn to recognize objects or patterns as one of *a general class*—such as another cricket.

To be sure, crickets manage. They've survived. They even do some apparently clever things, such as locating a potential mate at a distance. However, they do this unthinkingly. They rely on a hardwired biological trick, an anatomical detail evolved for this function alone. Similarly, a frog locates its food by relying on perceptuo-motor reflexes linking cells in its retina and brain to muscles that make it jump to just the right spot (Chapter 14.iv and vii).

People, too, sometimes use such biological tricks—for instance, in locating the source of a sound (14.viii.c). But their perception and learning involves much, much, more. Even without language, mammals (and birds) can do what crickets cannot: they can learn to recognize new stimulus patterns, and can generalize those patterns over different class members (see Chapters 12 and 14).

Moreover, some of the detailed brain structures that enable mammals to *see*, or to *hear*, arise by spontaneous self-organization in the womb. (So the fact that a newborn baby, or kitten, already has a certain perceptual ability *doesn't* prove that it was specifically coded in the genes.) This may seem surprising, even magical. But computer modelling has shown how such anatomical self-organization is possible (Chapter 14.vi.b and ix.c).

You may be sceptical. You may feel, for instance, that this general approach is merely an example of what Donna Haraway (1944–) calls "cyborg science", more a mark of the times than of the truth (Haraway: 1986/1991).

As she puts it, the many sciences currently informed by the concepts of information and computation involve a "reinvention of nature". They express a pervasive world-view, or "lived social reality", in which human minds and human beings "are constructed as [jointly] natural–technical objects". In her opinion, this view couldn't have arisen without post-Second World War military technology and its aggressive political background.

That last charge is true (see Chapters 4.vi.a, 11.i, and 12.vii.b). To a large extent, the "reinvention" charge is true also. Whether it follows that cognitive science is, as Haraway claims, deeply suspect and epistemologically compromised is quite another matter (Section iii.b–d, below).

Even if you're not an admirer of Haraway's writings, which include many provocative claims about the late twentieth-century Zeitgeist, you may nevertheless be sceptical about cognitive science. You may simply suspect that cognitive scientists have been seduced by the technology. Perhaps they're like the proverbial 'hacker' (11.ii.e), or those people from all walks of life who sit hunched over their bedroom computer for hours on end? Computers, after all, are only too capable of enticing people to waste their time. (Although, sadly, "waste" isn't always the right word: taking control over computers, which are usually much more predictable than other human beings, provides some devotees with their main source of ego strength and contentment—Shotton 1989: chs. 8 and 10.)

However, this type of research, vulgarly trendy though it may appear, is driven by a philosophical view of understanding and explanation that has deep and ancient roots. That is, it's an expression of the "maker's knowledge" *(verum factum)* tradition. This holds that in order to understand something properly one has to be able to make it. In other words, observation and abstract argumentation aren't enough.

One leading proponent of *verum factum* was Giambattista Vico (1668–1744), who famously argued that only the humanities can provide us with genuine knowledge, because they study the creations (not of God but) of human beings (Perez-Ramos 1988: 189–96; Miner 1998). Specifically, history—for Vico, the key to the understanding of human minds and cultures—involves an active re-creation of the thoughts of the people being studied. The eighteenth- and nineteenth-century Romantic philosophers used essentially similar arguments to prioritize art over science (see 2.vi.c and 9.iv). Others were more inclusive, applying *verum factum* to the natural sciences too. So for many of the early modern scientists, a scientific experiment was seen as a *construction*, and theory-based technology was an intellectual justification of 'pure' science.

In short: if you can build it, you can understand it. Cognitive scientists would agree—and their constructions include not only theories but computer models too.

c. Never mind minds?

There's an even more difficult question, one which threatens to undermine the rationale of cognitive science as a whole. Namely: Maybe we'd be better off if we avoided talk of 'mind' altogether?

One way of doing that would be to avoid psychological language entirely (cf. Chapter 5.i.a). But at what cost? Gossip would be impossible—an advance in morality, perhaps, but not in the gaiety of nations. And scientific studies of the topics that fascinate gossips would be impossible too. We could describe the bodily movements, but couldn't say what *action* was being performed, or what *purpose* was being followed. Similarly, we could say how the brain cells are responding, but not what they're *doing*.

Less radically, one could retain psychological language but gloss it in purely behavioural terms, or perhaps in the abstract, functionalist, terms of information processing. Then, scientific studies (of these types) would be justified. The second of these is the position taken by the vast majority of cognitive scientists.

Or—an option that's recently grown increasingly popular *within* cognitive science, as we'll see—one could say that 'mind' was *invented* by Descartes, not just described by him (Rorty 1979: 17–69). On that view, this Cartesian fiction (*sic*) separates the individual both from their own body and from other human beings—and the physical environment, too (see 2.iii.a–b). The implication is that cognitive science should stress embodiment rather than intellectualist reasoning, and social engagement rather than individualistic action and thought.

A much more radical approach would be to argue that both 'mind' *and* 'body' are concepts constructed on some deeper philosophical base, and are highly misleading when taken—by scientists, for example—as fundamental realities (see 14.xi and 16.vi–viii). That would eliminate many puzzles, but only by dismissing hope of *any* scientific explanation of psychology (and any naturalistic account of meaning). Cognitive science, on this view, wouldn't just be difficult: it would be non-sense.

You don't have to be a rocket scientist to guess that I don't share that last view. Perhaps you don't share it, either. But I'm not going to counter it yet. Indeed, we shan't consider it at length until Chapter 16.vi–viii (although it will push its nose above the surface in Section iii.b below, and also in Chapter 14.xi).

Even there, I shan't be able to give a knockdown argument against it: it's perhaps the deepest division in philosophy. To make things worse, it's often closely allied with a form of relativism that would undermine *all* scientific knowledge (see Section iii.b, below). However, many people—including many scientists—aren't even aware of this division, and don't take it seriously if they are. Throughout most of these pages, then, I'll continue to speak of 'mind' and 'mental' phenomena as though such talk were relatively unproblematic. That is, I'll assume that minds and mental phenomena do exist, even while admitting that there are many disagreements about just how they should be described.

Similarly, I'll assume (until Chapter 16) that *some* scientific psychology is in principle possible. Even if it isn't, the first fourteen chapters won't be irrelevant. For a science-denier should offer an alternative interpretation of the facts discovered by science—for which task, they need to know something about what these facts are (see 16.viii.a).

All the examples I've mentioned in this section fall under the centuries-old questions about the mind that were listed at the outset. As remarked there, most of us have mused on these at some time or other. For cognitive scientists, they're a prime concern. And as we'll now see, they ask them in a particular way, which was inconceivable before the late 1930s.

1.ii. The Scope of Cognitive Science

Cognitive science is a catholic field, in three ways:

* First, it covers all aspects of mind and behaviour. (That was illustrated by the wide range of questions listed above.)
* Second, it draws on many different disciplines in studying them.
* And third, it relies on more than one kind of theory. Broadly speaking, it's the study of *mind as machine*—a definition that covers various types of explanation, as we'll see.

a. Of labels and cans

In a neat and tidy world, where every label fitted what's inside the can, cognitive science would be the science of cognition (knowledge). Indeed, it's often defined that way. However, things aren't so simple.

In fact, cognitive science deals with all mental processes. Cognition (language, memory, perception, problem solving...) is included of course. But so are motivation, emotion, and social interaction—and the control of motor action, which is largely what cognition has evolved *for*.

You may feel that these types of psychological process aren't clearly distinguishable. If so, you're in good company. The 'holistic' belief that they're intimately intertwined is both very old-fashioned and very new. Its heyday was 200 years ago (see 2.vi). It never died out entirely: the cybernetic psychoanalyst Lawrence Kubie, for instance, said that "the various areas of psychic life are so interdependent that no one of them can be [experimentally] assayed alone and apart from the others" (1953: 48). However, it did go out of favour with the scientific community, resurfacing only very recently (Chapters 14.x.c, 15.vii.c, and 16.vii.c). Even now, it's an unorthodox view. For the moment, then, let's go along with the common assumption that cognition, motivation, emotion, social interaction, and bodily action can be considered separately.

Given that cognitive science isn't focused only on cognition, the label is highly misleading. Why, then, were these words chosen in the first place?

Today, they don't trip off Everyman's tongue. At the time, however, they were less arcane than one might think. Both had recently been popularized by social psychologists discussing "cognitive dissonance" (Festinger 1957). That terminology had even entered the media. Many journalists had summarized their explanations of the power of advertising, and of high pricing, on consumer behaviour. And the newspapers had had a field day in rehashing the social psychologists' reports about a recent cult in the mid-West of the USA (Festinger *et al.* 1956). These people had expected to be rescued from The End Of The World on a certain day by aliens in spaceships—only to see no EOTW, no spaceships, and *no* diminution of their faith in the cult leader (see 7.i.c).

In any event, professional psychologists were perfectly familiar with "cognition" as a technical term. But it had originally been coined, some two centuries earlier, specifically to *exclude* motivation and emotion. So, again, why choose it?

One of the two men mainly responsible—George Miller and Jerome Bruner—has explained it like this:

> In reaching back for the word "cognition", I don't think anyone was intentionally excluding "volition" or "conation" [aka motivation] or "emotion" (Hilgard 1980). I think they were just reaching back for common sense. In using the word "cognition" we were setting ourselves off from behaviorism. We wanted something that was *mental*—but "mental psychology" seemed terribly redundant. (Miller 1986: 210)

In short, they intended "cognitive" science to address cognition *and more*.

A glance through *Plans and the Structure of Behavior* (G. A. Miller *et al.* 1960) confirms this. That inspirational book (see Preface, ii, and Chapter 6.iv.c) discussed animal behaviour, instinct, and learning, as well as human memory, language, problem solving, personality, mental illness, and hypnosis. Social and cultural matters were touched on also. No aspect of mental life was excluded.

(It was left to another early volume, however, to highlight political, bureaucratic, and economic behaviour: Guetzkow 1962. The editor, Harold Guetzkow, had been a close friend of Simon when they were both graduate students at Chicago. At that time, Simon's interests—like Guetzkow's—had been in economics and management science, not psychology: see 6.ii.a.)

The breadth of coverage in *Plans and the Structure of Behavior* was seen—by readers who were sympathetic at all—as being just as it should be. Virtually all the founding fathers of cognitive science (Noam Chomsky excepted) had asked how motives and emotions interact with cognition, and several had also mentioned psychopathology. In short, these more sexy matters (literally!) were often discussed in the early days.

That didn't last. Because motivation, emotion, and social interaction (whether in small groups or in societies) are even more difficult to study—and to simulate—than cognition is, they were soon put onto the back burner. They were left there for thirty years, while cognition got almost all the attention. Vast amounts of research were done on perception, language, problem solving, concepts, belief, memory, and learning. The name of the field reflects this.

In fact, the label on this particular can was changed several times. In the early 1960s, the field was known by the more neutral "computer simulation". A Harvard graduate course was run under this rubric, and books appeared with titles such as *Computer Simulation of Personality* (Tomkins and Messick 1963), *Simulation in Social Science* (Guetzkow 1962), and *Computer Simulation of Behaviour* (M. J. Apter 1970). As research on cognition became more dominant, however, three new names emerged: cognitive studies, cognitive sciences, and cognitive science.

"Cognitive *Studies*" was chosen in 1961 by Bruner and Miller (the lead author of *Plans and the Structure of Behavior*) to name their new research centre at Harvard. This comprised a wide variety of psychologists, leavened by a few linguists and computer specialists and the occasional interdisciplinary philosopher. Nelson Goodman (1906–98), who co-founded Harvard's "Project Zero" studying representation and education in art, was in house when I was there, for example. These psychologists didn't do simulation as such, although Miller had co-published with Chomsky on mathematical models of language (9.vi.a). But they typically used ideas drawn from early AI, and from information theory, in their seminars and experiments (6.iv.d).

The "Cognitive" in the centre's title reflected the two co-founders' main interests: perception, language, memory, and problem solving. Even when Bruner studied values, he focused on their effect on *perception* (6.ii.a). Nevertheless, Miller later admitted that "conative" and "affective" phenomena (i.e. motives and emotions) should also be mentioned in the definition (see the quotation above, and also G. A. Miller 1978: 9).

"Studies" had become "sciences" by 1973. Already used in everyday chat by an Edinburgh research group for a couple of years, "cognitive sciences" first appeared in print in a defence of AI-based psychology, then under attack by a world-famous mathematician (Longuet-Higgins 1973: 37; cf. 11.iv). And the singular version—cognitive *science*—appeared soon afterwards, in two widely read collections of papers (Bobrow and Collins 1975, p. ix; Norman and Rumelhart 1975: 409).

Now, that's the label which is used most often. Even so, the editors of the recent *MIT Encyclopedia of the Cognitive Sciences* (R. A. Wilson and Keil 1999) chose the plural version—which highlights the fact that several very different disciplines are involved in the field.

The singular form, by contrast, highlights the intellectual links between them. That's why I've chosen to use it here. For as we'll see, there have been countless instances of work in one discipline being radically influenced by work in another. That's not surprising. To understand the mind (mind/brain) properly, one doesn't only need to look at it from all directions: one must also *integrate* the various views.

b. Two footpaths, many meadows

The field would be better defined as the study of "mind as machine". For the core assumption is that the same type of scientific theory applies to minds and mindlike artefacts. More precisely, cognitive science is *the interdisciplinary study of mind, informed by theoretical concepts drawn from computer science and control theory.*

These concepts change, as time passes. (Many examples of such change are described in later chapters.) So cognitive scientists don't believe that today's computer-related concepts suffice to explain the mind. Rather, they believe that they're a good beginning, and that later explanations will use concepts drawn from what then happens to be the best theory of what computers do (see Chapter 16.ix.f).

My "two-footpaths" definition, above, carries a health warning. As we'll see later, one highly influential alternative definition of the field specifically *excludes* control theory. It allows only explanations in terms of formal symbol manipulation (see Chapters 12.x.d and 16.iv.c–d). Cybernetics, and even connectionism, is therefore said to lie outside cognitive science.

For reasons which I hope will become clear throughout the narrative, I regard that definition as much too narrow. It's true, however, that cognitive science has seen—and is still seeing—competition, as well as cooperation, between computer science and cybernetics as ways of thinking about the mind. Indeed, the pendulum-swings between these two intellectual sources are a central, and fascinating, aspect of the story (see especially Chapters 4, 10, and 12–15).

As remarked in the Preface, the main disciplines involved are psychology, linguistics, AI, A-Life, neuroscience, and philosophy—and, though it's relatively rarely mentioned, anthropology. Certain areas of biology, such as ethology and evolutionary theory, are also included (and, at the fringes, some aspects of biochemistry are relevant too: 15.x.b). Moreover, the many examples in Section i.a (above) imply that the relevant research ranges all the way from mate finding in crickets to grammar, and even grief, in human beings.

The history of cognitive science is marked by a deep, and continuing, interdisciplinarity. This is a more intellectually intimate relationship than mere multidisciplinarity. Again and again, researchers in one area have borrowed *theoretical ideas*, not just *data*, from another.

Certainly, many specialist sub-areas (and sub-sub-areas . . .) have emerged over the past half-century. Each has its own conferences, journals, and textbooks. Moreover, their personnel rarely communicate. "Fair enough!" you may say. "If someone's interested in depth vision or learning, why should they bother with English grammar?" Well, they don't need to, in order to tackle their current problems. It remains true, nevertheless, that stereopsis and learning are studied in the way they are today partly because of mid-century work on syntax and mid-1980s research on past-tense verbs (see Chapters 7.vi.a and 12.vi.e). In short, even the most 'separate' specialisms share life-giving historical roots.

They share some central assumptions, too. All areas of cognitive science are informed by computational concepts, and driven by computational questions. In other words, all cognitive scientists use such concepts as core theoretical terms. This isn't the same thing as using computers. Biochemists or geologists, and non-computational psychologists too, often use computers as research tools (to do statistics, for example)—but their *theories* aren't computational. (Nor is it the same thing as building computer models: many cognitive scientists do this—but many don't.)

Broadly speaking, computational concepts are of two main types. On the one hand, they're drawn from computer science, AI, and software engineering. On the other hand, they hail from information theory and control engineering—in a word, cybernetics.

This dual definition, like my catholic definition of the field (above), carries a health warning. 'Computation' is often understood as Alan Turing defined it (see Chapter 4.i.b–c). Indeed, his definition remains the only rigorous one. And it *doesn't* cover cybernetics, nor even connectionist AI. Nevertheless, many people today—including computer scientists—use the term more widely. In other words, ideas about what computation *is* have become more extensive. (We'll see in Chapter 16.ix that, despite the undeniable loss of rigour, there are good reasons for this.) I'm one of many who use the term more widely than purist symbolists do. In general, the context should show when I'm using the term to refer to ideas from only one of these two sources.

The two sides of the computational coin were first clearly distinguished in the mid-twentieth century (see Chapter 4), although each had been prefigured much earlier (2.vii–x). Over the years, the theoretical concepts involved have developed into a varied group, defining many different types of information processing, virtual architecture, and computer model. This development, which hasn't been without hiccups, has involved both competition and cooperation between the two sides—first seen as competing in the late 1950s (Chapter 4.viii).

For example, research based on 'dynamical systems' falls on the cybernetic side of the fence, and is typically peppered with disparaging remarks about symbolic AI (14.ix.b, and 15.viii.c–d and xi). In particular, dynamicists claim that they can explain the temporal aspects of cognition, which earlier approaches ignored. But many people who work in this area were trained in AI, and depend heavily on it—for instance, when using genetic algorithms as ways of evolving dynamical systems (15.v). Similarly, most connectionist AI is closer to cybernetics, and has often been fiercely opposed to GOFAI—that is, to Good Old-Fashioned AI (Haugeland 1985: 112). Nevertheless, some researchers have tried to combine these two approaches (7.i.e–f, 12.viii–ix, and 13.iii.c).

As for the hiccups, Chapter 12 describes the birth *and renaissance* of connectionism—and the Sleeping Beauty phase in between. A-Life, too, had its Sleeping Beauty phase, from which it awoke one year later than connectionism. Psychology and philosophy have reflected these changes, offering very different theories of mind at different times.

The two computational pathways wound through many disciplinary meadows. The meadows were close neighbours at the beginning. Indeed, in the 1940s and 1950s—when distinct disciplines were being deliberately, outrageously, juxtaposed—highly *inclusive* consciousness-raising meetings were important (see Chapters 4.v.b and 6.iv.a–b). From the mid-1960s, however, the specialisms reasserted themselves. A second phase of "outrageous" interdisciplinarity was launched in 1987, at a party in the New Mexico desert (see 15.x). But most of the party-goers, though newly enthused, went back home to work in their own specialist houses.

This affects how our tale can be told. Chapters 2 to 6, by and large, move along a single time line, taking us from antiquity up to the mid-twentieth century. In Chapters 7 to 16, the time line branches. Each discipline has its own chapter (although AI has three, and we'll backtrack about 500 years for linguistics). Even so, most of the important topics feature in *several* 'disciplinary' chapters. For the same two computational pathways are there throughout, connecting the different disciplinary meadows with each other.

c. Why computers?

Computers as such are *in principle* less crucial for cognitive science than computational concepts are.

To be sure, computer technology (both digital and analogue) is an important player in the narrative. Computer modelling has a prominent role because it's often needed *in practice* to confirm—or even to discover—the full implications of a computational theory. Indeed, advances in software design (especially high-level programming languages: 10.v) and computer engineering may be needed before such theoretical modelling can be attempted. Turing himself was unable to develop many of his ideas because of the primitive state of computers in his day (15.iv). As he put it:

> At my present rate of working I produce about a thousand digits of programme a day, so that about sixty workers, working steadily through about fifty years might accomplish the job, if nothing went into the wastepaper basket. Some more expeditious method seems desirable. (A. M. Turing 1950: 61)

But cognitive scientists don't always build computer models. Chomsky's linguistics, John von Neumann's cellular automata, and David Marr's early brain theories, for example, were formal models—not functioning simulations (see Chapters 9.vi, 15.v, and 14.iv, respectively).

Some highly influential discussions weren't even *formal*. Marvin Minsky's "society of mind" theory (12.iii.d) and his earlier account of "frames" (10.iii.a), Michael Arbib's schema diagram for control of the hand (14.vi.c), and Robert Abelson's work on the structure of belief systems (7.i.a) are all cases in point. Indeed, the two seminal papers co-authored by Warren McCulloch and Walter Pitts (4.iii.e, 12.i.c, and 14.iii.a) were published in the early to mid-1940s, before the first modern digital computer had been built. It was another ten years before computer simulations of psychology were feasible. Some people would argue that *serious* simulation wasn't possible before the late 1980s—if then (see Chapter 14.vi.d).

Because computational concepts are essential, AI—or AI/A-Life—is a central discipline. Not all of AI is germane, however. Workers in AI—and A-Life—can have either of two motives. (Some have both.) The first is to build computer systems that are useful in some way. These range from automatic translators and financial networks to robot toys and remote-controlled surgeons. The second is to use software and/or robotics to help us understand human and animal minds (or life), or even *all possible* minds (or life). Let's call these 'technological' and 'psychological' (or 'biological') AI/A-Life, respectively. Only the latter project falls squarely within cognitive science.

Occasionally, this project has a further motive: to build, or anyway to start on the road towards building, a *real* intelligence, or a *real* living thing. These aims have driven some of the most well-known AI/A-Life research. And they've been very widely discussed for over fifty years—in terms (for example) of the Turing Test, strong AI, and strong A-Life (Chapter 16.ii, v.b–c, and ix.b). Nevertheless, they're minority tastes.

Despite the sensational quarter-truths peddled by the media ever since the 1950s, most researchers in AI/A-Life haven't argued—and probably haven't believed—that an AI program could actually be an intelligent mind, or that a merely virtual 'creature' could really be alive. Some have even denied it, claiming that *embodiment* is needed for life and intelligence (see Chapters 15 and 16.vii and x.) In short, this third motive has sometimes played a role in psychological AI/A-Life, but it isn't essential to it.

Technological AI is usually irrelevant to cognitive science—so is only rarely mentioned in my narrative—because it seeks to do something *irrespective* of how the mind/brain does it. The developers of IBM's Deep Blue, which beat the world chess champion Gary Kasparov in New York on 11 May 1997 (winning a prize of $100,000 dollars in the process: see 16.ii.c), were happy to use dedicated computer chips. These enabled the program, processing 200 million positions per second, to rely on exhaustive look-ahead over eight moves. Anyone studying how human beings play chess would avoid this biologically unrealistic hardware.

There's one type of situation, however, where even purely technological AI is relevant: namely, if someone believes that certain tasks simply *cannot* be done by computers.

For instance, Hubert Dreyfus's judgement (in 1965) that no program could play even "amateur" chess was falsified only a year later, when he himself was defeated by a program (H. L. Dreyfus 1965: 10; Papert 1968, para. 1.5.1). And his claim that no computer would ever play chess at a human level unless it could distinguish perceptually between "promising" and "threatening" areas of the board (H. L. Dreyfus 1972, pp. xxix–xxxiii, 208) was decisively refuted by the performance of Deep Blue. The fact that its exhaustive "counting out" strategy, to use Dreyfus's term, isn't one that humans can use is irrelevant.

Admittedly, the distinction between the two types of AI isn't clear-cut. For instance, I said in the Preface that Margaret Masterman's pioneering work on machine translation and classification was technological, but also guided by strong intuitions about how people process language. It's not that she wasn't interested in how the mind works—although, as a post-Fregean philosopher, she was wary of 'psychologism' (Chapter 2.ix.b). But detailed psychological questions would have been premature. There was no experimental evidence enabling her to decide, for example, that one of two thesauri was the more realistic. She had to rely on other, more intuitive, criteria.

Even today, fifty years later, most technological AI is grounded in intuitions about human thinking. What the writers of expert systems call "knowledge engineering", for example, includes a method of questioning human experts, to help them make their expertise explicit (10.iv.c). Sometimes, data from experimental psychology and/or neuroscience influence the program too. Some industrial applications even use special-purpose hardware chips modelled on the mammalian visual system (12.v.f). But the interpretation of the AI systems as models of actual mental processes isn't the object of the exercise.

d. What's in, what's out

Not all of *psychology* is germane to cognitive science, either. All psychological *data* are relevant, in the sense that cognitive science, if it is to succeed, must one day explain them. But many psychological theories aren't computational. This history narrates how some theoretical psychology became computational—and the dramatis personae are selected accordingly. (The same applies, *mutatis mutandis*, to anthropology, linguistics, and neuroscience.)

I'll say relatively little, for example, about the behaviourists, or Sigmund Freud—despite their importance for the history of psychology in general. With respect to computational psychology, behaviourism was significant as something to be reacted against, not developed (see Chapters 5 and 6.i–iii).

As for Freud, his psychodynamics (Chapter 5.ii.a) inspired some early AI simulations of neurosis, and a model of the effects of anxiety on speech (7.i.a and ii.c). In a broad sense, it informed Minsky's work on the society of mind (12.iii.d), and spread from there to Daniel Dennett's philosophy of consciousness (14.xi.b and 16.iv.a–b). It contributed to Arbib's schema theory, and especially to the application of schema theory to religion (7.i.g). And it provided examples and ideas that fed into Aaron Sloman's work on the architecture of grief (see Chapter 7.i.f). So I mention Freud in all those contexts—but I don't focus closely on him.

Even "cognitive psychology" doesn't always fall within cognitive science (Chapter 6.v.b). For instance, David Clark's (1996) explanation of—and therapy for—anxiety disorders (such as phobias, panic attacks, and post-traumatic stress disorder) analyses them in terms of the person's underlying beliefs about danger: it's cognitive, but doesn't involve explicit reference to computational concepts or theories. Admittedly, the term "cognitive psychology" was first defined in a computational context (Neisser 1967). But some cognitive psychologists, including that author himself in later years, specifically reject computational theories (7.v.e–f).

You may be surprised to see neuroscience included in this account. For a recent dictionary of psychology states that "cognitive scientists rarely pay much attention to the nervous system", and that cognitive science and neuroscience are "almost mutually exclusive" (N. S. Sutherland 1995: 83). The explanation given there is that "cognitive science deals with the brain's software, neuroscience with its hardware".

As a quick summary, that's correct. The cognitive scientists of the 1960s and 1970s adopted, or defined, an *abstract* (functionalist) philosophy of mind, and most still do (see 16.iii–iv). But facts about the brain have inspired various forms of AI, and of its cousin, cybernetics (Chapters 4.iii–vii and 12). Moreover, the intellectual traffic is increasingly two-way. Computational ideas inspired one of the most famous papers in neurophysiology, 'What the Frog's Eye Tells the Frog's Brain' (Lettvin *et al.* 1959), and they've been used for thirty years to model the brain (see Chapter 14). Neuroscientists today regularly use computational ideas, asking not only which cells and neurochemicals are involved but also what functions they're computing and/or modulating.

You may not expect A-Life to be counted as a member discipline, either. For A-Life researchers usually make a point of distancing themselves from traditional cognitive science, rejecting its representational ("Cartesian") view of mind. Even when they do so, however, they may admit that they "count as cognitive scientists" in so far as they try to create "working artifacts that demonstrate basic cognitive abilities" (I. Harvey 2005: first page).

Moreover, A-Life bears a new name on an old bottle. It dates back to mid-century cybernetics, and its three founding fathers—von Neumann, Turing, and W. Ross Ashby—were crucial also for the rise of other areas in cognitive science. In addition, many people believe that life and/or embodiment is *necessary* for mind (Chapter 16.vii and x). If they're right, then A-Life would be essential to our discussion even without the strong historical links.

The inclusion of anthropology may surprise you also. For it's rarely mentioned as being part of cognitive science. But we'll see in Chapter 8 that this is truer today than it was forty years ago, and that part of the reason is that much of the relevant work now appears under a different label—namely, evolutionary psychology.

As for philosophy, it plays a key role in the story. Indeed, it's rarely absent. Besides having its own chapter (Chapter 16), it has dedicated sections in several others (see Chapters 2.ii–iii, vi, and ix.b; 7.iii; 9.iv–v and viii; 10.iii.f and vi.a–d; 12.x; and 14.viii and xi). And it provides many passing remarks throughout the story.

This isn't merely because, before the scientists got their act together, only philosophers were saying anything about the mind. It's also because the scientists themselves have repeatedly raised philosophical questions—and they still do.

That's why there are so many different definitions of cognitive science, and so many competing camps (and changing fashions) within it. My pages don't tell a happy tale of consensual colleagues harmoniously seeking the same sort of truth. Personal rivalries and self-regard have played their part, of course: we're talking about human beings here, after all. But the deepest divisions have been at philosophical fault lines, so that *what could possibly count* as the truth, or as an explanation, isn't always agreed.

For example, cognitive scientists differ profoundly on how the mind–body distinction should be understood. A few even reject the distinction entirely. They also differ on what can count as a *representation*, a term that's often used within the field—and in some influential definitions of it.

Moreover, some philosophers outside the field argue that cognitive science can't illuminate *any* aspect of mind—not even cognition (or representation) itself. Such fundamental disputes have their own history, recounted in Chapter 16.iii–viii. Meanwhile (as explained in Section i.c, above), I'll continue to use "mind" as a useful shorthand, setting these basic philosophical questions aside until then.

The existence of philosophical disagreements between cognitive scientists may partly explain N. Stuart Sutherland's acid comment that cognitive science is "[an] expression [that] has come into being mainly in order to allow workers who are not scientists to claim that they are" (N. S. Sutherland 1995: 83). "Surely", someone might say in Sutherland's defence, "a *science* must have some fundamental philosophy on which all its practitioners agree?" This objection is too confident: even quantum physics would be excluded, given the notorious disagreement among physicists about its interpretation. With respect to cognitive science, it's partly because the theoretical foundations are still in dispute that it's so fascinating.

In sum, my view of cognitive science is relatively broad. It covers seven disciplines, and various forms of explanation. In particular, it includes both symbolic and cybernetic theories, despite the differences between them. Many people define the field in a much narrower way (see 12.x.d, 14.ix.b, and 16.iv.c–d). To do that, however, is both philosophically controversial and historically misleading. Cognitive science is a rich intellectual tapestry, woven over the years from many different threads.

1.iii. *Caveat Narrator*

Before embarking on my story of how cognitive scientists came to think as they do today, eight caveats are needed—for me as author, and for you as reader. (So it's *caveat lector* as well.)

The first three concern the temptations of Whig history (Butterfield 1931) and of over-idealized views of science. The fourth warns against the seductiveness of heroic accounts of creativity. The next two point out the difficulty in identifying an idea as a discovery—or even as new (which isn't the same thing). The seventh focuses on the role of rhetoric and publication in gaining a place in history. And the last reminds us that whether a new idea is favoured can depend on what sort of thing people with influence are prepared to count as an 'explanation'.

These are very *general* dangers. They make history-telling inherently problematic, a matter of more than just good plain fact. Indeed, a pessimist might argue that *Joe Bloggs was born in year x* is as near to plain fact as one can possibly get. I don't think the situation is quite as hopeless as that (see subsection b, below). But the historical claims in the following chapters, whether made by me or quoted from someone else—in their personal recollections, perhaps—are all prey in principle to the dangers listed here.

I'll mention some examples in this section as illustrations—but only very briefly: each one is discussed at greater length later. And if you return to this chapter after having read all the others, you'll be able to add many further instances.

There's a ninth, higher-order, warning too: the accepted history of the field is itself *part of the field*, helping to shape researchers' judgements and aims. However, because of the difficulties noted in the other caveats, there's no such thing as 'the' accepted history. For cognitive scientists with different theoretical agendas often differ about who did what, why they did it, and what the consequences were.

One illustration concerns the twenty-year connectionist 'winter': was it due to the undeveloped state of technology, to theoretical weakness, to the self-serving activities of two highly combative men, or to the fact that one of them had a very old friend in very high quarters? I'll ascribe it in some measure to all four (Chapter 12.iii.e). But that's not what it looked like at the coal face—and even there, it looked different depending upon which seam one was working on.

In brief: myths matter. History gets told over the coffee cups, as well as (sometimes) being made there—and the telling can affect the making.

a. Beware of Whig history

I don't assume nor, I hope, imply the Whiggist view that everything gets progressively better and better until we reach today's date.

Still less do I assume that it gets better smoothly. I've already mentioned the Sleeping Beauty phase of connectionism, for instance. Hotly disputed theoretical alternatives have surfaced, disappeared, and resurfaced over the years, and one aim of this book is to help readers understand why.

The Whiggist assumption, within the history of science, leads to two common types of anachronism.

On the one hand, ideas that don't fit into the progressivist narrative are ignored. This is usually because, from the historian's own viewpoint, they're incorrect or misguided. They may even be overlooked because they've become near-unintelligible. It can be difficult to understand what questions scholars and scientists of the past were asking, and why, if their concerns didn't align with what counts today as a scientific enquiry (N. Jardine 1991).

On the other hand, past ideas may be tendentiously misrepresented. This may be done by stating/implying that earlier thinkers were trying to solve *our* problems, and/or by using today's terminology in describing their work. It's all too easy to project our own concerns onto ancient writings that bear some superficial resemblance to ours, in order to make a progressivist story appear more plausible.

It's also possible, of course, to be correct in attributing something very like our interests to a past thinker, but incorrect in assuming—because of Whiggist presuppositions—that their work was actually influential. Conversely, if someone was unrepresentative of their age, we may fail to look closely enough to find interesting parallels between their thought and today's ideas.

Within cognitive science, there are specific examples that should warn us against all these varieties of Whiggism. They include arguments relating to fundamental issues, such as disputes about whether Descartes's account of mind was an advance on Aristotle's (Chapters 2.iii and 14.x–xi), or whether Goethe's biological views were well understood, and rightly scorned, by most of his scientific successors (2.vi.d–f, and 15.iii.a and vii.c). They also include narrower issues, such as whether Humboldt's views on language were as similar to Chomsky's as Chomsky himself has claimed (9.v.d–g). In addition, some work should have been more influential than it has been (7.i.e–f). An uncritically Whiggist approach would either exaggerate its early influence or ignore it entirely.

As for backwards projection of our own concepts onto the past, Descartes—contrary to what is now widely believed—did not deny (what we call) conscious experience to animals. Nor did he ascribe it to them, either. This wasn't because he was agnostic about the facts of the matter, but because he lacked the particular concept of consciousness used in formulating the question (see Chapter 2.ii.d–e). I argue also that we shouldn't attribute any version of the modern idea of 'man as machine' to those early Greek philosophers who posited materialism, nor to any ancient engineers inspired by them (Chapter 2.i). It's similarly anachronistic to assume that Charles Babbage had any intention of likening minds to machines, for he didn't: he was *not* an early cognitive scientist (see 3.iv).

Some historical figures with (some) ideas remarkably close to our own had a negligible—perhaps even regressive—influence on the later development of those ideas. Usually, this was because their ideas were so far ahead of their time that their contemporaries couldn't appreciate them.

Babbage, again, is (arguably) an example. He's often described as a crucial player in a Whiggist story about the development of computer technology, but some experts believe that knowledge of his work actually delayed the invention of modern computers (Chapter 3.vi.a). And Jacques de Vaucanson is often wrongly assumed to have had no scientific intent in building his automata, largely because most of his fellow artificers didn't—and because he committed the sin of exhibiting them for money (2.iv). Most of his contemporaries were under the same illusion, so even they didn't go to him for scientific inspiration.

As for more recent examples, Petr Smirnov-Troyanskii's pioneering ideas about machine translation, part-patented in 1933, were ignored even in his native Russia (9.x.a). John Clippinger's intriguing mid-1970s model of the effects of anxiety on speech fell into a black hole from which it hasn't yet emerged (7.ii.c). And even Karl Lashley's now famous discussion of 'The Problem of Serial Order in Behavior' (K. S. Lashley 1951*a*) caused barely a ripple when it was first delivered in 1948, or first published in 1951 (5.iv.a). It came into its own in 1960, when—thanks to late 1950s work in AI and linguistics—cognitive scientists were at last in a position to appreciate it.

Nevertheless, there are many areas where one can point to definite progress. And the value I have in mind here—given that "progress" can't be defined in a value-neutral way—is *closeness to the truth in describing and explaining aspects of the world.* This includes both how real people *actually* manage to think about real and imaginary worlds, and how it's *possible* for any creature—man, mouse, or Martian—to do so.

The growth of our understanding of cellular automata is one clear example of progress (15.v–viii). The advance of neuroscience is another (2.viii, 7.v.b–d, and 14), and of interactive human–computer interfaces yet another (9.xi.f and 13.v–vi). A fourth (related) example concerns the idea of mind-expanding 'cognitive technologies'. These were prefigured by a physicist in the 1940s, named by an experimental psychologist in the 1960s, and clarified/complexified over the next forty years by psychologists, computer scientists, and philosophers (10.i.h, 6.ii.c, 13.v, and 16.vii.d respectively).

Even when significant controversy remains, much may have been found out. That applies, for instance, to the degree of 'modularity' in human minds (7.vi.d–i); to the relation between autism and 'Theory of Mind' (7.vi.f); to how we understand religious concepts (8.vi); and to how we compensate for the fact that our rationality is limited (6.iii, 7.iv, and 8.i.b). Indeed, *every* chapter instances—quite apart from the amassing of new empirical data—the gradual clarification and development of theoretical ideas.

This constitutes scientific advance, even if the idea eventually turns out to be an explanatory cul-de-sac. (As Karl Popper put it, science grows by conjectures *and* refutations: K. R. Popper 1963; likewise Francis Bacon, in *The New Organon*, "Truth emerges more readily from error than from confusion": Bacon 1620.)

I make no apology, therefore, for the strong whiff of Whiggism that attends many parts of my text.

b. Losing the Legend

That claim—that we've already seen some progress in understanding the real world—may raise a few readers' eyebrows. Indeed, it may raise their blood pressure too. For, as evidenced by the 'science wars' that have raged for the last thirty years, social constructivists—supporters of 'the strong programme' in the sociology of knowledge—reject the possibility of scientific progress (as just defined) *in principle.*

That's because they deny that science is the study of some objective reality which exists, and which has certain properties rather than others, independently of human minds. In other words, they reject *realist* accounts of science, and *objectivist* accounts of truth.

One root of this position, which was officially inaugurated by Edinburgh's Science Studies group in the early 1970s (Bloor 1973), and backed by a quasi-anthropological study of the Salk Institute soon afterwards (Latour and Woolgar 1979), is the work of Thomas Kuhn. And cognitive science, arguably, offers examples illustrating Kuhn's remark that science "progresses" because old scientists die (1962: 150). I say "arguably" partly because most of the field's leaders are still alive. But partly, too, because the new ideas resisted to the end by some of those who've died—connectionism, in the case of the late Simon and Allen Newell for instance—are still *competing theoretical positions*, rather than new *paradigms* unquestioningly accepted by everyone else (see Chapters 12.viii–ix and x.d, and 15.viii.b–c).

Another root is neo-Kantian philosophy (introduced in Chapter 2.vi), or what's often called Continental philosophy—including the version of it known as postmodernism. For the science wars are a special case of what Simon Blackburn (2005) dubs "the truth wars", which oppose relativism to realism in respect of *all* areas of knowledge/belief. He sees this opposition as "arguably the most exciting and engaging issue in the whole of philosophy" (2005, p. xx). Certainly, it's fundamental to the philosophy of mind, as well as of science: that's why I flagged it very early on, in Section i.c above. Indeed, some disputes within and about cognitive science today are grounded in this philosophical debate (7.vii.c, 14.xi.a, and 15.viii–xi).

However, neo-Kantianism is so radically different from the Cartesian–empiricist Anglo-Saxon approach that the two are, to borrow Kuhn's term, virtually incommensurable. The one prioritizes sophisticated interpretation, or hermeneutics; the other relies on scientific objectivity, as understood in the experimental tradition. Consequently, they have very different ideas on what counts as a proper scientific attitude, especially with respect to the biological sciences—and a fortiori the human sciences of psychology and anthropology.

Both philosophical positions can be defended, to be sure (see 16.vi–viii). And historical evidence can be marshalled for each of them: the case for constructivism in seventeenth-century science, for example, has been made with some spirit, and much fascinating detail, by Steven Shapin and Simon Schaffer (Shapin and Schaffer 1985; Shapin 1994). But there's no clincher argument on either side. At some point, one must opt (*sic*) for one or the other.

For my part, I regard the constructivists' position as fundamentally irrational, even though they did have some important insights—which need to be remembered if one is to understand the history of this, or any other, scientific field (see below). The case against them was put in a nutshell nearly 400 years ago by Bacon:

> After the human mind has once despaired of finding truth, everything becomes very much feebler; and the result is that they turn men aside to agreeable discussions and discourses, and a kind of ambling around things, rather than sustain them in the severe path of inquiry. (Bacon 1620: 56)

Bacon himself, of course, was largely responsible for defining what is now familiar as well-honed scientific method. So today's constructivists are even more shocking than Bacon's contemporaries, in their rejection of "the severe path of inquiry".

The irrationality of their approach has been clearly summarized by philosophers Susan Haack (1996) and Noretta Koertge (2000), and put more tendentiously—but with panache—by the sociologist Ernest Gellner (1992). A serious philosophical rebuttal can't merely ask, "If science is mere cultural convention, why would anyone ever board an aeroplane?", and leave it at that. The arguments for constructivism, and also those against it, are more subtle. But this isn't the place to recount them (for an excellent discussion, see Blackburn 2005: esp. chs. 6–7). So I'll merely say that I see relativist and anti-realist philosophies of science as both self-defeating and fundamentally implausible. The relativism undermines every philosophical claim; and the rejection of realism, despite science's practical successes, ignores what philosophers of science call IBE, or inference to the best explanation (Harman 1965; Lipton 1991).

If realism is correct, as the vast majority of practising scientists (and most readers of this book?) assume, then scientific progress—in the sense defined above—is in principle possible. And in cognitive science, it has actually happened. In one discipline, it's happened *despite* opposition from highly influential constructivists in the area concerned (Chapter 8.ii.b–c).

So where's the caveat? Well, even where there has been progress, it hasn't always happened in a 'purely scientific' way. Personal and political interests sometimes influence the generation and/or acceptance of ideas in science, as they do in the arts. This is part of what the constructivists have been pointing out.

In other words, science in practice *doesn't* fit what the physicist, and science-policy expert, John Ziman (1925–2004) has called "the Legend". The Legend is "the stereotype of science that idealizes its every aspect", representing it as wholly objective and rational (Ziman 2000*a*: 2). If that mismatch isn't recognized, one's historical view of the field will be misleading.

In Chapter 2, for example, I describe the rise of Cartesianism within science in general and neurophysiology in particular. This took place very quickly: "From 1700 on, [such ideas] were taken to 'go without saying'; and, in practice, they often went unsaid" (Toulmin 1999: 108). The rapid spread of Cartesian ideas happened partly because they supported the sorts of experiment and theory already pioneered by Galileo Galilei and William Harvey, and taken up by the nascent Royal Society in London and Académie royale des sciences in Paris. (Three cheers for the Legend!) But there were sociopolitical reasons, too.

People weary of the religious disputes and bigotry that had been brewing in Europe since the late sixteenth century, and which had recently led to the ruinous—and destabilizing—Thirty Years War, welcomed Descartes's stress on clarity and agreement (even certainty) independent of religion. This was especially true in England and France, and in 'establishment' circles (Toulmin 1999: 119–25). And, of course, the gentlemen members of the two "Royal" societies were establishment figures par excellence. So they had more than purely scientific reasons for favouring Descartes.

That's not to say that these nebulous political considerations were consciously recognized by the individual thinkers concerned—still less, that the novel ideas were *generated* because of them. Even far less nebulous political influences may be invisible to the people whose creative work is being accepted largely because of them. For example, we'll see below (in subsection c) that Cold War politics encouraged the explosion of Abstract Expressionism in post-war New York, and fostered its acceptance by art critics around the world. However, this fact has come to light only relatively recently. At the time, it was hidden *even from the artists themselves.*

Sometimes, political influences are more easily visible. So, for instance, progress in cybernetics and AI has been (and still is) largely fuelled by military aims and funding, as opposed to the disinterested pursuit of truth (Chapters 4.vi.a, 9.x.a, 11.i, and 12.vii.b). That's a large part of what Haraway was saying (Section i.a, above), although she supplemented it with much more questionable claims—about the "need" for a feminist philosophy of science, for example (for a sensible rebuttal, see Koertge 2000).

Personal interests have entered the picture too. A hugely influential Report on AI prepared for the UK's Science Research Council in 1972 would probably never have been commissioned, and for sure would have been very different, if the personality of the UK's leading AI scientist had been other than it was (see 13.iv.a–b). A disinterested document this was not.

Much the same applies to an equally influential, and apparently highly abstract, attack on connectionist AI (12.iii). The published version pulled no punches, and the draft had contained many vitriolic remarks which friends persuaded the authors to remove.

Perhaps you're surprised, even shocked? The more one buys into the Legend, the more surprised one will be. The Legend has had powerful defenders—above all, among followers of Popper (1935, 1963). Popper himself was a sophisticate: he saw his theory as an idealization (a "rational reconstruction") of science, an account of how it should be done rather than a description of how it is in fact done. (Predictably, he dismissed Kuhn as a *philosopher* of science, seeing him as a sociologist/historian instead.) But cruder versions of the Legend still abound, often written by practising scientists (e.g. L. Wolpert 1992).

In such accounts, personal prejudices and sociopolitical factors are glossed—where they're even recognized—as unwanted subjective intrusions into objective study. While they may, and perhaps should, often determine which problems will be considered by science, they're held to be irrelevant to the content of the solutions offered. James Watson (1968), for instance, admits to the personal ambitions that drove him and Francis Crick in their search for the structure of DNA, but regards these as irrelevant to what they said in their scientific papers. The claim is that theory is value-free, even if choice of research topic isn't.

But that's too quick. For one thing, choice of research topic is hugely relevant in understanding the history of an entire scientific field. For another, even the driest theories are sometimes infected by personal/political values. (So a claim about the superiority of one programming language, or programming style, over another—*Boring*!—was typically stated, during the years of the Cold War, in terms pitting liberal democracy and egalitarianism against authoritarian regimes: Chapter 10.iv.a.) As for theories carrying implications regarding the nature of human beings, the scope for personal/political issues to prejudice clear judgement is significant (see Chapter 8.ii). Third, the core point of the constructivists' case, science is influenced also by *conceptual* assumptions, which can even affect researchers' *perception* of the 'data' (see 16.iv.e).

Ziman gives countless examples (mostly drawn from physics and chemistry)—and careful arguments—showing that the Legend not only isn't true, but couldn't be true. It underplays the sociology and psychology of science, and distorts its epistemology.

However, the Legend's polar opposite (constructivism) isn't true either—as I've said, above. Indeed, Ziman wrote his book to *defend* science against the most extreme of these attacks. Besides his many anti-Legend examples, he also cites a host of cases showing how science, as a body of theory and as a social institution, works as a relatively objective, rational, enterprise—despite the frailties of its human practitioners.

Why did Ziman feel the need to do this? After all, he wasn't a professional philosopher. Can't science be left to look after itself? Well, apparently not:

> Science is under attack. People are losing confidence in its powers. Pseudo-scientific beliefs thrive. Anti-science speakers win public debates. Industrial firms misuse technology. Legislators curb experiments. Governments slash research funding. Even fellow scholars are becoming sceptical of its aims. (Ziman 2000*a*: 1)

The "fellow scholar" mentioned here was the historian Gerald Holton, a long-time friend of science whose recent work had, perhaps surprisingly, given some credit to the "anti-science" position (Holton 1992, 1993).

Ziman, too, gave constructivism some credit. He even went so far as to declare himself a postmodernist:

> Our investigation thus arrives at a paradoxical conclusion. Academic science, the spearhead of modernism, is *pre-modern* in its cultural practices: and yet it turns out to be *post-modern* in its epistemology.
>
> Contrary to the Legend, science is not a uniquely privileged way of understanding things, superior to all others. It is not based on firmer or deeper foundations than any other mode of human cognition. Scientific knowledge is not a universal "metanarrative" from which one might eventually expect to be able to deduce a reliable answer to every meaningful question about the world. It is not objective, but reflexive: the interaction between the knower and what is to be known is an essential element of the knowledge. And like any other human product, it is not value-free, but permeated with social interests. (2000*a*: 327)

He hastened to add, however, that "terms such as 'modernism' and 'post-modernism' are very ill-defined . . . Most scientists only know of them as slogans, uttered wholesale by the partisans of the most diverse fashions and fads."

That's true. Even the "partisans" use the term postmodernism in differing ways. When it was used by the French philosopher Jean-François Lyotard (1924–98) in the late 1970s, in his study of "knowledge in computerized societies", he'd attacked all overarching philosophies ("meta-narratives" or *grands récits*) of human progress. As though his little (110-page) book weren't succinct enough, he put it in a nutshell:

"Simplifying to the extreme," he said, "I define *postmodern* as incredulity towards metanarratives" (1984, p. xxiv). So he attacked Marxism and Christianity, modernist aesthetics (which trumpeted the restorative 'transcendence' of art), and scientism too. His main target, indeed, was the "legitimacy" of scientific knowledge, and its "mercantilization" owing to "the computerization of society" (1984: 3–9). In other words, postmodernism was explicitly opposed to the growing role of science and computers in the late twentieth century.

Indeed, science had already been targeted by the literary scholar Roland Barthes (1915–80). In his essay on 'The Death of the Author' (written in May 1968, during the revolutionary events on the streets of Paris), he'd rejected its 'quasi-theological' claim to objectivity:

> Literature . . . by refusing to assign a "secret", an ultimate meaning, to the text *(and to the world as text)*, liberates what may be called an anti-theological activity, an activity that is truly revolutionary since to refuse to fix meaning is in the end to refuse God and his hypostases—*reason, science, law*. (Barthes 1968/1977: 147; italics added)

Renaissance scholars, too, had seen the world as a text, to be interpreted by us. But they'd supposed it to be God's text, expressing His hidden meaning. The scientific revolution had consisted, in large part, of rejecting these hermeneutic (intentional) accounts of the world in favour of empirical ones—hence the subtitle of Bacon's 1620 book: *or, A True Guide to the Interpretation of Nature* (and cf. Glanville 1661–76). Now, four centuries later, Barthes was describing the world as a text with *no* author, whose multifarious "meanings" aren't actual messages (or scientific laws) for us to discover but possible conceptualizations for us to construct.

Lyotard himself admitted that his thesis "makes no claims of being original"—"or", he added (as a relativist must), "even true" (p. 7). It revived the neo-Kantian contrast between scientific and humanist knowledge emphasized by Wilhelm Dilthey (1833–1911) and Max Weber (1864–1920), and by many philosophers today (see Chapter 16.viii.b). Where Dilthey (1883) had spoken of empirical/scientific versus historical/hermeneutic knowledge, and Weber of *Naturwissenschaft* versus *Geisteswissenschaft* (i.e. natural science versus human, or interpretative, science: Shils and Finch 1949), Lyotard spoke of science versus narrative. (And often of discourse rather than knowledge, for the legitimization of a given type of discourse *as* knowledge was being questioned.) Moreover, he was offering a political critique (of global capitalism) as much as an exercise in 'pure' epistemology.

Lyotard's influence on political and aesthetic discourse (and on the practice of art) was immense. But science, and its computerized offshoots, didn't escape. And Barthes, even though he'd focused primarily on literary texts, engendered hostility to science—more specifically, to science's claims to objectivity—in his readers.

As postmodernism flourished in the 1980s and 1990s, becoming (to traditional minds) ever more outrageous and bizarre, some followers of Lyotard and Barthes—often encouraged by the Science Studies work mentioned above—passed from deep scepticism to genuine incredulity, *even with respect to scientists' answers to 'properly' scientific questions.* They didn't confine their doubts to the social/behavioural sciences, which many philosophers believe lie outside science's reach (see 16. viii.a–b), for even the objectivity of physics was suspect. Calls were made, for example, for a specifically feminist philosophy—and practice—of science (e.g. Harding 1986; Longino 1990). As one illustration, the feminist anthropologist Emily Martin (1991) made a bitter sociopolitical attack on theories of reproductive biology (for a sanity-restoring reply, see P. R. Gross 1998). It was as though, in these postmodernists' eyes, the metanarrative was not just scientism but science itself.

That was many steps too far for Ziman. He was an opponent of scientism—but not of science. His version of postmodernism, then, is very different from that of the virulently anti-science faction in the science wars.

In short, the Legend is false. The disinterested pursuit of truth is rarer than many choose to believe, and the conceptually innocent ditto is impossible. On those points at least, the constructivists have had important things to say—things which must be borne in mind when thinking about the history of cognitive science.

c. The counter-cultural background

The science wars were part of a more general intellectual/political movement, which the sculptor Theodore Roszak (1969) dubbed the counter-culture. This was politically prominent in the 1960s and early 1970s, and still intellectually prominent in the 1980s and 1990s—by which time it had largely metamorphosed into the postmodernist movement. On the whole, counter-culturalism favoured non-scientific, and even explicitly anti-scientific, ways of thinking. (For examples of counter-cultural anti-science diatribes, see P. R. Gross and Levitt 1994/1998; Parsons 2003.) It wasn't only the arts that were celebrated. Religions in general, and 'New Age' spirituality, flourished. Crystals were more respected as amulets than as chemicals on the laboratory bench.

Computers were even less favoured than chemicals. Then available only to very large organizations, they were seen by most 1960s–1970s counter-culturalists not as a useful tool but as a threat. And the image of mind as machine was the deepest threat of all: "Technocratic assumptions about the nature of man, society, and nature [i.e. cybernetics]", said Roszak, had "warped" the experience of scientists, scholars, and policy-makers at source. "In order to root out those distortive assumptions, *nothing less is required than the subversion of the scientific world view*" (1969: 50; italics added).

This movement had philosophical roots in the nineteenth and early twentieth centuries (Romanticism, Marxism, and Continental phenomenology). In that guise, the culture it was countering was Cartesian modernism, and especially Enlightenment optimism about the reach of science.

The philosophical arguments were reinforced/overlain by mid-century disillusion about some of the intended (e.g. Hiroshima) and unintended (e.g. pollution) effects of science. Rachel Carson's environmentalist book *The Silent Spring* (1960), and her *New Yorker* articles on the ill effects of industrial chemicals, led to heated debates in the US Congress. It wasn't only decisions about technology that were questioned: explicit attacks were mounted against taxpayers' money being given for science education and research in general (R. C. Atkinson 1999). (Science education in the USA had been hugely boosted around 1960, because of the shock presented by the Soviets' Sputnik, a football-sized satellite orbiting the earth every 90 minutes: see 11.i, preamble.)

But the anti-science campaign also drew major strength and inspiration from political events—above all, the Cold War. This dominated Western, and especially American, politics and military planning from 1947 on. (There was some lessening of tension during the 1970s, but it was ratcheted up again by Ronald Reagan in the 1980s: see Chapter 11.i.c.) So the counter-culture was driven by a growing unease about the effects of technology (from nuclear weapons to agrochemicals), and about the arms race and MAD (Mutually Assured Destruction) calculations of the Cold War.

It was further inflamed by the passions involved in the mid-1960s civil rights movement in the USA. Not least, it was fostered by discontent/disillusion about the Vietnam war—in which more US bombs were dropped on that undeveloped country "than had been dropped by all combatants in *all previous wars combined*" (P. N. Edwards 1996: 137). Huge sums of money for scientific (including AI) research aiding the Vietnam adventure were made available from the mid-1960s to early 1970s, although lack of success in both the research and the war led to a tightening in the early 1970s.

Encouraged by intellectual/political leaders such as Chomsky (9.vii.a), Herbert Marcuse (1964), and Angela Davis, young people were enthused to protest against these political forces. Student activism in the USA began in 1964 with the Free Speech Movement at Berkeley, which led to overreaction by the authorities and eventually to riots and tear gas on campus (McGill 1982). Those confrontations, and many others across the world, were fuelled by the student-led *événements* in Paris in May 1968, which are still frequently mentioned today. Even those young people who didn't join in the violence were often broadly sympathetic.

Science and technology were seen as enemy forces. Both were crucial to Cold War rhetoric, and to Cold War military preparations, whether offensive or defensive—and generous funding was provided by the US (and UK) governments accordingly. (Some of it reached AI and other areas of cognitive science: 6.iv.f and 11.i.) Moreover, that funding was clearly visible to the public, and thus provocative to those whose politics might lead them to be provoked. Roszak himself was one of those, and his *Counter-culture* book provided a host of references to the angry and/or despairing writings of many others (1969; cf. 1986).

He wasn't concerned only—or even primarily—with bombs. Scorning the "commonplace contemporary idiocies which small minds are now busily elaborating into a *Weltanschauung*", he rued "the degradation of human personality" that he believed was resulting from the use of Wiener's cybernetic metaphors for mind.

This effect, he said, was no mere philosophical bagatelle. For it was influencing military policy too: "Not even Jonathan Swift could have invented such pernicious lunacy as the balance of terror or thermonuclear civil defense" (1969: 295). The "lunacy" was rooted in scientific and technological research in general, and especially in the work of RAND, the Stanford Research Institute, "and ever so many other military–industrial–university think-tanks". (Both RAND and SRI were crucial to the rise of AI, as we'll see in Chapters 10.ii.a and 11.i.)

Sciences having no direct military relevance were despised too, especially if they were seen to imply a non-humane image of mankind. Even at the outset of the Cold War, social science in general, and behaviourism in particular, was already being lambasted in the popular press. For example, an editorial in *Life* magazine commented viciously on Burrhus Skinner's utopian (or dystopian?) novel *Walden Two* (1948). Its author, the cultural commentator John K. Jessup, also reviewed the book for *Fortune*, where he said:

> If social scientists share Professor Skinner's values—and many of them do—they can change the nature of Western civilization more disastrously than the nuclear physicists and biochemists combined. (quoted in Skinner 1979: 369)

In the counter-culture's opinion, then, the sciences were deeply compromised (and technology, even more so). The arts, and religion, were seen as the saviours of civilization. For in this neo-Romantic version of the *verum factum* tradition (i.b above), the creative arts were both truly free and essentially intelligible. In particular, they were untouched by the contaminating fingers of the military–industrial complex.

That's a historical irony, since Cold War influences on the arts in the USA were in fact even deeper, if less costly, than those on science. But they were much less provocative, because apart from Senator Joseph McCarthy's early 1950s banning of "subversive" authors from government-funded libraries (and his committee's protests at a US drama group's staging "two productions of some Russian guy called Anton Chekhov"—Caute 2003: 62), they were deliberately—and for many years, successfully—hidden.

Or rather, they were hidden in the USA. In the Soviet Union, comparable pressures on artistic style and content were clearly explicit. The dramatist Konstantin Simonov declared that "in defense of Communism, most sacred of all things, all powers must be employed, including art" (Caute 2003: 108). This was official policy. Lenin himself had banned Russian modernism (e.g. Marc Chagall and Vasily Kandinsky) in the early 1920s, insisting on "revolutionary realism" instead; by mid-century, painters had to join the Union of Soviet Artists, whose rules prescribed adherence to socialist realism (Caute 2003: 510, 519). And in 1947 the Director of the Academy of Arts had announced: "In educating young artists we must make it absolutely clear that penetration of the walls of Soviet art schools by this or that decadent influence from the capitalist West is absolutely out of the question" (Caute 2003: 514). He meant modernist painting, from Paul Cézanne (and even the Impressionists) onwards—and especially Abstract Expressionism, which was making its mark in New York at that time.

The rapid rise of Abstract Expressionism in the 1950s, and the displacement of Paris by New York as the 'hub' of modern art, weren't due to an excess of creative genius in the studio lofts of SoHo. To be sure, Mark Rothko and Jackson Pollock painted as they did (and Roszak sculpted as he did), even before the Cold War started, for their own reasons: aesthetic, not political. (Similarly, scientists developed their theories for their own reasons: scientific, not political—although some of the *applications* were specifically military.) These artists were drawn to abstraction, and to individual expression, by pressures internal to art and their own psychology. But their huge public success wasn't due only to aesthetic values, nor even to enthusiastic art critics/collectors like the notorious Clement Greenberg. It was hugely encouraged by US government investment. For the politicians, it wasn't art for art's sake but art for America's sake.

The government's interest wasn't mere cultural imperialism, a wish to be top dog in the ateliers of the world. Admittedly, Americans would take pride in the fact that painters were increasingly looking to New York, not Paris, for inspiration. The art critic Robert Hughes says, "It would be foolish to claim that 1945–70 in New York rivaled 1870–1914 in Paris. America has never produced an artist to rival Picasso or Matisse, or an art movement with the immense resonance of Cubism" (Hughes 1991: 3–4). Nevertheless, he remembers:

> In the early 1960s, when I was a baby critic in Australia, it seemed that faraway New York had become a truly imperial culture, heir to Rome and Paris, setting the norms of discourse for the rest of the world's art. . . . One saw this triumph from afar. . . . In Australia one's response to it came out as a sigh—resignation to one's own cultural irrelevance. (Hughes 1991: 3, 4)

> Thirty years ago, Abstract Expressionism was pretty well a mandatory world style. We in Australia looked at it with awe. . . .
>
> This act of unwonted humility was made by thousands of people concerned with the making, distribution, teaching, and judgment of art, not only in places like Australia but throughout Europe and—not incidentally—in America in the mid-1960s. They resigned themselves to an imperial situation. (pp. 5–6)

> When Americans in the fifties and sixties eagerly claimed that their art had superseded that of Europe, their eagerness itself was a period phenomenon. . . . The idea that Europe was culturally exhausted was an important ingredient of American self-esteem. (p. 7)

However, there was more to it than that. Specifically, there was also a Cold War dimension. If Rothko and Pollock had painted tractors, or even meticulous still lifes, the government money wouldn't have been forthcoming. For there were two unspoken political messages. On the one hand, there was a clear distinction and a choice: Abstract Expressionism was the aesthetic opposite of the Realist representational art favoured in—or allowed by—the Soviet Union. On the other hand, the American artists' freedom to shock, to counter accepted artistic canons and to express their individuality in doing so, was a visible sign of the freedom generally available in the West.

But such messages, to be truly effective, had to be unspoken. If it was only the Soviet Union who directed their artists, individual thought in the West being no affair of government, then US government support had to be invisible. (A second reason was that modernist art wasn't popular with the US taxpayer. An ex-CIA man later admitted that "It had to be covert because it would have been turned down if it had been put to a vote in a democracy"—Caute 2000: 550.)

Accordingly, much of the investment came via supposedly independent bodies, such as the Museum of Modern Art (MOMA) and the National Endowment for the Arts (NEA), founded in 1965 (Guilbaut 1983). These institutions had access to Rockefeller largesse. (Governor Nelson Rockefeller, who had been supporting modernist painting since the mid-1940s, had set the ball rolling in 1960 by initiating the New York State Council on the Arts, having failed to persuade Congress to sponsor a national body.) They not only bought the New York School's canvases, but also sponsored Expressionist exhibitions across the USA and abroad.

Europe was the prime target, because of its proximity to (and sympathies for) communism. But non-aligned India and Japan were targeted too. For example, a huge exhibition of American art was sent to New Delhi in 1967, and a two-day seminar, led by Greenberg, was held in the hope of encouraging young Indian artists to study in New York rather than Paris—or, worse, Moscow (D. Guthrie, personal communication). The intention was to impress a political moral on an international audience, by contrasting the rigid social realism of Soviet painting with the liberated artistic expression of 'the Free World'. (After the Cold War was declared 'won' in the early 1990s, NEA funding for the visual arts plummeted accordingly: D. Guthrie, personal communication.)

Analogous activities went on in the literary and musical worlds and in film, theatre, and dance too (F. S. Saunders 1999; Caute 2003, pts. II–IV). Large sums of money were silently channelled through supposedly neutral bodies such as the Rockefeller and Ford Foundations, and the Congress for Cultural Freedom—which was set up for these specific purposes. The National Endowment for the Humanities (NEH) supported work by writers whose left-leaning politics would have led to their being spurned in the McCarthyite period a decade earlier.

(Over the years, the NEA and NEH were hugely beneficial to the arts in America. A recent NEA chairman reported that, since 1965, the number of state-supported arts agencies had grown from 5 to 56, and local ones from 400 to 4,000; non-profit theatres had burgeoned from 56 to 340; symphony orchestras had grown from 980 to 1,800, opera companies from 27 to 113, and dance companies had multiplied eighteen times—Ivey 2000: 3.)

Where literature was concerned, spreading the message internationally was a problem. The NEH couldn't send exhibitions of writing around a multilingual world. Nor, as a "National" body, could they support foreign writers directly. So the CIA stepped in: the transatlantic literary journal *Encounter*, and several 'progressive' European magazines including Germany's *Der Monat* and France's *Preuves*, were funded in large part by the CIA. CIA money ($750,000) was used also for the New York Metropolitan Opera's tour of Europe in 1956, and the Agency sponsored the Boston Symphony Orchestra's 1956 European visit.

What's relevant for our purposes here is that the covert political agenda in the arts wasn't fully uncovered until the mid-1980s. Its discovery caused outrage, among the people who had long been suspicious only of science.

To be sure, *Encounter*'s political stance had already been questioned in the 1960s, largely because it "dealt gently, if at all, with topics such as race and Vietnam" (Collini 2004: 10). Conor Cruise O'Brien, who then had no inkling of the financial arrangements, commented in 1963 that the editorial policy seemed consistently designed to support the US government. As he said later, when the truth had come out: "the beauty of the operation... was that the writers of the first rank, who had no interest in serving the power structure, were induced to do so—unwittingly" (quoted in Collini 2004: 10). The magazine's funding-cover was conclusively blown in 1967, leading the English co-editor Stephen Spender, who'd been innocent of the CIA involvement, to resign (J. Sutherland 2004). But the *visual* arts remained relatively untouched by such rumours.

Only "relatively": in the early 1970s, Max Kozloff (1973) highlighted the political implications of the sudden success of the New York school, and the American muralist Eva Cockcroft (1974) had some even more pungent things to say—linking MOMA, the CIA, and the Rockefellers. But the cat was well and truly let out of the bag in the 1980s by the French art historian Serge Guilbaut, in his fascinating book *How New York Stole the Idea of Modern Art* (1983). The major scandal arose from his July 1984 article of the same title in the art magazine *Commentary*, and from back-up articles by others, including Cockcroft, in the August 1986 number (several are reprinted in Frascina 2000). Since then, evidence of comparable—indeed, cooperative—cultural/intellectual meddling by Britain's intelligence services has come to light (Dorril 2000).

Given that the counter-culturalists of the 1960s and 1970s didn't realize that art was (unknowingly) acting in the service of capitalism, they couldn't be enraged by this. Their political animus was reserved, rather, for *science*. And after all, it was the scientists and technologists, not the artists, who were more directly involved in the governmental war machine.

The history of cognitive science was coloured as a result. It wasn't to be all hostility, however. As we'll now see, late-century theoretical shifts within the field led the attitude of the counter-culture to change: from being firmly against cognitive science, to being (to some extent) in favour of it.

d. The counter-cultural somersault

Even an anti-science movement can have preferences *within* science. For example, the mentalistic aspect of the cognitive revolution was more attractive to the early counter-culturalists than behaviourism was. However, computers—given their place in military–industrial technology, and in the growth of global communication—were a special focus of suspicion. (We've already seen, for instance, that Lyotard's key text was explicitly aimed at "knowledge in computerized societies".) In particular, mind-as-machine was anathema (Roszak 1986).

That's an over-simplification. For instance, it doesn't apply to one of the most influential members of the counter-culture, the biologist–artist Stewart Brand (1938–). Brand's first *Whole Earth Catalog* (1968), a compendium of "tools" for environmentally friendly living, inspired a host of back-to-nature projects worldwide. (The 1972 edition sold over one and a half million copies, and won a US National Book Award.) He had criticisms of technology aplenty, but he wasn't an enemy of technology as such. Far from it. He'd helped Douglas Engelbart to demonstrate the first computer mouse at a now-famous meeting in San Francisco (see Chapter 10.i.h). And he shared Engelbart's faith that personal computers, and IT in general, could help towards more ecologically viable lifestyles. Computer software was featured in the first *Catalog*, and increasingly in the following ones. So one of the early heroes of the counter-culture was also a computer enthusiast and computer visionary.

Moreover, a few brave artists in the mid-1960s were beginning to experiment with computer art (see Chapter 13.vi.c). Indeed, a major international exhibition (the first) was held in London in 1968 (Reichardt 1968). Many of the visitors were excited—though not the art critic of *The Guardian*, who described it as a "frivolous activity" (A. Sutcliffe, personal communication). But it was the relative *weakness* of the counter-culture in late 1960s Britain that had enabled the show to go ahead. One of the artists involved recalls that it "could at this time not have taken place in Paris. The revolutionary students would have swept it away" (Nake 2005: 59). The organizer herself has said: "The same venture in Paris would have needed police protection" (quoted in Klutsch 2005: 109). And a historian has commented:

> Could it be that the ICA's "happy accidents" flourished so well because they were staged in an atmosphere of breathtaking *naïveté*? Only a few lone voices seem to acknowledge the more serious and inevitably unhappy accidents that litter the history of cybernetics.... [In Great

Britain] the subversive momentum of 1968 never unfurled in the same way, with the same force, as it did in continental Europe or the United States. . . . Against this backdrop, [the London exhibition] offered a light-hearted view of the modern world without raising too many (if any) objections or stirring fears. (Usselmann 2003: 391 ff.)

AI/cybernetics, by contrast, did stir fears—and was a prime target for criticism accordingly. From the late 1970s on, it suffered high-profile counter-cultural attacks by Dreyfus and Joseph Weizenbaum (11.ii). And cognitive science as a whole, which has AI at its intellectual core, was targeted too (Roszak 1986).

Chomsky's specifically political attacks on orthodox social scientists, and on their policy advice to the US government, were partly responsible: after all, he was highly respected *as a cognitive scientist.* But since the field had unfashionable—and apparently threatening—things to say about human beings and human nature, it would have been attacked by the counter-culture even without him. And, significantly, it wasn't only the intellectuals who disapproved of AI and cognitive science. These powerful cultural forces led US politicians to approve a temporary drop in military funding for AI, and a skewing towards civilian applications (see 11.i.b).

The component disciplines were each affected. So in the philosophy of mind and of psychology, neo-Kantianism (including the newly published Kuhn) soared in popularity in the UK and USA. Cognitive anthropology was nipped in the bud in the early 1970s, and all but destroyed by the 'literary' turn in the discipline (8.ii.b–c). At much the same time, many social psychologists reacted strongly against experimental, computational, and information-theoretic approaches (6.i.d). A feisty best-seller written by a well-known professional psychologist in Great Britain was called *The Cult of the Fact*, a title that says it all (Hudson 1972). One leading cognitive scientist was so worried by these professional developments that he wrote a book, and organized a high-visibility conference, to protest against them (see 6.i.d).

In short, cognitive science in its early years faced hostility from counter-cultural critics. In its later years, however, they *welcomed* certain intellectual changes that took place within the field.

The change of heart was initiated by the public availability of the personal computers that had been dreamt of by Brand and Engelbart, and by the development of the Internet, which allowed for what Haraway called "network identities" (cf. Chapter 13.v.d and vi.e). But this intellectual somersault was soon reinforced by specific theoretical aspects of late-century cognitive science.

For instance:

* Computational psychology highlighted situatedness, embodiment, and epigenesis (Chapters 7.iv.g, v.b–f, and 15.vii–viii).
* In the late 1970s, the AI researcher Terry Winograd left MIT spiritually as well as geographically to join Fernando Flores and Hubert Dreyfus in California (9.xi.b), and
* Minsky and Seymour Papert started drafting their theory of the decentralized "society" of mind (12.iii.d).
* Dennett, following Minsky, described the "self" not as a unitary thing but as a laboriously constructed—and not fully coherent—"narrative" schema that helps to guide our choices (7.i.e and 14.xi.b).

* In the 1980s Rodney Brooks and Randall Beer rejected GOFAI in favour of situated robotics (13.iii.b, 15.vii and viii.a).
* GOFAI researchers started studying "distributed cognition", wherein coherent behaviour emerged from the action of many separate "agents"—with no central controller (13.iii.d).
* Similarly, PDP connectionism flourished, and seduced the public at large—and even some postmodernist philosophers (Globus 1992; Canfield 1993; E. A. Wilson 1998)—largely because of its decentralized approach (12.vi and x).
* Much the same happened when A-Life hit the scene, offering what some saw as a near-magical alternative to traditional cognitive science (15.x.a, and S. R. L. Clark 1995).
* The magic was seemingly underlined by the emergence of unexpected properties through computerized evolution (15.vi).
* Some anthropologists used the ideas of situated action and distributed cognition to analyse the behaviour of human groups (8.iii).
* Brian Cantwell Smith started work on a radically new, and admittedly highly eccentric, "participatory" account of the nature of computation as such (16.ix.e).
* The availability of personal computers encouraged a wide range of experiments in computer art, including interactive and/or evolutionary art (13.vi.c).
* And, by the end of the century, the technology of 'virtual reality' enabled people to experiment with the presentation, and perhaps even the construction, of *self* in very 'non-Cartesian' ways (13.vi. d–e).

These ideas fitted well with certain aspects of the late-century counter-culture. For example, postmodernists in literary and aesthetic circles—who'd already proclaimed "the death of the author" (Barthes 1977)—welcomed the implications of non-hierarchical control, and of user-directed computer technologies such as hypertext and interactive art (Chapters 10.i.h and 13.v.d and vi.c).

Alongside their undermining of the authorial signature had been their undermining of the unitary self. So, likewise, they had some sympathy with theoretical work that presented the mind and/or self as a virtual machine, consisting of many interacting agents as opposed to a unitary Cartesian centre (Chapters 7.i.e–f, 12.iii.d, and 16.iv.a–b). By the same token, they welcomed the playfulness in self-presentation that was made possible by the Internet (13.vi.e).

Connectionism was explicitly favoured by some postmodernist writers. Several claimed that Jacques Derrida's notions of deconstruction and *différance* were largely homologous with PDP ideas, even suggesting that his approach was scientifically authorized by them (Globus 1992; Canfield 1993). The feminist philosopher Elizabeth Wilson rejected that last paradoxical claim but she, too, dressed connectionism in deconstructionist clothes (1998, esp. 14, 24–30, 196 ff.). She saw PDP as providing an opportunity "not merely to rethink cognition, but also to rethink our [i.e. counter-culturalists'] reflexive [self-aware, not automatic] critical [i.e. postmodernist] recoil from neurological theories of the psyche" (Wilson 1998: 14). Connectionism couldn't solve the philosophical/political questions she and like-minded colleagues were interested in—but, despite its provenance in (normally suspect) biology and cybernetics, it merited their attention:

> Can we [i.e. postmodernists, and feminists in particular] think the subtlety of neurology and cognition on their own terms? Can we read the internal machinations of traditional empiricism in ways that do not return us to the routinized accusations [from the counter-culture, against biological science] of essentialism, reductionism, and political stasis? Specifically, does connectionism offer a political reading of psyche, cognition, and biology not despite its neurocomputational inclinations, but *because* of them? (Wilson 1998: 14)

Her answer was *Yes!* Connectionism, she argued (and she might have added "epigenesis"), offered feminists a way to accept scientific findings about the body without being trapped in a simple-minded biological determinism. On the contrary, it attributed "a fundamental mobility" to the mind/brain (p. 203).

A-Life, in particular, drew interest from some previously hostile sources. The feminist Sarah Kember approvingly described A-Life as "a discipline which developed precisely at the end of the cold war and which rejected the militarist top-down command and control and the masculinist instrumental principles of AI" (Kember 2003, p. vii). Explicitly contrasting cultural practices and self-images influenced by A-Life with "the previous race of cold-war cyborgs" (i.e. GOFAI-based models), she declared: "Posthuman identity, informed by the discourses of artificial life, centres symbolically on the humanization of HAL..." (p. 116)—where the emotionless HAL of *2001: A Space Odyssey* was said to exemplify the "failure" of the AI project. (Whether AI, or even GOFAI, has indeed failed is another question: see 13.vii.b.)

Kember also said that A-Life is "a cultural discourse [describing] posthuman life", where

> The posthuman is cyborgian in the sense of its enmeshment, at all levels of materiality and metaphor, with information, communication and biotechnologies and with other non-human actors. (Kember 2003, p. vii)

The pervasive "information" and "communication" she had in mind included telecommunications in general, but especially the Internet. And the "other non-human actors" ranged from semi-autonomous software agents (13.iii.d), through robot surgeons and automated mechanics (13.vi.b), to computerized companions (13.vi.d) and humans-as-avatars (13.vi.e).

But Kember wasn't attacking this millennial form of man–machine identity. Unlike her predecessor Haraway, whose critique of GOFAI-based "cyborg" culture had *predated* the rise of A-Life, she didn't see the post-human cyborg as a largely destructive "product of cold war AI". On the contrary, she saw the cultural legacy of situated AI and A-Life as liberating, in its stress on autonomy and emergent organization.

She even believed that it offered "the resolution of the science wars", because it enabled one "to become independent of the distinction between nature and culture which forms [their] 'epistem-onto-logical' ground" (p. 216). On that view, the battles raging around the Legend were fated not to be won or lost, but to die away. (They haven't died yet: see Blackburn 2005.)

Philosophically, the flight from centralization and the abstract was supported not only by feminism but also by the phenomenologists' notion of "situated" intelligence and "embodiment" (16.vii.a). However, it seeped into the intellectual air being breathed by people who'd never read a word of phenomenology—and who had scant sympathy for feminism.

Even US army generals were affected, despite their scorn for the explicit pronouncements of the counter-culture (P. N. Edwards 1996: 72, 111). They protested against the centralization brought about by computer-based approaches to military matters, as a result of which officers on the ground were, literally, losing control. Their strategic, and sometimes tactical, decisions were pre-empted by game-theoretic simulations, and their logistic decisions taken over by cost–benefit analyses devised by statisticians. Besides threatening their status and self-esteem, this Pentagon-driven trend favoured formal theory over hands-on application. That is, it substituted often unrealistic abstractions for discriminating responses to specific situations in the real world. (What's more, official reports of what actually happened in the Vietnam war were hugely unreliable as a result: P. N. Edwards 1996: 137–40.)

As for the counter-culture's attitude to computer technology, that somersaulted too. Roszak's complaints about "distortive assumptions" had been explicitly aimed at Norbert Wiener, despite the cyberneticist's attempts to humanize his approach (Wiener 1950). And with the advance of computing—and digital computers—in the late 1960s and 1970s, which depended crucially on military funding (11.i), counter-culturalists had become even more disaffected. By the early 1980s, however, these cultural doomsayers were being explicitly countered by people whose views vied with them for popularity.

Alvin Toffler, for instance, published a widely serialized best-seller, which achieved twelve printings and twelve translations within two years, that challenged "the chic pessimism that is so prevalent today" (1980: 2). Remarking that "Despair—salable and self-indulgent—has dominated the culture for a decade or more", he argued at length that this attitude was "unwarranted". The reason for optimism, he said, lay largely in the computer-based technologies, including new forms of communication, to which Roszak had been so hostile.

As we've seen, certain applications of personal computers were viewed by late 1980s counter-culturalists as philosophically liberating. The terminology of "emergence", and/or "life", made New Age souls—and journalists—even more receptive (15.x.a). Visual, performance, and installation artists, for instance, were quick to respond (Whitelaw 2004).

Granted, computers remained something of an embarrassment—and they still are. One art critic recently defended his approval of interactive art like this:

> [These new aesthetic theories] propose personal and social growth through technically mediated, collaborative interaction. They can be interpreted as aesthetic models for reordering cultural values and recreating the world. As much as these theories depend on the same technologies that support global capitalism, they stand in stark contrast to the profit-motivated logic that increasingly transforms the complexion of social relations and cultural identity into a mirror-reflection of base economic principles. (E. A. Shanker 2003: 6)

In short, the computer hasn't been all-conquering. Theories resting on computer technologies still arouse philosophical/sociopolitical suspicion in certain quarters.

So far, we've focused on how the counter-culture responded to cognitive science. But what about influences in the opposite direction?

The theoretical developments mentioned above weren't primarily caused by the sociopolitical values of the counter-culture. Indeed, the intellectual seeds had been sown long before its rise, and nurtured out of public view for twenty years (4.viii, 12.i–ii and iv–v, and 15.iii–vi). However, some of the relevant concepts may have occurred to the scientists concerned partly as a result of it. For new ideas are often hugely overdetermined, in the sense that *many* influences and associations play a part in their generation (Boden 1990*a*: 186–98, 244–8).

For example, consider what Stephen Toulmin has called the postmodernist "revaluation of the concrete", and the closely associated distaste for top-down hierarchical control—even in hard-headed business management (Toulmin 1999, ch. 5). This was reflected in a number of ways within cognitive science:

* by connectionist work on distributed control and bottom-up emergence in anthropology, AI, and psychology (8.iii and 12);
* by GOFAI work on 'agents' (13.iii.d–e), interactive interfaces (13.v–vi), and even AI programming languages (10.vi.b);
* by A-Life and situated robotics (13.iii.b–d, 15.vii and viii.a–b);
* and by some aspects of philosophy (12.x).
* It was explicitly endorsed by Papert, in relation to various late-century trends in psychology (Turkle and Papert 1990).
* Arguably, it was even indicated by the new willingness of laboratory neurophysiologists to hobnob with practising physicians (see 14.i.a).

Irrespective of whether these late-century scientific ideas were part-caused by sociopolitical influences, such influences helped determine whether they met a receptive audience. They were more readily accepted, even *within* the field, because of this particular Zeitgeist.

In intellectual history in general, remarks of that type are common. Indeed, the pioneering, and punctilious, historian of psychology Edwin Boring (1957) often cited the Zeitgeist in his work. On the very first page of his *magnum opus*, he said:

> Discovery and its acceptance are [limited] by the habits of thought that pertain to the culture of any region and period, that is to say, by the *Zeitgeist*: an idea too strange or preposterous to be thought in one period of western civilization may be readily accepted as true only a century or two later. Slow change is the rule—at least for the basic ideas. On the other hand, the more superficial fashions as to what is important, what is worth doing and talking about, change much more rapidly . . . (Boring 1957: 3)

Later chapters featured the Zeitgeist too (his index lists a dozen entries, some pointing to several pages).

Boring is sometimes mocked as a result, by critics who feel that he wheeled in this Hegelian notion almost as an extra character in his script. (For instance, he spoke of individual scientists becoming "the means by which the *Zeitgeist* prevails": p. 23.) Perhaps he was tempted by the animistic flavour of the term, which literally means "the spirit of the times", concepts/assumptions that inform virtually all aspects of a given culture. But whichever word one chooses to use, the point is that—as a pervasive intellectual background, informing virtually all areas of life—the Zeitgeist is a real phenomenon. In the second half of the twentieth century, then, counter-cultural ideas and values had a significant effect on the history of cognitive science.

In sum, although my narrative of cognitive science is predominantly 'internalist', dealing with the data and theoretical ideas as such, it's partly 'externalist' too. We'll see repeatedly that social and personal factors played a role within the scientific community (and a fortiori in the public media: P. N. Edwards 1996). Sometimes, these influences were explicitly acknowledged by the scientists concerned (e.g. Resnick 1994: 6–19). More often, they were implicit—betrayed by the choice of theoretical terminology and/or illustrative metaphor (e.g. AI workers' descriptions of heterarchical programming: see 10.iv.a). But in either case, they were there. Since the Legend—though highly attractive to many scientists—is false, that's only to be expected.

e. Hardly hero worship

The fourth warning concerns the Romantic myth of the creative prodigy. The thinkers I've chosen to discuss weren't intellectual heroes solely responsible for the ideas I attribute to them. Sometimes, it's even doubtful whether they *could* have had the idea independently. Herbert Simon, for instance, has said as much about the three-man origin of what's often (wrongly: 10.i.b) called the first AI program:

> We [three] were in closest communication during the whole period, through long association had developed an extraordinary capacity to communicate even our subtleties to each other, and *the whole product must be regarded as joint and inseparable. I am firmly convinced that none of us alone had much chance of accomplishing [it]*. (quoted in McCorduck 1979: 139; italics added)

Quite apart from other *individuals*, the writers discussed in this narrative were influenced and/or encouraged by the social context at the time. Indeed, one historian of science has said that the "great man" view of history might well be replaced by a "great opportunities" view, "with the emphasis on the socially given possibilities rather than on the people who exploit them" (Fleck 1982: 217). In AI, for instance, socially driven changes in funding policies have offered encouragement and discouragement alike (Chapters 11.i and v.b, and 12.iii.e and vii.b).

Occasionally, of course, people's self-serving rhetoric suggests the contrary. I've not come across any example so extreme as James Watson's *The Double Helix*, whose unfairness to several other DNA pioneers, including Rosalind Franklin, led Harvard University Press to drop its plans to publish it (Maddox 2002: 312). But a few famous names in cognitive science are guilty to a lesser degree (see 14.v.d). Certainly, some individuals were exceptionally fertile thinkers. Turing, McCulloch, von Neumann, Marr, and Simon are examples—which is why each of them features at length in *several* chapters. But even geniuses aren't lone geniuses.

An important idea is rarely, if ever, due to only one mind. Indeed, it often arises near-simultaneously in several. Given that creativity involves either novel associations of familiar (and shared) ideas, or the exploration and transformation of structured conceptual spaces acquired from one's culture, this is only to be expected (Boden 1990*a*, 1994*b*).

It's made even more likely by the fact that science has been an increasingly communal enterprise since the mid-seventeenth century (2.ii.b–c). Today, that's often flagged by multi-authorship: most of the scientific papers cited in the References have more than one author (one lists twenty-five). But even the most prolific credit listings are likely to omit the names of people who were, in fact, part of the communicative network that made the discovery possible.

One of the most important techniques in connectionist AI, for instance—namely, back propagation—was independently discovered by at least four people, and prefigured by several others (12.vi.d). And the group who actually got the credit for it were just that: a *group*, who collaborated for several years to improve one member's initial idea before publishing it.

Sometimes, this has bizarre, even comic, consequences. The Nobel Prize committee, when considering Ivan Pavlov in 1903, were much discomfited by his frequent declarations in his *Lectures on the Work of the Main Digestive Glands* (1897/1902) that his discovery of the conditioned reflex was "the deed of the entire laboratory". It's clear from the discussions recorded in the Nobel archives (Todes 2001) that Pavlov's modesty nearly prevented his winning the prize.

But perhaps "modesty" is the wrong word here? For Pavlov was (rightly) proud of his role as a pioneering—and visionary—laboratory manager. The more that scientific research depends on complex equipment and/or multidisciplinary cooperation, the more important this role becomes. Indeed, J. Robert Oppenheimer became world-famous as the scientific manager of the Manhattan Project, not as its leading researcher. Of course, he had enough scientific knowledge and imagination to have a good nose for new ideas—even (when heading the Princeton Institute after the war) some in the nascent cognitive psychology (Bruner 1983: 96, 121). So did Sydney Brenner. When Brenner was director of the Molecular Biology Unit in Cambridge, he offered an unused cubbyhole to the young Marr—who had scant interest in molecular biology, but was exploring highly abstract ideas about the brain (see 14.v.b).

In that particular case, there's no ambiguity: Brenner provided the space, not the intellectual content. In general, however, assigning individual responsibility for creative ideas isn't straightforward. And the more that people are interacting, the more this is true.

(Discovery and invention are in much the same boat, so far as group influences are concerned. Seymour Cray, the charismatic inventor of the supercomputer, couldn't have designed 'his' machine without a rich network of technical and social relationships—even including his child's willingness to accept help with her algebra from invisible "elves", working their magic long after her bedtime: MacKenzie and Elzen 1991.)

This fact lies behind the many horror stories about supervisors stealing their research students' ideas. No names, no pack-drill—except to pay tribute to the psychologist Edward Tolman. He opened his major book by thanking his research students not merely for doing most of the experimental work, but for having many of the theoretical ideas—"which ideas I have often no doubt quite shamelessly appropriated as if they were my own" (Tolman 1932, preface). To be sure, they might not have had them without Tolman's prompting. And they might not have been able to develop them as coherently as he did. Nevertheless, the ideas recounted in Tolman's book weren't just *Tolman's*. His own view was that "If it had not been for those students, this book could not have been written."

Finally, people may be unable to recall just who said what to whom—and even just who wrote what down. For instance, George Mandler was named in two different reports of a 1959 conference on language learning as the author of a provocative "manifesto" attacking both behaviourist and early-cognitive approaches (Mandler 2002*a*: 348, 350). (Specifically, it criticized the "glib invocation of 'schemas,' structures,' and 'organization'" and the "mere postulation" of new mechanisms and processes.) Mandler now—over forty years later—admits to being *one of three* initiators, and "probably" one of three authors. However, another member of the trio says that he himself didn't take part in the actual writing (which took place one evening), even though he'd contributed some of the ideas.

This isn't a case, for once, of an unpleasant priority dispute: the person who was given the public credit is loath to take it undeserved. Rather, the people concerned simply can't remember.

And what if they could? Even then, their memories couldn't necessarily be taken at face value. For as Mandler points out (2002*a*: 348), memory is construction, not recall (see Chapter 5.ii.b). If—as is usual—they were motivated to *claim* priority, instead of shrugging it off, the constructive process might have been biased accordingly. The same applies, of course, to all the other personal recollections quoted in this history.

f. Discovering discoveries

The next three warnings concern judgements of originality—of ideas, as well as people. The attribution of an original idea to one person or group rather than another, and even its recognition as important, or as a 'discovery', are complex social processes (Brannigan 1981). These judgements aren't subject to cut and dried criteria. They involve social negotiation and rivalry, as well as historical enquiry and theoretical argument. Let's consider *discovery* first.

Discovery is a highly loaded term (Sturm and Gigerenzer 2006). An idea deemed by some people as a discovery may not be so regarded by others—even by those whom one might expect to appreciate it.

A famous example is Charles Darwin's paper on natural selection, which—alongside one by Alfred Wallace on the same topic—was read at the Linnaean Society on 1 July 1858. (Whether this counts as 'simultaneous' discovery is doubtful: Darwin had been working on this idea for many years before Wallace came up with it; however, both had been inspired by Thomas Malthus's *Essay on the Principles of Population.*) This first public presentation of the theory of evolution failed to impress the Linnaean's president, Thomas Bell. His Presidential Address some ten months later declared:

> The year which has passed has not, indeed, been marked by any of those striking discoveries which at once revolutionize, so to speak, the department of science on which they bear.

Only those little words "at once" can save Bell, today, from ridicule. As Janet Browne (2002: 42) has remarked, his verdict, though "accurate enough in the short term", has become known as "one of the most unfortunate misjudgments in the history of science".

Some cognitive scientists—though by no means all—would say much the same of Minsky and Papert's unrelenting dismissal of connectionist AI (Chapter 12.iii). And, almost exactly 100 years after Bell's misjudgement in London, a similar blindness occurred at MIT (C. G. Gross 2002: 85). In 1959 Jerome Lettvin, the lead author of what would become perhaps the most famous paper in neuroscience (see 14.iii.a), wasn't invited to a two-week local seminar closely related to his (MIT-based) research. In the event, a visitor from England (Horace Barlow) arranged an invitation for him, and a second—much less famous—paper by Lettvin's team was tagged on at the end of the official Proceedings (Lettvin *et al.* 1961).

Sometimes, someone makes a discovery they don't count as a discovery—not because they don't realize its interest (although this happens too), but because they can't explain it. Lettvin still hasn't published a remarkable finding of the late 1950s, concerning a discrimination made by the frog's retina which seems to be far too complex for a retina to make (see 14.iv.a). Because it's a mystery in theoretical terms, Lettvin has never reported it officially. In his eyes, then, he's noticed something but discovered nothing.

In other cases, an idea is used to great effect in a specific context, but without anyone's recognizing its wider implications. One example is the Watt governor (Chapter 4.v.a). This was copied in countless machines of the steam age, but it was seen as a mechanical trick not a theoretical principle. Its importance as an example of a general type of control mechanism wasn't recognized for ninety years—and even so, it wasn't fully appreciated for another seventy. Indeed, James Watt wasn't the first to use such a mechanism: it's been found in some ancient Greek automata and fourteenth-century clocks. One might say, then, that feedback was *invented* long before it was *discovered*.

Some judgements about what counts as a discovery are grounded in explicitly heroic assumptions. So people may say: "If so-and-so thought of it, it must be important", or "If so-and-so was involved, he must have been the leader" (very rarely "she", of course). Sometimes they don't even realize that they're doing this: the name of the Nobel prizewinner Simon was often wrongly put first in references to papers co-authored by his younger colleague Newell (see Chapter 6.iii.b).

By contrast, some individuals are treated as anti-heroes. One example is the French mathematician Louis de Branges, recently described by a reviewer as personally "cranky", but "not a crank" (Sabbagh 2004). His latest work, in which he claims to have solved a famous mathematical problem (the Riemann hypothesis), is being systematically ignored by his peers, and science journalists are being told not to bother with it—by people who themselves haven't actually read it. (Apparently, it is fiendishly difficult even for professional mathematicians, and would require "a team" of experts working for at least six months.) De Branges is known to have done fine work in the past, but he's very unpopular (there's some suggestion that he may have Asperger's syndrome: see Chapter 7.vi.f). In short:

> It may be that a possible solution of one of the most important problems in mathematics is never investigated because no one likes the solution's author . . . The entire mathematical profession is turning its back on what could be the most important development in the last hundred years of mathematics. (Sabbagh 2004)

I can't think of such an extreme case in cognitive science. But it's certainly true that personal animosities often hinder, and sometimes even prevent, proper consideration of new ideas. For example, AI people in the late 1960s failed to take proper account of Dreyfus's criticisms, because they resented his savage attack on them—not to mention his technical ignorance (11.ii.b).

Such heroic/anti-heroic assumptions are often buttressed by social snobberies of various kinds. These include the superior trust accorded to the word of a "gentleman" (Shapin 1994), which was crucial to the emergence of scientific communities in the seventeenth century (Chapter 2.ii.b).

On the negative side, they include suspicion of uneducated 'country bumpkins' (such as the champion of the neurone theory, Santiago Ramón y Cajal: 2.viii.c), and systematic undervaluing of the contribution of technicians (Shapin 1989; Schaffer 1994). For example, Richard Gregory's influential work on visual illusions (6.ii.d) owed much to the ingenuity of his technician Stephen Salter (later, a professor of engineering who invented an ingenious way of harnessing wave power). Gregory is generous enough to acknowledge this, in print as well as conversation. Many others wouldn't.

Group loyalties enter the picture, too. Judgements of originality and/or value can be strongly influenced, for instance, by chauvinistic nationalism. In his inaugural address in January 1996, President Clinton called the electronic computer an American invention—to the chagrin of compatriots of Turing, Max Newman, Thomas Flowers, Frederick Williams, and Maurice Wilkes (3.v.b–d).

Such judgements can be skewed also by the 'not-invented-here' syndrome. This systematically distorts the reminiscences and the bibliographies of workers in at least one leading AI laboratory, and at least one leading department of linguistics (Chapter 9.viii.a and ix.a).

It even prevented proper recognition being given to the only working AI program to be presented in the final days of the famous Dartmouth Summer School in 1956 (Chapters 6.iv.b and 10.i.b). Although it electrified a few people present there, the program wasn't taken up as the paradigm for AI. Far from it. One of the originators could afford to be 'philosophical' about this years afterwards, but it had rankled at the time:

> [The new field of AI] was going off into different directions. They [i.e. Minsky and John McCarthy] didn't want to hear from us, and we sure didn't want to hear from them: we had something to *show* them! . . . In a way, it was ironic because we already had done the first example of what they were after; and second, they didn't pay much attention to it. But that's not unusual. The "Not Invented Here" sign is up almost everywhere, you know. (H. A. Simon, interview in Crevier 1993: 49)

As for *which* groups are involved in evaluating a discovery, that can rest largely on chance. Frank Rosenblatt's work on "perceptrons" became very widely known, very quickly, because he'd been communicating with physicists as well as psychologists (12.ii.e). This was more accidental than intellectual. He'd been working alongside physicists (in his university's Aeronautical Laboratory) because he'd been borrowing their computer: the psychologists didn't have one.

An idea may be hailed as a discovery largely because the context in which it's put forward makes it highly significant. For example, Goethe is commonly credited with having discovered a particular similarity between the bones of the rabbit's jaw and ours. In fact, someone else had noticed it before him. Goethe, however, placed this anatomical fact in the context of an ambitious philosophy of the unity of nature. The rabbit's jawbone was of interest because he related *all* vertebrate skulls to a single archetype, and because he saw morphological archetypes in the structure of flowers as well as skulls (2.vi.e).

g. So what's new?

To view something as a discovery is to see it both as *valuable* and as *new*. The Goethe example (above) shows that an idea may be more or less valued depending on its intellectual context. Judgement, even negotiation, must enter in. But what about the newness? One might think that novelty is a more cut-and-dried matter. In fact, however, it's a minefield for the unwary.

There are three very different reasons for this. One is obvious, even boring: namely, unavoidable ignorance. No one can be *sure* of knowing about every previous thought that's relevant to a given topic, nor even of having read everything that's been published about it (especially if some of the texts exist only in a language one doesn't read).

Moreover, publication is sometimes long delayed, and/or long unread, so that later work is mistakenly believed to have been the pioneer. One example is Paul Werbos's algorithm defining what was later called "backprop" (12.iv.d). Not only did this lie buried for many years in a computer manual (unread by psychologists), but wider publication was at first deliberately suppressed by a governmental committee, because other results in the same report were politically embarrassing. It was only later that Werbos's priority could be recognized. Another example is Konrad Zuse's work on computers, which remained unpublished for many years—and wasn't translated from the German for several years after that (3.v.a, 10.v.f). In short, judgements of historical novelty can only be provisional. (Maybe they should carry a government health warning?)

The second difficulty in assigning priority is more interesting: no creative idea is entirely new. It's either a novel combination of familiar ideas, or an exploration or transformation of a culturally accepted style of thought (Boden 1990*a*, chs. 3–5). Even in the latter case, where the new idea can be so surprising as to seem *impossible*, there will be some intelligible relation with the previous way of thinking. It may take a long time, and considerable persuasion, for someone/everyone to recognize this in a particular instance (such as the holographic theory of memory, or the travelling-depolarization theory of nervous conduction: 12.v.c and 2.viii.e). But the point stands, nevertheless. It follows that there are differing *degrees* and varying *types* of novelty, never absolute newness.

It's the third reason which is especially relevant here, however. Namely, people are human—often, all too human. Priority claims can be grounded in various kinds of moral frailty: from deliberate deception, through uncritical self-deception, to lazy (i.e. avoidable) ignorance. Instances of each of these can be found in cognitive science.

For a start, there's the unfortunate habit of representing other people's work as one's own. Occasionally, this boils down to outright *theft*. Two examples, both utterly trivial in the grand scheme of things but pretty annoying to me: Many years ago, I was sent a book for review that was 75 per cent lifted, largely word-for-word, from parts of one of my own—cited only twice. (I felt unable to point this out, and merely said that the book was "highly derivative"; but another reviewer, Donald Michie, did so—thanks, Donald!) And recently, while googling on the Web, I found some "Detailed Comments" on heterarchy (10.iv.a) comprising pp. 125–42 of the same book—which wasn't even mentioned. (On regoogling in April 2005, I could no longer find this entry; it must have been removed.) The limit of this person's originality was to make a few cosmetic changes, including one—adding the words "I'm afraid" to a critical remark of mine—specifically intended to strengthen the impression that he was the author of the stolen comments.

I don't know of any non-trivial examples quite as shameless as this, although I've been told (in confidence) of several that come very close. But it's pertinent to note that the pseudonymous Father Hacker includes the theft—and judicious renaming—of old ideas as a key item of advice in his spoof 'Guide for the Young AI-Researcher' (*AISB Quarterly* 2003; see Chapter 13.vii.a).

More common than such shameless theft is the deliberate suggestion that one's new ideas are more original, and/or more independent, than they actually are.

An exceptionally dishonourable example was directed against a cognitive scientist—let's call him Dr X—who worked in the hard sciences before turning to matters of the mind. During that period, he'd had idea A about topic B. When he was preparing a report for publication, his colleague Dr Y—who was thinking along the same lines, on idea A′—persuaded him to delay publication until his own thought was more advanced, so that they could co-author an even stronger paper. Dr X agreed, and waited for Dr Y to contact him. In vain: soon afterwards, Dr Y published alone—and later received a Nobel Prize for this work.

Within cognitive science as such, there are a number of cases of people exaggerating, even deliberately misrepresenting, their own originality. Sometimes, the earlier author isn't even mentioned—in which case only evidence from informal conversations (e.g. between the two individuals involved) can support the charge of deliberate plagiarism. Since this evidence is (*a*) unpublished and (*b*) often given in confidence, I haven't specified any such charges in my story: but I'm sorry to say that I could have done.

More commonly, the original author *is* mentioned—but in a misleading way. As one colleague remarked to me: "The art of plagiarism is to make a marginal and inaccurate citation of earlier work, but to give the impression that it is new and yours." For example, Marr (who has been accused several times of exaggerating his priority: 14.v.d) used this ploy. His early work on stereopsis has been described as "a minor variation" of a model published by someone else, who was cited dismissively by Marr in a low-profile footnote (J. A. Anderson and Rosenfeld 1998: 231–2). "Low-profile" not just because it was a footnote, but because it came almost at the end of a long list of footnotes. A generous acknowledgement this was not.

A more general way of exaggerating the novelty of one's own work is to describe it not as an exciting new development within an *existing* field, but as something quite different. Many pioneers mentioned in later chapters were guilty of this, in that they ignored/denied the close links—historical, sociological, methodological, and even philosophical—between their work and earlier forms of cognitive science.

Consider, for instance, a 'historical' paper by the situated roboticist Rodney Brooks (1991*b*). In the opening paragraphs he mentioned "traditional" AI, implying (correctly) that his new approach was part of the AI enterprise as a whole. In the main body of the text, however, he repeatedly used the term "Artificial Intelligence" to mean only symbolic AI—implying that his approach was even more new, even more revolutionary, than it was (cf. 13.iii.b). Other examples of this self-glorifying rhetorical strategy include the frequent attempts to distance connectionism from AI as a whole (Boden 1991), and to exclude A-Life from "cognitive science" because it doesn't share the Cartesian assumptions of the traditionalists (15.viii.b–c).

Even more common than deliberately overplaying one's originality is the *ignorant* suggestion that one's ideas are new. George Miller was bitterly accused by Newell and Simon of not giving them proper recognition as the originators of many ideas in *Plans and the Structure of Behavior* (Chapter 6.iv.c). However, as Miller said later, "they were old familiar ideas; the fact that they had thought of it for themselves didn't mean that nobody ever thought of it before" (1986: 213). In the event, he redrafted the text and added zillions of footnotes "so they would no longer claim that those were their ideas". They had some excuse, since they hadn't been trained as psychologists. Nevertheless, and despite Simon's knowledge of Gestalt research, they were evidencing the shameful ignorance of past work that pervaded American psychology from the 1920s to early 1960s (see Chapter 5.i.b).

A more excusable example concerns logic programming (Chapter 10.v.f). Its first occurrence (so far as is known) was due to Zuse, in 1946. But his ideas weren't taken up by the early computer scientists. Moreover, the description of his Plankalkul language wasn't published until 1972 (nor translated from the German until four years later). So "the early computer scientists", who were based in England and the USA (3.v.b–e), didn't know about them. The people who "pioneered" logic programming in the early 1960s may therefore be forgiven for believing that they were the very first to think along these lines.

Some cases of sincere-but-mistaken claims to originality rest not on ignorance so much as on *failure to recognize* the similarities between one's own ideas and others'. This failure, in retrospect, can be very surprising. Thus Simon recalled in his autobiography that he and his father—an early servo-engineer who'd designed gun-turret controls for battleships in the First World War—didn't appreciate their shared interests until it was almost too late:

> It wasn't until [the end of the Second World War] that I realized that his whole life had been spent in what you might call protocybernetic work, and that it was just a direct ancestor to this whole business. And until the last year or two of his life—he died in 1948—we never had a conversation about this. *He used to tell me about his work, but that was about his work, and I used to tell him about what I was doing, but that was about what I was doing,* and I don't think the thought crossed either of our minds, certainly not until about 1947 or 1948, that these had any relation to each other. And I don't really understand that now. (Simon, quoted in McCorduck 1979: 130; italics added)

Perhaps the explanation was a form of what the Gestalt psychologists had called "functional fixedness" (5.ii.b). Much as someone may be unable to tie two far-separated strings together because they see a nearby pair of pliers only as pliers, not as *a weighty object suitable for use as a pendulum bob*, so Simon (and his father) apparently thought in terms of "his work" and "my work"—labels which prevented them from noticing the conceptual links. Someone with an emotional and/or professional investment in classifying an idea as "my discovery" may be especially likely to fall victim to this type of blindness.

Failing to recognize the similarity can depend on failing to be consciously aware of the other person's idea. Paul McCartney was accused (in July 2003) of having drawn the superb melody, and some of the words, for *Yesterday* from a song released when he was only 11 years old. (The older version was played on BBC Radio, and the likeness was striking.) McCartney had already reported that he'd been so surprised at waking up with the tune (but no lyrics) "all there" that he'd thought at the time that something like this might have happened. He'd even asked his friends and fellow Beatles if they'd ever heard the tune before, but no one had.

This case of unrecognized memory wasn't identified as such until some fifty years after the original memory was formed. So priority claims, however sincere, can only be provisional. For instance, the neuroscientist Michael Arbib sincerely believed that he'd thought up the name *Rana computatrix* for his computerized frog all by himself, but realized years later that he'd been inspired by a similar name used by someone else (see 14.vii.c). Ideas about names for frogs, computerized or not, are of course relatively trivial. But who can know how many analogous cases concern ideas much more significant than that?

Sometimes, people deliberately suggest that they've done relevant previous work when in fact they haven't. The Nobel prizewinner David Hubel has confessed to a deception of this kind:

> [Torstein Wiesel and I had gone to a lecture by Vernon Mountcastle in the late 1950s] in which he had amazed us by reporting on the results of recording from some 900 somatosensory cells, for those days an astronomic number. We knew we could never catch up, *so we catapulted ourselves to respectability* by calling our first cell No. 3000 and numbering subsequent ones from there. When Vernon visited our circus tent we were in the middle of a three-unit recording, cells number 3007, 3008 and 3009. We made sure that we mentioned their identification numbers. (Hubel 1982: 516; italics added)

This was just a youthful prank, of course. But less innocent examples sometimes appear in print. They ride on the fact mentioned above, that by and large scientists trust each other's empirical reports even if they disagree on their theoretical interpretation. ("By and large", because in important cases the reported experiments must be replicated by someone else.)

Another prank, confessed years after it happened, was meant to suggest not just originality but effortless superiority:

> Alan Kay, Marvin Minsky, and I got together and did some back-of-the-envelope calculations—we actually killed about five minutes to find an envelope so we could later say we did back-of-the-envelope calculations—on how much knowledge would be required [for an AI system embodying common sense: see 10.viii.c] and how much time it would take. That was a million frames over ten years. (D. B. Lenat, interviewed in Shasha and Lazere 1995: 233)

Pretty harmless, admittedly . . . but some attempts to influence people's judgement by providing misleading information are more blameworthy than this one.

Even if someone honestly cites inspiration by, or similarity to, a previous worker, it's often not straightforward to decide who "discovered" the idea—because it's not clear *what*, exactly, the relevant "new" idea is.

The development of hypertext is an illustration (Chapters 10.i.h and 13.v). Individuals commonly cited as crucial originators include Vannevar Bush, Douglas Engelbart, Joseph Licklider, Theodor (Ted) Nelson, and Alan Kay. And Licklider, a leading influence on library science, was well aware that librarians had identified some of the core problematic issues long before—if not quite as early as Gabriel Naudé (see

Preface). To understand the history of hypertext, then, one must distinguish carefully between statements of:

* a general conceptual framework;
* abstract organizational principles;
* desirable/conceivable end-results;
* in-principle-possible technical methods;
* feasible computational techniques;
* commercially efficient designs;
* and specific improvements of pre-existing technologies.

Without doing this, one can't say just what was "new" about an individual's (or a research group's) contribution. The same applies, of course, to many other discoveries besides hypertext.

h. Rhetoric and publication

The seventh warning concerns not what ideas people come up with, but how they present them. For the history of science is not only about what people thought, but also about what they were *thought* to have thought. And that, in turn, depends on how—and where—they chose to express them.

The identification of discoveries (both as *new* and as *valuable)* can depend heavily on the discoverer's rhetorical skills, or lack of them. We'll see in Chapter 9.vii.b, for instance, that Chomsky's review of Skinner's *Verbal Behavior* became famous largely because of its sparkling, stinging wit. By contrast, several classic papers in cognitive science—even including *the* classic paper (Chapter 4.iii.f)—were all but ignored on first publication, because of their notational and/or mathematical difficulty.

Arguably, one case in point is the early work of the highly creative cognitive scientist Stephen Grossberg. He was perhaps the first to formulate three ideas that are influential today under the names of other people: Hopfield nets, the Marr–Albus model of the cerebellum, and Kohonen self-organizing maps.

(These things are tricky. Although Kohonen 1982 didn't cite Grossberg directly, he did cite four papers by Christoph von der Malsburg. The earliest of these had clearly acknowledged Grossberg as its prime inspiration: see Chapter 14.vi.b. But without following that particular paper-trail, readers might be unaware of the kinship between Kohonen's ideas and Grossberg's. This is, of course, a general point: the fact that Bloggs didn't cite Squoggins, or even hadn't read Squoggins, *doesn't* mean that he wasn't influenced by Squoggins.)

Grossberg also pioneered many more notions—including back propagation—that are commonly attributed to others, if not actually named after them. As he puts it:

> This has, all too often, been the story of my life. It's tragic really, and it's almost broken my heart several times. (J. A. Anderson and Rosenfeld 1998: 179)

> Trying to live with so many false [priority] claims has been difficult for me at times. If I try to get credit where it is due, then people who want the credit for themselves often mount a disinformation campaign in which they claim that all that I think about is priority. Because I have been a very productive pioneer, who innovated quite a few ideas and models, that can create quite a chorus of disinformation! If I don't try to get credit for my discoveries, then

I am left with the feeling that eventually most of my ideas may become attributed to other people . . . (pp. 186–7)

Grossberg's own diagnosis of his "tragic" life history identifies three causes:

The problem is that, [1] although I would often have an idea first, I usually had it too far ahead of its time. *Or [2] I would develop it too mathematically for most readers.* Most of all, [3] I've had too many ideas for me to be identified with all of them . . . [From the mid-1960s on] many things that I discovered started getting named after other people. (J. A. Anderson and Rosenfeld 1998: 179; italics added)

The key sense in which his work was "too far ahead of its time" was the unfamiliarity of the mathematics. He was talking about brain and behaviour as a complex dynamical system (as he put it: "nonlinear, nonlocal, and nonstationary"—1980: 351), a theoretical approach that didn't become popular in cognitive science until the late 1980s. But it's the second point that's of special interest here: the rhetorical style.

His early work was largely unintelligible even to the few psychologists who took the trouble to read it (see Chapters 12.v.g and 14.vi.a). He combined intellectually demanding (and unfamiliar) mathematics with a host of interdisciplinary details, most of which would be unfamiliar to any individual reader. They were there because he was trying to show the unsuspected theoretical *unity* behind hugely diverse data. His writing was unusually voluminous too: 500 pages for his first-year graduate report (1964), and many long and richly cross-referenced journal articles. Faced with this challenge from a youngster they'd never heard of, most people gave up before reaching the end, if they could summon up the courage to start reading at all.

Scientifically speaking, the weakness was theirs—not his. It's now taken for granted (which it wasn't then) that the problems he was discussing demand a fair degree of mathematical literacy on the researcher's part. Specifically, they require non-linear mathematics. (Broadly speaking, this is a type of mathematics in which a very small change can bring about a much larger, and global, change.) In general, indeed, one can't expect cutting-edge ideas to be easy to understand. Humans being human, however, new scientific claims have to compete for attention. In a cynical mood, one might even say that they have to compete *in the marketplace* for attention. So any rhetorical obstacles put in readers' way will tend to obscure the science, however worthy it may be. Grossberg wasn't easy to read.

By contrast, the group who (wrongly) got the credit for discovering back propagation went to exceptional lengths to describe their work in an intelligible way (12.vi.a). If they didn't emulate the sparkle of Chomsky's famous review, they did achieve clarity. This doesn't count as a *scientific* achievement. But, for good or ill, it does—and did—make a difference in bringing scientific ideas into historical prominence.

Sometimes, the rhetorical obstacles are so great that a primary author gains recognition largely through the efforts of a secondary one. For instance, Richard Montague's theory of semantics was nigh opaque to most cognitive scientists, even including linguists accustomed to dealing with formalism. Without Barbara Partee's (1975) more accessible introduction, his ideas might have been virtually ignored except by his fellow logicians. In fact, they became hugely influential in theoretical linguistics—and underlay the most important challenge to Chomsky (9.ix.c–d).

If one thinks of rhetoric as *presentational* skill, verbal or otherwise, one can even find examples in robotics. We'll see in Chapter 4.vii.b, for instance, that whereas William Grey Walter's mobile "tortoises" caused a sensation in the early 1950s, his—even more interesting—learning machine did not. The problem, apparently, was that this was built as an ugly metal box sitting boringly on a bench. Even though it was sometimes connected (wired) to a mobile robot, its significance wasn't recognized at the time. Another example is the robot arm built by Minsky when he was a boy: no one paid any attention—until he put the sleeve from a flannel shirt on it (Newquist 1994: 62).

The line between rhetorical and theoretical difference, in turn, is often unclear. I implied (above) that the people who independently 'discovered' backprop, Hopfield nets, or the Marr–Albus rule each discovered *the very same thing*. But this is debatable—which is why I said that Grossberg is "arguably" a case in point, and had "perhaps" discovered many ideas later attributed to others.

In general, it's easier to decide such matters when (as in these examples) the idea is expressed mathematically. Two authors might use an identical mathematical equation, or two equations that can be *proved* equivalent. That's not possible for verbally expressed theories, where the many subtleties of interpretation and association of ideas complicate matters hugely. Indeed, showing that two verbal theories or concepts are equivalent is analogous to translating from one natural language to another—a deeply problematic enterprise (9.iv.b and x).

Even with mathematical examples, however, there can be difficulties. A third person, Shun-Ichi Amari, is sometimes said to have "discovered" Hopfield nets (also known as additive nets). But whereas Amari gave only fourteen lines and a mathematical equation, John Hopfield drew out the implications at length, showing how they could be explored in computer simulations. Largely because it was so much easier for cognitive scientists to understand and be inspired by Hopfield, he got the credit. Yet Amari had defined the identical mathematical function—and Hopfield had cited him (see 12.v.f).

As for Grossberg, he'd defined additive nets even earlier, while he was still a student. But he hadn't given a succinct account of them (although he did so later, as we'll see: 12.v.g). Nor had he presented this new idea alone. He'd combined it with many others in describing the processes within a dynamical brain–behaviour system, so that even mathematical psychologists were unable to appreciate it. The scientific value was greater than that of Hopfield's paper (see 14.vi). But the impact, at the time, was less.

Besides the difficulties in getting recognition after a discovery has been published, there's the problem of getting it published in the first place. Strictly, official publication isn't necessary for an idea to be recognized. Chomsky's mimeographed research notes, like Ludwig Wittgenstein's lectures, were read by some experts long before appearing 'in press' (Chapter 9.vi.a). So were Brian Smith's research notes on the nature of computation (16.ix.e). But publication certainly helps.

If Bertrand Russell and Alfred North Whitehead hadn't had some cash to spare in 1910, cognitive science might have arisen much later than it did. For they had to help subsidize the publication of their epochal *Principia Mathematica* (see 4.iii.b). Work on cellular automata, by contrast, might have burgeoned earlier, had von Neumann's lectures not remained unpublished for many years after his death (15.v.b). Occasionally, wider publication is actively prevented. For example, the pioneering work on computing done at Bletchley Park in the Second World War remained an official secret for over thirty years (3.v.d). And the 'true' discoverer of backprop was unable to circulate his idea more accessibly because of the political sensitivities of a US government department (12.vi.d).

In addition, the source of publication matters. One could publish an idea in *The Beano*, and its scientific value wouldn't be affected—but its reception by other scientists, and therefore its place in history, would. In the seventeenth century acceptance, or even respectful consideration, required the word of a gentleman. Today's equivalent is the peer-reviewed journal. And the perceived relevance and quality of the peers make a difference. In the 1950s it was acceptable, even if not the authors' preferred choice, for an important neurophysiological paper to be published in the *Proceedings of the Institute of Radio Engineers*, because of the influence of cybernetics (see Chapter 14.iii.a). Today, no biological reader would see it. And the people popularly associated with backprop made a point of publishing this idea simultaneously in the high-prestige *Nature* and in a more accessible, easily affordable, but well-respected source: an MIT Press trade book, aimed at students and professionals in all areas of cognitive science.

Last but not least, it follows from what's just been said that the editors and reviewers of the professional journals have the power to block, or to censor, the ideas sent into them. This can affect even well-known scientists, if they're presenting ideas not favoured by the editor concerned. Bruner himself has come up against this problem:

> The human sciences are about human beings in specific situations, and why should I hide that fact? You know, the editor of the *Journal of Experimental Psychology* typically *deletes anything of that sort* from my papers. That's probably why I don't write as much for the *Journal of Experimental Psychology* any more. (quoted in Shore 2004: 157; italics added)

And stories are rife about as-yet-unknown writers being spurned by their chosen journals. Indeed, work which eventually won a Nobel Prize was initially rejected by an "elder statesman" reviewing it for the editor of the *Psychological Review* (Chapter 7.iv.f).

i. An explanatory can of worms

A new field of enquiry typically spawns new journals (see 6.v.c and 10.ii.b). It's easy to assume that this is because there are so many discoveries, of an increasingly detailed nature, that there wouldn't be room for them in the existing journals—even if the editors happened to be interested in them.

That's true, to be sure. But it's also true that the existing editors may reject *the general approach* underlying all the discoveries—in which case, they'll probably refuse to recognize them *as* discoveries. (Similarly, researchers are usually loath to offer jobs to youngsters who disagree with them: Newell was a refreshing exception, as we'll see in Chapter 10.v.b.) The problem, then, isn't a mere lack of space, or even a mere lack of interest, but a (supposed) lack of intellectual respectability. New journals are founded, accordingly, which do regard these new ideas as respectable.

Bruner's current publication difficulties (mentioned above) illustrate this point. For they're due less to specific, and unacceptably shocking, ideas on his part than to a general disagreement about what counts as explanation in psychology. Having started out in the 1950s as a straightforward, if highly creative, experimentalist (6.ii.a–c), he's now prepared also to consider *interpretative*, hermeneutic, accounts (8.ii.a). In a nutshell, this means that he relies on his intuition for "narratives" (a critic might say "gossip"), to understand "human beings in specific situations". But whether *human beings in specific situations* is, as he claims, a proper subject for a theoretical psychology is problematic.

The eighth caveat, then, is that Bruner's little story about his problems with the *Journal of Experimental Psychology* has opened a philosophical can of worms that we'll encounter at various points in later chapters. And there are two species of worm in the can.

On the one hand, it's highly controversial whether theoretical (as opposed to therapeutic) psychology should hope to deal with *particularities*. Some people, such as Bruner (and his long-time Harvard colleague Gordon Allport: 1942, 1946), argue that to understand specific events in individual human lives we need empathetic interpretation, not naturalistic science (Chapter 7.iii.d). Indeed, some philosophers argue that psychological phenomena *in general* (not just particularities) can't, in principle, be explained in scientific terms (14.x.c and 16.vi–viii). To the editors of the *Journal of Experimental Psychology*, such a view is an abomination. Naturally, then, they won't be willing to publish papers expressing it.

On the other hand, it's also disputed whether showing how certain things—either particularities (e.g. Joe Bloggs's saying "It wasn't me!") or general phenomena (e.g. language)—are *possible* really counts as 'explaining' them (7.iii.d). Since that's precisely what the computational approach enables us to do, cognitive scientists have a stake in this abstract philosophical dispute.

You'll have noticed that the FAQs listed in Section i.a weren't about what Joe Bloggs did or didn't do last Saturday, nor about what he'll do next week. Rather, they concern *how it's possible* for him—or anyone else—to do the sorts of things that he does every day of his life. If cognitive science can explain that (and if it really is an 'explanation'), it will have done what it set out to do.

1.iv. Envoi

As Naudé recognized (Preface, preamble), others would have told this tale differently. For as we've seen, people disagree about *just what counts* as cognitive science. They also disagree about what message they'd want to send in writing about it. Instead of relaying (as I've tried to do) the enormous interest and promise of the field, and its necessary interdisciplinarity, some would pen an account of the futile pursuit of a philosophical illusion (Chapters 11.iv.a and 16.vi–viii).

What's more, the many factors mentioned in Section iii make historical attribution within cognitive science a difficult, and delicate, matter. Such judgements can turn out in more than one way.

If one worried too much about all that, however, no history could even be attempted. So here goes. Now that we know what we're talking about, let's begin the story.

2

MAN AS MACHINE: ORIGINS OF THE IDEA

'Machine as man' is an ancient idea, and an ancient technical practice. Ingenious android machines whose movements resembled human behaviour, albeit in highly limited ways, were already being built 2,500 years ago. 'Man as machine' is much more recent.

Admittedly, its birth date can be disputed. Some might even say that the two ideas are equally old. That claim, that 'man as machine' is also an ancient idea, could be argued in two very different ways.

One could say that Plato sowed the seeds of a formalist account of knowledge which later inspired the rationalist philosophical tradition in general, and (symbolic) AI and cognitive science in particular (H. L. Dreyfus 1972). This is true. In that sense, intellectualism goes way back to Plato. But to describe 'man as machine' as a concept favoured by Plato would be anachronistic. It's plausible only in retrospect, only after the association of the idea of formalism with man-made calculating machines and digital computers.

The second argument for the antiquity of 'man as machine' as a guiding idea is stronger. For uncompromisingly materialist, even mechanistic, philosophies—of soul, as well as body—were proclaimed five centuries before Christ. However, these were highly programmatic. They didn't, and couldn't, guide systematic empirical research.

With hindsight, some Presocratic speculations seem remarkably apt. For instance, Democritus (*c.*460–370 BC) suggested that discrimination by taste and smell is possible because atoms (or combinations thereof) of different shapes enter the mouth and nostrils, where they fall into a variety of suitably shaped depressions in the sense organ. In essence, modern science teaches something similar. But our concept of atom is also very dissimilar. And whereas we can add theoretically detailed and empirically tested chapter and verse, Democritus couldn't. Nor did he suggest, as René Descartes was to do some 2,000 years later, that his mechanistic ideas be systematically tested by experiment. Before the rise of modern science, the experimental method wasn't an option.

Accordingly, much as atomic theory in chemistry is normally dated to John Dalton, not to Democritus, so 'man as machine' is best seen as an idea arising four centuries ago (with Descartes), not twenty-five. Indeed, if 'man' denotes the mind as well as the body, we have to pass on to the eighteenth century to see this vision developed as a philosophical theme, and to the late nineteenth to see it explored by scientific research.

Even at that time, talk of *mind* as machine was taken to mean merely that the mind, like the body, works according to scientific principles. Nineteenth-century marvels such as the telegraph or telephone were sometimes used as analogies in describing how the brain works, as earlier technologies had been before them (Fryer 1978; McReynolds 1980). Such comparisons lasted well into the twentieth century, and were still prominent in popular writing in the 1950s. But they were applied to brains, not to minds or thought processes as such.

Today, by contrast, 'man as machine' is usually interpreted more strongly (see Chapter 1.ii). It's now taken to mean that there are scientifically intelligible principles capable of controlling not only human (and animal) bodies and minds, but also *significantly lifelike or mindlike artefacts that we might actually try to build.* Understood in that way—which excludes clocks, and even telephones—the idea is very new. Although some assume that Charles Babbage originated it 160 years ago, it's more securely dated to around the time of the Second World War.

But that's to get ahead of our story: Babbage and his post-war intellectual successors are discussed in Chapters 3 and 4. In this chapter, I outline how *machine as man* was eventually joined by *man as machine*, explaining the remarks about alternative birth dates made above. I sketch how mind was later added to body in the interpretation of this slogan. And I mention some pioneering attempts to study 'man as machine' in science.

These efforts, initially in general physiology and later in neurophysiology and psychophysics too, prefigured ideas that are influential, and sometimes highly controversial, in cognitive science today. Virtually every topic mentioned in this chapter, then, will resurface in a later one. In short, nothing is discussed here simply because it *happened*, but because it's still relevant now.

The first six sections are ordered chronologically. Section i discusses some ancient examples of 'machine as man'. Sections ii–vi sketch the rise of mechanistic science—and especially of 'man as machine'—in the seventeenth and eighteenth centuries, and the neo-Kantian reaction to it.

The remaining sections focus mainly on the nineteenth and early twentieth centuries, and cover distinct themes. Sections vii and viii concentrate on general physiology (including embryology) and neurophysiology, respectively. Section ix describes early logic machines. And Section x turns to scientific psychology: mind as mechanism, but not yet as machine.

2.i. Machine as Man: Early Days

Automata building was hale and healthy in ancient times, but weakened in the Dark Ages. In the late Middle Ages, it rallied. By the early modern period, from about 1500, automata were increasingly common in Europe—and by the seventeenth century, they were among the technological marvels of their time. Only at the end, however, did these changes reflect changes in how scholars were thinking about *human beings*.

a. Ancient automata and Dark Age decline

'Machine as man' was exemplified in ancient Egypt (a 'talking' wooden head of the Jackal God of the Dead now sits in the Louvre), and was a familiar engineering practice in ancient Greece. Aristotle's *De Anima*, in the fourth century BC, described a number of moving statues built by Daedalus. One was a figure of Venus, worked by being filled with mercury, which had to be tethered to prevent it from running away.

By the first century BC this ingenious practice had blossomed further. Hero of Alexandria (roughly AD 10–62) wrote many treatises on "automata", or self-moving things—the word is his coinage. He described machines worked by gears, levers, valves, pistons, and pulleys, mostly powered by water, falling weights, or steam (Drachmann 1963: 19–140). Some of them involved feedback devices, first used by the Egyptians over 300 years earlier.

Some of the most interesting, in engineering terms, bore no relation to the human body: a handheld mechanical calendar of about 87 BC, which computed the solar equinoxes and the phases of the moon, used thirty intermeshing toothed gear wheels like those later used in clocks—and in Babbage's calculating engines, described in Chapter 3.ii–iii (de Solla Price 1975). But many early automata mimicked bodily actions, whether of men or animals.

A few ambitious engineers even attempted both of these within the same piece. For instance, among the constructions listed by Hero was a statue of Hercules with a snake:

> On a pedestal is a small tree, around which is coiled a snake. Nearby stands the Archer Hercules. An apple also lies on the pedestal. If the apple be raised a little by hand Hercules shall then discharge his arrow at the serpent, and the serpent shall hiss. (from Hero's *Pressure Machines and Automatatheatres*, quoted in Klemm 1957: 36)

Yet more lifelike behaviour was seen in Hero's construction of the birds and the owl. When the water from a fountain reached a particular level in a hidden tank (with a syphon mechanism in it), the birds were caused to sing, the owl to turn 'threateningly' towards them, and the 'frightened' birds to fall silent once more.

An automaton even featured in the political machinations of ancient Rome. Automated extravaganzas often formed part of the circuses in the Coloseum, but this device was built for a more specific political purpose. Mark Antony's funeral oration following the assassination of Julius Caesar, and his passionate collapse in front of Caesar's bier, had aroused the masses gathered in the Forum to fever pitch. Then, this happened:

> An unendurable anguish weighed upon the quivering crowd. Their nerves were strained to the breaking point. They seemed ready for anything. And now a vision of horror struck them in all its brutality. From the bier Caesar arose and began to turn around slowly, exposing to their terrified gaze his dreadfully livid face and his twenty-three wounds still bleeding. It was a wax model which Antony had ordered in the greatest secrecy and which automatically moved by means of a special mechanism hidden behind the bed. (G. Walter 1953: ii. 237)

This automaton did as much as Mark Antony's speech to whip up the "collective frenzy" and "wild ecstasy" at the funeral. As Caesar's dead body was burnt, soldiers hurled their swords and insignia into the flames, and women threw in "their jewels and the sacred amulets taken from the necks of their children". The history of Rome was decisively affected as a result.

(Fearing that this story was too good to be true, I tracked down the original sources and asked a classicist friend to translate them from the Latin, as well as asking a medievalist whether she believed the story. It turned out that the automated avatar had been described by Appian, and that simple wax models of the deceased were typically presented at important funerals in Roman, medieval, and Renaissance Europe.)

But being sponsored by Mark Antony, reported by Aristotle, or even analysed by Hero didn't grant respectability. These projects weren't seen, nor (with one qualification remarked below) were they intended, as having any philosophical or scientific interest. Indeed, machine building of any sort was scorned in antiquity by the high-born and the educated, even including some engineers themselves.

Plato expressed the fourth-century (BC) Greek's attitude to engineers, even to one whose skills could save a whole city. He reported Socrates' saying to Callicles (in the *Gorgias* 512bc): "You despise him and his art, and sneeringly call him an engine-maker, and you will not allow your daughter to marry his son, or marry your son to his daughter." This attitude was fully justified, said Plato—not for reasons of social snobbery, but because only a life guided by purely intellectual activity is truly worth living.

Contempt for practical skills lasted into the technologically sophisticated Roman Empire—and through into seventeenth-century England and the British Empire too (see Section ii.b below, and Chapter 3.ii.b). So Seneca (*c*.55/1932, letter xc) dismissed all handicrafts and useful technologies—glass, piped central heating, architecture, and even shorthand—as "the inventions of the meanest slaves". (Little did he know that science would develop in the West, not China, largely because of the invention of transparent glass: Macfarlane and Martin 2002.) Techniques of mining, and domestic tools, he declared, "were invented by some one whose intellect was nimble and keen, not massive and sublime: so was everything else the quest of which involves bowed shoulders and earthward gaze" (ibid.)

Likewise, the first-century historian Plutarch eulogized Archimedes, who 400 years earlier had designed many military engines, including a crane which picked up enemy ships and dashed them against the rocks, and a compound pulley that enabled him to draw a fully laden merchant ship across the beach by a mere hand movement. All very impressive, said Plutarch—but most admirable of all was Archimedes' judgement of this work as unworthy of written record. (This, from a historian!) As he put it:

> [Archimedes] would not deign to leave behind him any commentary or writing on such subjects; but, repudiating as sordid and ignoble the whole trade of engineering, and every sort of art that lends itself to mere use and profit, he placed his whole affection and ambition in those purer speculations where there can be no reference to the vulgar needs of life . . . (from Plutarch's *Lives*, quoted in de Solla Price 1975: 51)

In Europe's Dark Ages, even such "sordid and ignoble" knowledge was largely lost, and automata building went into abeyance. (In China, however, where a mechanical orchestra had been built for the Emperor in the third century BC, similar practices flourished—until they waned in the fifteenth century.)

b. In fashion again

At the time when China's automata were disappearing, automata building had resurfaced with a vengeance in Europe. It did so via the Arabs, who had translated Hero's treatises in the ninth century and who were enthusiastic practitioners of the art. (And of algebra, too: our word comes from the Arabic *al-jabr*: reunion, or connection; and our word algorithm recalls al-Khwarizmi, a ninth-century Arab mathematician.)

For instance, the twelfth-century mechanical engineer Ibn al-Jazari wrote *The Book of Knowledge of Ingenious Mechanical Devices*, which was soon translated into Latin. This discussed the design, manufacture, and assembly of fifty automata powered by hydraulics, pneumatics, and gears. Some were workaday instruments, such as clocks or pumps. But some were androids, including one—which looked like a 5-year-old boy—called the "boon-companion". Described by its inventor in drily meticulous detail, so that it could be understood—and perhaps rebuilt—by others (D. R. Hill 1974: 115–17), its "outside appearance and purpose" was recorded by al-Jazari as follows:

> It is a kneeling figure made of jointed copper. He holds a goblet in his right hand with fingers extended along its stem, and in his left hand he holds a waterlily by its stalk. It was one of the customs of the king in those days, when they were drinking, to leave some [of the wine] in the goblet and this, when it had collected, was drunk by a boon-companion designated for that duty. This boon-companion [i.e. the model] is placed in front of the head of the carousal. When he drinks a goblet the steward takes it, pours what is left in it into the boon-companion's goblet, and stands aside. When left by himself he [i.e. the boon-companion] lifts the goblet in his hand until its rim is between his lips [where it stays] for a while. Then he lowers the goblet from his hand and nods his head several times. This happens every time wine is poured into the goblet. His left hand moves and is observed by the head of the carousal until after a while it reaches a certain position. (al-Jazari, trans. D. R. Hill 1974: 115; all phrases in square brackets are Hill's)

At that point, the automaton would do something which his human equivalent would never do—and which few members of the "carousal" would ever forget:

> Then the head of the carousal says to someone he wishes to make fun of, and who does not know the purpose of the boon-companion, "So-and-so, take this boon-companion, who drinks wine and hides a secret. Put him on your knee, drink, and give him [wine] to drink." So he takes him without argument, puts him on his knee, drinks and gives him drinks. He does not finish two or three goblets before [the boon-companion] pours on to him all he has drunk since the beginning of the carousal, wetting his clothing. The wine flows beneath him, making him a target for laughter. (ibid.)

But the boon-companion could also be used less mischievously:

> This [practical joke: M.A.B.] is appropriate at certain times, but it is more usual for the king, when he knows the boon-companion is about to discharge what he has drunk, to have him carried outside the company and given two or three goblets, so that he discharges what he has drunk. [Then] he is brought back into the company. (ibid.)

News of such machine-as-man delights travelled from Islam to Sicily, then to the court of Frederick the Great, and eventually throughout Europe. The "news" often concerned technicalities as well as gossip. For instance, al-Jazari's writings provided a 1,600-word explanation of just how the boon-companion's hidden syphon worked, and similarly detailed accounts of his other automata. So Western engineers were not only inspired but also instructed. Countless European automata were built as a result.

Medieval automata in Europe included a device built by the Spanish theologian and missionary Ramón Lull (*c*.1234–1316)—and inspired by his experience of Islamic culture in North Africa. This wasn't an android toy to delight the senses, but a practical device to support the mind. That is, it was an early attempt to mechanize logical (non-numerical) reasoning. In his writings on the *Ars Generalis Ultima* or *Ars Magna (Ars Inveniendi Veritatum)*, Lull claimed to have schematized the content of all natural and theological knowledge, and to have formulated rules for reasoning within it. These were, in effect, forerunners of Venn diagrams (M. Gardner 1958: 28–59).

As Lull's first Latin label declared, his system was intended as a truly *general* problem-solver, capable of generating all knowable truths. He believed that it embodied "an Art of thinking which was infallible in all spheres because based on the actual structure of reality, a logic which followed the true patterns of the universe" (Yates 1982: 12). His justification, given in *The Hundred Names of God*, was that: "If understanding followed no rule at all, there would be no good in the understanding nor in the matter understood, and to remain in ignorance would be the greatest good." (This was specifically aimed at the Arab theologian Averroes, who taught that something could be false in philosophy but true in theology—a special case of the intriguing cultural–cognitive phenomenon discussed in Chapter 8.vi.b–e.)

Lull built several machines to embody this knowledge and reasoning. They're remembered today largely because—along with the *Ars Magna* itself—they inspired Gottfried Leibniz to build his calculating machine four centuries later (Section ix.a, below).

They'd also inspired the famous/infamous Cornelius Agrippa (1486–1535), and the equally renowned and controversial John Dee (1527–1607). Whether these two men should be seen as (respectable) "mathematical" scholars or "alchemists", or as (dangerous) "magicians" was a question that exercised Leibniz's near-contemporary Gabriel Naudé (1600–53). As remarked in the Preface (preamble), Naudé favoured the former description.

Jonathan Swift was less impressed, and put his famed talent for mockery into top gear. In *Gulliver's Travels*, he described a professor in Laputa's Lagado Academy who'd built a forty-handled frame that filled the room. The frame contained many squares of wood, linked by slender wires and carrying bits of paper displaying "all the words of their language, in their several moods, tenses, and declensions, but without any order". The order was provided by the professor's forty students, who turned the iron handles at random. Whenever they found three or four words juxtaposed which could form part of a sentence, they told four students assigned as scribes to write the word strings down. The professor's intention, said Swift, was that:

> the most ignorant person at a reasonable charge, and with a little bodily labour, may write books in philosophy, poetry, politics, law, mathematics and theology, without the least assistance from genius or study. (Swift 1726: 227)

(With the benefit of nearly 300 years of hindsight, one might see Swift as a soulmate of Senator Proxmire, who complained bitterly that the US government was funding apparently trivial—but ultimately useful—research on the sexual behaviour of the screw-worm fly: Chapter 6.iv.f.)

Some of Lull's machines were worked by levers or cranks (hence Swift's image of the many iron handles). But the most famous was a set of fourteen concentric discs of metal, wood, and cardboard (M. Gardner 1958). Each disc bore up to sixteen different symbols (denoting different types of knowledge) on different segments. The discs could be individually rotated, and their possible alignments allowed for a huge variety of symbol combinations—in other words, a huge variety of propositions, or 'truths'.

The machine's usefulness (for Lull) included the fact that it showed the dogma of the Trinity to be logically possible, so that Jews and Muslims—who shared the core beliefs of monotheism—might eventually be persuaded to accept it as true. However, his work was condemned by Pope Gregory XI, who saw it as confusing faith with reason. (This was understandable: *no* religion is expressible in strict logical form: see Chapter 8.vi.b–d.) Nevertheless, the Church eventually blessed Lull, who'd also said that faith as well as reason was needed to appreciate the highest Christian beliefs.

Rotating discs, of course, aren't so immediately engaging as android devices. Most of the many medieval examples of machine as man were less austere—and less 'useful'—than Lull's, bearing some visible likeness to human or animal forms and movement. Among these was a talking iron head constructed by Albertus Magnus in the thirteenth century—which was deliberately destroyed by his pupil Thomas Aquinas.

Some four centuries later, one writer said Aquinas had done this "because he thought it the Deuil, whereas indeede it was a meere Mathematical inuention" (quoted in A. Marr forthcoming). Another, namely Naudé, said he'd done it "meerely because he could not endure its excesse of prating" (Naudé 1625/1657: 254). As these deflationary remarks suggest, some seventeenth-century scholars understood automata as legitimate examples of the engineering arts. But others "who are so easily carryed away with the slender assurance of a common opinion" (Naudé 1625/1657: 254) saw them, still, in more threatening terms: not as mathematical wonders, but as dangerous magic. The "common opinion" to which Naudé referred, in his impassioned defence of the automata makers, included many of his literate contemporaries—not just the unlettered masses.

One of the hundreds of self-moving machines that had graced the Renaissance was a mechanical lion built by Leonardo da Vinci. Intriguingly, it combined (contrasted?) life with engineering. The lion "being brought into a large Hall before Francis the first, King of France . . . after he had a while walked vp and downe, stoode still opening his breast, which was all full of Lillies and other flowers of diuers sortes" (Marr forthcoming, n. 66).

In the late sixteenth century, added inspiration came from Hero's writings, which were translated into European languages from 1575 onwards (Boas 1949; Marr forthcoming, nn. 21–41). Although some copies reached the Bodleian soon after publication, they didn't circulate widely in England or France. Indeed, as late as 1648 John Wilkins (who would soon help to found the Royal Society) complained that "discourses [on automata] are for the most part . . . of great price and hardly gotten". And Naudé himself didn't possess any of Hero's works in his library (Marr forthcoming, n. 32). Nevertheless, they'd already been cited by various other authors, and had been highly influential in Italy.

The result was a plethora of "Androides" (Naudé 1625/1657: 254) and animal-like machines in the palaces and public spaces of Europe. (The word "android" was used by Naudé's English translator in the 1650s; but the *OED* dates its coinage to 1728, in the *Chambers's Cyclopaedia*'s description of Albertus Magnus' ill-fated talking head:

see Section ii.c.) By the middle of the seventeenth century, Europe could glory in two artificial armies of 100 men, besides the flesh-and-blood variety—sadly, more numerous. By the eighteenth, it was graced by countless moving ornamental figures and fountains. Some of these even toured the Continent. Visitors to an exhibition held in the Opera House in London's Haymarket in 1742 marvelled at Jacques de Vaucanson's mechanical musicians playing flute, tabor, and pipe, and at his all-too-lifelike digesting duck (see Section iv).

But, but... These wonders were focused on finding ingenious tricks to produce observable movements—or, sometimes, sounds: the hissing of snakes, the banging of drums, the blowing of trumpets. There was no attempt to model the mind as well as the body. (Strictly, this remark applies only to the later examples, for the mind–body distinction as we know it didn't exist before the mid-seventeenth century: see Sections ii–iii.)

Indeed, it's not clear that the automata builders were trying to model the body with a view to *understanding* it. With the interesting exception of the gifted engineer Vaucanson (1709–82), discussed in Section iv, even the post-Cartesians were challenging their practical ingenuity rather than their biological curiosity. In terms of the distinction drawn in Chapter 1.ii.b, these were technological, not scientific, enterprises.

The reason was that, prior to the seventeenth century, there was no philosophical tradition encouraging people to think about bodily function or behaviour in detailed mechanistic terms. There was Democritus, of course—but his atomic materialism could be pursued in practice only in the most general terms.

Some automata, even in the early days, may have been intended—like some early models of heavenly bodies—not as trivial toys but as tests of materialism. That is, they may have been artefacts "whose very existence offered tangible proof, more impressive than any theory, that the natural universe of physics and biology was susceptible to mechanistic explanation" (de Solla Price 1964: 9). But if so, they were general existence proofs rather than specific demonstrations of bodily mechanisms. As such, they weren't without interest in a pre-scientific and/or largely anti-materialist culture. However, if one wanted to know how our bodies actually work, more was required.

2.ii. Descartes's Mechanism

The emergence of 'man as machine' as a motivating theme for experimental science was due primarily to Descartes (1596–1650). As we'll see, his approach was *mechanistic* in two different, though closely related, senses:

* On the one hand, he believed that the principles of physics can explain all the properties of material things, including living bodies.
* On the other hand, he often drew explicit analogies between living creatures and man-made machines, seeing these as different in their complexity rather than their fundamental nature.

His approach was also *mechanical*: its principle of activity was the movement of one body on collision with another. But a philosophy can be mechanistic without being mechanical. Indeed, the mechanical philosophy died only a generation later, with Isaac

Newton (1643–1727). For Newton, the key principle of activity wasn't the familiar causation-by-contact but the "occult" force (action-at-a-distance) of gravity.

a. From physics to physiology

Moves towards mechanization (in the first sense) were already happening when Descartes was a young man. For others at the time had similar ideas. Galileo's advances in physics, and his insistence that "the great book of the universe . . . is written in the language of mathematics", had initiated what was later called "the mechanization of the world picture" (Dijksterhuis 1956). And comparable intellectual moves were being made in biology, too.

Even before Galileo (1564–1642), the structure of the human body had been greatly clarified by Andreas Vesalius (1514–64). His anatomical atlas of 1547 drew on careful dissection, and some animal vivisection, to correct many traditional beliefs due to Aristotle and Galen. (Leonardo's still earlier anatomical illustrations were as yet unknown.)

Moreover, the sixteenth-century physician Jean Fernel (1497–1558), whose textbook of physiology—it's his word—was still widely used in Descartes's time, had rejected Aristotle's view that all human bodily movement is informed by the rational soul (Sherrington 1946). He'd pointed out that many movements—such as breathing, shifts of position in sleep, and some movements of the eyes and eyelids—are independent of both will and sensation.

But Fernel *didn't* draw the conclusion that the body is a machine. Even these involuntary movements, and all the actions of animals, he thought to be due to some vital principle informing the material body—namely, the Aristotelian animal soul. The first person to put machine analogies to good effect in physiology was Descartes's contemporary William Harvey (1578–1657), widely regarded as the father of experimental physiology.

Like Vesalius before him, Harvey corrected some ancient anatomical beliefs, notably about the heart and blood vessels. And in relating anatomy to physiological function he took an explicitly mechanistic approach, describing the circulatory system on the analogy of water pipes fed by a pump (W. Harvey 1628, 1649).

Harvey supported his claims by a systematic programme of physiological experiments. His ingenious studies included careful quantitative measurements of blood volume, and close observations of blood flow in many different animals. These included cold-blooded creatures such as bees, wasps, shrimps, slugs, snails, mussels, frogs, newts, and eels . . . whose hearts beat much more slowly than those of mammals.

Significantly, this work attracted violent hostility from medical professors and practitioners. In part, this was mere social snobbery: like engineering, slugs and snails weren't considered fit topics for the attentions of gentlemen (Shapin 1991: 304–12). More to the point, many scholars weren't ready to accept the comparison of human physiology with that of lowly animals, or—Vesalius notwithstanding—to favour observation and experiment over ancient anatomical authorities. One leading opponent mentioned Harvey's work on "slugs, flies, bees, and even squill-fish" and wrote sarcastically: "We congratulate thee upon thy zeal. May God preserve thee in such perspicacious ways"; and he continued, "Dost thou declare, then, that thou knowest what Aristotle did not?" (quoted in Chauvois 1957: 222).

But even Harvey didn't regard the body as merely a machine: he believed that medicine must recognize the influence of the soul as well as the body in explaining disease. To be sure, biologists' talk of "souls" didn't necessarily imply anything supernatural. The souls posited by Aristotle to explain the properties of living things were a series of increasingly powerful organizing principles informing matter. Only the "rational" soul was—perhaps—metaphysically distinct (Matthews 1992; cf. Chapter 14.xi.a). But since these principles couldn't be detailed, it was all too tempting to see the lower degrees of soul as metaphysically mysterious also.

By contrast, Descartes aimed to banish talk of any type of soul from medicine and biology. His own physiological experiments, on the circulation of the blood, were crude by Harvey's standards. But it was he, not Harvey, who—at a time when radical scepticism was rampant, and occultism and magic rife—provided the philosophical rationale for a programme of physiological research that continues today. That's why he, too, is sometimes described as the father of scientific physiology.

His first publication, *Discourse on the Method of Rightly Conducting the Reason and Seeking for Truth in the Sciences* (1637), was a brief essay laying out the principles of his lifetime's work. It contained references to a much longer (scientific) treatise, whose publication he'd suppressed in 1634 on hearing of the Church's condemnation of Galileo; and its main points were elaborated in his later writings. (The quotations below are taken from Parts V and VI of the *Discourse*, unless otherwise noted.) It was written in French, not Latin, to invite those not blinded by the scholastic study of traditional authorities to consider how new knowledge might be discovered.

There, and in other works, he argued that a systematic experimental physiology was philosophically respectable, practically necessary, and even financially feasible. This scientific optimism, as we'll see, was grounded in four Cartesian claims.

The core claim was that the body is a mechanistic system, in the *first* sense distinguished at the opening of this section. That is, it functions according to the laws of physics (for Descartes, mechanics: see above), like everything else in the material world. This, he said, is true of all living things. It's therefore reasonable to expect that we might learn something about human physiology from studying animals, newts not excepted.

Descartes had no qualms, therefore, in suggesting in his *Dioptrics* that a tube of water fixed to the front of the eye could improve someone's sight, since "Vision will occur in the same manner as if Nature had made the eye longer than it is." He even said that the pupil of the biological eye could then be removed entirely, since it would become "not only useless but even deleterious, insofar as it excludes, by its smallness, rays that could otherwise proceed toward the edges of the back of the eye". As a recent scholar has remarked, this would provide "a hybrid, a fusion of machine and organ, superseding the eye God gave us, but no less 'natural' " (Des Chene forthcoming).

No one could actually do that yet, of course. But the point is that—on Descartes's view—'integrated' internal prostheses, and even artificial sense organs, were possible *in principle*. Now, over three centuries later, they exist. For examples of the former, consider plastic heart valves or electronic pacemakers. As for the latter, instead of Descartes's tube of water consider Paul Bach-y-Rita's 'visual' tickling pads, located on the back or on the tongue (Bach-y-Rita 1984, 2002). Even artists have got in on the act: the Australian artist Stelarc deliberately pushes/questions the boundary between man and machine in his extraordinary performances involving specially designed robots 'wired' into his own nervous system (Stelarc 1984, 2002; see Chapter 13.vi.c).

It was this rigorous mechanism which led Descartes, mistakenly, to explain the heart's action in a way very different from Harvey. One of the first to acclaim Harvey's work, he had much to say, in Part V of the *Discourse*, in praise of "the English physician". Besides his many post-mortem dissections (and his frequent visits to abattoirs and gallows), he followed Harvey in vivisecting a few animals—fishes, eels, and a hare—to study the action of their hearts (Lindeboom 1979: 106–22). And he, too, related anatomical structure to physiological function. For example, he explained the different number of membranes, or "little doors", in the mitral and bicuspid valves in terms of the shapes of the heart openings they closed (oval or round, respectively). But he disagreed on a crucial—and philosophically relevant—point.

For Harvey, the heart was a mechanical pump, whose muscular contraction (systole) pushed the blood out into the arteries. For Descartes, it was a heat engine, and diastole was the active moment. The heart, he said, was the hottest part of the body. The expansion of the heart wall, and so the flow of blood out of the heart, was caused by the rising pressure of the venous blood, which will "promptly expand and dilate" on entering the heart ventricles, "as liquids usually do when they are allowed to fall . . . into some very hot vessel".

This (mistaken) theory, you'll notice, relied on purely physical principles, not on some unexplained vital property: the heart's inherent power of muscular contraction. "In supposing the heart to move in the way Harvey describes, we must imagine some faculty or inherent power that causes the movement," said Descartes—and this he was not prepared to do. (It wasn't until very much later, in my own student days, that the contractile power of muscle began to be understood: see Preface, ii, and Section viii.e, below.)

b. Science as cooperation

As a metaphysician, Descartes celebrated the solitary ego. As a scientist, he stressed the need for cooperation. From the vantage point of the twenty-first century, we can appreciate his prescient claim that *collective* empirical study would be necessary to discover the actual mechanism of the material world.

Francis Bacon (1561–1626) had said much the same already. He'd argued in *The Advancement of Learning* (1605) and *The New Organon* (1620) that "natural philosophy" (*science*, in its current meaning, is a late nineteenth-century word) should be built on facts, not theories. And he gave detailed advice on how to do inductive reasoning systematically—and reliably. This covered both the observation of "Nature free and at large (wherein she is left to her own course and does her work her own way)" and also experimentation, wherein "by the art and hand of man [Nature] is forced out of her natural state, and squeezed and moulded".

Significantly, he added that scientists needed not only the real world but also each other. Whereas the medieval scholastics had resembled spiders, weaving fragile webs of words out of their own innards, scientists should emulate the bees, collecting nutriment from the real world *and working on it collectively in the hive.* Shortly before his death, he'd imagined a country with a scientific community called Solomon's House, "dedicated to the study of the works and creatures of God" (Bacon 1624).

Solomon's House, however, was still but a dream. As Lord Chancellor of England, Bacon had had access to the king. He'd tried repeatedly to persuade James I to found a college for experimental science, to encourage learned societies fostering scientific research, and to establish chairs of experimental science at Oxford and Cambridge. But none of those things happened in his lifetime—which ended just before Descartes wrote the *Discourse.*

Descartes was no Baconian. Whereas Bacon had exulted in his vision of scientific fact gathering, Descartes saw this activity as a regrettable practical necessity. His rationalism led him to say that every scientific law follows deductively from the first principles of physics—which, in turn, follow from the nature of God. Ideally, then, science could be conducted from the armchair—and he himself inferred various physical laws (such as the conservation of "motion") a priori.

But unfortunately, this could be done (by human minds) only up to a point. Only God, he said, has the intellectual power to carry out the deductions to the most detailed level; and only God knows which of the many theoretically possible worlds He has freely chosen to create. Human beings therefore have to rely not only on logical deduction but also on empirical observation.

Moreover, for Descartes (unlike Bacon) observation must be theoretically guided. That's why, in discussing the heart, his a priori mechanism had won out over Harvey's careful experiments and his claim to have "publicly shown, that arteries increase in volume because they fill up like bags or leather bottles, and are not filled up because they increase in volume like bellows" (W. Harvey 1628: 13). In the *Discourse*, Descartes made no mention of Harvey's telling observations of blood flow in newts and the like, giving more weight to machine analogies. He declared that "those who do not know the force of mathematical demonstration and are unaccustomed to distinguish true reasons from merely probable reasons" may not realize that the heart action as described by him "follows as necessarily from the disposition of the organs . . . as does that of a clock from the power, the situation, and the form, of its counterpoise and of its wheels".

But he did believe that the painstaking experimental research needed to apply Galilean principles to biology would involve the active cooperation of many people over many years. This (second) claim led him to publish the *Discourse* in order that:

> the best minds would be led to contribute to further progress, each one according to his bent and ability, in the necessary experiments, and would communicate to the public whatever they learnt, so that one man might begin where another left off; and thus, in the combined lifetimes and labours of many, much more progress would be made by all together than any one could make by himself. (Part VI)

This would be feasible, he remarked, if their time was funded by wealthy philanthropists. As for his own work, he closed the *Discourse* with thanks to "those by whose favour I may enjoy my leisure without hindrance", as opposed to "any who may offer me the most honourable position in all the world".

Descartes's hopes (and Bacon's) for the cooperative development of experimental science were soon to be realized. Informal communication networks of scholars—what

Robert Boyle called "the invisible college" (de Mey 1982: 133 ff.)—had been developing in Europe from around 1630. For science to move forward as Descartes had hoped, these invisible colleges would have to be made visible (cf. de Solla Price and Beaver 1966). In the 1660s they were officially chartered by Charles II and Louis XIV as the Royal Society of London and the Académie royale des sciences in Paris (Hunter 1981, 1994).

The two monarchs may have been part-moved by scientific curiosity, but it doubtless helped, in prompting their generosity, that the people on the receiving end of their sponsorship were drawn from the highest levels of society. All the early Fellows and Academicians were "gentlemen", and a goodly few of these were high-born aristocrats. Boyle, for instance, was a son of the Earl of Cork, and brother to a crony of the king (Shapin 1994: 138).

Significantly, this fact made *science*, not just *sponsorship*, possible. Cooperation could readily take place, because people in that social stratum shared a cultural expectation of trustworthiness (Shapin 1994, chs. 3 and 5).

This wasn't mere wishful thinking or complacent self-congratulation. All children naturally develop trust, including discrimination between more and less reliable informants (Harris and Koenig forthcoming). If they didn't, cultural sharing of non-observational knowledge would be impossible. But in certain social groups, being seen as trustworthy is especially important.

In Descartes's time, men of gentle birth faced huge social costs on being judged dishonourable. The code of chivalry had encouraged deadly duels in defence of one's honour, a practice that continued—though much abated—into the nineteenth century. By the seventeenth, the gentry had learnt, as a result, to practise probability.

That is, they'd realized the importance of making it clear when one *wasn't* committing oneself to a fact, but merely expressing a challengeable opinion. Books of gentlemanly etiquette urged them to protect themselves by using expressions such as "almost". Indeed, two popular sayings were: "Almost was never hang'd" and "Almost and very nigh, save many a lie" (Shapin 1994: 114). They were even counselled not to relate astonishing facts which they *knew* to be true, for fear of tarnishing their reputation—a form of self-denial still practised by scientists today (see Chapter 14.iv.b). This distinction, originally developed largely to avoid the social disaster of being branded a liar or incompetent, could be directly transferred to the new scientific discourse of 'empirical data' and 'theoretical speculation'.

In short, the gentleman's culture (and language) of credibility eased the acceptance of other scientists' observational reports. Everyone didn't have to repeat everyone else's experiments, even if they disagreed—politely—on the theory.

Initially, this was true *provided that* the observers were leisured amateurs, with no pecuniary interest in 'observing' one thing rather than another. So the observations of the Royal Society's Robert Hooke, who was officially a waged technician, were often subjected to doubt, or to verification by others, because of his lowly social status. His lack of social standing overshadowed his intellectual brilliance.

("Brilliance" is no exaggeration: L. Jardine 2003. Among many other achievements, it was Hooke who, in a letter to Newton, first suggested the inverse square law of gravitational attraction; and besides originating the eponymous Hooke's Law of elasticity, he also helped in the discovery of what's now called Boyle's Law.—None of that, unfortunately, endeared him to Newton: see Chapter 5, preamble.)

Gradually, these social conventions were being developed into relatively explicit canons of acceptable scientific behaviour, or "epistemological decorum" (Shapin 1994: 193). New conventions concerning how to do experiments, how to report them, and how to handle the scientific conversations provoked by them, were painstakingly established. (Whether it follows, as two historians of seventeenth-century science have argued, that the "facts" and "laws" accepted as a result are mere social constructions, not objective truths, is another matter: see 1.iii.b, and Shapin and Schaffer 1985.)

As a result, Descartes's hope was realized: one man could begin where another left off. Indeed, the Royal Society saw itself as following in Descartes's footsteps—and Bacon's. It took their advice not only on how to make new empirical discoveries (hence its motto, *Nullius in verba*: "On no man's word"), but also on fostering fruitful communication. It was Solomon's House, on London ground.

c. Cartesian cooperation develops

But how was the entry to the House to be widened? In Descartes's time, that wasn't clear. When he expressed the hope that scientists "would communicate to the public whatever they learnt", what he meant was very different from what we'd mean by those words now. It even differed from what would be meant a mere hundred years after him.

In his time, the early seventeenth century, there were only a few philosophical/scientific books, and only a few readers too—experimental philosophy was still very much a minority taste. The books were expensive, and published in the scholar's lingua franca, Latin. (Although the *Discourse* was—shockingly—written in the vernacular, so as to reach outside the charmed circle of scholars, Descartes's other works appeared in Latin as usual.) Moreover, scientific discussion was carried on primarily by individual correspondence, not in public places or in generally accessible print. The Royal Society's *Journal-Book* provided a permanent record, but it wasn't a public document. Descartes's dream of a growing community of cooperating scientists would require significant cultural changes in methods of communication.

Over the eighteenth century, the situation was transformed. Gutenberg finally came into its own, as western Europe—especially, England—became a "print culture" (Porter 2001, ch. 4). England was prominent here because of legal changes associated with the political revolutions of the 1640s and 1688, which not only reduced censorship but also enabled presses to be set up across the land. (Previously, not even advertisements or theatre tickets could be printed outside London, York, or Oxbridge.)

The reading public burgeoned—one might rather say, it *began*—as books became commonplace and Latin took a back seat. And not only books: newspapers and periodicals (including magazines *about books*) were founded, provincial as well as metropolitan, and ephemeral pamphlets abounded. The numbers speak for themselves:

> About 6,000 titles had appeared in England during the 1620s; that number climbed to almost 21,000 during the 1710s, and to over 56,000 by the 1790s. . . . Between 1660 and 1800 over 300,000 separate book and pamphlet titles were published in England, amounting perhaps to 200 million copies all told. (Porter 2001: 73)

These writings in the vulgar tongue reached way beyond the gentry: "Prodigious numbers in inferior or *reduced* situations of life", as one 1790s bookseller put it, could potentially benefit—thanks in part to his own pioneering of cheap remaindering (Porter 2001: 75). A visiting Irish clergyman in 1775 had wondered at seeing workmen reading the gazettes: "a whitesmith in his apron & some of his saws under his arm, came in [to the tavern], sat down and called for his glass of punch and the paper, both of which he used with as much ease as a Lord" (Clifford 1947: 58). And Samuel Johnson had remarked in the late 1770s that "General literature now pervades the nation through all its ranks," virtually every household being "supplied with a closet of knowledge" (Altick 1957: 41).

Most of these publications concerned religious, political, and moral–personal matters—the latter addressed not only in printed sermons but also in the new literary genre of the novel (J. M. Levine 1992). But some dealt with scientific topics, such as the exciting discoveries of Newton (whose *Principia* had appeared in 1687, and *Opticks* in 1704), and the empiricist philosophy of John Locke (1690).

Locke was championed several times by Joseph Addison in 1712, in the pages of his recently founded *Spectator*. And Addison wasn't alone: the demanding volumes of Locke, and especially of Newton—both early members of the Royal Society, and one its long-time President—were repeatedly rewritten by other hands. They even appeared as brief abstracts in the *Young Students' Library*, the *Reader's Digest* of the 1690s. So too did Thomas Sprat's (1667) account of the growing *community* of science, namely the Royal Society.

By the early 1700s, readings from these popularizing books, and Newtonian lectures with ingenious practical demonstrations, were regularly given in London coffee houses. By mid-century, this was happening in provincial cities too (Porter 2001: 142–5). In a word, science was now fashionable (Schaffer 1983). At last, most literate young men (of course!) had the opportunity to learn about it and to be excited by it.

No doubt, the popularizers' writings weren't wholly reliable. (*Plus ça change*...) But they whetted the appetite. (And in Sprat's case, they even pointed to the source community *qua* community.) Publications started to appear whose main aim was to educate, even to entice. The first modern English encyclopedia, the *Lexicon Technicum* of 1704, was slanted towards science, which was prominent also in Ephraim Chambers's two-volume *Cyclopaedia, Or an Universal Dictionary of Arts and Sciences* of 1728. This not only gained Chambers election to the Royal Society and burial in Westminster Abbey, but spurred the French Philosophes to embark on the hugely influential *Encyclopédie* (see Chapter 9.iii.d).

Shortly afterwards, books written specially for children appeared for the first time (Porter 2001: 348). Moreover, the publisher of the phenomenally successful *Little Goody Two-Shoes* (sixty-six editions between 1766 and 1850) also produced a popular children's book on science in 1794. Its official title was *The Newtonian System of Philosophy* (and its encouraging subtitle *Adapted to the Capacities of Young Gentlemen and Ladies* [*sic*]), but it was popularly known as *Tom Telescope* (Secord 1985). It depicted a young man of that name, invited by some children into their nursery to entertain them with six lectures/demonstrations on Newtonian physics. *Tom Telescope* was a staple on children's bookshelves for nearly a century.

By the beginning of the twentieth century, such scientific primers were regularly intriguing young minds. We'll see in Section viii.f, for instance, that one of these was read by Alan Turing as a child, who credited it in later life with having opened his eyes to science. Perhaps you, too, remember being drawn in long ago by brightly coloured pages on *Pond Life*, or by *The Story of a Molecule*? (And perhaps your eyes were opened wide by the devilish illustrations in the first undergraduate textbook of cognitive science? See 6.v.b.)

In the late twentieth century, the researchers themselves were sometimes doubling as popularizers, deliberately writing in such a way as to encourage budding scientists, aka graduate students, to work with them (Chapter 12.vi.a). Indeed, 'trade' books (and TV programmes) were increasingly focused on science, infuriating the purists but enlarging the interested public. In 1991, for instance, two influential books on consciousness—one by a science journalist (Nørretranders 1991), the other by a professional cognitive scientist (Dennett 1991)—appeared from non-academic publishers. Both sold widely, arousing interest in the topic and, in the latter case, contributing provocative original ideas (14.x.a).

Meanwhile, scientific journals had become an accepted tool—and trumpet—of professionalism. The Royal Society's pioneering *Proceedings* and *Philosophical Transactions* continued, spanning all areas of science—including cognitive science. And new research areas had spawned new journals, as we'll see, from the *Bulletin of Mathematical Biophysics* in the 1930s to the *Journal of Consciousness Studies* in the 1990s. There was even a revival of science-as-correspondence, as email blossomed. Eventually, important draft manuscripts became openly available on the World Wide Web (Chapter 7.i.e): Solomon's House in cyberspace.

Besides these examples of scientific collegiality, late twentieth-century cooperation had achieved a new form: extensive research collaboration. Multi-authored work had become the scientific norm (and email allowed people based in several different countries to work together). Indeed, one of the papers cited in Chapter 15.x.b lists no fewer than twenty-five names. Those twenty-five collaborators were biochemists and geneticists, working in "wet" A-Life. And a nine-author paper cited in Chapter 10.iv.c involved medical clinicians and AI scientists, collaborating in order to test an "expert system" designed for prescribing antibiotics.

In cognitive science proper, multi-authoring is common—but much less luxuriant, rarely rising above four. Long strings of names typically occur when the use of complex instrumentation means that expert technicians are cooperating with the theoreticians. So, for instance, two seven-author papers cited later concern a brain-imaging study of schizophrenic hallucinations (7.i.h), and recordings from "face-and-gaze" cells in visual cortex (14.iv.b). But sometimes, multi-authorship reflects an especially wide interdisciplinarity (as in Bedau *et al.* 2000, cited in Chapter 15.xi.c).

This research practice makes it even more difficult than usual to attribute a discovery to this person or that one (see 1.iii.e–f). Readers of the "manifesto" for cognitive science (G. A. Miller *et al.* 1960), which listed a triumvirate on the title page, joked that one author thought of it, another wrote it, and the third believed it (6.iv.c and 17.i). They were making an important intellectual point about the book, but the joke was funny partly because such inter-author distinctions are rare. Only the cognoscenti know just which person was mostly responsible for just which aspects of a multi-authored publication. (The journal *Nature* recently tried to rectify this situation, by inviting authors to spell out their individual contributions at the end of the paper: "After a slow start, more and more authors are responding to this situation": *Nature* 2005*a*.) And an eminent name may swamp a lesser one: an ingenious way of reading the minds of very young babies was largely due to a graduate student, but because his co-author was a very famous man his contribution is often forgotten (6.ii.c).

Multi-authorship wasn't what Descartes had had in mind. He'd foreseen the need for collegial communication: scientists would talk to each other, across the table or in letters and (rarely) publications. But then they'd each go off and do their own thing. By the millennium, however, both types of cooperation were in full swing.

To cap it all, jet travel enabled the oldest method of all: talking face to face. Chatting over the coffee table is especially helpful when those holding the coffee cups differ in outlook and expertise, since explanation and enthusiasm can readily be combined. It's no accident, then, that several key interdisciplinary gatherings are recognized as having been crucial for the birthing of cognitive science and of specific areas within it (Chapters 4.v.b, 6.iv.b, 12.v.b and vii, and 15.ix.a). And established conference series, like established journals, helped to keep the conversations going.

In brief, the scientific cooperation that Descartes had dreamed of was thriving within cognitive science. Admittedly, it encompassed more than a little ill-tempered rivalry too (see Chapters 4.viii, 9.viii.a and ix.a, 11.ii.a–e, and 12.iii–vii). But the communication channels were there.

As for the wealthy philanthropists he'd asked for, they were doing their bit. A few were precisely that: wealthy individuals, personally encouraging research they regarded as interesting. Edward Fredkin and Hugh Loebner both offered prizes of $100,000 for specific AI achievements (see Chapter 16.ii.c), and some of the UK research mentioned in Chapter 7.i.f was supported by Gerry Martin's Renaissance Trust.

Others were commercial firms (such as Bolt, Beranek & Newman), private organizations (like the Rockefeller and Alfred P. Sloan Foundations), public charities (like the Imperial Fund for Cancer Research), or privately funded universities. Some of these put millions of dollars into cognitive science, but even so small a sum as $7,500 could have a huge effect (Chapter 6.iv.f).

The most generous sponsor of all was the taxpayer. The Royal Society had long used its private money to sponsor research by non-Fellows: so in 1768 it had funded Lieutenant James Cook's voyage to Tahiti on the *Endeavour*, to observe the transit of Venus on behalf of the astronomers. (Later, they'd elect Cook as a Fellow.) Now, 200 years later, the Society was partly government-funded—and it supported some endeavours in cognitive science. For instance, Christopher Longuet-Higgins and Horace Barlow were both Royal Society research professors for many years (see Chapters 6.iv.e, 12.v.c, and 14.iii.b); and several relevant symposia were sponsored by the Society—and by its 'twin', the British Academy (e.g. Longuet-Higgins *et al.* 1981; Boden *et al.* 1994; Parker *et al.* 2002).

Other tax-funded work was paid for by governmental departments—notably, the USA's Department of Defense, or DOD (Chapter 10.ii.a, 11.i, and 12.vii.b). Although much of the DOD money was targeted on specific applications, or on unsolved scientific problems chosen for their practical relevance, some was set aside for speculative, blue-sky, research. And some of that was given with no strings attached: laboratory directors could spend it on whatever they liked, including high-risk maverick projects. This resembled the attitude of (some) aristocratic patrons three centuries earlier. Neither Descartes nor Locke (who was generously supported by Lord Ashley,

Earl of Shaftesbury) ever had to submit specific research proposals for vetting by lesser minds.

So, for instance, Paul Armer, director of RAND's computer science when NewFAI was getting off the ground, was told "Here's a bag of money, go off and spend it in the best interests of the Air Force" (interview in McCorduck 1979: 117). Similarly, Joseph Licklider at the DOD's Advanced Research Projects Agency (ARPA) not only funded the first university AI labs but let them decide what to spend the money on. Without that intellectual leeway, the early history of cognitive science would have been very different (Chapters 6.iii.b, 10.ii.a, and 11.i.a–b).

In addition, taxpayers' money came via national universities and research councils, and from centralized research organizations—such as the UK's National Physical Laboratory, France's CNRS (Centre national pour la recherche scientifique), and several of Europe's Max Planck institutes. If one of these agencies reversed its funding policy, this could have huge repercussions on cognitive science research (see 9.x.f, 11.i and iv, and 12.iii.e and vii.b).

We take this state of affairs for granted. So much so, that one paper contains the joking plea: "More research is needed, send money!" (Mayhew 1983: 215). Long before the days of jet aeroplanes or income tax, Descartes couldn't assume that international get-togethers or public funding would be available. He did, however, point out the need.

d. Descartes on animals—

A programme of systematic physiological, including neurophysiological, research was presaged by Descartes's third claim: for animals, the body is all there is. And his repeated descriptions of animals as machines show Descartes to be a "mechanist" in both senses defined at the opening of Section ii.

Descartes didn't merely state that animals are machines. He offered hypotheses about just what sort of machines they might be, and encouraged experimentalists to test them. And he repeatedly stressed the analogy between animate motion and the movements of the marvellous lifelike devices then visible in Europe's public places, some of which were triggered into motion by deliberate or unintentional movements on the spectator's part (in effect, an early form of interactive art: see 13.vi.c). They included the moving statues of Diana and Neptune, nymphs and satyrs, in the spectacular grotto fountains of the royal palace at Saint-Germain.

He first saw these fountains when he was 18 years old, and recovering from a breakdown. One commentator has speculated that "these glistening, fizzing statues in their eerie torchlit world became the surrogate friends of a brilliant intellect unable to cope with people" (Jaynes 1970: 224). However that may be (Descartes as the ancestor nerd?), Descartes later insisted that animals, in effect, are even more marvellous automata: "if there were machines with the organs and appearance of a monkey, or some other irrational animal, we should have no means of telling that they were not altogether of the same nature as those animals".

This applied to animals' behaviour, as well as to their physiology. Descartes made this clear in a letter to the Marquess of Newcastle:

Doubtless when the swallows come in spring, they operate like clocks. The actions of honeybees are of the same nature, and the discipline of cranes in flight, and of apes in fighting, if it is true that they keep discipline. (Descartes 1646: 207)

The nervous system, he insisted, is no less mechanical than the circulation of the blood. It functions by means of "the animal spirits, which resemble a very subtle wind, or rather a flame which is very pure and very vivid, and which [proceed from the brain] through the nerves to the muscles, thereby giving the power of motion to all the members". He even dreamt up a mechanical explanation of why the muscle stops contracting at the right time. When it has reached the appropriate length, he said, it pulls on a tiny thread inside the hollow nerve, which closes the valve in the brain through which the animal spirits were flowing to the muscle. (This may have been the first proposal for a feedback mechanism in the nervous system: cf. Chapter 4.v.c.)

In animals, Descartes wrote, there's nothing more than this. In other words, their behaviour is based in what we now call reflexes (see Section viii, below). Specifically, they have no soul of any kind. There is therefore no reasoning faculty, and no judgement or self-conscious thought behind any animal behaviour. This applies even to what we'd normally describe as action based on perception, as when we say (to take one of his own examples) that the lamb saw/smelt the wolf, and ran away to escape from it.

(Some of Descartes's mechanistically inclined contemporaries were willing to say that animals are machines, but *not* that they lack souls. Alphonsus Borelli (1608–79), for instance, suggested a purely mechanical explanation for how the muscles can "thicken" without acquiring any extra matter, and how they can move the bones to which they're attached. The first question wouldn't be satisfactorily answered until the mid-twentieth century: see viii.e below. But the *source* of movement, said Borelli, lay in the animal's soul.)

e. —but just what did he mean?

Here, we must avoid a subtle misunderstanding whose roots go back 300 years—and for which cognitive scientists in general have recently been castigated (Baker and Morris 1996: 69–100, 124–38).

Descartes is widely believed to have taught that animals have no consciousness, no experiences of vision, hearing, or pain. Indeed, his follower Nicolas Malebranche (1638–1715) explicitly argued that animals can't feel pain. Significantly, however, Malebranche's argument wasn't that animals are machines and therefore can't feel pain. Rather, it was that animals (being machines) aren't moral beings, so can't sin—and pain is God's punishment for sin.

That animals are machines meant, to Descartes, that they have no rational soul—nor any Aristotelian sensitive soul either. (St Thomas Aquinas had said that animals are machines too, but he'd been denying only the rational soul.) It *did not* mean that they weren't "sensitive", that they couldn't "see", "hear", or "smell", or even (*pace* Malebranche) have "pain". Nor did it mean that they couldn't endure the passions of "fear", "hope", and "joy" (Cottingham 1978). However, all these terms were interpreted by Descartes differently from how we understand them—hence my use of scare quotes (cf. Malcolm 1973).

Descartes himself put it like this:

> I should like to stress that I am talking of thought, not of... sensation; for... I deny sensation to no animal, *in so far as it depends on a bodily organ.* (quoted in Cottingham 1978: 557, italics added)

In other words, by calling animals "sensitive", or "sentient", Descartes seems to have meant what we mean by 'responsive'—which (for us) leaves it open whether or not there is (what we call) conscious experience. (More accurately, I should have said "for *most* of us": see Chapters 14.xi and 16.iv.)

Similarly, by "pain" Descartes meant an abnormal state of the body that is potentially harmful to the creature and which, when (mechanically) discerned by the body's "inner sense", leads it to try to avoid it in various ways. And by "hunger", he meant a physical state, caused by an empty stomach, which (when sensed) prompts food seeking. In all these cases, Descartes insisted that animal "sentience", including animal "passions", can be wholly explained in terms of mechanical processes in the body. He even posited seven different types of sensory nerve, for the five external and two inner senses.

Crucially, Descartes could neither have asserted nor denied what most philosophers today (but not all: see 16.viii) regard as obvious: that many animals have sensory experiences, or "raw feels", or "non-conceptual content", or a subjective feeling of "what it is like to be...". These concepts weren't in his repertoire, and they couldn't have been added by a ten-minute tutorial. A more judgemental way of putting this point is to refer to his "notorious confusion... between mere consciousness and reflexive [self-referential] consciousness" (B. A. O. Williams 1978: 286).

For Descartes, there was the responsive ("sensitive") body, and there was the rational soul—the origin of judgement, concepts, free choice, and self-consciousness. Arguably, he posited a third kind of property (though not a third ontological category), namely "confused perceptions", possessed only by humans (Cottingham 1986: 122–32). He believed that human perception is grounded in the sense organs, but also involves judgement. For example, we say we see men walking down the street—but, according to Descartes, all we really *see* is cloaks and hats (or even moving colour patches). Because perception involves judgement, it may be mistaken. He instanced someone's claiming to have a "sensation" of pain in a phantom limb: this isn't sensation but perception, he said, since it involves (erroneous) judgement. Animal 'perception', by contrast, is mere responsiveness, with no admixture of thought or judgement.

Suppose that a time traveller from our own century had asked him whether animals, besides being responsive, also have "raw", mental but non-judgemental, conscious experiences—which seems to be what we now mean by "sensations". They'd have been disappointed. He wouldn't have understood the question.

In short, Descartes didn't—couldn't—*deny* that animals have such experiences. But in saying that they're sensitive, or that they see, hear, smell, and have pains, he wasn't *asserting* it, either.

It follows that one must be very careful in juggling the cross-centuries ambiguities when one asks whether cognitive scientists—and, for that matter, the early physiologists—have radically misunderstood Descartes. Consider this comment, for instance:

Since consciousness is standardly seen as including sentience, the suggestion [made by "analytic philosophy" in general and "cognitive science" in particular] that the force of the *Bête-Machine* Doctrine was to deny consciousness to animals is thus 180 degrees off course. (Baker and Morris 1996: 91)

The use of the term *sentience* is problematic here. As we've seen, Descartes ascribed sentience to animals. But—unlike the Aristotelians—he explained sentience mechanically, not in terms of a "sensitive" soul, an organizing principle not found in animate nature. And—unlike his twentieth-century successors—he didn't interpret it as a form of non-judgemental consciousness. Given that these very points were made by the authors of the criticism just quoted, their description of cognitive science's view of Descartes's position as "180 degrees off course" is bizarre.

Part of the problem, of course, is that we don't know just what *we* mean by consciousness, experience, sensations—in other words, what today's philosophers call *qualia*. Indeed, some philosophers of cognitive science explicitly *deny* the existence, even in humans, of experiential qualities over and above the functional properties of the brain. The comforting, and widespread, notion that the concept of consciousness—if not its explanation—is clear to everyone whose IQ is larger than their shoe size melts away when put under a philosophical spotlight (see Chapters 14.xi and 16.iv).

In my judgement, it's nearer the truth to say (anachronistically) that Descartes denied animal consciousness than to say (also anachronistically) that he admitted it. Certainly, if—as many people believe—sensory experience is something over and above brain states, inexplicable by objective science, then he would not have ascribed it to animals. In that sense, the misunderstanding isn't so much of a misunderstanding after all.

Moreover, it's been the usual interpretation of his work since the late seventeenth century, when the empiricist notion of 'pure' experience (ideas, sense data) untainted by judgement or reflection became widely current. In discussing anyone's historical influence, one must normally focus on what people thought they said, even if this differs from what they actually did say. For these two reasons, I'll sometimes assert, without further qualification, that Descartes denied animal consciousness.

f. Vivisection revivified

Besides his mechanism (in both senses), Descartes had a theological reason for denying that animals have rational souls. He argued in the *Discourse* that the soul was "separable" from the body (see Section iii)—and therefore immortal. And in a letter of 1649, he wrote to a doubter: "It is less probable that worms, gnats, caterpillars, and the rest of the animals should possess an immortal soul than that they should move in the way machines move."

But there was, and still is, an alternative. Perhaps animals do indeed have souls—some principle that "animates" the body in some way, perhaps even involving a lowly form of consciousness—but, unlike Christian souls, these die with them.

This broadly Aristotelian position seems to have been held by the Jansenist priest Antoine Arnauld (1612–94), who doubtless shared Descartes's scepticism about the immortality of caterpillars:

I fear that this belief [that animals have no soul] will not carry persuasion into men's minds, unless supported by the strongest evidence. For at the first blush, it seems incredible that there is any way by which, without any intervention of the soul, it can come to pass that the light reflected from the body of a wolf into the eyes of a sheep should excite into motion the minute fibres of the optic nerves and, by the penetration of this movement to the brain, discharge the animal spirits into the nerves in the manner requisite to make the sheep run off. (trans. Haldane and Ross 1911: ii. 85–6)

Descartes's reply was unyielding. Citing the example of someone who, in falling, automatically puts out their hand to protect their head, he asked: "why should we marvel so greatly if the light reflected from a body of a wolf into the eyes of a sheep should be equally capable of exciting in it the motion of flight?" (trans. Haldane and Ross 1911: ii. 104).

Although he didn't go on to say so, it was clear that the "strongest evidence" requested by Arnauld would necessitate neurophysiological research—including experiments on living animals. As we've seen, Descartes did some such experiments himself, and recommended others to do them.

Many before him, including Vesalius and Harvey, had done so too. And although the experimental animals' suffering could, if one wished, be minimized, pain relief was impossible (awaiting the discovery of anaesthesia in the late nineteenth century).

What's not clear is whether the Cartesians in general did wish to minimize the suffering, or whether they thought there was no suffering to minimize. As we've seen, Malebranche argued—on *theological* grounds—that animals don't feel pain. But we've also seen that (despite what's commonly claimed) we can't be certain whether Descartes himself believed this. Nor is it certain whether his animal–machine hypothesis specifically encouraged physiologists towards increased cruelty, or even studied indifference.

There's one 'eyewitness' report that Cartesians associated with the Port-Royal school (Chapter 9.iii.b–c) did unpleasant experiments with an attitude of what looked like appalling callousness, confident in the belief that animals could feel no pain:

They administered beatings to dogs with perfect indifference, and made fun of those who pitied the creatures as if they felt pain. They said the animals were clocks . . . They nailed poor animals up on boards by their four paws to vivisect them and see the circulation of the blood which was a great subject of conversation. (Fontaine's *Mémoires de Port-Royal*, quoted in Leiber 1988: 313)

However, this report may not be trustworthy. Not only was it written years after the supposed events, but its author "held Cartesianism to be scientistic blasphemy" so had every reason to malign Descartes's followers (Leiber 1988: 314).

There are also reports of late seventeenth-century experiments which troubled one—but apparently not all—of their perpetrators:

Besides all this, I have also imitated Bilsius' experiment on the movement of chyle, when I was at Amsterdam; but I did not find that diversity in the blood, although I kept the dog alive up to the third hour who would have lived even through the whole day in this torment; but since the first attempt did not provide any certain result, I may admit that I tortured them with such lengthy cruelties with horror. The Cartesians boast hugely of the certainty of their philosophy: I wish they could persuade me as certainly as they are persuaded themselves, that there is no soul in animals and that there is no difference between the nerves of a live animal and the wires in a machine which is set in motion when you touch them, dissect them, burn them: then I'd be more willing to probe the entrails and vessels of an animal for a number of hours, since I see that there are many things which ought to be investigated, which one cannot expect to do in any other way. (Lindeboom 1979: 64 n., trans. S. Medcalf)

This comment—left in the Latin in the 1979 edition, so as not to upset the readers—seems to imply that Cartesian experimenters, confident that animal suffering is an illusion, didn't even try to minimize it. *Zero* is low enough to salve anyone's conscience.

However that may be, there were indeed many things—physiological questions suggested by the mechanistic approach—that could be investigated in no other way. Vivisection therefore increased hugely as a result of Cartesianism (Rupke 1987; Daly 1989). It was practised by the nascent Royal Society: Hooke, for instance, vivisected a dog in 1667, to study respiration (Shapin 1994: 393). And it was defended towards the end of the century by Locke's hero Boyle (1627–91), who saw its critics as superstitious sentimentalists:

The veneration wherewith men are imbued for what they call nature has been a discouraging impediment to *the empire of man over the inferior creatures of God*: for many have [described it as] something impious to attempt. (Boyle 1744: 363; italics added)

Dr Johnson, one might think, was no sentimentalist. Yet, a hundred years after Boyle, he complained bitterly about the "arts of torture" practised by some medical men, whose "favourite amusement" seemed to be "to nail dogs to tables and open them alive" (Wiltshire 1991: 125–9). However, such strictures had little or no effect at the time: Royal Society members such as Stephen Hales continued their experiments on live frogs, dogs, and horses (Daly 1989). Not until the 1870s would official action be taken to limit these practices (see Section v.a, below). Meanwhile, Cartesian mechanism—and *perhaps* Cartesian views on animal consciousness—encouraged them in the service of science.

g. Human bodies as machines

The fourth claim made in the *Discourse* went even further. Human bodies, too, are machines—in *both* of the two senses distinguished above. Because they're governed by the laws of physics, medicine must be based on scientific physiology. One can even discover certain anatomical/physiological facts a priori, by considering the body as a mechanical, even a machine-like, system.

Thinking of muscles as levers of varying length, for example, Descartes argued (in the *Treatise on Man*) that many movements—such as the movements of the eyes—*must* involve the relaxation of certain muscles at the same time as the contraction of others. And this reciprocal action, he said, must involve active nervous inhibition of some muscles, as well as excitation of their antagonists. We now know that his armchair hypothesis was correct (see Section viii.d). But for two centuries it was thought absurd, even by physiologists who had deserted the armchair for the laboratory: "Several who witnessed the fact did not report it, hesitating to accept it as true" (Sherrington 1940: 166).

Moreover, said Descartes (in the *Discourse*), human bodies are comparable to automata:

And this will not seem strange to those who [know] how many different *automata* or moving machines can be made by the industry of man... From this aspect the [human] body is regarded as a machine which, having been made by the hands of God, is incomparably better arranged... than any of those which can be invented by man.

Indeed, he himself discussed how one might build an android activated by magnets. And a historian of technology has remarked:

Legend has it that [Descartes] did build a beautiful blonde automaton named Francine, but she was discovered in her packing case on board ship and dumped over the side by the captain in his horror of apparent witchcraft. (de Solla Price 1964: 23)

Sadly, this story is probably no more than "legend". But Descartes was adamant in seeing the body as no more than a wonderfully complex machine. Thus in reply to another objection from Arnauld, he wrote:

[the body alone is responsible for] the beating of the heart, the digestion of our food, nutrition, respiration when we are asleep, and even walking, singing, and similar acts when we are awake, if performed without the mind attending to them. When a man in falling thrusts out his hand to save his head he does that without his reason counselling him so to act, but merely because the sight of the impending fall penetrating to his brain, drives the animal spirits into the nerves in the manner necessary for this motion, and for producing it without the mind's desiring it, and as though it were the working of a machine. (trans. Haldane and Ross 1911: ii. 103–4)

'Man as machine' was now being unambiguously argued—and even supported by references to 'machine as man'.

2.iii. Cartesian Complications

The *Discourse* contained two further claims also, each more ambiguous—as regards the prospects of a future cognitive science—than the other four.

One of these licensed psychophysiology and psychophysics, but excluded 'pure' psychology from the reach of science. It also laid philosophical tripwires for people studying the relation of thought and perception to the external world. The other ruled out many, though not all, sorts of AI.

a. The mind is different

Descartes's fifth claim was that a human being has a mind (a rational soul) as well as a body—and that these are radically different. The mind is not a machine, and is nothing like a machine. Indeed, in so far as the human body is guided by the mind, it isn't a machine either:

And as a clock composed of wheels and counter-weights no less exactly observes the laws of nature when it is badly made, and does not show the time properly, than when it entirely satisfies the wishes of its maker, and as, if I consider the body of a man as being a sort of machine so built up and composed of nerves, muscles, veins, blood, and skin, that though there were no mind in it at all, it would not cease [even in sickness] to have the same motions as at present, *exception being made of those movements which are due to the direction of the will, and in consequence depend upon the mind*... (Descartes 1641/1642: 195; italics added)

The mind, Descartes argued, doesn't exist in the material world, but in the realm of self-conscious substance. Consciousness and intelligence pertain to the mind, not the body. Indeed, a mind just *is* a thinking thing, a psychological subject essentially distinct from the body. That, by contrast, is an extended, or material, thing. So whereas in animals sensory processes are linked *directly* to motor actions, in people there is what one might call a sensori-motor sandwich: sensory input and motor output (both aspects of the body) separated by conscious thought (the mind). (As we saw in Section ii.d, Descartes—by our lights—conflated consciousness and reason, or thinking, since he didn't differentiate consciousness as awareness, including *qualia*, from self-reflexive experience.)

This was a novel claim—and, many philosophers would argue, a disastrous one (14.xi and 16.vi–viii). Descartes was saying something very different from Aristotle, who had taught that living things are informed by various levels of soul, or *psyche*, these being animating principles that aren't separable from the body (Matthews 1977, 1992). He was saying something different, too, from his compatriot Michel de Montaigne (1533–92), whose scepticism had spurred Descartes's search for *certainty*, and who'd insisted on "the equality and correspondence between ourselves and beasts" (1580: 167). In particular, Montaigne had ridiculed the belief that we possess a special faculty ("a reasonable soul") able to see the truth, pointing out that our mental faculties, like those of animals, are inseparable from our bodies:

> It's certain that our apprehension, our judgement, and the faculties of our mind [*âme*] in general suffer according to the movements and alterations in the body, which alterations are continual. Isn't our spirit more lively, our memory more available, and our speech more lively in health than in illness? And don't joy and gaiety make us receptive to ideas that would present themselves to our minds in quite a different way in unhappiness or melancholy? (Montaigne 1580: 269; my trans.)

Descartes allowed these facts to be so (how could one deny them?). But for him, they were *psycho-physiological* facts in the strict (i.e. Cartesian) sense, whereas for Montaigne they weren't.

This is why some people argue that Descartes wasn't so much describing the mind, as *inventing* it: why had no one else noticed the inner realm of consciousness (Rorty 1979: 17–69; Putnam 1999)? As we'll see in Chapter 16.viii, such people reject Descartes's separation of mind and body—though a form of dualism usually persists, in the guise of *cause* and *reason*.

One implication of Descartes's fifth claim was that a purely psychological science is impossible. Our conscious choice, and our reason, is not only not subject to the laws of physics, but is unconfined by any law. On the contrary: "It is only will, or freedom of choice, that I experience in myself [as unlimited]; so that it is in this regard above all, I take it, that I bear the image and likeness of God" (1641/1642, Meditation IV). Despite the many limits on our understanding, he said, we can always avoid error by freely refusing assent to a doubtful idea. (As a strict rationalist, "satisficing" wasn't in his vocabulary: see 6.iii.a.) Admittedly, he described the "passions", and complex conflicts between "natural appetites" and "the will", in terms of interacting streams of movement of the animal spirits (1649). However, he didn't suppose the will itself to be determined by the body, nor subject to psychological laws or systematic explanation.

Psychophysiology, by contrast, was tacitly allowed. Our daily experience assures us, Descartes said, that mind and body are closely related. Not only does my arm rise when I will it to do so, but brain injuries—and drunkenness—lead to various changes in consciousness. Indeed, he argued that each distinct mode of consciousness is accompanied by a different state of the brain.

In principle, then, the path lay open for some future experimental science in which these mind–body linkages might be mapped. One example would be the late twentieth-century use of brain scans by cognitive neuroscientists, aiming to discover which areas of the brain are active when someone thinks of one kind of thing rather than another (Chapter 14.x.b).

This argument, you may notice, took for granted that it *makes sense* to suggest that mental events and brain states may be correlated. Some early twentieth-century philosophers, Maurice Merleau-Ponty for instance, rejected Cartesianism at such a fundamental level that it followed that this wasn't so: his language of "mind" and "brain" wasn't acceptable, except as a (philosophically misleading) everyday shorthand. Today, their followers argue that cognitive science is radically flawed as a result (Chapters 14.xi and 16.vi).

But that was for the future. In the mid-seventeenth century, Descartes's assumption became widely accepted. Indeed, it still is. The overwhelming majority of the readers of this book, I've no doubt, assume that hypotheses about mind–brain correlations obviously make sense—and that many are very likely true.

"How can be there be any question about it?" you may be wondering: "Isn't brain-scanning discovering more instances every day?" Well, perhaps. I shan't ask you to consider the counter-arguments until we discuss brain and consciousness in Chapter 14.x–xi. Meanwhile, I'll go along with the prevailing view that talk of mind–brain correlations not only makes sense, but is very often true. (To anticipate: I'll *still* go along with it, after the discussion; but at least we'll have seen why it is that intelligent people have sometimes denied it.)

b. Birth of a bugbear

Quite apart from the radical anti-Cartesian challenge, which we're ignoring here, there was—and is—a catch. On Descartes's view, body and mind are so different, one material the other immaterial, that there can be no intelligible relation between them.

Their systematic association, therefore, can be due—said Descartes—only to God's benevolent fiat. This reference to divine fiat was interpreted in three ways. One of these ignored the divinity while keeping the systematicity, and—though battered—survived to modern times.

The first interpretation was the doctrine of occasionalism. Briefly suggested by Descartes himself, occasionalism was developed by his contemporary Géraud de Cordemoy (of whom, more in Chapter 9.iii.a) and his successor Malebranche, among others. It claimed that, on the occasion of my willing to move my arm, my arm is caused by God to move; and on the occasion of my hand being injured, God causes me to feel pain. An act of will is thus only the occasional cause of a voluntary action, not the real cause.

The notion that God intervenes in the human world—performing miracles, punishing communal sin by famine, and so on—was in those days a commonplace. What occasionalism added was systematicity. The doctrine was approved in theological circles, because of its stress on the unceasing activity of God in sustaining our everyday life.

The second interpretation, which Descartes usually favoured, was subtly different. It saw God as establishing a two-way causal interaction between mind and body. This was called "Natural" causation, to contrast it with "efficient" (mechanical) causation. Though metaphysically unintelligible, Natural causation is sanctioned by God as an enduring part of human Nature, which is a unique "union" of mind and body. (Sometimes, Descartes even described human Nature as a third philosophical 'primitive'—Cottingham 1986: 127–32.) This type of causation doesn't require continual divine intervention, but relies on the Natural union between mind and body. The will really does cause the arm to move, by a mysterious type of causation provided to human beings by God.

This metaphysical miracle is highly detailed, providing countless brain–mind correlations—in principle, open to experimental study. However, there's no intelligible reason why *this* brain state rather than *that* one should accompany a particular mental state. As Descartes put it:

> God could have constituted the nature of man in such a way that this same movement in the brain [physically caused by a certain movement of the nerves in the foot] would have conveyed something quite different [from a sensation of pain-in-the-foot] to the mind . . . [indeed] it might finally have produced consciousness of anything else whatsoever. (Meditation VI)

This implies that psychophysiology can provide only correlational data, not theoretical intelligibility (see 14.x–xi). Philosophically, its findings can be understood only in the context of divine benevolence.

However, Descartes was often read in a third way. Here, his references to God's benevolence in establishing (Natural) mind–body causation were ignored. He was supposed instead to have posited a (natural) causal interaction between mind and body, one that we might hope to understand. Descartes himself invited this interpretation, for he sometimes answered objections in terms apparently referring to some sort of natural causation, citing familiar analogies such as rudders and gravity.

Princess Elizabeth of Bohemia, for instance, wrote to him in 1643: "I beg of you to tell me how the human soul can determine the movement of the animal spirits in the body so as to perform voluntary acts—being as it is merely a conscious substance." And she added that movement is something that can be caused only by material contact between extended things. Descartes (1643) admitted that "what your Highness is propounding seems to me to be the question people have most right to ask me in view of my published works". His reply, which appealed to the analogy of gravity, didn't convince Elizabeth. She wrote back saying she found it unintelligible, and that she "could more readily allow that the soul has matter and extension than that an immaterial being has the capacity of moving a body and being affected by it".

Two years later, and partly as a result of this correspondence, Descartes wrote *The Passions of the Soul* (published in 1649, a few months before his death). There, he suggested that mind–body interaction is mediated by movements, or (in recognition of "the conservation of motion") by changes in the direction of movement, of the stalk of the pineal gland—which he compared to the rudder of a ship. His choice of the pineal gland wasn't arbitrary: since it was a single organ in the midline of the brain, he thought it might provide the basis for the integration of the different senses—for example, sight and touch. (His choice wasn't entirely original, either: Fernel had said that the pineal gland controls the passage of the animal spirits between distinct ventricles in the brain.)

The causal interaction, he said, works both ways. As regards voluntary movement:

> the whole action of the soul consists in this, that solely because it desires something, it causes the little gland to which it is closely united to move in the way requisite to produce the effect which relates to this desire.

Purely physical causation can then take over, because the gland is in contact with the animal spirits flowing through the superior ventricle of the brain. Likewise, movements of the pineal gland caused by the animal spirits coming from the sense organs lead—in human beings only, not in animals—to consciousness. In speech, for example, a person's lip movements are (originally) caused by their conscious thoughts, and (eventually) lead to conscious understanding in people whose ears are physically affected by the air disturbances produced by speech.

For the next 300 years, many of Descartes's readers thought that this was a respectable scientific hypothesis. That's not to say that the pineal gland was universally regarded with awe. To the contrary, the *Spectator* magazine in 1712 published a spoof report on this "lover's gland", saying that it "smelt very strong of Essence and Orange-Flower Water", and on dissection was found to contain a cavity "filled with Ribbons, Lace and Embroidery" (Porter 2001: 502 n. 35). But if the notional anatomy was widely ridiculed, the metaphysics wasn't.

However, perhaps it should have been. For it was subject to the same philosophical objections that Princess Elizabeth, among others, had already raised.

To be sure, the animal spirits might move, and be moved by, the pineal gland. But how could such movements—or any other purely physical changes—cause, or be caused by, consciousness? Correlations between mind and body there may be, and many of them. (Remember, we're here ignoring the later suggestion that this 'obvious fact' isn't even mistaken, but fundamentally confused.) But dualistic interaction, in either direction, is unintelligible *in Descartes's own terms.*

Even if we drop Descartes's commitment to mental substance, there's still a causal/metaphysical gap between bodily processes on the one hand and conscious states, intentionality, and reason on the other. (The still-unsolved problems of how to explain mind–body correlations and intentionality are discussed in Chapters 14.x–xi and 16.iv–x, respectively.)

This fifth Cartesian claim, the metaphysical separation of mind and body, presents a further ambiguity relevant to cognitive science. For Descartes's writings were to be influential in two radically different philosophical traditions.

On the one hand, his mechanism and recommendation of experimental science were accepted by empiricists, from his contemporary Thomas Hobbes onwards.

For Hobbes, even the idea of *infinity* (in mathematics or theology) was an extrapolation from experience. He took mechanism fearlessly into psychology and political philosophy. For instance, he likened society—"the Commonwealth"—to "*Automata* (Engines that move themselves by springs and wheeles as doth a watch)" (1651: 1). He even saw *thinking* as a form of *computation*. So, for instance, he said:

When a man *Reasoneth*, hee does nothing else but conceive a summe totall, from *Addition* of parcels; or conceive a Remainder, from *Substraction* of one summe from another: which (if it be done by Words,) is conceiving of the consequence of the names of all the parts, to the name of whole; or from the name of the whole and one part, to the name of the other part. And though in some things, (as in numbers,) besides *Adding* and *Substracting*, men name other operations, as *Multiplying* and *Dividing*; yet they are the same; for Multiplication, is but Adding together of things equal; and Division, but Substracting of one thing, as often as we can. These operations are not incident to Numbers onely, but to all manner of things that can be added together, and taken out of one another. For as Arithmeticians teach to add and substract in *numbers*; so the Geometricians teach the same in *lines*, *figures*... degrees of *swiftenesse*, *force*, *power*, and the like; The Logicians teach the same in Consequences of words; adding together *two Names*, to make an *Affirmation*; and *two Affirmations*, to make a *Syllogisme*; and many *Syllogismes* to make a *Demonstration*... [Writers of Politiques and Lawes, similarly.] In summe, in what matter soever there is a place for *addition* and *substraction*, there is also place for *Reason*; and where these have no place, there *Reason* has nothing at all to do.

Out of all which we may define, (that is to say determine,) what that is, which is meant by this word *Reason*, when wee reckon it amongst the Faculties of the mind. For REASON, in this sense, is nothing but *Reckoning* (this is, Adding and Substracting) of the Consequences of general names agreed upon, for the *marking* and *signifying* of our thoughts; I say *marking* them, when we reckon by our selves; and *signifying*, when we demonstrate, or approve our reckonings to other men. (1651: 18–19)

Moreover, this "reckoning" was effected purely by movement within the body—as were sensation, imagination, and voluntary choice too:

All [the] qualities called *Sensible*, are in the object that causeth them, but so many motions of the matter, by which it presseth our organs diversely. Neither in us that are pressed, are they anything else, but divers motions; (for motion, produceth nothing but motion). (p. 3)

For after the object is removed, or the eye shut, wee still retain an image of the thing seen, though more obscure than when we see it. And this is [*Imagination*, or *Fancy*, which] therefore is nothing but *decaying sense*, and is found in men, and many other living Creatures, as well sleeping, as waking. (p. 5)

There be in Animals, two sorts of *Motions* peculiar to them: One called *Vitall* [the heart beat, breathing, digestion, etc.]. The other is *Animal motion*, otherwise called *Voluntary motion*; as to *go*, to *speak*, to *move* any of our limbes, in such manner as is first fancied in our minds. [I have already shown] That sense, is Motion in the organs and interiour parts of mans body... And that Fancy is but the Reliques of the same Motion... And because *going*, *speaking* and the like Voluntary motions, depend alwayes upon a precedent thought of *whither*, *which way*, and *what*; it is evident that the Imagination is the first internall beginning of all Voluntary Motion. And although unstudied men, doe not conceive any motion at all to be there, where the thing moved is invisible; or the space it is moved in is, (for the shortness of it) insensible; yet that doth not hinder, but that such Motions are. These small beginnings of Motion, within the body of Man, before they appear in walking, speaking, striking, and other visible actions, are commonly called ENDEAVOUR. (p. 23)

[If a man should talk to me of] *A free Subject*; *A free-Will*; or any *Free*, but free from being hindered by opposition, I should not say he were in an Errour; but that his words were without meaning; that is to say, Absurd. (p. 20)

In other words, our thoughts, goals, and decisions aren't merely correlated with invisible bodily motions: they *consist in* bodily motions. Such a radical expression of mechanism was a step too far for Descartes, as we'll see.

Empiricism and scientific progress were soon explicitly linked by the philosopher Locke, an early Fellow (from 1668) of the Royal Society. He described himself as merely "an under-labourer in clearing the ground a little, and removing some of the rubbish that lies in the way to knowledge". And he identified the current "master-builders of the commonwealth of learning" as the scientists Boyle, Thomas Sydenham, "the great Huygenius" and "the incomparable Mr. Newton" (J. Locke 1690, Epistle).

On the other hand, Descartes's emphasis on the epistemological primacy of the individual consciousness (the *cogito*) led first to Kantianism and then to neo-Kantian idealism and Continental phenomenology (see Section vi, below). Although empiricism is much the stronger tradition in modern cognitive science, phenomenological critiques—and research programmes—are increasingly being promoted (see Chapters 14.xi and 16.vi–viii). Indeed, this is one aspect of the general cultural challenge to Cartesian modernism that blossomed in the 1960s and is still in bloom (Roszak 1969; Toulmin 1999).

Ironically, a leading phenomenologist has recently used the technology of virtual reality, or VR (see 13.vi), to revive—if only to dismiss—Descartes's hypothesis of the evil demon, or *malin génie*. Because Descartes believed that he was trapped inside his own thoughts and perceptions, he said it was metaphysically possible that he was dreaming, or that some evil demon might be deceiving him into thinking that his sensations, including his bodily sensations, were caused by real material objects when in fact they weren't. He mentioned the phenomenon of the phantom limb, for instance, in which an amputee feels pain seemingly caused by an injury in their non-existent leg. He was relieved of this anti-realist doubt only by his argument that God, whose existence he had now proved to his own satisfaction, wouldn't allow us to be deceived in that way. Today, Hubert Dreyfus regards VR in general, and the film *The Matrix* in particular, as constituting "Descartes' Last Stand" (H. L. Dreyfus 2000; cf. H. L. Dreyfus and Dreyfus 2002). The realist/anti-realist implications of VR are discussed in Chapter 16.viii.c.

c. The prospects for AI

If Descartes's fifth claim had mixed implications for the biological and psychological aspects of cognitive science, the sixth carried mixed messages regarding the prospects for AI.

Bearing in mind the many ancient and post-Renaissance automata, Descartes allowed that there might be machines like monkeys, having all their behavioural capabilities. But, he argued, there could never be a convincing android automaton. A humanoid robot capable (for instance) of fleeing from an approaching wolf or tiger, and of using its hands to prevent itself from falling, was certainly possible. But one with language, reason, or general intelligence was not:

> If there were machines resembling our bodies, and imitating our actions as far as is morally possible, we should still have two means of telling that, all the same, they were not real men. First, they could never use words or other constructed signs, as we do to declare our thoughts to others. [A machine, if touched at a certain spot, could cry out that it was hurt, but it could not] be so made as to arrange words variously in response to the meaning of what is said in its presence, as even the dullest men can do. Secondly, while they might do many things as well as any of us or better, they would infallibly fail in others, revealing that they acted not from knowledge but only from the disposition of their organs. For while reason is a universal tool that may serve in all kinds of circumstances, these organs need a special arrangement for each special action . . .

Descartes wasn't speaking here about mere technological difficulties. He believed that these were principled philosophical reasons why intelligent language-using automata could never be built. These reasons also forbade any ascription of language to animals, no matter what sounds they might be taught to utter (see Chapters 9.ii.b and 7.vi.c).

It would follow that psychological AI is doomed to failure. The mind is nothing like a machine, and even brain events have no intelligible relation to consciousness. The notion that a man-made automaton could help us understand human psychology (thinking and consciousness) is therefore absurd.

By contrast, some—but not all—sorts of technological AI are, on this view, achievable. Computer vision, for example, is possible. So is verbal data processing, including so-called expert systems, wherein the word strings are composed beforehand by the programmer. Robots simulating animal behaviour, no matter how complex, are feasible. And brain-inspired modelling, such as connectionist pattern recognition, could be successful.

However, natural language processing in general—good machine translation, for instance—is impossible, as is the automation of common sense. Even with the help of non-human strategies or tricks, the Turing Test (16.ii.c) couldn't even *seem* to be passed (Gunderson 1964*a*). As we'll see (in Chapters 9.x and 16), these anti-AI claims still have committed supporters today—some of whom cite Descartes as an intellectual ancestor.

In sum, then, Descartes's vision of 'man as machine' was both ambitious and limited. It concerned only the body, not the mind. Nevertheless, it assured scientists that they could hope to discover (though not to understand) the detailed correlations that exist between human minds and human brains. And, since bodies are machines, science could explain how the nervous system works, and how it interacts with other bodily organs. Indeed, the entire range of animal behaviour could be understood in this way.

2.iv. Vaucanson's Scientific Automata

In the late seventeenth century, two very different projects gained inspiration from Cartesian mechanism. One was experimental physiology itself: man—or rather, man's body—as machine (see Section v). The other was automata building: machine as man.

a. Fairs and flute-players

The post-Cartesian designers of automata were concerned with what Naudé had called Mathematicks, not Magick (see Preface). But, like their predecessors whom he had defended so strongly, they were more interested in Mathematicks than in Man.

In other words, their efforts were technological rather than scientific. At best (as remarked in Section i.b), they were existence proofs of mechanism in the broadest sense. Most twentieth-century roboticists dismiss them as scientifically uninteresting. One such, for example, complained that "these eighteenth-century gadgets were developed purely for their entertainment value" (Raphael 1976: 258).

This modern critic specifically included Vaucanson's automata in his complaint. Describing these as "the entertainment sensations of the courts of Europe", he compared them to the "audioanimatronics" of Disneyland and contrasted them with modern robots built "to test our current theories" about "how certain biological systems behave". In part, this is correct: though there was no dedicated theme park, Vaucanson's robots were exhibited to general delight in a Parisian fair, and in London's Haymarket too (see Section i.b, above). But if the comparison is just, the contrast is not.

To be sure, the young Vaucanson's scandalous attempt to build automatically flying angels, which caused his hurried departure from the religious order in which he was then studying, was probably a mere fancy (Bedini 1964: 36). It may even be fanci-ful: an alternative story has it that he built androids to wait at table for some distinguished visitors, and was dismissed the next day despite their having been favourably impressed (G. Wood 2002, ch. 1).

However, his famous automata were neither fancies nor fanciful. Nor were they undertaken "for the sole purpose of making money", as one historian of technology has claimed (Bedini 1964: 37). Indeed, it was common in those days for scientists to demonstrate their work to the public for money (Schaffer 1983). But Vaucanson was unusual, perhaps even unique, in often prefacing his demonstrations with an explanation of how his automaton worked, before setting it in motion to the wonder of the audience. (He did this, for instance, when his flute-player was shown to a select audience in a Paris mansion, for which an entrance ticket cost the equivalent of a workman's weekly wage.)

Unlike most other eighteenth-century gadgets, Vaucanson's were intended to show, in relatively specific terms, how our bodies work (Doyon and Liaigre 1956; Fryer and Marshall 1979). That is, they were theoretically motivated simulations of actual bodily processes. An early example (now lost, and only briefly mentioned in the contemporary documents) apparently involved several "anatomies": a model of a group of animals of different species. But the species that would be remembered were duck and man.

Sometimes, Vaucanson focused on chemical processes. His famous mechanical duck, made of gilded copper, accepted food into its mouth; (seemingly) broke it down inside its stomach; passed it through its rubber-tube intestines; and then mimicked the usual digestive denouement. (Although the denouement was genuine, the digestion wasn't. After his death, it was discovered that the duck contained two hidden chambers: one pre-loaded with evil-smelling mash, the other used to store the swallowed corn—Landes forthcoming, n. 4.)

Vaucanson's primary aim, here, wasn't to exhibit an amusing clockwork toy, but to illustrate his theory of digestion as dissolving ("Dissolution") and "to represent the Mechanism of the Intestines" (Vaucanson 1738/1742: 21). Indeed, he sometimes removed the duck's outer covering so as to let its innards show: "My Design is rather to demonstrate the manner of the Action, than to show a Machine" (p. 22).

But Vaucanson seems to have been even more interested in observable behaviour—or rather, the anatomy that made behaviour possible. His duck did a number of engaging tricks, such as quacking, splashing about on water, stretching its neck to peck corn from people's hands, rising on its feet, and flapping its wings—each of which contained over 400 articulated parts. As he put it, in a letter to the Abbé de Fontaine:

> I don't believe the Anatomists can find any thing wanting in the Construction of the Wings. The Inspection of the Machine will better shew that Nature has been justly imitated, than a longer [written] Detail, which wou'd only be an anatomical Description of a Wing... Not only every Bone has been imitated, but all the Apophyses or Eminences of each Bone. They are regularly observ'd as well as the different Joints... (Vaucanson 1738/1742: 21–2)

Sadly, the duck is now lost, as is a copy made in Germany in 1847 (however, a master automata maker is now trying to rebuild it as best he can: A. Marr, personal communication). Even 200 years ago, in 1805, it was already a sorry sight. Owned at that time by the Duke of Brunswick's doctor, in Helmstadt, it was sitting in "an old garden house", "utterly paralyzed" and "mute" (G. Wood 2002, ch. 1; Riskin forthcoming). It could still eat, but had lost its feathers. In short, it—and the flute-player alongside it, whose "playing days were past"—was in "the most lamentable state".

These descriptions came from the pen of its august visitor Johann von Goethe. As we'll see in Section vii.c, Goethe had little sympathy with the analytic/scientific motives that had energized Vaucanson. His visit to the dilapidated duck was an exercise in sceptical curiosity, not a pilgrimage of honour. But even he seemed somewhat depressed by its aged decline.

More significant, in this context, were Vaucanson's automatic musicians built in the late 1730s: the flute-player and the tabor-and-pipe player. These were life-size manikins, each with mobile lips, tongue, and fingers (made/padded with leather). The flute-player took the form of a shepherd boy, while its drummer cousin was a seated faun—half-man, half-goat—closely modelled on a statue in Louis XIV's Tuileries gardens.

To sound the German flute and the three-hole pipe, air was impelled through the mouth by three sets of bellows, which could deliver air at different pressures. (Vaucanson chose to use the German flute because of its reputation for difficulty.) The tabor-and-pipe player held a drumstick in its right hand, with which it could play single or double strokes, drum rolls, or time-keeping beats synchronized with the pipe held in its left hand.

Both machines could play a range of melodies: the flautist eleven, the piper–drummer over twenty (a score of minuets, plus some other dance tunes). Features such as pitch, speed, timing, echo, and crescendo could be varied by Vaucanson at will. These variations weren't made by interfering with the performance on the fly, but by adjusting mechanical parameters before the performance began: "[my] Machine, when once wound up, performs all its different Operations without being touch'd any more" (p. 23).

The flute-player was first exhibited (in 1737) at the fair of Saint-Germain, amidst ribbons, rattles, and freaks of various kinds. The wondering Parisians could be forgiven for thinking it a mere toy. But Vaucanson's scientific aims were made clear in a paper (describing the flute-player) addressed to the French Académie royale des sciences and in a personal letter to the Abbé de Fontaine (describing the drummer and the duck). These were translated in a pamphlet accompanying the London exhibition (Vaucanson 1738/1742).

No doubt, most of the general public visiting the exhibition didn't read it. Had they done so, they'd have realized that Vaucanson was even more ambitious than appeared at first sight.

b. Theories in robotic form

It was clear from Vaucanson's (largely unread) writings that his automata weren't mere gimmickry, but test-models of theories of instrumental playing. That is, they were intended as studies of how people move their lips, tongue, and fingers, and how they regulate their breathing, so as to make specific musical sounds with these wind instruments. (These "hows" should be interpreted charitably: unlike the A-Life work described in Chapter 15.ii.a, Vaucanson's machines weren't modelling individual muscles, but were focused on the movements of observable body parts.)

As Vaucanson put it, they were attempts to simulate "by Art all that is necessary for a Man to perform in such a Case" (1738/1742: 21). Having opened his paper by describing the structure of the flute, and (in great detail) the various movements involved when it's played by a human musician, he continued:

> These, Gentlemen, have been my Thoughts upon the sound of Wind-Instruments and the Manner of modifying it. Upon these Physical Causes I have endeavour'd to found my Enquiries; by imitating the same Mechanism in an *Automaton*, which I endeavour'd to enable to produce the same Effect in making it play on the German Flute. (p. 12)

A few pages later, he said:

> [My theory suggests that if certain movements are executed] it will then follow... according to the Principle settled in my First Part, the flute will give a low Sound: and this is confirmed by Experience. (p. 17)

One can almost hear his sigh of satisfaction and relief. But such reports, as is usually the case in scientific papers, underplayed the amount of work, and frustration, involved in getting the desired result. Vaucanson's descriptions of the movements executed by the automaton were highly detailed. Even so, they omitted a great deal:

> the fear of tiring you, GENTLEMEN, has made me pass over a great many little Circumstances, which tho' easy to suppose are not so soon executed: the Necessity of which appears by a View of the Machine as I have found it in the Practice. (p. 20)

This comment is an eighteenth-century version of the computational modeller's experience today: one doesn't know what bugs the system contains, nor what crucial processes have been omitted from the theory, until the robot is tested or the program is actually run.

As for the scientific value of Vaucanson's simulations, his translator John Desaguliers—a Fellow of the Royal Society of London, and inventor of the planetarium—had no doubts. In his Preface, he said: "this Memoire . . . in a few Words gives a better and more intelligible Theory of Wind-Musick than can be met with in large Volumes".

How, "better"? Well, Vaucanson had carefully described how each note, within three octaves, was produced by the flute-player. For the higher octaves, he found that extra mechanical variability must be added to the four basic operations. As well as studying single movements, he asked what combinations of movements are needed for certain effects. Pointing out the possibility of "a prodigious Number of Mechanical Combinations", he used his machines to find out what the results of some of these combinations might be. And he came up with some surprises ("Things which could never have been so much as guessed at").

For instance, he discovered (what Karl Lashley was to stress 200 years later) that the way in which a note is played—the bodily movements involved—may depend on the previous note. The need for context sensitivity had actually been discovered by him earlier, with respect to his mechanical duck. He had found that if the duck were sometimes to flap its wings, and sometimes to rise on its legs, a given mechanical part had to serve different functions in either case:

> Persons of Skill and Attention [will observe that] . . . what sometimes is a Centre of Motion for a Moveable Part, another Time becomes moveable upon that Part, which Part then becomes fix'd. (pp. 21, 23)

Again, it turned out that the three-holed pipe needed air pressures ranging from 56 pounds down to only 1 ounce to play the notes in its repertoire. And very fast tunes were performed better by the pipe-player than by human beings, because human tongue movements are too slow (compare the distinction between *competence* and *performance*: Chapter 7.iii.a).

Vaucanson wasn't the only eighteenth-century simulator. Others included two of his compatriots, the surgeons François Quesnay and Claude-Nicolas Le Cat—both of whom built simple models of the circulatory system, intended to teach anatomy and possibly to aid in therapy. In addition, various clumsy efforts were made to build prosthetic hands/limbs for amputees (Riskin 2003).

But these examples *didn't* lead to a proliferation of imitators in the following century. Partly, that was because Vaucanson's engineering skills outclassed those of all but the most ingenious of automata makers. More importantly, the Zeitgeist of Enlightenment rationalism gave way to that of Romanticism (see vi.c–d, below).

This favoured holistic over analytic science, and vitalism over mechanism. Accordingly, the countless nineteenth-century automata were *not* further exercises in scientific simulation, but—as in pre-Enlightenment times—mere toys, or engineering challenges. For instance, hidden pedals and levers were more common than moving body parts as such. Often, people went out of their way to deny that the bodily movements of human speech, namely the movements of tongue, larynx, epiglottis . . . , could be simulated artificially, and to insist that they could occur only in a living organism (Riskin 2003). Even limb movements weren't considered fit for *scientific simulation*: rather, they were superficially copied, for purposes of entertainment or prosthetics. It wasn't until the mid-twentieth century that automata for theoretical simulation became prominent once more (Chapters 4.viii.a–b and 10.iii.c).

c. Robotics, not AI

By that time (the mid-twentieth century), an extra—psychological—dimension was being added, over and above the aims of Vaucanson. For he'd been concerned with *only one half* of the Cartesian metaphysical divide.

Descartes himself would have enjoyed the flute-player, and would have endorsed Vaucanson's scientific hopes. For Vaucanson was studying the body, not the mind. *Psychological* aspects of musical performance and appreciation weren't at issue. Strictly, then, Vaucanson—the first scientific roboticist—wasn't doing psychological AI, *even though* he was modelling motor behaviour.

(Not until very recently could machine models be used to suggest how it's possible for people to understand the structure of music, and to apply this understanding both in listening intelligently and in performing expressively: Longuet-Higgins 1979, 1994; Longuet-Higgins and Steedman 1971. Even now, such studies can arouse scepticism. The editor of *Nature* rejected a paper that used the example of *Colonel Bogey* in describing how a computer could transcribe melodies into musical notation, suggesting—seriously? sarcastically?—that the computer be asked to transcribe something from Wagner instead. It was; it did; and publication followed—H. C. Longuet-Higgins, personal communication; 1976.)

To be sure, a curious quirk of history links Vaucanson with IBM. Vaucanson was a pioneer in the development of machine tools, such as metal-cutting lathes. While working as an inspector in silk factories in France, he had the idea—improved in 1801 by Joseph Jacquard (1752–1834)—for an apparatus using punched cards for weaving brocades automatically. Punched cards had already been used in France to control weaving, but they had to be hand-fed into the loom one by one. Vaucanson suggested stringing them together and feeding them in automatically, in sequence (Hyman 1982: 166). He even built a loom controlled by punched cards moved by a perforated cylinder.

One and a half centuries later, Herman Hollerith used electromechanical punched-card machines for analysing the results of the US Census of 1890. Manual analysis of the 1880 data had taken almost ten years (T. I. Williams 1982: 348), and the population had grown since then.

Hollerith left the Census Bureau in 1896 to start the Tabulating Machine Company, which in 1924 formed the core of IBM. And IBM eventually became the world's leading computer manufacturer—despite the early prediction of its Director, Thomas Watson, that only a handful of such machines would be needed worldwide. Even in the 1960s, state-of-the-art computers (the prototype IBM 360, for instance: see Preface, ii) were still programmed by ordered packs of punched cards.

But intriguing linkages of this sort aren't the same thing as real historical influence. Some members of the French Académie and the Royal Society recognized the scientific interest of Vaucanson's work on automata, but most of his contemporaries didn't.

His successors didn't recognize it either. One hundred years later, the committed mechanist Hermann von Helmholtz (1821–94) praised Vaucanson's "inventive genius" in aiming "to imitate living creatures", mentioning the flute-player "which moved all its fingers correctly". But he saw it as directed towards the "great problem" of "practical mechanics", as an exercise significant to technology, not biology (Helmholtz 1854: 137–8). Even now, as we've seen, Vaucanson's work is dismissed as "entertainment" by people doing research of an essentially comparable type.

Nor did the technological possibilities in the mid-eighteenth century encourage engineering projects of this—scientifically oriented—kind. Even the engineering wizard Vaucanson probably couldn't have done much more. If one wanted to further Descartes's vision of man as machine, the best method wasn't building automata, but studying physiology.

2.v. Mechanism and Vitalism

From the mid-1600s onwards, increasingly many scientists tried to explain physiological processes—respiration, for example—in terms of physics and chemistry. It's no accident that this project didn't look really persuasive until the late nineteenth century (see Section vii.a). Even then, some life scientists remained unconvinced.

a. Animal experiments: Are they needed?

Biology couldn't be mechanized without important advances in non-biological sciences, such as the theory of gases and the chemistry of combustion. It also required an extensive programme of experiments involving live animals—some of which included vivisection. (Only "some", because one can study some of the effects of a drug or a gas on a living animal *without* doing vivisection, which involves surgery.)

The need for experimentation on animals wasn't universally accepted at the time, however. This isn't a comment about the opinions of outsiders—such as Dr Johnson, whose qualms were cited in Section ii.e, above. Rather, it concerns doubts among the scientists themselves.

There were three different reasons for holding back. One was a doubt about the relevance of *animals*, another a doubt about the relevance of *experiments*, and the third a worry about the *morality* of certain sorts of animal experiment.

Many people still regarded human beings as so special that nothing useful could be learnt from work on animals. And this attitude survived into the late nineteenth century. Fully 200 years after Harvey had been scorned for investigating slugs, snails, and squill-fish, physiologists still felt bound to defend the study of "all animals", because "without such comparative study of animals, practical medicine can never acquire a scientific character" (Bernard 1865: 122–6).

Some influential biologists even regarded live experimentation as irrelevant in understanding lowly snails or squill-fish, because it destroyed the holistic character of life itself. The zoologist Georges Cuvier (1769–1832), for example, argued that:

> All parts of a living body are interrelated; they can act only in so far as they act all together; trying to separate one from the whole means transferring it to the realm of dead substances; it means entirely changing its essence. (quoted in Bernard 1865: 60)

And in 1823 the distinguished anatomist Charles Bell (1774–1842) declared:

> Experiments have never been the means of discovery; and a survey of what has been attempted of late years in physiology, will prove that the opening of living animals had done more to perpetuate error, than to confirm the just views taken from the study of anatomy and [the observation of] natural motions... (quoted in Robert M. Young 1970: 47)

(Such holistic qualms are still with us: many researchers have similarly criticized work, whether in neuroscience or in computer modelling, that neglects the *whole* animal—see Chapters 14.vi–vii, 15.vii, and 16.x.b.)

Other critics, unconvinced by Descartes's (supposed) denial of animal consciousness, saw such work as morally questionable—sometimes, *irrespective* of its scientific value. To be sure, in the eighteenth and early nineteenth centuries the view that nature (or even Nature) was created by God primarily for humans' use and delight was still widespread. Clergymen and scientists alike (often, the two were the same) commonly made this view explicit. For instance, the Quaker geologist William Phillips declared that mankind is "the Lord of Creation" and that "everything is intended for the advantage of Man" (Phillips 1815: 191, 193). However, such remarks were more likely to be made in the contexts of chemistry, geology, or botany than of zoology.

Eventually, the law declared zoology to be a special case. In 1876 the British Parliament's Cruelty to Animals Act set up the first statutory committee to legalize and regulate the treatment of experimental animals. Coinciding with the foundation of the Physiological Society of London, this Act encountered considerable opposition from the general public (R. D. French 1975). For many non-scientists felt that animal experiments shouldn't be regulated so much as banned.

Similar feelings are regularly aimed today at research carried out under the UK's Animals (Scientific Procedures) Act passed in 1986. Some high-profile physiologists—and their families—have endured threats, and even violence, accordingly. And several commercial companies for breeding and/or experimenting on animals have recently been closed down by a combination of imaginative financial pressures and physical assaults. Indeed, an entire village has been attacked in various ways, in an attempt to prevent the inhabitants having anything to do with the staff and owners of a nearby guinea-pig farm; the final straw was the theft of a relative's remains from the graveyard of the village church, and the farm was closed down a few months afterwards (*Nature* 2005*b*).

The Animals Act requires experiments on living animals to be justified by a detailed cost–benefit analysis before a licence is granted. The numbers of animals, the species selected, and the severity of the procedure (the potential for suffering) are all carefully considered, as is the likely benefit to science and/or human or veterinary medicine if the experiment goes ahead. Due in part to the findings of ethologists such as Jane Goodall (Chapter 7.vi.b), additional rules were introduced in the 1990s to protect wild-caught primates from experimentation, and to tighten the rules with regard to captive-bred primates.

In all cases, the government's inspectors have to be convinced that no less severe means are available. They consider "the three R's": Reduction, Replacement, and Refinement (J. A. Smith and Boyd 1991). The second of these is relevant because the possible alternatives may not involve any animals at all. Sometimes, experiments can be done *in vitro* using tissue cultures, or perhaps even *in silico*, using computer simulations of metabolic functions (Balls 1983; and see 15.viii.b). A charity called FRAME (Fund for the Replacement of Animals in Medical Experiments) has been sponsoring the development of such techniques for several decades.

Some questions, of course, can be answered only by vivisection—and some of these are considered sufficiently important, in scientific and/or medical terms, to be allowed under the law. Most British physiologists today, despite occasional grumbles about the bureaucratic delays and paperwork involved (more burdensome than in any other country), are content to have an Act of Parliament that protects animals from unjustifiable suffering. As explained in Section ii.d, however, we'll never know for sure whether Descartes himself would have seen any need for the Act, or for FRAME.

b. Holist chemistry

In the early days of scientific physiology, even the committed experimentalists weren't all unremitting mechanists, as Descartes was. Many of them doubted whether physics and chemistry could suffice to explain living processes.

This doubt was an intellectually respectable position in the seventeenth century, and wasn't unreasonable even in the nineteenth. The chemistry of respiration would take 200 years to clarify. And the familiar phenomenon of 'animal heat'—the ability of living animals to conserve a body temperature different from (and often higher than) that of their environment—appeared actually to contravene the laws of physics (Goodfield 1960, chs. 6 and 7). In general, the self-organizing properties of biological creatures were highly mysterious (and they still aren't fully understood: see Chapter 15).

Small wonder, then, that several versions of vitalism flourished long after Descartes's death (Lenoir 1982). Some were derived from philosophical idealism, gaining popularity (in Germany and elsewhere) in the form of *Naturphilosophie* (see Section vi). These posited a vaguely specified vital principle, animating the living body to give it capacities different from those of the inorganic world. Occasionally, chemistry was even said to be utterly irrelevant to biology.

Other forms of vitalism posited an additional (vital) force fully consistent with physics and chemistry (perhaps even obeying something like a Newtonian inverse-square law), but not reducible to them. Sometimes, a mysterious power of self-organization was ascribed to *all* matter, by analogy with crystal growth (T. Brown 1974). Although this was a retreat from Descartes's mechanism, it wasn't intended as a rejection of materialist science in general. It was comparable, rather, to late twentieth-century attempts to 'informationalize' matter and/or to imbue it with some primitive form of consciousness (Chapter 14.x.d).

With advances in chemistry and experimental physiology, vitalism of both types (and especially the first) became less common. Nevertheless, it survived—even among people using detailed chemical knowledge to advance the study of living things.

For example, in 1842 the organic chemist Justus von Liebig (1803–73) published a ground-breaking book on *Animal Chemistry*. This hugely clarified the processes of digestion and respiration. In the same book, however, he argued that life involves a "vital force", one which—like the known forces in physics and inorganic chemistry—must be regulated by certain laws. This "independent force", he said, is responsible for the growth and maintenance of the living creature's bodily matter and form:

> The vital force *causes a decomposition* of the constituents of food, and *destroys the force of attraction* which is continually exerted between their molecules; it *alters the direction of the chemical forces* in such wise, that the elements of the constituents of food *arrange themselves in another form*... it forces the new compounds to assume forms *altogether different* from those which are the result of the attraction of cohesion when acting freely... [By] its presence in the living tissues, their elements *acquire the power of withstanding* the disturbance and change in their form and composition, which external agencies tend to produce; *a power which, simply as chemical compounds, they do not possess.* (quoted in Goodfield 1960: 137; italics added)

Many vitalists were fundamentally sympathetic to mechanism, but faced apparently insuperable problems in applying it to life. With respect to living bodies, they would have been undiluted (or Newtonized) Cartesians if they could. And, after the work of the physiologist Claude Bernard, discussed in Section vii.a, they *could*—at least with respect to fully developed adult animals. But Liebig was different.

For all his superb experimentation, Liebig's view of organic chemistry, and of its place in nature, was holistic rather than analytic. For instance, he stressed its agricultural and even ecological aspects, rather than following the majority of his professional peers in studying a host of increasingly specialized laboratory reactions (Merz 1904–12: ii. 394–6). And, as we've seen, he was convinced that some vital principle could actually override chemical mechanisms.

Liebig's holism, and his suspicion of purely chemical accounts of life, dated from his youth. As a young man, he was one of many German scientists profoundly influenced by a fundamentally non-mechanistic philosophical movement, neo-Kantian idealism. As remarked above, this was one of the two opposing world-views that eventually developed from Descartes's work.

2.vi. The Neo-Kantian Alternative

Neo-Kantian ideas were hugely influential in *humanistic* circles in the early nineteenth century. But even at the height of their popularity, they were less prominent in physiology than Cartesian mechanism was.

That's even more true today. Nonetheless, they've surfaced in various areas of cognitive science. For example, some cyberneticians—for instance, Pierre de Latil (4.v.a)—recommended the Kantian account of "organisms" as systems following *internal* laws of order, or organization, so functioning at the same time as an end and as a means. Linguistics (and NLP—natural language processing) has engaged with, and to some extent supported, Wilhelm von Humboldt's neo-Kantian account of language (Chapter 9.iv and x). And A-Life includes a number of people sympathetic to neo-Kantian views of biology (Chapters 15.iii and vii, and 16.x). Moreover, such views underlie a fundamental critique of orthodox cognitive science that's now arising from several sources (see 13.ii.b–e, 15.viii.c, and 16.vi–viii).

For our purposes here, the crucial ideas concern epistemology, the philosophy of science, and biological holism. All three topics were discussed by the key writer in this philosophical tradition, Immanuel Kant (1724–1804).

a. Kant on mind and world

Kant was rooted in Descartes, but grew radically critical of him. For he'd tried to solve a central problem in Cartesian philosophy, one that had become clearer with the development of empiricism through the eighteenth century—especially, with the work of David Hume (1711–76).

According to Descartes, our conscious perception represents, and to a significant extent corresponds with, external (material) reality. Its disciplined application in science helps us to see where it's reliable (for instance, when it represents material things as having size and shape) and where it's misleading (for instance, when it represents them as having colour). Some non-correspondences, such as colour, have systematic links with the real ("primary") qualities of underlying reality. Science is thus a *realist* enterprise: it tells us about the external world, which exists independently of human beings—and, more to the point, of human perceptions.

But, given this representational theory of mind, our senses can give us only indirect knowledge of the material world. In other words, we can't get beyond our conscious thought, to see the world directly—or to check the correspondence between our empirical observation and the real world. As Addison put it (in explaining Locke's version of this doctrine) in *The Spectator*:

> Our Souls are at present delightfully lost and bewildered in a pleasing Delusion, and we walk about like the Enchanted Hero of a Romance, who sees beautiful Castles, Woods, and Meadows . . . but upon the finishing of some secret Spell, the fantastick Scene breaks up, and the disconsolate Knight finds himself on a barren Heath, or in a solitary Desart. (Addison and Steele 1712: 546–7)

For Descartes, there was only one way of avoiding this bleak denouement. Ultimately, he said, we have to rely on God's goodness in not misleading us:

> For since He has given me . . . a very great inclination to believe that [my sensory experiences] are conveyed to me by corporeal objects, I do not see how He could be defended from the accusation of deceit if these ideas were produced by causes other than corporeal objects. Hence we must allow that corporeal things exist.

This position was unstable. It led eventually to the empiricist Hume's scepticism about the existence of the external world and the reliability of causal relations. Kant, saying that Hume had "roused me from my dogmatic slumbers", tried to restore philosophical respectability to science. Specifically, he tried to rescue Newtonian physics.

The rescue of science from Hume's scepticism was to be achieved by a "Copernican revolution" in epistemology (Kant 1781). Kant claimed that the objectivity of physics lies not in its veridical representation of an external reality, but in the agreement among scientists that reality is to be described in certain (basically Newtonian) ways. In other words, its "objectivity" is actually inter-subjectivity.

But this agreement, according to Kant, isn't a matter of convention, or choice. The "intuitions" of space and time, and the basic "categories" (such as identity and causation) of science and of empirical perception in general, aren't arbitrary. On the contrary, they're built into the human mind a priori. We cannot help but perceive the world as structured by these principles. As Kant put it: "Categories are concepts which prescribe laws *a priori* to appearances, and therefore to nature, the sum of all appearances" (1781: 102).

A single-paragraph digression, not important at this point in our story but crucial for later chapters: To explain how intellectual concepts and abstract categories can be applied to experience, Kant posited "schemata". A schema isn't a phenomenal image, but *a capacity to form images* of a certain type. So someone who can apply the concept of a triangle to a variety of triangular things does so by way of a schema. Even the highly abstract categories—cause, substance, necessity—he said, are applied to experience by way of (transcendental) schemata. For example:

> The schema of cause . . . is the real upon which, whenever posited, something else always follows. It consists, therefore, in the succession of the manifold [experience], in so far as that succession is subject to a rule. (1781: B183–4)

Kant's contemporaries and immediate successors paid less attention to his notion of "schema" than to his position on objectivity and the real world. But it was destined to play a central role in cognitive science. It was picked up, and richly developed:

* by the neurologist Sir Henry Head in 1911 (see Chapter 4.vi.a);
* next, by the psychologist Sir Frederic Bartlett in 1932 (5.ii.b);
* by his student Kenneth Craik, who renamed it *model*, in 1943 (4.vi);
* by the neurophysiologist Warren McCulloch (*neural nets representing universals*) in 1947 (12.i.c);
* and by cognitive scientists such as Robert Abelson (*script*), Marvin Minsky (*frame*), and Michael Arbib (back to *schema* again) in the 1960s (7.i.c, 10.iii.a, and 14.vii.c respectively).
* It was also resurrected within 1960s philosophy of mind, which depicted the mind as a set of *functional capacities* (16.iii–iv)—making "experience as such" (*qualia*) philosophically problematic as a result (14.x–xi).

In short, people in cognitive science today who speak of mental models, schemas, or representations, and especially those (e.g. the evolutionary psychologists) who regard these as inborn, are—knowingly or unknowingly—echoing Kant (7.vi.d–f, 10.iii.a, 12.x, and 14.vii.c and viii; cf. Brook 1994). Indeed, many *do* do so knowingly: Kant's name is often mentioned.

As just remarked, however, Kant's contemporaries were more concerned with his claims about the metaphysical and epistemological status of the real world. According to him, there is indeed something—the "thing-in-itself"—which really exists outside us, completely independent of our thinking. But we can know nothing about it as it exists in itself:

> Things in themselves would necessarily, apart from any understanding that knows them, conform to laws of their own. But appearances are only representations of things which are unknown as regards what they may be in themselves. As mere representations, they are subject to no law of connection save that which the connecting faculty prescribes. (Kant 1781: 103)

In short, all our concepts of the external world are informed by the empirical intuitions and categories with which the mind is imbued. Other rational creatures, if such there be (Martians? angels?), might have a different set of empirical principles; but these are of necessity inconceivable to us.

Kant had intended his philosophy to save Newtonian science from Humean scepticism, and from arbitrary relativism. But in extending Descartes's stress on the epistemological primacy of consciousness, he offered hostages to absolute idealism.

If the thing-in-itself is unknowable, why posit it at all? And if our concepts of reality are inescapably mind-based, why not focus on the constructive activities of the mind—instead of on the 'real' world? This epistemological/metaphysical volte-face was the seed of the counter-cultural anti-realism that beleaguers science today (see Chapter 1.iii.b–c). But before it could flower in the form of today's social constructivism, it had nearly 200 years to develop.

Inevitably, the late eighteenth and early nineteenth centuries saw the rise of a variety of neo-Kantian idealist philosophies. These all stressed the creative power of mind (and life) and the epistemologically secondary nature of materialist science.

Most neo-Kantian philosophers reacted also against the atheistic rationalism of the Enlightenment. They were drawn more to religion and spirituality than to science, and more to (their interpretation of) biology than to physics (see below). They favoured intuition, imagination, and the specific individual—whether person or culture—over logic, experiment, and universal law. And many viewed the whole of Nature as a sort of organism, imbued with something akin to intelligence. Many, too, saw Nature as a system akin to a developing embryo, expressing the self-creative power of the divine.

Accordingly, even Newtonian physics, with its picture of inert matter pushed here and there by external forces, became suspect. Electricity—with its capacity to repel as well as to attract—was thought to be a more crucial, more philosophically promising, phenomenon than billiard balls. A fortiori, none of these idealistically inclined thinkers would have allowed that the nature of any living thing could be illuminated by comparing it to a man-made machine.

An example of neo-Kantianism focused on the creative power of language will be discussed in Chapter 9.iv. Others feature in subsections c–f, below. Before we can understand where they were coming from, however, we have to consider what Kant said about biology.

b. Biology, mechanism, teleology

A hundred years after its publication, Newton's *Principia* was still unrivalled as a theory of the inanimate world. As we've seen, Kant himself regarded it as an a priori aspect of human understanding. But he insisted that the animate world had to be understood differently (Korner 1955: 180–2, 196–214; Grene and Depew 2004: 92–127). Indeed, he discussed them separately. Whereas he justified physics in his *Critique of Pure Reason* (1781), he discussed biology only later, in the *Critique of Judgment* (1790, pt. 2)—which was concerned with aesthetics and teleology.

The life sciences, according to Kant, weren't on a par with physics. On the one hand, there could never be a scientific (mechanistic) *psychology*, because of the "autonomy" of human consciousness and voluntary action. On the other, there couldn't be a wholly mechanistic *biology*, because the purposive self-organization of living things—although effected by Newtonian processes—can't be analysed in Newtonian terms. Living things are purposive wholes, which means not that they are designed for a purpose but that their parts can't be properly understood without being (teleologically) related to the whole organism.

Kant didn't deny the value of the currently rising science of physiology. To the contrary, he declared as a guiding maxim that "All production of material things and their forms *must be considered as* being possible in accordance with merely mechanistic laws" (italics added). Nevertheless, he also declared a second maxim, that "Some products of material nature *cannot be considered* in accordance with *merely* mechanistic laws (their consideration requires an altogether different law of causality, namely, that of final causes)" (italics added). In other words, living organisms function in accordance with physics, but can't be explained solely by physical causes. They are *purposive* beings, requiring teleological explanation. Their purposes aren't relative to those of some other creature (like the purpose of a tool) but are inherent in the organism itself, which is therefore not a means to an end but an end-in-itself.

In saying this, Kant *wasn't* claiming that, at the metaphysical level, there are both mechanistic and teleological causes. Rather, he was saying that although the whole of biology runs by purely mechanistic causation, *our thinking of it* must also bear biological purposes in mind: mechanism as metaphysics, teleology as method (Korner 1955: 209). Partly, he said, this was helpful because thinking about the purpose of an organ might lead to mechanistic hypotheses about how it works ("For where purposes are considered as the conditions of the possibility of certain things, means have to be assumed . . ."). More importantly (for him), viewing living things as ends-in-themselves demands a certain—non-exploitative—moral attitude towards them.

Significantly, he wasn't sure just how far a mechanistic biological science could reach. He made two remarks about this which would later resonate strongly, both with his followers in the early nineteenth century and with modern biologists of various stripes (see below, and Chapter 15.iii, viii, and ix.c):

> When we consider the agreement of so many genera of animals in a certain common schema, which apparently underlies not only the structure of their bones, but also the disposition of their remaining parts, and when we find here *the wonderful simplicity of the original plan, which has been able to produce such an immense variety of species* by the shortening of one member and the lengthening of another, by the involution of this part and the evolution of that, there gleams upon the mind a ray of hope, however faint, that *the principle of the mechanism in nature, apart from which there can be no natural science at all*, may yet enable us to arrive at some explanation in the case of organic life. (Kant 1790: 418; italics added)

> [This hypothesis, which has probably occurred to most acute scientists, is daring but not absurd—like] the generation of an organized being from crude inorganic matter.... [It's not absurd to suppose that] certain water animals transformed themselves by degrees into marsh-animals, and from these after some generations into land animals. In the judgment of plain reason there is nothing *a priori* self-contradictory in this. But experience offers no example of it. (Kant 1790: 420)

He might have added that no one had managed to suggest a mechanism by which such evolutionary changes could take place. That, of course, is what Charles Darwin (and modern genetics) eventually provided.

Meanwhile, prior to Darwin, Kant's followers—especially Goethe (d–f, below)—focused less on the second quotation above than on the first. (And few people, if any, disagreed with him that the self-organization of life from crude matter was "absurd": see Chapter 15.x.b.)

c. Philosophies of self-realization

One of the neo-Kantian movements mentioned at the end of subsection a was the anti-rationalist *Naturphilosophie*, a form of German idealism espoused by Friedrich von Schelling (1775–1854), and others.

Kant himself, as we've seen, had argued that there could never be a mechanistic biology, because living things—being both causes and effects of themselves—are holistic teleological systems that can't be understood in purely physical terms. As he put it, they are self-organizing systems that propagate their own organization, and generate their own purposes, or ends. The followers of *Naturphilosophie* agreed, and took this view even further.

Their positions differed, some being more consonant with experimental physiology than others (Lenoir 1982; Cunningham and Jardine 1990). In general, however, they regarded life, adaptation, embryological development, and the origin of species as aspects of the self-creation of a divine mind immanent in nature. Most of them didn't believe in evolution, but those who did saw it also as due to some self-realizing creative force. And most of them argued for hylozoism: the view, now revived as the Gaia hypothesis (Lovelock 1988), that the earth as a whole is an organism.

The 'father' of this movement was Schelling, whose books *Ideas on the Philosophy of Nature* and *On the World Soul* were published at the close of the eighteenth century. Schelling saw the whole of Nature as infinite self-activity, an organic system constantly striving towards self-realization—an idea applied to *history* by his close friend Georg Hegel. As Schelling put it: "Nature is visible Spirit, Spirit is invisible nature."

He argued that all physical forces manifest the same "pure activity" of self-realizing Nature. Nature, he said, is—indeed, can only be—a balance of opposed forces (an idea inherited from Kant). So our "first principle" in studying it must be to "go in search of polarity and dualism throughout all nature" (quoted in Stern 1988, pp. ix–x). Since the dualism was a matter of balance, not absolutist metaphysics, he regarded knowledge as fundamentally identical with willing, and action with perception. And he saw art as superior to science. For the self-realizing aesthetic intelligence, being free of all abstraction, actually creates the world (Margoshes 1967). (This was a Romantic version of the *verum factum* tradition: see Chapter 1.i.b.)

Schelling was enormously influential in non-scientific circles. He is 'the' philosopher of Romanticism, and a crucial predecessor of existentialism—and of Freudian psychodynamics. But despite his prioritizing of art, he had some influence on science too.

To be sure, his more mechanistically minded contemporaries would have endorsed a much later judgement of his *Naturphilosophie*: that it was "fantastic to the verge of insanity" (Singer 1959: 385). Nevertheless, it had a significant impact on science in Germany, and elsewhere (Lenoir 1982; Merz 1904–12, vol. i, ch. 2; vol. ii, ch. 10). Darwin's contemporary Richard Owen was a zoologist in this neo-Kantian tradition. And the chemist Liebig, as remarked in Section v.b, adopted its holism and its stress on the uniqueness of life.

Not least, Schelling had a sympathetic hearing also from Goethe (1749–1832), who ensured that he was appointed to a university chair.

d. Goethe, psychology, and neurophysiology

Today, Goethe is remembered chiefly as a Romantic poet, or as yet another pantheistic neo-Kantian idealist. He would have demurred on two counts.

First, he didn't see himself as a Romantic: he regarded most Romantic artists as third-rate poseurs, and described the movement itself as a "disease" (I. Berlin 1999: 112). More to the point, he was keenly interested in scientific questions. Indeed, he regarded his scientific and literary works as equally important—and criticized the heavy arts-bias of Romanticism, accordingly.

One might almost say that, like Charles Snow (1959) after him, he lamented the split between "the two cultures". But that would be anachronistic. Certainly, Goethe had little patience with Romanticism's dismissal of science—understood as the systematic observation and study of the natural world. But he had even less with the analytical empiricism that Snow understood to constitute "science". (Hence his lack of intellectual sympathy for Vaucanson's duck.)

His own approach was relentlessly holistic. He was deeply suspicious of the Galilean–Cartesian programme of mathematizing science, and favoured careful—respectful—observation of the phenomena of Nature over analytic experimentation on them.

It's no accident, then, that the most detailed modern commentary on Goethe's science was written by an ex-student of the physicist David Bohm, whose *Wholeness and the Implicate Order* became something of a cult book in the 1980s. As Henri Bortoft (1996: 258 ff.) puts it, "the difference between a genuinely holistic perspective [such as Goethe's] and the analytical counterfeit" is that the former enables one to see "multiplicity in the light of unity, instead of trying to produce unity from multiplicity". We can recognize, indeed glory in, the rich diversity of the One instead of reducing all difference to uniformity.

So, for instance, in the anthropology of religion (my example, not Goethe's: see Chapter 8.vi.d–e), we can explore the many culturally different ways in which certain universally shared psychological principles come to be expressed, instead of reducing all finalized religions to the activation of one (or several) psychological module/s. Or, to take an example discussed at length by Goethe himself, we can see all the different individual plants as members of (better: expressions of) the one species, or class, instead of focusing on individual plants and seeking sets of shared properties in terms of which to define the class.

As for how we arrive at the concept of the species/class, if not by analytic property-counting, Goethe argued that this could be done by an act of intuitive perception in which we actually participate in the natural phenomenon we're studying. If that sounds overly mystical (and to orthodox scientific ears, it does), we'll consider some examples below.

Because of his holism, Goethe downplayed the relevance of experimental physiology much as Cuvier had done (see Section v). He anticipated some of Schelling's ideas about living organisms (and hylozoism). His insistence on the fundamental unity of natural phenomena was distilled in two philosophical concepts: "primal polarity" (similar to Schelling's explanatory dualism) and "primal phenomena". In applying these concepts to various scientific areas, he produced theories very unlike the empiricist norm.

Most cognitive scientists have read not one word of Goethe's science—and if they did, they'd probably recoil with horror. Nevertheless, his emphasis on painstaking observation, as applied to his own subjective experiences, made him an important forerunner of Gestalt psychology and of the phenomenological movement in Continental philosophy (see Chapters 5.ii.b and 16.vi–viii). He also anticipated a type of biological explanation (i.e. morphological explanation) that's gaining influence today (15.ix.c). And he's sometimes quoted by cognitive scientists and philosophers committed to the dynamical systems approach (15.viii.b–c and ix, and 16.x.a).

That's not to say that his detailed arguments were acceptable, or repeated by his twentieth-century successors. Important scientific and/or philosophical insights may well be introduced on unconvincing grounds, to be widely dismissed—even ridiculed—until better arguments are found.

For example, Goethe used his notion of primal polarity in attacking Newtonian physics—and excoriating Newton himself. He offered (in 1791) a fundamentally non-mechanistic account of the optics and perception of colour (Boring 1942: 112–19; Magnus 1906: 100–50). His optics was 'non-mechanistic' in being based on the qualitative aspects of visual perception, rather than mathematical analysis of the physical properties of light. A defender might say that he was addressing (psychophysiological) questions that Newton had not asked. But he didn't put it that way: he saw his optics as an alternative to Newton's, not a complement to it.

So he denied that white light is made up of the seven spectral colours. He argued instead that it's a fundamental, and unanalysable, aspect of the world. Newton's prism, he said, was a medium that led the polar opposites of light and dark to cooperate so as to produce colours. And this cooperation involves the activity of the eye: whenever the eye perceives a colour, it demands the complementary colour. He argued that there are two primary colours, yellow and blue, to which all the others are related. His polar distinction between yellow and blue was based on subjective criteria, not on optics. It was comparable (he said) to other dichotomies such as bright–dark, warmth–cold, active–passive, proximity–distance, and even acid–alkali.

Not surprisingly, mechanists who based their work on physics, or who aimed to do so, had scant sympathy with this subjectivist approach. But Goethe persisted, publishing (in 1810) a two-volume treatise on colour running to 1,411 pages. A leading historian of psychology has described this book as "an example of how personal pride distorts the use of evidence and how frustration induces scientific activity" (Boring 1957: 99).

Despite such charges of "distorting" the evidence, Goethe's treatise did have some influence in mechanistic physiology. For his extensive and subtle descriptions of colour experiences identified many important, and previously unnoticed, visual phenomena. These prompted some physiologists to work on the perception of colour.

One of these was Johannes Müller (1801–58)—yet another person, besides Harvey and Descartes, sometimes described as the father of experimental physiology. That's largely because of his monumental *Handbook of Human Physiology*, published between 1833 and 1840. As his handbook clearly showed, Müller was working within the mechanistic tradition.

But his mechanism, like that of so many others at the time, was qualified. Not only did he believe in mind-to-body causal interaction, but he was a vitalist. He couldn't endorse the project solemnly announced in 1845 by four of his students, to try to explain all living processes in terms of physicochemical laws (Boring 1957: 708). Two of those students were destined to have a huge influence in the gradual mechanization of mind. One was Helmholtz, who was soon to publish his paper on the law of conservation of energy. Another was Ernst Brücke, a future teacher of Sigmund Freud—who always insisted that there must be some mechanistic base to psychological phenomena (see Preface, ii, and 5.ii.a).

Müller's doctrine of "specific nerve energies" (see Section viii.a) was in part an attempt to deal mechanistically—and non-idealistically—with the subjective aspects of colour vision. The patterns in various types of nerve, Müller argued, cause different sorts of experience, and are themselves systematically caused by distinct types of stimulation originating in the external world. Here he was echoing Descartes, who had posited seven types of sensory nerve (see Section iii).

This approach was developed further by Helmholtz, who posited three types of colour receptor in the retina (Helmholtz 1860). Indeed, Helmholtz described Müller's law as comparable in importance—for psychophysiology in general—to Newton's law of gravitation. He also saw it as "the empirical exposition of the theoretical discussion of Kant on the nature of the intellectual process of the human mind" (from the *Handbuch der Physiologischen Optik* of 1896, p. 249; quoted in Merz 1904–12: ii. 483).

Nevertheless, a decade after Goethe's death, Helmholtz poured scorn on Goethe's ideas on psychophysics. He gave a lecture in which he echoed "the denunciation heaped by all physicists on [Goethe's] researches in their department, and especially on his 'theory of colour'" (Helmholtz 1853: 34).

Helmholtz justified this denunciation at length. He pointed out that, in his anti-Newtonian experiments (using a prism), Goethe had been hampered by impure colours and faulty apparatus. He described Goethe's account of the physical phenomena as "absolutely irrational" (p. 50). And he countered Goethe's subjectivist theory with a robust defence of man as machine:

> Even nature is, in the poet's eyes, but the sensible expression of the spiritual. The natural philosopher, on the other hand, tries to discover the levers, the cords, and the pulleys which work behind the scenes, and shift them. Of course the sight of the machinery spoils the beautiful show, and therefore the poet would gladly talk it out of existence, and ignoring cords and pulleys as the chimeras of a pedant's brain, he would have us believe that the scenes shift themselves, or are governed by the idea of the drama. (Helmholtz 1853: 50)

Helmholtz's point wasn't that *only* lever-and-pulleys explanations are allowed, for he himself posited "unconscious inferences" in perception (see Chapter 6.ii.e). These were needed, he argued, to compensate for the incompleteness of the stimulus. For instance, any given viewpoint will show only one face of a three-dimensional object, and one object may be part-hidden by another—yet we normally have no difficulty in recognizing them. However, Helmholtz took it for granted that those inferences are somehow grounded in "machinery", alias neurophysiology.

Scorning Goethe's "resolute hostility to the machinery that every moment threatens to disturb his poetic repose", he ended by saying:

we cannot triumph over the machinery of matter by ignoring it . . . We must familiarize ourselves with its levers and pulleys, fatal though it be to poetic contemplation.

One could be forgiven for inferring that Helmholtz thought Goethe worthless as a scientist, except perhaps as an observer of visual phenomenology or as an intellectual gadfly. But one would be wrong. In the same breath as dismissing Goethe's "egregious failure" in the psychophysiology of vision, Helmholtz referred to his "immortal renown" in another branch of biological science—namely, morphology.

e. The birth of morphology

Morphology, a word coined by Goethe, is the study of organized things. It concerns not just their external shape, but also their internal structure and development—and, crucially, *their structural relations to each other.*

Goethe's morphology was much wider in scope than the biological taxonomies of Carl Linnaeus and Cuvier. For he intended it to cover both living and inorganic nature, even including crystals, landscape, language, and art. But Helmholtz's interest, like ours, was with its application to biology.

Putting it that way, however, is anachronistic—and underplays Goethe's originality. For it wasn't yet clear that there's a unified subject, here. The inclusive term *biology*, meaning the life sciences in general, wasn't coined until 1796 (by which time Goethe was nearing 50), and was widely accepted only after Darwin. And not until the 1840s, well after Goethe's death, was it clear that *both* plants and animals are made of cells (Merz 1904–12: i. 193–5). In short, the fundamental unity of animals and plants wasn't yet clear in Goethe's lifetime.

Nevertheless, he applied similar ideas to both. In his *Essay on the Metamorphosis of Plants* (1790), he argued that superficially different parts of a flowering plant—such as sepals, petals, and stamens—are derived by transformations from the basic, or archetypal, form: the leaf. Later, he posited an equivalence (homology) between the arms, front legs, wings, and fins of different animals. All these, he said, are different transformations of the forelimb of the basic vertebrate type. And all bones, he claimed, are transformations of vertebrae.

In other words, he combined meticulous naturalistic observation with a commitment to the fundamental unity of nature. For instance, he's widely credited with a significant discovery in comparative anatomy. Namely, that the intermaxillary bone—which bears the incisors in a rabbit's jaw—exists (in a reduced form) in the human skeleton, as it does in other vertebrates. (Strictly, he *rediscovered* this fact (Sherrington 1942: 21–2), and *restated* the claim that sepals are a type of leaf (Goethe 1790: 73): cf. Chapter 1.iii.f.) The issue was "significant" because some people had used the bone's seeming absence to argue that God created a special design for human beings, marking them off from the animals. Goethe, by contrast, related human skulls to the archetypal vertebrate skull, much as he related sepals to the archetypal leaf.

Goethe himself didn't think of morphological transformations as temporal changes, still less as changes due to Darwinian evolution—which was yet to be defined. Rather, he saw them as abstract, quasi-mathematical, derivations from some Neoplatonic ideal in the mind of God. But these abstractions could be temporally instantiated.

So in discussing the development of plants, for instance, he referred to actual changes happening in time as the plant grows. He suggested that sepals or petals would develop under the influence of different kinds of sap, and that external circumstances could lead to distinct shapes, as of leaves developing in water or in air (see 15.ii). Similarly, some of the other Naturphilosophen spoke of temporal transformations in the embryo, as it ascends through the scale of animal being from animalcule, to mussel, to fish, to mammal. But they weren't evolutionists, so didn't—as the fervent mechanist Ernst Haeckel (1834–1919) later did—gloss this as "ontogeny recapitulating phylogeny".

The point of interest for our purposes is that Goethe focused attention on the restricted range of basic forms (primal phenomena) in the organic world. He encouraged systematic comparison of them, and of the transformations they could support. He also suggested that only certain forms are possible: we can imagine other living things, but not just *any* life forms. In a letter of 1787, he wrote:

> With such a model [of the archetypal plant (*Urplanz*) and its transformations] ... one will be able to contrive an infinite variety of plants. They will be strictly logical plants—in other words, even though they may not actually exist, they could exist. They will not be mere picturesque and imaginative projects. They will be imbued with inner truth and necessity. And the same will be applicable to all that lives. (quoted in Nisbet 1972: 45)

Similarly, in his essay on plant metamorphosis (1790), he said: "Hypothesis: All is leaf. This simplicity makes possible the greatest diversity."

Critics soon pointed out that he overdid the simplicity. He ignored the roots of plants, for instance. His excuse was telling:

> It [the root] did not really concern me, for what have I to do with a formation which, while it can certainly take on such shapes as fibres, strands, bulbs and tubers, remains confined within these limits to a dull variation, in which endless varieties come to light, but without any intensification [of archetypal form]; and it is this alone which, in the course marked out for me by my vocation, could attract me, hold my attention, and carry me forward. (quoted in Nisbet 1972: 65)

To ignore apparent falsifications of one's hypothesis, or even challenges to one's general approach, so shamelessly seems utterly unscientific in our Popperian age. And some of Goethe's contemporaries complained about it, too.

But his attitude stemmed from his idealist belief in the essential unity of science and aesthetics. He even compared the plant to a superb piece of architecture, whose foundations—the roots—are of no interest to the viewer. More generally: "Beauty is the manifestation of secret laws of nature which, were it not for their being revealed through beauty, would have remained unknown for ever" (quoted in Nisbet 1972: 35). For Goethe, this language had an import much richer than the familiar appeals to theoretical simplicity, symmetry, or elegance. It underlay, for example, his rejection of Newton's optics and his stress on the details of visual phenomenology.

Questions about such abstract matters as the archetypal plant were very unlike those being asked by most physiologists at the time. If a body is not just a flesh-and-blood mechanism, but a transformation of an ideal type, how it happens to work—its mechanism of cords and pulleys—is of less interest than its homology.

Indeed, for the holist Goethe the mechanism may even depend on the homology. Perhaps a certain kind of sap, a certain chemical mechanism, will induce a primordial plant part to develop into a sepal rather than a petal. But what's more interesting—on this view—is that sepals and petals are the structural possibilities on offer. How one describes the plant or body part in the first place will be affected by the type, and the transformations, supposedly expressed by it. It's not surprising, then, that Goethe was out of sympathy with the analytic, decompositional methods of empiricist experimentalism.

Initially, Goethe's morphology attracted scepticism even from descriptive (non-experimental) biologists. This was partly because of the poetical manner in which he wrote. And his close association with *Naturphilosophie*, not to mention his bizarre account of optics, didn't help.

But in 1830, two years before his death, his morphological ideas were publicly applauded by a highly respected biologist, Étienne Geoffroy Saint-Hilaire (Merz 1904–12: ii. 244). Geoffroy agreed with Goethe that comparative anatomy should be an exercise in "rational morphology", a study of the successive transformations—rational, not temporal—of basic body-plans.

Largely because of Geoffroy's influence, Goethe's ideas on morphology were cited approvingly after his death by a number of leading scientists. These included Haeckel and Thomas Huxley (1825–95), and even the self-proclaimed mechanist Helmholtz. Indeed, in the lecture quoted above, Helmholtz credited Goethe with "the guiding ideas [of] the sciences of botany and anatomy... by which their present form is determined", and praised his work on homology and transformation as "ideas of infinite fruitfulness" (Helmholtz 1853: 34, 30).

f. Goethe's eclipse

"Infinite fruitfulness" isn't on offer every day. So why were Goethe's ideas largely forgotten by the scientific community? Surely, such an encomium from such a high-profile scientist—and committed mechanist—as Helmholtz would be enough to guarantee close, and prolonged, attention?

Normally, yes. However, only six years after Helmholtz spoke of Goethe's "immortal renown" in biology, Darwin (1809–82) published *On the Origin of Species by Means of Natural Selection* (1859). This radically changed the sorts of enquiry that biologists found relevant. One might even say that they changed the sorts of enquiry that biologists found *intelligible* (see N. Jardine 1991).

Biological questions were now increasingly posed in ways that sought answers in terms of either mechanistic physiology or Darwinian evolution—that is, "descent with modification". (Darwin avoided the term "evolution", which was widely used to posit Lamarckian change in biology or to refer to embryonic development.) After the turn of the century, genetics soon became an additional source of biological enquiry. This might have happened earlier, but the work on hereditary "factors" of Gregor Mendel (1822–84) remained unknown until it was rediscovered, in 1900, by the founders of modern genetics.

The mix of physiology, evolution, and genetics was a heady brew. It quickly became the biological orthodoxy, eclipsing *Naturphilosophie* in all its forms—Goethe included.

One might wonder at the speed with which this happened. After all, Darwin himself was unsure of the mechanism of heredity, and accepted the evolutionary theory of

Jean-Baptiste Lamarck (1744–1829), who taught that acquired characteristics can be inherited. The mentalistic form of Lamarckism, which suggested that creatures can successfully strive for greater adaptation, was broadly consonant with *Naturphilosophie*. It fitted well with its vitalism, if not with its non-evolutionary structuralism. But even some of Goethe's sympathizers rejected Lamarck's theory. Geoffroy, for instance, criticized not just the mentalism but the adaptationism too. Moreover, Lamarckism was discredited ten years after Darwin's death, when August Weismann (1834–1914) attributed heredity to the cell nucleus, or "germ plasm", as opposed to the cell body. If acquired changes to the cell body weren't inheritable, Lamarck must have been mistaken.

In the late nineteenth century, then, *Naturphilosophie* as a form of biology was first cast into the shadows, and soon virtually eclipsed. To be sure, a recent bibliography lists over 4,500 titles on Goethe *as a scientist* (Magnus 1906/1949, pp. xiii, 249–53). But many of these were old and/or written by literary scholars rather than scientists. The life sciences were widely reinforced or reinterpreted as neo-Cartesian, *not* neo-Kantian, projects. Darwin encouraged systematic comparisons between different organs and organisms, as Goethe had done. But Darwin posited no ideal types. He explained morphological similarity in terms of contingency-ridden variation and selective descent, or coincidental likeness between environmental constraints.

In short, morphological—as opposed to metabolic—self-organization largely *disappeared* as a scientific problem.

That's not to say that the problem had been solved. It still survived in embryology (see Section vii, below). Moreover, Goethe's morphology was to be revived by the biologist D'Arcy Thompson in 1917, and is sympathetically regarded by some developmental biologists today (see Chapter 15.iii, and Webster and Goodwin 1996, esp. chs. 1 and 5). So Charles Sherrington's comments that "were it not for Goethe's poetry, surely it is true to say we should not trouble about his science", and that metamorphosis is "no part of botany today" are less true now than they were when he made them, some sixty years ago (Sherrington 1942: 23, 21).

Nevertheless, Goethe is still only a minority taste. From the 1860s onwards, his scientific work was sidelined, even scorned. Instead, Darwin and Bernard were paramount.

2.vii. The Self-Regulation of the Body

At much the same time as Darwin was preparing to publish on evolution, some other central mysteries regarding biological self-organization were being solved in essentially Cartesian terms. The leading figure, here, was the medic and physiologist Claude Bernard (1813–78).

a. Automatic equilibria

Bernard was hugely influential in both the practice and the philosophical justification of experimental physiology. Indeed, he was elected to both French Académies, of science and humanities. He argued that the science of physiology should underlie biology in general, and medicine in particular.

Like Descartes, he described an organism as "a machine which necessarily works by virtue of the physico-chemical properties of its constituent elements" (Bernard 1865: 93). But unlike Descartes, he took seriously the striking differences between the autonomous flexibility of living organisms and the passive predictability of stones, gases, and clocks.

His solution to this seeming paradox wasn't to deny biological determinism, as some of his contemporaries did. Rather, it was to attribute the (low-level) physical variation seen in organisms to (higher-level) adaptive powers of self-regulation found only in living things. Physiology, he argued, requires new concepts not needed in physics or chemistry.

In that sense, and *only* in that sense, he agreed with those who felt that there must be some sort of vital principle. But these new concepts, he argued, are ultimately grounded in those more basic sciences, knowledge of which is essential:

> In a word, biology has its own problem and its definite point of view; it borrows from other sciences only their help and their methods, not their theories. This help from other sciences is so powerful that, without it, the development of the science of vital phenomena would be impossible. (Bernard 1865: 95)

Animal heat, for example, is due—he said—to self-equilibrating mechanisms that continuously regulate the body's response to changing environmental conditions such that body temperature is conserved. In suggesting (and experimentally demonstrating) how such mechanisms might work, Bernard clarified many physico-chemical, and anatomical, details. He explained major aspects of the generation and regulation of body heat, and of digestion and nutrition. Among his many significant discoveries were the function of the vasomotor nerves in varying the rate of blood flow in different parts of the body, and the complex role of the liver in regulating blood sugar levels.

Just as important, and especially relevant for our story, was his general concept of the internal (or microcosmic) environment:

> I believe I was the first to insist upon this idea that there are for the animal really two environments: an external environment in which the organism is situated, and an internal environment in which the tissue elements live ... The invariability of the internal environment is the essential condition of free independent life: the mechanism which permits this constancy is precisely that which insures the maintenance in the internal environment of all conditions necessary to the life of the elements [the tissue cells] ... Far from being indifferent to the external world, the higher animal is, on the contrary, narrowly and wisely attuned to it in such a way that, from the continual and delicate compensation, established as if by the most sensitive balance, equilibrium results. (Bernard, quoted in Bodenheimer 1958: 415)

Adding that "these conditions are the same as needed for simple organisms", he suggested that plants and lowly animals have less "freedom" from environmental pressures because their compensating mechanisms are less well developed. But they, too, show the spontaneity characteristic of life.

Bernard was uncompromising about the relevance of physics and chemistry in understanding how such biological autonomy is possible. Unlike von Liebig, he insisted that it can be explained in fundamentally mechanistic terms:

> We cannot, therefore, admit in living organisms a free vital principle struggling against the influence of physical conditions. *The opposite has been proved*, and thus all of the contrary

conceptions of the vitalists are seen to be overthrown. (Bernard, quoted in Bodenheimer 1958: 416; italics added)

b. The embarrassing embryo

In truth, however, the opposite had not been fully proved. Mysteries remained, especially with regard to embryology.

To be sure, some complex, and previously mysterious, physiological phenomena had been explained by Bernard. And it was reasonable to expect that others would eventually be understood in much the same way. Indeed, Walter Cannon's influential work on "homeostasis" in the 1920s would describe a wide range of metabolic processes in terms very similar to Bernard's (Cannon 1926, 1932). Later still, the cybernetic movement of the 1940s would apply similar ideas to control engineering, too (Chapter 4.v–vii).

This last amplification wouldn't have discomfited Bernard. He'd been more than willing to apply his concept of the internal environment to man-made machines:

> We easily understand what we see here in the living machine, since the same thing is true of the inanimate machines created by man. Thus, climatic changes have no influence at all on the action of a steam engine, though everyone knows that exact conditions of temperature, pressure, and humidity inside the machine govern its movements. For inanimate machines we could therefore also distinguish between a macrocosmic environment and a microcosmic environment. In any case, the perfection of the machine consists in being more and more free and independent, so as to be less and less subject to the influence of the outer environment. (Bernard 1865: 98)

But some of Bernard's biological claims were mere statements of mechanistic faith, as most of Descartes's had been. If they now look more plausible than they did when he made them, that's not because of ideas that can be traced directly back to him.

In particular, this was true of his remarks about embryological development:

> But the term "vital properties" is itself only provisional; because we call properties vital which we have not yet been able to reduce to physico-chemical terms; but in that we shall doubtless succeed some day. So that what distinguishes a living machine is not the nature of its physico-chemical properties, complex as they may be, but rather the creation of the machine which develops under our eyes in conditions proper to itself and according to a definite idea which expresses the living being's nature and the very essence of life.
>
> When a chicken develops in an egg, the formation of the animal body as a grouping of chemical elements is not what essentially distinguishes the vital force. This grouping takes place only according to laws which govern the chemico-physical properties of matter; but the guiding idea of the vital evolution is essentially that of the domain of life and belongs neither to chemistry nor to physics nor to anything else. In every living germ is a creative idea which develops and exhibits itself through organization. (Bernard 1865: 93)

Perhaps so—but the nature of this "creative idea" was unknown. Self-organization in general was still not well understood, and biological morphology in particular was a mystery (see Chapter 15).

Consequently, the forms of vitalism that survived Bernard's work—some of which endured well into the twentieth century—drew their plausibility largely from difficulties within embryology and/or evolutionary theory.

The experimental embryologist Hans Driesch (1867–1941), for example, posited teleological "entelechies" guiding the development of the embryo. His theory attracted attention not least because he'd started out as a mechanist but, just before the turn of the century, rejected mechanism for vitalism.

This change was prompted by his experimental results showing (for instance) that each half of the two-cell stage of a sea-urchin egg could, if separated, develop into a whole embryo. Driesch soon abandoned experimental science for philosophy, and gave the first systematic statement of his vitalism as the Gifford Lectures, a series concerned with issues in the philosophy of "natural" theology (Driesch 1908).

But in doing this, he hadn't abandoned *science*. Lord Gifford's brief in founding the Gifford Lectures in 1885 had been clear:

> *I wish the lecturers to treat their subject as a strictly natural science*, the greatest of all possible sciences, indeed, in one sense, the only science, that of Infinite Being, without reference to or reliance upon any supposed special exceptional or so-called miraculous revelation. I wish it considered just as astronomy or chemistry is. (Deed of Foundation, 1885; italics added)

(It's no accident, then, that several scientists cited in later chapters delivered the Gifford Lectures too—including Sherrington, J. Z. Young, Donald MacKay, Christopher Longuet-Higgins, and Arbib. See also Chapter 8.vi.b–d.)

Driesch wasn't cheating. Although he'd rejected mechanist science, he hadn't rejected science as such. Or rather, he hadn't rejected what he understood science to be: the systematic study of natural phenomena, guided by experiment and observation. To describe his vitalism as unscientific, or even anti-scientific—as is often done, today—is anachronistic. For it presupposes a *modern* (neo-Cartesian) concept of science, and of biological science in particular. As we'll see (in Chapter 15.iii and viii), this concept itself is now being questioned. If some vitalists were enemies of science, broadly understood, Driesch was not.

c. Creative evolution

The most influential anti-mechanist account of evolution was developed by the late Romantic philosopher Henri Bergson (1859–1941). In 1907 he posited a mysterious creative principle, or spirit (*élan vital*), which he argued must play a directive role in evolution (Bergson 1911*a*).

In part, Bergson was driven (like Goethe and Driesch) by a recognition of holism. He didn't believe that Darwinian variation and natural selection could explain the wholeness of evolved organisms. In addition, he was puzzled by the appearance of novelty and especially of ever-increasing complexity. If ammonites and barnacles lived, and reproduced, successfully then why should fish and mammals ever arise? His answer was that there's a creative principle present in all living things, which relies on material processes for its expression.

So instead of the Cartesian mind–body dualism, he offered a life–matter dualism. The *élan vital* was something different from matter, but closely interdependent with it.

Mind–body dualism was reinterpreted accordingly (Bergson 1911*b*). Memory, for example, was seen by Bergson not as a store of material traces in the brain, but as enduring patterns of activity constituted by self-organizing associative processes—finding expression in the matter of the brain, but not to be identified with it. As for human consciousness, he often described this as the highest expression of the vital principle—and as the way in which human beings can become aware of it.

We ourselves, he argued, have some experience of the *élan vital* through the consciousness of our existence and self-creation through time. (That sentence is deliberately ambiguous, because Bergson himself wasn't clear about whether consciousness and/or "real duration" were supposed to be separate from the *élan vital*, or distinct aspects of it. Many passages, however, suggest the latter.) We know from experience that human life is becoming, not mere being. It is change, not stasis—nor even repetition. The same applies, he said, but in a lower degree, to other living things. The *élan vital* is "a current of consciousness" that has infiltrated living matter and which passes from one generation to the next by way of the reproductive cells. On this quasi-mystical view, the *élan vital* enables/directs the laws of matter and physical energy to produce novel structures which aren't necessitated by them but are somehow contingent. Complexity increases as biological evolution proceeds because the vital impulse itself consists in a creative potential that constantly strives to find expression in new, and fuller, ways.

Human consciousness, for Bergson, was the highest expression of this impulse so far. And despite repeatedly denying (as Darwin did, too) that evolution is directed to specific, pre-selected, goals he did say that humanity was "prefigured" in the creative evolutionary process at its start. He even said, some years later, that the appearance of human beings is the *raison d'être* of evolution on earth (1946, introd.). Moreover, he believed that essentially (though not superficially) similar beings are prefigured in the evolution that's surely happening on other planets.

Bergson's form of vitalism was far from clear, even when compared with other vitalistic philosophies. We've seen, for example, that the relation between life (*élan vital*) and mind (consciousness) was obscure: they were certainly very closely related, but just what sort of relation this was supposed to be is arguable. Again, he was unclear, even self-contradictory, about whether evolution is or isn't directed towards some end.

In general, his work contained more rhetorical flourishes than careful arguments. It was hugely popular in the early decades of the century, especially because of his attacks on "closed" societies and authoritarian religions (Bergson 1935). But it rarely convinced people who weren't initially sympathetic.

In particular, it didn't convince those committed to orthodox science or materialism. Bergson's neo-Romantic view was unpopular with most professional biologists, even before they knew just what a "gene" was. Once genes had been identified with DNA in the 1950s, the synthesis of molecular biology with Darwinism eclipsed Bergson much as Goethe had been eclipsed by Darwinism itself.

As is the way with eclipses, however, he eventually reappeared—if only as a minority taste. His mysterious dualism was replaced towards the end of the century by a definition of matter itself as intrinsically active, so enabling creative self-organization in life, mind, and evolution (Chapter 16.x.a).

2.viii. The Neurophysiological Machine

Bernard's work was situated within a tradition of *general* physiology, concerned with widespread biological processes such as respiration, nutrition, and thermo-regulation. Descartes himself had written on such matters, in connection with the circulation and functions of the blood. But, as we've seen, he'd also done the philosophical groundwork for an experimental *neurophysiology*, whose main focus would be the role of the central nervous system in mediating between perception and behaviour.

(This distinctly Cartesian way of posing the problem informs most neurophysiology, and most cognitive science, today. Nevertheless, it hasn't gone unchallenged: see 15.vii and 16.vi–viii.)

a. Getting on one's nerves

Neurophysiology took longer to develop than other sub-areas of physiology. Nevertheless, by the end of the nineteenth century the mechanistic basis of nervous function was already evident. For many decades after Descartes, the "animal spirits" of the nerves were thought to exert a force found only in living things (sometimes termed *vis viva*). The discovery of electricity in the mid-eighteenth century raised the question whether this force, then termed "animal electricity", was the same as the physicists' variety.

In a series of experiments on artificial electric fish, Henry Cavendish—whose ancestor the Marquess of Newcastle had been assured by Descartes of the automatism of swallows in spring—proved in 1776 that these two electrical forces were one and the same (Chapter 15.ii.a). And at about the same time, in 1780, Luigi Galvani showed that passing an electric current through a frog's leg (detached from the body) causes the muscles to contract. Since nerves were already known to cause muscular contraction in whole animals, the implication was that the nervous impulse itself is electrical—in the sense understood in physics.

Throughout the first half of the nineteenth century, this idea spread among the educated populace. One person enthused by it was Alfred Smee (1818–77), surgeon to the Bank of England and self-styled "electro-biologist". In several popular books written in the 1840s, he went way beyond frogs' legs: he argued that the full range of human thought and behaviour could be deduced from the electrical properties of the nervous system (e.g. Smee 1849, 1850). Sensory properties such as redness or roundness, he said, cause specific electrical events in the brain, which in turn lead to electrical activity in the appropriate muscles. And an idea, on his view, is a collection of such properties. Smee's writings introduced many readers to electro-physiology as a way of seeing the brain as a biological machine. (He also had something provocative to say about "brainlike" artificial machines: see ix.a, below.)

The increasingly mechanistic view of nervous action was brought to a head by Helmholtz, who in the mid-nineteenth century measured the speed of the nervous impulse (Helmholtz 1850). His experiment was intellectually as well as physically electric, galvanizing neurophysiology and much educated opinion alike. Its demystifying effect was compounded by the fact that nerve action turned out to be surprisingly slow. Many physiologists had assumed it to be instantaneous, or comparable to the speed of light, but Helmholtz clocked the nervous impulse at less than 100 miles an hour.

Clearly, the case for regarding animal—and human—nervous systems as scientifically intelligible mechanisms was strong. Strong, but not yet universally agreed. Helmholtz's own father refused to accept his son's results, saying, "I regard the idea and its bodily expression, not as successive, but as simultaneous, a single living act, that only becomes bodily and mental on reflection" (Boring 1957: 48).

As for what the nerves were doing, an eighteenth-century physiologist might have said, in anticipation of Gertrude Stein, that a nerve is a nerve, is a nerve . . . But early in the next century (in 1811), the anatomist Bell and the physiologist François Magendie (in 1822) independently discovered that where a spinal nerve connects with the spinal cord, it splits into two parts—one sensory, one motor. (Later, Bell proved that the nervous impulse can pass down a nerve in one direction only. So sensory nerves were afferent, and motor nerves efferent, with respect to the brain.)

Bell saw the difference between the dorsal and ventral roots of the spinal nerves as a special case of a general principle concerning the central nervous system:

> the nerves which we trace in the body are not single nerves possessing various powers, but bundles of different nerves, whose filaments are united for the convenience of distribution, but which are distinct in office, as they are in origin from the brain. (Bell 1811: 114)

Some time later, he showed that although some cranial nerves are mixed, as spinal nerves are, others are purely sensory or purely motor.

He also suggested that the nerves subserving the six senses are somehow physiologically distinct. Six senses, not five: it was Bell who discovered the need for kinaesthetic feedback—and the reciprocal innervation of flexor and extensor muscles which Descartes had hypothesized 200 years earlier (see Section ii.f). For example, he pointed out that when the eyeball is pressed one experiences not pressure, but light. He inferred that changes in the optic nerve, even if originally caused by pressure, can lead only to visual sensations. In general, he said, electricity powers the nerves—but we don't perceive *electricity*.

This suggestion seemed to offer a non-idealist account of the secondary qualities, such as colour and warmth, in so far as it explained them in relation to distinct physiological mechanisms. More accurately, it posited a systematic link between specific sensory qualities and distinct physiological mechanisms. From a Cartesian point of view, anatomists' perceptions of the nerves, whether of their secondary or primary qualities (their whiteness or their size), are themselves mere representations of the material world.

Bell's idea about the multiple specificity of nervous function was echoed in 1826 by Müller, partly influenced by Goethe's phenomenological work but leaning also on his own experiments on the spinal nerves. Müller developed the idea at length (in 1838) as the doctrine of specific nerve energies—which, as we've seen, Helmholtz took to be the pivotal law of neurophysiology.

b. Reflections on the reflex

The functions of the spinal nerves were investigated also in studies of reflex movement. The word "reflex" is Marshall Hall's (1833), but Thomas Willis (then Professor of Medicine at Oxford) in 1664 had spoken of "reflexion" in the brain, and Descartes himself had described what we'd now call reflex action (see Section ii.d, above). So it's an old idea.

The first relevant experiments were done in the mid-eighteenth century, by Robert Whytt in Edinburgh. He showed that reflex movements can occur as a result of sensory stimulation even when a frog's spinal cord is separated from its brain.

Almost 100 years later, similar work by Müller and Hall on newts and snakes aroused controversy. The debate concerned whether reflex movements are ever mediated by the brain (as well as by the spinal cord), whether they're involuntary, and whether they're conscious. Some physiologists argued that all nervous action is conscious to some degree, others that consciousness requires activity in the brain.

In Russia, the early reflexologist Ivan Sechenov (1829–1905) turned the question on its head. In *Reflexes of the Brain* (1863), he argued that "all acts of conscious or unconscious life are reflexes". Sechenov's book was officially banned as a result, and he was (unsuccessfully) taken to court by the Petersburg Censorial Committee for undermining public morals (Boring 1957: 635). Clearly, reflexology was both radical and risky.

By the turn of the twentieth century, it was more widely accepted. Indeed, Ivan Pavlov (1849–1936) was awarded the Nobel Prize in 1904 for his work on digestion and conditioned reflexes.

Pavlov's prime interest was physiology, not psychology. (He founded the physiology laboratory at Russia's Imperial Institute for Experimental Medicine.) He started by studying how dogs come to produce saliva and gastric juices on hearing a bell previously associated with the presentation of food (Pavlov 1897/1902). This work, unlike Sechenov's earlier pronouncements, *didn't* obviously threaten ideas of consciousness: even humans don't salivate consciously. But in the new century, he broadened his interests and studied behaviour (Pavlov 1923/1927, 1925/1928). The issue of consciousness became problematic again when his students and—especially—the behaviourist psychologists applied his theory of conditioning to types of motor behaviour often attributed to conscious intentions (see 5.iii.a–c).

But even Sechenov hadn't denied animal consciousness. Few, if any, late nineteenth-century neurophysiologists shared Descartes's opinion that, irrespective of what happens in their brains, non-human animals are never conscious. (Perhaps one should rather say, "what was widely believed to be Descartes's opinion": see Section iii.d.) This belief had become doubly problematic after the publication of Darwin's theory of evolution. To regard animals as mere (non-conscious) automata not only went against nineteenth-century common sense, but implied a stark discontinuity between *Homo sapiens* and other species.

Darwin's colleague Huxley, in his paper 'On the Hypothesis that Animals Are Automata, and its History' (1874), offered an account of mind–body relations that specifically admitted animal consciousness. His view was later named "epiphenomenalism" by James Ward (1902).

Epiphenomenalism was intended by Huxley—and gratefully received by most experimental biologists—as a philosophical justification of research in neurophysiology (and psychophysiology). However, it inherited major metaphysical problems from Descartes, while also being counter-intuitive on two crucial points. Indeed, twentieth-century neurophysiology was bedevilled by similar problems—as the new century's research is also (Chapter 16.i.b).

Accepting the Cartesian mind–body split, Huxley denied that consciousness can cause physical change (which it seems to do, when someone performs a voluntary action), because such mind-to-body interaction was inconsistent with Helmholtz's law of conservation of energy. He even implied, without actually saying so, that there's no mind-to-mind causation either. In a telling technological metaphor, he compared consciousness to the smoke emitted by a steam-engine passing behind a hedge: the only thing visible, but merely an inconsequential side effect of the powerful machinery underneath.

However, he agreed with Descartes that conscious feelings and sensations are somehow caused by brain states—leaving unsolved the puzzle of how body-to-mind causation is possible. (Being the first self-confessed "agnostic", Huxley could hardly rely on Descartes's theological account of Natural causation.) This neo-Cartesian puzzle is still unsolved—some argue that it's insoluble (see 14.x–xi).

Although most early neurophysiologists were concerned with the peripheral nervous system, some anatomists worked also on the brain. Bell, for example, generalized his idea about functional specificity from the nerves to the brain.

He suggested that different parts of the brain do different things, and distinguished various aspects of brain specialization—such as motor and sensory centres. He didn't consider only the brain's gross anatomy, the distinction between cerebrum and cerebellum for example. On the contrary, he stressed the importance of discovering the pathways and interconnections of individual nerves:

> My view about the differentiation of nervous function will explain the apparently accidental connection between the twigs of nerves . . . and *it will give an interest to the labours of the anatomist* in tracing the nerves. (Bell 1811: 114; italics added)

The labours of the anatomist, at the time Bell was writing, depended primarily on careful dissection. Where the brain was concerned, the aim was to distinguish, and to trace, bundles of nerve fibres differing only slightly in colour, texture, or density. Given that techniques for firming up the porridge-like consistency of the brain were still primitive, this was a tall order. Thirty years later, the method of nerve sectioning became available also, whereby the anatomist could trace the degeneration of nerve fibres separated from their cell bodies. But this method came too late for Bell.

c. From nerves to neurones

A "nerve", for Bell, was *not* a neurone. It was a structure visible to the naked eye, a thread forming part of the nervous system.

Microscopy wasn't well enough advanced to be very helpful in identifying what we know to be neurones. Not until 1833 would it show that there were both cells and fibres in the brain: the grey and white matter, respectively. Besides the relatively low optical resolution of the instruments, cell staining in general wasn't discovered until the mid-nineteenth century. And it was even later, in 1873, that Camillo Golgi (1843–1926) originated 'Golgi staining', which was needed to make individual brain cells visible under the microscope.

Even after the advent of Golgi's method, it wasn't evident that the brain contains neurones. In other words, it was still unclear whether the brain is a continuous network, with cell bodies positioned along the nerve fibres like isolated beads on a reticulated string, or whether the network is discontinuous. Up to the final years of the century, most neuro-anatomists assumed the former. By 1906, to be sure, neurones were gaining favour. But Golgi himself took pains to reject "the current opinion"—namely, the neurone doctrine—in his Nobel Prize acceptance speech that year (Shepherd 1991: 261).

Golgi's unrelenting speech was especially ironic, for he was sharing the Nobel Prize with the person who had provided the first proofs of the neurone theory: the Spanish histologist Santiago Ramón y Cajal (1852–1934). In 1888 he had shown that the brain contains a host of discrete units (Ramón y Cajal 1901–17: 321–41). Using an improved version of Golgi staining (but an out-of-date microscope), he observed units differing in size and shape, some of which are found only in specific parts of the brain, such as the cerebellum.

The details of his drawings—microscopic photography was impossible—were astonishing (see Figures 2.1–2.3). In many cases, the cell bodies sprout a mass of tiny dendrites at one end and a single fibre (the axon) at the other—which divides into dendrites at its extremity. The dendrites from different cells intermingle at the (newly observed) synapses, but aren't physically continuous.

In short, the brain is made up of individual cells, many of which are neurones. (The others include the glial cells.) Ramón y Cajal hypothesized that only neurones carry messages, that each neurone does this in one direction only (from the cell body, through the axon), and that nerve impulses are transmitted only by contact between neurones. I say "hypothesized", but in the final case that was too weak:

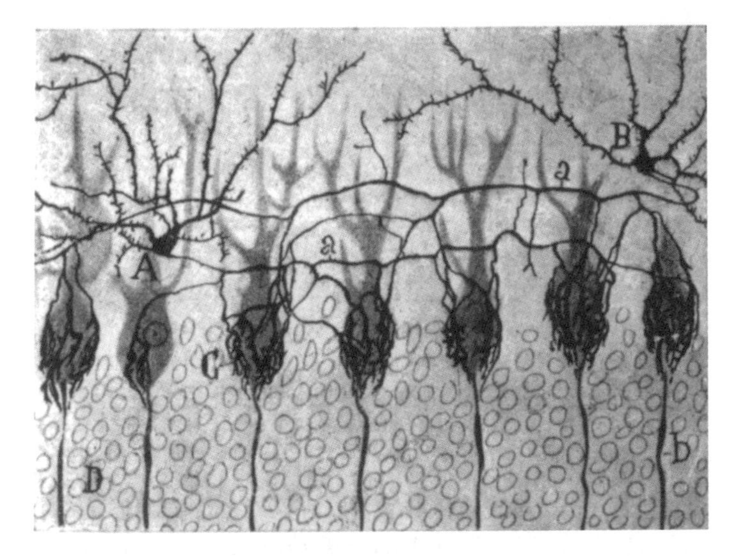

Fig. 2.1. Drawing of neurones, done in 1888, by Ramón y Cajal. Original caption: Transverse section of a cerebellar lamella. Semidiagrammatic. A and B, stellate cells of the molecular layer (basket cells), of which the axon (a) produces terminal nests about the cells of Purkinje (C); b, axon of the Purkinje cell. Reprinted from Ramón y Cajal (1901–17: 330)

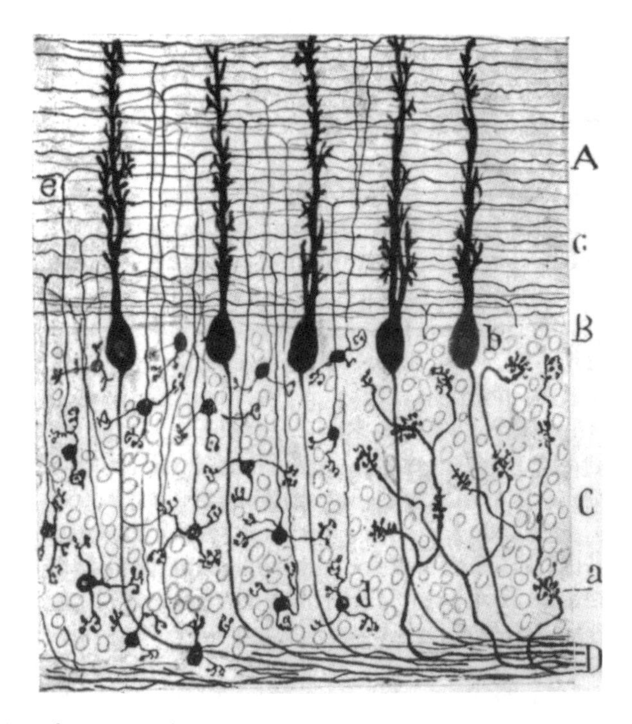

Fig. 2.2. Drawing of neurones, done in 1888, by Ramón y Cajal. Original caption: Longitudinal section of a cerebellar convolution. A, molecular layer; B, layers of Purkinje cells; C, granular layer; D, white matter; a, tuft of a mossy fibre; b, body of a Purkinje cell; c, parallel fibres; d, granule cell with its ascending axon; e, division of this axon. (Semidiagrammatic). Reprinted from Ramón y Cajal (1901–17: 331)

This fortunate discovery [of the cerebellar climbing fibres], one of the most beautiful which fate vouchsafed to me in that fertile epoch, formed *the final proof* of the transmission of *nerve impulses by contact.* (Ramón y Cajal 1901–17: 332; first italics added, second in original)

Did his fellow neuro-anatomists accept this "final proof"? Not at all. Largely because of the hold of the reticular theory on their minds—and also because they couldn't immediately replicate his findings (it turned out that there was something special about the tap water he used to rinse his slides for thirty minutes or so)—these discoveries were initially ignored. It didn't help that Ramón y Cajal was an untutored country boy from Zaragoza, not a university researcher (see Chapter 1.iii.f). As he put it in his (fascinating) autobiography, his papers were greeted with "silence", "excessive reservedness", and even "contempt" (Ramón y Cajal 1901–17: 352–3). One contemporary neurologist later recalled (in 1913):

The facts described by Cajal in his first publications were so extraordinary that the histologists of the time—fortunately I did not belong to the number—received them with the greatest scepticism. The distrust was such that, at the anatomical congress held in Berlin in 1889, Cajal, who afterwards became the great histologist of Madrid, found himself alone, exciting around him only smiles of incredulity. (quoted in Ramón y Cajal 1901–17: 356)

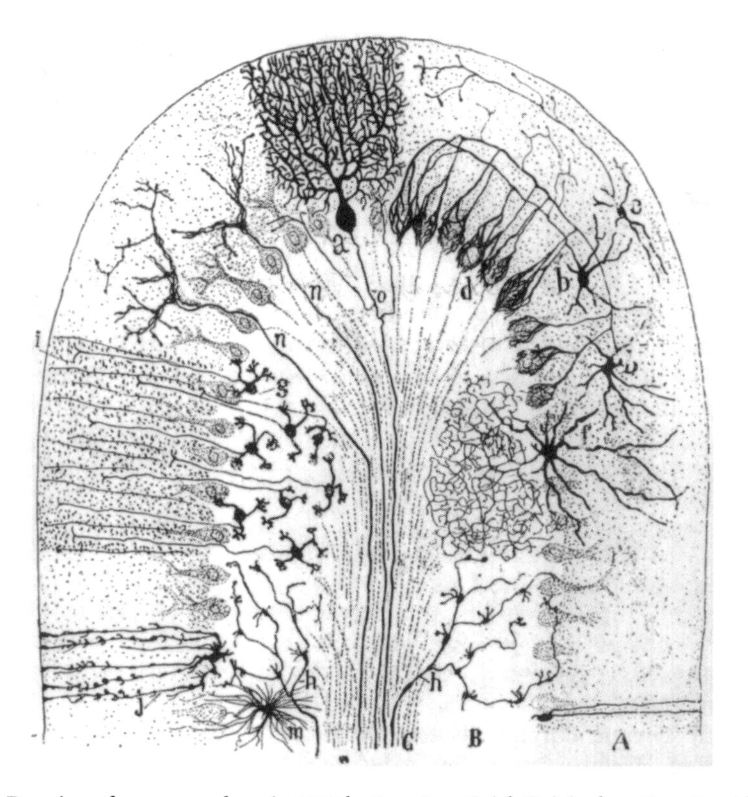

Fig. 2.3. Drawing of neurones, done in 1888 by Ramón y Cajal. Original caption: Semidiagrammatic transverse section of a cerebellar convolution of a mammal. A, molecular layer; B, granular layer; C, layer of white matter; a, Purkinje cell with its dendrites spread out in the plane of section; b, small stellate cells of the molecular layer; d, descending terminal arborizations embracing the cells of Purkinje; e, superficial stellate cell; f, large stellate cell of the granule layer; g, granules with their ascending axons bifurcating at i; h, mossy fibres; j, tufted neuroglia cell; n, climbing fibres; m, neuroglia cell of the granule layer. Reprinted from Ramón y Cajal (1901–17: 333)

The smiles of incredulity soon faded. The same observer also recalled that the microscopic preparations that Ramón y Cajal showed at the Berlin congress were "so decisive" that his claims were soon confirmed by Albrecht Kolliker, "the unquestioned master of German histology".

But to fade isn't necessarily to disappear. The physiological experiments of Sherrington, an early champion of the neurone theory, provided further evidence (see Section viii.d, below). Most people found this evidence compelling. Nevertheless, doubts remained even up to the mid-twentieth century, when disagreements between "neuronists" and "reticularists" ended at last (Changeux 1985, ch. 1).

In the early twentieth century, Ramón y Cajal's microscopic studies showed also that there are many different kinds of neurone, located in different parts of the brain. The structure of the cerebellum, for example, differs from that of the cerebral cortex. And the striate (i.e. striped) cortex contains several distinct layers, whose neurones differ in both shape and spatial arrangement.

Half a century after that, computational neuroscientists would use updated versions of the Zaragozan's pioneering work to try to puzzle out just what the cerebellum is doing, and how (Chapter 14.iv.c–d and viii.b).

d. Integration in the nervous system

In 1894 the champion of the neurone theory was invited to speak to the Royal Society in London. There he met the leading English neurophysiologists, including Sherrington (1857–1952) and his old friend Michael Foster.

Foster was the first professor of physiology at Cambridge, and the founder of the *Journal of Physiology*. (He'd been appointed, in 1870, on Huxley's recommendation.) At the time of Ramón y Cajal's visit, he was acting also as Secretary of the Royal Society. At the banquet held in the Spanish visitor's honour, Foster declared that "thanks to [his] work, the impenetrable forest of the nervous system had been converted into a well laid out and delightful park" (Ramón y Cajal 1901–17: 421).

Soon, Sherrington (at the University of Liverpool from 1895 to 1913, when he moved to Oxford) would show that the park was laid out on a number of interconnected levels, involving subtle hierarchies of control. And in the final years of the century, Sherrington added the new neurone theory to his study of mammalian reflexes. In 1897, while editing a chapter of Foster's *Textbook of Physiology*, he realized that the all-important nerve junction needed a name. He coined the term "synapsis", soon shortened to "synapse", from the Greek word for *to clasp* (Sherrington 1937).

It was already clear that the man-machine was a self-equilibrating system in its motor behaviour, no less than its blood temperature. As early as 1817, Magendie had defined a reflex not as a mere 'string' of events, but as a circular activity initiated at some bodily location, passing via the central nervous system, and returning to the initial point to cause some bodily reaction there. And he and Bell, in their studies of sensory and motor nerves, had identified two arcs of the circle.

But the detailed neural mechanisms were still unknown. Thus in his textbook of 1879 Foster had remarked:

> The spinal cord, and indeed the whole central nervous system, may be regarded as an intricate mechanism in which the direct effects of stimulation or automatic activity are modified and governed by the checks of inhibitory influences; but we have as yet much to learn before we can speak with certainty as to the exact manner in which inhibition is brought about. (quoted in Granit 1966: 41)

In Sherrington's hands, the neurone theory promised to help provide the answers.

Sherrington made precise time measurements, allowing not only for nervous conduction but also for synaptic delay (and refractory periods); and he kept precise records of the forces exerted by antagonistic muscles. Using these experimental data, he was able to explain gross patterns of movement in terms of neural circuits involving inhibitory and excitatory action. Moreover, he showed that inhibition didn't happen only at the nerve–muscle junctions of inhibitory nerve fibres, but also at (nerve–nerve) synapses within the brain.

In 1892, for example, he showed that—despite its very low latency—the knee jerk is indeed a reflex, which can be actively inhibited by electrical stimulation of the antagonist muscle. This was an early example of his lasting interest in the reciprocal innervation of opposing muscle groups—a principle anticipated by Descartes (see Section ii.f). Sherrington described it in several papers in the *Journal of Physiology*, and in the Royal Society's *Philosophical Transactions*, around the turn of the century.

A few years later, he showed that decerebrate cats, in which the cerebrum is separated from the brain stem, respond to a touch on the skin of one paw by an integrated pattern of movements involving all four limbs (Sherrington 1898). These animals suffered a characteristic rigidity (previously observed by Bernard), because the proprioceptive feedback mechanisms regulating their muscle tone had been destroyed. In investigating this, Sherrington pioneered the study of the sensory organs (muscle spindles) involved.

In 1906 he published an influential book citing even more complex automatic movements. For instance, a decerebrate animal will continue to 'step' on a moving belt, walking fast or slowly in perfect adjustment to the speed of the belt. He explained these movements in terms of specific sensori-motor mechanisms (Sherrington 1906). In this case, and others, he tried to relate particular neuronal circuits to the "integrative" sensori-motor action of the organism as a whole. His work on "circular" reflexes and "integration" in the nervous system would later help inspire the cybernetics movement (see Chapter 4.v).

Although Sherrington's theory of 1906 treated the neurone as the analytical unit of the nervous system, he couldn't then record from individual nerve fibres. The electrical signal was too faint. His hypotheses about the functional properties of neurones had to be tested indirectly. Direct confirmation, and an explanation of how these functions actually work, was to come from research done over the next half-century (see below).

e. How do neurones work?

Meanwhile, some sceptics still doubted whether 'mental' functions like associative learning and inhibition could be grounded in *any* physical mechanism. Neurophysiologists assumed that they could—but they weren't able to explain how.

Just before the First World War, the engineer S. Bent Russell set out to prove that this is in principle possible, by building a hydraulic compressed-air model of nervous conduction (Bent Russell 1913). He even suggested that engineers, using mechanical simulation as a general method, might do interdisciplinary research with neurophysiologists and psychologists:

> It is thought that the engineering profession has not contributed greatly to the study of the nervous system ... *As the cooperation of workers in different fields of knowledge is necessary in these days of specialists* it may be argued that engineers can consistently join in the consideration of a subject of such importance to man. (1913: 21; italics added)

With hindsight, his suggestion looks prophetic. At the time, it looked unconvincing—or anyway, not readily feasible. Bent Russell himself soon abandoned simulation (Cordeschi 2000: 321).

(The early interest in *physical* models of neural learning soon waned: Cordeschi 2002. But it revived in the early 1930s, largely due to Clark Hull's behaviourist 'robot' approach: see Chapter 5.iii.c. It blossomed in the 1940s, thanks to cybernetics: Chapter

4.v–vii. By the 1950s, most simulations of the nervous system were computational rather than physical: Chapters 4.iii–iv and 14.iii.)

By the middle of the twentieth century, Foster's question about *how* conduction and inhibition actually happen had been largely answered. The answer was found by work done at Trinity College, Cambridge—with the help of technical equipment pioneered in the early 1920s by neurophysiologists at Washington University, St Louis (Finger 2000: 245–8). (All the prime movers involved ended up with Nobel Prizes.)

Trinity's Edgar D. Adrian (1889–1977)—later, Lord Adrian—was able to amplify the nerve impulse so that the action potential, or spike, in a single axon could be recorded. He first did this in experiments on muscle (Adrian 1926; Adrian and Zotterman 1926), which confirmed Sherrington's ideas about proprioception due to muscle spindles (see Section viii.d, above). But very soon afterwards he was able to do so by working directly with motor-nerve fibres (Adrian and Bronk 1928). His technique became more widely used when the London anatomist John Z. Young discovered 'giant' axons in the squid, a millimetre across and several centimetres long.

This method vindicated the "all-or-none" principle of nervous conduction, which had been posited—and demonstrated in muscle fibres—by Adrian's Trinity supervisor Keith Lucas (Lucas 1905, 1909) and confirmed *indirectly* by Adrian himself (Adrian 1914). It also provided precise data on the temporal properties of the neurone, such as the refractory period and the timing of bursts of impulses in different stimulus conditions.

Adrian also discovered that neurones are destined to fire spontaneously, without being stimulated to do so by other neurones. As he put it, "The moment at which [firing] occurs can be greatly altered by afferent influences, but it cannot be postponed indefinitely" (Adrian 1934: 1126). He assumed that this unpredictable noise in the nervous system was due to unknown chemical changes at the synapse.

The mechanism of nervous action was further clarified in 1939 by Adrian's fellow Fellows Alan Hodgkin (1914–98) and Andrew Huxley (1917–1998). They suggested, against the all-or-none orthodoxy, that the spike travels down the axon by successive *local* depolarization of the cell membrane, with a swing in potential from inside-negative to inside-positive and back again. However, their hypothesis was too heretical to be accepted—and for the next few years, they were diverted into war work.

By the late 1940s they were again free to study the biophysics of nerve conduction. In 1949 they discovered the "sodium pump", by means of which the electrical impulse travels along the axon. Finally, at mid-century, they published their research in the form of the fundamental 'Hodgkin–Huxley equations' (Hodgkin and Huxley 1952).

And that, for a while, was that. As Huxley later put it:

> When we had completed the work on the [giant] squid fibre that we published in 1952, we could not see what could be done next to take the understanding of the excitation process to a deeper level. Huge advances have been made since, but all have depended on technical improvements or on advances in other branches of biology—notably molecular genetics—that were unforeseeable in 1952. (A. Huxley 1999)

Hodgkin, accordingly, switched his interests to other aspects of nerve physiology. As for Huxley, he turned to study the mechanism of contraction of striped muscles. By the mid-1950s, he was awing Cambridge's medical students with hot-from-the-lab reports of his discoveries (see Preface, ii).

Many questions remained, however, about just which nerve cells do what. Hodgkin had used miniature electrodes to record intracellular changes in the *axon.* John Eccles (1903–97)—who shared the Nobel Prize with Hodgkin and Huxley in 1963—developed their technique to stimulate and record from individual *cell bodies.* This heralded a new era: by the early 1960s, neuroscientists had fallen prey to what one of their number would term "a virtual obsession with unit recording" (see Chapter 14.iv and ix.b). "Obsession", because although a huge amount was learnt through this technique, certain questions were systematically sidelined.

It would be some time yet before neuroscientists focused on the properties of large networks of neurones, or on the detailed functionality of anatomically identifiable areas of the brain. Some of that work—examples of *computational* neuroscience—will be described in Chapter 14. (For a broad summary of post-1960s advances in general neuroscience, including neurochemistry, see Kandel and Squire 2000.)

f. Brains and machines

Eccles's new knowledge of neural mechanisms didn't commit him to mechanism as a philosophical position (see 16.i.a). On the contrary, he argued that, thanks to quantum indeterminacy, human free will can cause physical changes in "critically poised" neurones in the brain (Eccles 1953, ch. 8). He specifically described the cerebral cortex as a "detector" whose sensitivity is of a different kind and order from that of any physical instrument. (Some thirty years later, he was still arguing essentially the same position: see Eccles 1986.)

Most mid-twentieth-century neurophysiologists ignored that speculation. They valued Eccles, rather, as one of the first in a growing line of researchers using single-cell techniques to demystify the brain. His techniques would soon lead to the discovery of feature detectors (Chapter 14.iv). And his own experiments would clarify the detailed structure of the cerebellum—and, ironically, help ground a computational account of the role of the cerebellum in "voluntary" action (14.v.b–d). Even in 1950, those results still lay in the future. But 'man as machine' now had firm neurological underpinnings.

Before Darwin, Bell's robust defence of his practice of "examining the human body as a piece of machinery" had included the argument that the "perfection of the instrument" exemplifies the Creator's "plan universal" and "prospective design" (Bell 1834: 3, 15 ff.). Indeed, Bell's pioneering study of the anatomy and functioning of the hand, including its sensory and motor nerves, was published as one of the eight *Bridgwater Treatises.*

Appearing between 1833 and 1836 (and very soon reprinted), these had been established by the will of the eighth Earl of Bridgwater, who died in 1829. An ordained minister, he was more theologically active after his death than before it, being described as "a noble clergyman who had always neglected his parish assiduously" (Gillispie 1951: 209).

His executors carried no little clout: the Archbishop of Canterbury, the Bishop of London, and the President of the Royal Society. These three gentlemen were required by the terms of the will to choose eight scientific writers supporting the general theme:

"On the Power, Wisdom, and Goodness of God as Manifested in the Creation". (The "Ninth" *Bridgwater Treatise*, as we'll see in Chapter 3.i.b, was unofficially added by Charles Babbage.)

But the will also suggested more specific topics, one of which was "the construction of the hand of man". Having picked up that baton, Bell's preface explicitly condemned:

> the futility of the opinions of those French philosophers and physiologists, who represented life as the mere physical result of certain combinations and actions of parts, by them termed Organization. (Bell 1834, p. ix)

In short, he was just one of many scientists, at that time, who saw the functional organization of living things as a proof, or even as a self-expression, of divine intelligence.

After Darwin, the notions of natural machines and God's design were divorced in many—though not all—biologists' minds. And after Bernard (with his not-so-futile French opinions), and especially Sherrington, some of the most mysterious workings of the body were being compared to inanimate machines. But instead of clocks and fountains, people now cited the steam-engine, the telegraph, and the telephone. The last two of these would feature prominently for the next fifty years, occasionally accompanied by the accordion, harpsichord, or jukebox. (Jukebox, because it seemed, in the late 1920s, that a particular signal from the brain evokes *an entire pattern* of innate or learnt behaviour: see Chapter 14.v.c.)

These ideas didn't spring merely from turn-of-the-century technophilia. For they all exemplified what Gerd Gigerenzer (1991*b*) has called the "tools-to-theories" heuristic. This is a widespread creative strategy in science, of which *man-as-machine*, *mind-as-computer*, and *heart-as-pump* are all special cases. Thinkers guided by it use a pre-existing physical artefact—or perhaps a mathematical calculus, such as statistics (Chapter 7.iv.g)—as the inspiration for a new scientific theory.

The telephone-exchange analogy was undermined in the 1930s, by Adrian's observations of spontaneous neuronal firing ('noise' in the nervous system) and by his and others' work on EEG waves. But it wasn't finally dropped until later. Its demise was partly due to further work on random activity in the brain. "Random", here, didn't mean uncaused: rather, a neurone's firing was not, as had previously been thought, wholly determined by the input—or lack of it—from other neurones (Delisle Burns 1968). (Later still, Horace Barlow would argue that this seeming unreliability of individual neurones had been an illusion: Chapter 14.x.e.)

What replaced the telephone exchange, as we'll see in Chapter 4, was the computer. (Whether the computer itself will some day be replaced by some other machine is discussed in Chapter 16.ix.f.)

By the early to mid-twentieth century, then, the biological version of 'man as machine' had blossomed luxuriantly. Interpreted in the scientific sense, there was a powerful consensus that the body, including the nervous system, is scientifically intelligible. Interpreted in the technological sense, man-made machines were commonly used as analogies in theorizing about the body, including the brain.

Even exciting new toys such as motor boats and aeroplanes were sometimes mentioned, if only to inspire the minds of the young. In 1912 Edwin Brewster's book on physiology, enticingly named *Natural Wonders Every Child Should Know*, confidently declared:

For, *of course*, the body is a machine. It is a vastly complex machine, many, many times more complicated than any machine ever made with hands; but still after all a machine. It has been likened to a steam engine. But that was before we knew as much about the way it works as we know now. It really is a gas engine; like the engine of an automobile, a motor boat, or a flying machine. (quoted in Hodges 1983: 13; italics added)

For all his confidence in mechanism, Brewster didn't pretend to have all the answers. For example, he asked how the "living bricks" of the body "find out when and where to grow fast, and when and where to grow slowly, and when and where not to grow at all"—something, he said, which "nobody has yet made the smallest beginning at finding out".

Brewster's book earns a footnote in history by being the one which, so Turing told his mother, opened Turing's eyes to science (Hodges 1983:11). Indeed, Turing later made "a small beginning" in answering Brewster's question about the living bricks (see 15.iv). Even more to the point, he made a *large* beginning in extending the man-machine analogy from body to mind (Chapters 4.ii and 16.ii). But, as the next two sections explain, that extension was long awaited.

2.ix. Strictly Logical Automata

The wheels and pistons, the switchboards and electric wires, and even the motor boats and jukeboxes were aspects of the *bodily* machine: analogies for the brain, that material stuff inside the skull. If 'man as machine' is taken also to mean 'mind as machine', little had changed. At the beginning of the 1900s there was still scant temptation to think of the mind by analogy with automata.

a. Early gizmos

It may seem strange that no one was suggesting the mind–machine analogy. For would-be logic machines had been attempted in the Middle Ages (Section i.b, above), and primitive calculating machines had existed since the early seventeenth century. Gottfried Schickard, a friend of Johannes Kepler, invented one of these in 1623 (his letters were lost for over 300 years, so very few people have heard of him: Rojas 2002). However, these machines were seen as aids to human thought, not models of it.

So when Blaise Pascal (1623–62) built his cogwheeled adding machine in 1642, he was attempting to embody the laws of mathematics that the mind must respect when calculating, rather than the activities of the mind itself. Nor did he value logic above all other types of thinking. Quite the reverse: a forerunner of existentialism, he praised the leap of faith involved in religious belief (cf. Chapter 8.vi) and insisted that "the heart has reasons that Reason does not know".

Pascal's near-contemporary Leibniz (1646–1716) did believe that every human problem could in principle be settled by logical thought. He dreamt of a universal language in which all properties could be accurately mapped. This, he said, would enable future statesmen and philosophers to settle disputes by formal reasoning, instead of resorting to rhetorical persuasion or even abuse. Disagreements between philosophers or politicians need be no greater than those between accountants:

> For it would suffice for them to take their pencils in their hands, to sit down each to his abacus, and (accompanied if they wished by a friend) to say to each other: "Let's calculate! [*calculemus*]". (Leibniz 1961: 200; my trans.)

This vision had been part-inspired by the writings—and the logical devices—of Lull, 400 years earlier (Section i.b, above). Aged only 20, Leibniz wrote a *Dissertatio de Arte Combinatoria* whose scope and style were clearly inspired by Lull and his followers. In his maturity, he remembered:

> When I was young, I found pleasure in the Lullian art, yet I thought also that I found some defects in it, and I said something about these in a schoolboyish essay called *On the Art of Combinations*... I have found something valuable, too, in the art of Lully and [a scholar who] pleased me greatly because he found a way to apply Lully's generalities to useful particular problems. (quoted in Fauvel and Wilson 1994: 55)

Like Lull, Leibniz believed that "the understanding" follows "rules". But his own calculating machines, some of which could multiply and divide as well as add, were concerned only with mathematics—not logic. An early model was shown to the Royal Society in 1672, and several other designs were described later (Leibniz 1685). Surprisingly, perhaps, Leibniz *was not* influenced by Pascal's gizmo:

> When, several years ago [about 1670], I saw for the first time an instrument which, when carried, automatically records the numbers of steps taken by a pedestrian, it occurred to me at once that the entire arithmetic could be subjected to a similar kind of machinery so that not only counting but also addition and subtraction, multiplication and division could be accomplished by a suitably arranged machine easily, promptly, and with sure results.
>
> *The calculating box of Pascal was not known to me at that time. I believe it has not gained sufficient publicity*. (Leibniz 1685: 173; italics added)

If we wanted to produce "a more admirable machine", he said, "things could be arranged in the beginning so that everything should be done by the machine itself" (1685: 178). But he didn't see that as cost-effective, or even of much practical use. All the human operator had to do, after all, was to turn the wheels, or (for multiplication) move the machine from one operator to another.

What's most relevant here is that Leibniz saw his calculators as specialized tools for compiling mathematical tables (for use in commerce, estate management, navigation, and astronomy: p. 180)—not as attempts to mechanize thought in general. He did envisage mechanizing *logic*, using calculating machines based on binary numbers. But he could actually build only (decimal) calculators (see Chapter 4.iii.d).

One of the eighteenth-century machines inspired by Pascal and Leibniz was a calculator built in 1784 by the engineer Johann Müller (1746–1830) (Pratt 1987: 99; Swade 1991: 21–2). Modelled on Leibniz's ingenious stepped-cylinder design (Lindgren 1990), this machine was intended not to do single sums but to perform several calculations in series.

Its central mathematical, and mechanical, principle was to be employed nearly forty years later in Babbage's Difference Engine (see 3.ii). Indeed, the principles embodied in Leibniz's machine were used in almost every subsequent mechanical calculator. But much as the Difference Engine was concerned with arithmetic, not logic (and still less with thought: 3.iv), so Müller's machine was a mathematical device, not a logical one.

The nineteenth century saw the design of two calculating machines enormously more powerful than anything previously available (Chapter 3). One of these—Babbage's Analytical Engine, initiated as early as 1834—was capable in principle of being adapted for logical use. But the Analytical Engine was (and still is) no more than a design. At mid-century, then, there was still no actual mechanical device that would do logic, as opposed to arithmetic.

There was, to be sure, a maverick vision of a logical machine—intended, unlike the Analytical Engine, to "reason" *in the very same way that people do*. This was due to the popular science-writer Smee (1851). Drawing on his ideas about electro-biology, he designed a Relational machine to embody just one idea at a time. ("Relational", because he saw an idea as a set of interrelated properties: see viii.a, above.) Based on George Boole's just-published logic (see below), it would consist of a large metal plate successively divided into two parts by (a hierarchy of) hinges. Their position, open or shut, would represent the presence or absence of the relevant properties. Then, he said, two Relational machines could be combined to make a Differential machine, whose task would be to compare two different ideas.

Smee saw this implementation of Boolean logic as sufficient, in principle, to simulate all human thought. But even he had to admit that what's possible in principle may not be achievable in practice:

> When the vast extent of a machine sufficiently large to include *all words and sequences* is considered, we at once observe *the absolute impossibility* of forming one for practical purposes, inasmuch as it would cover *an area exceeding probably all London*, and the very attempt to move its respective parts upon each other, would *inevitably cause its own destruction*. (Smee 1851: 43; italics added)

By the end of the nineteenth century, things had changed. A device for handling "all words and sequences" was still a pipe-dream. But a few machines now existed for doing simple logic.

One had been designed in 1869 by the economist and logician Stanley Jevons (1835–82). He was part-inspired by Babbage, but relied on electricity instead of cogwheels (Hyman 1982: 255). His device used binary relays to model the logical relations—conjunction, disjunction, if–then, equivalence, and negation—between up to three propositions (Jevons 1870). The state of each relay was signalled by a lamp, all of which would light up to indicate a necessary truth, or tautology.

This device prompted lively correspondence in the 1870s (Mays and Henry 1953). As noted already (Preface, ii.a), it was still influential almost 100 years later. The first modern electrical logic machine was built on Jevons's principles, lamps and all, in 1949 (Mays and Prinz 1950). It could generate logical proofs made up of a sequence of steps. A special store accepted the results of applying one of the five operations to the two propositions in the two lower-level stores. These results (thanks to two auxiliary stores) could then be fed back into the lower level—so modelling a "chain" of inferences.

b. Logic, not psychology

None of those nineteenth-century inventions, however, was thought of as a machine for thinking, or even for simulating thought. Smee's never-built, and self-confessedly impractical, device had been intended as a simulation of reasoning, but these logic machines were not. Rather, they were devices for illustrating/following the formal principles that must be satisfied if thinking is to be logically valid or mathematically correct.

This may seem surprising. For logic was traditionally concerned with argumentation, whether analysed in terms of statements (propositions) or concepts (classes). Lull's *Ars Magna* had supposedly concerned reasoning, knowledge, and truth. The Port-Royal logicians of the seventeenth century (discussed in Chapter 9.iii.c) had described logic as "the art of thinking". And one of the nineteenth century's most distinguished logicians, Boole (1815–64), described logic as "the laws of thought" (Boole 1854).

However, that little word "of" was ambiguous. It could be given *either* of the two interpretations distinguished in the opening paragraph of this subsection. Not until the 1880s was the ambiguity clearly recognized (see below).

Boole is remembered for his development of Boolean algebra, or Boolean logic—which is fundamental to modern computers (Chapters 3.v and 4.iii). These alternative labels mark his demonstrations that algebra can be expressed in terms of a binary notation, which in turn can be mapped onto statements that are either true or false (Boole 1847). By using variables to stand for whole propositions, he showed that the various types of syllogism, familiar since Aristotle, can be expressed in this symbolism and thereby explicitly proved to be valid. (Previously, their validity had been taken as intuitively obvious.) Jevons's logic machine was based on Boole's work.

Babbage, who regarded Boole as "a real *thinker*" (Hyman 1982: 244), noted that there could be a machine for doing logic, but he never designed one. Even if he had, he wouldn't have seen it as a psychological model (see Chapter 3.iv). As for Boole himself, his book title *An Investigation of the Laws of Thought* was misleading: his interest was in the logical foundations of mathematics, not the facts of psychology.

The ambiguous position of logic with respect to psychology was clarified towards the end of the nineteenth century by the mathematician Gottlob Frege (1848–1925). Frege, like Boole, believed that mathematics could be grounded in logic (Frege 1884; Beaney 1997). In trying to prove this, he defined a long-standing logical evil, psychologism—saying that his guiding principle was to avoid it (Frege 1884, Preface).

Psychologism is any approach which confuses formal logic (or *norms* of rational thinking) with empirical facts about how people think. Frege's point was not that people don't—or don't always—think logically. It was that whether they do or not—and how they do, when they do—is of no interest to the logician. The logical should always be distinguished from the psychological, since the (normative) laws of logic are not the (empirical) laws of thought. In modern philosophical jargon, Frege's position was that logic, or rationality, can't be "naturalized". (Kant had said much the same thing, and neo-Kantians today criticize cognitive science accordingly: Chapter 16.vi–viii.)

In his critique of psychologism, Frege introduced new standards of logical rigour, and contributed a host of new ideas to logic and the philosophy of language (Chapter 9.ix.c). For instance:

* He defined the notion of a "truth value".
* He pioneered the truth-functional propositional calculus.
* He provided a formalism, the predicate calculus, that could represent the internal structure of propositions, and deal with quantifiers such as *all* and *some*.
* He defined higher-order functions, or functions made up of functions (an idea that would later feed into LISP: Chapter 10.v.c).
* And he made important distinctions between various meanings of "meaning"—such as *sense* and *reference*. The phrases 'the Morning Star' and 'the Evening Star' have different senses, so would be translated differently; but, as astronomers eventually discovered, they have the same reference: namely, the planet Venus.

As a result of Frege's work, which became widely influential with the publication of Bertrand Russell and Alfred North Whitehead's *Principia Mathematica* in 1910, logic and psychology were pushed even further apart. Not until the mid-twentieth century would people see logic machines as having psychological relevance. Ironically, they'd do so largely because of Frege—via Russell (1872–1970) and Russell's student Rudolf Carnap (1891–1970): see Chapter 4.i–ii.

Still more shocking, cognitive scientists would also suggest the reverse: that psychology is relevant to logic. The philosopher of science Paul Thagard, for instance, suggested a way of "*revising* normative (prescriptive) logical principles in the light of descriptive psychological findings" (1982: 25; italics added). This was based not merely on the outward facts of psychology: people's patterns of error, for example. It was based also on computational principles explaining how it's possible for people to reason at all.

In general, cognitive scientists—even if they didn't seek to "revise" logic—rejected the orthodox assumption that logical norms are ideal for guiding actual thought. One illustration is Herbert Simon's stress on "satisficing", not optimizing (Chapter 6.iii). Others include pleas for a more realistic philosophy of inductive reasoning (e.g. Boden 1980; Stich and Nisbett 1980), and a stress on "fast and frugal" heuristics for problem-solving (7.iv.g). The seeming imperfections ("boundedness") of human rationality, some said, were what made it possible for humans to act rationally at all (7.iv.h).

2.x. Psychology as Mechanism—But Not as Machine

If, 200 years after Pascal and Leibniz, people still weren't ready for 'mind as machine', that's not to say that they weren't ready for a mechanistic psychology. By the end of the nineteenth century, some psychologists had started to study the mind by the methods of science.

a. Visions of a scientific psychology

These fledgling psychologists were grounded in empiricist philosophy, which from its inception in the mid-seventeenth century had always been very close in spirit to the science of the time (see Section iii.b, above). So Locke, for instance, although he *didn't* pick up Hobbes's suggestion that thinking is essentially a form of computation, did think that philosophy must be fully consistent with science. As he put it, it was "ambition enough to be employed as an under-labourer in clearing the ground a little" for scientists such as Boyle and "the incomparable Mr. Newton" (1690, Epistle). Indeed, Locke himself assisted Boyle on various occasions, and saw his *General History of the Air* through the press after his death (Shapin 1994: 398 n.).

In his philosophical writings, Locke discussed many questions which today we'd call psychological. One famous example, posed to him in a letter from William Molyneux, was whether a man blind from birth would be able to recognize things if his sight were suddenly restored:

> Suppose a man born blind, and now adult, and taught by his touch to distinguish between a Cube and a Sphere of the same metal, and nighly of the same bigness, so as to tell, when he felt one and t'other, which is the Cube, which is the Sphere. Suppose then the Cube and Sphere placed on a Table, and the Blind Man to be made to see. *Quaere*, Whether by his sight, before he touch'd them, he could now distinguish, and tell, which is the Globe, which the Cube. (J. Locke 1690: II. ix. 8)

Locke's view (and Molyneux's) was that the man *would not* be able to distinguish them. Shape-learnt-by-touch, he believed, is quite distinct from shape-learnt-by-sight, so the association between a seen cube and a touched one couldn't be inferred (by "reflection"). Instead, it would have to be freshly learnt. The 'Molyneux question' would fascinate philosophers and psychologists for centuries (Morgan 1977), and was largely answered by a cognitive scientist in 1959 (Chapter 6.ii.e).

Another example of a fundamental psychological puzzle that exercised Locke was how we can think about "universals". For instance, how can we understand the concept of "a triangle" in the general case? As he put it, how can a triangle be thought of as "neither oblique nor rectangle, neither equilateral, equicrural, nor scalenon; but all and none of these at once"? (1690: IV. vii. 9). One of the earliest papers in connectionism, some 250 years later, would ask the very same question—and sketch a not-implausible answer (Chapter 12.i.c).

Even in Locke's time, the study of the mind (or, as it was often put, the soul) was still classified as an example of "pneumatology": the philosophy of incorporeal substances—such as angels, and even God Himself. As such, it was associated with theology rather than science (then called natural philosophy) or medicine. By the 1720s, however, it was being termed "psychology"—and in Chambers's *Cyclopaedia* (see Section ii.b above), although it was defined as "a Discourse Concerning the Soul", it was regarded as part of anthropology and linked with Locke's views on physiology (Vidal 1993). In brief, it was moving away from theology and towards science.

Locke had followed Descartes in assuming regular correlations between brain and mind (although, unlike Descartes, he left open the possibility that matter might be inherently active, so capable of thought). But he'd described ideas in atomistic terms. On his view, the primitive ideas of "sensation" were passively received by the sense organs, and combined into complex ideas (in the brain) by "reflection".

Hume carried this approach further. In his vocabulary, the data of the senses were *impressions*, and the copies of them in memory were *ideas*. The mind, then, was made up of a host of impressions and ideas. He described it as a kind of theatre, where several perceptions successively make their appearance; pass, repass, glide away, and mingle in an infinite variety of postures and situations. (Hume 1739: I. iv. 6)

And the script, according to him, was written by Newton—or someone very similar: namely, himself.

Newton had suggested (in Query 31 of the *Opticks)* that if experimental science were to be perfected then "the Bounds of Moral Philosophy [i.e. psychology] will be also enlarged". Hume took up this suggestion in the introduction to his *Treatise of Human Nature*. The book was tellingly subtitled *Being an Attempt to Introduce the Experimental Method of Reasoning into Moral Subjects*, and if he didn't do what we'd now regard as properly controlled experiments, he did base much of what he said on careful observation and introspection. (His assumption that perception is basically passive prevented him from realizing that even introspection isn't pure, but theory-laden: see Chapter 16.iv.e.)

The mental associations he described in the *Treatise* were atomistic, automatic, and based on spatio-temporal contiguity—and on similarity, too. Indeed, he declared mental association to be "a kind of ATTRACTION, which in the mental world will be found to have as extraordinary effects as in the natural" (1739: I. i. 4). He explicitly suggested that the laws of thought were closely analogous to Newton's laws of gravitational attraction (Battersby 1978, 1979).

And the respect due to them was no less: "we may hope to establish . . . a [psychological] science *which will not be inferior in certainty, and will be much superior in utility*, to any other of human comprehension" (Hume 1739: 8; italics added). (Notice the mid-eighteenth-century assumption that physics, for all its intellectual glory, has scant "utility".) Similarly, in his *Enquiry Concerning Human Understanding*, Hume said:

But may we not hope, that philosophy, if cultivated with care, and encouraged by the attention of the public [see ii.c, above], may carry its researches still farther, and discover, at least in some degree, *the secret springs and principles, by which the human mind is actuated in its operations?* (Hume 1748: I. 15; italics added)

Strictly, Hume's wish to model psychology on *physics* can be distinguished from his commitment to *associationism*. In other words, someone might hold (1) that psychology should be modelled on physics, as the most fundamental of the sciences, without holding (2) that psychology should be atomistic and associationistic (that is, 'Newtonian': see Chapter 5.i.a).

Hume himself didn't make that distinction, because in his day physics and Newton were virtually synonymous. But some cognitive scientists today, who also wish to model psychology on physics, stress thermodynamics rather than Newtonian mechanics. Thought and behaviour, they argue, must be understood in terms of state transitions in dynamical systems (14.ix.b, 15.viii.x, ix, and xi, and 16.vii.c). Moreover, they must be understood as essentially *embodied*—so physics is relevant not because it's the fundamental science but because it's the science *of bodies*. Significantly, one leading proponent of this view describes himself as vindicating Hume's dream of a scientific psychology based on mathematical laws like those of physics (van Gelder 1992/1995). The specific "mathematical laws", however, are of course very different.

Newtonian associations, for Hume, weren't the only things to be considered in explaining thought: the emotions were crucial too. Reason, he said, is and ought to be "the slave of the passions"; it "can never pretend to any other office than to serve and obey [our emotions]" (1739: II. iii. 3. 4). He even declared: "'Tis not contrary to reason to prefer the destruction of the whole world to the scratching of my finger" (ibid.). In short, purely intellectual thinking can't tell us what to do, only how to do it.

In the years that followed, many scientific psychologists would forget Hume's stress on emotions. Not until the late twentieth century would it be widely realized that even *the most rational* thinking can't be understood without taking them into account (Chapter 7.i.d–f).

Hume's contemporary David Hartley (1705–57) also took inspiration from Newton—specifically, from his theory of light as involving vibrations, or waves. But whereas Hume had focused on ideas, Hartley (1749) concentrated on what might be going on in the *brain.*

Hartley posited (large and small) vibrations of infinitesimal particles in the nerves and brain, assumed to be "of the same kind with the oscillations of pendulums". These, he said, are "the physiological counterpart of ideas". He was a dualist—"Man consists of two parts, body and mind"—but not an interactionist. He believed, instead, that mind and brain involve parallel laws. So, just as physical waves take some time to die away, the experience of heat persists after the hot object has been removed and the image of a brightly lit window remains after the eyes are shut.

Moreover, mind and brain involve similar sorts of combination. Hartley's explanation of learning held that if only one part of an oft-repeated association recurs, it will be reconstructed as a whole both physiologically and mentally:

> Any sensations A, B, C, etc., by being associated with one another a sufficient Number of Times, get such a Power over the corresponding Ideas, a, b, c, etc., that any one of the Sensations A, when impressed alone, shall be able to excite in the Mind, b, c, etc., the Ideas of the rest. [And, analogously:] Any [nervous] Vibrations A, B, C, etc., by being associated together a sufficient Number of Times, get such a Power over a, b, c, etc., the corresponding Miniature Vibrations, that any of the Vibrations A, when impressed alone, shall be able to excite b, c, etc., the Miniatures of the rest. (Hartley 1749: i. 65, 67, Propositions 10, 11)

It's not surprising that Hartley is now thought of as an early connectionist, for this passage rings distinctly Hebbian bells in modern readers' minds (see Chapters 5.iv.b and 12). It might even be paraphrased as *Cells that fire together, wire together* (today's 'ft/wt' rule).

The eighteenth-century French physician Julien Offray de La Mettrie (1709–51) went even further—and earned a highly scandalous reputation as a result. Indeed, after publishing his book anonymously (in Holland), he had to seek the protection of Frederick the Great when his authorship was discovered—and he was never able to return to his native country (Vartanian 1960). Denying any distinction between mind and matter, he insisted that *matter itself* is active and feeling. He interpreted these terms in a strictly materialist way. As he put it, "the brain has its muscles for thinking, as the legs have muscles for walking". It was one of his followers, Pierre Cabanis (1757–1808), who famously said:

> [The brain is] a special organ whose particular function it is to produce thought just as the stomach and the intestines have the special work of carrying out the digestion, the liver that of filtering the bile, etc. (Cabanis, quoted in Brett 1962: 475)

In a book tellingly entitled *L'Homme machine* ('Man a Machine', 1748), La Mettrie cited Vaucanson, whose automata were touring Europe as he wrote it. He referred to various conditions (including cataract, drunkenness, and fever) wherein people's senses, thinking, and willpower, or self-control, are clearly dependent on the body. While he praised Descartes for being "the first to prove completely that animals are pure machines", he argued that the same applies to human beings:

Since all the faculties of the soul depend to such a degree on the proper organization of the brain and of the whole body, that apparently they are but this organization itself, the soul is clearly an enlightened machine. For finally, even if man alone had received a share of natural law, would he be any less of a machine for that? A few more wheels, a few more springs than in the most perfect animals . . . any one of a number of causes might always produce this delicate conscience so easily wounded, this remorse which is no more foreign to matter than to thought, and in a word all the differences that are supposed to exist here. (La Mettrie 1748: 48)

To be a machine, to feel, to think, to know how to distinguish good from bad, as well as blue from yellow, in a word, to be born with an intelligence and a sure moral instinct, and to be but an animal, are therefore characters which are no more contradictory, than to be an ape or a parrot and to be able to give oneself pleasure . . . I believe that thought is so little incompatible with organized matter, that it seems to be one of its properties on a par with electricity, the faculty of motion, impenetrability, extension, etc. (p. 64)

In short, human beings are *nothing but* complex automata, just as Descartes had declared animals to be.

Having cited Vaucanson's duck and flute-player with admiration, La Mettrie declared that "another Prometheus" might make a mechanical man that could talk (1748: 100). And, he added, there's no reason why one shouldn't teach apes to speak, for their physical organs appear to be perfectly suitable. (It took another two centuries for biologists to realize that this isn't true; a host of inter-species differences in the anatomy of the mouth, larynx, and respiratory muscles enable us to speak and prevent other primates from doing so: Lenneberg 1964: 34–51, 75–98.)

La Mettrie, and most of his readers too, believed that he was saying something which Descartes himself could easily have said. The implication was that only theological squeamishness, or what today might be termed "species-ism", had prevented Descartes from extending his doctrine of animal automatism to human beings. But this was a mistake.

The problem was that La Mettrie hadn't realized the special place of language (in particular, its endless generativity) in Descartes's argument that there could never be a plausible mechanical man (see Section iii.c, above). He thought, in effect, that the only problem facing the future Prometheus was to make the flute-player's mobile lips and tongue pronounce phonemes instead of blowing air. But Descartes would have readily conceded that possibility: a flute-player's body, like yours and mine, is indeed a machine—and, as such, might be copied by a superb engineer. His metaphysical problem wasn't with the hypothetical android's articulation of speech, but with its generation of language—something La Mettrie didn't discuss. In short, the seeming continuity between Descartes's *bête-machine* and La Mettrie's *homme machine* is "one which involves the inconsistent adaptation of half-understood views of one's predecessor" (Gunderson 1964*a*: 220).

La Mettrie's theory was comparable to behaviourism: the mind hadn't been explained, so much as explained away. Hardly anyone at the time went that far. Another French physician, the surgeon Claude-Nicolas Le Cat (1700–68)—inventor of the first ventricular shunt for hydrocephalus—apparently did so, even before La Mettrie. Most of his writings are now lost, including his intriguingly named treatise *Description d'un homme automate dans lequel on verra exécuter les principales fonctions de l'économie animale*, but his *Traité des sens* (1744) survives and gives a rigidly mechanistic account of the senses.

However, Le Cat and his fellow physician compatriot were unusual. In particular, they were unlike their philosophical neighbours across the English Channel. The British empiricists were interested in the mind *as distinct from* the body, even though they believed that it, too, followed mechanistic (quasi-Newtonian) laws.

The nineteenth century saw the rise of associationist psychologies aimed at detailing those laws. It also provided novel research programmes (such as psychophysics) aimed at discovering systematic mind–body correlations.

The most influential theories of the association of ideas included those of the philosophers James Mill (1773–1836) and his son John Stuart Mill (1806–73), and the pioneering psychologist Wilhelm Wundt (1832–1920). Wundt's influence was enormous because he founded professional psychology. He established one of the first two laboratories in 1879 (William James started the other), founded the first journal, and wrote extensive textbooks defining the field. In effect, it was defined there as the study of consciousness—so the prime experimental method wasn't the observation of behaviour, but introspection.

Wundt concentrated on sensation and perception, regarding the "higher mental processes", such as thinking, as lying beyond the reach of associationism. (That doesn't mean he didn't discuss them: see Chapter 9.v.a.) However, Hermann Ebbinghaus (1850–1909) generalized associationism to memory, by inventing the experimental method of nonsense syllables. Whereas *dog* has an unknown and uncontrollable number of previous associations, *dob* doesn't. It followed, of course, that Ebbinghaus had got rid of *meaning* too—for which he would eventually be reproached (see Chapters 5.ii.b and 6.iv.c).

Other nineteenth-century research tried to link psychology with a mechanistic neurophysiology—or anyway, with physical mechanism. For example, Ernst Weber (1795–1878) and Gustav Fechner (1801–87) initiated psychophysics. Their aim was to quantify the relations between specific physical stimuli and conscious sensations—something with which Descartes would have been content.

b. Non-empiricist psychologies

Although the empiricist paradigm was the dominant form of experimental psychology in the late 1800s, not all psychologists were drawn to it.

Animal psychologists, such as George Romanes (1848–94) and Conwy Lloyd Morgan (1852–1936), clearly couldn't use an introspective methodology. Indeed, Lloyd Morgan formulated his famous "canon" to reduce anthropomorphism as much as possible:

In no case may we interpret an action as the outcome of a higher psychical faculty, if it can be interpreted as the outcome of the exercise of one which stands lower in the psychological scale. (1894: 53)

He wasn't forbidding "psychical faculties" outright: in that sense, he and Romanes were more alike than they're commonly thought to be (R. K. Thomas 2001). But neither of them could study their non-human subjects by making use of introspection. Moreover, they were interested in the functions—not just the underlying mechanisms—of animals' behaviour.

Some students of *human* psychology, too, took a different approach. Their criticism was more often directed against associationism than mechanism in the broader sense (one exception is mentioned below).

The introspectionists of the Würzburg School, and their successors the Gestalt psychologists, criticized Wundt's atomism. And William James (1842–1910), despite his general commitment to what we now call connectionism, famously remarked that Wundtian associationism "could hardly have arisen in a country whose natives could be *bored*" (James 1890: 192).

Never one to brush difficult questions under the carpet, James discussed a host of psychological conditions that couldn't be described, never mind explained, in the atomistic terms of empiricism. They included many examples drawn from psychopathology and hypnosis.

These phenomena had been studied (for instance) by the clinician Jean Charcot (1825–93) at the Salpêtrière hospital in Paris. (It was Charcot who identified the puzzling phenomenon of hysterical paralysis: Preface, ii.b.) And, of course, they were studied also by Freud (1856–1939). He'd worked on hypnosis and hysteria with Charcot for one year (1885–6), and with the physiologist Joseph Breuer for many more (from 1882 to 1895)—Ellenberger (1970, ch. 7).

These turn-of-the-century psychologists were less aggressively mechanistic than those in the empiricist tradition. To be sure, bodily mechanisms weren't ignored. But interpretative (hermeneutic, intentionalist) theorizing was also prominent.

The philosophical relation between these two forms of theorizing was controversial—and it still is (see Chapter 16). The two extremes were represented by Freud (1856–1939) and William McDougall (1871–1938).

Perhaps not surprisingly for a pupil of Brücke (who'd taken an oath to counter vitalism: see Section vi.d), Freud insisted that his interpretative theory was compatible with a strictly mechanistic neurophysiology (S. Freud 1895). McDougall, by contrast, explained human action and personality—and much animal behaviour too—partly in terms of psychic energy (*horme*) intrinsically directed to specific purposes, a view he defended at length in his book *Body and Mind: A History and a Defense of Animism* (1911).

McDougall's theories of motivation in everyday social life and clinical psychopathology (for instance, multiple personality disorder) showed great insight into mental architecture (see 5.ii.a, and Boden 1972, chs. 6–8). And his pioneering textbooks on physiological psychology and social psychology were hugely successful (1905, 1908). But his hypothesis of teleological energy was widely spurned. Even psychologists out of sympathy with associationism avoided such a radically anti-mechanistic position.

By the early 1900s, then, a scientific psychology grounded in mechanistic assumptions had got off to a strong start. But that was the only sense in which 'man as machine' encompassed the mind. Psychologists at that time *did not* draw on artefacts to find analogies for mental processes.

As for building a specifically psychological automaton, this wasn't in sight: 'machine as man' was still restricted to mimicry of the body. The primitive calculators and logic toys weren't thought of as psychological models (see Section ix, above), and no nineteenth-century engineer seriously suggested that a man-made device could produce behaviour comparable to the workings of the mind. Even the maverick Babbage didn't think of his ambitious project in that light (see 3.iv). Still less did people imagine that an artefact might be intelligent *in just the same sense* as we are. Those ideas didn't arise until later. Specifically, 'mind as machine' was first taken seriously in the 1940s (Chapter 4).

3

ANTICIPATORY ENGINES

Men of monumental achievement are often credited with even more than they deserve. That's true of Charles Babbage (1791–1871). Indeed, some readers may have been surprised by the conclusion of the previous chapter: that the mind-as-machine hypothesis hadn't emerged by the turn of the twentieth century. For in 1834 Babbage had designed his Analytical Engine: the first program-controlled, essentially general-purpose, digital computer. Because later examples of such machines were a crucial intellectual source for cognitive science, and also because of a handful of famous (but misinterpreted) quotations, his work is often assumed to be an early step in the field.

That's a mistake. One must resist the temptation to see Babbage as a cognitive version of Jacques de Vaucanson (see Chapter 2.iv), an ingenious engineer whose automata modelled not lips and fingers but thoughts. For Babbage didn't attach any psychological significance to his project.

It's not even clear that he was an important historical influence for the development of electronic computers. His Analytical Engine involved startling anticipations of these modern machines, to be sure. To anticipate, however, isn't necessarily to influence. Babbage's effect on computing technology 100 years later is controversial, being variously described as positive, absent, and even negative.

As for his historical role in the development of *ideas about mentality*, this was zero. I don't mean that he was unjustly neglected. Rather, I mean that he had no thought of modelling minds. Mind-as-machine was first taken seriously in the middle of the twentieth century, not the nineteenth (see Chapter 4).

This chapter, then, is a necessary digression. Before resuming our main theme, we need to understand Babbage's paradoxical relation—superficially close, yet fundamentally irrelevant—to cognitive science.

Babbage's general philosophy is outlined in Section i, and his two remarkable Engines in Sections ii and iii. In Section iv, I explain why, contrary to widespread belief, his work was irrelevant to mind-as-machine.

Section v describes the invention of the modern computer. This process involved two of the 'founding fathers' of cognitive science, and informed its early theories in various ways. Finally, Section vi asks what influence Babbage's research had on computer technology. We'll see that some people regarded his work as over-optimistic techno-hype, with counter-productive effects like those which would blight various types of AI in the 1970s and late 1980s.

3.i. Miracles and Mechanism

Babbage was interested in many other things besides machines, and many other machines besides those for which he's famed today. Moreover, he drew many parallels between machines and other matters—including factory organization, and even theology.

a. Babbage in the round

Born only two years after the French Revolution of 1789, Babbage was widely recognized during his lifetime as one of the most progressive thinkers of his day. He was at the hub of England's intellectual society, constantly interacting with outstanding political, literary, and scientific figures. And he had close contacts with intellectuals in France and, to a lesser extent, Italy. He'd still be regarded, today, as a historically significant figure even if he'd never given a thought to automatic computation.

Babbage's strengths were personal as well as intellectual. As his friend and literary executor Harry Buxton put it: "No man ever enjoyed society with a greater relish than he, and few, indeed, were better calculated to adorn it" (Buxton 1880: 356). (Sadly, this became less true as he grew older, and burdened with disappointments: see below.)

The early feminist Harriet Martineau was one of his many friends. In her autobiography, she remarked on his "exemplary patience" and "genuine good nature" when a "lady examiner" visiting his salon, on being shown his miniature Difference Engine (see Section ii.b), naively asked: "Now, Mr. Babbage, there is only one thing that I want to know. If you put the question in wrong, will the answer come out right?" (quoted in Hyman 1982: 129).

(Such a query isn't always foolish. Babbage's later Engine had a self-correction facility, as we'll see in Section iii.a. Moreover, user-friendly AI interfaces, not to mention spell-checkers, enable today's machines to compensate for users who put the question in wrong: Rich *et al.* 2001; cf. Chapter 13.v.)

Nor was this naive "lady examiner" Babbage's only fan. Martineau, again: "All were eager to go to his glorious soirées; and I always thought he appeared to great advantage as a host." Evidently, the attractions of his soirées weren't only intellectual: the young Charles Darwin, recently disembarked from the *Beagle*, was advised by a friend to attend Babbage's house to meet some "pretty women" (Swade 1996: 44). In short, Babbage was for many years the toast of London's intellectual community, admired by everyone who was anyone.

Yet the title page of his autobiography, published seven years before he died, carried a curious snippet about an oyster-scientist, who "instead of countenance, encouragement, and applause ... was exposed to calumny and misrepresentation". And in the closing pages he described himself as "a father whose name in his own country has been useless to himself and to his children" (Babbage 1864: 364).

The reason for this bitterness, contrasting with his previous ebullience and sociability, was his failure to win continuing financial support for the Analytical Engine, his prime interest during the last forty years of his life (see Section iii). It may be that an earlier lapse of his usual sociability had been largely to blame. Already frustrated, he behaved very badly—and very stupidly—when meeting the Prime Minister in 1842 to ask for more funds (Swade 1996: 47–8).

Professionally, Babbage was a mathematician, one who—with friends such as George Peacock—contributed to important advances in 'analytical' algebra (Pratt 1987: 93–7). He used his mathematical skills to advance cryptology too, an interest he'd nurtured from his schooldays: he was widely recognized as an authority, and was regularly sent coded messages to decipher (Babbage 1864: 173–9; Buxton 1880: 346–7; Kahn 1996: 204–7). He held the Lucasian Chair of Mathematics at Cambridge for eleven years, a post formerly filled by Isaac Newton—who, unlike Babbage, had actually delivered some lectures. Nevertheless, he scandalized many of his compatriots by championing the 'French' (originally, Leibnizian) version of the differential calculus, rather than Newton's.

He was also a highly accomplished engineer, and—like Vaucanson—a widely consulted expert on machine tools. His work on automatic calculators resulted in significant advances in precision engineering. He was deeply involved in the accelerating industrialization of the time, and played a prominent part in the economic and political debate associated with it.

His fascination with moving gadgets had started early:

> During my boyhood my mother took me to several exhibitions of machinery. I well remember one of them in Hanover Square, by a man who called himself Merlin. [Remarking my great interest, the exhibitor] proposed to my mother to take me up to his workshop, where I should see still more wonderful automata. We accordingly ascended to the attic. There were two uncovered female figures of silver, about twelve inches high.
>
> One of these walked or rather glided along a space of about four feet, when she turned round and went back to her original place. She used an eye-glass occasionally, and bowed frequently, as if recognizing her acquaintances. The motions of her limbs were singularly graceful.
>
> The other silver figure was an admirable *danseuse*, with a bird on the forefinger of her right hand, which wagged its tail, flapped its wings, and opened its beak. This lady attitudinized in a most fascinating manner. Her eyes were full of imagination, and irresistible.
>
> These silver figures were the chef-d'oeuvres of the artist: they had cost him years of unwearied labour, and were not even then finished. (1864: 12)

Many years later, after the wizardly "Merlin" had died, his effects were sold at auction. Babbage bought the lady-with-the-bird. Seeing that "No attempt appears to have been made to finish the automaton; and it seems to have been placed out of the way in an attic uncovered and utterly neglected" (1864: 273), he took the mechanism to pieces, and restored it.

He asked "one or two of my fair friends" to "supply her with robes suitable to her station". But it wasn't only the fair friends who took an interest in her wardrobe: Babbage himself fixed "a single silver spangle" to each of her small pink satin slippers, and "a small silver crescent in the front of her turban". He'd even made the turban, out of pink and light green crepe and "a plaited band of bright auburn hair". Satisfied with her appearance at last, he lodged the Silver Lady in a glass case in his drawing-room (p. 274). In the next room, as we'll see, he placed an even more amazing—not to say miraculous—mechanism (and the juxtaposition intrigued his many friends: Schaffer 1996).

If Babbage respected the social conventions in clothing the Silver Lady so decorously, he didn't do so in his writings, or his politics. In those areas, his radical—and polemically expressed—ideas aroused attention and controversy (Hyman 1989). His polemics were often ill-judged: tact wasn't one of Babbage's strong points. But many of his challenging suggestions, scorned in his lifetime, are now commonplace.

He recommended the introduction of decimal currency, flat-rate postage, and life peerages. He outlined new principles of taxation and life insurance. He suggested that lithography might somehow be used to reproduce facsimiles of out-of-print books. And he argued that science should be included in the general university education, criticizing the aristocracy and industrialists for their ignorance of scientific matters.

But science was one thing, scientists something else. Naming names with abandon, Babbage (1830) publicly scorned the intellectual mediocrity—and, he claimed, the venality—of the Royal Society at that time. (His attack prompted some rethinking in the Society's rooms: from 1847, Fellows were elected solely on the basis of scientific merit.) Naturally, if Babbage was prepared to be slanderous in public he was equally ready to compose calumnies in private. One of his autograph letters (16 June 1854) contained an "epigram" scorning two former Royal Society presidents:

> Methinks I've seen three things look wondrous small:
> A penny loaf in Davies Gilbert's hall;
> A tiny flee upon a lion's hide,
> And Banks' marble block by honoured Newton's side.
>
> (quoted in J. M. Norman 2004: 56)

Davies Gilbert, president in the late 1820s, had been explicitly named and shamed in Babbage's book. He wasn't a scientist, but a littérateur and historian (of Cornwall). Although he'd used his money and his political contacts to support fine scientists such as Sir Humphrey Davy, Babbage had little time for him. As for the aristocratic Joseph Banks (1743–1820), who'd explored the South Pacific with Captain Cook on the *Endeavour*, he was a fine botanist (and founder of Kew Gardens), and a champion of science in general. Hence the marble bust, placed next to Newton's portrait in a Royal Society meeting room by some of his admirers. But he was also vain, arrogant, and ostentatious—all qualities exacerbated by his forty-two-year presidency (from 1778 to his death), and all guaranteed to arouse Babbage's contempt.

When he wrote that mocking squib, Babbage wasn't a disinterested critic. For in 1831, soon after his book appeared, he'd co-founded the British Association for the Advancement of Science. He constantly compared the Royal Society unfavourably with the British Association, not least because of its snobbish exclusivity. The "British Ass" has since gone from strength to strength, and today is hugely effective in bringing science to the public. It's just one example where Babbage helped to bring about important reforms.

In many other cases, however, his provocative ideas gathered dust for decades. Some are still gathering. At the turn of the third millennium, the House of Lords in the British Parliament still retained nearly 100 hereditary peers—and they're not yet ousted, in 2005.

Babbage's most widely read book, which influenced both Karl Marx and John Stuart Mill, was *On the Economy of Machinery and Manufactures* (Babbage 1832/1835; Hyman 1982, ch. 8). This discussed the organization of a factory or other large institution. In the Preface, Babbage said it had grown from the ideas he'd had while designing his "calculating engine", and remarked that it was based on several chapters he'd written for the "mechanical" section of the *Encyclopaedia Metropolitana* of 1829. The thirty-volume encyclopedia may not have been a best-seller, but Babbage's book was. It ran to four editions within three years, and was translated into six foreign languages.

It gave a general recommendation of the division of labour, illustrated (for example) by nine eye-opening pages on the manufacture of pins (Hyman 1989: 132–40). In addition, the book included detailed recommendations on costing and marketing; on industrial relations and profit sharing; on 'clocking-in', for which he invented a suitable machine; and on the application of science in industry. One chapter, later published separately, advised on how to invent machinery, and how to combine machine tools for manufacturing purposes. He foresaw a time when machines would take over the basic repetitive tasks, while the human workers concentrated on "mental" labour.

As if the book weren't already wide-ranging enough, he closed his review of 'The Future Prospects of Manufactures, as Connected with Science' by venturing into theology. Science, he declared, has given us "resistless evidence of immeasurable [divine] design", and reason to believe in extra-terrestrial intelligence:

> [It] would indeed be most unphilosophical to believe that those sister spheres . . . should each be no more than a floating chaos of unformed matter;—or, being all the work of the same Almighty architect, that no living eye should be gladdened by their forms of beauty, that no intellectual being should expand its faculties in decyphering their laws. (Hyman 1989: 200–1)

As these remarks indicate, Babbage was an unorthodox but committed believer, who regarded religion as "the highest calling of man". This didn't prevent his arguing that bishops were educationally unfit to govern in the House of Lords—"My Lord Bishop" wasn't, and still isn't, just an empty phrase (Hyman 1989: 207). Nor did it forbid his remarking that the Athanasian Creed appeared to be "written by a clever, but most unscrupulous person, who did not believe one syllable of the doctrine" (Babbage 1864: 302). But despite these characteristically 'contrary' opinions, his religious faith was secure.

b. Religion and science

Babbage was much concerned by the growing tendency for people to see science as fundamentally opposed to religion—a tendency encouraged by Romanticism (see Chapter 2.vi). He was equally concerned by an influential book arguing that his particular type of science could contribute nothing positive to theology.

William Whewell (1794–1866) had distinguished two sorts of scientific reasoning: deduction and induction (Whewell 1833). A deductivist science, he said, tends to irreligion, but the inductive method doesn't. Whewell was no less keen than Babbage that science should pose no threat to religion. Indeed, his religious scruples would lead him, a quarter-century later, to prevent *On the Origin of Species* from being placed in the library of Trinity College, Cambridge, where he was Master (Browne 2002: 107).

Deduction leaves us no choice, and implies that everything that exists had to be as it is. By contrast, Whewell argued, induction requires a visionary faith in the possibility of deciphering nature so as to glimpse the mind of God. The reason is that scientific laws can show how the physical world is sustained from day to day, but not how it—or its fundamental forces, such as gravity—came to exist in the first place. Biological phenomena, he said, require a distinctive (teleological) type of explanation, not expressible in deductivist terms. He even argued that "deductive" scientists lacked "any authority" in natural theology, and that "we have no reason whatever to expect from their speculations any help, when we ascend to the first cause and supreme ruler of the universe".

Babbage, whose research—on the manufacture of pins, as well as on calculating machines—took a highly deductivist approach, didn't agree. What's of interest for our purposes is the apparently 'modern' nature of his reply, presented in *The Ninth Bridgwater Treatise.* This title cheekily attached his volume to the eight official *Bridgwater Treatises* on natural theology, of which Whewell's had been the first—and Charles Bell's another (see 2.viii.f). (Why there was a need for natural theology in the first place is discussed in Chapter 8.vi.)

Babbage held that the scientific laws we have discovered "converge to some few simple and general principles, by which the whole of the material universe is sustained, and from which its infinitely varied phenomena emerge as the necessary consequences" (Hyman 1989: 209). This comment may have come as no surprise from someone who was on visiting terms with the notorious determinist Pierre Laplace (1749–1827). What distinguished it from a mere post-Laplacian banality was Babbage's view on the origin of those "infinitely varied phenomena".

He argued, in effect, that God is a cosmic programmer whose foresight enabled Him to ensure the emergence of new, and humanly unpredictable, types of natural structure. Examples of such emergence include the appearance of novel species (believed at that time to be sudden), and biological metamorphosis:

> The laws of animal life which regulate the caterpillar, seem totally distinct from those which...govern the butterfly...[These changes] were equally foreknown by their Author: and the first creation of the egg of the moth...involved within its contrivance, as a necessary consequence, the whole of the subsequent transformations of every individual of [its] race. (Hyman 1989: 213)

Read out of context, this passage might seem to be groundless intellectual hand-waving. After all, the physiology of the 1830s couldn't even explain body temperature, never mind butterflies or embryology (2.vii). It's not surprising that Whewell, along with many others, saw such phenomena as requiring a special form of explanation. But Babbage claimed, amazingly, that a machine he'd built himself could do essentially the same sort of thing.

The machine was designed to calculate a series, or table, of numbers according to some arithmetical rule: for example, successive squares, or the iterated addition of 7 to the previous number. In such a device, the eventual appearance of periodic and/or apparently random changes could be built in from the start, by arranging that one rule of progression would automatically be replaced by another when a certain numerical value was exceeded. This type of behaviour is intelligible only deductively, not inductively. Induction alone can be seriously misleading: even if "one unbroken chain of natural numbers [passed] before your eyes, from *one* up to *one hundred million*", the next number, or the one after that, could be totally unexpected. The sudden change in the machine's behaviour would seem utterly mysterious to observers ignorant of the underlying rationale:

In contemplating the operation of laws so uniform during such immense periods, and then changing so completely their apparent nature, whilst the alterations are in fact only the *necessary* consequences of some far higher law, we can scarcely avoid remarking the analogy which they bear to several of the phenomena of nature. (Hyman 1989: 213)

By "several of the phenomena of nature", Babbage meant biological metamorphosis, embryological development, and historical 'leaps' from one set of biological forms (fossils) to another. Indeed, the latter suggestion was taken up by Robert Chambers (1802–71) in his *Vestiges of the Natural History of Creation* (1844: 206–11).

Chambers's book, which created a sensation on its (anonymous) publication in 1844, was a provocative defence of evolution—and, like Darwin's some years later, it had to account for the visible discontinuities in the fossil record. After all, in our experience of flora and fauna, "like produces like". Having quoted Babbage at some length, Chambers continued:

[The] gestation (so to speak) of a whole creation is a matter probably involving enormous spaces of time . . . All, therefore, that we can properly infer from the apparently invariable production of like by like is, that such is the ordinary procedure of nature in the time immediately passing before our eyes. Mr. Babbage's illustration [of how entirely unexpected numbers could be produced by his Difference Engine] powerfully suggests that this ordinary procedure may be subordinate to a higher law, which only *permits* it for a time, and in proper season interrupts and changes it. (Chambers 1844: 211)

(Darwin, of course, would substitute natural selection for Chambers's 'computational' explanation of species change.)

Not content with explaining only biological wonders such as butterflies and fossils, Babbage went even further. He suggested that analogous rule changes could explain religious miracles too. On his view, a miracle isn't some last-minute interference by the hand of God. Rather, it's a divinely preordained singularity falling between two similarly preordained rule changes, the second of which restores the laws of nature that were in effect before the occurrence of the miraculous event.

This logical possibility was mentioned, but without theological comment, in the first published description of Babbage's mathematical machine:

[The] very nature of the table itself may be subject to periodical change, and yet to one which has a regular law . . . [Tables] are produced, following the most extraordinary, and apparently capricious, but still regular laws. Thus a table will be computed, which, to any required extent, shall coincide with a given table, *and which shall deviate from that table for a single term, or for any required number of terms, and then resume its course*, or which shall permanently alter the law of its construction. (Lardner 1834: 95–6; italics added)

The author—not Babbage himself, but a supporter—remarked that it might be impossible for the observer to predict the machine's behaviour inductively, or to discover "any function . . . capable of expressing its general law". In other words, like systems termed chaotic in modern terminology, this machine was deterministic but unpredictable.

3.ii. Differences that Made a Difference

The miraculous machine mentioned in Section i.b was Babbage's Difference Engine. Designed in the early 1820s, this was the first of his two automatic calculators. The second, his Analytical Engine—on which he worked from 1832 until his death—was even more marvellous. It's the Analytical Engine (to be described in Section iii) which anticipated the modern computer.

a. Division of labour, again

The Difference Engine was designed not to do individual sums ($2+2=4$), but to carry out long series of interconnected calculations. These were necessary for preparing mathematical tables "of infinite extent and variety" for use in astronomy and mathematics, and in practical activities such as navigation and commerce. They included, for instance, "tables of the moon's place for every hour, together with the change of declination for every ten minutes" (Lardner 1834: 55–61).

Such tables existed already. But they were known to be riddled with thousands of errors. Indeed, the published lists of errata would often introduce new, and sometimes worse, errors (Lardner 1834: 61–70).

The basic reason for this wasn't that the calculations were difficult, but that they were repetitive and boring. Indeed, when Babbage had been verifying some tables for the recently founded Royal Astronomical Society in 1820–1, he'd said in exasperation, "I wish to God these calculations had been executed by steam" (Buxton 1880: 46). Moreover, the printers often made mistakes when reading handwritten figures, or when converting them into movable metal type.

About 140 years earlier, Gottfried Leibniz had longed for a machine—not, of course, executed by steam—to compile mathematical tables (Chapter 2.ix.a). And he too had referred to the advantage of their providing "sure results". But he'd designed/built only a highly minimalist version. Indeed, even that description is overly complimentary: what he'd designed was no more than a small hand calculator. Babbage's Difference Engine was quite another kettle of fish.

The mathematical method that Babbage used in the Difference Engine was inspired by the method used by the French mathematician Gaspard de Prony to prepare tables of logarithms (Hyman 1982: 44). De Prony's aim had been to devise a method that could be executed by relatively uneducated clerks. A highly accomplished mathematician would identify the relevant formula (governing changes in the moon's declination, for example), expressing it in terms of mathematical functions that could be numerically calculated by simple steps. Next, a few competent mathematicians would insert the relevant numbers (data) into the equations. Last, a large number of clerks would then do the repetitive arithmetic.

This division of labour, suggested to de Prony by his reading of Adam Smith's *The Wealth of Nations* demanded very little from the clerks. All they had to do was to add and subtract. (Over 100 years later, de Prony's approach was still in use. The untrained people doing the sums for the Los Alamos implosion bomb "were assigned different tasks—adding, multiplying, cubing, and so on—in a kind of reconfigurable arithmetical assembly line": MacKenzie and Spinardi 1995: 225.)

Babbage approved of the division of labour, here as elsewhere. As he put it in his book on factory organization,

> The master manufacturer, by dividing the work to be executed into different processes, each requiring different degrees of skill or of force, can purchase exactly that precise quantity of both which is necessary for each process; whereas, if the whole work were executed by one workman, that person must possess sufficient skill to perform the most difficult and sufficient strength to execute the most laborious, of the operations into which the art is divided. (Babbage 1832/1835: 226)

His Difference Engine was a mechanical embodiment of this manufacturers' maxim. Specifically, his "method of finite differences" used in the Engine was a variation of de Prony's approach. (The idea came full circle, from economics and manufacturing to mathematics and back again: the 1833 edition of *Economy* added a long footnote describing how de Prony's division of labour had been implemented in the machine.)

It's possible that Babbage's method of differences was part-inspired also by Johann Müller, who had employed a similar principle in 1784 (see 2.ix.a). Müller had suggested also that a machine might print its results automatically—which Babbage's machine did. Babbage became aware of Müller's work, some of which was translated for him by his close friend John Herschel, the astronomer. However, it's not known when this translation was done, and it's unclear whether Müller's ideas, as well as de Prony's, influenced Babbage's earliest design (Pratt 1987: 103; Swade 1991: 21–2).

The Difference Engine was basically an adding machine. Used iteratively, it could also do multiplication. It performed a series of additions, where the result of one addition was used as the starting point for the next. The number that was added to the starting point at each step could either remain constant, or change at predetermined points. (Hence the miracles.) The machine checked the accuracy of its results, and printed them as tables or graphs—thus eliminating clerical and printers' errors.

As a practical engineer, Babbage discussed various ways of doing the printing. His first plan involved 30,000 pieces of movable type, to be fed one at a time on the instructions of the calculating part of the engine. Another used type fixed onto the rims of wheels, similar to the printing wheels commonly used today. But the real interest of the machine lay in the principles involved.

Adding machines, of course, weren't new. Besides early forays by Hero of Alexandria and others, many commercial designs had been developed since Blaise Pascal sent his cogwheels summing in 1642 (see Chapter 2.ix.a). But they were less widely used than one might think. Most people wanting help in calculation preferred to use logarithmic tables, slide rules, or Napier's bones as aids (Pratt 1987: 38, 83–5).

Moreover, all 'automatic' calculators prior to Babbage had required the continual intervention of the human operator. For example, separate 'carry wheels' might store information that had to be manually fed into the calculation by the user. Or the result of one calculation would have to be newly entered by a human, so as to function as the input for the next. Using the most recent output as the new input was a crucial aspect of multiplication-by-addition, and of some other mathematical procedures—including the method of differences itself. So removing the human from the iterative loop would save enormously on effort.

The Difference Engine achieved this: it was the first machine that could carry out a sequence of calculations in a fully automatic way. As Babbage put it, it was the first to be "self-acting" (Babbage 1864: 30).

b. Design and disappointment

The calculating part of the Difference Engine was constructed from geared metal wheels mounted on columns. Each number wheel had ten teeth (for the digits 0–9), the wheels for units, tens, hundreds, etc. being placed above one another on the relevant column. The first column stored the starting number, the others the (hierarchy of) numbers to be added. The machine was prepared by setting the wheels to the relevant starting positions by hand, and was operated by turning a handle.

A few wheels took no part in the actual calculations, but provided helpful "memoranda" to the user. An ingenious "anticipatory carry" mechanism enabled the Engine to add 1 to 9,999,999 fairly quickly, without enduring the mechanical equivalent of a nervous breakdown. And bells informed the user when the calculations were completed (much as my new microwave instructs me to OPEN THE DOOR when the food is cooked).

Babbage himself gave a brief explanation of the Difference Engine, imagining how "papa" and "mamma" might comment on a game of marbles (Babbage 1864: 30–50). But the details don't concern us here. A clear discussion, written with Babbage's guidance, can be found in Lardner (1834).

(Readers interested in the engineering aspects can find reproductions of some of Babbage's fine mechanical drawings in Hyman 1982, 1989. For his mechanical notation, an "algebra" of mechanism capable of describing any conceivable machine, see Babbage 1826; Lardner 1834: 99–106; Hyman 1989: 312–17; Buxton 1880, ch. 9. For further details, see Lovelace 1843; Buxton 1880; Bromley 1990, 1991; Swade 1993. For its reception by mathematicians and machinists at the time, see Schaffer 2003*a*: 266 ff.)

Thus far, I've described the Difference Engine as though it were an actual machine. This is misleading in two ways. First, the Difference Engine—and the Analytical Engine too—was in fact a *class* of machines.

The more wheels per column, the larger the numbers that could be represented. And the more columns, the higher the powers (squares, cubes, etc.) that could be computed. Babbage's most ambitious engineering design would accept numbers of up to eighteen digits. But considered as an abstract mathematical device, like the Turing machine described in Chapter 4.i, the Difference Engine could compute *any* polynomial equation.

A further complication is that Babbage's first machine was superseded in 1847–9 by the Difference Engine-2. This was a more powerful, elegant, and much smaller version, whose design had benefited from his work on the Analytical Engine.

Second, the Difference Engine-1—again, like the Analytical Engine—was never actually built. Or rather, the full-scale Engine was never completed. It would have had over 25,000 parts, weighing several tons, and would indeed have needed "steam" to power it.

Babbage did, however, construct a miniature version. Having built a working model illustrating the basic arithmetical principle in 1820–2, he instructed his (superb) toolmaker Joseph Clement to build a small Difference Engine ten years later. This hand-cranked device, constituting about a seventh portion of the full-scale Engine, had nearly 2,000 parts. Its three columns, each with six wheels, could deal with five-digit numbers and square-powers (and cubes up to 9). And it included the anticipatory carry and an automatic printer. This was the machine exhibited, to universal amazement, in Babbage's drawing-room at the "glorious soirées" described by Martineau. (It's now in London's Science Museum—rescued for posterity much as his Silver Lady had been.)

A year later, in 1833, the construction of the Difference Engine was abandoned—to Babbage's lasting distress. When "Babbage's Engine" was demonstrated at London's International Exhibition in 1862, it was the small version completed thirty years earlier that was put on show.

In a just world, it should have been shown at the Great Exhibition of 1851, held in the specially built Crystal Palace in Hyde Park. Indeed, Babbage should have been allowed to accept the organizers' invitation to head the exhibition's Industrial Commission. But the government refused him permission to have anything to do with the event, including showing his machine—probably the finest product of precision engineering to date. (He got his revenge, if not his just deserts, by publishing a vitriolic account of the event and of its organizers—and by including his ideas about how it *should* have been run: Babbage 1851.)

As for calculating machines inspired by Babbage and built by his contemporaries, two small-scale versions of the Difference Engine were manufactured in Sweden by George and Edward Scheutz in the 1850s (Scheutz and Scheutz 1857). These coped with four orders of difference, and up to fifteen digits. Although logically similar to Babbage's Engine, they were mechanically different—and engineered at a much lower level of precision. But they were generously praised by Babbage at the time, and also in his autobiography later (Hyman 1982: 239–40; Babbage 1864: 35). They were featured in the *Illustrated London News*, which suggested that a small dog on a treadmill would be able to turn the handle. And one was bought by the General Register Office in London, to calculate tables for life insurance. Nevertheless, Scheutz's machines weren't easy to use, and weren't a commercial success. (The Register Office soon reverted to manual calculations with logarithms, which they used until switching to mechanical calculators in 1911.) George Scheutz died almost bankrupt in 1873, and his son achieved that dismal fate a few years after him.

Other people—in London, Sweden, and America—also tried to build Difference Engines, and one or two succeeded (Swade 1991: 14–21). But they, too, lost money as a result. Their machines were soon forgotten, to be rediscovered only with the computer-age renaissance of interest in Babbage. Scheutz's first engine, for instance, lay ignored in a Stockholm museum until 1979.

Just why Babbage failed to build his full-scale Difference Engine (or his Analytical Engine, either) is a matter of dispute. Six reasons are commonly suggested.

Arguably, he was betrayed by the politicians—after having received significant financial support, in the hope of improving navigation. In the 1850s the Chancellor of the Exchequer Benjamin Disraeli ruled: "The ultimate success [is] so problematical, the expenditure certainly so large and so utterly incapable of being calculated, that the Government would not be justified in taking upon itself any further liability." Babbage tartly noted, later, that his Difference Engine could have been used to "calculate the millions the ex-Chancellor of the Exchequer squandered" (Babbage 1864: 81).

Conceivably, his ideas were too close to the workbench to be quite respectable. His associate Dionysius Lardner had judged it "more a matter of regret than surprise" that his mechanical notation received little attention: "In this country, science has been generally separated from practical mechanics by a wide chasm" (Lardner 1834: 104). I'm reminded of Plato's assumption, quoted in Chapter 2.i.a, that one wouldn't wish an engineer to marry one's daughter.

Possibly (and thirdly), Babbage's outspokenness, and increasing bitterness, didn't endear him to those who might have been able to help. For instance, his reference to Disraeli's "squandered millions" was followed by his sneer that the novelist–Chancellor's mathematical understanding was surely surpassed by "any junior clerk in his office"—and many similar examples could be cited. Probably, he was let down by Clement's intransigence, and perhaps dishonesty. And apparently, the potential of his engines was unappreciated even by most of the scientific community.

All these five factors, very likely, contributed in preventing the building of his machine. The sixth reason often cited is more questionable: that building a full-scale Engine (as opposed to the mini-versions manufactured by Scheutz, for instance) was then impossible—or, at least, not practically feasible. The British Association may have been right in deciding, a few years after his death, that building a Difference Engine wasn't an affordable project. But we now know that it wasn't absolutely impossible.

A full-scale Difference Engine-2 was built to Babbage's original designs in his bicentenary year, 1991, and the printer was added in 2000 (Swade 1993; 2000: 252–307). The core Engine and the printer were each made of about 4,000 parts. They were constructed using only materials and methods available to Babbage, except that modern techniques were used to make often repeated parts. Even these 'mass-produced' parts showed variations that had to be filed down by hand: otherwise, the twentieth-century Engine wouldn't have worked.

This work of reconstruction was undertaken by London's Science Museum, where the large Difference Engine is now on permanent exhibition. But it wasn't all plain sailing. Like Babbage himself, the scientists responsible for this recent project suffered many vicissitudes. These included the sudden bankruptcy of the engineering company originally contracted, and a crucial deadline met with only minutes to spare (Swade 1993: 67). If one believed in jinxes, the Difference Engine would surely be a candidate.

3.iii. Analytical Engines

Intriguing though the Difference Engine was, it was surpassed by the even more amazing Analytical Engine—the cause of Babbage's greatest pride and bitterest disappointment.

From 1832 onwards, most of his formidable energies were devoted to the invention of this mechanical calculator. He saw it as "wholly independent" of its predecessor, not as a development of it (Lovelace 1843: 275). In his view, the Analytical Engine embodied foresight, whereas the Difference Engine had embodied only memory.

a. From arithmetic to algebra

Babbage's first design for the Analytical Engine dated from 1834, and he improved it throughout the rest of his life. It was highly ambitious as an engineering project, requiring a large number of near-identical parts—for which he experimented with die casting and sheet-metal stamping. Various small portions were built, and he was working on a small trial model when he died. But an entire machine was never completed.

Again, the details needn't concern us (but see Menabrea 1843; Lovelace 1843; Buxton 1880, chs. 7–10). The important point is that, whereas the Difference Engine had been confined to arithmetic, the Analytical Engine could deal with algebra.

It was extremely general, for it could find the value of almost any algebraic function. This generality alone would make it interesting to us, not least in the light of Alan Turing's work on universal machines (see Chapter 4.i). But its interest is greatly increased by the fact that *in essence* it was a general-purpose, program-controlled, digital, symbol-manipulating computer that included many specific features found in modern machines.

Although general-purpose *in essence*, the Analytical Engine was designed to deal only with numbers. Babbage knew, from Boole's work in the 1840s, that algebra can be mapped onto logic (see Chapter 2.ix.b). That is, he realized that the general *class* of possible Analytical Engines included machines capable of dealing with logic as well as mathematics. But he didn't try to design such a machine. As an engineering project, his Analytical Engine was a special-purpose device dedicated for mathematical use.

It was like its predecessor, the Difference Engine, in being constructed from columns of interlocking toothed wheels. Having considered using a binary mechanism, Babbage again chose decimal. However, the data and instructions were entered not by manual wheel setting, but by punched cards. As noted in Chapter 2.iv.c, Joseph Jacquard—following Vaucanson—had used punched cards for weaving fancy textiles. (The technique is described in Menabrea 1843: 254–5; Lovelace 1843: 283–4.) This wasn't a technology to be sneezed at: in Babbage's drawing-room there hung a portrait of Jacquard, woven in silk on a loom controlled by 24,000 cards, each with the capacity for 1,050 holes.

It was the use of sets of punched cards, automatically read one-by-one by the machine, which justifies our describing the Analytical Engine as 'program-controlled', and which often leads people to describe it also as a 'stored-program' machine. Babbage's method enabled an entire program to be provided in one step, as a single batch of cards, to be followed without further human intervention. To make it follow a different program, he didn't have to tinker with the wheel settings, as he'd done for the Difference Engine. Instead, he merely had to exchange one batch of cards for another. But the Engine's program wasn't stored internally, as in modern computers: see Section v.b. Since the card reader operated at much the same speed as all the other components, there wouldn't have been any advantage in this.

The Engine's cards had three main purposes. Some stored the numerical value of a constant (a number of up to fifty digits). Some represented a variable (determining the column on which the number would be placed). And some defined the operations to be performed (addition, division, etc.). Several operational cards could be combined and treated as a unit, allowing for repetitive looping and nested subroutines.

There was a separate "store" and "mill", for numbers being held and numbers being worked on. As in the Difference Engine, there was a facility for making (inoperative) "memoranda" reminding the user of what was going on. And there was even a way of ensuring self-correction if some of the number wheels were set to wrong figures during the calculation (Buxton 1880: 249–50).

b. Programs . . . and bugs

The startlingly modern flavour of the Analytical Engine, including the constraints involved in using it efficiently, is evident from these contemporary descriptions (taken, unless otherwise noted, from Lovelace 1843):

The Analytical Engine is therefore a machine of the most general nature. Whatever formula it is required to develop, the law of its development must be communicated to it by two sets of cards. When these have been placed, the engine is special for that particular formula . . .

Every set of cards made for any formula will at any future time recalculate that formula with whatever constants may be required.

Thus the Analytical Engine will possess a library of its own. (Babbage 1864: 90)

The use of the cards offers a generality equal to that of algebraical formulae, since such a formula simply indicates the nature and order of the operations requisite for arriving at a certain definite result, and similarly the cards merely command the engine to perform those same operations . . . In this light the cards are merely a translation of algebraical formulae, or, to express it better, another form of analytical notation (Menabrea 1843: 264)

. . . the collection of columns of Variables may be regarded as a *store* of numbers, accumulated there by the mill, and which, obeying the orders transmitted to the machine by means of the cards, pass alternately from the mill to the store, and from the store to the mill, that they may undergo the transformations demanded by the nature of the calculation to be performed. (Menabrea 1843: 257)

[We] may, if we please, retain separately and permanently any *intermediate* results . . . which occur in the course of processes [having] an ulterior and more complicated result as their chief and final object . . . (p. 282)

The Operation-cards merely determine the succession of operations in a general manner. They in fact throw all that portion of the mechanism included in the *mill*, into a series of different *states*, which we may call the *adding state*, or the *multiplying state*, etc., respectively. In each of these states the mechanism is ready to act in the way peculiar to that state, on any pair of numbers which may be permitted to come into its sphere of action. Only *one* of these operating states of the mill can exist at a time; and the nature of the mechanism is also such that only *one pair of numbers* can be received and acted on at a time. (p. 281)

There are certain numbers, such as those expressing the ratio of the circumference to the diameter . . . which frequently present themselves in calculations. To avoid the necessity for computing them every time they have to be used, certain cards may be combined specially in order to give these numbers ready made into the mill, whence they afterwards go and place themselves on those columns of the store that are destined for them . . . [By such means] Mr. Babbage believes that he can, by his engine, form the product of two numbers, each containing twenty figures, in *three minutes*. (Menabrea 1843: 264)

The mode of application of the cards, as hitherto used in the art of weaving, was not . . . sufficiently powerful for all the simplifications which it was desirable to attain in such varied and complicated processes as those required [for] an Analytical Engine. A method was devised of what was technically designated *backing* the cards in certain groups according to certain laws. The object of this extension is to secure the possibility of bringing any particular card or set of cards into use *any number of times successively* in the solution of one problem. (p. 283)

It is desirable to arrange the order and combination of the processes with a view to obtain them as much as possible *symmetrically* and in cycles [or even in cycles of cycles: see p. 308], in order that the mechanical advantages of the *backing* system may be applied to the utmost. It is here interesting to observe the manner in which the value of an *analytical* resource is *met* and *enhanced* by an ingenious *mechanical* contrivance. We see in it an instance of one of those mutual *adjustments* between the purely mathematical and the mechanical departments, [which is] a main and essential condition of success in the invention of a calculating machine. (p. 298)

The engine is capable, under certain circumstances, of feeling about to discover which of two or more possible contingencies has occurred, and of then shaping its future course accordingly. (Lovelace, Menabrea 1843: 252 n.)

Figures, the symbols of *numerical magnitude*, are frequently *also* the symbols of *operations*, as when they are the indices of powers. Wherever terms have a shifting meaning, independent sets of considerations are liable to become complicated together, and reasonings and results are frequently falsified. Now in the Analytical Engine the operations which come under the first of the above heads, are ordered and combined by means of a notation and of a train of mechanism which belong exclusively to themselves; and with respect to the second head, whenever numbers meaning *operations* and not *quantities* (such as the indices of powers), are inscribed on any column or set of columns, those columns immediately act in a wholly separate and independent manner, becoming connected with the *operating mechanism* exclusively, and re-acting upon this. They never come into combination with numbers upon any other columns meaning *quantities*; though, of course, if there are numbers meaning *operations* upon *n* columns, these may *combine amongst each other*... It might have been arranged that all numbers meaning *operations* should have appeared on some separate portion of the engine from that which presents numerical *quantities*; but the present mode is in some cases more simple, and offers in reality quite as much distinctness when understood. (p. 270)

There are frequently several distinct *sets of effects* going on simultaneously; all in a manner independent of each other, and yet to a greater or less degree exercising a mutual influence. To... perceive and trace them out with perfect correctness and success, entails difficulties whose nature partakes to a certain extent of those involved in every question where *conditions* are very numerous and intercomplicated; such as for instance the estimation of the mutual relations amongst *statistical* phenomena, and of those involved in many other classes of facts. (p. 288)

Last, but unfortunately not least:

[It may be objected] that an analysing process must... have been performed in order to furnish the Analytical Engine with the necessary *operative* data, and that herein may also lie a possible source of error. Granted that the actual mechanism is unerring in its processes, the *cards* may give it wrong orders. This is unquestionably the case... (p. 274)

The vocabulary of computer science wouldn't be coined until over a century later (see Section v). But Babbage's cogwheel design for the Analytical Engine had exemplified a wide range of computational phenomena that are implemented electronically today.

It's clear from the quotations given above that these included programmed control; stored programs (in the sense previously remarked); data and operations; central processor; memory store; addressing; hierarchically nested subroutines; microprogramming; looping; conditionals; and comments.

They even included the programmer's nemesis, bugs. (These are so called by extension: the first bug wasn't a software bug at all, but a moth crushed on a relay switch. It was found in the Harvard Mark I by Grace Hopper, who stuck it into the log book, noting: "First actual case of bug being found". For a photo of this famous animal, see Kurzweil 1990: 179.) True, Babbage didn't foresee that bugs would later be regarded as a "powerful idea", used both in education and in programmed planning (Chapter 10.iii.c and v.g). But he was alive to their possibility.

In sum, Babbage not only designed the first stored-program, general-purpose digital computer. He also anticipated many specific computational functions found in modern machines.

3.iv. Had Wheelwork Been Taught to Think?

In the year of the Great Exhibition, the surgeon Alfred Smee (1818–77)—who'd already argued that "instinct" and "reason" could be "deduced" from electrobiology (Smee 1850)—suggested that a logic machine could be built to operate like the human mind (Smee 1851). He allowed that it would fill an area larger than London (see 2.ix.a), but insisted that it was possible *in principle*. And a close friend of Babbage wrote in the 1870s: "The marvellous pulp and fibre of the brain had been substituted by brass and iron, he had taught wheelwork to think" (quoted in Swade 1993: 64).

Was that Babbage's view? And did he believe that he'd provided evidence for the 'mind as machine' hypothesis? (These are different questions: substitution is neither simulation nor replication.)

Babbage himself said very little on these matters. But what he did say—and also what he didn't say—suggests that the answer to both questions is *No*. In other words, he wasn't a precocious supporter of strong AI, nor of weak AI either (see Chapter 16.v.c). In that sense, he was firmly located in his own century, not the next.

a. For Lovelace read Babbage throughout

Direct evidence from Babbage's own hand is scant. He wrote very little about the general implications of his work, and nothing that one can confidently interpret as addressing the question we're raising here.

In his autobiography, for example, he described "calculating machines" not as devices that can do arithmetic, but as "pieces of mechanism for assisting the human mind in executing the operations of arithmetic" (Babbage 1864: 30). Possibly, he intended there to make the distinction (mentioned in Chapter 2.ix, with respect to Blaise Pascal and Stanley Jevons) between thinking as such and the formal principles that thinking must satisfy if it is to be valid. On other occasions, his language was more ambiguous, as when he called his Difference Engine "a substitute for one of the lowest operations of human intellect" (Hyman 1989: 44).

On most occasions, however, his comments about mind in relation to machine weren't so much ambiguous as non-existent. He published very little about the Analytical Engine, even considered as a mathematical tool or engineering design. When he did, he didn't discuss the mind–machine relation. Nor did he do so in his unpublished notebooks. Moreover, even when—towards the end of his life—he wrote that *the most challenging scientific question* is how inventive thinking works, he mentioned his writings on induction and analogy but said not a word about either of his Engines (Babbage 1864: 321–2).

Occasionally, he did express a passing interest in what a cognitive scientist would call psychological mechanisms. So in his reminiscences he said: "It has often struck me that an analysis of the causes of wit would be a very interesting subject of enquiry. With that view I collected many jest-books . . ." (1864: 272). But "fortunately", he continued, he had the "resolution to abstain from distracting my attention from more important enquiries".

What his analysis of jests might have been like is indicated in the few ensuing remarks. For instance, he recalled one of his friends saying to another, "I am very stupid this morning: my brains are all gone to the dogs"—to which the instant reply was "Poor dogs!" The "wit" here, he said, arose from "sympathy expressed on the wrong side". And he pointed out that "jokes formed upon this principle [*sic*]" depend on the meaning of the words but not on their sound or arrangement, so are "rare" in being translatable into all languages. Puns—which he considered "detestable"—depend, he said, on double meanings of one and the same word, or on similar pronunciation of words that are differently spelt. As an example of "a triple pun", he gave this:

A gentleman calling one morning at the house of a lady whose sister was remarkably beautiful, found her at the writing-table. Putting his hand upon the little bell used for calling the attendant, he enquired of the lady of the house what relationship existed between his walking-stick, her sister, and the instrument under his finger.

His walking-stick was { cane / Cain }, the brother of { a bell / a belle / Abel.

(Babbage 1864: 273)

Then, instantly, he changed the subject (to talk about his Silver Lady).

It's clear from those examples (and from his use of the word "principle") that he'd have been sympathetic to cognitive analyses of jokes. But nowhere did he suggest that one of his Engines might implement such an analysis. He'd probably have been intrigued by a recent joke-generating program that uses some of the punning principles he identified here (Binsted and Ritchie 1997; Binsted *et al.* 1997; Ritchie 2001, 2003*a*; see Chapter 13.iv.c). But whether he'd think it a model of "the causes of [human] wit" is quite another matter.

This absence of psychological comment on the Engines is highly indicative. Babbage wasn't a blinkered mechanic, blind to the wider implications of his work. He was quick to draw theological morals from his Difference Engine, for instance, as we've seen. Nor did he fear controversy—indeed, he relished it. It's hard to believe that someone prepared to link wheelwork with miracles wouldn't have linked it also with minds, if he'd thought the comparison to be of interest. And if he had, he'd have shouted it from the rooftops.

Something more, however, can be said. If Babbage didn't describe the Analytical Engine in print, his collaborator Ada, Lady Lovelace (1815–52), did. And, much as the spoof history book *1066 and All That* famously advises "For pheasant read peasant throughout", so we can read Lovelace as Babbage—pretty well throughout.

Lovelace, wife of the Earl of Lovelace, was the daughter of the poet Lord Byron (D. L. Moore 1977; D. Stein 1985). It was Byron who was famously described on first meeting—by Lady Caroline Lamb (later, his lover), in her diary of 1812—as being "mad, bad, and dangerous to know". It's a delicious historical irony that this leader of the Romantic movement, with its contempt for order and for science (see Chapter 2.vi.c), fathered Babbage's closest co-worker and confidante—and chose the name that would eventually be given to a widely used programming language, Ada (Chapter 13.i.c).

But Byron's daughter didn't betray her Romantic father so far as to liken minds to machines. On the contrary, in her comments on the general nature and potential of the Analytical Engine she too, like Babbage, avoided drawing any psychological moral.

Lovelace's comments occur in one important document. She translated a paper on the Analytical Engine written by an Italian scientist, Luigi Menabrea—later, Prime Minister of Italy (Menabrea 1843). This paper was based on a lecture given by Babbage himself in Turin, the only occasion on which he spoke at length about his Analytical Engine to fellow scientists. And, significantly, she added lengthy 'Notes', and a few footnotes, giving her own corrections and comments (Lovelace 1843). We may read Menabrea's (translated) paper as being Lovelace at one remove, for her scrupulousness suggests that any claim that she translated without comment was one with which she didn't disagree.

Likewise, we may read Lovelace as being Babbage at one remove—and not just because it was Babbage's lecture that Menabrea was reporting. It's not known just how much Babbage contributed to the first draft of these 'Notes' (nor to the translation of Menabrea's paper), and how much is 'pure' Lovelace. Recent historical research has shown that "most of the technical content and all of the programs in the *Sketch* were Babbage's work" (Campbell-Kelly 1994: 27). But at one point in the 'Notes' (p. 271), Lovelace mentions an idea and says "we do not know" whether Babbage agrees with it: presumably, that passage was drafted by her. However, Babbage did oversee both papers before publication, even if he didn't supply the answer to her implied question. It's therefore most unlikely that he disagreed strongly with anything she said there.

Moreover, the two were very close, both intellectually and personally. They first met, at a party, when she was only 17. He addressed her as "my dear and much admired interpreter", and perhaps called her "the Enchantress of Numbers". I say "perhaps", partly because the context would allow this phrase to mean not Ada but mathematics, or even his beloved Engine (Swade 2000: 165), and partly because her mathematical abilities are actually in doubt. One sour biography ungenerously downplays them (D. Stein 1985: 89–120 *passim*). Another Babbage historian, Bruce Collier, endorses this judgement (1990: [5]), having already complained in exasperation:

> There is one subject ancillary to Babbage on which far too *much* has been written, and that is the contribution of Ada Lovelace. It would be only a slight exaggeration to say that Babbage wrote the 'Notes' to Menabrea's paper, but for reasons of his own encouraged the illusion in the minds of Ada and the public that they were authored by her. It is no exaggeration to say that she was a manic depressive with the most amazing delusions about her own talents, and a rather shallow understanding of both Charles Babbage and the Analytical Engine.... Ada was as mad as a hatter, and contributed little more to the 'Notes' than trouble. (Collier 1990: [4])

Declaring "an open mind on whether Ada was crazy because of her substance abuse ... or despite it", Collier ended tetchily: "I am somewhat sorry I did not debunk her role more vigorously.... But, then, I guess *someone* has to be the most overrated figure in the history of computing" (1990: [5]).

Overrated? Well, perhaps. It seems pretty clear, however, that Ada understood the principles of the Analytical Engine. If she didn't grasp every mathematical jot and engineering tittle, she nevertheless appreciated its general nature and implications. It's even possible that she had a better idea of some of these implications than did Babbage himself: non-specialists may be better able to see the wood, even if they can't see all of the trees. His epithet "admired interpreter", unless it was utterly empty flattery, suggests that she did add something to his own view of his work.

However that may be, and even if some of Lovelace's comments wouldn't have come spontaneously from Babbage, it's hardly credible that he disagreed with her on any important point. So we can read Lovelace as Babbage, albeit at one remove. (Even Collier would agree with that: see his remark about the "illusion", quoted above.)

b. What Lovelace said

Lovelace took pains to stress the generality of the Analytical Engine:

The distinctive characteristic of the Analytical Engine, and that which has rendered it possible to endow mechanism with such extensive faculties as bid fair to make this engine the executive right-hand of abstract algebra, is the introduction into it of the principle which Jacquard devised for regulating, by means of punched cards, the most complicated patterns in the fabrication of brocaded stuffs. It is in this that the distinction between the two engines lies. Nothing of the sort exists in the Difference Engine. We may say most aptly that the Analytical Engine *weaves algebraical patterns* just as the Jacquard-loom weaves flowers and leaves. (Lovelace 1843: 272–3)

[Although the Engine was built so as to deal with numerical data and results] it must be easy by means of a few simple provisions and additions in arranging the mechanism [to enable it also to bring out] *symbolical results* [being] the necessary and logical consequences of operations performed upon *symbolical data*. (Lovelace 1843: 271)

By the word *operation*, we mean *any process which alters the mutual relation of two or more things*, be this relation of what kind it may. This is the most general definition, and would include all subjects in the universe. (Lovelace 1843: 269)

She pointed out that there could be a similarly general machine for doing logic. She also remarked that the Analytical Engine could in principle compute all the laws of science, for "Mathematics . . . constitutes the language through which alone we can adequately express the great facts of the natural world . . . ." And she (and Menabrea) foresaw a time when scientists would need to call on mechanical calculation, to deal with otherwise unmanageable amounts of data. (How right she was! Today's scientists regularly depend on computers to do their sums. A few have even used computers to help make scientific discoveries: see 10.iv.c and 13.iv.c.) So, whereas the Difference Engine was to be used for more mundane tasks (helping navigators and arithmeticians), the construction of the Analytical Engine would mark "a glorious epoch in the history of the sciences" (Menabrea 1843: 266). This prophecy was the culmination of Menabrea's paper, and perhaps—who knows?—the triumphant ending of Babbage's lecture.

Even musical composition, she said, might turn out to be explicable by science, and amenable to (programmable for) this machine:

The operating mechanism . . . might act upon other things besides *number*, were objects found whose mutual fundamental relations could be expressed by those of the abstract science of operations, and which should be also susceptible of adaptations to the action of the operating notation and mechanism of the engine. Supposing, for instance, that the fundamental relations of pitched sounds in the science of harmony and of musical composition were susceptible of such expressions and adaptations, the engine might compose elaborate and scientific pieces of music of any degree of complexity or extent. (Lovelace 1843: 270)

However, for Lovelace (and presumably for Babbage), the Engine's generality *didn't* make it capable of thought.

Some evidence for this comes from her comments, or lack of them, on Menabrea's paper. She didn't demur from Menabrea's opening remark, that "mathematical labours", which seem at first sight to be "the exclusive province of intellect", actually comprise two types: "the mechanical [involving] precise and invariable laws, *that are capable of being expressed by the operations of matter*", and "reasoning, [which belongs] to the domain of the understanding" (Menabrea 1843: 246; italics added). Nor did she contradict his judgement that "although [the Analytical Engine] is not itself the being that reflects, it may yet be considered as the being which executes the conceptions of intelligence" (Menabrea 1843: 265). Indeed, she wrote a long note on this very sentence, in which she warned against over-enthusiastic descriptions of the machine:

It is desirable to guard against the possibility of exaggerated ideas that might arise as to the powers of the Analytical Engine. In considering any new subject, there is frequently a tendency, first, to *overrate* what we find to be already interesting or remarkable; and secondly, by a sort of natural reaction, to *undervalue* the true state of the case, when we do discover that our notions have surpassed those that were really tenable. (Lovelace 1843: 300)

One is reminded of the 'hype' sometimes attached to various forms of AI, and the see-sawing public responses to it (see Chapters 9.x.f, 13.iv–vi, 12.iii and vii, and 15.ix). This passage, especially in the context of Menabrea's remark, implies that Lovelace didn't think of the Engine as a "being that reflects".

Further evidence that she saw it merely as a machine, not as anything like a mind (neither simulation nor replication), came from her own pen:

In enabling mechanism to combine together *general* symbols, in successions of unlimited variety and extent, a uniting link is established between the operations of matter and the abstract mental processes of the *most abstract* branch of mathematical science... [The] mental and the material... are brought into more intimate and effective connexion with each other. We are not aware of its being on record that anything partaking in the nature of what is so well designated the *Analytical* [algebraic] Engine has been hitherto proposed, or even thought of, as a practical possibility *any more than the idea of a thinking or reasoning machine.* (Lovelace 1843: 273; final italics added)

That last sentence seems to mean that, whereas the idea of an algebraic machine is wrongly supposed to be nonsense, the idea of a thinking machine obviously is nonsense. Nor was the notion of a creative machine acceptable to her:

The Analytical Engine has no pretentions whatever to *originate* anything. It can do whatever we *know how to order it* to perform. It can *follow* analysis; but it has no power of *anticipating* any analytical relations or truths. Its province is to assist us in making *available* what we are already acquainted with. (Lovelace 1843: 300)

She would have regarded the music machine, had it ever been built, as an existence proof that "scientific" music can be composed by machine, not as a simulation of how human musicians actually compose ("originate") it.

Possibly—indeed, probably—part of her scepticism here related to the assumption that no mere machine could be conscious. However, she didn't actually say so. Nor did Babbage, when referring to the difficulty in explaining inventive thought (1864: 321–2). Rather, they seemed (to me) to assume that *processes* of some kind are involved in human thinking—but of an utterly mysterious type.

Of course, Lovelace (and Babbage) used psychological language to describe the operations of the Analytical Engine. She even added a footnote describing it as "feeling about" to discover which of several possible events have occurred, in rebuttal of Menabrea's claim that it "must exclude all methods of trial and guess-work, and can only admit the direct processes of calculation" (Menabrea 1843: 252). But she didn't quibble when Menabrea declared, in the very next sentence: "[The] machine is not a thinking being, but simply an automaton which acts according to the laws imposed upon it." Presumably, then, she agreed.

c. Babbage and AI

In his *Ninth Bridgwater Treatise*, Babbage declared: "I was well aware that the mechanical generalizations I had organized contained within them much more than I had leisure to study, and some things which will probably remain unproductive to a far distant day."

Bearing in mind his searing disappointment in failing to complete any full-scale Engine, this remark—with hindsight—may seem poignantly prophetic. In fact, it was more poignant than prophetic.

Babbage was neither a successor of Vaucanson nor a precursor of Allen Newell and Herbert Simon (see Chapters 6.iii, 7.iv.b, and 10.i.b). Unlike them, he had no biological or psychological aims in designing his automata. His machines embodied mathematical principles, not psychological theories.

Given his Boolean assimilation of logic to mathematics (and science), one might say that, *in spirit*, he anticipated Turing's claim that anything computable can be computed by some machine (see 4.i.c). But if his Analytical Engine was the first truly general automatic problem-solver, he didn't see it as such—still less, as a model of human problem solving. He designed a powerful, technologically motivated, calculator. But he got no further along the AI road than to propose a noughts-and-crosses (tic-tac-toe) machine—never built—to raise money for his research (Swade 1991: 32). Even that, had he built it, would have been a technologist's toy, not a psychologist's model. In the terminology of Chapter 1.ii.b, his life's ambition was to do technological, not psychological, AI.

In short, Babbage didn't believe he'd taught wheelwork to think. Nor, despite a still-widespread assumption to the contrary, did he imagine that wheelwork might have anything to teach us about our own thinking. In that sense, he was irrelevant to cognitive science.

Some people would argue, however, that he merits mention in the history of the field in virtue of his role in the invention of computers—without which, cognitive science wouldn't exist. After all, it was he—or so this claim goes—who started us on the road to general-purpose computing. Before we can weigh that argument (in Section vi), we first need to consider briefly how it was that today's machines came on the scene.

3.v. Electronic Babbage

The technological advances in computing that occurred in the mid-twentieth century were crucial for cognitive science. They not only made it potentially possible in practice, but also influenced its theory.

I say "potentially" because the very earliest computers weren't usable by anyone but a few dedicated boffins. Indeed, with one exception they weren't even seen by more than a few hand-picked boffins (plus the servicewomen who 'fed' them, night and day). They were literally a state secret, and remained so for many years. What's more, they were being used to solve problems much more urgent than modelling minds.

Their immediate successors, in the late 1940s, were almost as difficult to use. By the mid-1950s, the machines were somewhat—only somewhat—more manageable. But for several more years, they weren't actually available except within a tiny handful of computing research labs—whose "service clients" were mostly physicists and mathematicians (plus occasional pioneers in machine translation), not psychologists or biologists.

That's why the cognitivist ideas that were exciting a number of professional psychologists in the 1940s and early 1950s still couldn't be implemented as computer models (see Chapters 4.ii, 5.ii–iv, 6.ii.a, and 12.i). Even if—which wasn't always so—the theorists concerned were thinking about psychology in broadly computational terms (perhaps because they'd read a paper in an obscure journal of 1943: see 4.iii.e–f), they had no opportunity to translate their thoughts into computer programs. That could be attempted only by hands-on computer aficionados, who were still very thin on the ground—and most of them shared Babbage's unconcern for matters mental.

Most . . . but not quite all. As it happens, modern computing was largely initiated by two men who, unlike Babbage, *were* interested in core questions about life and mind. Both are now counted among the founding heroes of cognitive science: namely, Turing (1912–54) and John von Neumann (1903–57). Their theoretical ideas will be considered in later chapters (4.i, 10.i.f, 12.i.b–d, 15.iv–v, and 16.ii). Here, the focus is on their practical work on the design of computing machinery. For our purposes, the interest isn't in the engineering details but in the functions, or tasks, which the electronic nuts and bolts enabled mid-century computers to perform.

a. A soulmate in Berlin

The "one exception" to the claim that the earliest computers were known only to a few boffins was a machine ensconced in a family flat in pre-war Germany. In 1935, just over 100 years after Babbage started work on the Analytical Engine, the German inventor Konrad Zuse (1910–95) gave up his job and set up a computer workshop in his parents' flat in Berlin.

Their living-room was near-filled by his machines (the first of which was the size of an eight-place dining table), and the floor littered with the debris from his fretsaw.

Lacking Babbage's independent means, he was financially supported by his father (who went back to work, having been retired for a year), by his sister (who donated much of her wages), and by a dozen friends who provided financial and/or technical help.

For the next six years, until he was called up into the German army, he devoted all his time to designing and building a series of increasingly powerful calculating machines. These had begun even earlier: in 1933–4 (well before Turing's 1936 paper), he'd snatched spare moments from his salaried work to design the first version. They culminated in the first fully automated, program-controlled, computer. Now known as the Z3, this was sketched in 1938 and operational by 1941.

Zuse himself later described the Z3 as "a Babbage machine" (Zuse 1993: 50). That's not to say that it was based on Babbage's work, for it wasn't (see Section vi.a). Nor was it exactly similar in composition or in function. Z3 was built of electrical switches and mechanical parts, not columns and cogwheels, and it used a binary instead of a decimal base. Moreover, it did floating-point arithmetic, and could cope with mathematical matrices as well as series of 'individual' sums. Nevertheless, it was similar in principle to the Analytical Engine. And it was driven by very similar motivations: Babbage and Zuse were soulmates.

Both, for instance, were 'natural' engineers. Zuse had been an inventor of mechanical and electro-mechanical gadgets ever since his schooldays. He repaired no silver ladies. (Not because he lacked aesthetic appreciation: he was an inventive photographer and film-maker, and *art or engineering* was a difficult career choice for him—1993: 9.) But as a schoolboy, he'd managed to mend the broken second gear on his bicycle when the bike-dealer couldn't. And as an engineering student he invented an automatic dispensing machine, which not only accepted different coins and different prices, but provided change. He even built a snow machine for use in theatres, and designed the first split-level (clover leaf) intersection for street crossings. In short, "All types of smaller and larger inventions kept me busy" (p. 16).

Much as Babbage eventually abandoned the Silver Lady, so Zuse eventually allowed "all types" of invention to be overtaken by just one. This happened in his mid-twenties: "If I remember correctly, my thoughts [about the computer] first took concrete shape in 1933" (p. 28), and "In 1935 I decided to become a computer developer" (p. 34). In 1947 he would found Germany's first computer company, with the aim of developing CAD/CAM applications as well as 'mere' calculation. (The company succeeded partly because there was no competition from IBM. He'd suggested a joint effort, but the negotiations came to nothing: IBM wouldn't guarantee that he could continue to work on his computer: p. 114.)

Like Babbage, too, Zuse saw his machine as a *general* computer. Indeed, he'd seen this right from the start. (Babbage hadn't: the Difference Engine wasn't a universal machine, but a dedicated one.) For gadgets weren't Zuse's only passion. He was enthused also by mathematical logic—and the logic was as important as the mathematics. (Surprisingly, he was unaware of the propositional calculus: p. 44.) Indeed, the Z3 was based not on adding wheels, but on logic gates: implementations of *conjunction*, *disjunction*, and *negation*, used to define conditional statements. By 1946 Zuse had even developed a "logical" programming language, the Plankalkul (Chapter 10.v.f). In short, he was well aware that his machines had the potential for computing symbolic expressions such as chess moves and propositions, as well as numbers.

This was revolutionary, despite Babbage's (and/or Ada's) prescient remarks on the same theme and despite the notoriety of *Principia Mathematica* (Russell and Whitehead 1910). And it was resisted accordingly. Even his close collaborators doubted the possibility of non-numerical computing (p. 57).

That was par for the course in the 1940s. When logic gates were independently defined by Warren McCulloch and Walter Pitts two years later (1943), they caused a sensation in a few well-prepared minds but weren't understood by most readers (see 4.iii.e–f and iv). Zuse's lecture on symbolic computing to the Society for Mathematics and Mechanics in 1948 was ignored. And a full twenty years after that, MIT's AI group still felt the need to make a song and dance about the possibility of "semantic" information processing (Minsky 1968; see 10.iii.a).

An added similarity between Zuse and Babbage was their lack of interest in cognitive science. It's true that Zuse's diary for 20 June 1937 declared: "For about a year now I have been considering the concept of a mechanical brain . . ." (p. 44), and that "[It] was clear to me that one day there would be computing machines capable of winning international chess matches. I estimated that it would take about fifty years before this would happen" (pp. 49–50). It's also true that by 1944 he'd sketched a version of associative memory, and that he'd considered parallel processing almost from the start. But these projects were more technological than psychological. Zuse's autobiography says nothing about modelling human thought processes as such.

Like his English soulmate, he wasn't afraid of being outrageous. He originated the concept of "the computing universe" (to be described by a digital physics), discussed self-reproducing systems, and analysed attractor cycles and other abstract properties of cellular automata (Zuse 1969/1970)—all topics that are still highly controversial. At the time, he reported that he'd found it "rather difficult to find a publisher" for his work, "which stands somewhat outside the presently accepted method of approach" (1969/1970: 2). That was an understatement: he later admitted that his hypothesis of the computing universe appeared to be "crazy", while hanging on to it nonetheless (1993: 174). So he surely wouldn't have baulked at a 'digital' psychology. Since he didn't claim to have illuminated the mind, it's pretty certain that he wasn't interested in doing so.

The greatest difference between the two men was that Zuse did—eventually—receive recognition in his lifetime. In 1957 he was awarded an honorary doctorate by the Technical University of Berlin. Universities and professional societies in other countries later followed suit (US recognition first occurred in Las Vegas in 1965). His own comment puts the point clearly:

> [As regards professional recognition and honours] I have experienced much of both and was just as happy to receive both. Sometimes I cannot help but think of Babbage, who was denied this as well as the inventor's greatest reward: the successful realization of one's idea. (1993: 166–7)

Why did Zuse have to wait so long for professional recognition? The main reasons why his ideas on symbolic computing weren't taken up earlier were the intellectual isolation and physical devastation caused by the Second World War. Zuse himself was forced to divert his energies onto weapons development (Chapter 11.i.a), and his prototype machines and most of his early papers were destroyed in Allied bombing raids. By the end of the war, he still knew nothing of Claude Shannon or Turing. And from 1947 on, he had to concentrate on running his business, not on advancing or publicizing his theory.

By the time his work became known in Anglo-American circles, the computer industry had already been established there. But whether he'd have received a sympathetic hearing in Allied countries immediately after the war is doubtful in any case.

b. Call me MADM

In 1944 a novel calculating machine was built that was soon hailed in *Nature* as 'Babbage's Dream Come True' (Comrie 1946). That machine was Howard Aiken's (1900–73) Harvard Mark I, an electromechanical (fixed-point) calculator constructed (by IBM, and part-funded by the US Navy) by combining several Hollerith-type statistical machines.

But the salutation was premature. Aiken's machine (originally proposed in 1937) wasn't a general-purpose computer, but a dedicated calculator. Moreover, it lacked the full conditional branching allowed for in the design of the Analytical Engine (I. B. Cohen and Welch 1999; Pratt 1987: 148–50). Aiken himself later said, "If Babbage had lived seventy-five years later, I would have been out of a job" (quoted in Swade 1991: 34). Even so, Babbage's dream was fully realized only by the electronic, stored-program, general-purpose, digital computer.

As we've seen, Zuse's electromechanical Z3 was the first stored-program general computer. But it remained unknown, owing to the chaos of post-war Germany. It had no influence on the Anglo-American developments that led to today's machines.

In that historical context, the 'first' computer was the Manchester Mark I, or MADM (Manchester Automatic Digital Machine). The small-scale prototype (affectionately named "Baby" by the team who built it) was finished in June 1948, and a commercial version was marketed by Ferranti in 1951 (Kilburn and Williams 1953). A simple demonstration computer, called Nimrod, was exhibited by Ferranti at the Festival of Britain in the same year.

MADM's intellectual ancestry was highly distinguished. The design team, led by the electronic engineers Frederick Williams and Thomas Kilburn, was largely inspired by Maxwell Newman (1897–1984)—the recently appointed Professor of Mathematics at Manchester:

> Neither Tom Kilburn nor I knew the first thing about computers when we arrived in Manchester University... Newman explained the whole business of how a computer works to us. (Williams, quoted in Copeland and Proudfoot 1998: 6)

Newman had taught Turing in Cambridge in the mid-1930s—indeed, it was his lectures which led Turing to write the seminal paper on the Turing machine (see Chapter 4.i.c). He was familiar with Turing's early construction (in 1937–8) of a simple binary multiplier. And he'd designed and used pioneering computers at Bletchley Park, where he'd headed the cryptanalysis unit (see below).

Turing himself joined Newman at Manchester in 1948, just after the full Mark I had become operational. But he had no direct contact with the MADM engineers until after their first program had been run (Copeland 1999). His one contribution to the hardware of the Manchester machine was in 1949, when he helped to design a random number generator (Hodges 1983: 402).

Williams later described MADM as "pure Babbage", the "sole difference" being that it used subtraction rather than addition as its basic operation (quoted in Randell 1972: 9). Its code allowed for eight functions, including STOP.

Its first program, entered by way of a 5 × 8 array of push-button switches, was a 'highest factor' mathematical routine, only seventeen instructions long. (A facsimile is shown in Lavington 1975: 11.) The program tested about 130,000 numbers (generated by repeated subtraction), by means of 3.5 million operations, in a run of fifty-two minutes. A quarter-century on, Williams recalled the achievement:

> A program was laboriously inserted and the start switch pressed. Immediately the spots on the display tube entered a mad dance. In early trials it was a dance of death leading to no useful result, and what was even worse, without yielding any clue as to what was wrong. But one day it stopped, and there, shining brightly in the expected place, was the expected answer. It was a moment to remember. This was in June 1948, and nothing was ever the same again. (quoted in Stracey 1997: 17)

The programs for the Manchester machine were written by a number of people, including the mother and father of Tim Berners-Lee, inventor of the World Wide Web (Berners-Lee 2000: 3), and Turing himself. Turing was employed for a while as MADM's software writer—officially, as "deputy director of the Computing Machine Laboratory".

One of his efforts was a programmer's joke: it used random numbers to choose words making up "love-letters" (Chapter 9.x.c). Another, which could have been used for a serious purpose, did long division (it's reproduced in Stracey 1997). A third was a much-improved (faster) version of a prime-number program written by Newman (Lavington 1975: 12). And a long section of Turing's manual discussed how to write code to enable MADM to play tunes on its "hooter". This advice was heeded by Christopher Strachey, one of whose programs—the first ever, of any significant length—ended its activity by playing *God Save the King* (B. J. Copeland, personal communication).

Only a few of these very early programs remain. Indeed, not many were written in the first place. The reason was that MADM's programs, expressed in an unfamiliar notation full of %s and $s and £s and ////s, could be written or understood only by someone fully conversant with the basic structure of the machine itself. Turing's *Programmers' Handbook* was the only aid available, and it sometimes presupposed advanced mathematical knowledge (Hodges 1983: 398–401).

Consequently, this epochal machine could be dealt with directly only by accomplished mathematicians entirely familiar with its construction. That's not to say that others couldn't benefit from it: the Manchester Computing Laboratory offered a consultancy service to both academics and industry from about 1950. But the "others", themselves, couldn't use it.

The first 'usable' stored-program electronic computer was the EDSAC, built by Maurice Wilkes (1913–) and colleagues at the University of Cambridge (M. V. Wilkes 1953). This did employ mercury delay lines (the D stood for 'Delay'), which Turing had wanted for the ACE, the Automatic Computing Engine built in London after the war—see below (Campbell-Kelly 2002). It ran its first program on 6 May

1949, when it calculated a table of squares and printed the result. It provided a regular computing service for members of the University (including CLRU: Preface, ii) from 1950 until 1958, when EDSAC-2 was ready.

One reason why EDSAC was easier to use was that it accepted an alphabetic shorthand instead of binary numbers, converting the one into the other automatically. (Even so, it was still fiendishly difficult: genuine ease of use had to await the development of high-level programming languages: Chapter 10.v.) Moreover, Wilkes's colleagues provided its users with a growing library of commonly used subroutines—for doing division, for example. Similarly, you will remember, Babbage had used a special type of card to express "numbers which frequently present themselves in calculations".

c. Intimations of AI

But where were the "elaborate and scientific pieces of music"? They were nowhere in sight, and almost nowhere in mind. For if these British machines were, in principle, the realization of Babbage's (or Ada's) ambitious dream, that's not to say that most of their users saw them in that way.

They—and their transatlantic cousins the ENIAC and EDVAC—were usually thought of merely as powerful numerical calculating machines, not as general symbol manipulators. (The letter C stands straightforwardly for 'Calculator' in EDSAC, and for the ambiguous term 'Computer' in ENIAC and EDVAC.) Indeed, the ENIAC wasn't a general-purpose machine but was used mainly for calculating bombing tables. (The wartime security surrounding the ENIAC was less fierce than at Bletchley. Even so, when it was in the initial stages of construction for the US Army in 1943, those involved deliberately encouraged the rumour that it was a "white elephant"—J. Norman 2004: 16.)

One person involved with computer design in the mid-1940s has admitted that "in general, the idea of universality of a general purpose digital computer took some grasping" (quoted in Randell 1972: 15). It was obvious that Babbage's algebraic ambitions had been realized, if "algebra" was interpreted in its usual (mathematical) sense. But very few saw these early computers as applying to "other things besides number".

Turing was among the few who did see their more general potential. For instance, we've already seen that he used MADM for non-numerical tasks such as generating "love-letters". He also programmed it to play "music" on its hooter—though not to compose melodies, which is what Lovelace had had in mind. Indeed, he'd already employed the electromechanical Bombes to break the German military codes. And he knew that the Colossus (see below) had been used, from December 1943, for cryptography based on Boolean logic (Hodges 1983, ch. 5).

In 1945 he wrote a report outlining a logical design for a stored-program computer to be built at the National Physical Laboratory (NPL), where he worked for a few years immediately after the war (A. M. Turing 1946; Carpenter and Doran 1977). He called it the Automatic Computing Engine (ACE)—in homage to Babbage. The ACE report was "perhaps the first written discussion of software since A. A. Lovelace" (Carpenter and Doran 1986: 13). Among its many original programming suggestions was a stack allowing nested subroutines (10.v.b), pushed and popped by "BURY–UNBURY" instructions (A. M. Turing 1946: 36, 75–9).

It also included some engineering suggestions, such as the use of mercury for the delay lines. (Sometimes, Turing mischievously recommended gin—M. V. Wilkes 1967: 199.) Turing even attempted to cost the labour that would be required to build it. However, his costings have been described as "amateur" and "unrealistic", and blamed on "the traditional and absurd British academic contempt for engineering" (Carpenter and Doran 1986: 11)—see Chapter 2.i.a. In the event, construction was long delayed. The pilot model worked by 1950, the full version—which differed significantly from the original proposal—by 1957 (Copeland 2005).

ACE was intended from the start as a general-purpose machine (Hodges 1983: 318–24). Indeed, Turing described it as a "practical version" of the universal machine he'd defined in the 1930s (see Chapter 4.i.c) (A. M. Turing 1947*a*: 107). He might have proposed it even earlier, had he not had more pressing matters on his mind: an early history of British computers points out that work on machines based on Turing's 1936 paper was "delayed by the war" (Bowden 1953: 135).

Before its construction was completed, Turing and his NPL colleagues had written a number of "sophisticated" mathematical programs for it (Copeland 2004: 367). And Turing suggested in his report that ACE might be used, for instance, for playing chess, or for information retrieval. He even said:

> the machine should be treated [when programming it] as entirely without intelligence. There are indications however that it is possible to make the machine display intelligence at the risk of its making occasional serious mistakes. By following up this aspect the machine could probably be made to play very good chess. (A. M. Turing 1946: 41)

This suggestion wasn't taken seriously by everyone in the emerging computing community. According to Turing's biographer, the idea of a universal machine—more strictly, of a general-purpose machine with finite storage capacity—"was stoutly resisted well into the 1950s" (Hodges 1997: 28). Probably, Turing's remarks about machine intelligence didn't help, for they—deliberately—raised philosophical hackles. Though only a few people saw the ACE report itself, such provocative remarks soon multiplied in Turing's writing (see Chapters 4.ii.b and 16.ii.a).

Nevertheless, some people—including a group in a Cambridge apple orchard (Preface, ii)—were intrigued, even persuaded. Turing offered some specific guidance, for instance in an early discussion of chess programming (not an actual program) (A. M. Bates *et al.* 1953: 288–95). Within a few years, papers on chess and other board games were being published by the first generation of AI programmers, some of whose letters and publications acknowledged their debt to Turing (see Chapter 10.i.a and b, and Copeland 2004).

d. Turing's invisibility

Turing's practical work with computers, as opposed to his theoretical founding of the field, is less widely known outside his own country than it deserves. Even in the UK, his contribution wasn't known until relatively recently.

It didn't begin with ACE, but dated back to the 1930s. He'd built a gear-wheel calculator in the late 1930s, and an electronic speech scrambler in 1944. He'd designed and used Bletchley Park's electromechanical Bombes, which broke the Nazis' Morse-based

Enigma codes in August 1940. (For an overview of the Enigma project, see Copeland 2004: 217–64; for Turing's description of the machines employed see Turing 1940/2004; an account of the code-breaking written by a Bletchley colleague, and kept top-secret until 1996, is Mahon 1945/2004.) And although he wasn't directly involved with the Colossus machine (which broke the non-Morse Fish codes), some of his methods for automatic cryptanalysis were used in it (Hodges 1983: 230, 266).

Colossus was the world's first large-scale programmable, special-purpose, electronic digital computer. It was outlined by Turing's ex-teacher Newman (who'd joined the Bletchley code-breakers in 1943), and designed and built by the Post Office engineer Thomas Flowers, who'd been recommended to Newman by Turing. Its name was bestowed by the Wrens (WRNS: Women's Royal Naval Service) who operated it: it was the size of a small room, and weighed about a ton.

After building several simple "Heath Robinson" machines (named after the English designer of joke machines, like the USA's "Rube Goldberg machines"), Flowers came up with the Colossus in 1943. It took Flowers and his team at the Post Office Research Station ten months to construct this, "working day and night, pushing themselves until (as Flowers said) their 'eyes dropped out' " (Copeland 2001: 344). Ten examples were eventually built—whose 1,500 valves were supplemented by mechanical pulleys.

Flowers had the revolutionary idea of storing *data* (the Fish key patterns) internally, using electronic valves to do so (Hodges 1983: 267). This was *so* revolutionary, indeed, that Newman and others at Bletchley simply didn't believe that a machine with about 2,000 valves (Colossus had 1,600) could be reliable (Copeland 2001: 360–1). They ignored Flowers's suggestions and concentrated on the Heath Robinson machines instead. Left to himself at the Post Office, Flowers developed the Colossus anyway—at his own expense (his bank account was in the red by the end of the war).

But Flowers hadn't needed the demands of war to originate the idea. Long before the war, he'd explored the possibility of using valves for controlling telephone connections. By 1934 the Post Office had approved his design for an automatic controller (built of 3,000–4,000 valves) for 1,000 telephone lines, and they'd even started using it by 1939. Moreover, in 1938–9 he was already experimenting with a high-speed electronic data store for use in telephone exchanges (Copeland 2001: 352). So the technology he'd intended as an aid to people's everyday chats ended up being used for desperately urgent code breaking.

However, the Colossus *program* wasn't stored internally. To fit Colossus for a new task, the operators had to rewire it using plugs and switches. Internally stored programs would be independently suggested by Turing and von Neumann in 1945.

By the time of the D-Day landings in June 1944, Colossus had been devoted to code breaking for six months. Until as late as 1975, however, its very existence was unknown in the computer community, except by the half-dozen British computer pioneers who'd worked at Bletchley Park. (For writings on this period by Turing and others, see Copeland 2004: 217–352.) Every detail of their wartime activities was covered for thirty years by the government's Official Secrets Act.

Eight of the ten Colossus machines were destroyed at the end of the war, on Winston Churchill's orders. It was revealed over fifty years later that the other two had been secretly installed at the intelligence centre GCHQ, where one was still operating in the early 1960s (Enever 1994: 38; cf. Sengupta 2000).

Thirty years after the war ended, Turing's contribution to the war effort could be acknowledged at last. It's now known that Churchill himself regarded it as so important that he ordered that all Bletchley's requests for equipment and personnel should be met immediately. (That was in response to a letter from Turing and his colleagues, complaining not only that their work was being held up by a lack of typists but that their requests for help had been repeatedly ignored by Whitehall: A. M. Turing *et al.* 1941.) And Turing's chief statistical assistant at Bletchley, I. Jack Good—later, a leader in AI chess—has remarked: "I won't say that what Turing did made us win the war, but I daresay we might have lost it without him" (McCorduck 1979: 53).

Even so, some of the relevant technical information was still secret at the end of 1999. Academic publications on Colossus were being delayed as a result (B. J. Copeland, personal communication). The full details of the Bletchley code-breaking exercise weren't finally released until the new century, when GCHQ's 500-page report on *The History of Newmanry* was made available. That was largely due to Donald Michie, a youngster at Bletchley (and co-author of the report) and later an AI pioneer in the UK (6.iv.e and 11.iv.a). He'd badgered the authorities for years to declassify it. However, a replica of Colossus was constructed in the late 1990s, and is on show at Bletchley—now, a museum.

Quite apart from the thirty-year secrecy blacking out Bletchley and the Colossus, Turing's visionary ACE report written in 1945 remained largely unread for forty years, especially outside the UK. There were only fifty or 100 mimeograph copies of the original report (issued in 1946), and even the designers of the EDSAC never saw it (Carpenter and Doran 1986: 16). A limited edition was printed for Babbage's centennial in 1972, in connection with an NPL Open Day. But it was widely published only in 1986, almost half a century after Turing wrote it.

Moreover, knowing full well (from his Bletchley experience) that working computers could indeed be built, Turing soon lost interest in the practical details of how to improve them. He turned instead to the theoretical problem of self-organization in biology (see Chapter 15.iv).

In short, much of the early British work on computing was either deliberately suppressed or circulated only narrowly. Small wonder, then, that the contribution of Turing and his British colleagues to the invention of the modern computer—as opposed to the notion of formal computation—isn't always recognized, especially outside the UK. As we saw in Chapter 1.iii.f, a US President recently claimed the computer for America—and he's not alone in this illusion.

e. Von Neumann's contribution

By contrast, von Neumann's 'Draft Report on the EDVAC' (von Neumann 1945) was circulated fairly widely, though not officially published, as soon as it was written in June 1945. (Turing was one of the first people outside the USA to see a copy—Hodges 1983: 307.)

Although MADM would be the first functioning example of a stored-program general-purpose electronic computer (EDVAC's debut was in 1951), von Neumann's document was the first 'published' account of such a machine. It was part-inspired by the ideas of McCulloch and Pitts, who in 1943 had described neural nets in logical-computational terms (see Chapter 4.iv.e). And it was largely inspired, too, by Turing's

1936 paper (4.i). Indeed, von Neumann "repeatedly emphasized that the fundamental conception was Turing's" (Copeland and Proudfoot 2005: 114; see also Copeland 2005: 21–7). Even so, "Many books on the history of computing in the U.S. make no mention of Turing," probably because—despite internal evidence that its ideas had been used—there was no *explicit* reference to 'On Computable Numbers' (A. M. Turing 1936) in the EDVAC Report (p. 10).

Von Neumann's paper immediately aroused a good deal of interest. For example, Turing cited it in his discussion of the ACE, written a few months later (A. M. Turing 1946: 21). His own stored-program design, however, *wasn't* inspired by von Neumann—and was very different (Carpenter and Doran 1986: 6). By contrast, Cambridge's EDSAC—although completed earlier than von Neumann's machine—*was* largely inspired by the plans for the EDVAC. Wilkes had discussed these on a visit to Pennsylvania's Moore School of Electrical Engineering in 1946, and he acknowledged his intellectual debt by the similarity in names.

Von Neumann became even more influential after developing the IAS machine at the Institute of Advanced Study, Princeton (Aspray 1990: 52–72, 92–4). This was officially inaugurated in 1952, but was running large programs for Los Alamos in mid-1951. (It would be heavily used for the H-bomb project, in which he was a participant: MacKenzie 1991*a*.) Von Neumann's IAS machine is commonly taken as the prototype of most modern computers. It was indeed the most influential of the early examples—but whether it resembled modern computers more significantly than MADM or ACE did is debatable (B. J. Copeland, personal communication).

How much von Neumann owed to Turing for his ideas on computer design is debatable also. Certainly, he was deeply impressed by Turing's theoretical work of the mid-1930s, and was doing pencil-and-paper experiments with it by 1938 (Aspray 1990: 178). Moreover, he acknowledged that the fundamental conception of modern computing was Turing's (Randell 1972: 10). But it's doubtful whether he was influenced also by Turing's practical experience and designs (Pratt 1987: 169; Randell 1972).

Von Neumann probably first met Turing in 1935, and offered him a job in 1938—which was declined. Some people believe, though this is disputed (B. J. Copeland, personal communication), that he and Turing met again in Princeton during the war (Aspray 1990: 100, 177–8). They had long, and not fully recorded, discussions on practical matters in 1947 (Randell 1972)—but both NPL and IAS were well advanced in significantly different designs by that time.

As for indirect influences, von Neumann became committed to computing on a visit to England in 1943, and wrote his first program (for a tabular calculating machine) at the Nautical Almanac Office there (Aspray 1990: 27, 231). He immediately wrote to friends that he'd "developed an obscene interest in computational techniques", and later told his English host that he'd "received in that period a decisive impulse which determined my interest in computing machines" (quoted in Aspray 1990: 27, 28).

(This "interest" was soon developed, in the spring of 1944, at Los Alamos. Interestingly, his frustrating experiences with their plugboard parallel machine "led him to reject parallel computations in electronic computers and in his design of the single-address instruction code where parallel handling of operands was guaranteed not to occur"—MacKenzie 1991*a*: 113.)

However, there's no evidence that von Neumann met Turing in person on his 1943 visit (and he certainly wouldn't have known of the Bletchley project), although he may have talked to British scientists influenced by him. Nor is there any evidence that he was influenced by Turing's ACE design. Possibly, the two pioneering designs were conceived independently. Turing's biographer sees the British and American initiatives as having only a "tenuous" connection, alongside a "very marked independence" (Hodges 1983: 304). In short, just who owed what ideas to whom may remain forever opaque (but see Copeland 2001).

This particular priority question is intriguing (though not crucial) here, because the protagonists are so important for our general discussion. Both Turing and von Neumann raised basic theoretical issues in cognitive science, and both feature prominently in several later chapters.

By contrast, the historical priorities in computer design *as such* aren't relevant for our purposes. Perhaps this is just as well—for a discussion of them would not only involve many names besides the few mentioned above, but would be highly contentious to boot.

For instance, the design of the EDVAC used ideas also from J. Presper Eckert and John Mauchly, whose wires-and-plugboard electronic calculator ENIAC was functional by November 1945. Indeed, von Neumann was a latecomer to the Moore School's EDVAC team, which already included Eckert and Mauchly (Arthur Burks was also a member of the team: see 15.v.b). A bitter dispute over priorities, and patents, ensued. Even after a court ruling in 1972, which attributed the core idea to someone else entirely (John Atanasoff 1940), priority battles continued for many years (Aspray 1990: 34–8). Other priority arguments may have escaped the law courts, but are hotly disputed nonetheless (Bowden 1953; Randell 1972, 1973; Hodges 1983, chs. 5 and 6; Aspray 1990).

Such complexity, and contention, isn't surprising. As remarked in Chapter 1.iii.f, several people often contribute to a discovery or invention, especially one involving both theoretical and technical developments. And the more complex the origins of the idea, the more difficult it may be to identify what counts as "the" idea under discussion.

Just what, for example, counts as a *computer*? Our attribution of priorities must be influenced by how we answer that question. (Hence the need to distinguish different sorts of machines: analogue or digital, calculators or symbol manipulators, (electro)mechanical or electronic, dedicated or general-purpose, rigidly "wired" or program-controlled, hand-instructed or stored-program.) Moreover, a highly valued and/or commercially successful idea will often be claimed after the fact by many people, some more disingenuous than others.

In the case of computer design, matters are even more complicated than usual. Historical details have been, and many will forever remain, hidden by wartime—and post-war—secrecy in the USA and the UK. Even the development of supercomputers, designed well after the Second World War for the nuclear weapons industry, is still part-veiled in secrecy (MacKenzie 1991*a*).

3.vi. In Grandfather's Footsteps?

Babbage's work has some historical interest for cognitive science, even though he himself wasn't a budding cognitive scientist. It doesn't follow that he actually had any historical *influence* on the field (see Chapter 1.iii.a). If he did, it was in virtue of the part he played in the development of the modern computer. However, that "if" wasn't an empty rhetorical device: there's genuine disagreement over Babbage's role in this matter.

To be sure, his work had no influence on the development of analogue computers. In such machines, quantities are represented not by discrete (digital) states but by continuously varying physical features, such as mechanical rotation or electrical voltage.

Simple analogue calculating devices had been used since the early eighteenth century (Pratt 1987: 139). If any nineteenth-century writer prompted 'modern' developments in analogue computing, it wasn't Babbage but the physicist William Thomson, later Lord Kelvin (1824–1907).

In the mid-1870s Thomson's elder brother James built an ingenious device wherein *any desired fraction* of the motion of a revolving disc could be communicated to a cylinder mounted above it—and he used it to do simple integration (J. Thomson 1876). William realized that (combinations of) machines based on his brother's idea might be applied to more complex problems, such as analysing harmonics or solving differential equations (W. Thomson 1876; Bowles 1996). So he built another analogue machine, intended "to substitute brass for brain in the great mechanical labour of calculating the elementary constituents of the whole tidal rise and fall". However, although his tidal harmonic analyser worked in practice, his design for a differential integrator didn't.

When the first useful differential analysers were built in the USA in the late 1920s some authors, such as MIT's Vannevar Bush, cited the Thomsons' research—but they didn't cite Babbage (Bush 1931). That's not surprising. If Babbage was important in the development of computers, it's digital machines which are at issue. But, as we'll now see, that "if" is genuinely iffy.

a. Conflicting evidence

In 1851 Albert Smee had announced his imaginary machine to be an "absolute impossibility... for practical purposes" (2.ix.a). But twenty years before that, Babbage had declared his Analytical Engine to be feasible: it might fill his workshop, but it wouldn't need—to borrow Smee's words—"an area exceeding probably all London".

By the mid-1950s, when functioning modern computers had appeared on the scene, it was clear that Babbage's vision had been vindicated. Late twentieth-century machines incorporate many logical and procedural principles enunciated in the early nineteenth century by Babbage (see Section iii.b). In short, the Analytical Engine and the modern computer are essentially equivalent.

However, as we saw in the case of Vaucanson (Chapter 2.iv), theoretical equivalence isn't the same thing as actual historical influence. Babbage's biographer describes him as "Pioneer of the Computer" (Hyman 1982). In the sense that he was the first to design a symbol-manipulating machine whose fundamental principles are, in essence, those of a general-purpose computer, this appellation is just. Indeed, there's probably no one who'd deny that he had "vision verging on genius" (M. V. Wilkes 1991: 141). But the implication that where he led, others followed (as with the pioneers who opened the American West), is disputed.

Broadly, there are three views on this matter. Some say that Babbage was unknown to or disregarded by computer scientists. Others declare that he was an inspiration to them. Yet others believe that his failure to complete his Analytical Engine, which so embittered his own life, acted as a serious disincentive to comparable work a century later. If that's true, then Babbage's over-confidence delayed computer engineering much as the over-optimism of early machine translation, symbolic AI, and connectionism hindered their later development (see Chapters 9.x.e, 11.iv, and 12.iii).

These very different views on Babbage's influence are illustrated in turn, below. But adjudicating between them isn't easy. I'll be quoting various reminiscences, but one must be wary of taking these at face value. Someone may want—consciously or not—to claim an influence because of the eminence of the historical figure concerned, or to deny it in order to imply their own originality. This caveat will need to be borne in mind in later chapters, too (for example, in Chapter 9.ii–iii).

The first view is that "Babbage was neither influenced by what had gone before nor influential upon what followed him" (Collier 1970, p. v). Collier's not the only historian to have come to that conclusion.

A press notice prepared (by Doron Swade) for the Babbage bicentenary exhibition at the Science Museum in 1991 (see Section ii.b), while acknowledging the closeness of Babbage's ideas to today's computers, said that "he had no effect on the development of large-scale calculating machines". Similarly, Allan Bromley, an expert on Babbage's papers (and a computer collector whose finds are now in Sydney's Powerhouse Museum), judged that "Babbage had effectively no influence on the design of the modern digital computer," and that "Babbage's papers cannot have influenced the design of modern computers in more than the most superficial manner" (Bromley 1991: 9; 1982). That wasn't said from any lack of respect for Babbage: it was Bromley who suggested that London's Science Museum build the Difference Engine for the bicentenary.

These dismissive historical judgements are supported by specific disclaimers from some of the modern computer pioneers themselves. For instance, Wilkes, as a Cambridge mathematician, was well aware of his eminent predecessor. But at the Memorial Meeting in 1971, on the bicentenary of Babbage's death, he remarked: "In writing of Babbage as a computer pioneer one must at once admit that his work, however brilliant and original, was without influence on the modern development of computing" (M. V. Wilkes 1971: 1).

Or again, John Brainerd—the Director of the Moore School, and project manager for the ENIAC—declared in 1965 that "The development of the ENIAC was in total ignorance of Babbage's work... Babbage's direct influence was *nil*" (quoted in an unpublished Ph.D. thesis by Alex Arbel: D. Swade, personal communication). And Wilkes has said that "It was not until after the project was completed that [the ENIAC team] heard about the work of Charles Babbage," adding (what is not the same thing: see 1.iii.h) that "indeed it is obvious that the ENIAC shows no influence from that source" (M. V. Wilkes 1982: 55).

Babbage was certainly mentioned in lectures at the Moore School in 1946 (Metropolis and Worlton 1980). By that time, however, ENIAC was complete. And he may well have been discussed by Flowers, who knew of him from Bletchley Park, when Flowers visited the Moore School in 1945 (Copeland 2004). Again, however, ENIAC was by then more than a sparkle in the Moore School's eye: it was fully functional by November of that year.

If we count Zuse as a (even *the*) computer pioneer, we have an entirely clear case of a 'modern' computer being built *in ignorance* of Babbage. For Zuse learnt about his Victorian predecessor only when he applied for a US patent in 1939:

> When I began to build the computer [in 1933], I neither understood anything about computing machines nor had I ever heard of Babbage. It was only many years later, when my constructions and switches were basically set, that an examiner from the American Patent Office showed me Babbage's machines. The otherwise extremely thorough German examiners had not been acquainted with Babbage. (Zuse 1993: 34)

The remark about the German examiners is telling: even they, despite their usual thoroughness, knew nothing of Babbage. (Or if they knew of him, they didn't see the connection.)

Others imply, to the contrary, that Babbage's work was a positive influence in the construction of modern computers. A few say this explicitly. For instance, a friend of some members of the ACE team—who often spoke of Babbage during the mid-1940s—recalled in 1965 that:

> In 1913, I turned back to a boyhood interest in Babbage's 1834 dream of an Analytical Engine, a self-operating, self-recording calculating machine—and during the 1914–18 war I was still thinking in terms of gear wheels . . . In 1934 [I had developed a] plan of [an] electronic computer working in binary, but with octonary (digits 0 to 7) input and output completed to make the human operator's task easier. Babbage's 1834 sleeping beauty had awakened—after the proverbial hundred years. (E. W. Phillips, quoted in Randell 1972: 14)

Most express it less directly. The editor of an early volume on digital computers, which included the first reprint of Lovelace's paper of 100 years before, declared: "This book is devoted to an account of the construction and use of the machines which [Babbage's] vision inspired" (Bowden 1953: 7). Twenty years later, another computer historian wrote: "Thus the saga of Babbage's Analytical Engine came to an end, although its fame lingered on *and inspired several other people to attempt what Babbage had failed to achieve*" (Randell 1973: 12; italics added).

Babbage's Analytical Engine was praised by Turing in 1950 as the first universal digital computer, but without any explicit suggestion that it had inspired his own work (A. M. Turing 1950: 16, 26). However, Turing's biographer points out that he certainly knew of Babbage's "universal" Engine by the mid-1940s, and discussed it on several occasions with a friend (Hodges 1983: 297). Others at Bletchley later described Babbage's ideas *and* his proposed machine as "on occasion a topic of lively mealtime discussion" there (Copeland 2001: 348). And Babbage's biographer endorses the implication: "It seems likely that [Babbage's work] was one of the sources for Alan Turing's [theoretical] Turing Machine" (Hyman 1982: 255).

Babbage's biographer also remarks that: "Amongst the mathematical elite of Cambridge Babbage's work was never forgotten and remained almost synonymous with the idea of mechanizing computation" (Hyman 1989: 327). That elite included not only Turing himself, but also Newman and Wilkes—yet Wilkes insisted, as we've seen, that Babbage's research was "without influence" on the development of modern computers. (Aiken was led to Babbage's work in the late 1930s by seeing the model Difference Engine donated by Babbage's son to Harvard—Swade 1991: 36; however, as noted in Section v.b, his Mark I was a calculator, not a general-purpose computer.)

These remarks suggest the possibility, though hardly the probability, of Babbage's influence. There's no unambiguous acknowledgement here, as had earlier been given by the Spanish engineer Leonardo Torres y Quevedo (1852–1936).

Torres y Quevedo constructed a wide range of (analogue and digital) calculating machines and automata, including the first chess automaton in 1911, which played endgames using only three chessmen and only five possible moves (*Scientific American* 1915). He credited Babbage at length as the inspiration for his own design for an electromechanical "analytical engine", and for his conception of "automatics" in general (Torres y Quevedo 1914: 91, 101–2). However, his work didn't directly influence the design of the electronic computer. Rather, it was rediscovered at mid-century largely as a consequence of that advance.

Yet others (thirdly) see Babbage's legacy as a hindrance, not a help. Evidently, the "calumny and misrepresentation" which he—and his oyster-scientist—endured cast its shadow for 100 years. Wilkes (1971: 1) again:

> More important was the fact that Babbage's projected image became one of failure, with the result that others were discouraged from thinking along similar lines and eventual development of computers was delayed.

This casts a new light on the remark about Cambridge's "mathematical elite" (quoted above), whose intended implication was that Babbage's work played a positive role in the Cambridge achievements. Wilkes suggests, to the contrary, that the widely shared knowledge of Babbage's example was a largely negative influence.

Again, the very man who hailed the Harvard Mark I as the realization of Babbage's dream also said: "This dark age in computing machinery, which lasted one hundred years, was due to the colossal failure of Charles Babbage" (L. J. Comrie, quoted in I. B. Cohen 1988: 180). He himself, an expert on the production of mathematical tables, was specifically discouraged (in the 1930s) by Babbage's example from trying to design a machine for calculating them automatically (Swade 1996: 40–1).

Such commentators argue that Babbage's failure cast its shadow not only over the intellectual optimism of (some) individual scientists, but also over the readiness of funding agencies to support their more ambitious projects.

The earliest recorded example of this effect occurred in the early 1840s. A calculating machine "in certain respects vastly more promising than Babbage's", designed by a Devonshire printer called Thomas Fowler, was refused governmental support, or even consideration. The reason—so Fowler's son wrote—was that they'd already "spent such large sums, with no satisfactory result, on Babbage's [Engine]" (quoted by Swade 1996: 40). And the father's fate was eerily similar to Babbage's. In his son's words, again:

> It is sad to think of the weary days and nights, of the labour of hand and brain, bestowed on this arduous work, the result of which, from adverse circumstances, was loss of money, loss of health, and final disappointment. (quoted in Swade 2000: 311)

Given the drying-up of money, following the early high hopes (of Babbage and of Fowler), one's reminded of two twentieth-century phenomena. Namely, the setbacks in research funding, *and* in intellectual respect from the wider scientific community, that have beset AI as a result of its failure to live up to various promises—some of which had been carelessly made (see Chapters 11.iii–iv and v.b, and 12.iii and vii).

b. So what's the verdict?

The three positions just outlined aren't so starkly inconsistent as they may appear. For example, some social groups—elite or otherwise—may have known about Babbage's example while others didn't. Moreover, familiarity with Babbage's work could have encouraged someone to believe that machines doing highly general computations are in principle possible, while *also* discouraging them from trying to build computers using his approach—or even from building computers at all.

Nevertheless, conflicts in evidence remain. The truth will probably never be clear. Given this situation, the judgement offered by the director of the Science Museum's Difference Engine project seems to me to be fair:

> There is no unbroken line of development between Babbage's work in the nineteenth century and the modern computer. His Analytical Engine was a developmental cul-de-sac. His efforts represented an isolated episode, a startling and magnificent one, but an episode nonetheless. There is a great gap: the movement that led to the modern computer did not resume until the 1940s when pioneers of the electronic age of computing rediscovered many of the principles explored by Babbage, largely in ignorance of his designs.
>
> However, there is more owing to Babbage than a respectful and perhaps awed salute across a barren gulf of time. His exploits and his aims were an integral part of the folklore shared by the small communities of scientists, mathematicians and engineers who throughout remained involved with tabulation and computation. Babbage's failures were failures of practical accomplishment, not of principle, and the legend of his extraordinary engine was the vehicle not only for the vision but for the unquestioned trust that a universal automatic machine was possible. The electronic age of computing was informed by the spirit and tradition of Babbage's work rather than by any deep knowledge of his designs which have attracted detailed attention only in the last few decades. (Swade 1991, pp. ix–x)

Let's give Babbage himself the last word. Nearing the end of his long life, he said—proudly, if sadly:

> If, unwarned by my example, any man shall undertake and shall succeed in really constructing an engine embodying in itself the whole of the executive department of mathematical analysis upon different principles or by simpler means, I have no fear of leaving my reputation in his charge, for he alone will be fully able to appreciate the nature of my efforts and the value of their results. (Babbage 1864: 338)

4

MAYBE MINDS ARE MACHINES TOO

Maybe minds are machines, too! As late as 1930, this shock-horror thought hadn't even attained the status of a heresy. It wasn't a heresy, because no one believed it. Indeed, no one had even suggested it.

The brain, to be sure, had long been compared to machines—in the Roaring Twenties, even to jukeboxes—and nervous conduction had been crudely mimicked by physical models before the First World War (see Chapter 2.viii.f). But *mental* processes lay untouched. Many psychologists avoided all mention of mind, referring to behaviour and/or brain instead (5.i.a). Others granted that mind *as such* is a fit subject for science (2.x). But no one was defending the strong interpretation of 'mind as machine': that *the same type of scientific theory could explain processes in both minds and mindlike artefacts.*

To put the point another way, by 1930 no one had yet argued that mind and/or mental processes, *conceptualized as somehow distinct from matter*, could be understood in *machine-based* terms. Two hundred years earlier, of course, Julien de La Mettrie had provocatively spoken of 'Man, a Machine' (2.x.a). But he'd conflated mind and matter. And by "machine", he meant mechanistic physics in general, not specific types of artefact.

This situation changed in the years around 1940, with the emergence of two new ways of conceptualizing the mind—based on two novel types of machine. The newly minted mind-as-machine hypothesis drew on ideas from logic and/or physiology whose beginnings lay in the late 1800s (see 2.vii–ix). But the logic was now associated with computer science, and the physiology with cybernetics—or "circular systems" (the term *cybernetics* wasn't used until 1947).

By mid-century, and largely because of the invention of the modern digital computer (3.v), the stage was set for a flowering of research on this general theme. That efflorescence had a lasting effect on psychology, anthropology (up to a point), linguistics, AI/A-Life, neuroscience, and philosophy: see Chapters 5.iv.b–f and 6–16.

Some early devotees—notably, Warren McCulloch—drew on formal computation and cybernetics more or less equally. And most of them were sympathetic to both. But the theoretical loyalties, and the sociological groupings, gradually diverged. In general, however, and with the arguable exception of dynamical theorists (see 14.ix.b and 15.viii.c), mind and mental processes were distinguished from their material base, being glossed in abstract (non-physical) terms.

In short: by 1960 things had changed. *Mind as machine* was now a major heresy. ("Heresy", because most people still thought it incredible: see 16.i–ii.) Indeed, McCulloch himself used this very word: "Our adventure is actually a great heresy. We are about to conceive of the knower as a computing machine" (1948: 144). This chapter narrates how that heresy was born, and how it gave rise to two different—sometimes, passionately competing—sects within the field as a whole.

We'll begin by seeing that Alan Turing was the first proponent of this new way of placing mind in nature. Having defined computation formally for the first time (see Section i), he soon argued that minds and computing machines employ *the same general type* of operation—even if they may also employ others (see Section ii). Section iii describes how Turing's ideas were taken up by McCulloch, who suggested that *the same specific computations* are involved. "The whole of psychology", he said, boils down to the definition of particular logical networks. And Section iv shows how McCulloch's work, in turn, influenced John von Neumann (1903–57) to use binary logic in designing his electronic computer.

The cybernetics movement saw the mind as controlled by feedback processes like those in the bodies of living things—and in the machines then being developed by control engineers. The core theoretical idea is explained in Section v, as is the closely related concept of "information". Section vi outlines a cybernetic theory based on the notion of cerebral models, or representations. A wide range of mid-century self-regulating machines is described in Section vii, and two especially influential—and recently resuscitated—examples in Section viii.

Finally, Section ix notes the professional schism that eventually separated the two ways of thinking about mind as machine.

4.i. The Turing Machine

First and foremost, Turing was a mathematician. His ideas about mind (discussed in Section ii) were grounded in his work on the nature of computation *as such*. The famous 'Turing machine' was an abstract mathematical concept, not a clanking or sparking device. And he described other abstract "machines", with no suggestion that they could ever be implemented.

Nevertheless, some of the work discussed in this section influenced the development of modern computing (which *isn't* to say that this wouldn't have developed without it: see Chapters 3.v.a and 16.ix.a). More to the point, it was used by others to help ground a persuasive vision of the possibility of computational psychology and AI (see Section iii below, and Chapter 16.iii.b).

a. Turing the man

As a person, the relatively solitary Turing was very different from the gregarious Charles Babbage (Hodges 1983; Babbage 1864; Hyman 1982). The "lady examiner" who asked Babbage an ignorant question and was treated with gentleness by him (see 3.i.a) would have received a far less gentle response from the prickly Turing.

He didn't bear fools gladly—and, in his eyes, most people were just that. On one famous occasion in 1943, when he was visiting Bell Labs in wartime New York, a sudden lull in conversation in the executive lunchroom enabled all to hear his high-pitched voice saying:

> No, I'm not interested in developing a *powerful* brain [in a computer]. All I'm after is just a *mediocre* brain, something like the President of the American Telephone and Telegraph Company. (Hodges 1983: 151)

That's not to say that he was impatient with people whose intellect he did respect. After his death, the zoologist John Z. Young wrote this to his mother:

> My impression of your son is of his kindly teddy-bear quality as he tried to make understandable to others, ideas that were still only forming in his own mind. To me, as a non-mathematician, his exposition was often difficult to follow... (S. S. Turing 1959: 105–6)

Young's memories here were probably correct, but his reference to Turing's "kindly teddy-bear quality" may have been a diplomatic exaggeration aimed at a grieving mother. For she'd lost her brilliant son when he was only 41. Turing—again, like Babbage—ended his life in unhappiness. Indeed, he endured tragic desperation (Hodges 1983, ch. 8).

He was a homosexual at a time when homosexual activity was illegal in England. (He'd proposed marriage to Joan Clarke, another Bletchley cryptologist, in 1941, and had given her a ring; but he broke off the engagement a few months later—Hodges 1983: 206 ff., 216–17.) In January 1952, on the very day when his BBC debate about mind and machine was repeated on the radio (see 16.ii.a), he made the extraordinarily unworldly mistake of informing the police of a trivial burglary he already suspected might have been committed by a young man he'd invited into his flat (Hodges 1983: 449–55). He was arrested, charged, and tried. There was a whispered word to the judge, to the effect that he'd done great service to his country and should be spared prison accordingly. So he was offered probation—provided that he accepted treatment with female hormones. After a while, the effects on his physique became highly embarrassing.

As though that weren't dispiriting enough, a major spy scandal involving homosexuals (and the Director of MI5, Kim Philby, discovered to be a double agent whose prime loyalty was to the Soviets) prompted the Civil Service in 1952 to embark on positive vetting of everyone who already had security clearance, not just those on the waiting list to get it. Because of his work in the Second World War (see below, and 3.v.c–d), Turing had the highest security clearance possible. Very likely, he would have to suffer the humiliation of having it withdrawn—possibly, with a farewell fanfare in the nastier newspapers. (The decidedly nasty *News of the World* had already featured his court case.)

On 7 June 1954 he committed suicide by taking potassium cyanide.

That, at least, was the verdict of the coroner's court. His mother disagreed. This was perhaps only to be expected, given that suicide was then a crime—and especially as she knew nothing of the background mentioned above. (None of those matters, nor even the already public trial, were mentioned at the inquest—Hodges 1983: 488.) Moreover, she'd often warned him about his carelessness in playing around with poisons (ibid.).

In her memoir of her son, Sara Turing reported not only that neighbours had said he was very cheerful a few hours before he died, but also that he'd been looking ahead:

On his writing table ready for the post were acceptances of invitations for the near future, as well as tickets for the theatre, to which he was to take friends that very week. . . . [And his close friend Dr Gandy told me]: "When I stayed with him the week-end before Whitsun, he seemed, if anything, happier than usual; we planned to write a joint paper and to meet in Cambridge in July." (S. S. Turing 1959: 118)

Her explanation was that his lifelong habit of doing experiments at home had led to "some unaccountable misadventure". Robin Gandy, on his last visit, had found him surrounded by many such experiments, most dealing with electrolysis. One of these, he told Sara,

almost certainly did produce potassium cyanide, and may have been intended to produce it: but this was not the aim of the experiments as a whole, which was to produce a wide range of chemicals from the simplest and most easily obtainable ingredients [i.e. in "desert island" conditions]. (p. 116)

But, said his mother, "No poison was found in his bedroom. There was just a partly-eaten apple by his bed, for, as a rule, he used to eat an apple at night" (p. 117).

The apple, strangely, was never analysed. If, "as seemed perfectly obvious" (Hodges 1983: 488), it had been dipped in cyanide (two jars of which were found in the house), this was never conclusively proven. Even if it had been, the reasons behind the tragedy would have remained a mystery. Turing's biographer said: "Like Snow White, he ate a poisoned apple, dipped in the witches' brew" (p. 489). "But what", he asked, "were the ingredients of the brew? What would a less artificial inquest [have] made of his last years?" The answer, discussed over his next thirty-eight pages, remains unclear to this day. Whatever the truth of it, Turing was dead at half Babbage's age: Babbage had reached 80.

Let's pass on to happier comparisons.

As mathematical intellects Turing and Babbage were remarkably similar. Both were superb cryptologists, for instance. In the late 1930s, Turing built a very small machine (using electromagnetic relays) to use binary multiplication for relatively secure encoding; and during the war he developed a general theory of cryptanalysis.

More to the point, for our purposes, Turing was the first person "fully able to appreciate the nature of [Babbage's] efforts and the value of their results" (see the closing quotation of Chapter 3).

He did so a good twelve years before modern computers were built (3.v). For he'd proved in the mid-1930s that a machine defined in purely abstract terms could—like the Analytical Engine (which was itself only a *design*: 3.ii.b and iii)—compute anything that's computable (A. M. Turing 1936). This proof is so important for the history of computing that a collector recently paid $19,000 for a copy of the journal in which it first appeared (J. M. Norman 2005). For this abstract machine, soon dubbed a "Turing machine" by Alonzo Church (1937), was essentially identical to a general-purpose digital computer—whether built from "wheelwork" or electronic circuits.

b. Playing the game

Babbage, like everyone else before Turing, had assumed that we know intuitively *what it is* for something to be mathematically deducible (computable). But Turing offered an explicit definition. Indeed, the new "machine" was described specifically in order to make this definition clear.

Before discussing his seminal definition of computation in any detail, let's consider an informal version given by Joseph Weizenbaum (1976: 51 ff.).

Weizenbaum asks us to imagine a game played with a roll of toilet paper, many white stones, five black stones, and an ordinary die. To set up the game, one does this:

1. Roll out the toilet paper on the floor.
2. Put down stones as follows:
   (i) on an arbitrary square, one black stone;
   (ii) on successive squares to the right of the square holding the black stone, as many white stones as you please, one to each square;
   (iii) on successive squares continuing to the right, one black stone, skip one square, one black stone;
   (iv) on successive squares continuing to the right, an arbitrary number of white stones;
   (v) on the square to the right of the last white stone, one black stone;
   (vi) finally, one black stone, the "marker", [placed on the floor] above the square holding the rightmost white stone.
3. Turn the die so that its one-dot side is facing upward, i.e., so that it is showing "1."

Having set up the game, one starts to play it. The rules of the game are called "transformation rules". In general, they work like this:

The marker stone is moved either one square to the left or one to the right on each move. However, before each move, the stone under the marker stone is replaced or removed according to the applicable rule. The die may be turned to a new side after each move.

Each of the eighteen rules mentions a starting position defined in terms of an orientation of the die and a certain number of stones (which may be zero) on the square under the marker. To play the game, the player looks at the die and the toilet roll to see what the existing position actually is, and finds the appropriate rule accordingly. Then, he does what that rule tells him to do:

Each rule says to do three things:

1. Turn the die so that it reads the stated number.
2. Replace the stone under the marker by the kind of stone specified—possibly by no stone at all.
3. Move the marker one square in the indicated direction.

Doing this, of course, creates a new game situation. Accordingly, the rule which is *now* appropriate is found, and obeyed. When a rule tells the player to turn the die to read "0", the game stops.

As for just what Weizenbaum's eighteen rules are, they're defined by the table shown in Table 4.1. Boring, they may be. This game won't rival soccer, or even tiddlywinks. But unclear, they're not. At every move of the game, the player knows precisely what must be done. There's no ambiguity, and no choice—except in setting up the game in the first place. What happens depends entirely on (*a*) the rule being followed, (*b*) the position of the marker, and (*c*) the stones, or lack thereof, at present under the marker.

For anyone who doesn't already know what a Turing machine is—in other words: how, at base, a digital computer program works—this trivial game should help

TABLE 4.1. The rules of the toilet roll game

IF THE DIE READS	AND THE STONE UNDER THE MARKER IS	THEN TURN THE DIE TO	REPLACE THE STONE BY	MOVE MARKER
1	none	3	white	left
1	black	2	no stone	left
1	white	1	white	left
2	none	2	no stone	left
2	black	3	no stone	left
2	white	5	no stone	right
3	none	3	no stone	left
3	black	4	no stone	right
3	white	5	no stone	right
4	none	4	no stone	right
4	black	1	black	right
4	white	6	white	left
5	none	5	no stone	right
5	black	1	black	right
5	white	1	white	left
6	none	0	no stone	right
6	black	0	black	right
6	white	3	white	left

Source: Weizenbaum (1976: 53)

ease them into some very non-trivial mathematical ideas. For Turing's 1936 paper defined *computation* in terms of transformations (moves, rules) of the same general type. His definition was both mathematically rigorous and relatively easy to grasp. In addition, it was evidently possible to implement computation (so defined) in a *physical* machine—though we'll see in Section ii.a that Turing's informal example featured pencil and paper, not toilet roll and stones.

c. What computation is

Turing's central aim in his 1936 paper was to clarify the intuitive notion of computability. Initially, he thought about this concept purely as a mathematical exercise. He'd been prompted to do so, early in 1935, by a remark of Max Newman's, in his lectures on the foundations of mathematics (Hodges 1983: 90 ff.).

Newman had been discussing one of three fundamental questions about mathematics that he'd heard David Hilbert (1862–1943) raise a few years before, at a conference in 1928. The meeting had been puzzling over the type paradoxes described by Bertrand Russell and Alfred Whitehead in their *Principia Mathematica* (1910). These paradoxes threatened to scupper their goal of basing mathematics on logic—a goal previously recommended by Russell's logical muse, Gottlob Frege (2.ix.b). Hilbert had asked whether one could state in general terms just what could, and what could not, be proved by their theory. In other words: is there some way of deciding, for *any* mathematical statement, whether that statement is provable? (Hilbert himself believed the answer must be *Yes*. "In mathematics", he said, "there is no *ignorabimus*"—that is, there's no "We shall never know".)

This question (known as the *Entscheidungsproblem*) didn't seek the proof of a particular mathematical assertion. Rather, it sought a reliable and precisely definable method for deciding whether or not, for any such assertion, some proof is in principle available. As Newman put it in his lecture: is there a "mechanical procedure" for proving any assertion to be decidable?

Turing (1936) took Newman's expression seriously. But (like Newman himself) he interpreted "mechanical" to mean *precisely definable*, *deterministic*, or even *mindless*—not *physical*. That is, he asked not how such a procedure could be implemented, but how one could define—in abstract terms—just what sort of procedure it would have to be.

Gödel-bells will now be ringing in many readers' minds. For Hilbert's three queries about mathematics had soon been addressed by Kurt Gödel (1906–78), who'd proved a highly counter-intuitive result: that a mathematical statement may be incapable of proof, even though it can be seen to be true (Gödel 1931). (Later, this would be cited in various arguments trying to prove the impossibility of "strong AI", or formal–computational intelligence: see 16.v.a.) This result contradicted Hilbert's hunch that every mathematical statement must be decidable. But it didn't resolve the *Entscheidungsproblem*, for it didn't show whether there's any method of telling just which statements are/aren't decidable—still less, what that method might be.

One might expect that Turing's first thoughts about computability would have been informed by Gödel's theorem. But they weren't. It's not clear that he even knew about it when he wrote his paper in 1935. Certainly, Newman's final lecture had mentioned it—but perhaps Turing hadn't been there? There's good evidence that he was introduced to Gödel's work not by Newman, but by the philosopher Richard Braithwaite (Preface, ii).

Turing's biographer says merely that "He had a few words with Richard Braithwaite at the High Table one day on the subject of Gödel's theorem" (Hodges 1983: 108–9). But Braithwaite once remarked to me on "Turing's complete ignorance of Gödel's work when he wrote his 'computable numbers' paper", adding: "I consider I played some part in drawing T's attention to the relation of his work to Gödel's" (letter from R.B.B. to M.A.B., 21 Oct. 1982). However that may be, Turing's paper made no reference to Gödel. He took the *Entscheidungsproblem* neat, without Gödelian ice.

In his paper, Turing conjectured that a statement is decidable if and only if there exists a definite method for computing it. And "computation" and "definite method" (i.e. algorithm) were, for the first time, precisely defined.

In addition, he proved that some statements *aren't* decidable in this sense, and that some numbers *aren't* computable—even though they can be mathematically defined. (Hilbert's hunch assailed a second time.) In particular, whether or not *any* given statement is decidable (computable) is not itself decidable: the goal motivating the *Entscheidungsproblem* is unattainable. It follows—though he didn't say so, for computers weren't yet on the horizon—that there's no *general* way of knowing, for just any computer program, that it will eventually halt. (*Ignorabimus*, after all.)

Computability (decidability) was defined by Turing in terms of an abstract machine—a fully deterministic system—now called a Turing machine. More accurately: the Turing machine, like the Analytical Engine, was an infinite class of machines. Similarly, there are an infinite number of toilet roll games. That's no accident, of course, for Weizenbaum's game was closely modelled on the ideas in Turing's paper.

In general, a Turing machine consists of: (1) a tape divided into squares, on each of which a symbol can be written, (2) a "scanning-head" that can consider one, and only one, square at a time, and (3) a set of rules (a "table of behaviour") specifying what the machine will do next in this or that circumstance. The machine can read the symbol on the scanned square, and—if the rules so determine—erase it or replace it with another. The tape can be moved one square backwards or forwards, or the machine can stop.

"Moving backwards and forwards" refers to predecessors and successors in some ordered sequence, and is often interpreted in terms of discrete moments in time. For any moment t, the state ("configuration") of the machine at the following moment $(t + 1)$ is fully determined by two things: the state of the machine at time t and the symbol on the square being scanned at time t. The system has a finite set of rules, and a finite set of allowable symbols. However, the tape itself is infinite, so some computable numbers (such as *pi)* are infinite in length.

Turing showed how such abstract machines could be described in a standard logical form, and how they could be used to do elementary computations out of which all standard arithmetical operations could be constructed. He proved—or rather, he claimed—that anything that is computable can be computed by some Turing machine. This is circular, if "computable" is defined as Turing-computable. Turing's claim, rather, was that anything that is computable in the mathematicians' previous (intuitive) sense of the term is also computable in his own, more rigorous, sense.

These were "claims" rather than proofs, since the intuitive sense of computation isn't definable. Indeed, this is why people speak of the Church–Turing "thesis".

The logician Church reached an essentially similar conclusion about the *Entscheidungsproblem* in the same year as Turing, though he defined computability in terms of a novel mathematical formalism, the lambda calculus, not in terms of a potentially physical process. His claim was that intuitive computability boils down to what mathematicians call recursive functions (which would later be implemented by LISP: Chapter 10.v.c). Turing soon proved that these pick out the same set of mathematical operations as Turing-computability does. But because it can't be strictly proved that they map onto the intuitive concept of computability, the claim that they do is known as the Church–Turing *thesis*, not the Church–Turing *proof*. Nevertheless, because this thesis is widely believed to be true, people—including computer scientists—often say that Turing "proved" that anything which is computable is Turing-computable.

Even more significantly, Turing proved (*sic*) that one abstract machine—the "universal" Turing machine—is capable of imitating any other, if the standard description of the second machine (in modern parlance, its program) is written onto the tape. More accurately, there's an infinite class of universal machines, each defined by a different set of rules enabling it to imitate other Turing machines. Any universal machine is capable of computing anything that is computable.

d. Only programs, not computers

You may be muttering that the statement I've just made is—strictly—false, because it ignores the issue of run-time input. A Turing machine that accepts data from the environment while the computation proceeds can compute functions which a universal Turing machine cannot.

True. But this talk of "accepting data from the environment" inevitably suggests a context of *physical* computing machines—for vision, perhaps, or natural language processing. And that's not surprising. For Turing, in his 1936 paper, was (in effect) talking about a computer program: a set of rules defined by software, but implementable in hardware.

Babbage had already had this idea, as we've seen. Moreover, Ada Lovelace had specifically pointed out that numbers could symbolize either data or operations (see 3.iii.b). And her observation that, in the Analytical Engine, "numbers meaning quantities" and "numbers meaning operations" were not—though they could have been—stored on different cards and executed by different columns applied *pari passu* to Turing's device. In a universal Turing machine, data and operations are stored on the same tape, read by the same scanning head, and acted on by the same set of rules.

Indeed, one and the same sequence of tape symbols can be interpreted sometimes as data and sometimes as instructions—a fact which was later crucial for von Neumann's account of self-reproduction (see 15.v). To take a homely analogy, suppose that your sister writes the words "Scratch your left foot" on a piece of paper. Then she asks you to fetch a paintbrush and a tin of red paint, and to copy those words above her front door. In that case, you're treating the written words as *data* (and someone who can read only the Cyrillic alphabet could do the job almost as well as you can). Now, suppose that she asks you, instead, to read the words on the paper and to do what they tell you to do (namely, scratch your left foot). In that case, you're treating the words as *instructions*—specifically, as instructions to carry out the *operation* of foot-scratching. (And the solely Cyrillic reader would be at a loss.)

Unlike Babbage, however, Turing (in his *Entscheidungsproblem* paper) didn't ask how, if at all, the primitive operations could be implemented. Even had he done so, he couldn't have said how to implement an entire Turing machine, because these are defined as having an infinite tape—which no actual machine can have.

He might have asked, instead, how one could build an *approximation* to a Turing machine, or even to a universal machine. The latter would be a general-purpose device, in the sense that it could do many different things, depending on which non-universal Turing machine (which program) was described on its tape at any given time. But, in the 1936 paper, he didn't ask this question either.

According to Newman, he'd been privately speculating about it right from the start (Copeland and Proudfoot 2005: 109 n. 5). But it's not clear that he then thought a quasi-universal computer to be a practical possibility. (I say "quasi-universal", because *in practice* no computer, and no programming language, is actually universal: see Chapter 10.v.) Indeed, he remarked dismissively to a friend that such a machine would have to be as big as the Albert Hall (Copeland 1998*a*). (Even this would be an advance on Alfred Smee, whose notional machine would have exceeded "probably all London": 2.ix.a.)

However, that was before he learnt about electronics. From the mid-1940s, the developments in computer technology outlined in Chapter 3.v were to build on Turing's first paper. He himself, for example, designed the ACE as a "practical version" of his abstract machine of 1936; and MADM—the first stored-program general-purpose electronic digital computer—was based on his theoretical ideas. It was now clear that a special-purpose digital computer is equivalent to some specific Turing machine, and a general-purpose computer approximates a universal Turing machine. In short, if one avoided cogs and wheels (and toilet paper?), programs could be usefully implemented after all.

Turing's initial lack of concern for practical implementation was evident also in his doctoral thesis of 1938 (supervised by Church). There, he defined another abstract machine, called an O-machine: O stands for 'oracle' (Turing 1939; Copeland 1998*a*,*b*). An O-machine is a Turing machine with one or more extra primitives. Each of these is an operation that returns the values of some mathematical function which is *not* Turing-computable. In other words, it "computes the uncomputable"—where the first of these terms is understood intuitively, and the second in the Church–Turing sense (see Copeland and Proudfoot 1999, 2000).

As for how these extra primitives do it, Turing said that—like the Delphic oracle—they work by "some unspecified means", and that we need "not go any further into [their] nature" (Turing 1939: 166–7). He was concerned only with what sorts of mathematical powers such systems would possess, if they existed. (Half a century later, people would start asking in earnest whether, and how, computers going beyond Turing-computation could be built: see 16.ix.a and f.)

4.ii. From Maths Towards Mind

If someone asks, "Did Turing believe the mind is a machine?", the answer is a definite "Yes". If they ask, "Did he believe the mind is *a Turing machine*?", the answer is "Yes, but . . .". However, if they ask whether he already believed this in the mid-1930s, his 1936 paper gives only a hint, not a definitive reply. It wasn't until the 1940s that he made his commitment clear—and even then, it was to theme rather than detail.

a. Computers and computors

The hint in the *Entscheidungsproblem* paper lay in how Turing led up to his abstract definition of computability. He did so by describing, in the simplest possible terms, how a human "computer" does, or could do, arithmetic.

In the quotations below (taken from pp. 135–40 of the 1936 paper), I've explicitly marked the fact that, for Turing—and everyone else—at that time, a "computer" was a human being. For instance, the "computers" in London's Scientific Computing Service, set up in 1936 by Leslie Comrie (1893–1950), were the people who operated the mechanical calculators—not the machines themselves (J. M. Norman 2004: 97).

Initially, the new devices (i.e. small calculators) were sometimes called "computing machines". And that language was carried over to (what we would call) the first real computers: in a meeting held in 1948, von Neumann himself referred repeatedly to computing machines, never to computers (Jeffress 1951: 37 ff.). But "computers" eventually won out—not without some misgivings. Even as late as 1955, most newspapers put the word *computer* in scare quotes, if they used it at all (Ceruzzi 1991).

Indeed, the early electronic computers were sometimes called "computors" (*sic*), to preserve the man–machine distinction. That applied, for instance, to the title of "the founding document in the electronic computer industry" (J. M. Norman 2004: 224), the 1946 business plan co-authored by J. Presper Eckert and John Mauchly: "Outline of Plans for Development of Electronic Computors [*sic*]". The o-spelling has now fallen out of use, but it was routinely favoured by computer scientists in the early days (R. L. Grimsdale, personal communication). And by philosophers, too: at mid-century, Wolfe Mays—or perhaps the editor and/or copy-editor of *Mind*—was writing about "modern digital computors" (*sic*) (Mays and Henry 1953: 484; cf. Mays *et al.* 1951: 262).

Having described how the human arithmetician computes, using pencil and paper in the familiar way, Turing then proposed that we imagine the person being replaced by a machine:

> Computing is normally done by writing certain symbols on paper. We may suppose this paper is divided into squares, like a child's arithmetic book... [Turing next pointed out that it needn't be two-dimensional: it could be a one-dimensional paper-tape.]
>
> The behaviour of the computer [by which Turing meant the person] at any moment is determined by the symbols which he is observing, and his "state of mind" at that moment...
>
> Let us imagine the operations performed by the computer [i.e. the person] to be split up into "simple operations" which are so elementary that it is not easy to imagine them further divided. Every such operation consists of some change in the physical [*sic*] system consisting of the computer and his tape....
>
> The operation actually performed is determined... by the state of mind of the computer [i.e. the person] and the observed symbols. In particular, they determine the state of mind of the computer after the operation is carried out.
>
> We may now construct a machine to do the work of this [human] computer....
>
> We suppose that the computation is carried out on a tape; but we avoid introducing the "state of mind" by considering a more physical and definite counterpart of it. It is always possible for the computer [person] to break off from his work, to go away and forget all about it, and later to come back and go on with it. If he does this he must leave a note of instructions (written in some standard form) explaining how the work is to be continued. This note is the counterpart of the "state of mind". We will suppose that the computer [person] works in such a desultory manner that he never does more than one step at a sitting. The note of instructions must enable him to carry out one step and write the next note. Thus the state of progress of the [person's] computation at any stage is completely determined by the note of instructions and the symbols on the tape.

These quotations are undeniably suggestive. But they don't establish beyond doubt that Turing already accepted 'mind as machine' as his credo. For Babbage, too, had designed computing machines by analogy with human beings (Gaspard de Prony's clerks) doing simple arithmetic. And Turing's "simple operations which are so elementary that it is not easy to imagine them further divided" were comparable to the detailed calculations embodied by Babbage as movements of toothed wheels. Yet Babbage, as explained in Chapter 3.iv, *didn't* see thought in mechanistic terms. Other evidence is needed, then, to show that Turing did.

There's no such claim in Turing's 1936 paper. In Babbage's terminology, the paper dealt with calculation and logic, not reason. Indeed, soon afterwards Turing implied that some thoughts, such as seeing the truth of an assertion Gödel had shown to be unprovable, are *not* computable. (Hence my answer: "Yes, but . . ." above.)

He attributed these thoughts to people's unconscious "intuition", which sometimes results in "correct" judgements—"leaving aside the question what is meant by 'correct'" (Turing 1939: 209–10). Some intuitive judgements later turn out to be provable; but some do not. He evidently believed that intuition, although not a computational process, is sometimes unavoidable:

> In consequence of the impossibility of finding a formal logic which wholly eliminates the necessity of using intuition, we naturally turn to "non-constructive" systems of logic with which not all the steps in a proof are mechanical, some being intuitive . . . [A non-constructive logic should] show quite clearly when a step makes use of intuition, and when it is purely formal. (Turing 1939: 210)

Similarly, his discussion of O-machines left it open whether the mind may be, in part, an O-machine—so able to think thoughts that aren't Turing-computable.

In sum, he certainly thought that much, perhaps even most, of what the mind does could be done by a Turing machine. But he allowed for the possibility that some mental powers may lie beyond such machines.

An anachronistic aside: This historical fact undermines a type of argument that appeared about twenty years later, and that is still frequently directed against cognitive science in general. It goes like this:

* Turing proved that *all* machines are restricted to Turing computation;
* so *any* mind-as-machine approach must be similarly restricted;
* if even one example of non-Turing-computable thought can be found (take your pick . . .),
* then cognitive science must be at best inadequate, at worst fundamentally flawed.

Despite its popularity, this argument misses its target—for the first premiss is false. It's false not because Turing tried but failed to prove that all machines are, in effect, Turing machines, but because he didn't even claim to have proved this (see Copeland 1997).

Moreover, by the end of the century various other definitions of "computation" were—and still are—being explored (16.ix). So those people are mistaken who claim that cognitive science, given its commitment to computational theorizing, *must* see the mind as a Turing machine. (I plead guilty: some years ago, I said that computational psychologists think of the mind "in terms of the computational properties of universal Turing machines"—Boden 1988: 5.) It's more accurate to say, as I did in Chapter 1.ii.a, that cognitive science aims to describe the mind by *concepts drawn from the best theory of what computers do—whatever that theory turns out to be.* (These somewhat cryptic remarks will be clarified in Chapter 16.ix.f.)

b. Commitment to the claim

Soon, however, Turing appears to have decided that the psychological relevance, if any, of O-machines was so limited that it could be ignored. (Presumably, that's why so many people believe that he claimed to have proven that minds are Turing machines.)

His remark (quoted in Chapter 3.v.c) that the ACE might display intelligence, though flawed by "serious mistakes", implies that he'd done so by 1945. And in 1947 he wrote a report in which he referred to man as a machine, and outlined the prospects for "making thinking machinery" (A. M. Turing 1947*b*: 12). This was the first AI manifesto—or rather it would have been, if it had seen the light of day. In fact, it wasn't officially published until 1969.

This paper made no mention of intuition, or uncomputable thoughts. On the contrary, it explicitly rejected the idea that Gödel's proof showed that intelligent computers are impossible. This would be so, said Turing, only if intelligence precluded the possibility of ever making a mistake (A. M. Turing 1947*b*: 4; cf. 1947*a*: 123–4).

In that same paper (A. M. Turing 1947*b*), he defined a different type of machine—what we would now call a connectionist system, or neural network—that could be induced not only to perform, but to learn (see Chapter 12.i.b). Indeed, he showed how this could happen in an initially random, or unorganized, network. And he pointed out that such networks could be simulated by a digital computer. From the historical point of view, this is interesting rather than significant, for the paper remained largely unknown for many years. Written as an internal report for the National Physical Laboratory, it wasn't communicated to the outside world until 1969.

Turing's discussion of neural networks was an aspect of his core argument for the possibility of intelligent computers: that "it is possible to make machinery to imitate any small part of a man". For instance, he said, TV cameras imitate the eye, and unorganized networks imitate (parts of) the brain. In other words, the body is the causal ground of thinking, so body-mimicking artefacts could think. He'd been musing on this topic for some years, having been excited by machine analogies for the body ever since childhood (see 2.viii.f).

Donald Michie, who founded the UK's first university department of "machine intelligence" (6.iv.e), worked in Newman's group at Bletchley in the early 1940s. He recalls that "many of the ideas [about thinking machines and artificial brains] in [Turing's] 1947 essay . . . were vigorous discussion topics during the war. Some of his younger associates were fired by this, although most regarded it as cranky" (quoted in Randell 1972: 6).

Cranky or not, in his assessment of the most promising areas for building intelligent machines, Turing highlighted most of the topics to be forefronted over the next twenty years by early AI—whether symbolic or connectionist.

For instance, he said that "intellectual activity consists mainly of various types of search" (Turing 1947*b*: 23). And he named the following as promising research areas: "Various games, e.g. chess, noughts and crosses, bridge, poker; the learning of languages; translation of languages; cryptography; mathematics" (p. 13).

He also mentioned vision programs, speech analysers, and robots—including a science-fiction version ("allowed to roam the countryside") of the Cog project (see Chapter 15.vii.a). To build that, one would "take a man as a whole and try to replace all the parts of him by machinery [such as] television cameras, microphones, loudspeakers, wheels and 'handling servo-mechanisms,' as well as some sort of 'electronic brain' ". These projects, he said, would be more difficult to achieve than chess, or even translation. For one would need to engineer sensory and motor organs interfacing with the outside world.

In that paper, and in his lecture on ACE to the London Mathematical Society in the same year (Turing 1947*a*), Turing had been asking scientific and practical questions—however visionary his answers may have been. He was considering just which sorts of computer-engineering methods, applied to just which sorts of thinking, would be most likely to afford progress. He identified exhaustive search as a promising computational technique, and implied that it might suffice for all types of thought: "we should not go far wrong for the time being if we assumed that all problems were reducible to this form. It will be time to think again when something turns up which is obviously not of this form" (Turing 1947*b*: 22).

The most plausible candidates for problems "obviously not of this form" were specifically included three years later. In 1950 he published a paper in *Mind*, in which he argued that there's no philosophical reason for denying the possibility of intelligent machinery (A. M. Turing 1950). If a computer artefact were to show performance indistinguishable from that of a human being, he said, we'd have no good reason for denying that it was *really* thinking, and *really* conscious. That argument, involving the so-called Turing Test (which he'd mentioned briefly already: Turing 1947*b*: 23), will be discussed in Chapter 16.ii. We'll see, there, that in his 1950 paper Turing was being part-mischievous—and that his mischief attracted far more philosophical attention than his serious points.

More relevant here, the *Mind* paper explicitly suggested that even the highest flights of human thought—not only reason, but also creative imagination—could be achieved by a computer. A computer might be able, for example, to compose Shakespeare's sonnet 'Shall I compare thee to a summer's day?', and to defend its choice of imagery in terms of rhyme, scansion, and cultural associations—such as seeing an affinity between Mr Pickwick and Christmas Day (Turing 1950: 53).

Mathematicians might have asked—as Roger Penrose (1989) later did—how we can see the truth of Gödel-unprovable statements. But Turing specifically rebutted Gödelian attacks, on the same grounds as in his 1947 paper (making no mention of "intuition", or of O-machines). In short, no sort of thinking was excluded.

c. But what about the details?

However, to say that Turing believed the mind to be a Turing machine isn't to say just which Turing machine he believed it to be.

Even if one assumes (with Turing in 1947) that all thoughts are computable, the mind may not compute them in exactly the same way that a particular program or connectionist network does. As Lovelace had long ago pointed out, there are always different ways of instructing (or designing) a machine to do 'the same' computation, different ways of getting from the same input to the same output. So the fact that a machine has achieved a particular (input–output) computation leaves open the question of precisely how the mind does it.

The psychologist—as opposed to the mathematician, the computer scientist, or the philosopher of mind—will want to know *what actually went on* in the mind (or brain) when someone did something. It's not clear that Turing cared about such questions—or perhaps he thought them premature. As we've seen, he sometimes argued that a certain type of behaviour (learning or vision, for example) could in principle be achieved by the general type of mechanism found in the brain and sense organs (Turing 1947*b*). But he didn't ask, and couldn't say, *just which* detailed computations the brain is executing when a person learns a language, recognizes a visual image, or picks up a coffee cup.

In other words, Turing wasn't doing computational psychology—nor even psychological AI. His use of computers (the Bombes, MADM, and ACE) tried to get the machines to do something reliably (code cracking, long division . . .), not to model the way in which people do it. His remarks about simulating the brain were programmatic, not neurophysiologically detailed. And his philosophically provocative arguments about machine intelligence, including his discussion of the Turing Test (see Chapter 16.ii), said almost nothing *specific* about just how human intelligence works.

In short, this was *mind as machine* only in a very general sense. It was necessary for the cognitive revolution, but not sufficient. Turing was propounding what John Searle (1980) would later call "strong AI", the thesis that a suitable AI system would *actually be* intelligent. But he wasn't seriously doing "weak AI", which uses AI programs to clarify and develop specific psychological theories.

His work was approaching psychology, but hadn't quite reached it. Computational psychology proper had to await the intellectual stimulus of McCulloch (and the empirical slant of Allen Newell and Herbert Simon: see Chapters 6.iii and 7.iv.b).

4.iii. The Logical Neurone

If Turing wasn't a psychologist, McCulloch was. Or rather, psychology—in particular, psychiatry—was one of the many interests he followed in his long life (1889–1969). Another was logic. Yet another, neurophysiology. His place in the history of cognitive science is due mainly to his bringing these three disciplines together.

When McCulloch said, "Everything we learn of organisms leads us to conclude not merely that they are analogous to machines but that they are machines", the *mind* was explicitly included:

> Brains are a very ill-understood variety of computing machines. Cybernetics has helped to pull down the wall between the great world of physics and the ghetto of the mind. (1955: 163)

His first clear statement (in the early 1940s) of *mind as machine* relied heavily on Turing, as well as on cybernetics. But he'd been moving in this direction long before encountering either.

a. McCulloch the Polymath

Initially destined for the Church, McCulloch read mathematics at a theological college before studying philosophy, and also psychology, at Yale. During the First World War he joined Yale's Officers' Training Program for the US Navy—of which, more anon (in Chapter 14.v.a). But in 1919 he began to work "chiefly" on logic (W. S. McCulloch 1961*a*: 2).

Later, he qualified in medicine, specializing in neurology and physiological psychology—and worked for some years with brain-injured people and psychiatric patients. He did experimental as well as clinical work, and was a gifted technician, who designed and made many of his own experimental instruments. In addition, he was a keen reader, and writer, of poetry.

He hadn't abandoned logic and mathematics for the brain, but tried to combine them—as we'll see. He spent a year doing graduate work in mathematical physics, and in 1939 was a member of Nicholas Rashevsky's mathematical biophysics group at the University of Chicago—whose pioneering 'manifesto' had just appeared (Rashevsky 1938).

After eleven years as a psychiatrist at Illinois Medical School, during which time he wrote his two most important papers (with Walter Pitts), he moved (in 1952) to MIT's Electronics Laboratory "to work on the circuit theory of brains" (W. S. McCulloch 1961*a*: 3). He applied his knowledge of the brain to robotics at the end of his life, using recent discoveries about the reticular formation in designing an autonomous Mars robot capable of switching appropriately from one basic behaviour pattern to another (Chapter 14.iv.a).

This rich intellectual background enabled McCulloch to play a highly significant role as an intellectual catalyst, bringing countless disparate ideas and people together. Many novel ideas were sparked off by McCulloch himself, who was nothing if not mentally adventurous. Turing, on meeting him in 1949, thought him a "charlatan" (Hodges 1983: 411)—but Turing wasn't known for generosity of spirit towards intriguing, though still half-baked, ideas. (Also, he may have been annoyed by McCulloch's claim that his neural nets were fully equivalent to Turing machines: in fact, they weren't—Papert 1965, p. xviii.)

Despite his frequent hand-waving and his often purple prose, McCulloch was what William James (1907) had called a "tough-minded" thinker. As such, he was a prime mover in founding both streams of mind-as-machine research: formal–computational on the one hand, and cybernetic–probabilistic on the other (see Section v.b, below).

His intellectual breadth was matched by his sociability. He initiated or encouraged many meetings of minds by his open and generous hospitality. For example, the neurophysiologist Jerome Lettvin and the mathematician Pitts lodged with him when they were impecunious students in the early 1940s. And the influential AI researchers Marvin Minsky and Seymour Papert met (for only the second time) in the early 1960s at a vibrant party in his house, on Papert's return from Jean Piaget's laboratory in Geneva (see 12.iii.a).

Besides his important role behind the scenes, McCulloch co-authored three papers now acknowledged as classics in the history of cognitive science. It's indicative of his intellectual breadth that these fit best into three different chapters of this book.

One, discussed below, was a powerful abstract statement of the mind-as-machine hypothesis (W. S. McCulloch and Pitts 1943). Another, just four years later (Pitts and McCulloch 1947), was a neurophysiologically informed connectionist model of fault-tolerant processing (see Chapters 12.i.c and 14.iii.b). And the third, published in his sixtieth year, identified the first 'feature detector' cells in the frog's retina (Lettvin *et al.* 1959: see 14.iv.a). (To be accurate, the first two authors of this four-author paper were the ones who were primarily responsible for it: McCulloch's name, and Pitts', was added as a "courtesy": C. G. Gross, personal communication.)

A fourth paper, less well known but influential nonetheless, outlined a Mars robot based on brain-stem anatomy (Kilmer *et al.* 1969). Very recently, interest in this long-dormant piece has been revived (see 14.v.a).

b. Experimental epistemology

McCulloch often described his work as "experimental epistemology" (e.g. W. S. McCulloch 1961*a*: 3). This may sound like a contradiction in terms, but it's not.

He *didn't* mean that physiology could provide a normative theory of knowledge (or rationality), which is what philosophers typically mean by epistemology. Whether any scientific theory could ever achieve this is highly controversial: it had been declared impossible by Gottlob Frege (see Chapter 2.ix.b), and is still widely rejected (16.vi–viii). (Indeed, it's a main reason, perhaps *the* main reason, for scepticism about cognitive science among philosophers.) Hence the whiff of contradiction in McCulloch's typically provocative phrasing.

What he did mean was that he sought to understand "the physiological substrate of knowledge". Specifically, he was concerned with the mechanisms that enable us to have concepts, and to use them in thinking about (and acting within) the world, and possible worlds.

He thought about such issues in broadly Kantian terms. That is, he distinguished innate structuring principles from specific environmental stimuli (see 9.ii.c and 14.iii.b). And besides declaring loyalty to Immanuel Kant, he saw himself as engaged in the same general enterprise as Hermann von Helmholtz (2.vii.c).

Two colleagues of the 1930s, too, influenced him deeply because of their general approach to knowledge. Both considered the wood as well as the trees, trying to relate the detailed data to the overall functions of the mind/brain. One was the psychiatrist Eilhard von Domarus, the other the neurophysiologist Joannes Dusser de Barenne (1885–1940), a Dutchman working at Yale.

Von Domarus had been advised by a philosopher (Filmer Northrop) for his doctoral thesis of 1934, which related the "logic" (the language and grammar) of schizophrenia to its neurophysiology. As for de Barenne, he also tried to relate high-level functions to their neurophysiological base. He collaborated with McCulloch on experiments to identify functional pathways in the monkey's brain (Dusser de Barenne and McCulloch 1938).

McCulloch's guiding epistemological theme was well established by the time he was 17. In old age, he recalled:

> In the fall of 1917, I entered Haverford College with two strings to my bow—facility in Latin and a sure foundation in mathematics. I "honored" in the latter and was seduced by it. That winter [the Quaker philosopher] Rufus Jones called me in. "Warren", said he, "what is thee going to be?" And I said "I don't know." "And what is thee going to do?" And again I said, "I have no idea; but there is one question I would like to answer: What is a number, that a man may know it, and a man, that he may know a number?" He smiled and said, "Friend, thee will be busy as long as thee lives." (W. S. McCulloch 1961*a*: 2)

Before this conversation took place, McCulloch had already been enthused by the mathematical logic of Russell and Whitehead (1910). Ironically, given its enormous later influence on logic and computing, the first publication of *Principia Mathematica* had to be subsidized both by the Royal Society and by the authors themselves (Monk 1996: 194). The *Principia* was in part a development of Frege's work (see Chapter 2.ix.b). It was heavy going in more senses than one: there were 1,929 pages, in three volumes (appearing between 1910 and 1913). And initially, as the need for financial subsidy suggests, not everyone could see the point.

But the young McCulloch would. Indeed, he was entranced by it. Nor was he to be the only one: much later, both Pitts and David Marr were bowled over by it while still at high school (see below, and 14.v.b). Moreover, it had an honoured place on Simon's bookshelf—and inspired his Logic Theorist, one of the very first AI programs (10.i.b). Besides all that, it was the logicians' bible for many years, and the foundation of important movements in philosophy. (Not bad, for a privately published tome!) Without the *Principia*, as we'll see, cognitive science would have been very different.

The book wasn't the first mathematical treatment of symbolic logic. That place was taken by Whitehead's earlier work *A Treatise on Universal Algebra* (in 1898). His novel approach to logic had excited the young Russell greatly. Not least, it liberated him from his commitment to Hegelian idealism:

> [Whitehead] said to me once "You think the world is what it seems like in fair weather at noon-day. I think it is what it seems like in the early morning when one first wakes from deep sleep." I thought his remark horrid, but could not see how to prove that my bias [towards clarity] was better than his. At last he showed me how to apply the technique of mathematical logic to his vague and higgledy-piggledy world, and dress it up in Sunday clothes that the mathematician could view without being shocked. (Russell 1956*a*: 41)

McCulloch, here, would have empathized with Russell.

(The source of Whitehead's "horrid" remark was his commitment to process philosophy, of which he was/is perhaps the outstanding proponent: Sibley and Gunter 1978. This sees matter as consisting essentially in self-organizing processes, not inert stuff subject to external influences. Recently, such ideas have been gaining ground in certain areas of cognitive science—but more orthodox colleagues still think them fairly horrid: see 15.viii.c–d and ix, and 16.x.a.)

In particular, as implied by his remark to his Quaker tutor, McCulloch accepted Russell's Fregean definition of a number as "the class of all of those classes that can be put into one-to-one correspondence to it".—But how could this highly abstract and general definition be understood by a finite human brain?

How can a material brain compute numbers, or do formal logic, in the first place? And how can it enable us to understand any particular number (2, 7, 27, etc.)? After all, for any given number we can't actually comprehend all the classes that could be put in one-to-one correspondence with it. It was said that we know "the rule of procedure by which to determine it on any occasion". But that rule of procedure, one-to-one matching, takes for granted our ability to recognize that different things fall into the same class (cats, cups, carpets). How do we manage to do that? And how can we recognize sets containing up to six things by visual perception, while larger numbers require (symbolic) counting?

As his tutor had predicted, this many-sided puzzle kept McCulloch busy for the rest of his life. And all three of his "classics" were part of his solution.

The first, published in the early 1940s, proved that neural networks are capable of computing any number, or logical proof, that a human can compute (see below). Next, in 1947, he offered a neurophysiological theory explaining both "How We Know Universals" (how we assign different, and slightly differing, things to distinct classes) and how the brain can recognize patterns reliably, despite faulty components and noisy data (see Chapter 12.i.c). As he later put it, this paper explained "[what is] a man, made of fallible neurons, that [he] may know [number] infallibly" (W. S. McCulloch 1961*a*: 18). These two papers together covered both deduction and induction, for universals can be projected as "expected regularities of all future experiments, [so the brain] can frame hypotheses in order ultimately to disprove them" (W. S. McCulloch 1948: 142).

Then, in the 1950s, the four-author paper 'What the Frog's Eye Tells the Frog's Brain' (Lettvin *et al.* 1959) showed that single cells, coding specific sensory features, can be involved in various types of ecologically relevant generalization (Chapter 14.iv.a). McCulloch (1965, p. v) later described that paper, co-authored with a mathematician and two neurophysiologists, as "our first major step in experimental epistemology".

c. Enthused by logic

Much as logic came before psychology for the young Jean Piaget (5.ii.c), so it did for McCulloch. His general approach, in his youth, was a psychophysiological version of logical atomism (Russell 1918–19; Wittgenstein 1922).

This philosophical movement, one of whose leading proponents was the early Ludwig Wittgenstein (1889–1951), applied Russell's mathematical logic to language, mind, and metaphysics. It saw natural language and propositional thought as grounded in logic, or as an approximation to the ideal language, logic. As a general position, this wasn't new: it had been held 200 years before by Gottfried Leibniz, for example (see Chapter 2.ix.a). But more powerful logical tools were now available: hence the excitement.

It was assumed that class concepts could be analysed in terms of necessary and sufficient conditions. And psychological verbs were thought to be definable in truth-functional (extensional) terms. As for everyday words such as *and*, *if–then*, *some*, and *the*, these were reduced to the terms of Russell's propositional and predicate calculus.

There were many prima facie problems here. For one thing, the truth of psychological statements such as *Mary believes that p*, or *John hopes that q* doesn't depend on, and nor does it guarantee, the truth of *p* or *q*. How, then, could such statements be dealt with by this type of logic?

For another, even seemingly boring English words like the four mentioned above weren't so simple as these logicians were suggesting. This had been pointed out soon after the publication of *Principia*, by Russell's colleague George Moore (1873–1958). The logicist view, said Moore (1919), was that the sentence *If Holmes is playing his violin, then Watson is writing about their adventures* means the same as *It is not the case both that Holmes is playing his violin and that Watson is not writing about their adventures.* But this isn't what we normally mean by "if–then": in ordinary English, some sort of meaningful/causal relationship between the two expressions is assumed. Or again, the logicist view that *Some tame tigers do not growl* means the same as *There is at least one tame tiger who does not growl* is problematic (G. E. Moore 1936). For we'd normally expect there to be at least two non-growling tame tigers.

The latter difficulty was easily ignored: *one*, *two*, *three*... didn't seem to matter much. But the difference between the "if–then" of logic ("material implication") and the "if–then" of English was more embarrassing. Even today, nearly a century after *Principia*, the relation between these two isn't clear. Nor is the logic of psychological verbs. (These problems set tough challenges for natural language processing, and for logicist AI in general: see 10.iii.e and 13.i.)

Despite such embarrassments, however, logicism flourished in Anglo-Saxon philosophy. This was largely because it was seen as a liberation from the ("horrid") holistic idealism that had been dominant earlier in the century (Ewing 1934; Boucher 1997). Various British versions of the spiritual/religious philosophies sketched in Chapter 2.vi were hugely influential, notably those of Thomas Green (1836–82), Francis Bradley (1846–1924), and John McTaggart (1866–1925). Even Russell, as a young man, had been seduced by the idealist movement.

A corollary of the new logic was that all aspects of meaning were to be made explicit, not left implicit. For instance, Russell claimed that sentences containing definite descriptions, such as "the present king of France", unambiguously assert (not just presuppose) the existence of one and only one thing fitting the description (Russell 1905). And as in language, so in metaphysics. The fundamental structure of the world was seen in terms of atomic facts modelled on logical primitives and related by truth-functional connectives.

Logical atomism was widely welcomed by philosophers with empiricist leanings, who were eager to escape the then prevailing idealism. Influential versions of it still persist today, notably in the empiricism of Willard Quine (1953*a*, 1960) and in the logicist approach to AI and cognitive science: see Chapters 10.iii.e, 13.i, and 16.iv.c–d. The centenary of Russell's 1905 paper was even marked by a special conference, organized by the universities of Padua and Bologna: 'One Hundred Years of "On Denoting"'. (However, the circle turns: idealism is much more fashionable now than it was at mid-century—see 16.vi–viii.)

After the early years, however, its assumption that the structure of world and language mirror the logical structures of *Principia Mathematica* would be increasingly questioned. At mid-century, Moore's doubts about "if–then" were joined by critiques of other specific aspects, including Russell's theory of descriptions (Strawson 1950). Most of these skirmishes were inspired by the later Wittgenstein's (1953) repudiation of the logicist approach in general—including his own previous work—as fundamentally misguided (Chapter 9.x.d). And around 1970, Wittgenstein's ideas were used by Herbert Dreyfus (1965, 1972) as the basis of an attack on the guiding assumptions of AI—and of McCulloch (see Chapters 11.ii.a–c and 16.vii.a).

In the first third of the century, these Wittgensteinian critiques lay far in the future. Even Moore's worries about "if–then" hadn't been published when the young McCulloch first decided to apply logic to language, as well as to number—and to psychology and neurophysiology, too.

In 1920 he started work on supplementing the predicate calculus with a logic of verbs that would cover both knowledge and action (W. S. McCulloch 1964: 391 ff.). But he found it too difficult. He distinguished transitive from intransitive (including reflexive) verbs, verbs of perception from verbs of motion, and verbs of action from verbs of sentiment (what Gilbert Ryle would later call "dispositional" verbs:

16.i.c). These distinctions, however, and their different temporal implications, were too subtle and complex to be easily formalized. Moreover, they weren't reflected in the subject–predicate form of English sentences.

In disgust, McCulloch gave up. Turning from verbs to propositions, he tried to develop a logical atomism of the mind/brain. In order to do this, he sought to define "a simplest psychic act", or "psychon".

A psychon "was to be to psychology what an atom was to chemistry, or a gene to genetics" (W. S. McCulloch 1964: 392). But, unlike atom and gene, it was an event, with a place in time and a temporal history. As a psychic unit, it had to be intrinsically capable of connecting knowledge and action (perception and movement, or even belief and inference)—much as a gene, as a hereditary unit, is intrinsically capable of connecting generations. It had to have a semiotic, or meaningful, aspect: that is, it had to correspond to a proposition. And, for McCulloch, it had to find some equivalent in the brain.

The then recent neurone theory (2.viii.c) offered a suitable candidate:

> In those days the neuronal hypothesis of Ramón y Cajal and the all-or-none law of axonal impulses were relatively novel, but I was overjoyed to find in them some embodiment of psychons. (W. S. McCulloch 1964: 392)

> In 1929 it dawned on me that these events [combinations of psychons] might be regarded as the all-or-none impulses of neurons, combined by convergence upon the next neuron to yield complexes of propositional events. During the nineteen-thirties [influenced by friends including von Domarus, Northrop, and Dusser de Barenne], I began to try to formulate a proper calculus for these events . . . (W. S. McCulloch 1961*a*: 9)

Chains of psychons were essentially comparable to compound propositions, which could in principle be mapped in logical terms. McCulloch tried to put this principle into practice:

> [In teaching physiological psychology] I used symbols for particular neurons, subscripted for the time of their impulse, and joined by implicative characters to express the dependence of that impulse upon receipt of impulses received a moment, or synaptic delay, sooner. (W. S. McCulloch 1964: 393)

Throughout the 1930s, then, McCulloch was already thinking of "the response of any neuron as factually equivalent to a proposition which proposed its adequate stimulus". And he was already trying to express the behaviour of complicated nets in a notation similar to propositional logic (W. S. McCulloch and Pitts 1943: 21).

But his nascent neural calculus couldn't deal with circular (feedback) networks. These were difficult to describe in formal terms, because excitation can continue reverberating around a circuit for an indefinite period of time. Moreover, McCulloch was troubled by the seeming paradox that, if something is negated as it goes around the loop, then an input must be equal to its negation (Arbib 2000: 200).

Circular networks were a key concern of Rashevsky's group. They'd been postulated, for instance, by McCulloch's psychiatrist colleague Lawrence Kubie (1930), when he was working with Charles Sherrington (see Chapter 2.viii.d). He saw them as the neurological basis of disorders of memory, including the "repetitive core" of psychoneuroses (Kubie 1941, 1953). McCulloch was strongly influenced by Kubie's ideas on circular memories (but would later vilify his psychoanalytic explanation of neurosis, as we'll see).

Others had made similar suggestions. For instance, the notion that *normal* memory, also, might depend on "self re-exciting" circuits had been put forward in 1922 by Rafael Lorente de No (1902–90), who gave anatomical and then experimental evidence for them later (Lorente de No 1933*a*, 1934, 1938). And "reverberative" chains of neurones were being discussed in the 1930s, too (e.g. Ranson and Hinsey 1930). In short, talk of looping circuits was prominent among neurophysiologists.

All these suggestions, however, were beyond the scope of McCulloch's calculus at the time. More generally, his logical notation couldn't represent feedback—the core notion of cybernetics and, under other names, of reflexology and Sherringtonian neurophysiology. Circular causality in general required cyclic (looping) nets.

d. The young collaborator

It was Pitts (1923–69) who provided the crucial insight, and the mathematical power, needed to deal with circular nets. The insight was that the seeming paradox could be avoided if *delay* was represented in the theory. And the power was evident in the use of an arcane logical formalism to express it (see below).

McCulloch and Pitts, and assorted guests (including Lettvin), spent "endless evenings sitting around the McCulloch kitchen table trying to sort out how the brain worked, with the McCullochs' daughter Taffy sketching little pictures which later illustrated [the key paper]" (Arbib 2000: 200). Pictures aside, it's highly doubtful whether McCulloch could ever have achieved the theory unaided.

Pitts was a close friend of Lettvin, who lived alongside him for some time in McCulloch's house. Pitts was largely self-educated, and had been accepted as a member of Rashevsky's group when he was only 14. (He was still only 18 when he and McCulloch published their 'Logical Calculus of Ideas'.) He'd already devoured *Principia Mathematica*, which he found in a library stack when he was 12. He'd even been invited to England by Russell, as a result of sending him some criticisms.

This was typical: Pitts acquired detailed knowledge of new domains extraordinarily quickly, and had an exceptional ability to see unsuspected links between—and logical faults in—other people's ideas. McCulloch used to tell a story (apocryphal? embellished?) of how one of these critiques came about:

> Walter Pitts was forced to drop out of high school by his father, who wanted him to go to work and earn money. Rather than do this, young Pitts ran away from home and ended up in Chicago, penniless. The fifteen-year-old boy spent a lot of time in the park, where he met and began to have conversations with an older man he knew only as Bert. When Bert detected the boy's interests, he suggested that young Pitts read a book that had just been published by a professor at the University of Chicago by the name of Rudolf Carnap. Pitts did, and showed up at Carnap's office. "Sir", he said, "there's something on this page which just isn't clear." Carnap was amused, because when *he* said something wasn't clear, what he meant was that it was nonsense. So he opened up his newly published book to where young Pitts was pointing, and sure enough, it wasn't clear; it was nonsense. *Bert turned out to be Bertrand Russell.* (McCorduck 1979: 73–4; final italics added)

Irrespective of the reliability of this particular story, there's no doubt whatever about Pitts' searching intelligence. Norbert Wiener (1894–1984) declared: "He is without question the strongest young scientist whom I have ever met" (quoted in Heims 1991: 40). Given that Wiener himself had been mixing with fine minds ever since his entry to Tufts University at the age of 11, this was no faint praise.

Admittedly, Pitts' strengths didn't include proof-reading. The first edition of Wiener's *Cybernetics* was riddled with errors, because it had been typeset by a small French press: MIT Press had come in at the last minute, insisting on co-publication because Wiener was on 'their' MIT faculty (M. A. Arbib, personal communication). Pitts was supposed to have checked it, but it even included pages printed out of order.

McCulloch's interest, however, was in his ability to cope with proofs of a different kind. Although he was very different from the highly sociable polymath McCulloch, Pitts' wide interests—and exceptional logical acumen—soon made him a valued collaborator.

He was to become a tragic figure, after a personal betrayal in 1952. Wiener suddenly turned violently against McCulloch, and cut himself off from McCulloch's associates too (14.vii.a). Having regarded Wiener as a father figure, the psychologically fragile Pitts ceased his research. He burnt his manuscript (he'd been working on probabilistic and three-dimensional neural networks), eventually succumbed to delirium tremens, and "became a ghost long before he died" (Lettvin 1999).

In later years, McCulloch often declared that Pitts had taught him more than he'd ever taught Pitts. One of the things that Pitts taught McCulloch was Leibniz's work on the mechanization of logic. Leibniz had seen logic and arithmetic as essentially connected, and mechanizable too (2.ix.a). He'd even shown that a calculator using binary numbers (which couldn't be built at the time) was in principle capable of doing logic. Given his prior commitment to Russell's views on logic, McCulloch was highly receptive.

e. Mind as logic machine

Their joint paper 'A Logical Calculus of the Ideas Immanent in Nervous Activity' (W. S. McCulloch and Pitts 1943) integrated three powerful twentieth-century ideas: the propositional calculus, Turing machines, and neuronal synapses. (According to McCulloch, it was Turing's 1936 paper which had initially inspired them— Hodges 1983: 252 n.) This brief text was an abstract manifesto for computational psychology, and—as soon became apparent—for both symbolic and connectionist AI.

The paper described networks of units modelled on what was then known about real neurones. But the units were highly simplified. A McCulloch–Pitts neurone was an all-or-none device, which fired only when its threshold was reached. The threshold was defined in terms of the number of excitatory impulses coming in from other neurones. Several known neural complexities were deliberately ignored. For instance, the activity of any inhibitory synapse in their abstract model absolutely prevented excitation of the (modelled) neurone at that time, whereas real inhibition isn't so simple.

McCulloch and Pitts used a novel notation to define specific types of net. Their formulae didn't denote timeless logical relations, as Russell's did, but events in time (compare: psychons). They called them TPEs, or temporal propositional expressions, and their notation showed the number of synaptic delays, or relays, involved.

They didn't attempt to represent precise time intervals—partly to avoid excessive complication, and also because neurophysiologists knew so little about such matters. Accordingly, their calculus applied only where "the exact time [needed] for impulses to pass through the whole net is not crucial" (W. S. McCulloch and Pitts 1943: 24). This blindness to timing was carried over into the digital computer by von Neumann, and still characterizes most connectionist AI (see 14.ix.g).

This notation wasn't merely a new manner of speaking, but a new *calculus.* The authors used it to prove various theorems about the computational power of their neural nets. For example:

> The behaviour of any non-cyclic net can be expressed in terms of the propositional calculus.
>
> Every logical function of the propositional calculus is realizable by some net.
>
> All nets are recursively definable out of four basic types: precession (comparable to identity), disjunction, conjunction, and negation (see items a,b,c,d in Figure 4.1).
>
> Every net computes a function that is computable by some Turing machine.
>
> And a universal Turing machine can compute anything that is computable by a McCulloch–Pitts net. (The converse isn't true: see below.)

This work was an early instance of what John McCarthy called "meta-epistemology" (see Chapter 10.i.g). Considered purely as mathematical logic, it was imperfect. (For example, we saw above that McCulloch–Pitts networks *are not* fully equivalent in computing power to Turing machines, and other flaws were identified by Stephen Kleene: 1956.) This potentially embarrassing fact was announced in the opening pages of the hugely influential volume on *Automata Studies*—but it hardly affected the authors' reputation. Having *tried* to use logic to address neurological or psychological questions was more significant than managing to produce utterly flawless proofs. Similarly, mathematicians' proofs may be tidied up by other mathematicians, but recognized nonetheless as important achievements—and as we saw in Section i.c, even a *non-proof,* such as the Church–Turing thesis, may be hugely important.

However, networks—unlike universal Turing machines—are finite. A McCulloch–Pitts neurone isn't a purely mathematical abstraction, but a simplified representation of a real thing. So, for any class of logical function, we need to calculate just what type of net could actually compute it. It's not enough to say that we know that *some* net could do so. Indeed, until we know what sorts of net can generate a psychological phenomenon, we don't really know what that phenomenon is: "If our nets are undefined, our facts are undefined" (W. S. McCulloch and Pitts 1943: 37).

The 'Logical Calculus' paper was hugely ambitious. Logic and language were merely part of it. As they put it: "if any number [any function] can be computed by an organism, it is computable by these definitions, and conversely" (p. 37). In other words, the authors' theoretical reach extended across the whole of psychology. For instance:

* They argued that learning could be embodied (in any net) as a threshold change resulting from the simultaneous excitation of two neurones. This idea was a mid-twentieth-century version of suggestions made in the 1740s by David Hartley (see Chapter 2.x.a).

* They proved that, in cyclic nets, threshold change can be mimicked by loops of fixed-threshold neurones.

* They said that purpose was grounded in cyclic networks that reduce the difference between two afferents. This was their way of expressing the cybernetic idea that goal-seeking behaviour uses feedback to reduce the distance from the goal (Rosenblueth *et al.* 1943).

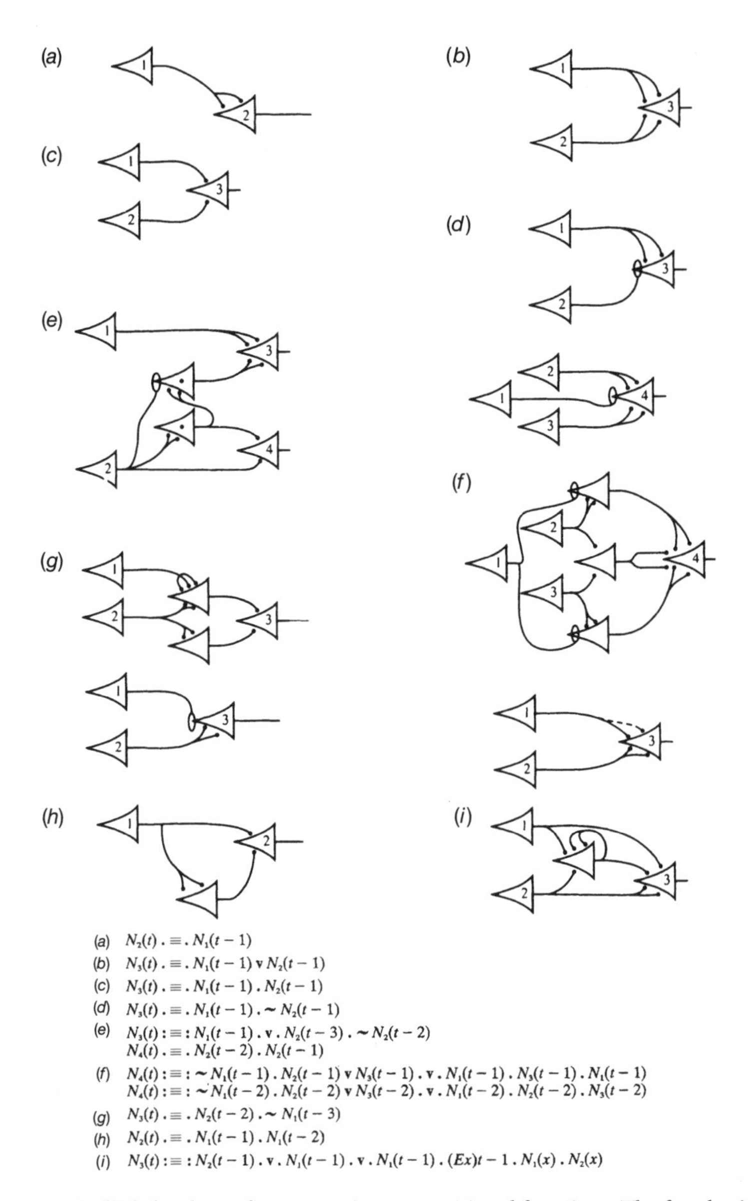

FIG. 4.1. Recursively defined nets for computing propositional functions. The four basic nets are (*a*) precession (identity), (*b*) inclusive disjunction, (*c*) conjunction, and (*d*) conjoined negation. In the diagrams, the threshold is a two-knob input; inhibition is shown as a circle at the tip of the inhibited neurone. In the TPE formulae, the neurone c_i is always marked with the numeral i upon the body of the cell, and the corresponding action is denoted by N with i as subscript. Adapted with permission from McCulloch and Pitts (1943: 36–7)

* And they attributed hallucinations and delusions in psychiatric patients to alterations in someone's normal networks whose causes are other than he/she supposes.

In short, they saw the mind as a Turing machine. And they weren't thinking only of cognition. For *all* psychological processes, "the fundamental relations are those of two-valued logic" (p. 38). Even in psychiatry, they said, "Mind no longer goes 'more ghostly than a ghost'."

This remark was a tacit reference to Sherrington's recent confession, in the 1937–8 Gifford Lectures (see 2.vii.b), that his neurophysiology couldn't explain mind—by which he didn't just mean consciousness:

> Mind, for anything perception can compass, goes therefore in our spatial world more ghostly than a ghost. Invisible, intangible, it is a thing not even of outline; it is not a "thing". It remains without sensual confirmation, and remains without it for ever. All that counts in life. Desire, zest, truth, love, knowledge, "values", and, seeking metaphor to eke out expression, hell's depth and heaven's height. Naked mind . . . It will sit down and watch life acquiescent, or on the other hand take life and squeeze it like an orange. (Sherrington 1940: 266)

All these phenomena, and 'madness' too, were now grist for their Turing-neuronal mill.

f. Initial reception

To say that the 1943 paper is a classic isn't to say that it was widely hailed as important at the time. On the contrary, it initially faced "a hostile or indifferent world" (W. S. McCulloch 1965, p. xvii). The authors themselves weren't surprised: they'd feared that the paper might never be noticed, or even published (Heims 1991: 20).

One reason for this was its fierce logical technicality, unfamiliar to psychologists and physiologists alike. Even the expert mathematician Michael Arbib (1940–) found it "almost impenetrable" (J. A. Anderson and Rosenfeld 1998: 217). Indeed, Arbib abandoned his attempt to check/correct the original proofs, finding it easier to derive new ones instead (Arbib 1961).

The impenetrability was due to the paper's rhetorical form (see Chapter 1.iii.h). This may seem surprising, for anyone who's read McCulloch knows that he was a fine, if flowery, wordsmith. But here, he needed to be a mathsmith too. By most people's standards, he was—but not by Pitts'.

Unfortunately, Pitts persuaded him that they should express their theorems in a rebarbative notation based on that of his ex-teacher the logical positivist Rudolf Carnap (1891–1970) : see Carnap (1934). Technicality was inevitable, of course. But this formalism was unnecessarily difficult, and the fact that it was supplemented by symbols drawn from their beloved *Principia Mathematica* didn't really help. McCulloch later admitted that Rashevsky's group in general was "under the spell" of their Chicago colleague Carnap, and "employed his terminology, although it was not most appropriate to our postulates and hypotheses" (W. S. McCulloch 1964: 393).

The notational difficulty obscured the paper's real-world relevance. Twenty years later, McCulloch complained about highly abstract interpretations of it, protesting that:

> [Our] temporal propositional expressions are events occurring in time and space in a physically real net. The postulated neurons, for all their over-simplifications, are still physical neurons as truly as the chemist's atoms are physical atoms. (W. S. McCulloch 1964: 393)

At the time, however, it was all too easy for people to assume that the new neural calculus was merely a mathematical game.

That's assuming that they even encountered it (Chapter 1.iii.h, again). The paper was published in a relatively obscure journal, Rashevsky's *Bulletin of Mathematical Biophysics*. This didn't achieve a high profile, even though it did raise interest in biophysics, because the Chicago group were not only highly abstract in their interests, but socially isolated too. They rarely got involved with the meetings of the newly founded Mathematical Biophysics Society, and in particular, they didn't interact much with experimenters working on the nervous system—or on metabolism in general (Harold Morowitz, personal communication). It's understandable, then, that most experimental neurophysiologists didn't regard the abstract arguments of the McCulloch–Pitts paper as relevant to their concerns.

Another reason for hostility was the paper's imperialistic claim that "to psychology, however defined, specification of the net would contribute all that could be achieved in that field" (W. S. McCulloch and Pitts 1943: 37). Most experimental psychologists in mid-century America were rampantly behaviourist (see Chapter 6). Since they not only avoided reference to meanings and thought processes, but also treated the nervous system as a black box, they were hardly likely to welcome McCulloch and Pitts' definition of psychology.

As for clinical psychologists and psychiatrists, their therapeutic methods had nothing to do with detailed neural circuitry. 'Talking cures' ignored the brain entirely. And even physical methods, such as lobotomy and electro-convulsive therapy, weren't directed to specific neural networks. On that count, the paper seemed irrelevant to many clinicians—even those who shared McCulloch's general view that psychiatry requires physiological understanding and therapies (W. S. McCulloch 1949).

Moreover, the paper argued that—because of disjunction and circularity—retrospective (as opposed to predictive) description of neural nets is in principle impossible (W. S. McCulloch and Pitts 1943: 35, 38). It followed, the authors said, that historical diagnoses of mental illness are unavailable, and would be therapeutically irrelevant anyway. Clinicians committed to 'child-based' theories wouldn't look kindly on that.

Nor did McCulloch look kindly on most of *them*. Ten years later, he would deliver a vitriolic attack on psychoanalysis (W. S. McCulloch 1953; see also Heims 1991: 115–46). He didn't mince his words. He accused Freudians in general of intellectual futility, dishonesty, and greed.

His cybernetic colleagues were less florid in denigrating Freudian influences in psychiatry. But most of them shared his suspicion, even his contempt. This led to some searing interchanges at interdisciplinary meetings set up to bring physics and psychology together (Dupuy 2000: 85). Even McCulloch's old friend Kubie, for whom Freud had now joined Sherrington as a major influence, received some taxing questions—from McCulloch and others—about how to measure the emotions he was concerned with in his psychoanalytic work (Kubie 1953: 62–72).

What eventually brought the paper to the attention of psychologists capable of appreciating it was its role in the design of the modern computer (which also explains why a copy fetched $6,000 some sixty years later: J. M. Norman 2005). As McCulloch himself admitted, "as far as biology is concerned, it might have remained unknown" had it not been picked up by von Neumann and used to design computers (W. S. McCulloch 1961*a*: 9).

4.iv. The Functionalist Neurone

McCulloch and Pitts' first joint publication made no reference to computer modelling, nor to control engineering either—both to become key influences on cognitive science (see Chapter 1.ii.a). Nevertheless, it was soon seen as the pivotal moment in the foundation of the field. The person who made this possible was von Neumann, wearing his hat as a computer designer.

a. From calculus to computer

The 'Logical Calculus' paper, in effect, had provided an abstract manifesto for computational psychology and AI (both symbolic and connectionist). It was "abstract" in two senses.

First, it didn't mention computer modelling. A few years later, McCulloch would make the declaration quoted at the outset of this chapter: "Our adventure is actually a great heresy. We are about to conceive of the knower as a computing machine." But he followed it by this:

> My problem *differs* from that of the men who build computing machines *only in this*—that I am confronted by the enemy's machine [i.e. the brain]. I have not been told and must learn what it is, what it does, and how it does it. (W. S. McCulloch 1948: 144; italics added)

But his 1943 paper, although it relied on the concept of a Turing machine, made no reference to man-made Turing-equivalent computers.

This is hardly surprising, since no such machines existed at the time—or rather, none was known outside the wartime secrecy of Bletchley Park (3.v.d). McCulloch was familiar with early cybernetic analogies between control in animals and machines. But that approach—rooted in physiology, not logic—didn't focus on language or reasoning as such (although it sometimes referred to systems of belief: see Section v.e, below). The 1943 paper, therefore, *didn't* raise the possibility that a man-made machine might compute in the way it described.

However, the paper's potential for computer design was very soon recognized by von Neumann. He was alerted to it by the cyberneticists Wiener and Julian Bigelow (1906–2003). Both men were members of Rashevsky's biomathematics group, and Bigelow would later be the chief engineer for the Princeton IAS machine (see Chapter 3.v.e).

In particular, von Neumann was impressed by McCulloch and Pitts' use of binary, or Boolean, elements. This contrasted with the early electronic calculator ENIAC, already designed by his colleagues Presper Eckert and John Mauchly (3.v.c and e), which was a decimal device. He was impressed also by the definitions of basic computational mechanisms—conjunction, disjunction, negation—which, he realized, could be embodied electronically as logic gates. (They'd already been embodied electromechanically by Konrad Zuse. But no one outside Germany knew that: see 3.v.a.)

That possibility seems not to have occurred to McCulloch and Pitts. If the logic enthusiast McCulloch was aware of it, he hadn't discussed it in any detail with his young collaborator. This is evident from Wiener's report of a conversation he had with Pitts in 1943:

At that time Mr. Pitts was already thoroughly acquainted with mathematical logic and neurophysiology, but had not had the chance to make very many engineering contacts. In particular, he was not acquainted with Dr. Shannon's work [on information theory], and he had not had much experience of the possibilities of electronics. He was very much interested when I showed him examples of modern vacuum tubes and explained to him that these were ideal means for realizing in the metal the equivalents of his neuronic circuits and systems. From that time, it became clear to us that the ultra-rapid computing machine, depending as it does on consecutive switching devices, must represent an almost ideal model of the problems arising in the nervous system. The all-or-none character of the discharge of the neurons is precisely analogous to the single choice made in determining a digit on the binary scale, which more than one of us had already contemplated as the most satisfactory basis of computing-machine design. (Wiener 1948: 14)

The "more than one of us" included Wiener himself. In 1940, as he recalled in his introduction to *Cybernetics*, he'd outlined what we'd call a digital computer (though with no stored program). Specifically, he'd sent a letter to Vannevar Bush, the inventor of the differential analyser, enclosing an unpublished 'Memorandum on the Mechanical Solution of Partial Differential Equations' (J. M. Norman 2004: 15).

More to the point, the "more than one of us" also included von Neumann. He had immediately seen the relevance of McCulloch and Pitts' work for the design of computing machinery, and his proposal for the EDVAC cited them explicitly:

Every digital computing device contains certain relay-like *elements*, with discrete equilibria. Such an element has two or more distinct states in which it can exist indefinitely... In existing digital computing devices various mechanical or electrical devices have been used as elements: [wheels, telegraph relays]... and finally there exists the plausible and tempting possibility of using vacuum tubes... It is worth mentioning, that the neurons of the higher animals are definitely elements in the above sense... Following Pitts and McCulloch [i.e. McCulloch and Pitts 1943], we ignore the more complicated aspects of neuron functioning... It is easily seen, that these simplified neuron functions can be imitated by telegraph relays or by vacuum tubes. (von Neumann 1945: 359–60)

A few paragraphs later, he pointed out that when vacuum tubes are used by engineers as current valves, or gates, they function as all-or-none devices. And he continued:

Since these tube arrangements are to handle numbers by means of their digits, it is natural to use a system of arithmetic in which the digits are also two-valued. This suggests the use of the binary system. The analogs of human neurons, discussed [above, in reference to McCulloch and Pitts], are equally all-or-none elements. It will appear that they are quite useful for all preliminary, orienting considerations on vacuum tube systems. It is therefore satisfactory that here too, the natural arithmetical system to handle is the binary one. (von Neumann 1945: 5.1)

The architecture of the most widely used modern computer was designed accordingly. McCulloch and Pitts' brain-inspired definitions of logical functions were embodied in electronic circuitry as and-gates, or-gates, and the like.

(Von Neumann wasn't the only one to adopt a version of McCulloch and Pitts' notation in designing the logic gates for his computer. Turing was already doing so too, in his design for the ACE computer: see 3.v.c. In fact, he extended their original notation significantly—Hartree 1949: 97, 102.)

This engineering development highlighted McCulloch and Pitts' insight that their networks (like Babbage's Analytical Engine) could compute propositions as well as numbers. In short, von Neumann's use of their ideas made it even clearer than it was already (in regard to MADM and the other early computers discussed in Chapter 3.v) that computers aren't mere number-crunchers, but general-purpose symbol-manipulating machines. That's why they seemed so promising, to early cognitive scientists interested in simulating (and explaining) thought.

b. Function, not implementation

The 1943 paper was abstract in a second sense, also. Although determinedly materialist ("Mind no longer goes 'as ghostly as a ghost' "), it ignored the physical details.

McCulloch–Pitts neurones represented real things, as we've seen, and were part-inspired by research on the brain. But those "real things" could be of many different types. The defining properties of the basic computational units had been deliberately chosen so as to avoid commitment with respect to current neurophysiological controversies.

For instance, neuroscientists disagreed about whether the mechanism of inhibition in real brains is direct or indirect. McCulloch and Pitts showed that this was irrelevant for their purposes, because they ignored precise time intervals. Similarly, they argued that the functional effects of experimentally observed facilitation, extinction, and learning could be represented without knowing the underlying electrochemical mechanisms (W. S. McCulloch and Pitts 1943: 20–2).

It followed, though they didn't explicitly say so, that the material embodiment of a McCulloch–Pitts neurone is irrelevant. That is, their computational psychons allow "multiple realizability" (Chapter 16.iii.b).

Von Neumann agreed. Indeed, his EDVAC proposal attracted scepticism from some engineers, precisely because it focused on the logical properties of abstractly defined computing units instead of detailing the hardware issues. Von Neumann explicitly refused to get bogged down in discussions of the switching elements *considered as electronic mechanisms*:

> The ideal procedure would be to treat the elements as what they are intended to be: as vacuum tubes. However, this would necessitate a detailed analysis of specific radio engineering questions at this early stage of the discussion, when too many alternatives are still open, to be treated all exhaustively and in detail. Also, the numerous alternative possibilities for arranging arithmetic procedures, logical control, etc., would superpose on the equally numerous possibilities for the choice of types and sizes of vacuum tubes and other circuit elements from the point of view of practical performance etc. All this would produce an involved and opaque situation in which the preliminary orientation which we are now attempting would be hardly possible.
>
> In order to avoid this we will base our considerations on a hypothetical element, which functions essentially like a vacuum tube—e.g. like a triode element with an appropriate associated RLC-circuit—but which can be discussed as an isolated entity, without going into detailed radio frequency electromagnetic considerations. We re-emphasize: This situation is only temporary, only a transient standpoint, to make the present preliminary discussion possible. (von Neumann 1945: 29–30)

The situation was indeed only temporary. Von Neumann's "preliminary discussion" led to the EDVAC and the JOHNNIAC, up and running in 1952 and 1953 respectively, and to all subsequent 'von Neumann' computers. In the 1950s, and largely due to the availability of this computer technology, McCulloch and Pitts' work prompted early research in both connectionist and symbolic AI (see Chapters 10 and 12) and in computational neuroscience too (Chapter 14).

In sum, the abstractness (in this second sense) of McCulloch and Pitts' networks was significant. It licensed von Neumann to design electronic versions of them. It allowed him, when doing so, to ignore "detailed radio frequency electromagnetic considerations". It permitted computer scientists, including AI workers, who followed him to consider software independently of hardware. It enabled psychologists to focus on mental (computational) processes even while largely ignorant of the brain. And it led to *functionalism* in the philosophy of mind and cognitive science (Chapter 16.iii–iv).

4.v. Cybernetic Circularity: From Steam-Engines to Societies

Cybernetics is the study of "circular causal systems". These are self-regulating systems, in which information about the results of the system's actions is fed back so as to cease, adjust, or prolong the original activity. In the eyes of the early cyberneticians, they ranged from steam-engines to human societies.

Many such systems are highly constrained, having a relatively small number of possible states; but some can achieve self-organization from a random starting point. And whereas some have only one end point (only one state of equilibrium), others have many. There may be an unending succession of different 'circles', as when perceptual input leads to motor action, which leads to a new perception prompting a new bodily action . . . and so on (see Chapter 14.vii and ix.a–b). The cyberneticians of the 1940s, however, focused more on 'circles' leading to single end points.

In any given case, the self-regulation is done in a particular way: for instance, by a metal contraption in a steam-engine, or by chemical metabolites in a cell, or by the nervous system in behaviour. And the early cyberneticians (unlike some AI successors: see Chapters 10 and 12) were seriously interested in the underlying physical processes that made such phenomena possible.

Even so, the focus of cybernetics was on the flow of *information*, as opposed to the matter or energy involved. Because information is an abstract notion, it could be applied to many different types of system—even including minds.

a. Feedback, way back

The central principle—informational feedback—wasn't new, even if the *term* was. It was first mathematically developed in 1868 (see below), and had been quickening within biology since the 1860s (see Chapter 2.vii.a). But engineers had used it long before that, to enable machines to reach the same end in varying conditions.

It had featured, for instance, in some ancient automata (see Chapter 2.i), and in the escapement of a fourteenth-century clock, now in London's Science Museum, that alternated between being slightly fast and slightly slow (J. O. Wisdom 1951: 2–3). It had been exploited also in the early seventeenth century by the chemist–engineer Cornelius Drebbel, whose thermostatic incubator was controlled by the expansion and contraction of alcohol, raising and lowering a rod connected by levers to a damper (Bedini 1964: 41). By the mid-eighteenth century, it was common in water mills, and in the grain distributors of windmills. And it had entered the libraries: feedback devices were described in a book written by an Italian military engineer in the 1580s, and in the windmills chapter of an encyclopedia of 1786 (de Latil 1953: 116 ff.).

However, it was first used widely (and visibly) in the late eighteenth century—notably, as the ingenious "governor" designed by James Watt (1736–1819). This was developed to control the speed of a steam locomotive.

Watt's original device was a pair of heavy metal spheres suspended from an upright rotating rod, whose rotation was caused by cogwheels linked with the main shaft of the engine. Centrifugal force made the suspended spheres move away from the rod: the faster the rod turned, the higher they rose. As they did so, their weight dragged on the main shaft, so slowed the engine down. When the slowed-down rotation of the rod caused them to fall again, the drag on the shaft was decreased and the engine speeded up again. (In later versions, the speed of rotation of the rod controlled the steam pressure more directly, by being linked with the throttle.)

For Watt, this was an elegant practical solution to a specific engineering problem. Indeed, that's how it struck most people. The same was true of the other examples: the windmill gizmo was described as *a way of controlling grain feed*, not as an instance of *feedback*. The French cybernetician Pierre de Latil (1953) put it like this:

> Now, this mechanism [the Watt governor] has been in existence since round about 1780; and, throughout more than a century and a half, *no one* has seen that, far from being merely "ingenious", this classic device contains the makings of a revolution.
>
> In all the various attempts at classification of machinery, the Watt governor has never been rated worthy of any very prominent position. [The author of a well-known *Traité des mécanismes* of 1864] drew up what he called: "a systematic index of mechanical devices" in which he lists governors simply as "accessory apparatus". (de Latil 1953: 47–8; italics added)

He went on to complain about his own contemporaries, many of whom were irritated by the fuss attending this "new science":

> Aware of the fact that cybernetics borrows from their technique and vocabulary, certain electronic engineers discount the evidences of the new science, claiming that they know all about it and have been working along these lines for a very long time. That is as may be!—but are they not merely applying the principles without having grasped the inherent theoretical possibilities, in the same way as Watt and his successors did? For it is a fact that, unwittingly, they held within their grasp the key to problems that had arisen in many other fields, even in metaphysics. (de Latil 1953: 48)

De Latil's references to "metaphysics", here and elsewhere (e.g. pp. 195–204), was less a promise to illuminate the Cartesian mind–body problem than a reminder—and a recommendation—of Kant's distinction between "organism" and "mechanism" (see 2.vi.b).

However, de Latil's "no one" wasn't quite right. James Clerk Maxwell (1831–79) had seen Watt's invention—admittedly, some ninety years after Watt's use of it—as an example of an important general principle. In 1868 he read a paper 'On Governors' to the Royal Society, in which he discussed negative feedback in mathematical terms (J. C. Maxwell 1868). And it was in honour of Maxwell that Wiener named the field in 1947, for the word "cybernetics" has the same Greek root as the word "governor" (Wiener 1948: 12). (André Ampère had coined the term some seventy years earlier, using it in a political sense in his *Essay on the Philosophy of the Sciences* of 1884.)

The mid-twentieth century saw countless novel engineering applications, some driven by the need for weapons and defence in the Second World War. Many of these involved computer—but not Turing-computational—technology. These automated systems monitored various physical parameters of the processes they controlled. Most were analogue devices, such as the Bush differential analyser. Computers were here being used as number-crunchers, not as general-purpose symbol manipulators.

b. Infant interdisciplinarity

Our interest isn't in 'pure' engineering, however, or in analogue computers as such. Cybernetics is relevant to cognitive science because it was a self-consciously interdisciplinary project that studied organisms *as well as* artefacts, and took 'man as machine' as covering mind *as well as* body.

This was made explicit by Wiener, who defined cybernetics (when the US version was already about 10 years old) as the science of "control and communication in the animal and the machine" (Wiener 1948). (In fact, in terms of page-counts, his book was more about animals than about machines.) The British pioneers, too, applied the principles of automatic control—described in terms of physical "models" and/or self-organization—to organisms as much as engineering (see vi and viii below and Chapters 12.ii.c, and 15.i.b and vii).

Moreover, cyberneticians on both sides of the Atlantic were interested in the social/psychological aspects of man–machine systems. Besides the pioneering experimental research done in Cambridge, England (Section vi, below), this topic was highlighted in Wiener's non-technical book *The Human Use of Human Beings* (1950).

It's not surprising, then, that the infant cybernetics was nurtured by an intellectually diverse group—of whom McCulloch was one of the most prominent members. Indeed, he became the first President of the American Cybernetics Society. This was due not only to his own unusually wide-ranging research, but also to his gift for encouraging others to cooperate across disciplinary boundaries. Fired up by the seminal Josiah Macy Foundation meeting on hypnosis and reflexology in 1942, it was he who proposed the post-war Macy seminars on cybernetics—which he organized and chaired from 1946 to 1951.

These hugely influential international mini-meetings—only twenty-to-thirty core members—were held in New York from 1944 to 1953, with published proceedings from 1950 on (von Foerster 1950–5). With hindsight, it's interesting that the very first speaker was von Neumann, who regaled the small audience with a description of the digital computer—but expressed grave doubts about its usefulness for modelling the thoughts generated by human *brains* (see Chapter 12.i.d).

This emphasis on the life sciences was one factor distinguishing cybernetics from the closely related approach of Ludwig von Bertalanffy's (1901–72) general systems theory, developed at much the same time (von Bertalanffy 1933/1962, 1950). Whether von Bertalanffy himself regarded cybernetics as an example of systems theory isn't clear: sometimes he said that it was, but many of his remarks about the nature of "systems" implied that it wasn't (see Dupuy 2000: 129–35). However that may be, his general systems theory studied closed systems as well as open ones. Open systems exchange matter and/or energy with their environment, whereas closed systems don't—and the paradigm case of an open system is a living thing.

Another distinguishing factor was the differing treatment of *purpose*. Like cybernetics, general systems theory focused on self-organization or equilibration. One of von Bertalanffy's core ideas was "wholeness", a property of complex systems where the changes in every element depend on all the others. Another was "equifinality", in which a system would reach some equilibrium state from indefinitely many different initial conditions. He remarked that this principle covered the startling examples of embryological development that Hans Driesch had seen as incontrovertible evidence for vitalism (2.vii.b). So far, so familiar: most cyberneticians sympathized with all that. But von Bertalanffy insisted that embryos—and, by implication, other non-human organisms—*don't* show "true finality, or purposiveness". This, he said, involved foresight of the goal by means of the symbolism of language and concepts (von Bertalanffy 1950). Most cyberneticians, by contrast, underplayed the goal-setting role of language and conceptual thought (see Sections vi and viii, below).

Cybernetics pre-dated digital computers (3.v). Its totem artefacts, instead, were analogue computers and devices used in control engineering. But, like the symbolic computationalists inspired by Turing and by McCulloch and Pitts' 'Logical Calculus' paper, the early cyberneticians saw the mind as controlled by the sorts of processes embodied in (some) machines.

At that time, the two types of researcher—cybernetic and computationalist—were mutually sympathetic. They saw themselves as intellectual blood brothers, interacting at the international conferences of cybernetics and at various specially convened meetings.

There were disagreements, to be sure. For instance, Donald MacKay (1922–87) saw "little merit" in comparing the brain to a digital computer, as opposed to a probabilistic analogue mechanism (D. M. MacKay 1951: 105). But he was keenly interested in the digital hypothesis—and had earlier suggested building a hybrid computer: part-analogue, part-digital (D. M. MacKay 1949/1959). As this example illustrates, the various mid-century proponents of mind as machine were generally convivial. Relations didn't become strained until later.

As late as 1958, for example, a meeting on "the mechanization of thought processes" sponsored by Britain's National Physical Laboratory embraced people from both groups, including MIT's Minsky and McCarthy (Blake and Uttley 1959: see 6.iv.b). The same was true of the Macy conferences, whose participants were even more varied: the 1947 meeting, for instance, included I. A. Richards and Suzanne Langer, invited to talk on human communication and symbols. Torres y Quevedo's electromechanical chess automaton (3.vi.a) was exhibited by his son at the first cybernetics conference (in Paris) in 1951, and Wiener was photographed playing a game with it—which he lost (de Latil 1953: 33, 258). And the two opening talks of the 1945 meeting of the Teleological

Society (Section vi.a, below) were one on digital computers by von Neumann, and another on communication engineering by Wiener.

Even twenty years later, one of the cybernetic pioneers would mention *both* programs *and* feedback devices in a paper tellingly entitled 'A Discussion of Artificial Intelligence and Self-Organization' (Pask 1964). It's not surprising, then, that Minsky's early bibliography of AI cited a host of cybernetic items (Minsky 1961*a*).

Moreover, many individuals at mid-century—not least, McCulloch and Pitts themselves—had their feet firmly placed in both camps. Wiener, for example, wasn't concerned only with dynamical feedback. He'd been a pupil of Russell's. He stressed the role of mathematical logic in cybernetics. He described Leibniz as the cyberneticians' "patron saint" (Wiener 1948: 12). And at the first Macy conference, he discussed Russell's paradox and the computer's oscillatory *true–false–true–false–true–false* response to it.

Similarly, Claude Shannon's (1916–2001) information theory—his M.Sc. dissertation!—was a core intellectual resource for cybernetics, but was based in Boolean logic (Shannon 1938). When Turing showed Shannon his own paper in 1943, they agreed that their approaches had been essentially alike (Hodges 1983: 250–1). Rashevsky encouraged formal theoretical approaches *in general*, and included both statistical and logical models of the nervous system in his *Bulletin of Mathematical Biophysics*. And von Neumann, whose digital computer made logical programming possible, worked also on probabilistic parallel networks, and studied self-organization in organisms and artefacts (see 12.i–ii and 15.v).

Nevertheless, the two approaches were potentially divergent—and they soon diverged, as we'll see in Section ix.

c. Biological roots

General physiology had been concerned with self-equilibrating mechanisms since the seminal work of Claude Bernard in the 1860s (2.vii.a). By the 1930s, the living body was recognized as an open system, continuously interchanging matter and energy with its environment while keeping certain vital quantities constant. This was largely due to the Harvard physiologist Walter Cannon (1871–1945), whose recent experiments on "homeostasis" had investigated many biochemical examples of circular causation (Cannon 1926, 1932).

These metabolic mechanisms maintain the internal environment of the body cells, regulating levels of blood temperature, sugar, salt, fat, proteins, calcium, and so on. Cannon showed that homeostatic control operates at many (interconnected) levels, under the regulation of the autonomic nervous system. For example, loss of blood leads not only to constriction of surface blood vessels but also to the release of blood corpuscles from the spleen. The sympathetic and parasympathetic nervous systems were found to have broadly oppositional, or complementary, functions, together maintaining the body's equilibrium.

Cannon also suggested extending the idea of homeostasis to society. This wasn't the first time that natural-scientific concepts had been applied to social matters. As we saw in Chapter 2.iii.b, Thomas Hobbes had likened the Commonwealth to "engines that move themselves by springs and wheels as doth a watch" (1651: 19). But Cannon's exemplar was a self-adjusting biological system, not a piece of inanimate clockwork. His suggestion would be followed by others. Indeed, the official theme of the Macy conferences was 'Cybernetics, Circular Causality, and Feedback Mechanisms in Biological *and Social* Systems'.

Interest in computer models of social institutions endured even as cybernetics was being overtaken by AI. Examples of the 1950s and early 1960s included work on bureaucracies, international relations, economics, management, and the daily rush hour (S. Beer 1959; Guetzkow 1962). And one of the founders of symbolic AI (namely, Simon) won the Nobel Prize for *Economics* (see 6.iii.a).

However, much as Simon turned towards individual psychology from the mid-1950s, so cognitive science in general became less social in character. By the 1970s, the social dimension had been largely eclipsed (1.ii.a). It would become visible again with AI work on 'agents' and the rise of the "new cybernetics", aka A-Life (Chapters 12.ii.e, 13.iii.d–e, and 15, and J. M. Epstein and Axtell 1996).

Even in the 1930s and 1940s, social homeostasis was a minority taste. The biological versions were far more prominent. One of the founders of cybernetics, the Mexican physiologist Arturo Rosenblueth (1900–70), worked with Cannon at Harvard for eleven years investigating (for example) the regulation of the heartbeat. He did many physiological experiments with Wiener, who dedicated his book *Cybernetics* to him. So the 'novel' science of cybernetics was a close descendant of Bernard's nineteenth-century work.

Neurophysiology—especially reflexology—was prominent in cybernetics from the start. The concept of reflex action took self-regulation beyond mere self-equilibration. Charles Bell and François Magendie, long ago, had described the reflex as a sort of circular causation, capable (for instance) of removing a hand or paw from an irritating stimulus. In the first half of the twentieth century, Charles Sherrington had shown how reflex mechanisms can provide a hierarchy of neural self-regulation. For instance, bodily posture when standing is maintained by a continuous interaction between flexor and extensor muscles, and by continuous sensory feedback from the touch sensors and muscle spindles; these stabilizing feedback circles are modified whenever a body part moves (see Chapter 2.viii.d).

Besides these reflex circularities, neural loops were being found, or hypothesized, within the brain. By the early 1930s, as we have seen, various people were suggesting that memory is grounded in reverberative neural circuits.

Neuro-circularity was assumed also by Lorente de No's mid-century "law of reciprocity of connections" (Lorente de No 1947). This law stated that any group of neurones in the brain that sends fibres to another group also receives fibres from it (either directly or by means of one internuncial neurone). It was based, for instance, on his studies of the anatomical connections between eye and inner ear (Lorente de No 1933*b*), positing what's now known as the vestibulo-ocular reflex (14.viii.b). Lorente de No himself was a prominent member of the cybernetics community. He took part in the first cybernetics meeting (organized by Wiener and von Neumann in 1943), and spoke about "the computing machine of the nervous system" at the Macy seminars.

In this neuroscientific context, McCulloch found it natural to describe each individual reflex arc as having a "goal, or aim, or end": namely, the particular bodily state (for instance, the length of a muscle) which it first alters and then acts to restore (W. S. McCulloch 1948: 151). His teleological language was encouraged also by "goal-seeking" machines developed in England in the early 1940s, and by an influential cybernetic comparison of purpose with the behaviour of anti-ballistic missiles (see Section vi.a).

As for learning, reflexology treated this as the acquisition of conditioned responses by means of positive and negative reinforcement. This approach was originated in mid-nineteenth-century Russia by Ivan Sechenov, and elaborated around the turn of the twentieth century by Ivan Pavlov (see 2.viii.b). By the 1940s, it was flourishing in the USA as behaviourist psychology (5.iii.a–c). Wiener, like McCulloch, restated ideas about conditioning in terms of feedback. And he suggested that artificial synapses in computers might be similarly adjusted by "experience" to enable them to learn (Wiener 1948, chs. 4 and 5).

Psychopathology, too, was included. Tremors, convulsions, rigidities, pathological pain, and even various types of anxiety neurosis were explained in terms of disturbances in neurological feedback. So therapy should aim to 'break the circle' sustaining the pathological behaviour. Preferably, this should be done by neurochemical methods (W. S. McCulloch 1949), but otherwise by some form of psychotherapy (Wiener 1948: 174).

d. Information theory

In discussing this wide variety of neurological circles, cyberneticists often spoke of the "information" they carried. Similarly, cybernetic physiologists explained bodily homeostasis in terms of information flow. And cybernetic engineers glossed the physico-chemical measurements made by their machines, and fed into the control loops, as providing information to the system.

This term was drawn from Shannon's information theory, developed at Bell Labs to measure the reliability or degradation of messages passing down telephone lines (Shannon 1948; Shannon and Weaver 1949). But the "messages" were thought of not as meaningful contents, conveying intelligible information such as that Mary is coming home tomorrow. Rather, they were the more or less predictable physical properties of the sound signal. In Shannon's words:

> Frequently the messages have *meaning;* that is, they refer to or are correlated according to some system with certain physical or conceptual entities. These semantic aspects of communication are irrelevant to the information problem. The significant aspect is that the actual message is one *selected from the set* of possible messages. (Shannon and Weaver 1949: 3)

This statement was fully endorsed by Shannon's colleague Warren Weaver (1894–1978), even though he also expressed the hope that a theory of meaning might one day be expressed in informational terms (Shannon and Weaver 1949: 99–100, 116–17).

In short, "information" wasn't a semantic notion, having to do with meaning or truth. It was a technical term, denoting a statistical measure of predictability. *What* was being predicted was the statistics of a physical signal, *not* the meaning—if any—being carried by the signal.

As a technical term for something measurable, "information" needed a quantitative unit. This new unit was the *bit* (an abbreviation of "binary unit"). One bit of information was defined as the amount of information conveyed by a signal when only two alternatives are possible. (It was Shannon who defined the "bit" in the context of information theory, but the term had been coined in 1946 by the statistician John Tukey, who also thought up "software": Leonhardt 2000.)

Each alternative, of course, may cover many distinct possibilities. For instance, the answer to the *Yes/No* question "Does the missing letter appear in the first half of the alphabet?" conveys only one bit of information. After all, either it does, or it doesn't. However, each possible answer covers thirteen possibilities: the letters A to M (for "Yes") and N to Z (for "No"). If the unit of measurement was unfamiliar, the general principle wasn't: any practised player of the parlour game Twenty Questions was aware that it's foolish to ask "Is it a cat?" before asking "Is it an animal?"

Information theory, then, had nothing to do with semantics. Certainly, Shannon sometimes discussed natural language. Specifically, he used information theory to predict the next word in generating 'English' word sequences. But he used only statistics, not syntax or semantics, to do so (see Chapter 9.vi.c). For Wiener or McCulloch, the signal might be a physical process in a neurone, instead of a telephone wire. Or it might be a written memorandum passed by one person to another. But the fundamental notions of "information", "messages", and "decisions" didn't imply *meaning*. Cybernetic theories didn't specify conceptual or propositional content.

Admittedly, cybernetic *psychologists* assumed meanings to be relevant. For instance, the precise nature of a psychosis would depend on the person's linguistically mediated experiences and beliefs. But its sudden onset was explained by analogy to the breakdown of an overloaded telephone exchange, where the *semantic content* of the many incoming messages is irrelevant. Similarly, cybernetic psychiatrists such as Kubie and McCulloch took it for granted that the feedback mechanisms underlying an anxiety neurosis involve *intelligible* associations. But they didn't specify particular semantic contents (see Wiener 1948, ch. 7—and cf. Chapter 7.i.a).

This flight from meaning and subjectivity has a lot to answer for: it became an important source of the structuralist and deconstructionist assaults on humanism that originated in France in the 1970s (Dupuy 2000). But that was still to come. The main point here is that mid-century cybernetic psychology rarely represented specific meanings in its theories. That's one reason why the symbolic-AI alternative, when it sprouted in the mid-1950s, seemed—to many psychologists—to be more promising.

e. Bateson, Pask, and a sip of Beer

Gregory Bateson (1904–80) was an exception, for he *did* apply the cybernetic approach to psychological meaning. His hugely influential "double-bind" theory of schizophrenia, which became popular in the 1960s, attributed it to the reception of contradictory messages from family members. These messages, often though not always verbal, were supposed to connote permissible or impermissible forms of behaviour (Bateson *et al.* 1956). Similarly, his therapy for alcoholism presupposed a circular semantics involving the patient's self-image (Bateson 1971). He'd picked up the notion of directive ideas, or schemas, from his tutor Frederic Bartlett in the 1920s (5.ii.b and vi.b, below), but the notion of feedback came from cybernetics.

The circularity was important—but, to Bateson's disappointment, the popularized versions of his theory lost sight of it. Most people interpreted the double bind as either a *logical* contradiction between messages, or a situation in which one would be hurt/punished *whatever* one did, as if forced to choose the lesser of two evils. For Bateson, it was more complex: a communicative system was set up which would systematically prevent the 'desired' outcome, owing to paradoxical relations between explicit communications and the background context (Harries-Jones 1995: 134–9).

This focus on multilevel contradictory meanings wasn't new. In his first book, a study of kinship rituals in a New Guinea tribe, Bateson (1936) had described what was going on in terms of complementary conflicts in relationships, defined at three hierarchical levels. These conflicts, he'd said, largely balanced each other out, so that potentially divisive behaviour was contained within a stable cultural system. (He didn't yet have the concept of feedback, but—not surprisingly—immediately got the point when he heard about it from McCulloch a few years later.)

Bateson's concern, in these contexts, with specifically human meanings was belied elsewhere. He frequently claimed that "mind" is present in *every* example of circular causality, even including whirlpools. He even said that all living things and social institutions—"the starfish and the redwood forest, the segmenting egg, and the Senate of the United States—have knowledge" (Bateson 1972; 1979: 12). But the sense in which the starfish "knows how to grow into five-way symmetry" is very different from the sense in which the Senate knows about George Washington, or the child knows what is expected by the mother. So much so, that one might judge the term "knowledge" to be inappropriate (see the discussion of autopoietic biology in Chapters 15.vi–vii and 16.x.c).

Another exception was A. Gordon Pask (1928–97), ensconced in the 1950s in a back room at the Cambridge Language Research Unit (Preface, ii). In the early 1960s he set up his own company, Systems Research Limited. Reached by a set of creaky stairs, this was situated over a launderette in Richmond. But even the creakiest of stairs couldn't keep his admirers away. Both McCulloch and Minsky, for instance, often visited; and the frequent home-based visitors included Stafford Beer, William Ross Ashby, and Richard Gregory.

Pask was a highly imaginative—and highly eccentric—early cybernetician, esteemed by McCulloch as "the genius of self-organizing systems" (W. S. McCulloch 1965: 220). His eccentricities and imaginative experiments had started at an early age:

> [As a young boy] Gordon had been a very determined designer of things electromechanical and chemical. I have it from a contemporary of his at Rhydal School in North Wales that he reputedly designed a special kind of bomb, this would be in 1942 or 1943 when he was aged twelve or thirteen, and my informant recalls Gordon being whisked mysteriously away from school in a military staff car. (Mallen 2005: 86)

Originally trained as a chemist and a psychologist, Pask considered meaning contexts of various sorts.

He was nothing if not ambitious, nothing if not interdisciplinary, and nothing if not prescient. For example:

* He planned a cybernetic theatre, in which audience reaction would influence the unfolding of the plot on stage—an idea that's now being claimed as "new" by enthusiasts for interactive media and virtual reality (Chapter 12.vi.a–c).

* He led a team of scientists, mathematicians, and artists planning Joan Littlewood's "Fun Palace Project", intended to provide a host of challenging diversions not found in ordinary theatres (Ascott 1966/1967: 119).

* He suggested how an architectural space (room, building, public arena . . .) might, in effect, *learn* and *adapt to* some of its users' goals, and their physiological and

psychological reactions (Pask 1969). Some of this learning and adaptation could be automatic (using sensor technology, variable lighting, and movable walls/screens), some would require feedback-driven modifications by human architects. This work anticipated today's ideas on "intelligent buildings", and is regarded by some as one of the "critical writings for the digital era" (Dong 2003; Spiller 2002: 76–7).

* He built a "Colloquy" of mutually adaptive mobiles, descendants of the Musicolour system mentioned in the Preface. They communicated by both sight (colour and pattern) and sound (pitch, rhythm, and simple phrases). They also provided an "aesthetically potent" environment for interactivity, enabling humans to include meaningful interpretation in the loop (Pask 1971). The Colloquy was exhibited at the seminal Cybernetic Serendipity conference, the world's first international exhibition of computer art (Reichardt 1968) (see l.iii.d).

* He concocted simple evolutionary machines, not programmed but implemented as solutions of chemical salts (Pask 1961: 105–8).

* One of his 'craziest' ideas was among the electrochemical models (e.g. Pask 1959), of which the "evolutionary" machine was one. These grew threadlike crystalline structures—dynamically balanced between the deposition and re-solution of ions—supposedly analogous to concepts, since they could discriminate sounds of different pitch (50 or 100 Hertz), or the presence/absence of magnetism, or differences in pH level.

Although the crystalline "ear" wasn't useful *qua* technology, the principle was interesting—even deep. For these chemical threads weren't *designed in order to* discriminate pitch (or magnetism, or pH), but naturally arose in a way that made this discrimination possible. In other words, they were a primitive example of the self-organization of *new* perceptual abilities, creating *new* perceptual dimensions—which is what happens from time to time during phylogenetic evolution. (Mostly, evolution *improves* an ear, or an eye—but sometimes, it *originates* a primitive ear/eye where none had existed beforehand.) That's largely why McCulloch thought so highly of Pask's work. In his Preface to Pask's first book (1961), McCulloch credited him with having illuminated "the central problem of epistemology", namely: where do our concepts *come from*?

McCulloch's accolade didn't prompt Pask's contemporaries to follow in his footsteps. His electrochemical ear was seen as a mere curiosity, and was soon well-nigh forgotten. The world would have to wait for thirty-five years before its theoretical significance was realized (Cariani 1993—and see Chapter 15.vi.d), and for forty-four years before a comparable example occurred (Bird and Layzell 2002).

That post-millennial example involved the evolution of a primitive radio antenna (a "radio wave sensor"), which picked up and modified the background signal emanating from a nearby PC monitor. The researchers involved were using a recently developed technique for evolving circuits in hardware (15.vi.c), in order to evolve an oscillator—not a radio. But, unexpectedly, some of their devices exploited unconsidered an accidental aspects of the physical environment (the aerial-like properties of printed circuit boards, and the proximity of the PC) to produce radio reception. Indeed, the (rewarded) oscillatory behaviour depended largely on accidental factors, such as the order in which the analogue switches had been set, and the fact that the soldering-iron left on a nearby workbench happened to be plugged in at the mains. In short, Pask had been way ahead of his time.

As though all that weren't interesting enough, Pask also discussed the cybernetics of analogy and parable in human learning and creativity (Pask 1963). And he designed a number of pioneering teaching machines that employed feedback to adapt to different individual learners. (True to form, the feedback was initially embodied in dishes of iron-salt crystals—but he soon cobbled simple computers to implement the system instead.)

Pask used this wide variety of adaptive machines to illustrate a theory of "conversation" covering human communication (and brain processes) as well as man–machine interfaces (Pask 1975*b*). However, his theoretical notation was rebarbatively complex. His general theory didn't spread beyond the core cybernetic community, but his machines were more influential. The cooperating mobiles, for example, were exhibited at London's Institute of Contemporary Arts, and the teaching machines inspired further research (Mallen 2005).

Unlike Skinnerian teaching machines, which employed a fixed program and syllabus, Pask's devices used their interaction with the student to alter their own decision rules and data presentation. After his pioneering SAKI, which taught a simple motor skill (see Preface, ii), Pask developed computers enabling human learners (and problem-solvers) to take different routes through a complex knowledge representation (Pask and Scott 1973; Pask 1975*a*, ch. 11). The domain might be probability theory, or an imaginary taxonomy of Martian animals. But the route through it which was proffered to the user by the system depended on the individual's general cognitive style. Specifically, it depended on their preference for semantically "holist" or "serialist" strategies.

The holist learner needs to confront relatively high-level aspects of the unfamiliar domain at the outset, even though these can't be fully understood at that point and may lead to over-generalization. The serialist is put off by such 'premature' presentation of high-order relations, preferring to build the new knowledge step by step, each step being readily intelligible in relation to its predecessor.

In his experiments on human subjects, Pask had identified a further distinction also, between "irredundant" and "redundant" holism (IH and RH, respectively). IH people avoid apparently irrelevant semantic content, and may be confused by it if it crops up. RH thinkers, by contrast, often find it positively helpful. So RH-ers are typically aided by analogies, 'side stories', and parables. But IH-ers (and serialists, too) find them unhelpful, and perhaps even counter-productive.

Pask couldn't explain these differences in cognitive style. (There's still no detailed explanation, but it now seems that RH-ers engage in a type of information processing which is relatively tolerant in estimating 'relevance': it not only *costs* them less to link 'far-flung' ideas, but they're able to get added cognitive *benefits* out of doing so: see Chapter 7.iii.d.) He pointed out, however, that they implied that different communicative styles would be best for reaching different learners. So tutorial programs should ideally be able to present their material in any of the three ways—which a fixed text, such as a book, clearly cannot.

Accordingly, his CASTE machine (Course Assembly System and Tutorial Environment) would engage in a "conversation" with the learner, first identifying and then adapting to the individual's cognitive style. Some students, Pask reported, experienced "a sense of participating in a competition (some say a conversation) with a not dissimilar entity" (Pask 1961: 92). Today's AI work on user-friendly interfaces, and on how people can combine their experience of genuine and 'virtual' reality, explores some of the questions that Pask raised forty years ago (see 13.v–vi).

Pask described the criminal underworld, too, as a cybernetic system. He even saw criminals and the police locked in a communicative symbiosis, which he and his associate George Mallen simulated—at first by role playing, then on a computer—in his laboratory over the launderette. The role playing, which also involved a senior police officer from the Met, was so convincing that it drew a visit from the writers of *Z Cars*, the first police series to be shown on British TV; it also drew a raid by the local constabulary, who'd been alerted by a neighbour overhearing "plans for a bankraid" (G. Mallen, personal communication). Moreover, the SIMPOL computer program (which went through three incarnations) was so convincing that it was used for several years in training at the Bramshill Police College.

However, the particular semantic content was provided by the human beings. For Pask's theory stressed not specific meanings, or propositions, but general notions of feedback in communicative networks. The same was true of other examples of social cybernetics.

For instance, Beer (1926–2002), a close friend of Pask, applied such ideas to industrial organizations at the first meeting of the International Association of Cybernetics in 1956. Unlike Babbage, who'd also compared a factory to a complex machine, and who discussed many detailed logistic and managerial problems (Chapter 3.i.a), Beer developed a general theory of management and operational research that focused on circular causation as such (S. Beer 1959). And unlike the sociologist Max Weber (1864–1920), who'd analysed bureaucracy in terms of rigid administrative hierarchies, Beer described the role of individual decisions and information flow in the organization as a whole. (What he *didn't* do was consider the decision-making processes within individual minds: see Chapter 6.iii.a.)

In 1971 Beer was invited to put his ideas about communication into practice in Chile, for Salvador Allende's government (Beckett forthcoming). He designed a network (called Cybersyn), based on 500 unused telex machines bought and stored away by the previous government, which ran the whole length of the country. Driven by socialist principles, Cybersyn linked mines, factories, local social groups, and government ministers (one of whom was Fernando Flores: 11.ii.g), enabling them to share detailed economic and political information speedily.

Cybersyn had an intellectual thread attaching it to England. For Beer's first computer-based experiments on issues of management had been done with Pask at Richmond. The team at Systems Research had developed "Ecogame", the first management multimedia interactive game (Lambert 2005: 73). Players sat at computer terminals depicting various situations, and made business decisions linked to social costs. Besides random-access 35 mm slide projectors, it utilized realtime displays showing numbers and graphs for each player, updated according to the person's decisions. Ecogame was exhibited at the Computer Graphics 1970 exhibition at Brunel University, and later taken to Davos for a summit of world economic leaders.

(In Chile, however, Beer's Cybersyn was no game, but deadly serious. After Allende's assassination in June 1973, the Pinochet government deliberately destroyed the system because of its egalitarian potential. It had already been used to support resistance to right-wing opposition movements.)

Although a wide range of social scientists were invited by McCulloch to the Macy meetings (Heims 1991, esp. chs. 6 and 8; Dupuy 2000, ch. 3), many Macy participants perceived social cybernetics as speculative and marginal. They respected von Neumann's theory of games (von Neumann and Morgenstern 1944). But they saw the (idealized) psychology of 'rational economic man' as very different from—and theoretically superior to—(empirical) social psychology.

Wiener himself wrote at length about social issues (Wiener 1950). But even he regarded applications of cybernetics to everyday social phenomena as unreliable, because they could initiate circular causation that eliminated previously observed social trends (Wiener 1948: 190). That is, they could provide self-defeating prophecies as well as self-fulfilling ones. And this fact could threaten an entire discipline, some of whose more imaginative practitioners (Bateson and Margaret Mead, for example) were Macy attendees:

> With all respect to the intelligence, skill, and honesty of purpose of my anthropologist friends, I cannot think that any community they have investigated will ever be quite the same afterwards. (Wiener 1948: 190)

Most cyberneticists' attention was devoted not to social systems but to biological organisms—and to artificial self-regulating mechanisms. Some of these artefacts provided ideas that were fed back into theories of the biological "machine", both bodily and mental. And this was happening, simultaneously, on both sides of the Atlantic.

4.vi. Brains as Modelling Machines

Independently of the cybernetics movement in the USA, similar ideas were being developed in England by the psychologist Kenneth Craik (1914–45). His credo was published in 1943, five years before Wiener's book (though near-simultaneously with his co-authored paper on teleology: Rosenblueth *et al.* 1943).

Craik, too, was inspired by the idea that control in analogue machines might be similar to control in the nervous system. But in exploring this theme, he developed a new approach to the brain. He suggested that it's a system which constructs "models" representing the world (and possible worlds, as in fairy tales). Both psychology and neurophysiology would be deeply influenced by his work.

"How is such and such represented in the brain?" seems, today, an obvious question to ask. Even people who deny the existence of internal representations aren't surprised to hear other people ask it, or to hear answers that compare the brain to some kind of computational machine. This wasn't always so. Even as late as 1940, such questions weren't seriously raised—and they wouldn't have been answered in that way in any case.

Certainly, it had been assumed ever since René Descartes that there must be some brain state correlated with the "such and such"—or anyway, with the perception or the thought of it. But correlation isn't representation. What more is needed, over and above correlation (temporal simultaneity), for a brain state to be a representation?

* In what sense—if any—must it match, or correspond to, the external reality, in order to represent it?

* And is correlation really necessary? After all, some brain states, presumably, can represent something even in its absence; and others can represent non-existent things (like unicorns), where no correlations are ever available. How are memories, predictions, hypotheses, illusions, fictions, or hallucinations possible?

* Last but not least, what about the nitty-gritty? Can we say why *this* brain state, rather than some other, should be the one to represent *that* environmental feature? Or are such facts mere contingencies, to be discovered but not understood?

a. A Cambridge cyclist

The questions just listed are a mixture of philosophical and scientific puzzles. On the one hand, they concern what representation *is*, what "representation" means. On the other hand, they concern how, in fact, the human brain manages to represent various things, so as to control behaviour. (Let's assume, here, that it does do this; doubts will be discussed later, in Chapters 13.iii.b–c, 14.viii, and 16.vi–viii.) Craik was well aware of this interdisciplinarity. He described his theory as a new *philosophy*, not just as psychology or neurophysiology.

Before the 1940s, neuroscientists hadn't asked such questions in a rigorous way. If they spoke of representations at all, they did so uncritically, as though the concept were a clear one. And they didn't seriously consider the idea that a certain type of brain state might be *inherently apt* for symbolizing a certain type of thing. That is, they didn't try to match the physiological syntax onto the psychological semantics.

To be sure, the clinical neurologist Henry Head (1861–1940) had spoken of a "postural schema" thirty years earlier (Head and Holmes 1911). This, he suggested, represented one's (continually changing) bodily attitudes. He'd posited some sort of isomorphism between brain states and reality, which supposedly explained how it was possible for the schema to function as a guide for, not just a correlate of, specific bodily actions. He'd even generalized the notion of schema from the body itself to its clothing:

> Anything which participates in the conscious movement of our bodies is added to the model of ourselves and becomes part of those schemata: a woman's power of localization may extend to the feather of her hat. (Head and Holmes 1911)

More recently, Head's student Bartlett (1932)—a leading Gestalt psychologist—had generalized it still further, to cover conceptual memories as well as bodily ones (see Chapter 5.ii.b). But neither man had asked *just what* the schema's neural mechanism might be—for the very good reason that there was no way of knowing. What's more, there seemed to be no way of coming up with fruitful hypotheses.

This changed in the early 1940s, when Bartlett's young colleague (and ex-student) Craik wrote his seminal book *The Nature of Explanation* (1943). This was published, in wartime Cambridge, only two years before Craik's tragic death in a cycling accident on King's Parade, aged only 31. Some say he was knocked off his bicycle by an American jeep on VE Day (V. Stone 1978: 87), others (probably more reliable) that he was thrown under a lorry on the eve of VE day, by someone suddenly opening a car door (Mollon n.d.). Either way, Craik was no longer around to see the effects of his ideas.

He wouldn't have expected a quick payback, in any case. For the book reported no new discoveries. Rather, it set a research agenda. (The closing words: "And so, *Tentare*".)

Craik tried to express ideas about interpretative schemas in precise, and neurologically plausible, terms. He argued that the brain events implementing perceptions and memories are *models* of the things represented, which can be useful in psychological processing because they *work in the same way* as those things. And he used a machine analogy to explain what he meant.

He referred to cybernetic devices such as "an anti-aircraft predictor, Kelvin's tidal predictor", and "the Bush differential analyser" (1943: 51, 60). As analogue machines, it was their *physical* properties which embodied their representational power. The tide predictor, for example, was made of pulleys and levers (so didn't resemble a tide visually), but *it works in the same way in certain essential respects* (namely, "it combines oscillations of various frequencies so as to produce an oscillation which closely resembles in amplitude at each moment the variation in tide level at any place"). The same was true, he suggested, of models in the brain:

> By a model we thus mean any physical or chemical system which has a similar relation-structure to that of the process it imitates. By 'relation-structure' I do not mean some obscure non-physical entity which attends the model, but the fact that it is a physical working model which works in the same way as the process it parallels, in the aspects under consideration at any moment. (1943: 51)

> I have tried ... to indicate what I suspect to be *the fundamental feature of neural machinery*—its power to parallel or model external events—and have emphasized the fundamental role of this process of paralleling in calculating machines. (p. 52; italics added)

His fascination with models, and with machines, dated from his schooldays. As a boy, he'd built several pocket-sized steam-engines—one of which he showed to Bartlett on their very first meeting (Bartlett 1946). (Now, they're all in an Edinburgh museum.)

Those toy models were very obviously similar to the full-scale reality. But this needn't always be so—which is just as well, since neuroscientists can't expect to find miniature steam-engines inside our heads. Nor need they fear that the complexity of the model must always match that of the real thing:

> [If] the fluid pressure at perhaps half a dozen points within an ocean wave were measured, it might be found that no form of surface other than that of the actual wave would cause the pressure at those points to be just what it is. It is thus conceivable that [in the visual system] a highly complicated pattern of local stimulation may make a fairly simple "label" or symbol for itself, in the form of a pattern of stimulation travelling to a common centre. (1943: 75)

This reference to a pattern of neural stimulation as a "symbol" was deliberate. Craik described his theory as "a symbolic theory of thought" (p. 120). However, he wasn't thinking of the formal symbols used in machines based on Turing-computation, for these didn't yet exist. Nor had McCulloch and Pitts published their formal–symbolic account of psychology (although they'd do so very soon, in the same year that Craik's book appeared). For Craik, symbols symbolized—representations represented—in virtue of their *physical* features:

Without falling into the trap of attempting a precise definition, we may suggest a theory as to the general nature of symbolism, viz. that it is the ability of processes to parallel or imitate each other, or the fact that they can do so since there are recurrent patterns in reality . . . [The] point is that symbolism does occur, and that we wish to explore its possibilities . . . [We must ask] is there any evidence that our thought processes themselves involve such symbolism, occurring within our brains and nervous systems? (1943: 58–9)

[This] symbolism is largely of the same kind as that which is familiar to us in mechanical devices which aid thought and calculation. (p. 57)

It followed that there's nothing eerie about memory, prediction, or even imagination. If the cerebral model, considered as a dynamic physical process, works in the same way as the external reality being modelled, it will behave (develop) in the same way. It can therefore represent past and future events—by persisting and by running, respectively.

But it can also model purely hypothetical matters. The results can then be evaluated (as positive/negative reinforcers) by other brain mechanisms, just as veridical models are. If they're evaluated as negative, the animal can avoid carrying out the (hypothetical) actions whose internal models produced them. "In the same way", he said, a tiny model of the *Queen Mary* helps the shipbuilder, and a differential analyser that calculates strains helps the designer of a bridge (pp. 52 and 61). In short, prediction and imagination—in general, thought—are both physiologically possible and psychologically useful.

The general message was that engineering, neuroscience, and psychology—and philosophy too, as we'll see—must be integrated. For the human brain, Craik said, is "the greatest machine of all, imitating within its tiny network events happening in the most distant stars" (p. 99).

The stars in one's eyes, whether literal or idiomatic, may of course be illusory. Craik grasped the nettle of non-veridical representation. He argued that perceptual illusions, errors in abstract thought, false beliefs, and much psychopathology (and religion, too: cf. 8.vi) also depend on adaptive modelling of the external world. But in all these cases, the models are skewed, often rigid, and counter-productive. For example:

[Hysterical conduct is] a form of adaptation . . . achieved by narrowing and distorting the environment until one's conduct appears adequate to it, rather than by altering one's conduct and enlarging one's knowledge till one can cope with the larger, real environment. Dissociation and schizophrenia and repression are further mechanisms for attaining this state of splendid isolation and pseudo-adjustment and of excluding difficulties and awkward suspicions. (1943: 90)

In general, his approach was to ask how it's possible for *anything* to represent something else. In criticizing the Gestalt psychologists, for focusing on description not explanation, he said:

The interesting thing in perception, surely, is not just *what* happens, but how and why it happens, and what has failed in the case of illusion or insanity. (1943: 114)

Given that cerebral models aren't inserted piecemeal into our brains by some benevolent deity, how do they arise? They're *in principle* possible, according to Craik, because physical processes in general are so basically similar that any one can be modelled by some other. (Hence his reference, above, to "recurrent patterns in reality".) But which *specific* reality processes are modelled by various species, and *how*, are empirical questions.

So neuropsychologists should ask:

* What physical features does reality-A possess? (Any one of these might function as a stimulus for some species or other.)
* What physical features in the nervous system, *if* they existed there, would be sufficiently similar to enable it to model reality-A?
* And are those features in fact possessed by process-A* found in the brain of this species?

If the answer to that last question is "Yes", then process-A* is a model of reality-A.—Or rather, it *could be*, it is *apt to be*, a model of reality-A. Whether it *is* a model of it depends on other things, as we'll now see.

b. Similarity isn't enough

Whether a brain mechanism that physically resembles some external reality *actually is* a model of it depends on whether it's linked into the animal's sensori-motor processing in such a way as to be useful *to the whole animal.*

In other words (mine, not Craik's), *model*—like *representation*—is an intentional concept. As Minsky would put it some twenty years later, when defending references to models in AI:

> The model relation is inherently ternary. Any attempt to suppress the role of the intentions of the investigator B leads to circular definitions or to ambiguities about "essential features" and the like. (Minsky 1965: 45)
>
> We use the term "model" in the following sense: To an observer B, an object A* is a model of an object A to the extent that B can use A* to answer questions that interest him about A. (ibid.)

A computer simulation, for the researchers using it, is a model of the real-world phenomenon concerned. With respect to *internal* models, the "investigator" or "observer" is the system itself. If a computer—or an organism—uses some internal mechanism A* to answer questions about A which it wants answered, then the one is an internal model of the other. Let's grant that a computer can't really *want* anything. But an animal can—and, in general, the "questions that interest him about A" involve survival value.

Forty years after Craik's death, neurophysiologists discovered that monkeys have cells in the visual cortex that respond to faces (monkey or human) in particular orientations: full face, profile, looking upwards, or looking downwards—distinctions that are important in the animals' social interactions (Chapter 14.iv.d). Suppose, for a moment, that the monkeys had been found to possess a *continuous* range of eye-aversion detectors, but *without* any added functionality—such as greater discrimination in their social behaviour. Those cells would be *apt* to model many different angles of gaze, but they wouldn't actually *be* models of a continuous gaze range.

Such a discovery is of course unlikely, because such useless—or anyway, unused—cells probably wouldn't have evolved in the first place. Certainly, a neural structure apt for representing this or that might arise accidentally, through random mutation. And, as A-Life work on *neutral* (*sic*) networks suggests, it might survive uselessly for a while and then be exploited by later generations (15.vi.c). But without attracting adaptively relevant connectivities, it would very likely disappear eventually.

Today, the appearance and loss of potentially useful but actually useless neural circuits can be observed close-up, using the techniques of evolutionary robotics (Chapter 15.vi.c, and vii.d below). Mini-circuits may be randomly generated which are apt to represent some aspect of the robot's environment. But if they lack the neural connections that would *put them to use* in the robot's task environment, the evolutionary program (genetic algorithm) will soon prune them. Only if they do increase the robot's 'fitness' will they be retained.

Craik pointed out that selection pressures (survival values) add an extra dimension to the neuroscientist's task. His research strategy outlined above assumes that the *fundamental* requirement is a matter of physical similarities between environment and nervous system. And he believed that to be so. But nervous systems have also been shaped by survival value, which "complicates" matters enormously. "In general", he said, "it is much more illuminating to regard the growth of symbolising power from this aspect of survival value, rather than from the purely physical side of accordance with thermodynamics" (1943: 60). That is, *the whole animal*, and its "interest" in survival, must be considered (see 14.vii and 15.vii–viii).

c. Craik and cognitive science

As for actual examples of neural models, Craik discussed various aspects of (mostly visual) neurophysiology in these terms. Our nervous system, he said, clearly does contain examples of "states of excitation and volleys of impulses which parallel the stimuli which occasioned them" (1943: 60). He suggested several hypotheses about the neural processes that *could* represent this or that environmental feature (see especially pp. 64–77).

These hypotheses were related to (one or more) specific computational functions, and were typically explored with particular machine analogies in mind: from differential analysers and anti-aircraft devices to electron-scanning cameras and temperature-compensated barometers. For instance, in discussing the mechanism underlying vernier acuity (seeing a tiny lateral shift of one part of a nearly straight line), he said:

> We must, then, consider how the nervous system might achieve two essential steps in the process:
>
> (1) The summation of several points of stimulation, lying on a straight line, so as to give rise to a state of neural excitation or trace or pattern having *direction*, which may be called a "vector element"; and
>
> (2) How differentiation occurs, so that a series or line of such vector elements having the same direction possesses little or no perceptual significance, whereas a bend or discontinuity in it has a very marked one. (1943: 68).

He immediately described what he saw as "the easiest mechanical or electrical method" by which to do this—but then pointed out that it underperformed the visual system in various ways. Next, he asked how we should "set about designing a mechanism" to do *those* things. In considering that question, he made careful functional distinctions between superficially similar tasks—suited to the "calculations" (computations) effected by distinct types of physical machine.

Even in a book of only 120 pages, neuroscientists might have hoped to find more than thirteen devoted to nitty-gritty empirical questions, such as these. And indeed, relevant remarks were scattered throughout the volume. But neuroscience wasn't yet able to support many detailed speculations—not even about vision, never mind "hysterical conduct" (pp. 90 ff.).

Moreover, Craik was a psychologist, not a neuroscientist. This didn't prevent him from making prophetic and/or influential (though sometimes mistaken) suggestions about the brain: for instance, that a randomly connected cortex might self-organize with experience, and that the EEG might be a cortical scanner (see Craik 1943: 115, and Chapter 14.iii.a and ix.b). But neuroscience wasn't his prime concern.

In a sense, neither was psychology. His first love—and his first degree, at Edinburgh University—had been philosophy. Moreover, he intended his highly interdisciplinary book as a statement of a new *philosophy.* This fact was declared in the Preface, reflected in the chapter titles (such as '*A Priorism* and Scepticism' and 'On Causality'), and evident on almost every page. As we've seen, the text addressed the conceptual puzzles concerning representation, as well as the empirical ones. Kant was mentioned more often than Sherrington, although Craik gave him a less central role than McCulloch did. And well over half of the forty-item bibliography was devoted to epistemology or philosophy of science—plus the ubiquitous *Principia Mathematica.*

Unlike both McCulloch and Pitts, however, Craik hadn't been seduced in adolescence by this new logic. Or if he had, the love affair was a short one. He described *Principia* as "a garden where all is neat and tidy but bearing little relation to the untidy tangle of experience from which the experimentalist tries to derive his principles" (p. 3). But he did this in the context of a discussion of *epistemology*, not just psychology.

As it turned out, Craik's book had no *direct* influence on philosophy (see Chapter 16.iii.a). Even in psychology and neuroscience, its influence outside his immediate circle was somewhat delayed. Relatively few read it on publication: in 1943, and especially in England, people had other things on their minds. And the transatlantic convoys dodging the U-boats (with Turing's help: 3.v.d) weren't shipping books across the ocean.

Immediately after the war, however, Craik's ideas were much discussed. In the high-profile Waynflete Lectures of 1946, Lord Adrian praised him for (as he saw it) making mentalism redundant: "ideas" were simply organized patterns of electrical activation in the nervous system (Adrian 1947: 94). Bartlett's (1946) obituary stressed his role as a highly creative experimental psychologist. The minutes of the inaugural meeting of the Experimental Psychology Group, founded in Cambridge a year after his death, noted their belief "that the type of work in which the Group will engage would have secured his approval" (Mollon n.d.).

Soon afterwards, the newly formed Ratio Club in London picked up Craik's baton (see Section viii below, and Chapters 12.ii.c and 14.iii.a). Most club members had recent wartime experience of analogue machines like those he'd mentioned. And, spurred by his writings, they leant on that experience in thinking about the mind/brain.

The neurophysiologist Horace Barlow, for instance, was strongly indebted to Craik (Barlow 1959: 542; see also Chapter 14.iii.b). So was Gregory, a Cambridge psychologist close to Barlow and to Bartlett's group. He endorsed Craik's enthusiasm for engineering analogies (14.iii.a), and later developed Craik's approach to visual illusions (6.ii.e). Gregory wasn't a member of the club, but learnt many years later that he'd been about to be elected to it when it closed (personal communication).

His friend and colleague Christopher Longuet-Higgins, who'd later give cognitive science its name (and many of its ideas as well), missed out on the Ratio Club too. But he didn't miss out on Craik. Inspired by his ideas, he built a very simple wheeled model to illustrate the use of internal representations for prediction and gap-filling, as opposed to direct control from the inputs (Gregory 2001: 389).

By the early 1960s, Craik's followers felt themselves to be "at the forefront of a revolution" (Baddeley 2001: 348; see also Zangwill 1980). And those followers were now spread far beyond the magic circles of Cambridge and the Ratio Club. Countless people were drawn to Craik's work by his (posthumous) data-rich papers in the *British Journal of Psychology* (1947–8). These discussed the role of human operators dealing with—or rather, participating in—servo-control systems.

Craik declared, for instance, that:

As an element in a control system *a man may be regarded as* a chain consisting of the following items:

1. Sensory devices, which transform a misalinement [*sic*] between sight and target into suitable physiological counterparts, such as patterns of nerve impulses, *just as* a radar receiver transforms misalinement into an error-voltage.
2. A computing system which responds . . . by giving a neural response calculated . . . to be appropriate to reduce the misalinement . . .
3. An amplifying system—the motor-nerve endings and the muscles—in which a minute amount of energy (the impulses in the motor nerves) controls the liberation of much greater energy in the muscles . . .
4. Mechanical linkages (the pivot and lever systems of the limbs) whereby the muscular work produces externally observable effects, such as laying a gun [NB: the second military example]. (Craik 1947–8: ii. 142; italics added)

A clearer statement of man-as-machine would be difficult to find—unless one also mentioned verbal thought (which Craik himself had done in his book, and which symbolic AI would do later).

Craik's two papers hit their mark. They attracted attention partly as pioneering work in applied psychology (he'd been appointed Director of the newly founded Applied Psychology Unit in 1944), and partly as intriguing examples of the new approach of cybernetics.

Cybernetic discussions—and cybernetic machines—were now springing up on both sides of the Atlantic, and Wiener's widely read rallying cry was published in 1948. It was evident from Craik's (posthumous) papers, and from his book, that he'd anticipated the core themes. In brief: although he wasn't part of the cybernetics *movement*, he was an early, and highly significant, *cybernetician*.

d. Might-have-beens

If Craik hadn't died in 1945, he might have broadened his discussion to include digital as well as analogue devices. He might even have joined Ratio member MacKay (1949/1959) in urging the development of hybrid "analytical engines", part-analogue and part-digital. Also, he might have been a strong voice in the development of neural/psychological theories of *emulation* (Grush 2004: see Chapter 14.viii.c).

Despite his previous disparagement of *Principia Mathematica*, he might have been excited by Turing's (1950) paper in *Mind* and by the symbolic AI of the 1950s. After all, he'd already described thought—and mental modelling—as involving three "symbolic" processes:

> (1) "Translation" of external processes into words, numbers, or other symbols.
> (2) Arrival at other symbols by a process of "reasoning", deduction, inference, etc., and
> (3) "Retranslation" of these symbols into external processes (as in building a bridge to design) or at least recognition of the correspondence between these symbols and external events (as in realising that a prediction is fulfilled). (1943: 50)

So, thankful that he'd explicitly avoided giving "a precise definition" of symbolism in 1943, he might have widened his concept to include formal representations as well as physical ones. (From the late 1950s on, cognitive scientists would do precisely this—in one case, marrying Craik's theory to formal semantics: Chapter 7.iv.e.)

However, those are all might-have-beens. What actually happened was that McCulloch and Pitts published their logicist manifesto in the same year as Craik's book.

As we saw in Section iii.f, this didn't sweep the board immediately either. Its technical proofs were taxing, and there were no comfortingly familiar logic machines to back it up. On the contrary, John von Neumann's computer would take the form it did partly because of the ideas in that paper. But by the early 1950s, people in and around MIT were increasingly thinking of digital–formal models of thought as well as analogue ones. The "models" had come from Craik, the "digital–formal" from McCulloch and Pitts.

4.vii. Feedback Machines

Cybernetics would have thrived even in peacetime. After all, it had already been given a good start (biologically) by Bernard and Cannon, and (mathematically) by Maxwell. Moreover, engineers hadn't needed a war to make them work on feedback in clock escapements, or steam-engine governors. But there's no denying that the Second World War provided opportunities—practical challenges as well as money—for mid-century engineers to develop cybernetic machines.

Most cyberneticians were already thinking about biology, as we've seen, and many were looking towards psychology too. It's no surprise, then, that insights and techniques initiated in the military context were soon applied in a wide range of post-war artefacts built with psychology and/or neurophysiology in mind. (Which is *not* to say that no such artefacts had been built before: see Chapters 2.viii.e and 5.iii.c, and Cordeschi 1991, 2002; Valentine 1989.)

a. Purposes of war

The wartime machines of most interest to cyberneticians weren't code-breakers, which in any case were shrouded in secrecy, nor even number-crunching computers (such as ENIAC) used for calculating bombing tables. Rather, they were systems involving self-corrective servomechanisms. In particular, the cybernetics community was concerned with weapons such as radar-guided missiles.

Anti-ballistic missiles were capable of a form of self-equilibration going beyond Bernard's "internal environment" to involve the external environment, too. Anticipating the movement of the tracked target (by extrapolation of its path observed so far), they would aim at the point where the target would be at some time in the future. These continuous predictive adaptations were seen by cyberneticians as a paradigm case of "purposive" or "teleological" behaviour.

In a highly influential paper of 1943, Rosenblueth, Wiener, and Bigelow defined teleological behaviour as "behavior controlled by negative feedback". This was the written version of their talk at the 1942 Macy meeting on hypnosis, and the first published account of the new ideas (see also Rosenblueth and Wiener 1950). Affecting to link hypnosis (and other forms of psychopathology) with missiles, it's no wonder that their audience was excited.

More specifically, they described teleological behaviour as:

> behaviour controlled by the margin of error at which the [behaving] object stands at a given time with reference to a relatively specific goal... The signals from the goal are used to restrict outputs which would otherwise go beyond the goal. (Rosenblueth *et al.* 1943: 19)

A clear case of purpose, they said, is a machine designed so as to impinge on a moving luminous or heat-emitting "goal". And they assimilated such machines to human purposes—and thereby to "voluntary activity", since "what we select voluntarily is a specific purpose, not a specific movement". For instance, in discussing the inability of patients with cerebellar disease to bring a glass of water to their lips, they said:

> The analogy with the behaviour of a machine with undamped feed-back is so vivid that we venture to suggest that the main function of the cerebellum is the control of the feed-back nervous mechanisms involved in purposeful motor activity. (Rosenblueth *et al.* 1943: 20)

The notion that purpose is analysable in terms of "difference reduction" (between the current state and the goal) had already entered cybernetic discussions. It was mentioned, for example, in McCulloch and Pitts' logical calculus paper of the same year. But the teleology paper raised the profile of this discussion. Von Neumann was greatly impressed by it, and asked Wiener to arrange for him to meet Rosenblueth (Aspray 1990: 267).

Its influence resulted partly from its definition of *two* categories of teleology. The simplest type is controlled merely by the current distance from the goal state. If the goal happens to be a moving target, the system aims for the place where it now judges it to be. An extrapolative system, by contrast, predicts some future location of the target and aims for that point. Such systems include biological organisms (a cat chasing a mouse) and anti-ballistic missiles, which compute the expected future path of the target in real time.

In 1944, while most information-processing technology was still classified, Wiener, von Neumann, and Aiken (designer of the Harvard Mark I) organized a small gathering of people already largely in the know. This group, which also included McCulloch,

Pitts, and Lorente de No, called themselves the Teleological Society. The organizers of the first (and, as it turned out, the only) meeting declared:

> Teleology is the study of purpose or conduct, and it seems that a large part of our interests are devoted on the one hand to the study of how purpose is realized in human and animal conduct and on the other hand how purpose can be imitated by mechanical and electrical means. (quoted in Aspray 1990: 183)

Whether the "mechanical and electrical means" best suited for imitating purpose are cybernetic or computational, involving dynamical or logical-symbolic systems respectively, wasn't then much debated. As remarked in Section v.b, this distinction wasn't yet clearly made.

To be sure, von Bertalanffy distinguished equilibration from "true finality" involving symbolic foresight (see Section v.b). But most cyberneticians seemed to see no fundamental difference between pure self-equilibration (as in homeostasis), purposive behaviour directed to some observable object (as in guided missiles), and goal-seeking directed to some intentional end (as in human deliberation and planning).

In particular, the concept of mediation by *symbolic* representations wasn't widely current. As we've seen, Craik had suggested that behaviour involves internal models of the world, which could be paralleled in artefacts. But he saw these models as analogue physical mechanisms inside the brain, not as symbolic structures embodied in neural networks. And he focused primarily on sensori-motor control, not abstract problem solving or purposive behaviour directed to still-imaginary ends.

Similarly, the cyberneticians in general relied on quantitative measures, and on equations describing correlations between continuously varying system parameters. They didn't, and couldn't, describe complex hierarchical structures or detailed discontinuous changes in system structure, such as are involved in much human thinking, and language.

Problem solving and planning would come into centre-stage only in the mid- to late 1950s. Then, Newell and Simon attributed them to *symbolic* representation of the end state—the key notion of difference reduction being retained (see 6.iii.c). At that time too, Noam Chomsky's (1957) new theory of generative linguistics explained hierarchically structured behaviour in terms of internal models conceptualized as *formal rules* (6.i.e and 9.vi). So by 1960, the general concept of internal models, or representations, had been hijacked by the symbolists. (As we'll see in Chapters 12 and 14.viii, other types of representation are also possible.)

Meanwhile, the early cyberneticians had been modelling purposive behaviour in either analogue–dynamical or, less commonly, reflex terms.

b. Post-war projects

Anti-ballistic missiles had been built for practical, and deadly, use. (So, for that matter, had ENIAC: see Chapter 3.vi.a.) But some self-regulating artefacts at mid-century were developed to throw light on psychological and biological phenomena.

Unlike machines (from Pascal's onwards) focused on logic or mathematics, these didn't attempt abstract problem solving. Admittedly, Shannon, like Turing, discussed the possibility of chess programs, and argued that these should use symbolic concepts like those humans employ to avoid considering every possible move (Shannon 1950*a*,*b*). But his own chess machine (Caissac), which played a variety of endgames by comparing the advantages of various moves, was a special-purpose gizmo, not a program (McCorduck 1979: 103).

In general, the focus of cybernetic devices—most of which were analogue, not digital—was on adaptive behaviour. Some were intended to illustrate fundamental organizational principles of life in general and nervous systems in particular.

Shannon built a mechanical maze-runner called Rat, which he took along to the eighth Macy conference in 1951. It used an electrical contact "finger" to detect the walls of a twenty-five-square maze. Two electrical motors enabled the fingers to move either north–south or east–west: if the finger touched a wall, Rat started again from the centre of the relevant square. Exploring the maze in ignorance, this machine would store its errors (those choices which led to a blind alley) in its memory, and avoid making them on its second run (Shannon 1951). However, if the maze wasn't solved within a given number of moves, the memory was erased, so enabling the maze-runner to escape from a repetitive cycle.

Early British devices, some of which were exhibited at the Festival of Britain in 1951, included a maze-learning "rat" designed independently of Shannon's by I. P. Howard (1953). This electromechanical rat, equipped with three whiskers ("feelers"), could learn any maze, of reasonable size and suitable lane widths.

Always trying the right turn first at any new choice point, it marked its errors on a memory wheel used to set the switches for its subsequent runs. Unknown to Howard, an essentially similar memory wheel had been used in a maze-solving machine built many years earlier (T. Ross 1938). But Howard's work was of more general interest. His theory didn't require that the creature always turn right first, nor that there be only two possible paths. Moreover, it could be adjusted to cover probabilistic behaviour (Howard 1953: 56).

Howard's rat wasn't lonely: comparable artefacts were legion. In the 1940s and 1950s, so many machine models of life and mind were developed that Pask (1964) soon counted over 350 instances. Besides those of Pask himself, interesting examples—interesting, that is, to cognitive scientists—were built by MacKay (1956*a*), Albert Uttley (1956, 1959*a*,*b*), Tony Deutsch (1954), and John Andreae (1963) (see Chapter 12.ii.b–c, iv.a, and c). Their machines ranged from adaptive robots, through ball-dropping hoppers that learnt by reinforcement, to abstract models of adaptation in neural networks. (Many examples were described in Blake and Uttley 1959; for reviews, see George 1959, 1961, and Cordeschi 2002.)

We can't consider them all here—and we don't need to. For their basic principles were illustrated in two 'families' of device constructed in England from the late 1940s on. Together with McCulloch and Pitts' theoretical papers of 1943 and 1947, and Craik's book, these artefacts inspired most of the cybernetic machine makers of the 1950s.

Both were the brainchildren of neurologists who were also amateur engineers. If they hadn't been, they couldn't have put mind-as-machine into practical form, nor kept so close (*despite* the huge oversimplifications) to the biology. In so far as neurological data could be exploited at that time, these machines were driven by neurological interests, and knowledge. And as the next section (and Chapters 12 and 14–15) will show, they were in many ways far ahead of their time.

4.viii. Of Tortoises and Homeostats

The cybernetic families of Tortoises and Homeostats were constructed by William Grey Walter (1910–77) and William Ross Ashby (1903–72), respectively. (Neither of them used the "William".) Both men were early members of the interdisciplinary Ratio Club (Husbands forthcoming).

This was a small dining club that met several times a year from 1949 to 1955, with a nostalgic final meeting in 1958, at London's National Hospital for Neurological Diseases. The founder–secretary was the neurosurgeon John Bates, who had worked (alongside Craik) on servomechanisms for gun turrets during the war.

His archive shows that the letter inviting membership spoke of "people who had Wiener's ideas before Wiener's book appeared" (P. Husbands, personal communication). Indeed, its founders had considered calling it the Craik Club, in memory of Craik's work—not least, his stress on "synthetic" models of psychological theories. In short, the club was the nucleus of a thriving British tradition of cybernetics, started independently of the transatlantic version.

The Ratio members—about twenty at any given time—were a very carefully chosen group. Several of them had been involved in wartime signals research or intelligence work at Bletchley Park (3.v.d). They were drawn from clinical psychiatry and neurology, physiology, neuroanatomy, mathematics/statistics, physics, astrophysics, and the new areas of control engineering and computer science. (For a discussion of the neuroscientists and psychologists involved, see Chapter 12.ii.c.)

The aim was to discuss novel ideas: their own, and those of guests—such as McCulloch, who was their very first speaker in December 1949. (Bates and MacKay, who'd hatched the idea of the club on a train journey after visiting Grey Walter, knew that McCulloch was due to visit England and timed the first meeting accordingly.) Turing gave a guest talk on 'Educating a Digital Computer' exactly a year later, and soon became a member. (His other talk to the club was on morphogenesis.) Professors were barred, to protect the openness of speculative discussion. So J. Z. Young (who'd discovered the squid's giant neurones, and later suggested the "selective" account of learning: 2.viii.e and 13.ix.d) couldn't join the club, but gave a talk as a guest.

The club's archives contain a list of thirty possible discussion topics drawn up by Ashby (Owen Holland, personal communication). Virtually all of these are still current. Indeed, if one ignores the details, they can't be better answered now than they could in those days.

These wide-ranging meetings were enormously influential, making intellectual waves that are still spreading in various areas of cognitive science. Barlow (personal communication) now sees them as crucial for his own intellectual development, in leading him to think about the nervous system in terms of information theory (see Chapter 12.ii). And Giles Brindley (another important neuroscientist: 14.iii.c), who was brought along as a guest by Barlow before joining for a short time, also remembers them as hugely exciting occasions (P. Husbands, personal communication).

Our specific interest here, however, is in the machines built by two members of the Ratio Club: Grey Walter's tortoises and Ashby's Homeostat. As befitted the diverse interests of the club, these were very different in style. One was intended to model purpose and learning, the other to approximate general features of life. And they were very different as physical objects, too. One was an intriguing gadget that attracted enormous publicity (frowned on by some Ratio members), while the other was a laboratory contraption predictably ignored by the newspapers. Their theoretical interest, however, was significant in both cases.

More "interest", perhaps, than immediate influence. With hindsight we can now see how hugely insightful they were. But that 20–20 vision wasn't available to their contemporaries. Some people got the point, to be sure. De Latil (1953), for instance, regarded both as "revolutionary", and wrote about their wider scientific—and philosophical—implications at length. (His book was soon translated into English by a relation of Grey Walter's boss, the neurologist Frederick Golla.) However, the primitive state of electronic technology didn't enable those implications to be explored in practice.

That wouldn't be possible until the late 1980s, with the development of behaviour-based (situated) robotics and dynamical theory: see Chapters 14.vii and ix, and 15.vii–viii and xi.

a. Robots at the festival

Grey Walter was the first Director of Physiology (from 1939) at Golla's newly founded Burden Neurological Institute, Bristol. He was a highly influential electro-encephalographer. For instance, he discovered the delta and theta rhythms, and designed several pioneering EEG-measuring instruments. In addition, he founded the EEG Society (in 1943), organized the first EEG Congress (1947), and started the *EEG Journal* (also 1947). With his EEG expert's hat on, he focused on the overall effects of large populations of neurones rather than on specific cell connections. But his robots, as we'll see, used as few 'neurones' as possible.

His interest in the EEG dated from his time at Cambridge in the early 1930s, when he worked on muscle contraction with Edgar D. Adrian. Ever since the first paper on EEG (by Hans Berger) in 1929, Adrian was a key pioneer in the field. He discovered body mappings—of the limbs, for instance—in both cerebellar and cerebral cortex. And he predicted that improved brain-monitoring technologies would one day (fifty years later, as it turned out: 14.x.c) enable neuroscientists to study the cerebral changes associated with thinking (Adrian 1936: 199).

In his youth, Grey Walter also studied conditioning with a team of Pavlov's students visiting from St Petersburg. Indeed, he met Pavlov briefly. But his prime neurological interest was in the activity of the brain as a whole.

Besides these psychophysiological skills, he was a skilled speaker and writer in several languages. He was much in demand to talk on both professional and political issues. Initially a communist, he later veered towards anarchism: that is, the rejection of top-down control. But he was so full of ideas that, as his son Nicolas remembers, "he found it difficult to produce more sustained work, and both of his two books were actually written by his father from his notes and conversations" (N. Walter 1990).

From 1949 onwards, Grey Walter built several intriguing cybernetic machines. These were intended to throw light on the behaviour of biological organisms—although he did point out that they could be adapted for use as "a better 'self-directing missile'" (Grey Walter 1950*a*,*b*, 1953; see also O. Holland 1997, 2002). Unlike actual missiles, however, his machines displayed a range of different behaviours.

He'd been inspired, in part, by a wartime conversation with Craik. Craik was then working on scanning and gun aiming, and visited Grey Walter at the Burden Institute to use some of his state-of-the-art electronic equipment. During his visit he suggested that the EEG might be a cortical scanner, affected by sensory stimuli. This idea became influential in neuroscientific circles (13.ii.a). And it was later modelled by Grey Walter as a rotating photoelectric cell, whose "scanning" stopped when its robot carrier locked onto a light source.

Wiener's influence on him was less effective. Thus in a letter to Adrian written in June 1947 (after Craik's early death), Grey Walter said:

> We had a visit yesterday from a Professor Wiener, from Boston. I met him over there last winter and found his views somewhat difficult to absorb, but he represents quite a large group in the States . . . These people are thinking on very much the same lines as Kenneth Craik did, but with much less sparkle and humour. (O. Holland 2002: 36)

Wiener himself was more generous—or perhaps just more polite. In a letter thanking Grey Walter for his hospitality during this brief visit, he wrote, "I got a great deal out of our trip, and am certain that it will be possible to renew our contact at some future date" (Wiener 1947).

In particular, Grey Walter sought to model goal seeking and, later, learning. But he did so as economically as he could—in both the financial and the theoretical sense. Not only did he want to save money (the creatures were cobbled together from war-surplus items and bits of old alarm clocks), but he was determined to wield Occam's razor. That is, he aimed to posit as simple a mechanism as possible to explain apparently complex behaviour.

And simple, here, meant simple. His wheeled robots, or "tortoises", had two valves, two relays, two motors, two condensers, and one sensor (for light or for touch). In effect, then, a Grey Walter tortoise had only two neurones. For, crucially, the tortoises weren't mere toys but models of (very simple) nervous systems.

Robot toys with simple tropisms were already common at the time, at public exhibitions if not in the toyshops. For instance, a French-made "Philidog" at the 1929 International Radio Exhibition in Paris would follow the light from an electric torch—until it was brought too near to its nose, when it started to bark (de Latil 1953: 240–1). Ten years later, visitors to the New York World Fair in 1939 were sadly robbed of their chance to be enchanted by another robot dog. It had committed suicide a few days earlier:

> [The "electrical dog"] was to be sensitive to heat and was to have attacked visitors and bitten their calves, but just before the opening of the exhibition it died, the victim of its own sensitivity. Through an open door it perceived the lights of a passing car and rushed headlong towards it and was run over, despite the efforts of the driver to avoid it. (de Latil 1953: 241)

Great fun—except perhaps for the dog! But nothing to do with neuroscience. Grey Walter, by contrast, was pioneering biologically based robotics, an activity taken up many years later by (among others) Arbib, Valentino Braitenberg, Rodney Brooks, Randall Beer, and Barbara Webb (13.iii.b, 14.vii.c, and 15.vii).

One of his tortoises, the *Machina speculatrix*, showed surprisingly lifelike behaviour. "Lifelike" rather than (human) mindlike—the Latin word meant exploration, not speculation. But Grey Walter clearly had his sights on psychology as well as physiology. This was the first step in a research programme aimed at building a model having

> these or some measure of these attributes: exploration, curiosity, free-will in the sense of unpredictability, goal-seeking, self-regulation, avoidance of dilemmas, foresight, memory, learning, forgetting, association of ideas, form recognition, and the elements of social accommodation. (Grey Walter 1953: 120–1)

"Avoidance of dilemmas" and "free-will" were supposedly modelled by the tortoise's ability to choose between two equally attractive light sources (O. Holland 2002: 44–5). Unlike Buridan's ass, forever poised between two identical bundles of hay, the tortoise would unknowingly exploit its scanning mechanism to notice, and so to follow, one light before the other. "Learning", "association of ideas", and "social accommodation" came later (see below). Meanwhile, the Latin tag being too much of a mouthful, this early tortoise was quickly named ELSIE (from Electro-mechanical robot, Light-Sensitive with Internal and External stability). The prototype, which was very similar, was dubbed ELMER: ELectro-MEchanical Robot.

ELSIE soon became something of a celebrity. Much as Jacques de Vaucanson's flute-player had delighted visitors to London's Haymarket 200 years before (2.iv.a), so ELSIE amazed visitors to the Festival of Britain held—only a couple of miles away—in 1951. A few years later, it caused great amusement at a meeting of Babbage's brainchild, the British Association for the Advancement of Science. For it displayed an unseemly fascination with women's legs, presumably because of the light reflected from their nylon stockings (Hayward 2001*a*). (Much later, it was resurrected at the Science Museum for the Millennium Exhibition, and for the centenary of the British Psychological Society in 2001.)

Members of the public unable to reach the Festival site were soon able to read about ELSIE in a potboiler entitled *The Robots Are Amongst Us* (Stehl 1955). A more serious account—but including a photograph of ELSIE and her infant human 'brother' in Grey Walter's living-room—had already come off the press in France, and appeared in English in 1956 (de Latil 1953). Grey Walter himself became something of a celebrity too. He was often invited to speak by the BBC, sometimes appearing as a panellist on the popular radio programme *The Brains Trust*.

b. Of wheels and whiskers

The Festival robot ELSIE explored its environment. It used a scanning photoelectric cell (coupled to the steering wheels) to seek light and to guide the wheels towards the illumination. If the light wasn't too bright, the tortoise would stay in front of it, oscillating very slightly to left and right. But a strong light would cause the creature to continue scanning. In that case, its attention—and its movement—might be attracted by another, perhaps a weaker, light. As Grey Walter put it (in deliberately biological language), the tortoise showed a positive tropism to moderate light, but negative tropisms with respect to both strong lights and darkness.

Besides these basic tropisms, Grey Walter's robot displayed simple forms of approach and avoidance. Its (slightly movable) "shell" acted as a three-dimensional whisker, or pressure sensor: it closed an electrical circuit whenever it encountered a mechanical obstacle, causing the creature to back away from walls, furniture, or people's fingers. Since the robot also carried a pilot light, it would approach its own image in a mirror, or another light-bearing tortoise. On touching the mirror, or the mate, it would automatically move away—only to be drawn back again by the other's pilot light. The mechanical minuet that resulted was, for a while, fascinating to watch.

Another cybernetic device designed (in 1950) by Grey Walter was an electrical learning circuit named CORA (COnditioned Reflex Analogue), now kept in London's Science Museum. Largely 'cannibalized' from ELSIE, this functioned only as a static box on the workbench. It wasn't incorporated by Grey Walter himself into a mobile tortoise, although there's some evidence that he'd intended to do so. That may explain why "in spite of his efforts, CORA had only a small fraction of the impact of the tortoises, and made little lasting impression" (O. Holland 2002: 46). This shouldn't have mattered, at least outside the exhibition halls. For, quite apart from its theoretical interest, Grey Walter did *connect* it to the circuitry of a moving tortoise, so presumably was able to demonstrate learning in action. But the rhetorical effect was less dramatic (see 1.iii.h), and even the scientists failed to see the point.

The "point", here, was Pavlovian. CORA was based in neurophysiology, being a development of an earlier circuit named NERISSA (Nerve Excitation, Inhibition, and Synaptic Analogue). It was intended as a model of the sort of conditioning that Pavlov had reported in his bell-salivating dogs (2.viii.b)—and which had fascinated Grey Walter in his Cambridge days.

If CORA was repeatedly presented with two stimuli in quick succession, only the second of which 'naturally' caused a particular response, it would eventually produce that response even when the first stimulus occurred alone. Like Pavlov's dogs, CORA needed occasional reinforcement (wherein the first stimulus is again followed by the second) in order to maintain the conditioning. It was based on a probabilistic theory, and may have owed something to Grey Walter's fellow Ratio member Uttley (see Chapters 12.ii.c and 13.ii.b). In short, CORA reflected current views on neural communication and learning—and it was described by Grey Walter in his account of the *living* brain (Grey Walter 1953: 203–7).

Grey Walter pointed out that CORA could be combined with *Machina speculatrix* to produce a robot capable of learning: *Machina docilis*. A tortoise equipped with CORA and an auditory sensor would learn to approach at the call of a whistle, if the whistle was repeatedly blown just before a flash of light. Similarly, it would learn to move away from its current position when 'whistled', if the whistle had often been blown just before it touched an obstacle. One follower of Grey Walter built tortoises, called *Machina reproducatrix*, sensitive to various combinations of lamp, flute, and whistle (Angyan 1959).

Significantly, Grey Walter noted that his model of associative learning could "respond to a part of the significant association as if the whole were present" (Grey Walter 1956: 368). This, he said, was "essentially the same process" as pattern recognition—what Pitts and McCulloch (1947) had called knowledge of universals (Chapter 12.i.c). He didn't say, because he couldn't know, that part-to-whole generalization would be hailed thirty years later as one of the triumphs of parallel distributed processing (12.vi).

Grey Walter's last reported device, built in about 1953 and shown to the Ratio Club in 1955, was called IRMA: Innate Releasing Mechanism Analogue (Grey Walter 1956: 367–8). As the name implies, this was designed to model the ethologists' notion of an IRM: an innate propensity to respond in a specific way to a specific stimulus (see 5.ii.c). In developing IRMA, Grey Walter was especially interested in stimuli originating in the action of some other robot, so that the activity of two (or more) creatures could be coordinated in adaptive ways. Again, this robotic research would be largely forgotten, only to be taken up several decades later (13.iii.e).

If IRMA was the last device to be built by Grey Walter, it wasn't the last to be envisaged by him. Near the end of his working life (his research ceased when he was seriously injured in 1970), he remarked on "a new era" made possible by transistor technology:

> We are now envisaging the construction of a creature which instead of looking as the original did, like a rather large and clumsy tortoise, resembles more closely a small eager, active and rather intelligent beetle.
>
> There seems to be no limit to which this miniaturisation could go. Already designers are thinking in terms of circuits in which the actual scale of the active elements will not be much larger, perhaps even smaller, than the nerve cells of the living brain itself. This opens a truly fantastic vista of exploration and high adventure . . . (Grey Walter *c.*1968: 7)

Grey Walter's intriguing tortoises, despite their valve technology and clumsiness, were early versions of what would later be called Vehicles (Braitenberg 1984), autonomous agents, situated robots, or animats. They illustrated the emergence of relatively complex motor behaviour—analogous to positive and negative tropisms, goal seeking, perception, learning, and even sociability—out of simple responses guided and stabilized by negative feedback.

The "tropisms" of *Machina speculatrix*, for instance, emerged from a few core rules linking the speeds of the two motors to the level of illumination. In the dark, the drive motor would run at half-speed and the steering motor at full speed. In a moderate light, the drive motor would run at full speed while the steering motor was switched off. And in strong illumination, the drive motor and steering motor would run at full and half-speeds, respectively. These simple mechanisms gave rise to a wide range of observable behaviour.

Even more to the point, in an unpublished manuscript of about 1961, Grey Walter described a complete behaviour—finding the way past an obstacle to reach a light source—as being achieved by four reflex "behaviour patterns", some of which were "prepotent" over others (O. Holland, personal communication). The four basic patterns envisaged were exploration, positive and negative phototropisms, and obstacle avoidance. His analysis, in this case, was equivalent to that used in the 'subsumption architecture' of modern behaviour-based robotics.

During Grey Walter's lifetime, and for nearly thirty years afterwards, his tortoises—like Vaucanson's flute-player—were commonly dismissed by professional scientists as mere robotic "toys". The general verdict was that they were superficially engaging, but of little theoretical interest.

This largely negative reception was due partly to the vulgar publicity they'd attracted in the mass media. The brouhaha surrounding the tortoises put off even some of Grey Walter's fellow Ratio members, who were better placed than anyone to appreciate their significance. He wasn't the last to suffer the counter-productive effects of publicity. The connectionist Frank Rosenblatt would do so too, only a few years later, and symbolic AI would suffer similarly in the early 1970s (Chapters 12.iii and 11.iv, respectively).

But excessive media attention wasn't the only obstacle. As remarked above, Grey Walter himself never provided an extended account of his tortoises' theoretical implications. Having his father draft his books from his notes and conversations simply wasn't enough.

c. Less sexy, more surprising

The "mere toy" criticism was less often made of another interesting cybernetic gadget of the 1940s, Ashby's "Homeostat". For this *was* explicitly situated within a comprehensive theoretical context by its designer. Whereas Grey Walter kept most of his neuropsychological speculations to himself, Ashby didn't.

In addition, the Homeostat was less sexy, less media-friendly, than the tortoises. (Rhetorical considerations, again.) It didn't explore its environment, approach other Homeostats, or learn to come at the call of a whistle. Indeed, Grey Walter sarcastically dubbed it *Machina sopora* (sleeping machine), or "a machine designed to do nothing" (O. Holland, personal communication, and de Latil 1953: 308). Nor did it look like any sort of animal. It was intended not to replicate superficially lifelike behaviour, but to explore the most general features of life and mind: feedback and self-organized adaptation.

The Homeostat aroused interest, and scepticism too, when it was described at the ninth Macy meeting in 1952. Interest, and even stunned incredulity, because its behaviour wasn't carefully predesigned like ELMER's but emerged from a random starting point (see below). Scepticism, because Ashby insisted that it was to be understood as a purely *physical* mechanism. Even the hard-headed engineer Bigelow was nonplussed by this, declaring that whereas Shannon's Rat—which stored "memories" of its encounters with the maze—could be said to learn, the Homeostat could not. One might as well say, he complained, that a ball-bearing that falls through a hole in the bottom of a shaken box has "learned to find the exit" (Dupuy 2000: 150).

Despite the initial scepticism, the Homeostat would later inspire important research in both connectionism and A-Life. In the 1950s, however, it was too far ahead of its time to be followed up (improved on). Indeed, it was too radical even to be properly appreciated by most people.

Ashby's definition of cybernetics was even more inclusive than Wiener's:

> [Cybernetics] takes as its subject-matter the domain of "all possible machines", and is only secondarily interested if informed that some of them have been made, either by Man or by Nature. What cybernetics offers is the framework on which all individual machines may be ordered, related, and understood. (Ashby 1956: 2)

In principle, then, any physical system whatever is grist to the cybernetics mill. But Ashby, like Grey Walter, was a neurologist (and he too worked at the Burden Institute, becoming Director for a brief period at the end of the 1950s). As such, he was most interested in biological examples.

He was interested in everyday homeostasis, as all the cyberneticists were. But he was especially intrigued by the startling reorganizations implied by spontaneous adaptation to inverting spectacles, which had been reported by George Stratton in the 1890s (see 14.viii.b). (A recent computer model of such reorganizations, explicitly based on Ashby's work, will be described in Chapter 15.xi.a.) Similarly, he was intrigued by Roger Sperry's (1947) then recent discovery that animals could recover from the surgical 'cross-wiring' of antagonistic muscles.

From the outset of the 1940s, and independently of Wiener, he conceptualized biological adaptation in terms of feedback-based equilibrium. Adapted behaviour is "the behaviour of a system in equilibrium", which can be described in mathematically rigorous terms (Ashby 1947). But whereas Wiener treated feedback as a fundamental concept, Ashby defined it in terms of difference and change in dynamical systems:

> The most fundamental concept in cybernetics is that of "difference", either that two things are recognisably different or that one thing has changed with time... All the changes that may occur with time are naturally included, for when plants grow and plants age and machines move some change from one state to another is implicit... (Ashby 1956: 9)

> That a whole machine should be built of parts of given behaviour is not sufficient to determine its behaviour as a whole: only when the details of coupling are added does the whole's behaviour become determinate...
>
> Cybernetics is... specially interested in the case where each [of two parts] affects the other... When this circularity of action exists between the parts of a dynamic system, *feedback* may be said to be present. (Ashby 1956: 53)

> ... the concept of "feedback", so simple and natural in certain elementary cases, becomes artificial and of little use when the interconnexions between the parts become more complex. When there are only two parts joined so that each affects the other, the properties of the feedback give important and useful information about the properties of the whole. But when the parts rise to even as few as four, if every one affects the other three, then... knowing the properties of all the twenty [possible] circuits does *not* give complete information about the system. Such complex systems cannot be treated as an interlaced set of more or less independent feedback circuits, but only as a whole.
>
> For understanding the general principles of dynamic systems, therefore, the concept of feedback is inadequate in itself. What is important is that complex systems, richly cross-connected internally, have complex behaviours, and that these behaviours can be goal seeking in complex patterns. (Ashby 1956: 54)

Ashby specifically included the neural control of behaviour as an example of adaptation (Ashby 1940, 1947). As he put it:

> The brain is a physical machine... Adaptive behaviour is that which maintains it in physical equilibrium with its environment. (Ashby 1947: 58)

> "Adaptive" behaviour is equivalent to the behaviour of a stable system, the region of the stability being the region of the phase-space in which all the essential variables lie within their normal limits. (Ashby 1952: 64)

The kitten who learns to avoid a hot coal on the carpet, he said, does so by a neural mechanism embodying feedback, whereby its brain adapts to the unwelcome stimulation. Its trial and error learning is not merely (as the early behaviourists implied) a blind effort at success, but a way of gaining specific information about the environment and how to adapt to it. And errors aren't valueless failures, but essential parts of the information-gathering process.

That process could be thought of as information gathering or as self-equilibration. Similarly, physiological processes as such could be thought of in either way. Talk of "information" was the more likely when *mind*, or behaviour, was in question. But life and mind were essentially similar, comprising many dynamical systems coupled to, and nested within, each other. In the words of Ashby's French admirer:

> As soon as we introduce the idea of functional activity [as opposed to "event"-driven causality], we are able to enter into a realm of thought where all is relative and the concepts themselves are not absolute. Within this realm each of the various systems seeks to attain its own particular state of equilibrium; but all are involved in ever vaster systems which again are themselves in search of equilibrium. Such a complex in a continuous process of equilibratory activity is the central nervous system. (de Latil 1953: 353)

In his books *Design for a Brain* (1952) and *An Introduction to Cybernetics* (1956), Ashby applied his view of "The Organism as Machine" to behaviour as well as to homeostasis. In doing so, he defined a form of negative feedback ("ultrastability") that could account for self-organization from *random* starting points.

He'd illustrated this principle some years earlier, in his analogue machine, the Homeostat—first demonstrated at the meeting of the Electroencephalographic Society on May Day 1948. There was no crawling, no bumping and turning away, no coming at the call of a whistle—in short, nothing to delight the kiddies. This was a machine for theorists.

d. How the Homeostat worked

The Homeostat consisted of four square boxes, each with an induction coil and a pivoting magnetic needle on top (Ashby 1948: 380 ff.). Current changes in a coil would cause its needle to move. But, because of a feedback system, these movements would cause further changes in the current. Each needle carried a wire dipping into a water trough, and the needle movements were affected by the electrical potential of the water at the dipping point. Electrodes were placed at the ends of each trough, providing a potential gradient in the water. And these gradients were influenced by the current generated by the induction coil.

So far, so boring: four magnet–trough–coil units, each embodying feedback. The interest of the Homeostat arose from the fact that these four feedback units were interconnected in a flexible way so that it could maintain the set of needles in a stable equilibrium.

Whenever any one (or more) was deflected, a coordinated activity would return them all so as to point to the centre of the four-box device. This was possible because each box sent its output to the other three (and so received input from them also): "As soon as the system is switched on, the magnets are moved by the currents from the other units, but these movements change the currents, which modify the movements, and so on" (Ashby 1952: 96).

The four main variables of the Homeostat were the positions of the four magnetic needles. Its "essential variable", corresponding to the viable range of blood temperature or blood sugar (for instance), was a maximum value of the current being output by any box. If the outputs of all four boxes fell below this maximum, each would be pointing to the centre.

The feedback links between the boxes could be continually altered by a stepping switch, or "uniselector". If—and only if—the maximum allowable output current from some box was exceeded, the switch would close so as to move the box's uniselector to a different position. This changed the quantities defining the relevant feedback link, so diverted the machine onto a different path through the space of its possible states.

One might even say, instead, that the uniselector *changed one machine into another*, defined by a different feedback function. Ashby himself pointed out that, where there are various levels of dynamical coupling, what we decide to call "one" system is largely a matter of convenience (Chapters 13.vii.b–c and 16.vii.d).

If the parameter values of a single-level Homeostat were to be set inappropriately, it would embark on an uncontrolled "runaway" until the magnets hit the ends of the troughs (Ashby 1952: 96). But in the double-level Homeostat, where second-order feedback ensured that the first-level controls were reset so as to achieve stability, this didn't happen. The parameter values (represented by the positions of the uniselectors) could be randomized, and the Homeostat would still achieve stability. Ashby even outdid Sperry, by crossing the wires from the mains switches so that input became output, and vice versa: again, the machine reached equilibrium.

Since each box's uniselector could take up twenty-five different positions, the system could in principle reach its end point from any one of 390,625 distinct starting conditions. Or rather, it could reach one of the many acceptable end points: the stable state defined as "four needles pointing to the centre" could be implemented by many different states defined in detailed physical terms. However, the Homeostat didn't always equilibrate quickly. In practice, the convergence would sometimes take a very long time. And it couldn't learn. Having reached an acceptable position by chance it stayed there, but its past achievements didn't improve its future performance.

Ashby saw this principle of double feedback as essential to the dynamics of physiological equilibrium and (the brain states underlying) learning and purposive behaviour. And he developed a mathematical theory of ultrastability, pointing out that the Homeostat illustrated only one example of a wide range of possible forms, many of which are found in biological systems (Ashby 1952, ch. 9).

Strictly, he said, a living organism isn't an ultrastable system: rather, it is multistable. In other words, it's a complex system made up of many ultrastable subsystems whose main variables are dynamically linked to each other (Ashby 1952, ch. 16). Ashby distinguished various types of multistability differing (for instance) in the degree and frequency of these subsystem interactions. The body's subsystems, he said, adapt to each other "in exactly the same way as animal adapts to environment" (Ashby 1952: 174).

The biologically typical form of multistability—where subsystems affect each other only weakly, or occasionally, or indirectly—was illustrated by a particular version of the Homeostat. Ashby showed that stability would be achieved even when the number of units in active combination varied from time to time. And he considered the question of just how long it would take for various types of multistable system to adapt successfully to changes in their (internal or external) environment.

In brief, Ashby originated a systematic formal theory of adaptive self-organization in dynamical systems (see 12.ii and 15.vii–viii and xi). As for its explanatory potential, he made sweeping—not to say grandiose—claims about the capabilities of future versions of the Homeostat (Ashby 1948: 383). But he himself acknowledged that they might, in effect, turn out to be black boxes, their behaviour being "too complex and subtle for the designer's understanding" (ibid.).

His more down-to-earth contemporaries were sceptical. Some complained that his stress on communication within the body, and especially the brain (Ashby 1960: 218–24), couldn't immediately be applied to neurology. So the psychophysiologist Karl Lashley, also a Macy participant, pointed out that Ashby's (and McCulloch's) account of the brain involved "a very great oversimplification of the problem", and declared that doing research on the brain itself should take priority over "indulging in far-fetched physical analogies" (Lashley 1951*b*: 70, and quoted in F. A. Beach *et al.* 1960, p. xix).

Detailed application of cybernetic and computational theories to brain function would come only later. This required powerful computers, in place of simple devices cobbled together on a hobbyist's workbench, and much-improved knowledge of the brain (see Chapters 14 and 15.vii).

4.ix. Schism

By the end of the 1940s, there were two ways of thinking about mind as machine. One was symbolic–computational, the other cybernetic–dynamical. (Early connectionism lay on both sides of the fence: see Chapters 12.i–ii and 14.iii.) One might even say that the former focused on *computer and mind*, the latter on *computer and brain*. For as we've seen, the cyberneticists—besides being more interested in the brain than the committed symbolists were—had little or nothing to say about specific mental contents. Adaptation, self-organization, feedback, purpose, reverberating circuits . . . took precedence over (for instance) doing logic or using natural language.

* The two approaches focused on different types of control, favoured different types of mathematics (discrete or continuous–statistical), and took different wartime computers as their artificial exemplars.

* Their technological fruits differed. One led to (digital) symbolic AI, the other to (analogue) control engineering and dynamical forms of connectionism.

* And their historical sources were distinct, as we've seen (see also M. W. Wheeler *et al.* forthcoming). The symbolic approach was nurtured by propositional logic and logicist philosophy of language. Cybernetics was grounded in control engineering, general physiology, reflexology, and information theory—which drew, in turn, on the mathematics of probability.

Nevertheless, the scientists concerned were collaborators agreeing to differ, rather than rivals—still less, enemies. Wiener himself, the high priest of cybernetics, had outlined a digital computer in 1940 (see iv.a, above). A few individuals explored both approaches in detail: McCulloch, for instance, defined neural networks both logically and statistically. And most people communicated willingly, feeling themselves to be involved in the same general project.

This cooperative spirit was still evident in the 1958 meeting on 'The Mechanization of Thought Processes' at the UK's National Physical Laboratory, or NPL (6.iv.b). Indeed, it survived for a few years after that: "Through the early 1960s, all the researchers concerned with mechanistic approaches to mental functions knew about each other's work and attended the same conferences" (Newell 1983: 201).

That happy state of affairs didn't last. The initially convivial mind-as-machine community split into different, even opposing, camps. There were two engines driving this split. The first was fuelled by some of the less admirable features of Everyman, the second by genuine theoretical differences.

a. All too human

By the mid-1960s, two distinct groups had emerged, namely formal computationalists and (what are now called) cyberneticians. For many years afterwards, there was little love lost between them. In other words, the Legend was tarnished, again (1.iii.b).

This splitting was due partly to the availability of two methodologies, involving different types of skill: 'logical' programming, and building or simulating dynamical or probabilistic systems. Both were used to study circular causation, or feedback. And they could be seen as theoretically equivalent at some very fundamental level, for both could be described by Boolean logic and modelled by a Turing machine (or rather, by a Turing machine with time specifications added). But in practical terms they were different.

The importance of tacit practical skills in science is too often forgotten, as though theoretical knowledge alone were enough (but see Polanyi 1958, 1967). The historian Donald MacKenzie, for instance, has criticized the over-quick assumption that atomic weapons can't be "un-invented": a large part of the Manhattan Project involved practical engineering rather than scientific theory, and the former isn't fully preserved in the library books (MacKenzie and Spinardi 1995). Indeed, some critics argue that the importance of hands-on manipulation and other kinds of tacit knowledge is an insuperable obstacle to human-level AI (see 13.ii.b).

In any event, and quite apart from specific theoretical disagreements and different target topics (such as adaptation versus meaning: see below), people who feel comfortable with different methodologies typically feel uncomfortable with each other. They won't even share—or understand—the same private jokes. It's not surprising, then, that the two computational communities grew increasingly distinct in sociological terms too.

The natural territoriality of Everyman, the desire to carve out and defend one's own patch, did nothing to prevent the schism. Far from it. In the mid- to late 1960s Minsky contrasted 'Artificial Intelligence and its Cybernetic Background' like this:

> While work on artificial intelligence draws upon methods from other fields, this is *not* a significantly interdisciplinary area: it has *its own concepts*, *techniques*, and *jargon*, and these are slowly growing to form an intricate, organized *specialty*. (1968: 6; italics added)

He even went on, cheekily, to define Newell-and-Simon *out* of what he was calling "AI":

> The [digital] computer made it practical to be much more ambitious [than the cyberneticians had been]. As a result, cybernetics divided, in my view, into three chief avenues.
>
> The first was the continuation of the search for simple basic principles—most clearly exemplified by the precise analyses of Ashby...

> The second important avenue was an attempt to build working models of human behavior incorporating, or developing as needed, specific psychological theories . . . Work in this area—Simulation of Human Thought—has focussed rather sharply at the Carnegie Institute of Technology . . . in the group led by Newell and Simon.
>
> The third approach, the one we call *Artificial Intelligence*, was an attempt to build intelligent machines without any prejudice toward making the system simple, biological, or humanoid . . . Much of the earlier work on artificial intelligence was done by the group at MIT . . . (pp. 7–8)

Eventually, mere difference led to mutual ignorance, not to say emotionally loaded schism. The enmity didn't involve "blood-boiling animosity", as disputes over Noam Chomsky's linguistics later did (9.viii.a). And the leaders didn't "hate each other's guts", as some feuding connectionists would (12.v.b). Nevertheless, the atmosphere was far from congenial.

In technological contexts (such as industrial research laboratories, university engineering schools, and equipment manufacturers) the schism started early. In the late 1940s and 1950s, these institutions constituted "a major source of resistance to the emerging digital paradigm, especially when it came to using the new machines for purposes other than mathematical calculation" (Edwards 1996: 68). One servo-engineer recalled later that analogue computer experts felt "threatened" by digital computers, and that (partly because of the huge post-war demand for control engineers) "only a few" of his peers were able to make the switch. (Besides professional jealousy, part of the problem was the genuine doubt "that a machine containing vast numbers of vacuum tubes could ever function for more than a few minutes at a time without breaking down": Edwards 1996: 69.)

Within cognitive science, where such practicalities could more easily be ignored, the group rivalries and jealousies didn't emerge until the late 1960s. On my visits to the apple orchard in the late 1950s (Preface, ii), and even at early 1960s Harvard, I heard plenty of healthy disagreement—but never (what I often encountered later) champions of one theoretical style mocking, even vilifying, those of the other. At mid-century, there was one heterogeneous research group rather than two homogeneous ones.

The invective appeared later, especially after symbolic AI achieved public prominence—and generous research funding—in the early 1970s. The merciless attack on connectionism by two of the high priests of symbolic AI didn't help (Minsky and Papert 1969). Early drafts of this attack, described by one insider as dripping with "vitriol", had been circulating for several years before the (toned-down) final version was published (see Chapter 12.iii). So there was plenty of time for resentment to grow, followed by a twenty-year funding famine to make it fester.

When the worm finally turned in the 1980s (12.vii), and the funding—and the glamour—was reversed, the invective didn't disappear. Too often, resentment on the part of the 'cyberneticists' gave way to ill-considered triumphalism.

Moreover, the excessive media hype previously attached to symbolic AI was now transferred—more accurately: transferred back again (Chapter 12.ii.g and iii.a)—to the connectionist variety. At each swing of this journalistic pendulum, some group of researchers suffered chronic irritation that often spilt over into personal abuse of those in the other group. The explosion of A-Life in the 1990s also soured the intellectual atmosphere with ungenerous denigration of opponents, not to mention yet more media hype guaranteed to exasperate the 'enemy' (15.ix.a).

One indication of this sorry state of affairs can be found in a recent number of the newsletter of the UK/Europe's oldest AI society. The editorial bemoaned the "many cliques" in the field, and the "frankly insulting" names often used by researchers for approaches different from their own. As the editor put it: "The lack of tolerance [between different research programmes] is rarely positive, often absurd, and sometimes fanatical" (Whitby 2002*b*). Outside academia, he continued, where theory is of less concern than practical efficacy, "different approaches to AI cohabit not only in the same office, but often in the same program" (see Chapter 12.ix.b and d). (For example, the Clementine commercial data-mining system includes *both* the ID3 machine learning program *and* a Kohonen network, *and* some evolutionary computing too: Whitby 2004; Khabaza and Shearer 1995.) These self-serving attacks on other forms of AI have even contributed to today's widespread, but mistaken, view that AI has "failed" (13.vii.b).

Granted, people recommending new (or newly revived) ideas typically overplay their hands. It's a good way of getting attention. But it's liable to backfire. It can cause fruitless squabbling, and unnecessary blindness to what "the others" have seen.

The Harvard sensory psychologist Edwin Boring complained about such behaviour in chastising the New Look psychologists (Chapter 6.ii)—*and* in recalling "Titchener's in-group" of many years before. He'd observed the same "digs" against opposing schools in psychology, the same "unsophisticated and unscholarly and biased ... aggressions about what [the speaker] did not care to understand" because it didn't come from the relevant in-group (letter to J. S. Bruner, in Bruner 1983: 91). Certainly, he said, in-groups can work "integratively" as well as "disruptively". But reaching the truth would require cooler heads, and less self-serving rhetoric.

The same applies here. If, as I claimed in Chapter 1.ii.a, there are *two* fruitful pathways through the many meadows of cognitive science, then we shouldn't be barred from either of them.

b. Adaptation or meaning?

Besides the all-too-human reasons for schism described above, there were substantive differences too. The two methodologies seemed to line up with two different areas of interest: adaptation (and life) on the one hand, and meaning (and mind) on the other.

To put it another way, the main controversy concerned the interpretation of *purpose*, and the place—if any—for *meaning*. Could Ashby's theory, or even Rosenblueth's, really explain purpose? Perhaps—as Craik had suggested—the modelling of purposive behaviour required internal representations of the goal, understood as symbolic structures in the artefact concerned (10.i.d)? More generally, there was discontent over information theory's flight from meaning (see Section v.d, above). Surely, psychologists must often pay attention to specific semantic content, whether of purposes or beliefs?

For example, Ashby was soon criticized by MacKay (1956*b*) for ignoring "goal-guided" behaviour, which involves some sensitivity to the required direction of action. Ashby's exemplary kitten and the Homeostat were indeed acquiring information that enabled adaptive change, but their actions ("trials") occurred at random. In the Homeostat, for instance, the difference reduction (between actual and permissible levels of output current) was achieved by chance. By contrast, difference reduction in guided missiles offered some teleological guidance as to the most promising line of action.

And hierarchical means–end modelling, soon to be developed in symbolic AI, would provide even more. (Today, half a century later, it's still disputed whether any form of dynamical systems theory can suffice to explain purposive behaviour: see 16.vii.c.)

The division in the mind-as-machine community that was brewing throughout the 1950s was (and still is) due also to differing views on the relation between *life* and *mind.* The cyberneticists assumed that broadly similar sorts of explanation (based in circular causation) are required for biology and psychology. Not only is the mind a machine, but it is essentially the same sort of machine as the living body is.

The computationalists, by contrast, focused mainly on logical–linguistic thinking, problem solving, and purposeful planning. When they studied non-linguistic capacities, such as vision and locomotion, they assimilated them to logical thought (Chapter 10.iv.b). But they didn't assume that *metabolic* systems work by logic. On their view, then, the human mind is a machine that differs significantly from the biological machine in which it's embodied.

For the next four decades, most—though not all—theorists who took *mind as machine* as their guiding theme concentrated on the non-biological issues. Their ambition was to further AI and computational psychology, not A-Life or neurophysiology. The early cybernetic work was either ignored or assimilated to connectionist AI.

Even when Turing and von Neumann themselves turned (in the early 1950s) to discuss life, few mind-as-machine researchers listened. Partly, this was because the lack of computer power didn't enable their ideas to be explored much further (see 15.iv–v). But the prime reason was that such researchers were interested in more evidently *psychological* (including linguistic) problems, which they addressed by way of symbolic and/or connectionist AI.

Connectionism, in turn, soon faltered, not rallying until the 1980s (Chapter 12). Meanwhile, in the 1960s–1970s, symbolic AI held court (Chapters 7 and 10). Only in the 1990s would ideas of *life* and *mind* be prominently drawn together again (15.ix and 16.x).

In sum, the "logical" side of the mind-as-machine divide gained the ascendancy around 1960 because it was better fitted to deal with goal-directed action and propositional meaning. The newly available digital computers could symbolize an indefinite variety of thoughts. (Whether these were *real* "meanings" wasn't, then, much discussed: see 16.v.c.) In psychology, linguistics, AI, and philosophy, the 1960s were largely logicist. Even neurophysiology was emphasizing matters, and models, more logical than dynamical. Explicitly logicist definitions of 'cognitive science' flourished (Chapter 16.iv).

But the dynamical stream was still flowing, underground. It would resurface in the last twenty years of the century, first with connectionism (Chapters 12 and 14) and then with A-Life (Chapter 15). The dual theoretical heritage of cognitive science, and the see-sawing ascendancy of one side or the other, is a central theme in its history.

5

MOVEMENTS BENEATH THE MANTLE

Sir Isaac Newton was not a nice man. His personal unhappiness (Manuel 1968, ch. 3) often fuelled intemperate attacks on others, including his social inferiors. One such was his Royal Society colleague Robert Hooke. Although Hooke was a scientific rival (with a competing theory of light), and the first to suggest (in 1679) that the planets move under some influence inversely proportional to the square of their distance from the sun, he was menially employed as a technician (Pumfrey 1991; Shapin 1994: 395–403). Newton rarely minced his words in criticizing him, and many others of his acquaintance too.

Yet Newton is now widely regarded as a model of magnanimity, because of his oft-quoted remark: "If I have seen further it is by standing on the shoulders of Giants." He said this in a letter to Hooke, one of several written at that time in which he fulsomely complimented his long-time adversary (Manuel 1968, ch. 7).

These compliments weren't intended seriously, however. Indeed, Newton's apparent modesty conveyed a venomous personal insult. His remark was a commonplace, dating back five centuries to John of Salisbury—for whom it was already second-hand:

> Bernard of Chartres used to compare us to [puny] dwarfs perched on the shoulders of giants. He pointed out that we see more and farther than our predecessors, not because we have keener vision or greater height, but because we are lifted up and borne aloft on their gigantic stature. (John of Salisbury 1159: 167)

John's words had often been quoted—for instance by the translators of the 1611 King James Bible, defending their reuse of previous translations if these weren't clearly wrong (McGrath 2001: 176). So Newton's contemporaries, including Hooke himself, would inevitably be reminded of puniness and dwarfs.

—And the punchline? The unfortunate Hooke was a tiny hunchback, described by an acquaintance as "[physically] but despicable, being very crooked . . . [and] but low of Stature, tho' by his limbs he shou'd have been moderately tall" (Manuel 1968: 137; cf. L. Jardine 2003). Magnanimity, this was not.

But if Newton was no *moral* exemplar, perhaps he exemplified good *science*?

David Hume certainly thought so: he explicitly modelled his 'scientific' psychology on Newton's laws of physics (see Chapter 2.x.a). The nineteenth-century associationists apparently agreed. So, too, did the twentieth-century behaviourists, who dominated American academic psychology for over forty years.

Behaviourism acted as a rough-spun mantle, thrown hastily over the body of psychology (and neurophysiology) and hiding it from view.—*Very* hastily: the 35-year-old John Watson was elected President of the American Psychological Association (APA) only two years after publishing the first behaviourist paper (Watson 1913). To be sure, the mantle eventually became less rough-spun. By the 1950s, behaviourism was much more sophisticated. But it was still obscuring other psychological ideas.

Admittedly, that wasn't true of Europe: one European psychologist, recalling the 1960s, described behaviourism as "completely absent from my horizon" (see 6.iv.d). But in the USA, where most professional psychologists were based, things were very different. George Miller, interviewed a quarter-century later, said: "The power, the honors, the authority, the textbooks, the money, everything in psychology was owned by the behaviorist school" (G. A. Miller 1986: 203).

Cognitive science arose when the mantle was shrugged off. If there was a cognitive "revolution" in the late 1950s, it consisted in the overthrow of the core assumptions of behaviourism—listed in Section i, below.

That overthrow was due in part to new ideas. These came from mid-century psychoneurology, and from late 1950s AI, linguistics, and informational psychology (see Chapter 6). But it also involved older ideas, already moving more strongly beneath the mantle as mid-century approached. These psychological concepts, highly unfashionable during behaviourism's heyday, would later—in some cases, only much later—play a positive role in cognitive science.

Those first-half-century approaches are outlined here. So Section ii discusses relevant theories of personality, cognition, and biological aspects of behaviour. Some anticipatory "softening" of the behaviourist orthodoxy is described in Section iii. Finally, the entry of neurology is reviewed in Section iv.

In short, this chapter concerns pre-computational work. The only person discussed at length, namely Donald Hebb (in Section iv.b–f), would later be hugely important for three areas of cognitive science: psychology, neuroscience, and connectionist AI/A-Life. But he wasn't a computationalist. So, apart from a quick peep over the parapet in Section iv.f, man-as-machine won't raise its head until Chapter 6—which describes the excitement that ensued when a new choreography was offered to describe the movements going on beneath the mantle.

5.i. Newtonianism

The cognitive revolution can't be understood, or evaluated either, unless one's aware of what it was reacting against. In one word, this was behaviourism. But more than one word is needed to explain just what it was about behaviourism which was challenged by cognitive science.

a. The six assumptions

Behaviourism was 'Newtonian' in six ways:

First, it was atomistic. The key theoretical categories, *stimulus* and *response*, were thought of as separately identifiable and unstructured, like a point mass.

Next, it was associationist. Stimulus and response were linked by contiguity, not structure. (Think billiard balls.)

Third, it was universalist. Where Newton had assimilated falling apples to orbiting planets, behaviourism posited psychological laws applying across all domains and many species. So people were said to learn language in much the same way as laboratory pigeons learn to play ping-pong—and, a fortiori, in much the same way as they learn table manners (see 8.vi.b).

Fourth, it was externalist. Everything the creature does, apart from a few inbuilt reflexes—like showing fear at loud sounds, as baby Albert B. famously did (J. B. Watson and Rayner 1920), was attributed to environmental influences. Left to itself, nothing would change—like a physical mass with no external force to act on it. Nativism (the claim that some psychological abilities are innate) and self-organization, then, were alike anathema.

Fifth, it was meaning-less. Much as point masses have nothing to do with meaning, so "stimulus" and "response" were operationally defined as *non-intentional* (non-mentalist) categories.

And last, it avoided hypotheses about hidden causes. Newton famously refused to speculate on the cause of gravity: "Hypotheses non fingo." The previous "mechanical" philosophy had seen physics in terms of causation by contact (see Chapter 2.ii, preamble). Similarly, behaviourists treated the organism as a black box. (For the record, some scholars believe Newton meant not "I offer no hypotheses", but "I'm not inventing fictions": P. M. Williams, personal communication.)

The behaviourists didn't doubt that neural mechanisms cause observable behaviour. Indeed, Ivan Pavlov and his fellow reflexologists were adopted as patron saints. But they themselves said almost nothing about the brain. Very little was known about it, in any case (see Chapter 2.viii).

They also avoided talk of mental causes. This was due less to philosophical worries, such as those expressed by Princess Elizabeth (2.iii.b), than to the practical difficulties—already familiar by the turn of the century—in getting reliable introspective reports from laboratory subjects. And dumb animals, of course, couldn't offer such reports anyway. The notion that there may be non-introspectible (non-conscious) mental processes was given even shorter shrift. As for meaning, purpose, and intentionality, these were mentalistic notions to be avoided at all costs. Minds were out, brains were ignored, behaviour was all.

As a result, even perception—surely, or so one might think, a mentalistic concept—was studied in fully objective, non-cognitive terms. And thinking was *hors de combat.* Jerome Bruner remembers the mainstream "high status" textbooks thus:

> When I was a graduate student, Woodworth's *Experimental Psychology* [1939] was *the* book.... It boasts 823 pages of text. By a generous count, the topic of thinking is treated in two brief chapters... (including animal studies)... All told, 77 pages. When the prestigious *Handbook of Experimental Psychology* appeared in 1951... it had 1,362 pages. This time, the topic was disposed of in a chapter called "Cognitive Processes": 27 pages. Add a few to that, for George Miller had a chapter on "Speech and Language", the last few pages of which were devoted to the relation of thought and language. The mind was not doing well in psychology. The eye, ear, nose and throat fared far better: nine chapters, about four hundred pages. (Bruner 1983: 106–7)

It will be helpful (I hope!) to bear my sixfold sketch of behaviourism in mind over all of the following chapters. But it's no more than that: a sketch. It isn't "the whole truth".

It isn't even "the truth and nothing but the truth". For the behaviourists didn't always follow their own diktats, as we'll see in Section iii.b below. The Harvard psychologist Burrhus Skinner (1904–90) came closest. His departmental colleague Edwin Boring (1886–1968), in his magisterial *History of Experimental Psychology*, described him as dealing with "the empty organism", because his theory was empty of neurones and almost empty of "intervening variables" (Boring 1957: 650). But even Skinner was covertly non-Newtonian in some ways (iii.a below, and Chapter 9.vii.b).

All six behaviourist assumptions were questioned by the cognitive revolution. Computational psychology did offer causal hypotheses. These were couched in mental/computational terms. And even in the 1940s, some were also interpreted as brain processes, broadly defined (see Chapters 4.iii.e and 12.i.c, and Section iv below). Eventually, more detailed theories of neural computation would emerge (Chapter 14).

b. What sort of revolution was it?

One might say that the cognitive revolution was made "official" in 1960, when the President of the APA used the term in the title of his Presidential Address (Hebb 1960). But what sort of revolution was it?

One shouldn't describe this anti-Newtonian rebellion as a *Kuhnian* revolution, though people sometimes do (O'Donohue 1993). The strongest point in their defence is that, by the 1950s, the most influential behaviourist theories were seemingly lost in untestable speculations about hidden Ss and Rs (see iii.b, below). In other words, behaviourism was running into trouble *even in the eyes of its sympathizers*—a state of affairs which Thomas Kuhn (1922–96) saw as the characteristic prelude to a "paradigm shift" (Kuhn 1962).

Perhaps that's why, as one psychologist remembers:

> It was just a tremendously exciting time [in 1968]. The basic assumption was that things were boiling over, and of course there was going to be a lot of argumentation and debate, but a new day was coming. And of course, everybody toted around their little copy of Kuhn's *The Structure of Scientific Revolutions.* (Jenkins 1986: 249)

However, the "new day" wasn't bringing wholly new ideas. Many of the ideas central to cognitive science had been around for years in continental Europe and the UK (see Section ii). They'd been spurned by the behaviourists—but, as George Mandler (1924–) has pointed out:

> American psychologists sometimes fail to understand that behaviorism was a *very* parochial event... [The] reaction of the Europeans during the '30s, '40s, and early '50s when [Americans] suffered through behaviorist orthodoxy, was "What is going on over there?" They paid very little attention to it. (G. Mandler 1986: 259)

In other words, the "revolution" could almost be seen as a *counter-revolution*. Mandler, himself a refugee from the old world, was struck less by the newness of these ideas than by their familiarity: "The so-called 'cognitive revolution'... was not so much a revolution as a return to an *ancien regime*" (2002*c*: 256).

Just what Kuhn meant by a scientific paradigm is rarely considered with care. (And we can't do so here: there were no fewer than twenty-one senses in his book, not all mutually compatible—Masterman 1970.) Broadly, however, it refers to some unquestioned base of assumptions (and preferred methodologies), accepted—rather than argued—by everyone in the field concerned.

So, as Mandler pointed out, behaviourism wasn't a paradigm for mid-century European psychologists. Cognitive science isn't a paradigm either. Even *within* cognitive science, the "argumentation and debate" hasn't finished. There's still fundamental disagreement—which is why, in Chapter 1, we couldn't simply launch straight into our story. As for professional psychology as a whole, many researchers still start from assumptions very different from those of cognitive science (see 7, preamble, and 7.vii.e).

Early computational psychology reconnoitred *both* of the two mind-as-machine footpaths (Chapters 1.ii.a and 4.ix). In the late 1950s, a single volume of the *Psychological Review* carried seminal papers scouting each route (Newell *et al.* 1958*a*; Rosenblatt 1958). Initially, the logical footpath was explored more fully. This was partly because it seemed more apt for representing meaning (Chapters 4.ix.b and 16.v.c), and partly because the cybernetic scout soon met a major obstacle on the way (Chapter 12.iii). Later, as psychological theories of self-organization blossomed, the cybernetic pathway became well-trodden too.

So the passage from behaviourism to cognitive science *wasn't* a shift from one Kuhnian paradigm to another. (Psychologists aren't the only ones over-ready to use the popular concept of paradigm: I'm reminded of the newspaper cartoon showing a newly hatched chick crying "Wow! Paradigm shift!")

The term *revolution*, however, is more apt—and it's widely used (e.g. de Mey 1982; Bruner 1983: 277; H. Gardner 1985; Baars 1986; D. M. Johnson and Erneling 1997). To be sure, one of the prime revolutionaries—Miller, whose frustrations with behaviourism were quoted above—insists that it wasn't really a revolution, but an "accretion" of old ideas to newer ones (Miller 1986: 210). This is much the same point as I made in speaking of a counter-revolution. For that there was *significant change* is incontestable.

Besides the accretion of new (computational) ideas, the empiricist psychological assumptions that had guided Hume and his intellectual successors for over 200 years were abandoned *by some of those most sympathetic to them.*

That's not to say that all six core assumptions were dropped at once. Universalist associationism is still robust today—though recently dubbed "the phlogiston of psychology" (14.ix.g). But many empirically oriented people adopted a new way of doing psychological science.

5.ii. Psychology's House

In the late 1950s, when cognitive scientists first entered the house set aside for Psychology, there were already many tenants lodging there. The most comfortable suites held the behaviourists—often unaware of the other inhabitants ("For a while in the 1920s it seemed as if all America had gone behaviorist"—Boring 1957: 645). But the less-favoured rooms were long-occupied too.

The longest-established sitting tenants were the Freudians, Gestalt psychologists, Piagetians, and ethologists. More recent entrants included the first Gibsonian psychologists, and the Third Force personality theorists. The once-fashionable William McDougall had been banished to the attic, but his grumbling could still be faintly heard.

Some early cognitive scientists took the sitting tenants seriously from the start. Others didn't. Eventually, most came to recognize the insights of these older traditions.

The more we consider mid-century *American* researchers, the truer this is. For behaviourism (as already remarked) was an American malady, not a European one. Mandler remembers being "asked, in all seriousness", by a senior British psychologist in 1965 "whether anybody in America really believed any of the behaviorist credo" (2002*a*: 344). And he adds:

> The important aspect of European psychology of the time was that not only was Europe *essentially unaffected and uninfluenced by behaviorism*, but also that *the developments in Europe became part of the American mainstream* after the decline of behaviorism. (pp. 343–4; italics added).

Long before the late 1950s, when American behaviourism first came under concerted fire, ideas which today would be called 'cognitive' were flourishing—indeed, dominant—in Britain, Germany, France, and Canada. That's why psychologists from those countries feature prominently in this chapter.

a. Sitting tenants with personality

When the early cognitive scientists were first given house room, they had no wish to invite the behaviourists to tea. For behaviourism was defined as "the enemy" (see Chapter 6.iv.d). But many were also loath to invite the other inhabitants.

For instance, in the 1960s–1970s neither Gestalt psychology nor Piagetian theory was taught in the Experimental Psychology group at the University of Sussex—at that time, the best psychology department in the UK. The group was headed by N. Stuart Sutherland (1927–98), one of the first British psychologists to embrace cognitive science. There was no question, however, of his embracing earlier thinkers whom he regarded (rightly) as less than rigorous. So he chided me on my arrival in 1965 for "wasting my time" on McDougall, and said much the same twelve years later when I was writing on Jean Piaget (Boden 1972, 1979).

As for Sigmund Freud, his name was mud. Sutherland wrote a passionate, part-autobiographical, attack on Freudian analysis as a clinical tool (1976). This attracted huge attention: it was extracted in the Sunday newspapers, went into several revised editions over the next twenty years, and was even made the basis of a West End (and off-Broadway) play.

One may share (as I do) most of Sutherland's scepticism about Freudian theory and therapy. Not least, one may deplore Freud's outrageously unscientific "evidence" for the Oedipus complex. This was based largely on leading questions about a little boy's dreams about giraffes—which weren't even posed to the child by Freud himself, but fed to him through the father (S. Freud 1909). Nevertheless, five aspects of Freud's work are highly relevant for cognitive science:

* First, he focused on particular intentional phenomena, or propositional attitudes: highly specific meanings, purposes, beliefs, anxieties, hopes . . . and so on.

* Second, in doing this he indicated the hugely subtle complexity of the semantic networks and associative/inferential principles in human minds.
* Third, he integrated cognition with motivation and emotion, suggesting how beliefs affect our purposes and anxieties—*and vice versa.*
* Fourth, he taught that one can't understand adult cognition without knowing how it developed. (Note that this is different from saying that "how it developed" is interesting in itself.)
* And fifth, he tried to specify the mental structures and processes involved.

For examples of mental structure, consider his views of memory and motivation. In *The Interpretation of Dreams* (1900, esp. ch. 7) and *The Psychopathology of Everyday Life* (1901), he presented memory not as myriad associative strings of meaningless beads, but as a richly complex structure. It was organized, he assumed, in terms of semantic and phonetic dimensions (think puns—and see 13.iv.c), as well as of contingent associations specific to the individual (think madeleines). As for motivation, his distinction between id, ego, and superego was a broadly sketched account of what cognitive scientists would later call the architecture of the mind. Indeed, Marvin Minsky, among others, has drawn on Freud's ideas in formulating his computational theory of mental structure (Chapters 7.i.e and 12.iii.d).

Examples of mental processes included the unconscious "mechanisms of defence", first mentioned by Freud in the mid-1890s and elaborated later. These were hypothetical transformations of one belief into another. Which one would be brought into play at a given time depended on the emotional circumstances: the level of anxiety caused by the superego. They might be evidenced as verbal reports and/or 'free associations' by the patient. Or they might show up (for instance) as stuttering and/or hesitations in speech (see 7.ii.c), obsessional hand washing, or hysterical paralysis (see Preface, ii).

Just how many defence mechanisms Freud believed there to be isn't clear. (There's even disagreement about the relation between "defence" and "repression"—Madison 1961, esp. 3–30.) His daughter Anna listed ten, giving clear examples of each: repression, regression, reaction formation, isolation, undoing, projection, introjection, turning against the self, reversal, and sublimation (A. Freud 1937, ch. 4). Eight more, not specifically listed by her, are often cited: denial, displacement, splitting, fixation, condemnation, neutralization, intellectualization, and rationalization (Rycroft 1968).

Most of these concepts have entered into common parlance, and many examples will spring to mind. What's less likely to spring to mind, because it's not widely realized even within cognitive science, is that some of the earliest "computer simulations" modelled the selection and application of Freudian defence mechanisms (see Chapter 7.i.a and c).

The defence mechanisms were supposed to describe *how the mind works.* Freud was quite clear about this. He saw them as "the corner-stone on which the whole structure of psychoanalysis rests" (S. Freud 1914: 16). And he claimed, in turn, that "Psychoanalysis is a part of psychology; not of medical psychology . . . but simply of psychology" (1926, postscript). In other words, his account of *personality* had deep implications for *general psychology.*

Admittedly, there were strong whiffs of homunculism. For id, ego, and superego were treated as little minds, whose functioning was taken for granted in a way that's impossible today. But Freud did attempt to specify some of the purpose-directed processes involved in memory, belief fixation, and emotional equilibrium.

What's relevant here is that he tried. Whether what he said about repression was true is another, hotly contested, question. (So, too, is whether it was original. The debts Freud owed to Friedrich Nietzsche, in particular, were many: Ellenberger 1970, ch. 7.)

Some critics argue that Freud's use of 'repression' was unfalsifiable (Cioffi 1970), others that there's empirical evidence both pro and con (e.g. Kline 1972; Grunbaum 1984, 1986; Farrell 1981). Yet others hold that genuine "evidence" simply isn't obtainable (K. M. Colby and Stoller 1988, chs. 5–6). And some practising analysts reject Freud's bizarre "hydraulic" concept of mental energy, but retain his views on mental structure (Colby 1955). So those cognitive scientists today who draw inspiration from Freud's work certainly aren't committed to an uncritical acceptance of his theories.

Another theorist of personality who was seeking underlying psychological causes was McDougall (2.x.b). He was hugely influential for many years, because of his pioneering books on physiological and social psychology (1905, 1908), and on the philosophy of psychology too (1911). His work on purposive behaviour in animals would inspire some core theoretical concepts formulated by the ethologist Konrad Lorenz (see ii.c below). And he was well known also for his part in the first systematic cross-cultural study, the 1898 anthropological expedition to the Torres Strait (Herle and Rouse 1998, esp. ch. 6).

Clearly, then, he was a man of parts—and a significant thinker. Eventually, he was headhunted from across the seas. He moved from Oxford to accept the Chair of Psychology at Harvard in 1920.

Unfortunately, that was just a few years too late. The youngsters there, behaviourist to a man (*sic*) since Watson's hijacking of American psychology in 1913, didn't respect him. Nor he, them: he dismissed behaviourism as "a most misshapen and beggarly dwarf" (1923, p. ix). In addition, he made enemies because of his obstinacy and arrogance, and got into hot water as a result of "incautiousness in lectures on public affairs" (Eysenck *et al.* 1975: 637).

Disaffected, he soon moved to Duke University, where he founded the parapsychology laboratory, with Joseph Rhine as his first assistant. Later, he taught the young Bruner. Indeed, Bruner left Duke with his mentor's warning against the "corruptions" of behaviourist Harvard ringing loudly in his ears (see Preface.ii). "Only Oxford was worse, he said" (Bruner 1983: 31). The warning worked: Bruner never succumbed to the American orthodoxy (see Chapter 6.ii).

Despite the scorn heaped on him by the behaviourists, McDougall's theory of personality was evaluated at mid-century as

> a systematic treatment of conation and affection [i.e. motivation and emotion] that, in completeness and thoroughness, is without a rival, and in penetration is second only to the work of Freud. (Flugel 1951: 277)

No serious critic was prepared to accept his views on psychic energy, which were even more bizarre than Freud's (2.x.b). But his account of mental architecture (both normal and abnormal), and of the dimensions that distinguish one individual personality from another, was highly suggestive.

McDougall described the mind as a "hierarchy", or "colony", of *sentiments.* These were ultimately based in the dozen or so "instincts" common to all human beings, and were controlled—in normal personalities, at least—by the "master sentiment of self-regard" (McDougall 1926; Boden 1972, chs. 6–7). The cognitive and motivational unity of the mind depended not on some mysterious Cartesian "self", but on the control exercised by the master sentiment. By the same token, mental *dissociation* occurred when one or more of the lower-level sentiments became to some degree autonomous.

Minor cases of dissociation, he said, express themselves merely as temporary confusion or absent-mindedness. But deeper and/or longer-lasting dissociations give rise to pathological states of various kinds. These ranged from fleeting fugues to persisting personality disorders.

For instance, if the master sentiment temporarily loses control of some member of the mental colony, the result may be a clinical automatism (where the person has *no* consciousness of their strange behaviour). Alternatively, he said, if the master sentiment fails to develop properly, we may see what's called multiple personality. In such cases, there appear to be two or more persons inhabiting one human body, with different interests and beliefs—and, sometimes, non-reciprocal co-consciousness between them.

On the Cartesian view, multiple personality is simply impossible. Irrespective of *why* it arises, that it can occur at all is profoundly mysterious (N. Humphrey and Dennett 1989; Boden 1994*d*). Indeed, postmodernists today use the inner diversity of every 'normal' self, as well as its pathological extremes, as a stick with which to beat Cartesianism in general (see 13.vi.e). But while they acknowledge the diversity of the self, they don't explain it—still less, offer a scientific account of it, which is what McDougall was trying to do. With hindsight, his explanation can be seen as an anticipation of Ernest Hilgard's (1977/1986) neo-dissociation theory, and even of current computational views on this clinical syndrome (7.i.h).

A sentiment, he explained, is "a system in which a cognitive disposition is linked with one or more emotional or affective conative dispositions to form a structural unit that functions as one whole system" (1908: 437). Each sentiment is centred around some intentional object, whether concrete or abstract. For example: "the sentiment of love for a child, of love for children in general, of love for justice or virtue" (1908: 140). And that cognitive core could grow into "an extensive system of abilities (a system of knowledge or 'ideas' concerning that object)" (1932: 223). Most sentiments reflect the person's place in a particular social group or culture. This is especially true for the political and religious sentiments (see Chapter 8.vi).

In sum, the idiosyncratic purposes, life goals, cultural interests, and personal dilemmas of Tom, Dick, and Harry are generated by organized emotional/cognitive structures. These develop, grow, and sometimes decline or decay, as the individual life progresses. Abnormal psychological states—of which McDougall had seen a great many, as a psychiatrist in and after the First World War—are due to the unusual content and/or organization of these everyday mental structures.

From the point of view of cognitive science, it's interesting that McDougall saw the "structure" of the mind—what's now called its computational architecture—as being abstract/functional, not physical. It's comparable, he said, to the "structure" of a poem or of a society; and he described the internal processes of control and communication in non-physical terms. Specifically, he spoke of "telepathy" between homuncular "monads" comprising the "society" of mind. Predictably, his more tough-minded readers were put off, not to say appalled. But since the vocabulary of information-processing and/or programming wasn't yet available, these dubious concepts and Leibnizian language (Leibniz 1714) were as near as he could get (see Boden 1972: 248–55; 1994*d*).

Surprising as it may seem, given his contempt for behaviourism, McDougall was the first person to define psychology as the study of behaviour (1908: 13). Clearly, however, he was no Newtonian. For him, behaviour was *by definition* purposive. Even in his account of perception, he stressed the organism's activity, goal-directedness, and (largely innate) organization. And his theory of personality, like Freud's, posited hidden causes galore—all integrated within a complex psychological structure.

In the early 1960s, the most recent occupants of the wing devoted to personality theory were the Third Force psychologists. The most prominent were Abraham Maslow, Gordon Allport, Ronald Laing, Carl Rogers, and Rollo May. They had varied (mostly neo-Kantian) philosophical roots: existentialism, phenomenology, Gestalt, organismic, and humanist psychologies. Nevertheless, they comprised an increasingly "total, single, comprehensive system of psychology" (Maslow 1962, p. vi).

Their central theoretical commitment was to human freedom (see 7.i.g). Their interest was not to support a quasi-mystical picture of human beings, but to insist that conscious choices *matter*: free will *makes a difference*. Accordingly, they all favoured "proactive" rather than "reactive" approaches in psychology—that is, theories stressing spontaneous, self-directed, and future-oriented behaviour (G. W. Allport 1960). And they all rejected Freud, whom they saw as backward-looking and reductionist, and behaviourism too. (Hence the "Third" Force.)

They were much closer in spirit to McDougall. For they saw personal growth into the future as an important theoretical dimension, and they emphasized the role of purposes, hopes, aspirations, and creativity in the personal life. Every human being, they said, is a self-determining subject with an idiosyncratic experience of him/herself and of the lived-in world. We need to appreciate this if we're to understand people's lives, and especially if we're involved (on either 'side') in clinical therapy.

While computational psychology was still no more than a hope or aspiration, the Third Force was highly fashionable in clinical psychology—and in the world outside. Laing (1927–89), in particular, became a guru for the young and radical worldwide. From his relatively sober critique of orthodox psychiatry (R. D. Laing 1960), to his later—and far more questionable—recommendations of schizophrenia as a way of getting *closer* to reality (e.g. R. D. Laing 1967; Laing and Esterson 1964), his maverick approach was hugely popular within the counter-culture of the 1960s and 1970s.

The uncritical acceptance of his ideas sometimes led to tragedy, as when a husband—advised by Laing himself—refused to yield to his suicidal wife's pleas for electro-convulsive therapy, a method which (though admittedly invasive and ill-understood) had helped her in the past (D. Reed 1976). And, over the years, Laing's increasingly radical politics would frighten many people off. Nevertheless, the psychological importance of the self-image, and of the recursive layers of perception linking family members, was now evident. And it could even be captured in an ingenious multilevel questionnaire (R. D. Laing *et al.* 1966).

(Gregory Bateson, of course, had been thinking about such issues since his days at the Macy conferences: 4.v.e. But he'd had less influence than Laing, largely because his writings were less intelligible: see Harries-Jones 1995, esp. chs. 2 and 8.)

For the Third Force, it was no more false to think of people as machines than to think of them as biological systems. Laing himself, for instance, said, "I am not here objecting to the use of mechanical or biological analogies as such, *nor indeed to the intentional act of seeing man as a complex machine* or as an animal" (1960: 21; italics added). But they regarded machine analogies as likely to be even more misleading than biological theories, because most people assume that machines can offer no purchase for everyday psychological concepts—especially the concept of *freedom*.

If by "machines" one means cars or jet engines, that's true. Whether it's true for computers is discussed in Chapters 7.i.g and 16. But it's worth noting here that the first computationalists to enter Psychology's House included several who *did* believe that this new technology could help us understand personal life (see Chapter 7.i.a–c). It's worth noting, too, that the person who first named "cognitive psychology", then describing it in computational terms, was an ex-colleague of Maslow (see 6.v.b).

In short, and *despite* the counter-culture's love affair with the Third Force, this movement wasn't so deeply incompatible with cognitive science as most people assume.

b. Sitting tenants with knowledge

The other long-time lodgers in Psychology's House weren't much interested in personality. They were more concerned with the nature and/or development of knowledge—sometimes, with a strong leaning towards biology (see subsection c). (So a *cognitive* psychology doesn't have to be *computational.*) But they, too, were non-Newtonian in that they posited internal processes and structures, and emphasized meaning.

The reason why the Gestalt psychologists stressed structure, holism, and meaning is that they were hugely influenced by neo-Kantian philosophy (see 2.vi). The influences went in both directions: Maurice Merleau-Ponty's philosophy of mind relied heavily on their research (16.vii.a). But where the Naturphilosophen had merely theorized, they reported a host of intriguing experiments.

They showed, for example, that how people perceive the parts of a visual stimulus may depend on the structural organization of the whole. So in 'figure–ground' reversals, a stimulus part can be seen either as a chin or as the indentation at the base of a moulded vase (see Figure 5.1). In other cases of "restructuring", a stimulus part previously seen as a young woman's chin is seen as an old hag's nose (see Figure 5.2).

Structure, and restructuring, affected problem solving too. For instance, Max Wertheimer (1880–1943) asked children to find the area of the parallelogram, not by applying some pre-learnt—and perhaps ill-understood—formula but by their own creative ("productive") thought. He paid attention not just to the final results but also to their remarks along the way. And he explained their thinking in an explicitly non-Humean (i.e. non-Newtonian) way (1945: 9). It depended not on "habit" or "past experience", but on

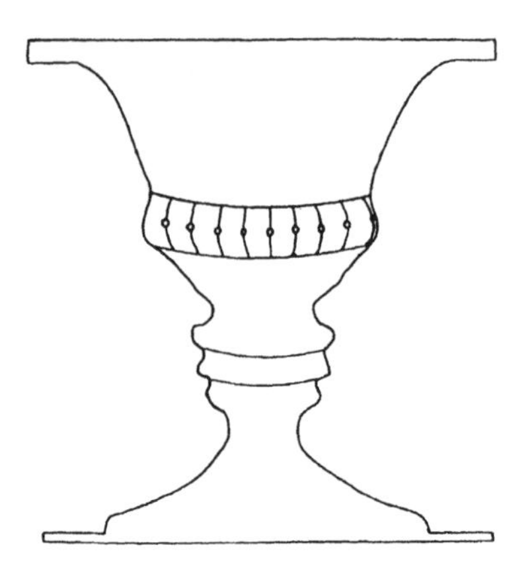

FIG. 5.1. Figure-ground ambiguity: one vase or two faces? Reprinted with permission from Gregory (1966: 11)

FIG. 5.2. E. G. Boring's object-ambiguous figure. Reprinted with permission from Gregory (1970: 39)

[spatial] *grouping, reorganization, structurization*, operations of dividing into sub-wholes and still seeing these sub-wholes together, with clear reference to the whole figure and in view of the specific problem at issue. (Wertheimer 1945: 9, 41)

(If this quotation reminds you of early AI work on problem solving, that's no accident: see 6.iii.) These Gestalt principles were applied to several historically famous problem solvers. One was Albert Einstein—with whom Wertheimer had spoken, from 1916 on, "for hours and hours" and "in great detail" about "the concrete events in his thought" (1945: 168).

Wertheimer's student Karl Duncker (1903–40) didn't try to emulate the interviews with Einstein. He excused himself, saying: "[Although] a thunderstorm is the most striking example of electrical discharge, its laws are better investigated in little sparks within the laboratory" (1945, p. v). So, in the early 1930s, he asked people to solve this mundane practical problem:

> Given a human being with an inoperable stomach tumor, and rays which destroy organic tissue at sufficient intensity, by what procedure can one free him of the tumor by these rays and at the same time avoid destroying the healthy tissue which surrounds it? (1945: 1)

He recorded the detailed "protocols" of his subjects, asking them to "Think aloud" and "emphatically [warning them] not to leave unspoken even the most fleeting or foolish idea".

Taking seriously every detail of these protocols, Duncker analysed the problem-solving process as a whole. He asked how his subjects managed to come up with a variety of tentative solutions; how they recognized impracticalities and dead ends; what made them switch to other alternatives; and how they finally arrived at the solution—which was to focus two or more *weak* rays on the tumour. (So, as he pointed out, the phrase "at sufficient intensity" in his statement of the problem was a huge clue.)

Duncker identified a number of general "heuristics", or solution procedures. These included adapting the solution to some similar problem solved in the past, and asking "*'Just why doesn't it work?'* or *'What is the ground of the trouble (the conflict)?'*" (1945: 21). Heuristic methods, he said, aren't properties of the solution itself, but "ways" to find it: "They ask: 'How shall I find the solution', not: 'How shall I attain the goal?'" (1945: 24). Moreover, he analysed the subject's thinking as a tree (with tree diagram) of goals and sub-goals, and compared the "family trees of two solutions" (1945: i. 6 and 13). He also marked different "functions" (operations) for achieving different things. For instance, a *lens* can be used to *focus* rays, and *swallowing* can *convey something to the stomach*.

All of this would eventually be reflected in early AI. What John Haugeland (1985: 112) has dubbed "Good Old-Fashioned AI", or GOFAI, not only adopted a particular form of computer modelling: sequential, symbolic, programming. Also, it included the psychological claim that our thinking *is* symbol manipulation of this general type. As we'll see in Chapter 7.iii, GOFAI was partly inspired by the findings and theories of the Gestalt psychologists.

One anti-associationist implication of Gestalt psychology was that the perceiver, as well as the problem solver, is *active*. The phenomenal percept is constructed by the mind, according to its own principles of organization, or "good form"—such as proximity, similarity, symmetry, and closure. I say "by the mind", for the Gestaltists mostly spoke in psychological terms. However, they believed that isomorphic physiological processes—continuous field-effects—were going on in the brain. (That theory was eventually disproved by inserting gold foil in the brain, to disrupt electric fields: Lashley *et al.* 1951.)

Another anti-associationist implication was the stress on *meaning*. They showed that ecologically important meanings can be constructed from highly artificial stimuli.

One example—described in 1912 by Wertheimer—was apparent movement: if two spots at nearby points are shown with a very brief time interval, they're perceived as *one spot moving*. Another was the perception of cause. Albert Michotte (1881–1965) found that if one spot moves to touch another, which then moves, people (given certain timing relations) see this as the first *causing* the second to move (Michotte 1963). A third involved social perception: Fritz Heider (1896–1988) showed that cartoons of moving triangles, squares, and circles were sometimes perceived as agents chasing, fleeing, threatening, fighting, and embracing each other (Heider and Simmel 1944). Similar effects, he said, occur in everyday life: we don't *infer* that someone snarling and clenching their fists is angry, but *see them as* angry (Heider 1958).

The "effort after meaning" was a prime theme of the British Gestaltist Frederic Bartlett (1866–1969), Kenneth Craik's mentor at Cambridge. A student of the neurologist Henry Head (Chapter 4.vi), he generalized Head's concept of the "postural schema" to *conceptual* memories. This work was immediately influential in British psychology (Oldfield and Zangwill 1942–3; Broadbent 1970*a*,*b*). Nevertheless, even his fellow countrymen later came to shun its vagueness. In 1978 a youngster at Sheffield University complained to an older colleague, an ex-student of Bartlett's, that "*Your* generation has virtually *betrayed* Bartlett. We know nothing about him"—and his senior had to admit that he was right (V. Stone 1978: 87). (The youngster clearly hadn't read the New Look psychologists, or anyway hadn't followed up their frequent references to Bartlett: see Chapter 6.ii.)

Like the other Gestaltists, Bartlett reported some remarkable experimental data. Following a suggestion from his "close friend" Norbert Wiener (Bartlett 1958: 144), he adapted the American parlour game Russian Scandal—in England, Chinese Whispers—to study memory. (It turned out that a French psychologist had done this first, in 1897—Bartlett 1932: 63.) He used inkblots, drawings, and stories, and got comparable results from all three.

For instance, he got people to read a story, and later asked them to recall it, again and again. He chose a folk tale ('The War of the Ghosts') from a native American culture, whose beliefs about the supernatural were unfamiliar to English people. (In other words, they differed from English people's notions of the supernatural: see 8.vi.d.)

What he found was striking. First, his subjects remembered only the gist of the story, not its details—still less, its words. They seemed to be reconstructing it rather than resurrecting it. Second, they subtly altered it (unconsciously) so that the unfamiliar concepts were assimilated to more familiar ones. That wouldn't have surprised Wilhelm von Humboldt, who'd taught that every culture has its own concepts, not fully intelligible to people from other societies (9.iv.b). Indeed, he'd have said that the original story was unknown even to its translator, the anthropologist Franz Boas. However, both findings surprised Bartlett's colleagues.

These things happened, according to Bartlett, because memories are stored as hierarchically organized, meaningful, *schemas*. "I strongly dislike the term 'schema,'" he confessed, adding that "pattern" and "organized setting" were better (1932: 200–1). But Head's term was already familiar—and Bartlett may have felt that its 'hard-headed' neurological provenance added respectability. (Originally, it had been Immanuel Kant's term, not Head's: it's no accident that the psychologists who first revived the concept had Kantian sympathies: see Chapter 2.vi.a.)

In recall—i.e. reconstruction—the higher levels, expressing the gist, determine the details at lower levels. This activity can't be explained as the re-collection of associationist strings:

"Schema" refers to *an active organization* of past reactions . . . which must always be supposed to be operating in any well-adapted organic response. That is, whenever there is any order or regularity of behaviour, a particular response is possible *only because it is related to other similar responses which have been serially organized*, yet which operate, *not simply as individual members coming one after another*, but as a unitary [holistic] mass. (1932: 201; italics added)

The implication was that experiments on memory using Hermann Ebbinghaus's method of nonsense syllables (Chapter 2.x.a) weren't dealing with typical memory at all. Moreover, the fact that people learning meaningless lists usually make up some mnemonic to help them—a fact noted later by Donald Hebb, and later still by Miller (iv.d below and 6.iv.c, respectively)—is entirely understandable accordingly.

Such feats of memory were special cases of the mind's *effort after meaning*, which characterizes "every human cognitive reaction—perceiving, imaging, remembering, thinking and reasoning" (1932: 44). (And, one might add, musical performance: see 13.iv.b.) It may involve psychological processes "of very great complexity", even though they appear, to the person concerned, to be "of the utmost simplicity". In a word, these processes are unconscious: not repressed (à la Freud), but hidden from introspection.

The Gestaltists, then, offered many hypotheses about holistic and/or central mechanisms. As though that weren't shocking enough to behaviourist ears, their hypotheses were always vague and sometimes (when describing the brain) implausible. Karl Lashley even said to Wolfgang Kohler (1887–1967), commenting on his studies of "insight" in chimps (W. Kohler 1925): "Excellent work—but don't you have religion up your sleeve?" (Heims 1991: 235). (He wouldn't have said this to the neo-Gestaltist Craik, who *did* try to describe interpretative schemas—"models"—in precise and/or neurologically plausible ways: Chapter 4.vi.)

The Gestaltists, other than Kohler (whose work had been translated in 1925), were largely unknown in America until the late 1930s. By that time, however, a number of them had fled the Nazis to the USA.

In 1933, when Adolf Hitler became Chancellor, the German Society for Psychology rapidly capitulated to the new regime in various ways—well before this was required by law, or even prudence (G. Mandler 2002*b*: 193). For instance, psychological journals instantly dropped Jewish members from their editorial boards; and the 1933 Congress featured several talks on racial purity, as well as many tributes to Hitler and his political project. Indeed, Gestalt psychology itself was appropriated by Nazi sympathizers to its intellectual cousin Ganzheitpsychologie, which emphasized "the unity of experience, transcending immediate or personal preoccupations and consistent with the notion of the new German *Volk* community" (G. Mandler 2002*b*: 192).

Psychologists who were Jewish, or married to Jews, resigned from the Board and/or lost their jobs over the next few years—and some emigrated. One who didn't was Otto Selz, whose work on thinking processes would eventually inspire Herbert Simon: he died in Auschwitz in 1943. (Another non-emigrator was the Aryan ethologist Lorenz. His address at the 1938 Congress warned against social types "whose dangerous virulent propagation threatens to invade the body of the nation", and urged further genetic research "to discover the facts that solidify 'our holiest racial, *völkisch*, and human heritage' "—G. Mandler 2002*b*: 198.)

The Gestaltists slowly gained a hearing in their newly adopted country. That's not to say that their foreign ideas were widely welcomed, for they weren't. It would have been amazing if that hadn't been so, given the predominance of behaviourism in the USA. They had to make do with spaces grudgingly made available on the periphery of American academia (J. M. Mandler and Mandler 1968). It didn't help that much of their work remained untranslated until the late 1930s, when a new "source book" made it more widely known (Ellis 1938). Even then, Gestalt psychology was a minority taste. However, it would eventually inspire important work in cognitive science (see Chapters 6.iii, 7.i.c, and 10.i.c).

To some extent, American appetites for Gestalt approaches had already been whetted by Norman Maier's intriguing work on "functional fixedness" in problem solving (Maier 1929, 1930, 1931, 1933). Maier showed that someone may persist pig-headedly in an unsuccessful strategy, despite the presence of (apparently 'invisible') clues.

For instance, he showed his subjects into a room with two strings hanging from the ceiling, and asked them to tie them together. Naturally, people would take hold of the end of one string and walk towards the other—only to find that the string they were holding was too short to reach it. The only way to solve this problem was to tie a heavy object to one string and then use it as a pendulum, swinging back and forth while the subject walked over to the 'empty' string and waited for the other one to arrive. The only things in the room which were suitable as pendulum bobs were scissors or pliers.

Maier found that: (1) many subjects couldn't immediately solve the problem; (2) failure was even more likely if, before the experiment (apparently) started, he had sat at his desk using the scissors or pliers for their normal purpose while chatting to the subject; (3) some previously frustrated subjects were able to solve the problem if he walked past one of the strings and 'accidentally' brushed it with his arm, so making it move slightly; (4) some of those subjects sincerely claimed that the idea of making a pendulum had occurred to them spontaneously, 'out of the blue'; and (5) some people who had the idea of the pendulum, spontaneously or otherwise, didn't think of using the scissors or pliers as a bob. In short, they were "fixated" on the usual function of these tools: that they might be used for a different purpose appeared to be inconceivable.

Maier's work had been well known since the early 1930s. By 1940, the exodus of German Gestaltists had taken place, and Ellis's *Source Book* had appeared. And by 1950, Wertheimer's and Duncker's research on problem solving had been translated. The movements beneath the mantle were becoming more robust.

c. Sitting tenants with biology

The sitting tenants with biology were the Piagetians and the ethologists. The former had moved into their rooms in 1923 (on publication of *Le Langage et la pensée chez l'enfant*), the latter in the mid-1930s. Piaget's people could communicate with all of their neighbours (although the behaviourists weren't interested), for four of his books were translated within ten years (1926, 1928, 1929, 1932). But the ethologists' occupancy was hardly noticed for twenty years by readers limited to English.

Piaget (1896–1980) is normally thought of as a psychologist. In his own mind, however, he was a biologist (Boden 1979, chs. 1 and 6). His first publication, at the awesome age of 11, reported his sighting of an albino sparrow; and he received his doctorate in 1918 for a thesis on molluscs. But this isn't a merely chronological point: throughout his life, his first intellectual loyalty was to biology—and his second, to philosophy. He took up *psychology*—in around 1919—only in order to integrate those two approaches, planning to spend no more than five years on it (Piaget 1952*a*: 255).

Accordingly, he called himself a "genetic epistemologist", not a psychologist. The label was borrowed from the child psychologist James Baldwin (1861–1934), now remembered by theorists of learning and A-Life through the 'Baldwin effect' (Baldwin 1915; Ackley and Littman 1992). And his epistemology/psychology was 'non-Newtonian' in various ways.

Piaget saw the child's mind not as a small-scale version of the adult's (just lacking a few facts and habits, so to speak), but as a fundamentally different system. Indeed, he said, children don't even have *concepts* until 4 years of age. The 3-year-old's words are "notions" attached to the first verbal signs, and "remain midway between the generality of the concept and the individuality of the elements composing it" (1950: 128).

The baby's mind (he said) differs from the 2-year-old's, the infant's from the 7-year-old's, and the 7-year-old's from the adolescent's—at long last, a mini-adult. In short, he posited four major "stages" of mental development (with up to six sub-stages within them): sensori-motor, pre-operational, concrete operational, and formal operational. Each one is necessary for the next, but the changes take place through holistic transformation rather than piecemeal addition. One of his key terms was "equilibration": a change that happens when a dynamical system is in conflict, ending in a more stable state.

Initially observing his own children, from the cradle onwards, he reported many startling facts. For example, a very young baby will lose interest in a toy once it's hidden under a cushion: as Piaget put it, it's as though the object has ceased to exist. An older baby will look for it under cushion-A *even though* he or she has just seen someone removing it from cushion-A and putting it under cushion-B (this is known in the trade as the A-not-B error). Or again, a 4- or 5-year-old will insist that there are "more" beads in a widely spread five-bead row than in a closely spread one, *even though* the longer row was visibly produced by someone's moving the beads further away from each other. Conservation of number—and, when playing with water-in-jars, of volume—isn't yet recognized. Similarly, apparently basic concepts describing the physical world, such as *movement* and *speed*—what AI workers would later call "naive physics" (13.i.b)—aren't developmentally basic at all, for they take time to emerge (Piaget 1970).

Piaget found also that it takes many years for the child to master apparently simple skills, such as "seriation": putting a set of items in order, according to some perceptible criterion. So a 2-year-old can't build a 'staircase' out of seven blocks. He or she may build one or two short sequences, but can't integrate them. Later, trial and error can be used to produce a perfect sevenfold staircase (and later still, an eighth block can be cleanly inserted into an already completed staircase). Eventually, the child constructs the staircase rationally, starting with the shortest or longest block and successively adding the correct neighbour. At adolescence, the child/adult can also express the general principle involved, and understand that seriating blocks is in essence *just the same thing* as seriating dolls, or pencils . . . or even numbers.

These qualitatively (observably) different stages of seriation, Piaget argued, are structurally discontinuous at the deeper psychological level. And they don't follow each other by magic, nor by inner determinism. The child's practical experience (clenching and unclenching the fists, shaking a rattle, playing with boxes and blocks, pouring water from one container to another . . .) is crucial in enabling stage transformations to take place. Even formal logic and mathematics—and language, too—can develop (he said) *only* from the seed of bodily action in the real world (Inhelder and Piaget 1958, 1964).

The word he used for this was "epigenesis", an ancient term for the unfolding of the embryo but used at that time by—and borrowed from—the biologist Conrad Waddington (Chapter 15.iii.b and Boden 1979: 93–103). Waddington's key claim, and Piaget's too, was that development—in embryo and/or behaviour—is neither a deterministic unfolding of a preformed 'germ' nor merely a response to the environment (whether internal, uterine, or external world). As Piaget put it, it's a self-regulatory "dialectic" between organism and environment, brain and behaviour.

Piaget's hypotheses were not only unfashionably non-Newtonian, but undeniably vague—hence Sutherland's disdain (see subsection a, above). (They were also overwhelming in quantity: he admitted he "could not think without writing"—1952*a*: 241.)

The vagueness remained even after the rise of cybernetics had put some scientific flesh onto dialectic bones. The imprecision, in some cases near-emptiness, of his core concepts was criticized even by his admirers. For example, Bruner—of all people—dismissed his concept of equilibrium as "surplus baggage". It contributed nothing to theory or experimental design, he said, save some "confusing imagery", and served merely to give Piaget "a comforting sense of continuity with his early biological apprenticeship" (Bruner 1959: 365).

Partly because of the vagueness, and partly the non-Newtonianism, Piaget wasn't taken seriously by American experimental psychologists (as opposed to educationists) for many years. Harvard gave him an honorary doctorate when he was 40, but he wasn't honoured by the APA until he was 72.

This also reflected the low status, at that time, of developmental psychology in the academic pecking order. It was seen as something one does if one loves babies—which more or less excludes the higher-status half of the human race. The notion that one might need to consider children in order to understand adults hadn't yet got across (see Chapter 7.vi.g). Granted, the first English textbook on Piaget (Flavell 1962) was *immediately* assigned for Bruner's graduate seminar. Indeed, Bruner had been aware of Piaget's work for over twenty years. But Bruner was no behaviourist, as we'll see (6.ii.a–c).

It would be misleading to describe Piaget as a non-Newtonian and leave it at that. For he went only half-way in rejecting the fourth tenet. Yes, he believed in self-organization. But—like the behaviourists—he *didn't* believe in innate knowledge of the external world. If he'd been told (what was discovered in the 1980s: Chapter 14.ix.c) that newborns can already recognize their mother's voice and the stress patterns of her language, he'd have been surprised—but not theoretically challenged. For the explanation is that some external sounds can reach the foetus while it's still in the womb. Other 1980s discoveries, showing that the neonate brain is prefigured to pay attention to specific aspects of the physical and social world, would have been far more embarrassing (see 7.vi.g–h).

In other words, he was equally drawn to the empiricist and neo-Kantian traditions, often describing his own theory as a "middle way". He agreed with the Gestaltists that we actively organize our perceptions in terms of categories such as *cause* and *object*. But he disagreed with them (and with Kant: see 9.ii.c) in arguing that these internal schemas aren't innate, but epigenetically constructed through the real-world activities of the baby/infant/child.

In the final decades of the century, after his death in 1980, epigenesis would gain ground in both psychology and A-Life (Boden 1994*a*, introd.). Indeed, the concept would become highly fashionable. Meanwhile, and with the important exception of Seymour Papert (Chapter 10.v.f), the Piagetians had been speaking more to each other than to the computationalists (Chapter 7.vi.g).

That's true of the ethologists, too. By mid-century, when cognitive scientists started knocking at the door of Psychology's House, the ethologists had taken up lodging there. But the newcomers had little to say to them. And the long-established behaviourists, even less. The early ethologists weren't so shocking as to speak of animal minds (or of animals' theories of mind: Chapter 7.vi.b and f). But they had a healthy respect for inherited capacities, and an unfashionable disrespect for laboratory experiments and universal laws.

Ethology began in continental Europe, thanks to Baron Jakob von Uexküll (1864–1944), Lorenz (1903–89), and Niko Tinbergen (1907–88). Tinbergen moved from Holland to Oxford in 1949, and collaborated with William Thorpe at Cambridge to establish ethology in England. He published an English-language textbook very soon (Tinbergen 1951), but his early work—and Lorenz's—wasn't translated until 1957 (Lorenz 1935–9; Lorenz and Tinbergen 1938; Tinbergen and Künen 1939).

In 1899 the Estonian biologist von Uexküll had co-authored a paper calling for a strictly objective vocabulary in neurophysiology (Boring 1957: 625). At that point, he was concentrating on the physiology of muscle. And with some success: his tough-minded muscular efforts won him an honorary doctorate from the University of Heidelberg, in 1907. But only two years after collecting that honour, he wrote a book on a very different topic: the inner world (*Umwelt und Innenwelt*) of animals.

A version of this appeared in English, much later, as a textbook on *Theoretical Biology* (1926). Despite being published in a highly respected series, alongside the hugely influential work on semiotics *The Meaning of Meaning* (Ogden and Richards 1923), it didn't make a splash. Today, von Uexküll is recognized as a pioneer of biosemiotics (his major paper was reprinted in the journal *Semiotica* in 1992), and his work has appeared in at least nine languages. However, that recognition came later. Thomas Sebeok, the American comparative psychologist who founded "zoosemiotics", read his book as a teenager when visiting England in the 1930s, but found the translation so poor that he got nothing from it (Paul Cobley, personal communication).

Soon afterwards, von Uexküll wrote a paper describing 'A Stroll Through the Worlds of Animals and Men: A Picture Book of Invisible Worlds' (von Uexküll 1934). That, too, remained untranslated for almost a quarter of a century—until it was included in a collection on instinct (C. H. Schiller 1957). It was this paper which eventually attracted most of von Uexküll's English-speaking admirers. ("Eventually", because among psychologists, as opposed to ethologists, "instinct" was a highly unfashionable concept in those behaviourist days.)

From "objectivity" to "invisible worlds": what on earth was going on? The difference was more apparent than real. This *wasn't* an exercise in neo-Kantian idealism, even though it's sometimes glossed in that way (see Chapters 15.viii.b and 16.vii and x.c). The point von Uexküll was making in his "picture book" was that different species have different perceptual and motor abilities, closely integrated in each case. So the animal's environment ("world") or *Umwelt* isn't best thought of as what we'd normally call the real world, but rather *that subset of it* to which the creature can respond and which it can affect.

To the members of one species, the *Umwelten* of other species are invisible—and, of course, unlivable. The best one can do is to rely on the charming, and unforgettable, pictures in von Uexküll's paper, portraying a living room as seen by a human being, a dog, or a fly. Different species, then, live in different worlds—or, as it's sometimes put, different "life worlds". Seventy years later, a leading MIT roboticist would cite von Uexküll in describing the "perceptual worlds" of his new-style robots (R. A. Brooks 1991*a*: sect. 3; see 15.viii.a). And as we'll see in Chapter 8, human beings from different cultures live in significantly different worlds, too.

However, the unavailability of good translations wasn't the only reason why von Uexküll was less famous than his fellow ethologists. For unlike them, he wasn't a field scientist. The first ethologists to base general theories on systematic study of animals in their natural habitats were Tinbergen and Lorenz.

They first met in London in 1936 (two years before that infamous 1938 Congress address: see subsection b). As a result, Lorenz invited Tinbergen to spend a few months at his family estate in Altenberg, near Vienna. (Today, it's the home of the Konrad Lorenz Institute for Theoretical Biology.) Together, they formulated theoretical concepts such as *imprinting, fixed action pattern* (FAP), and the *innate releasing mechanism*, or IRM (modelled in one of William Grey Walter's tortoises: see Chapter 4.viii.b).

One example of an IRM is the hawklike cross that causes a greylag gosling to react as if in fear (Tinbergen 1948). Not any cross will do: the crossbar, representing the predator's wings, needs to be slightly shorter than the axis-bar, and placed nearer to one end than the other (i.e. nearer the 'head'). The gosling's reaction, namely crouching, is the FAP that goes with this IRM. The concept of fixed action patterns was linked by Lorenz with that of "action-specific energy"—an idea taken over from McDougall, whom Lorenz acknowledged repeatedly (Hendriks-Jansen 1996: 208 ff.). But Lorenz took care to separate it from the notion of goal-directedness, and above all not to posit specifically *psychic* energies, as McDougall had done (2.x.b).

In general, the ethologist followers of Tinbergen and Lorenz explained animal behaviour in terms of interactions of simple reflex rules, not of inferences run on internal representations. In that sense, they were faithful "Newtonians".

They were non-Newtonian, however, in favouring nativism (the IRM and FAP) and rejecting universalism. Not only do various species behave differently, but animals of a given species conduct themselves differently in the laboratory and in the wild. The fact that cats in a behaviourist's puzzle box eventually learn to escape as a result of random thrashing around doesn't prove, they said, that cats always learn in that way (cf. Thorndike 1898).

We'll see later that some modern neurologists agree, even arguing that there's *no* universal learning mechanism but only "[many] different problem-specific learning mechanisms [which we may call] 'instincts to learn'" (Gallistel 1997: 82; see 14.ix.g). We'll see also that a new form of ethology, namely computational neuro-ethology (CNE), is now part of cognitive science (Chapters 14.vii.a and 15.vii). CNE helps explain species-specific adaptive behaviour, by modelling the neural mechanisms involved.

In the 1950s and early 1960s, however, the ethologists weren't invited to the cognitive tea parties. The first computational psychologists had happily dropped several behaviourist assumptions, as we'll see in Chapter 6.i–iii. But they weren't ready—yet—for nativism.

A very recent (1940s) entrant to the biological wing of Psychology's House was James Gibson (1904–79). Like the ethologists, he was much concerned with animals' overall lifestyle: how their sensory and motor repertoires fitted together. But he focused much more closely on the detailed *physics* of the environment, asking how it affects perception and behaviour. His approach wasn't popular, since it challenged Gestaltists and behaviourists alike. Nor did it become popular when, soon afterwards, cognitive science came on the scene. Having no time for "internal representations", Gibson had none for "computations" either.

Eventually, battles royal would be engaged between Gibsonians and Marrians (Chapter 7.v.e–f), and psychologists (and philosophers) interested in *embodiment* would recognize Gibson as a patron saint. Meanwhile, he and his disciples worked busily inside their self-contained flat.

5.iii. Soft Centres

The sitting tenants discussed in Section ii were all looking towards the centre of the mind/brain. (Even the ethologists were positing IRMs.) They believed in structural organization and/or central mechanisms integrating perception and action. However, they could say very little about *just what* these central influences were, or *just how* they worked. In other words, their theories had soft centres. (Sutherland's contempt for McDougall and Piaget was due to their softness, whereas I'd been intrigued by their centres.) The cognitive revolution would eventually revive important aspects of these older views, by turning soft centres into harder ones.

Hardness, however, can be overdone—as it was in orthodox behaviourism. If the ideas outlined in Section ii were to be salvaged, meaning and purpose would have to be salvaged too.

a. Mentalism goes underground

In behaviourism, as in most areas of life, propaganda and practice diverged. The propaganda was "Newtonian" (in the six ways listed above). But the practice wasn't. However—again, as in most areas of life—the divergence was mostly covert, more rarely overt. Although most behaviourists believed themselves to be avoiding mentalistic concepts entirely, they weren't. In a nutshell, the key notions of *stimulus* and *response* weren't as atomistic as they claimed.

John Watson (1878–1958) had borrowed these concepts from the reflexologists (Chapter 2.viii.b). In his first statement of behaviourism—*the* first statement of behaviourism—he concentrated on spurning "consciousness" and other mentalist terms (J. B. Watson 1913). Although he spoke of "the range of stimuli to which [the animal] ordinarily responds", he didn't explicitly define psychology as the search for S–R laws. By the mid-1920s, however, such definitions were common. Watson, again:

> We use the term *stimulus* in psychology as it is used in physiology. Only in psychology we have to extend somewhat the usage of the term . . . In a similar way we employ in psychology the physiological term "response", but again we slightly extend its use. (J. B. Watson 1924: 10–11)

> By response we mean anything the animal does—such as turning toward or away from a light, jumping at a sound, and more highly organized activities such as building a skyscraper, drawing plans, having babies, writing books, and the like. (J. B. Watson 1925: 7)

Though not all S–R psychologists were so crass as to call architecture or authorship a "response", they followed Watson in his atomistic assumptions concerning the two key terms.

This is true even of the arch-behaviourist Skinner (1938), who—strictly speaking—wasn't an S–R theorist. He focused not on classical conditioning (described by S–R laws) but on operant conditioning, described in terms of "contingencies of reinforcement". (It's often thought that he discovered operant conditioning, but that's not so—C. G. Gross 2002: 89. Ironically, it was first described by the Polish neurophysiologist Jerzy Konorski, whose highly anti-Newtonian ideas about "gnostic" units primed the discovery of brain cells for detecting a monkey's hand: see 14.iii.b.)

In classical conditioning, a response stably elicited by stimulus A becomes attached also to stimulus B, given repeated trials in which A and B occur simultaneously. Higher-order conditioning can "chain" reflexes, as yesterday's B is used as today's A . . . and so on. (That's true for the higher animals; it's much more difficult to achieve when working with animals having relatively simple brains.)

In operant conditioning, the experimenter waits for the animal to do something—why it does it, doesn't matter—and immediately rewards (or punishes) it. This raises (or lowers) the probability of that response being emitted again. The conditioned response can be "shaped" to a desired behaviour. For pigeons, perhaps batting a ping-pong ball; for babbling infants, saying "Mummy". This is done gradually, by rewarding only behaviour that's more like the desired pattern than yesterday's was. (Similarly, in evolutionary robotics one starts by selecting *any robot which moves*, next only robots which move *broadly* in the desired direction . . . and so on: see 15.vi.c.) But if Skinner didn't formulate S–R laws as such, he did speak of stimulus and response.

However, both the "S" and the "R" of the behaviourists were more soft-centred than they appeared. This point was made in the early days by the philosopher John Dewey (1859–1952). Indeed, Dewey wasn't early so much as anticipatory. Some twenty years before Watson's revolutionary paper, he gave a fundamental critique of the concept of the reflex arc (Dewey 1896).

He was glad that the reflexologists weren't focusing on "consciousness". That, he said, was a crucial advance. But he wanted them to appreciate—and study—the *teleological unity* in experience/behaviour. He was stressing what the cyberneticists, fifty years later, would call circular causation. For he complained of reflexology's "failure to see that the arc of which it talks is virtually a circuit, a continual reconstitution", and its offering us, instead, "nothing but a series of jerks, the origin of each jerk to be sought outside experience itself" (1896: 360). He hoped to persuade the new physiologists to his point of view by showing that they were endorsing it *covertly* already.

The reflexologists believed that their language of stimulus and response was associationist. But Dewey demurred. As he put it, the distinction between stimulus and response is a "teleological distinction of interpretation" with reference to an assumed *end*, not a "distinction of existence". (In modern terminology, stimulus and response are concepts used within the intentional stance, not the physical stance: Chapter 16.iv.b.) The reflexologists, like Hume and the two Mills before them (2.x.a), were ignoring the central—intentional—structure that generates the reflex and is served by it:

> [For the reflexologists, the] sensory stimulus is one thing, the central activity, standing for the idea [thought], is another thing, and the motor discharge, standing for the act proper, is a third. As a result, *the reflex is not a comprehensive, or organic unity, but a patchwork of disjoined parts, a mechanical conjunction of unallied processes*... (Dewey 1896: 358)

Instead of looking for "sensation-followed-by-idea-followed-by-movement", he said, physiologists should focus on the "organized coordinations" that unite them.

For over half a century, his pleas fell on deaf ears. Teleologically structured "coordinations" were shunned in favour of atomistic "connections". Lashley would challenge this non-structural orthodoxy in the 1940s, as we'll see (subsection iv.a). But not until the 1960s would neuroscientists start thinking in terms of the (teleological) *interests* of the whole animal (Chapter 14.iv and vii).

In short, Dewey held that reflexology seemed plausible only because it was covertly non-Newtonian. Six decades later, much the same point was made by Noam Chomsky, and by the philosopher Charles Taylor too.

Taylor (1931–) focused on the general philosophical points stressed by Dewey. But he also provided many detailed critiques of behaviourist experiments and S–R theories (C. M. Taylor 1964*a*). In addition, he argued that no *causal* account could possibly explain behaviour, which involves some non-causal "press of events" in the "essential nature" of behaving organisms (p. 24). Hence, he said, a mechanical dog, programmed to behave like a real one, could be *described* as "goal-directed", but not *explained* in that way (p. 20). Taylor's book was hugely successful in Anglo-American philosophical circles. But although his critiques impressed the philosophers, who felt that he'd obviously done his homework, they didn't impress the more knowledgeable Sutherland, who rebutted them forcefully (N. S. Sutherland 1965); and Taylor's recourse to the "essential nature" of organisms didn't impress me (Boden 1972: 119–37).

Chomsky (1928–) was more concerned with specific—in his view, outrageous—behaviourist claims about language. His savage review of Skinner's *Verbal Behavior* argued that Skinner, despite his anti-mentalist rhetoric, relied heavily on (covertly) intentional concepts, and on an intuitive recognition of linguistic structure (Chomsky 1959*b*). Chomsky's review is discussed at length in Chapter 9.vii.b. Here, the point is that it showed Skinner to be less Newtonian than he claimed.

In sum: although Bruner would complain of the behaviourists' "impeccable peripheralism" (Bruner *et al.* 1956, p. vii), their peripheralism was less impeccable than they believed. Even Skinner was covertly flirting with mentalism, meaning, and structure.

b. Behaviourism softens

If one were to ask "When is a behaviourist not a behaviourist?" the answer would seem to be "Always!" The previous subsection showed that, in a sense, *no* behaviourist was really a behaviourist. However, some diverged from the party line more openly. Ralph Perry, Edward Tolman, and Clark Hull all ignored the sixth tenet, and Perry and Tolman—self-styled "purposive behaviourists"—bravely flouted the first, second, and fifth as well.

Perry (1876–1957) was perhaps less brave than Tolman, for he wasn't a card-carrying psychologist. Rather, he was a philosopher. However, as a close friend and ex-student of William James, he was deeply interested in psychology. He described himself as a behaviourist, but had scant sympathy for the behaviourist propaganda.

Drawing on the writings of McDougall, who'd just been invited to Harvard, Perry argued around 1920 that *purpose* is an essential concept for psychology (R. B. Perry 1918, 1921*a*). He pointed out—what Continental philosophers would call the hermeneutic circle, and what Daniel Dennett would include within "the intentional stance" (16.iv.b)—that one can't identify a given purpose without assuming some specific belief, and vice versa (J. B. Perry 1921*b*). To infer, for instance, that someone carrying an umbrella *wants* to keep dry, is to assume they *believe* that umbrellas provide protection from rain. It followed, said Perry, that behaviour can't be described by the language of stimulus and response unless this is understood in intentional terms: "behaviour is incapable of being translated into simple relations correlated severally [i.e. atomistically] with external events" (J. B. Perry 1921*a*: 102).

Perry wasn't an experimentalist, so his views could safely be ignored by the more orthodox behaviourists. Tolman (1886–1959) was quite another matter. Working at UC-Berkeley from 1918 until his death (despite risking dismissal in the 1950s by refusing to take Senator Joseph McCarthy's loyalty oath), he reported many surprising experimental results. These cried out for attention even from those who rejected his explanations of them.

Many did reject Tolman's explanations, which sprang from his commitment (following Perry) to teleological structure. Even in the early 1920s, he called himself a "purposive" behaviourist (Tolman 1922, 1925*a*,*b*, 1932). Tolman had no use for atomistic "S" and "R": "I rejected the extreme peripheralism and muscle-twitch-ism of Watson" (1959: 94). He believed in "instincts", defined not as reflex movements but as inborn goals/interests (1920, 1923). And he posited many central organizing mechanisms, which he saw as "intervening variables" between episodes of observable behaviour (1935).

These included innate or acquired "exploratory impulses", or hunches, and meaningful "expectancies" linking different stimuli (which he often termed "Gestalts"). They also covered mechanisms of "means–end-readiness" linking perception to action, and "belief-value matrices" organizing the creature's choices. In the late 1940s, he even added "cognitive maps" and "hypotheses" (1948).

All these mentalistic concepts were attributed to rats in the first instance, and later to humans (1958). (I say "later", but Tolman's theoretical terms had mostly been drawn from common-sense psychology in the first place, as he himself admitted—1959: 98 n.)

Clearly, Tolman was no Newtonian. He preferred a very different approach: "I don't enjoy trying to use my mind in too analytical a way" (1959: 93), and welcomed very different ideas: "I was tremendously influenced by Perry... [and] much influenced by gestalt psychology" (1959: 94, 95). He'd interacted with the Gestaltist Kurt Koffka (1886–1941) while on an early visit to Germany, and later collaborated with Egon Brunswik (1903–55), who left Vienna for the USA in the 1930s (e.g. Tolman and Brunswik 1935).

In the 1920s–1930s in particular, Tolman's views were hugely unfashionable. To be sure, the young Simon, puzzling over human decision making, was primarily inspired by Tolman and William James (Simon 1947*a*, ch. 5). But Simon was then an economist. Professional psychologists were wary, not to say contemptuous, of Tolman's theories. But they couldn't ignore him—and in 1937, they even elected him President of the American Psychological Association—because of the startling data he reported.

Many of these concerned "latent" learning, in which rats appeared to learn mazes without any reinforcement. Allowed to explore a maze *without* any food in it, they later learned the maze more quickly than usual. Tolman explained this in terms of the "confirmation of expectancies": if the expected situation is reached, the probability of the action is increased. (Some psychologists posited a drive of curiosity, to provide hidden reinforcements; but this move was embarrassing for mainstream behaviourists.)

Later, Tolman described rats who seemed to have learnt internal models/schemas of spatial orientation, used to organize their behaviour. But instead of talking about the chaining of reflexes/responses, he spoke of the organization of stimuli (Gestalts) in "goal/sub-goal hierarchies", or "cognitive maps". Cognitive maps troubled the orthodox, for the first word was mentalistic and the second, structural. The other 'folk-psychological' terms troubled them too.

One S–R psychologist who tried to accommodate Tolman's data without using his explanations was Hull (1884–1952). Initially trained as a mining engineer, he did most of his psychological research at Yale. One of the people attending his seminars in the early 1930s was Warren McCulloch, and it wasn't only McCulloch whose attention was captured. Throughout the 1940s and early 1950s, Hull (not Skinner) was the most professionally influential—though not the most popularly famous—of all the behaviourists.

Hull was more firmly wedded to empiricism than Tolman was. He saw purposive explanations as illegitimate because, he argued, they assumed queer backward-working causes (a common misunderstanding which McDougall, for one, had countered clearly). Purposes, said Hull, are just "habit mechanisms" (1930), which are reducible to "colorless movements and mere receptive impulses" (1943: 25). He was an S–R theorist par excellence—except that the S and the R could happen *inside* the organism.

Moreover, Hull's philosophy of science was austere. He shared the logical positivists' view of scientific method (memorably mocked by Skinner: 1959). And he favoured their ambitious programme of the unification of science (see Chapter 9.v.a), even to the extent of axiomatizing his own theory (C. L. Hull 1937, 1943; Hull *et al.* 1940). Tolman, by contrast, confessed: "Apparently I have no scientific superego which urges me to be mathematical, deductive, and axiomatic... [My] system is based on hunches and on common-sense knowledge. It is certainly not 'hypothetico-deductive'" (1959: 97, 150).

So why mention Hull here? Surely, *he* was a behaviourist if anyone was?—Well, yes. But he was more open-minded than one might expect. His doctoral thesis (1920) had explored the topic of concept formation; he was interested in hypnosis, which he studied experimentally in some detail (C. L. Hull 1933); and he even invited Koffka to Wisconsin as visiting professor for a year. More to the point, he ignored the sixth tenet: his psychology bristled with intervening variables.

To be sure, these weren't overtly mentalistic. At worst (at best?), they were covertly mentalistic, in the sense explained in the previous subsection. But he attributed a complex functionality to the mind/brain—and in doing so, flouted the sixth tenet with abandon. As a computationalist critic would later put it: "stimulus–response theorists themselves are inventing hypothetical mechanisms with vigor and enthusiasm and only faint twinges of conscience" (Neisser 1967: 5).

The hypothetical mechanisms posited by Hull (1934) included pure stimulus acts (aka attention), persistent stimulus components (aka representations), reaction potentials, anticipatory responses, and habit-family hierarchies. The last of these was used to explain what McDougall had called "variation of means", and Tolman "means–end readinesses".

For Hull, each goal, and each of its anticipatory sub-goals, is associated with a habit-family, a set of alternative responses which share a common anticipatory response. The latter mediates the transfer of reinforcement from one family member to others. (Like human families, habit-families have members who vary in influence: a behaviour-based equation was provided for calculating "habit-strength".) Hull's chains of "fractional anticipatory goal responses" served much the same theoretical role as Tolman's "expectancies". So Hull felt that he'd allowed for the three crucial features of purposive behaviour: that it may involve a wide range of responses, that these may be aimed at a variety of sub-goals (and sub-sub-goals . . .), and that—up to a point—it persists until the goal has been achieved.

The remark (quoted above) about "only faint twinges of conscience" implied that *testing* Hullian theory was easier said than done. Axioms and deductions were all very well, but the increasingly baroque logic of Hull's hidden Ss and Rs seemed to bear little relation to specific experimental evidence. (At least Skinner reported observable phenomena, namely contingencies of reinforcement—which were reliable enough to be used to train animals for circuses and cinema films.) By the late 1950s, then, the leading psychological journals were publishing many critiques of Hull's (and similar) theories.

What such critiques didn't do, however, was to draw the explicit conclusion that *behaviourism in general* was doomed. That was left to Lashley in 1951 (see iv.a below), to Chomsky in 1959 (9.vii.b), and to the cognitive science manifesto of 1960 (6.iv.c).

c. Behaviourist machines

Hull quite often suggested neurophysiological underpinnings for the psychological functions he described. (These weren't always in the brain: the anticipatory goal responses were assumed to lie largely in the muscles and endocrine glands.) Tolman didn't do this, but he was well aware that his theories must allow for neural implementation.

Moreover, neither man had any philosophical qualms about artefacts being used to model their ideas. But when they discussed this, their approach was theory-to-machine, not machine-to-theory. In other words, they *didn't* use concepts that had been originated for describing machines to help them express their own psychological theories.

Tolman (1941), for instance, outlined a "schematic sowbug", a hypothetical creature broadly comparable to Valentino Braitenberg's "Vehicles" (15.vii). As for the ex-engineer Hull, he built a physical model of the conditioned reflex (having already built a logic machine and a device for calculating correlations in aptitude tests). He described his conditioning machine briefly in the pages of *Science* in 1929, and more fully soon afterwards (Baernstein and Hull 1931).

He wasn't alone: the psychological journals around 1930 reported several attempts to model basic behaviourist principles in machines (e.g. Stephens 1929; A. Walton 1930: 110–11; Ross 1938). In one of these, a doll would "naturally" raise her arms to approach a toy rabbit entering her field of view, and "naturally" tremble at a loud noise. But the wiring diagram of the machine (Figure 5.3) would ensure Pavlovian conditioning:

> If the loud noise be made first . . . and then the rabbit is presented, no conditioning takes place. Even if the doll is still trembling when the rabbit approaches, the trembling is somewhat reduced (probably by reason of the "distraction" involved), but there has been no conditioning. If the rabbit be again presented alone, the old approach reaction is made just as before. If, however, while the approach response is being called out, the loud sound be made [by banging together two blocks of wood], the arms fall to the side and trembling ensues. The rabbit being withdrawn and presented again now elicits no "approach response," but does bring on the fear reaction of trembling. Fortunately, unconditioning can be effected simply by pressing a push-button. (A. Walton 1930: 110)

This affecting vignette of the doll-and-rabbit was surprisingly engaging. But the existence of such machines wasn't surprising at all. We saw in Chapter 2.vii.e, for instance, that S. Bent Russell, who'd built a hydraulic model of nervous conduction as early as 1913,

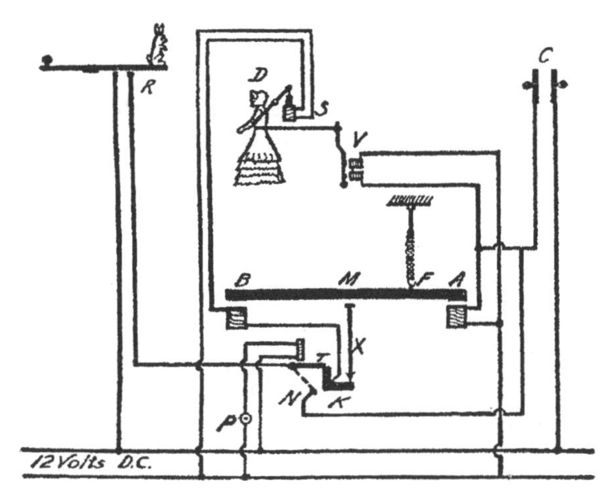

FIG. 5.3. Wiring diagram for a doll-and-rabbit conditioned reflex machine. Reprinted with permission from A. Walton (1930: 110)

had suggested then that *psychological* principles might be simulated too (see also Bent Russell 1917*a*,*b*). By Hull's time, quite a few people were playing this game (Boring 1946; Valentine 1989; Cordeschi 1991, 2002).

Boring approved. Indeed, he assumed that to be an S–R theorist *just is* to think of organisms as "robots" (Boring 1946: 177). He'd come to this conclusion after being "defied" by Wiener (in a letter of 1944) "to describe a capacity of the human brain which he [Wiener] could not duplicate with electronic devices". In his response, delivered as the Presidential Address to the Eastern Psychological Association, Boring described Tolman's schematic sowbug and said:

> The advantage of playing this kind of game lies *solely* in the fact that, if you talk about machines, you are more certain to leave out the subjective, anthropomorphic hocus-pocus of mentalism. There would be nothing wrong with mentalism if it used rigorous definitions of terms, but usually it does not. Hence the mentalistic concepts need first objective analysis into functions, and then a further test, the test of thinking about them as pertaining to a machine. (Boring 1946: 191; italics added)

He went on to discuss the design and construction of *actual* robots, like Hull's. He even mentioned the possibility of using electronic computers, if the result were thought worth the trouble and expense. And he cited a futuristic paper by Vannevar Bush (1945), full of hype about computers and how they might be used (and how, eventually, they *were* used: 10.i.h and 13.v–vi).

But this was 1946. MADM had yet to make her debut, the 'logical' neurone was still confined to the pages of the *Bulletin of Mathematical Biophysics*, and cybernetic robots were still hidden in infancy (see Chapters 3.v and 4.iii–v). Boring's word "solely" is a give-away: his assimilation of S–R organisms to robots *wasn't* an example of mind-as-machine. To be sure, he admitted that, challenged in Wiener's letter "to describe a capacity of the human brain which he could not duplicate with electronic devices", he couldn't do so (Boring 1946: 178). Nevertheless, lacking an "inventory" of psychological functions, he "could not be sure that there was not [at least one] which a nervous system could perform and an electronic system could not".

Similarly, Tolman and Hull weren't cognitive scientists, for they'd turned to consider machines only *after* formulating their theories. The artefacts could show whether their theories seemed to 'work'. But there was no suggestion that thinking about *the machines themselves* might generate psychologically interesting ideas. Computational psychology was still (just) below the horizon.

5.iv. Neurology Creeps In

Psychologists before mid-century had mostly ignored neurophysiology, not least because little was then known about it. McDougall was an exception, but by 1920 he'd fallen out of fashion.

The behaviourists, in particular, paid only lip-service to the brain—hence Boring's talk of "the empty organism" (see Section i.a). To be sure, they'd been inspired by Pavlov, and had little doubt that their psychological findings would one day be explained in neurological terms. But such explanations weren't their concern. Indeed,

Skinner (1938) had specifically warned psychologists not to digress into neurological speculation (subsection b, below).

The most important exceptions in the 1940s were Craik (4.vi), Lashley (1890–1958), and Lashley's student Hebb (1904–85). However, Craik's work was still relatively unknown outside the UK. The combination of Lashley and, especially, Hebb would encourage psychologists on both sides of the Atlantic to think about possible neural mechanisms.

Some people impressed by early cybernetics were encouraged to model them also (4.viii). Hebb's work soon set IBM's computers buzzing (subsection f, below). Plausible neurological models, however, wouldn't be available for twenty years or more (see Chapters 12, 14, and 15.vii). Meanwhile, most psychologists' talk about brain mechanisms was—like Tolman's theory—'soft' in the sense that it was *talk*, not *implementation*.

a. Hierarchies in the brain

The person who mounted the first *really* worrying attack on behaviourism, from the behaviourists' point of view, was initially one of their own: Lashley. To be sure, he'd never been one for *hypotheses non fingo*: as a psychophysiologist, his prime aim was to discover neural causes. But he'd given a behaviourist analysis of consciousness (Lashley 1923), and had been trained—by none other than Watson himself—in the canons of S–R behaviourism and reflexology.

Eventually, however, he undermined both. On the one hand, he questioned the *neural reality* of the reflex arc. On the other, he denied its *theoretical adequacy* as an explanation of behaviour.

By 1920, Lashley was already beginning to doubt 'pure' behaviourism. Before moving to Harvard in 1935 (and to the Yerkes primate laboratory in 1942), he'd commenced an epochal series of experiments showing that rats, after damage to their cortical cells and/or cuts in their brain circuitry, could still learn to negotiate mazes (1929*a*,*b*, 1931). They needed more practice after more extensive damage, and for the more difficult mazes—but they *could* still learn.

His conclusion was shocking:

> It is very doubtful that the same neurons or synapses are involved even in two similar reactions for the same stimulus . . . The results are incompatible with theories of learning *by changes in synaptic structure*, or with any theories which assume that particular neural integrations are *dependent upon definite anatomical paths specialized for them* . . . The mechanisms of integration are to be sought in the dynamic relations among the parts of the nervous system rather than in details of structural difference among them. (Lashley 1929*a*: 3; italics added)

There'd been a monumental failure to locate memory circuits in the brain. Conditioned reflexes were real enough, considered as learnt behaviours. But how they were embodied was suddenly a mystery. Twenty years later, he was still shocked:

> The series of experiments has yielded a good bit of information about what and where the memory trace is not. It has discovered nothing directly of the real nature of the engram. I sometimes feel, on reviewing the evidence on the localization of the memory trace, that the

necessary conclusion is that learning just is not possible. *It is difficult to conceive of a mechanism* which can satisfy the conditions set for it. (Lashley 1950: 477–8; italics added)

(In Chapters 12 and 14 we'll see how research in connectionist AI and computational neuroscience would eventually demystify this problem, at least up to a point.)

Even more relevant, here, is Lashley's lecture on "serial order" in behaviour (Lashley 1951*a*). This was delivered at Caltech's Hixon symposium, held in Pasadena in 1948. It caused a mini-sensation. (There were so many appreciative comments that he admitted: "I have been rather embarrassed by some of the flattering remarks made today"—Jeffress 1951: 144.) For Lashley explicitly rejected four of the six behaviourist assumptions: atomism, associationism, externalism, and black-boxery. Black-boxery, of course, was no great loss for people interested in the brain. But the other three were.

In Lashley's opinion, externalism was the major casualty. At the start of his lecture, he declared: "My principal thesis today will be that the input is never into a quiescent or static system, but always into a system which is already actively excited and organized" (1951*a*: 112). He ended with the same point:

Attempts to express cerebral function in terms of the concepts of the reflex arc, or of associated chains of neurons, seem to me doomed to failure *because they start with the assumption of a static nervous system.* (p. 136; italics added)

In addition, Lashley's paper posited top-down structural organization (so the first two tenets of behaviourism bit the dust). This, in a nutshell, was Lashley's answer to "the problem of serial order in behavior".

The phenomenon that was puzzling him was temporal order, or sequence. What controls the production of phonemes in speaking a word, or the movements of the fingers in playing an arpeggio? The orthodox answer was: a chain of S–R reflexes. This implied two things. First, that what happens can depend only on what's already happened, not on any future links in the (uncompleted) chain. Second, that each link is controlled separately, by sensory messages reaching the brain and triggering motor messages to the muscles. Lashley showed that neither implication fitted the evidence.

In rebutting the first, he started with seven pages discussing language: mainly speech, but also typing. Many everyday phenomena, he said, make it clear that what happens in speech *is* affected by what's going to happen later. Common linguistic errors such as 'spoonerisms', and anticipatory slips of the tongue (or typing-fingers) in general, show that speech (and typing) is actively organized by syntax, not merely describable by it. And the same applies, he said, to the *comprehension* of language (as in the 'garden-path' example given below).

By "syntax", he didn't mean Chomsky's NP–VP (which hadn't yet been defined: Chapter 9.vi), nor even the grammar-school child's *subject*, *predicate*, *noun*, *adverb*... and so on, but something much wider. He spoke of the "generality of the problem of syntax", by which he meant that all skilled action ("almost all cerebral activity"), whether in animals or in humans, is hierarchically organized. Examples of skilled action include speaking, piano playing, walking, jumping... even 'simply' reaching and grasping. Moreover, action errors comparable to those found in speech can be seen in other skills too—for instance, as Lashley pointed out, the many non-linguistic action errors described in Freud's *Psychopathology of Everyday Life*.

Hierarchical structure is most easily recognized in language. We all know that language involves several levels of organization: sounds, syllables, words, phrases, sentences . . . Moreover, the higher levels affect what happens below them. Lashley instanced the *spoken* sentence: "Rapid righting with his uninjured hand saved from loss the contents of the capsized canoe." Most people hear this as "Rapid writing with his uninjured hand . . ." until they hear the phrase "capsized canoe". At that point, they do a double-take and reinterpret the initial verb. (Garden-path sentences, such as this one, are discussed in Chapter 7.ii.b.)

In rebutting the second implication of orthodoxy, Lashley used not only qualitative arguments (concerning "syntax" and hierarchy) but quantitative ones as well. For instance, "whip-snapping" hand movements can be accurately controlled to less than an eighth of a second—yet that's the *minimal* reaction time for the arm to respond to touch or kinaesthesis. Moreover, the speed of nervous conduction doesn't allow enough time for messages to pass up to the brain and back again, so as to control every detail of the hand's movement. The same applies in playing music vivace, or in batting a cricket ball. It's clear, said Lashley, that "a series of movements is not a chain of sensory-motor reactions". On the contrary, there must be some central mechanism, working "in independence of any sensory controls", controlling the relevant muscles *as an organized whole.*

What might that mechanism be? Lashley had only the vaguest idea. Speech, for instance, involves a huge number of muscles and many different body parts, all subtly integrated whenever we utter a word. (A fascinating account of the anatomical coordinations involved, which draws heavily on Lashley's paper, is in Lenneberg 1967, chs. 1–3.) He posited "general schemata of action which determine the sequence of specific acts", defined as "elaborate systems of inter-related neurons capable of imposing certain types of integration upon a large number of widely-spaced effector elements". But the real problem was to discover their nature, "and to this problem", he confessed, "I have no answer". (Soft centres, again.)

Given the soft centres, one might expect that Lashley's ideas would instantly be taken up in a cognitivist spirit. They weren't. The mini-sensation didn't become a genuine sensation until about 1960, when his lead was acknowledged by Miller and Chomsky—and, afterwards, Eric Lenneberg (Chapter 6.iv.c). Indeed, the same applies to the 1948 Hixon symposium in general, which was a good ten years before its time (D. Bruce 1994).

Part of the reason was that, in criticizing behaviourism in this way, Lashley himself wasn't embracing the nascent cognitive science. He said, for example:

> The brain has been compared to a digital computer because the neuron, like a switch or valve, either does or does not complete a circuit. But at that point the similarity ends. [Among other differences, the] number of neurons involved in any action runs into millions so that the influence of any one is negligible . . . Any cell in the system can be dispensed with . . . The brain is an analogical machine, not digital. Analysis of its integrative activities will probably have to be in statistical terms. (Lashley 1958: 539)

Nevertheless, Lashley was a precursor of cognitive science, not just a predecessor of it (see 9.ii). For he sketched a physiological research programme to test von Neumann's ideas on cellular automata (Chapter 15.v.a; Aspray 1990: 182). And he spoke of "plans" and "structures": concepts that his student Karl Pribram would help to highlight in the cognitive science manifesto (6.iv.c). He even identified many of the questions addressed in that manifesto, and—in a very broad sense—anticipated some of the same answers.

b. Connectionism named

If Lashley's 'Serial Order' paper wasn't immediately recognized as important by all and sundry, neither was Hebb's book *The Organization of Behavior*. This was published in 1949, the year after the Hixon meeting. It galvanized some, to be sure—but not all. Mandler recalls:

> [As] a major alternative to mainstream behaviorism, [Hebb's book] was in part ignored to the eventual embarrassment of the conventional wisdom, but found enough support to become the core of a small counterrevolutionary movement. (G. Mandler 1996: 19)

However, it came into its own a few years later. Its rise to prominence was aided in 1958 by Hebb's hugely successful *Textbook of Psychology* (still in use today, in the fourth edition of 1987), and by the sudden arousal of interest in cognitive psychology to be described in Chapter 6. Thus Mandler, again:

> In the United States, it was not until another half-dozen years had gone by before the rejection of behaviorist dicta hit full stride. (G. Mandler 1996: 19)
>
> *A burst of activity rarely seen before* turned the field around between 1955 and 1960 and established a firm basis for the "new" cognitive psychology. (p. 20; italics added)

Shortly after that, in 1960, Hebb was elected President of the APA. (His Presidential Address promised that his non-behaviourist form of connectionism would throw light, at last, on higher-level cognition: Hebb 1960.) And in 1965, he was nominated for a Nobel Prize.

(Which, by the way, he didn't get. Perhaps the Committee felt that although the name was a new invention, the core idea wasn't?—see subsection e, below. And perhaps it was premature in any case. A quarter-century later, maybe he'd have got it—see 12.vi–vii, and 14.v–vi and ix.)

His 1949 book would turn out to be a crucial spur to cognitive science, in two ways. On the one hand, it soon encouraged work in connectionist modelling—already starting, thanks largely to McCulloch (4.iii and 12.i–ii). On the other, it encouraged a 'non-Newtonian' mentalism: specifically, a concern with *concepts* (verbal and non-verbal), and how they're learnt. "Mentalism", here, means the use of intentional vocabulary. Mentalism in the Cartesian sense was implicitly denied: "the task of the psychologist [is] understanding behavior and reducing the vagaries of human thought to a mechanical process of cause and effect" (Hebb 1949, p. xi). (He would reject it explicitly later, in his book on the mind–body problem: 1980.)

Hebb got away with this, in those predominantly behaviourist times, partly because—unlike Chomsky, ten years later (9.vii.b)—he didn't go out of his way to scorn the behaviourists. Quite the contrary: an ex-student has recalled that "Hebb always stated with complete conviction that he regarded B. F. Skinner as the greatest psychologist of the century" (Harnad 1985).

Mainly, however, he got away with it because although he was (from 1946) chairman of McGill's *Psychology* Department, he presented mentalism in the respectable form of a theory of *neural mechanisms*. Indeed, he pleaded that psychologists and neurophysiologists should cooperate more than they were then wont to do (1949, pp. xii, xix, 12).

That apparently unexceptionable plea was immediately spurned by his hero, who didn't see Hebb's theory as "respectable" at all. According to Skinner (1950), psychologists should stick to their last: describing changes in the probability of behaviour. They shouldn't meddle with neurophysiology, he protested, because studying actual structures and biochemical processes in the brain—including "synaptic connections [being] made or broken"—requires very different observational methods. Mere speculations aren't good enough.

Nor should they fog matters, as he'd tartly put it some ten years earlier (in criticizing Hull's theory), by treating the CNS (the common shorthand term for the central nervous system) as the "conceptual nervous system" (Skinner 1938: 421). Quasi-neurophysiological theories that treated the brain merely as "a system with a certain dynamic output", and postulated unobservables such as "concepts" and "expectancy", were useless. Nevertheless, he complained, "Theories of this sort are multiplying fast." He didn't mention Hebb by name, but the target was clear. (Hebb soon took up Skinner's challenge, with respect not only to "concepts" but also to "drives": 1955.)

As for what Hebb said about the nervous system, his position was broadly similar to David Hartley's two centuries earlier (Chapter 2.x.a):

> *The general idea is an old one*, that any two cells or systems of cells that are repeatedly active at the same time will tend to become "associated", so that activity in one facilitates activity in the other. The details of speculation [*sic*—and *pace* Skinner] that follow are intended to show *how this old idea might be put to work again*, with the *equally old idea* of a lowered synaptic "resistance", *under the eye of a different neurophysiology* from that which engendered them. (1949: 70; italics added)

In short, simultaneity of firing is the key: cells that fire together, wire together. A few pages before, he'd given a somewhat different version of this 'ft/wt' principle:

> Let us assume then that the persistence or repetition of a reverberatory activity (or "trace") tends to induce lasting cellular changes that add to its stability. The assumption can be precisely [*sic*] stated as follows: *When an axon of cell A is near enough to excite a cell B and repeatedly or persistently takes part in firing it, some growth process or metabolic change takes place in one or both cells, such that A's efficiency, as one of the cells firing B, is increased.* (1949: 62)

In fact, Hebb's two formulations of the ft/wt rule were neither "precise" nor equivalent. But that wasn't realized until a few years later (see subsection f, below).

This general idea, that synaptic resistance can be changed by activity, is nowadays 'common sense'. We've come a long way since Hartley spoke of unseen "Vibrations", and since Sherrington showed that synapses do actually exist. *Just how* synaptic change happens, at the neurochemical level, may still be controversial. But *that it happens at all* is taken for granted. In Hebb's time, that wasn't so.

The opening page of his chapter on the 'Growth of the Assembly' acknowledged as much:

> The first step in this neural schematizing is *a bald assumption* about the structural changes that make lasting memory possible. The assumption has repeatedly been made before, in one way or

another, and repeatedly found unsatisfactory by the critics of learning theory. I believe it is still necessary. As a result, I must show that in another context, of added anatomical and physiological knowledge, it becomes more defensible *and more fertile* than in the past.

The assumption, in brief, is that *a growth process accompanying synaptic activity makes the synapse more readily traversed.* This hypothesis of synaptic resistance, however, is different from earlier ones in [various] respects... (1949: 60–1; italics added)

He spoke of "a bald assumption" because no one knew just how synaptic resistance could be changed. (Lashley had stressed this fact, in his attack on theories based on the assumption: 1924.) The change might take place, Hebb suggested, through the growth of synaptic knobs, which would increase the area of contact between the two neurones concerned (1949: 62–6). But whatever happened at the level of (what we'd now call) the wetware, the need for an adequate higher-level, functional, explanation made the assumption "necessary".

The term "connectionism" is Hebb's coinage. But if someone had asked him whether he was a connectionist, he'd have dithered. He was certainly a connectionist as opposed to a field theorist:

Two kinds of formula have been used [to explain what goes on within the cortex], leading at two extremes to (1) switchboard theory, and sensori-motor connections; and (2) field theory... [The first claims that direct sensori-motor] Connections rigidly determine what [the] animal or human being does, and their acquisition constitutes learning... [The second] denies that learning depends on connections at all, and attempts to utilize instead the field conception that physics has found so useful. (1949, p. xvii)

(That's to say, Hebb accepted the second tenet of 'Newtonianism'.) However, he wasn't the then usual kind of connectionist, because the sixth tenet was being dropped:

[My] theory is evidently a form of connectionism, one of the switchboard variety, though it does not deal in direct connections between afferent and efferent pathways: not an "S–R" psychology, if R means a *muscular* response. The connections serve rather to establish *autonomous central activities*, which then are the basis of further learning... It does not, further, make any single nerve cell or pathway essential to any habit or perception. (1949, p. xix; first and third italics added)

The "no single cell" disclaimer was an obeisance to Lashley, whose search for the engram had suggested that memories aren't stored at specific locations.

A further ground for dithering was that, for Hebb, connections weren't the whole story. Timing was crucial too:

In a single system, and with a constant set of connections between neurons in the system, the direction in which an entering excitation will be conducted *may be completely dependent on the timing* of other excitations. *Connections are necessary, but may not be decisive in themselves*; in a complex system, especially, time factors *must always* influence the direction of conduction. (1949: 10–11; italics added)

Hebb admired Skinner greatly, as we've seen. But he also regarded Gestalt psychology as hugely important (1949: 23). His own approach was "based at least as much on *Gestalt* as on [behaviourist] learning theory" (p. 59). His major disagreement with both of them was that, in their very different ways, they each assumed "a sensory dominance of behavior" (p. 3). (This was more true of the early Gestaltists, such as Kohler, than of Bartlett.) By contrast, Hebb would emphasize largely autonomous *central* (cortical) processes. For Hebb's connectionism was defined in terms not of the linear single-cell circuitry of "switchboards", but of cyclical activity in *populations* of neurones—or "cell assemblies".

c. The cell assembly

Like the authors of the future cognitive science manifesto (6.iv.c), Hebb was concentrating not on the "S" and the "R" but on the hyphen between them. But whereas they would spell out the hyphen in computational terms, Hebb spelt it out in terms of (notional) neurophysiology.

Craik had done this before him, of course (Chapter 4.vi). But besides being unknown in the post-war USA, his account of "models" in the cortex was even sketchier than Hebb's. Indeed, Hebb's theory of cell assemblies could be used to put (speculative) flesh onto Craikian bones. So psychologists in the UK, already primed by Craik, were no less excited by *The Organization of Behavior* than their colleagues across the seas.

Because of the widespread mid-century doubts about the very possibility of learning-by-changes-in-synaptic-resistance (see subsection b, above), Hebb had to develop an unorthodox neurophysiology. Previous brain scientists (including McCulloch in 1943, though not in 1947) had tended to concentrate on single-cell circuitry. But Hebb focused on interacting *groups*, of neurones, or cell assemblies.

He proposed "two radical modifications" of earlier ideas about synaptic transmission. Transmission isn't simply linear, he said, but "always involves some closed or recurrent circuits". And a single impulse can't normally cross a synapse: "two or more must act simultaneously, and two or more afferent fibers must therefore be active in order to excite a third to which they lead" (1949: 10).

This led to a third neurophysiological idea with "revolutionary implications for psychology"—namely, the cell-assembly hypothesis:

> It is proposed first that a repeated stimulation of specific receptors will lead slowly to the formation of an "assembly" of association-area cells which can act briefly as a closed system after stimulation has ceased; this prolongs the time during which the structural changes of learning can occur and constitutes *the simplest instance of a representative process* (image or idea) . . . (1949: 60; italics added)

Considered as a physical structure, a cell assembly is "a closed solid cage-work, or three-dimensional lattice" (p. 72). It has "no regular structure", and connections are possible "from any one intersection to any other".

The implication was that only "stimulation of receptors" engendered significant structure in the cortex. Without it, cortical activity would be largely random. Whether such stimulation needed to be continued in order to *maintain* this structure was then an open question, although Hebb's belief was that it did. Eventually, it led to countless studies of "sensory deprivation" (Zubek 1969), starting with Woodburn Heron's hiring of college students to do virtually nothing for twenty-four hours a day (Heron 1957). (Heron also did the "brilliant" experiments on stabilized images which, Hebb said later (1980: 98), "abolished" his doubts about cell-assembly theory: Pritchard *et al.* 1960.)

In today's terminology, Hebb was describing a form of "unsupervised" learning (see 12.ii.b). There's no 'tutorial' feedback, to make the pattern of activity in the cortex become correlated with the input pattern. Mere temporal simultaneity/contiguity does the job.

A cell assembly is an example of what's now called distributed memory (Chapter 12.v–vi). The representation, or 'engram', isn't stored in any one cell; and the individual neurones making up an assembly needn't be spatially near one another. Nor need they all be firing when the assembly as a whole is activated: to that extent, a cell assembly is a "statistical" concept (1949: 77).

Moreover, because of the continuous (reverberating) self-excitation of the cell assembly, a *partial* stimulus can activate the assembly as a whole. (This includes cases where an object that stimulates several senses can eventually be recognized by one sense alone: hearing someone's voice, for example, brings many other aspects of that person to mind.) For the same reason, slightly *dissimilar* stimuli can activate one and the same assembly, thus allowing for stimulus generalization (McCulloch's "knowledge of universals": Chapter 12.i.c). And cell assemblies are noise-tolerant in another sense too, for they can continue to function even if some component cells are damaged.

It followed, Hebb said, that a concept isn't something cut-and-dried: "Its content may vary from one time to another, except for a central core whose activity may dominate in arousing the system as a whole" (p. 133). This important idea wasn't picked up by experimental psychologists until much later—and even then, it wasn't picked up from Hebb (see Chapter 8.i.b). Meanwhile, most psychologists continued to think of concepts as sets of necessary and sufficient conditions (6.ii.b–c).

(Hebb continued by saying: "To this dominant core, in man, a verbal tag can be attached; but the tag is not essential." This idea, too, would be resurrected later: see 12.x.g.)

Much the same was true of the way concepts were being envisaged in philosophy. Although Ludwig Wittgenstein had by then developed his notion of "family resemblances", his later work wasn't yet officially published. Only a tiny handful of the people doing empirical studies of language were taking any notice—and they, perhaps surprisingly, were working on machine translation (9.x.d).

Even horse-racing officials took the (common-sense) cut-and-dried view. So the UK's Jockey Club laid down supposedly strict rules for deciding automatically (by photography) which horse had won the race. However, Alan Turing himself possessed a print of a photograph showing 6 inches of spittle spewing out of a horse's mouth, which had forced the officials to reconsider the rules: was the spittle to be counted as part of the horse's head, or not? (Schaffer 2003*b*). He showed the photograph to his Manchester colleague Michael Polanyi, who delightedly used it as ammunition in his lectures defending "tacit" knowledge (see 13.ii.b). But Polanyi was a philosophical maverick. Most people, in or out of the Jockey Club, assumed that concepts were, or anyway should be—and therefore *could* be—defined (represented) strictly. Hebb's cell assemblies cast doubt on that assumption.

It also followed from Hebb's theory of cell assemblies that "there are two kinds of learning" (1949: 111). One is the learning of the newborn baby, or of the adult reared in darkness but now presented with light. This, said Hebb, is very slow—especially in primates. Indeed, the higher up the phylogenetic scale, the slower infantile learning is. (He cited evidence from experiments on insects and various mammals, and from 'restored-sight' patients such as those studied later by Richard Gregory: see 6.ii.e.)

That was an unorthodox claim: "We are not used to thinking of a simple perception as slowly and painfully learned" (p. 77). The reason was this:

A triangle then is a complex entity in perception, not primitive. As a whole, it becomes distinctive and recognizable *only after a prolonged learning period* in which there is a good deal of receptor adjustment—head-and-eye movement... [These] changes of visual fixation and some locomotion occur freely. The problem is to show how the variable stimulation which results from such movements can have a single effect, the perception of a single, determinate pattern. (p. 84; italics added)

Each grossly different pattern of stimulation, as the object is seen from one side or another, requires the establishment of a *separate* set of cell-assemblies... [The] various sets of assemblies would *gradually* acquire an interfacilitation—*if* sight of the object from one angle is often followed by sight of it from another. Arousing one would then mean arousing the others, and essentially the same total activity would be aroused in each case. (p. 91; italics added)

Only *Homo sapiens* can achieve this with full generality. A rat, for instance, can learn to perceive a triangle. But he fails to recognize it if it's rotated by 60 degrees, or if the familiar white-on-black is switched to black-on-white. The difference lies in the difference between the brains:

The possession of large association areas [in the cortex] is an explanation both of the astonishing inefficiency of man's first learning, as far as immediate results are concerned, and his equally astonishing efficiency at maturity. (p. 126)

The "equally astonishing efficiency" referred to the second type of learning, which is faster—even instantaneous. A human adult, said Hebb, can glance once at a face and remember it indefinitely. Yet a face is a highly complex structure, whose neurophysiological representation must code many different relationships.

Learning such complexity so quickly is far beyond the ability of a rat. Even human babies can't recognize individual faces until they're several months old. (Hebb assumed that faces are learnt *purely* by experience, since the newborn's cortex is randomly connected—see e.g. pp. 68–9, 71. Much later, it was discovered that innate neural mechanisms aid the baby's face learning, and seed the abilities of the adult—see 14.ix.c.) The adult human, however, has built up a host of mutually facilitative cell assemblies. Their associative structure is already so rich that the person can 'instantly' learn complex stimulus patterns: "The prompt learning of maturity is not an establishing of new connections but a selective reinforcement of connections already capable of functioning" (p. 132).

This aspect of Hebb's work defused some of the psychological disputes raging at mid-century. These had opposed "incremental" to "single-trial" learning, and learning to "insight". His distinction between early and mature learning integrated much of the data supporting these competing theories (pp. 109–20).

So Lashley's experiments on memory and learning were respected, and to an extent explained. Indeed, they were explained—or so Hebb claimed (Lashley disagreed: see subsection e, below)—in a way already intimated by the great experimenter himself:

[Lashley (1929*b*)] himself has pointed out [that his] evidence is consistent with the idea that the trace is structural but *diffuse*, involving, that is, a large number of cells widely spaced in the cortex, *physiologically but not anatomically unified*. (Hebb 1949: 13; italics added)

The physiological 'unification', according to Hebb (p. 41), was a matter of "connections and specialized conduction paths", not—as both Kohler and Lashley had thought—of neural "gradients and fields". The perceptual psychologist's classic problem of *stimulus generalization* was solved too—not by innate stimulus equivalence (favoured by Kohler and Lashley) but by learning.

d. Beyond perceptual learning

Hebb's vision for psychology went far beyond perception and sensori-motor learning. It covered thinking, attention, expectancy (set), purpose, emotion, and even psychopathology.

Psychopathology was one of his long-standing interests. He'd read James and Freud (and Watson) before enrolling as a psychology student. And, having worked with Wilder Penfield in Montreal after leaving Lashley and Harvard, he'd gained clinical experience of various types of brain damage and mental illness. Three chapters of his book were devoted to these issues—including a discussion of 'Mental Illness in Chimpanzees', such as "neurotic" and "psychotic episodes" and "depression" (1949: 245–50).

Mental illness? And in *chimpanzees*? Behaviourists' minds boggled. But the cell assembly was supposed to show how all of this is possible—hence Hebb's promise that his form of connectionism would be "more fertile than in the past" (p. 60).

He suggested, for instance, that temporally organized *thinking* is the sequential activation of a number of cell assemblies:

> Any frequently repeated, particular stimulation will lead to the slow development of a "cell-assembly"... capable of acting briefly as a closed system, *delivering facilitation to other such systems* and usually having a specific motor facilitation. *A series of such events* constitutes a "phase sequence"—the thought process. Each assembly action may be aroused by a preceding assembly, by a sensory event, or—normally—by both. The central facilitation from one of these activities on the next is the prototype of "attention". (1949, p. xix; italics added)

Besides temporal sequence, Hebb's theory allowed for structural complexity. Phases and phase cycles, like Lashley's motor skills and Bartlett's schemata, were assumed to be hierarchical. (Hebb often used the term "schema" as a near-synonym for "phase" and "phase sequence": e.g. p. 121.) It followed that "two ideas or concepts to be associated might have, not only phases, but one or more subsystems in common" (p. 131). This makes memory, and therefore thought, easier than in the case of "a simpler perception without [complexity or] meaning". Indeed, said Hebb (p. 132), that's why people trying to remember a list of unrelated things normally construct a meaningful mnemonic to help them do so (see ii.b above, and Chapter 6.iv.c).

Phase sequences—which, largely because of their hierarchical structure, can include alternative pathways—can be learnt from experience, just as individual cell assemblies can. So thinking can improve with practice. It can also deteriorate with practice: a maladaptive sequence, such as a neurotic's obsessional idea, will become more entrenched with repetition.

Like McCulloch before him (Chapter 4.v.c and d) and Kenneth Colby after him (7.i.a), Hebb was interested in neurosis. But like McCulloch (and unlike Colby) he didn't focus on the specific semantic content of individual neuroses. He was more interested in how *neurophysiological* factors could explain their origin and/or functioning:

It is assumed that the assembly depends completely on a very delicate timing which might be disturbed by metabolic changes *as well as by* sensory events that do not accord with the pre-existent central process. When this is transient, it is called emotional disturbance; when chronic, neurosis or psychosis. (p. xix; italics added)

He admitted that he couldn't fully explain the chimpanzees' mental illness—nor human neuroses either (p. 250). But the answer must lie in cell-assembly country:

Freudian theory has the credit of recognizing the existence of a kind of learning that causes, apparently, no immediate emotional disturbance and yet may contribute to one much later. Such learning undoubtedly occurs . . . [but] we must find some way of incorporating it into other learning theory—not as an *ad hoc* assumption specially made to deal with mental illness. (p. 250)

He suggested that emotional disturbances can prompt long-lasting "adaptive" responses of fear or aggression, associated with the thought patterns of the individual. The phase sequences concerned are assembled (learnt) differently, according to personal circumstances. So a child brought up in a punitive authoritarian regime learns that anger directed at the parents is ineffective, and becomes an adult whose anger is diverted onto those whom he/she believes can be influenced in some way and/or onto those who show (forbidden) insubordination (pp. 257–8).

In general:

. . . mental illness consists either of a chronic disturbance of time relations in the cerebrum, or a lasting distortion of the thought process from such a disturbance at an earlier time . . . [The illness may be "adaptive" in that it consists in] changes in thought (new phase sequences) that are not characteristic of the majority of the population but which avoid some conflict that makes for major disruption of [the] assembly or phase sequence . . . neurosis or psychosis is a product neither of experience nor of [neurophysiological] constitution, but a joint product of both. (p. 259)

It followed, as a matter of logic, that no mental illness can be caused *simply* by one's childhood experiences.

Although Hebb didn't use the term, nor belabour the point, this was an example of "epigenesis" (see Chapter 7.vi.g). Epigenetic development doesn't fit the nature/nurture dichotomy, because what happens at any given moment depends on *both* the genetic constitution (and past history) *and* the current environment—including the bodily tissues.

Where brain development is concerned, "environmental" influences are of several different kinds. They obviously include (1) sensory stimulation from the outside world after birth. Less obviously, they include (2) sensory stimulation from the external world before birth (newborn babies can recognize the intonation patterns of their mother's language). They also cover (3) sensory stimulation in the womb (as the foetus moves its limbs or sucks its thumb). And (4) non-sensory internal factors qualify too—for instance, random 'noise' as the unborn baby's neurones fire spontaneously.

Hebb didn't have specific behavioural (or histological) evidence of pre-natal learning. But he explicitly allowed for it, when discussing the lowering of synaptic resistance by the growth of synaptic knobs: "[this] is learning in a very general sense, which must certainly have begun long before birth" (p. 66). Similarly, he said that "the intrinsic organization of cortical activity" that's present at birth is (in that sense) "innate", but it "may or may not be unlearned": it could have been " 'learned'—established *in utero* as a result of the neural activity itself" (p. 121 n.).

That insight was relegated to a footnote. Half a century later, spontaneous self-organization in the cortex would be observed and modelled (Chapter 14.vi.b and ix.c). And innateness would be "rethought" much as Hebb had suggested (7.vi.g).

In short, Hebb's position on the development of mental illness was all-of-a-piece with his physiological psychology. As for whether psychoanalysis works better as a therapy than therapies based on conditioning, he doubted it. But no one knew. Moreover, "electroshock" and lobotomies were options too (pp. 271–4). The one sure point was that questions about therapeutic effectiveness were empirical questions—requiring experiments, not dogma (p. 260).

e. Hebb's originality?

Just how original was Hebb's theory? Connectionism, to be sure, was a new word. But it was a very old idea, and had already appeared in a number of neurophysiological guises. So what about Hebb's way of expressing it, in terms of cell assemblies?

According to Hebb himself (1949: 10, 61), the cell assembly was a development of Rafael Lorente de No's 1930s work on self re-exciting, or reverberatory, circuits (1933*a*,*b*, 1934, 1938). He'd found out about this, he said, from the new textbook on learning theory (Hilgard and Marquis 1940). Thirty years later, Hebb was still acknowledging his debt to Lorente do No:

> [The cell assembly theory] certainly looked improbable to its author—me—when it was first conceived [because it makes the ease of perception of common objects the result of a long process of learning].
>
> The problem of perception remained intractable for about five years (1939 to 1944) and as a result I made no progress in my attempt to understand concepts and thought. It seemed obvious that concepts, like images, must derive from perception, and I could think of no mechanism of perception that corresponded to my . . . preconceptions. *In fact, by 1944 I had given up trying to solve the problem.* What happened then was that I became aware of some recent work of Lorente de No in conjunction with some observations of Hilgard and Marquis (1940) [which led me to think about the problem] from a different point of view. . . .
>
> The essential basis of an alternative view was provided by Lorente de No, who showed that the cortex is throughout its extent largely composed of enormously complex closed or re-entrant paths, rather than linear connections only between more distant points. . . . When an excitation reaches the cortex, instead of having to be transmitted at once to a motor path, or else die out, it may travel round and round in these closed paths and may continue to do so after the original sensory stimulation has ceased. (Hebb 1980: 83 ff., 87; italics added)

That's surprising, since (as he noted—1949: 10) Lorente de No's "revolutionary" neural-feedback ideas were "well enough known by now to need no elaborate review". The cyberneticists had picked them up in the 1930s (see Chapter 4.iii.c). And Lashley had been writing about them for some years (Lashley 1938, 1942).

Half a century later, indeed, Jack Orbach (1998) would argue at length that Hebb owed a huge, and unacknowledged, debt to Lashley. Specifically, Hebb used four of

Lashley's ideas in defining the cell assembly: "non-sensory control of behavior, the central autonomous process, mechanisms of attention, and the importance of Lorente de No's reverberatory circuit" (Orbach 1998, p. xii). What's more, he adopted Lashley's focus on neurone groups—the assembly being his name for what Lashley called "neural lattices". (Hebb sometimes described the cell assembly as "a three-dimensional lattice", as we've seen.)

The notion that neural networks can be *learnt*, Orbach admitted, was indeed original to Hebb:

> Probably the most enduring idea embodied in the 1949 monograph, and for which the monograph is justly famous, is the *empirically assembled* nerve net that Hebb dubbed the "cell assembly." I would venture the opinion that the conception that functional nerve nets can be assembled *by experience* is one of the more important ideas in neuropsychological theory of the twentieth century. *Lashley never developed that idea, nor did he ever acknowledge it in print.* We have to credit the 1949 monograph for its dissemination. (Orbach 1998: 71; italics added)

Orbach's "controversial and startling if not downright outrageous" claim (1999) attracted significant peer commentary, most of it highly critical. For example, Peter Milner (1999)—who back in the 1950s had improved Hebb's learning theory by including inhibition (see below)—pointed out that Lashley's work on reverberatory circuits was very different from Hebb's. Whereas the former posited *innate mechanisms of stimulus generalization in perception*, the latter focused on *the learning of neural mechanisms representing concepts*.

Orbach knew that, of course. He'd even noted that on the only occasion when Lashley referred to Hebb in public after 1949, he disagreed with him. (The reason for Lashley's near-silence on Hebb, according to Orbach, was that the key ideas listed above had long been discussed in his own publications.)

In his last lecture, given in 1957 and never published before being summarized in Orbach's Prologue, Lashley criticized Hebb's account of stimulus equivalence as fundamentally behaviouristic: "[of all current learning theories, it's] the most in accord with conditioned reflex theory". He also dismissed it as mistakenly externalist, being "ruled out by a mass of evidence for innate [*sic*] discriminations and equivalencies". Moreover, in his four Vanuxem Lectures, given at Princeton in 1952 (and printed for the first time in Part II of Orbach's volume), he developed his long-standing view that learning *can't* be explained in terms of lowered synaptic resistance after cell firing—if only because unactivated neurones appear to learn too (Lashley 1924, 1952). (In fact, Hebb had explained this: neurones could be activated indirectly, through their membership of assemblies linked to the prime candidate.)

Orbach noted also that Lashley had been invited by Hebb to co-author *The Organization of Behavior*, but had declined. On scrutinizing the comments that Lashley gave to Hebb in early 1946, Orbach inferred that Lashley had read only the first ninety-six pages of the 300+-page draft. (These pages, however, contained the most sustained *neurophysiological* discussion.) He found that Hebb had made no changes as a result of Lashley's comments, not even when they concerned his descriptions of Lashley's own work.

This led Orbach to suggest that Hebb had asked Lashley to look at the manuscript not because (as he told others at the time) he thought the book would be more widely read with Lashley's name on the title page, but because he wanted to ensure that Lashley wouldn't claim the ideas as his own. Given that he evidently had no intention of making changes as a result of the comments, said Orbach, why else would he have asked for them?

This intriguing personal snippet doesn't prove much, however. Probably, Hebb had expected that some of Lashley's comments would be (from his point of view) worth taking on board. Possibly, he felt that Lashley—as his mentor—would be offended if he wasn't asked to comment. What's more, the story could be glossed *against* Orbach's main claim. For Lashley's failure to protest that his ideas were being either stolen or misused (as opposed to misdescribed) in Hebb's manuscript suggests that he didn't recognize them there.

That's hardly surprising. The two men's theories were very different—as Milner would point out later and as Lashley's own comments of 1957 (quoted above) confirmed. Moreover, by the 1940s no one connected with the cybernetics community—centred at MIT and Harvard, Lashley's base camp—could seriously have imagined that reverberatory mechanisms were "his" idea alone.

In sum, Orbach's ascription of priority to Lashley is, in my view, mistaken—albeit not "outrageous". Hebb is rightly credited with the cell assembly, even though the idea of neuronal reverberation (and neural lattice) was already being used by others.

For sure, his contemporaries saw him as original—and it was the cell assembly which excited them. His work had an enormous influence on experimental psychology, firing an explosion of research on early learning and/or sensory deprivation over the next decades.

Some people, for instance, asked how the development of vision depends on active, as opposed to passive, movement (Held and Hein 1963). Others observed the effects of rearing kittens in the dark, or in unnatural environments—such as having only vertical stripes to look at (Hirsch 1972). A few asked whether kittens could learn to overcome fundamental changes in the position of an eye (D. E. Mitchell *et al.* 1976). Yet others probed the effects of sensory 'silence' on mature human beings (Zubek 1969).

These studies weren't all purely behavioural: sometimes, brains were involved too. After the discovery of feature detectors and ocular "columns" in visual cortex (Chapter 14.iv), neurophysiologists soon asked whether these anatomical features could be modified by changes in early visual experience (C. Blakemore and Cooper 1970). In short, psychology was now being complemented by real neurophysiology, just as Hebb had hoped.

What's more, it was being complemented by computer modelling too.

f. Loosening the mantle

Hebb's book didn't achieve its phenomenal success with cognitive psychologists immediately (see the Mandler quotation, above). That's so *despite* the fact that parts of it had been read before publication by two of the manifesto writers featured in Chapter 6.iv.c: Karl Pribram, a fellow student of Lashley's, and Miller (Hebb 1949, p. vii). Not until the mid- to late 1950s would computer modelling become successful enough, or anyway promising enough, to shrug off the behaviourist mantle. For all that, Hebb had *loosened* the mantle, by offering a clear neurophysiological theory in support of mentalism.

However, there's clear—and then there's *clear*. Despite Hebb's claim to have stated the ft/wt (fire together, wire together) principle "precisely", there were three important unclarities. The first was that he supposed his two formulations (given at the beginning of subsection b, above) to be equivalent—but they weren't. Second, the earliest computer implementation of his work ran into unsuspected difficulties, forcing a redefinition of his learning rule. And third, when others came to define it later, they found that there were many different ways of doing so.

The first computer model of Hebb's rule was produced at IBM's research laboratory in Poughkeepsie, by a group led by Nathaniel Rochester, IBM's manager of information research (Rochester *et al.* 1956). It was begun soon after the appearance of *The Organization of Behavior*, and benefited from regular consultation with Hebb himself. It also built on network models being produced by Belmont Farley and Wesley Clark at MIT, and pioneered by Raymond Beurle at Imperial College, London (see Chapter 12.ii.b).

The Hebb simulation was one of the earliest examples of computer modelling being used to test/improve a specific psychological theory. And the effort was well spent, for Rochester's group discovered several previously unsuspected problems.

The first was that if Hebb's learning rule was implemented (as it was, initially) by McCulloch–Pitts on–off threshold units, a runaway positive feedback ensued—so that the strengths of all the synapses increased to saturation. To counter this, they introduced a "normalization" rule, according to which the sum of all the synaptic strengths remained constant. Strengthening *here* would be compensated by weakening *there*. They also mimicked neuronal fatigue, so that a unit that had just fired would be unlikely to fire again immediately.

Even so, the network (of 69 to 99 units) didn't develop recognizable cell assemblies. So the IBM group tried again. Now using a network of 512 units, they followed Peter Milner's (unpublished) suggestion that Hebb's theory needed to include *inhibition*. (This wasn't cheating. Sherrington had shown that the brain must contain inhibitory neurones, although at mid-century their manner of function still wasn't understood: see Chapters 2.viii.d and 4.iv.b.) The idea, here, was that cell assemblies compete. That is, they're not only self-exciting (reverberatory) but also other-inhibiting.

The units, or "neurones", were made less simple accordingly. Instead of being binary, unit-activity was now graded from 1 to 15. A mathematical measure of unit-activity (the ancestor of several such measures used today) took into account both synaptic strength and the average frequency of past firing. And Hebb's ft/wt rule was modified too. If the frequencies of the pre-synaptic and post-synaptic units were correlated then the synapse would be strengthened, and if one unit fired and the other one didn't then it would be weakened. That helped: Rochester was now able to say, "there is no doubt that cell assemblies did form, [even though the model] still needs improvements".

Rochester's paper had begun by making a historical point about the need for interdisciplinarity:

> As the neurophysiologist considers more and more complicated structures of neurons he gets into problems that are less and less related to his normal way of thinking. Curiously, however, some of these problems do not begin to resemble parts of psychology. *What is happening is* that the neurophysiologist is beginning to think about information handling machines that are too complex to be understood without the specialized knowledge of other disciplines. These other

disciplines are *information theory, computer theory, and mathematics.* People in these other fields need to augment the work of the neurophysiologists and psychologists before the brain can be properly understood. (Rochester *et al.* 1956: 80; italics added)

Across the pond, Horace Barlow, Giles Brindley, and the young David Marr were thinking along broadly similar lines (14.iii and v.b–e). In short, neurophysiology was beginning its long metamorphosis into computational neuroscience.

As for connectionist modelling, many people followed Rochester's example. Over the next forty years they found that there are many "Hebbian" rules, which give rise to different learning profiles (Chapters 12 and 14.v–vi).

Hebb himself hadn't pursued this methodology. When he was writing his 1949 book, it wasn't yet technologically possible (Chapter 3.v). The very first connectionist systems were wire-and-solder contraptions, not computer programs (see 12.ii.a–b). And even computers, in those days, could be used only by the technically minded. Hebb had originally aspired to be a novelist, settled sensibly for schoolteaching, and studied psychology part-time. He got his Ph.D., at 32, without having had time to learn engineering too.

It's not even clear that he was tempted towards modelling, whether material or mathematical. He hailed the work of McCulloch, and other members of Nicholas Rashevsky's "Mathematical Biophysics" group, as interesting. Indeed, he did this on the very first page of his book. But he valued McCulloch's neuro-anatomical research (cited several times in the text) at least as much as his mathematical studies: neither the 1943 nor the 1947 paper was mentioned, or listed in the fifteen-page bibliography. Indeed, he may not have known about the "Universals" paper:

For the present, Kohler and Lashley are the only ones who have attempted to say *where* and *how* perceptual generalization takes place. (1949: 38)

This was a mistake, for the 1947 Pitts–McCulloch paper had attempted to do just that (Chapter 14.iii.a).

Moreover, Hebb was suspicious of idealizations. Endorsing a recent review by the Washington University neurophysiologist George Bishop (1946), he complained that Rashevsky's group had been "obliged to simplify the psychological problem almost out of existence" (1949, p. xi).

It's intriguing, therefore, that one of the four people thanked for their "painstaking and detailed criticism" of his 1949 manuscript was the MIT (and Bolt, Beranek, & Newman) psychologist Joseph Licklider (1915–90). Licklider (1951, 1959) was already building simple connectionist models of perception, some of which implemented temporal correlations—stressed by Hebb, but largely neglected by computer modellers for many years afterwards (14.ix.g).

Looking back in the 1990s, he recalled: "I was one of the very few people, at that time, who had been sitting at a computer console four or five hours a day" (Edwards 1996: 264). This wasn't purely theoretical research. For in 1950 Licklider had outlined the interface design for the ambitious SAGE computer system, the US Department of Defense's (DOD's) early forerunner of the infamous "Star Wars" project of the 1980s (Chapter 11.i).

A few years later, Licklider would become a hugely important force in the development of computing, and of AI (Edwards 1996: 262–71). In 1962 he was appointed by ARPA, the DOD's Advanced Research Programs Agency (renamed DARPA in 1972: "D" for Defense), to found their information-processing section. In that role, he would enable—and influence—AI research for a quarter-century (Chapter 10.ii.a and 11.i.b).

(He sometimes acted also as a latter-day Gabriel Naudé, who had pioneered library science over 300 years earlier: see Preface, preamble. Licklider addressed advanced versions of Naudé's questions about librarianship by using new ideas about information technology, including timesharing, and by redescribing a library as a "thinking center": Licklider 1965.)

Apparently, Licklider didn't notice the unclarities in Hebb's formulation of ft/wt. If so, he wasn't alone. In the few years after 1949 it was the general approach which excited psychologists, not the nitty-gritty detail of just how ft/wt should be expressed. Or rather, they thought that he'd *provided* the nitty-gritty. The high standard of clarity enforced by computer modelling was then appreciated only by a tiny handful of aficionados. These, of course, included Licklider. But unless Hebb simply ignored his advice, as he did Lashley's, Licklider didn't pick up on the ambiguities either.

Rochester did. And the concluding paragraph of his paper drew a general moral:

> This kind of investigation cannot prove how the brain works. It can, however, *show that some models are unworkable* and provide clues as to how to *revise the models to make them work.* Brain theory has progressed to the point where *it is not an elementary problem to determine whether or not a model is workable.* Then, when a workable model is achieved, it may be that a definitive *experiment can be devised to test* whether or not the workable model corresponds to a detail of the brain. (Rochester *et al.* 1956: 88; italics added)

In short, computer modelling could help in the mind-as-machine project already taxiing along the runway.

A couple of years later, the project finally took off—fuelled partly by GOFAI and partly by cybernetics, including Hebb (Chapter 6.iv.b). The jet wash was strong. It blew away the behaviourist mantle, bringing all the non-Newtonian ideas sketched in this chapter into view.

6

COGNITIVE SCIENCE COMES TOGETHER

The 1950s saw the emergence of a self-consciously interdisciplinary cognitive science—though not, yet, under that name. Both the interdisciplinarity and the ideas had been prefigured in the cybernetics movement of the 1940s.

As for the interdisciplinarity, Warren McCulloch alone had combined no fewer than five areas: logic, computer science, philosophy, neurophysiology, and psychology—including clinical psychiatry (4.iii). His fellow cyberneticists had each integrated two or more of those, sometimes also weaving in anthropology, educational technology, or management/sociology (4.v.e). And, of course, biology and engineering had defined the overall frame.

As for the ideas, these came from both sides of the future divide in cybernetics (4.ix). Initially, the strongest influences in 1950s cognitive science were Claude Shannon's information theory and the concept of self-control by feedback. But thanks to the two epochal papers by McCulloch and Walter Pitts (1943 and 1947), plus the development of usable computers (3.v), there was an ever-growing interest in both symbolic and connectionist computation.

By the end of the 1950s, research using these ideas was more widespread and more confident. The confidence was expressed in 1960 in two highly visible ways: by the founding of an interdisciplinary centre, and by the publication of what turned out to be the manifesto of cognitive science. Crucial theoretical concepts were diffusing from one area to another.

The cognitive revolution began with psychology. Indeed, quite a few psychologists had opposed behaviourism well before 1950, and both Karl Lashley and Donald Hebb had recently hurled anti-behaviourist rocks into the water (see Chapter 5). Now, in the early 1950s, computational accounts of perception, concepts, and human decision making appeared on both sides of the Atlantic.

The other disciplines, at that time, were less easily visible. When the revolution picked up steam in the late 1950s, linguistics would provide much of its motive power (Section i.e, below). But only one linguist's work was important in this regard, and in the early 1950s he was still almost unknown. Neurophysiology was just beginning to feature informational/computational ideas (14.iii–iv), but connectionist AI was still in early infancy (5.iv.f and 12.ii). As for symbolic AI, this—apart from work in machine translation (9.x)—had hardly begun (10.i).

That being so, this chapter must begin by discussing psychology. How mind-as-machine first began to drive psychological experiments is outlined in Sections i–ii. These describe the development of informational psychology and the New Look, and sketch the first psychological work inspired by Noam Chomsky. Section iii recounts the electrifying effect of Herbert Simon's encounter with Allen Newell early in 1952, which prompted the first experimentally guided computer models of thinking.

The rest of the chapter relates how the various disciplines came together in the mid- to late 1950s, and how that led to the establishment of the first official research groups. It highlights the cooperation that's needed for science in general and interdisciplinarity in particular (2.ii.b–c). The three seminal meetings of cognitive science, the manifesto, and the early research groups are all described in Section iv. Finally, Section v says a little about how the word was spread in the following years.

(A note about nomenclature: This chapter is about what cognitive scientists were doing in the 1950s and early 1960s, and how they saw themselves at the time. As for what they *called* themselves, that varied—but it wasn't "cognitive scientists". The term "Cognitive Studies" was coined in 1960, and "cognitive science/s" had to await the early 1970s: see Chapter 1.ii.a. Even "cognitive psychology" wasn't named until 1967. So the current phrase is used here only for convenience: the people concerned weren't thinking of themselves under that label.)

6.i. Pointers to the Promised Land

The immediate forerunners of computational psychology—or, if broadly interpreted, its earliest examples—were based on information theory. At first, this was welcomed by psychologists because it offered quantitative measurements. Soon, its theoretical concepts were appreciated too. For it supported centralism without falling into homunculism: positing mysterious 'little men' in the mind/brain. (The soft centres were hardening: see Chapter 5.iii–iv.)

Information-processing psychology used ideas about machines (telephone systems) to conceptualize minds. But it didn't stress computer modelling, nor even programs. (The occasional exceptions included an early implementation of formal grammars—G.A. Miller and Chomsky 1963: 464–82.) The first person to use information theory to describe *the mind as a whole* was Donald Broadbent (see subsection c). But George Miller (1920–) was the first to apply it within (a more limited area of) psychology.

Miller was the first, too, to persuade psychologists that Chomsky's linguistic theory was relevant for them. Chomsky himself had deliberately remained silent on this point in the mid-1950s, for fear of frightening the horses (9.vi.e). In 1959, however, he published his notorious attack on behaviourism (9.vii.b). (His yet more provocative defence of nativism would be delayed even longer: see 7.vi.a and 9.vii.c–d.) Miller's interest in Chomsky was a key point in his own turn from "information" to "computation".

a. Informed by information

The mid-twentieth century saw an explosion of numerical/formal methods for describing mental processes, known collectively as "mathematical" psychology.

Some were developments of behaviourism, rendered axiomatic (by Clark Hull) or probabilistic (by Edward Tolman and Egon Brunswik): see Chapter 5.iii.b. For instance, William Estes (1919–), a founder of the Society for Mathematical Psychology, proposed a statistical theory of learning (1950, 1959). Eventually, he would develop mathematical models of many aspects of cognition.

Instead of treating the stimulus as an indivisible Newtonian atom (see Chapter 5.i.a), Estes saw it as *a set of stimulus elements*, only some of which will be "sampled" by the animal on any given trial. The conditioning that occurs involves those elements alone. So if the animal happens to sample a different subset on the next trial, the 'learnt' response won't ensue. Learning actually takes place in a single step; but it *seems* to be gradual, because time is needed for a response to be conditioned to all possible samples of the stimulus concerned.

Most mathematical psychologists were inspired by the cybernetic ideas outlined in Chapter 4.v–vii. Their new theories included John von Neumann's theory of games (von Neumann and Morgenstern 1944); signal-detection approaches (Wald 1950); Robert Luce's axiomatic theory of choice (Luce 1959); and Simon's (1957) account of decision making in social groups.

By the end of the decade, Chomsky's work on formal grammars was included too. Fully 40 per cent of the second volume of the *Handbook of Mathematical Psychology* (1963) was devoted to papers written or co-authored by Chomsky. As for the first volume, this carried a lengthy chapter on mindlike artefacts, ranging from the Homeostat, through GOFAI, to perceptrons (Newell and Simon 1963).

As well as new theories, mathematical psychology provided new methods of *measurement*. For instance, perceptual psychologists were offered a "geometrical" measure of similarity between structured stimuli (Eckart and Young 1936); each stimulus was located at a particular point within a multidimensional hyperspace (for an explanation of hyperspaces, see Chapter 14.viii.b). And psychologists in general were given an analysis of different types of measurement scale: some fully arithmetical, some dealing only with ratios, and some limited to the ordering of values (S. S. Stevens 1946). If psychology couldn't match the numerical precision of physics, it could use mathematical descriptions whose parameters were specified more vaguely.

So mathematical psychology was a broad church. But of all the new ideas fermenting in the 1940s, Shannon's were the most intoxicating. They were also the most important *as a lead-in to computational psychology*, for they dealt with the coding, or transformation, of information as well as its transmission. Psychologists theorizing about internal representations could look to Shannon in asking how—and why—these are constructed. Moreover, information theory, internal models, and computers were then considered all part of the same bag (Chapter 4.ix). Anyone interested in one was probably interested in them all.

Although information theory offered both explanatory concepts and numerical description, the initial feeding frenzy was due to psychologists' hunger for the latter. Measurement was widely believed to be essential for a psychological science.

In practice, not everyone agreed. Programs and generative grammars—both pioneered in this period—were formal systems having nothing to do with numbers. So were McCulloch and Pitts' logical networks, or TPEs (4.iii.e). (Indeed, science in general is best thought of as an empirically constrained search for *systematic structural possibilities*, of which numerical laws are a special case—Sloman 1978: chs. 2–3; see Chapter 7.iii.d.) But the initial attraction of information theory was its ability to provide numbers—that is, measurement in bits.

At base, Shannon's theory—like behaviourism—was a beads-on-strings affair, for it concerned sequences of events conceptualized as Markov processes. (It was therefore vulnerable to Chomsky's criticisms of Markovian theories: Chapter 9.vi.c.) But the informational psychologists used a non-Newtonian concept of the stimulus.

Instead of finding complexity inside the stimulus, as Estes had done, they found it outside. That is, a stimulus was no longer definable in isolation. Much as John Dewey and Ralph Perry had seen *stimulus* as a covertly purposive term (5.iii.a), so the information theorists saw it as covertly probabilistic. To be sure, information can be measured. However, a given input (or response) carries *this much* information if it's one of only two possible alternatives, but *that much* if it's one of three, or four . . . or more. In other words, *what the stimulus is* depends on what had previously been called environmental variance.

These psychologists asked how many bits a given sensory channel can carry. They looked to see whether some senses are more nicely discriminatory than others. And they studied how much (*sic*) it helps to have input from two or more senses simultaneously. Also, they asked how specific redundancies in the environment (in the signal) can improve the reliability of the message communicated.

For instance, Fred Attneave (1919–91) argued that perception relies heavily on redundancies (1959). A straight line, or a uniform curve, is highly redundant, highly predictable. But a sudden change of direction, a corner, or a break isn't. The perceiver should therefore pay more attention to those parts of the scene, or stimulus, which are low in redundancy. (A few years later, a remarkable Russian study of eye-movements would show that this is indeed what happens: Yarbus 1967.)

Eventually, psychologists woke up to the *theory*, not just the *measurements*. For example, the Oxford professor R. Carolus Oldfield (1954) used Shannon's concepts—plus ideas about computers—to formulate a theory of memory. He imagined a computer that codes redundancies, and stores stimuli in two parts: the core-schema and the deviances. Such a system, in principle, could represent facts like *All birds fly—but emus don't* (compare "frames" and "defaults" in GOFAI: 10.iii.a and 13.i.a). Oldfield envisaged hierarchies of coding schemas, much as Miller would later (see below).

As for problem solving, the typical information-theoretic approach was to remind their readers of the parlour game Twenty Questions (see 4.v.d). Only an idiot would ask *Is it a Manx cat?* before asking *Is it a cat?*, and some even wider questions—*Is it a mammal? Is it a domesticated animal?*—should come earlier still. The efficient problem-solver aims to halve the number of possibilities at each successive step. This assumes that all possibilities are equally likely: if they aren't, then other strategies must be calculated (Attneave 1959: 5–9). The core idea, that *some rational strategy or other* must be used, would later be picked up by the New Look researchers (Section ii.b, below).

Questions about the efficiency of different methods of coding were applied to neurophysiology too. By the end of the 1950s, Albert Uttley and (most notably) Horace Barlow were thinking about specific neural mechanisms in terms of economical coding (Chapter 14.iii.b).

Mathematical psychology didn't suit all tastes. Tolman, for instance, commented:

Psychology today seems to me to be carried away (because, perhaps, of feelings of "insecurity") into a flight into too much statistics and too great a mathematization . . . [To] me, the journals seem to be full of oversophisticated mathematical treatments of data which are in themselves of little intrinsic interest and of silly little findings which, by a high-powered statistics, can be proved to contradict the null hypothesis. (1959: 150)

However, he was talking primarily about behaviourist curve fitting. One could share Tolman's impatience with obsessively precise mathematical *descriptions* while being excited by formal–mathematical *explanations*. Among the first to take the explanatory concepts of information theory to heart was Miller.

b. Miller and magic

The Harvard luminary Miller was the first information-theoretic psychologist. (I say "Harvard", but for four key years he was at MIT: 1951–5. He'd been appointed by Joseph Licklider, who'd set up MIT's Psychology Department a few years earlier.) He was initially trained by the psychophysicist Stanley (Smitty) Stevens (1906–73), who ran Harvard's Psycho-Acoustic Laboratory.

This was "the largest university-based program of wartime psychological research" in the USA, and it continued to thrive in the post-war years (Edwards 1996: 212). Much of the Lab's effort went on studying communication in noisy conditions, and on investigating the psychological side of what Licklider called "man–computer symbiosis" (Chapter 11.i.a–b). In other words, it was the American equivalent of the MRC's Applied Psychology Unit at Cambridge (see 4.vi, and subsections c–d below).

Miller was always more interested in speech and language than in pure psychoacoustics. As a youngster intending to go into clinical psychology, he'd worked as a speech therapist at the University of Alabama in 1942 (G. A. Miller 1986: 201). The acoustics was added after the US entered the war, when he spent two years on a top-secret speech-jamming project for the Signal Corps.

In the mid-1940s, when he worked on auditory "masking" (see i.c, below), he realized that, in effect, he could turn this wartime research upside down in order to understand how speech is recognized in normally noisy conditions. (Military concerns were still relevant. Indeed, Miller's then colleague Licklider would later enable research on HEARSAY and on DARPA's "intelligent pilot's assistant", intended to interpret speech inside a jet's cockpit: 9.xi.g and 11.i.c.)

Eventually, the speech conquered the acoustics. Miller chose the young Chomsky as his assistant for his Stanford summer seminar on mathematical psychology in 1957, and they soon co-published on formal grammars (Chomsky and Miller 1958; see Chapter 9.vi.a and c).

Even at that point, Miller was already asking questions about the "psychological reality" of syntax (1986: 208). In 1960 he introduced Chomsky to psychologists in general. He did this both in his cognitive science manifesto (Section iv.c, below) and in an "elegant lecture on the psychological implications of Chomsky's grammar, given at many major universities, [which] became famous" (Baars 1986: 199). Later still, he'd make major contributions to psycholinguistics, concentrating more on vocabulary than on syntax (7.ii.d).

As for his interest in computers, this dated from the 1940s when he read about cybernetics and the McCulloch–Pitts neurone (G. A. Miller 1986: 205). But it wasn't until the late 1950s that computational concepts became central for him (Section iv.c).

Meanwhile, he'd made his name as the leading light of information-theoretic psychology. He'd pioneered this even before the seminal Shannon–Weaver book was published. By 1948 he was grilling his new graduate students on their knowledge—typically non-existent—of information theory and hill-climbing algorithms (McGill 1988: 8). And by 1949, when the Shannon–Weaver volume appeared, he was already teaching a course on phonemes, Shannon, and the redundancy of English (Bechtel *et al.* 1998: 45). Ulric Neisser (1986: 274) remembers this as being based on the manuscript of Miller's first book (1951), and "all full of 'bits' and 'phonemes' and the like" (Neisser 1986: 274).

Miller soon applied information theory in designing experiments. In the early to mid-1950s, he did a number of studies showing that redundancy (i.e. familiar context) helps people to recognize and/or remember (G. A. Miller 1951; Miller and Selfridge 1950; Miller *et al.* 1954). The stimuli he used were Shannon-inspired letter strings and word strings that approximated English more or less closely (see Chapter 9.vi.c).

The 1954 experiment helped kick-start the New Look (Section ii.a). And in 1956 his work suddenly leapt onto centre-stage for a wide range of psychologists (see below). Before then, he'd been seen as an austere specialist—some of whose papers were so symbol-ridden as to be unreadable by those not mathematically inclined (e.g. G. A. Miller 1952).

In fact, it was Miller who'd introduced information theory to psychologists in the first place. Almost immediately after Shannon's publication of it, he'd co-authored a paper with Frederick Frick in the *Psychological Review* (G. A. Miller and Frick 1949). This provided a method for *quantifying* the organization, or patterning, in sequences of events (including responses) where only a few alternatives are possible. Two years later, he and Frick would apply their analysis to the behaviour of rats in Skinner boxes (Frick and Miller 1951).

Miller's enthusiasm for information theory wasn't instantly persuasive. The psychophysicist Wendell Garner recalls that at first, even many *mathematical* psychologists (including Stevens) were sceptical, seeing Shannon's work as providing just another descriptive statistic. But that reaction didn't last long:

> The change in attitude occurred as soon as psychologists saw that information theory was much more than just another statistic and that *there was a fundamental theory behind the statistics, with a set of useful new ideas and concepts.* In retrospect, I would say that the wait-and-see attitude lasted for little more than half a year, and after that it was accepted [by psychophysicists] at what can only be called a phenomenal rate. (Garner 1988: 22; italics added)

Miller (1953) soon wrote a beginner's guide for psychologists in general, published in the main journal of the American Psychological Association (APA). Looking back, Garner now sees Miller and Frick's paper as "the birth date of cognitive psychology" (p. 32). The concepts it introduced (borrowed, of course, from Shannon)—*information*, *redundancy*, *channel capacity*, *coding*—are now part of the daily vocabulary of cognitive psychologists.

Miller's high visibility to psychologists *in general*, however, arose in 1956. In that year, he published 'The Magical Number Seven, Plus or Minus Two'—probably still the most widely read paper in his œuvre. A bibliographic study in the mid-1970s identified it as the most-cited paper in the whole of cognitive psychology (E. Garfield 1975). Although that may no longer be true, it continues to be cited today. Moreover, even when the "seven" is challenged, the core theoretical point is accepted (Cowan 2001).

This paper didn't report any new data, but surveyed a huge range of—individually, pretty boring—experiments in the literature. In so doing, it lit a blazing fire.

Miller opened with a strange confession: "My problem is that I have been persecuted by an integer." And he went on to suggest—on the basis of many different experiments, in distinct domains and involving various senses (and dating back to William James)—that, as in fairy tales, there really is something special about the number seven:

> For seven[!] years this number has followed me around, has intruded in my most private data, and has assaulted me from the pages of our most public journals. This number assumes a variety of disguises, being sometimes a little larger and sometimes a little smaller than usual, but never changing so much as to be unrecognizable. The persistence with which this number plagues me is far more than a random accident. There is design behind it, some pattern governing its appearances. (G. A. Miller 1956*a*: 81)

That "pattern", as declared in the paper's subtitle, concerned 'Some Limits on our Capacity for Processing Information'. Considered as an information-processing channel, the human mind is limited to about seven items, or 2.6 bits. That's why telephone numbers ("my most private data") typically have no more than seven digits, and why one-dimensional psychophysical discriminations (reported in "our most public journals") and arbitrary scales (found in *Woman's Own* as well as in journals of psychometrics) normally run between five and ten.

Of course, some telephone numbers have as many as fifteen digits. But they're grouped into a small number of "chunks" (international code, country code, area code, city code, individual number), none of which has more than eight elements.

Similarly, we can remember sentences of many more than nine words (and very many more than nine phonemes: a word is itself a chunk). But this holds only if they're genuine sentences, hierarchically structured (chunked) as noun phrase, verb phrase, etc. Asked to remember lists of randomly chosen words, we're constrained by the magical number seven again. The only way of overcoming this constraint is to use some mnemonic to add structure to the random signal.

In this way, the total amount of information remembered can be hugely increased. For it's the *number* of chunks on a given structural level, not their information content, which matters. (Given the usual size of an English-speaker's vocabulary, one English word carries about ten bits: G. A. Miller 1956*b*.) Another way of putting this is to say that information theory alone can't describe memory, although it can describe absolute judgement.

(Occasionally, we may appear to recall many more than seven items, as in reciting the alphabet. But no one learns the alphabet in the first place as an unstructured twenty-six-letter string. And even an adult, who can gabble the whole alphabet without taking breath, is usually aware of some sort of chunking of the letters.)

In short, said Miller, "There seems to be some limitation built into us either by learning or by the design of our nervous systems, a limit that keeps our channel capacities in this general range." (Conceivably, work in comparative and/or computational psychology might one day be able to explain why the magic number was seven—not ten, or four: Chapter 7.iv.h.)

He didn't jump to all the seemingly obvious conclusions. The seven wonders of the world, the seven seas, the seven deadly sins . . . and so on *may not* depend on this fact. The appearance of the magical number there, as well as in the psychophysicist's laboratory, *may be* "only a pernicious, Pythagorean coincidence" (1956*a*: 97). Nevertheless, the limits on human channel capacity underlie some of our most distinctively human properties. For hierarchical chunking, or recoding, which enables us to escape this basic limitation, is the general principle underlying schemas of all kinds.

Moreover, it was information theory we had to thank. "Recoding", Miller pointed out, was part of "the jargon of communication theory". The experiments he'd reported "would not have been carried out if information theory had not appeared on the psychological scene, and [so] the results are analyzed in terms of the concepts of information theory".

He'd already emphasized recoding-by-chunking as a memory mechanism some years earlier (S. L. Smith and Miller 1952). But that was in a private research report from MIT to the US Air Force—and, like most of his early papers, was highly technical. Now, he was expressing his ideas more intelligibly and circulating them more widely. Psychologists *in general* would prick up their ears when he said: "In particular, the kind of linguistic recoding that people do seems to me to be *the very lifeblood of the thought processes*" (1956*a*: 96; italics added).

c. Going with the flow

The "lifeblood" of psychology, or anyway its central problem, is often held to be consciousness. Miller himself said as much in later years: "I consider consciousness to be the constitutive problem for psychology" (1978: 233).

In the 1950s, such a remark—at least in America (see iv.d, below)—would have been too shocking, too out of kilter with the temper of the times. Whereas James had devoted several chapters to human consciousness (including attention) in his *Principles of Psychology* (1890), Stevens's *Handbook of Experimental Psychology* (1951) ignored it. Nevertheless, one of the formative texts of informational psychology was primarily concerned with consciousness. Not with *how consciousness is possible at all* (the theme of Chapter 14.x–xi, below), but with *the conditions in which it does or doesn't occur*—and with *why it's limited.* Even then, however, the theme wasn't identified as "consciousness", but as the more anodyne "attention".

The author of that text was Broadbent (1926–93). He was an English psychologist then working with Frederic Bartlett, and later—like both Bartlett and Kenneth Craik before him—Director of the MRC's Applied Psychology Unit (APU) at Cambridge. (Later still, he was the MRC Professor of Experimental Psychology at Oxford.)

His main interests were attention and perceptual vigilance. In particular, he was interested in their effects on how people—from airline pilots, through car drivers, to factory workers—interact with machines. Today, decisions about *what* information should be presented to the human operator, and *when*, and *how*, are often made (in real time) by computer programs. That applies to many medical monitoring systems, and to NASA's ground control for the space shuttle (see Chapter 13.iii.d). Fifty years ago, it wasn't even clear just what questions should be asked about this. Broadbent tried to find out.

In this, he was following in the footsteps of Craik (Chapter 4.vi.c). Alan Baddeley, a student at the APU in the early 1960s, remembers:

> There was little pressure to publish. There was, however, a feeling of real intellectual excitement. We felt that we were at the forefront of a revolution—a revolution inspired by Kenneth Craik's insights, developed into a broad information-processing approach to the human mind as reflected by Donald Broadbent . . . in his highly influential book *Perception and Communication*. (Baddeley 2001: 348)

Many people were enthused by Broadbent's ingenious and hard-headed—and practically important—research. In 1968 it earned him election to the Royal Society; for many years, he was the only psychologist so honoured. (William McDougall had been an FRS too, but he was elected for his work in the relatively high-status *physiological* psychology: 1897, 1901, 1902–6, 1904, 1905.)

As a young man, Broadbent developed a number of ideas drawn from Uttley's early connectionist models (12.ii.c–d). Like Uttley, he'd been impressed by Shannon's then recent account of information theory. He applied this not only to the nervous system but also to perceptual masking—including the interference between different speech signals presented to the two ears simultaneously (Broadbent 1952*a*,*b*).

Real-life masking/unmasking phenomena include the "cocktail-party effect", in which one suddenly hears one's own name above the general buzz of conversation (Cherry 1953; cf. Treisman 1960). They also cover the many distractions of noisy work environments—where "noise" means *any irrelevant information* (from loud sounds, to visual signals on a multi-dial dashboard or flight deck). Sometimes, of course, one *wants* to be distracted, because the so-called noise is an important warning signal. But how is it possible for such a signal to be consciously noticed, among so many others competing for one's attention?

In such cases, the *selective* nature of consciousness is evident. Broadbent argued that this depends on the information-processing properties of the mind/brain, and especially on the channel capacity of the perceptual system. That system, he said, involves several input pathways working in parallel, sensory buffers, memory stores, and a central processor. His ideas dominated research on attention (aka consciousness) for over twenty years.

The twelve general principles announced at the close of his book included several which rang magic-number bells, such as: "(A) A nervous system acts to some extent as a single communication channel, so that it is meaningful to regard it as having a limited capacity" (1958: 297). Others recalled the New Look's interest (see Section ii) in how motivation and emotion affect perception:

> (E) States of the organism which increase the probability of selection of classes of events are those normally described by animal psychologists as 'drives'. When an organism is in a drive state it is more likely to select those events which are usually described as primary

reinforcements for that drive. Thus food has a high probability of being selected if the animal has been deprived of food for 24 hrs . . . (p. 298)

(K) There is [or rather, may be] a minimum time during which information from one class of event is sampled before any action is taken about it.

(L) This minimum time is [or rather, may be] shorter in persons who are extroverted, by Eysenck's operational definition of that word. (p. 299)

Broadbent imagined someone complaining that *we already knew* that attention is limited, that noises distract us . . . and so on: "What gain is there from putting these everyday experiences into this stilted language?" (p. 300). He replied that previous psychologists hadn't expressed these matters adequately, even if they'd sometimes stated them. The precision of information theory was a significant advance.

Moreover, this new theoretical approach had some non-obvious implications, including: "The interference between two tasks will increase as the probability of the stimuli in each decreases: two highly probable stimuli will hardly interfere with one another" (p. 300). In short, the common-sense notion that "noise distracts" is correct, but it doesn't capture *just when*, *just how*, or (above all) *just why*.

One of Broadbent's most influential contributions was to use flow diagrams to express psychological theories. They'd been developed in the early 1950s for designing computer programs. But very few people knew anything about that. So the first example given in Broadbent's book had to be carefully explained: today, it seems laughably simple (see Figure 6.1).

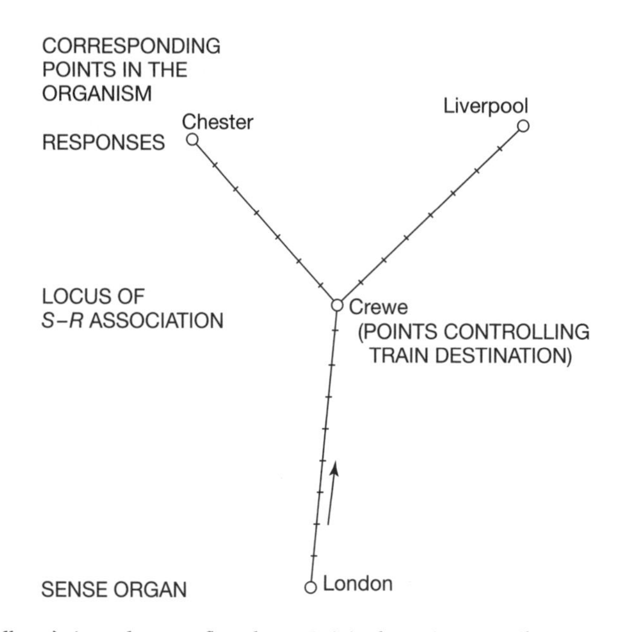

FIG. 6.1. Broadbent's introductory flowchart. Original caption: A railway system analogous to the flow of information through the nervous system in conditioning. The locus of 'inhibition' may be before that of S–R association: it is usually assumed that it is after it, for no good reason. Redrawn with permission from Broadbent (1958: 188)

Both flow diagrams and concepts derived from computer design were soon widely used, to describe not only perceptual information processing but also learning and memory:

* For instance, Earl Hunt's account of concept learning was profusely illustrated by flow diagrams and decision trees (see Figures 10.10–10.12).
* Memory stores for pronouncing or recognizing words were distinguished in John Morton's flow-diagrammed "logogen" model (1969, 1970). (For a recent development of that model, see Coltheart *et al.* 2001.)
* Thanks to the programming analogy, short-term memory was presented as an active "working" memory, not just a temporary data store, by Broadbent's student Baddeley (Baddeley and Hitch 1974).
* Also, it was found that recognition and (especially) recall are improved when their environmental conditions match those at the time of the initial "coding" (Tulving and Thompson 1973).

A more ambitious example of a flow diagram was given at the close of Broadbent's book (see Figure 6.2). Broadbent claimed that this single diagram included contemporary views on immediate memory, learning, anticipation, refractoriness, noise, multi-channel listening, and prolonged performance (p. 299). In other words, the organism was here being presented as an integrated system. And this, in a sense, was the whole story: "Nervous systems", he declared, "are networks of the type shown in Figure [6.2], *and of no other type*" (p. 304; italics added).

For all his hard-headedness, Broadbent was no behaviourist. Besides allowing one to emphasize "the relationship between the stimulus now present and the others which might have been present but are not", information theory fostered many hypotheses about internal mechanisms. But this wasn't mystery mongering:

> [Mine] is a non-positivistic approach, and attempts to find out what happens inside the organism . . . [Statements] about unobservables are not necessarily mystical or scientifically useless . . . [We] have tried to make our hypothetical constructs of such a kind that they could be

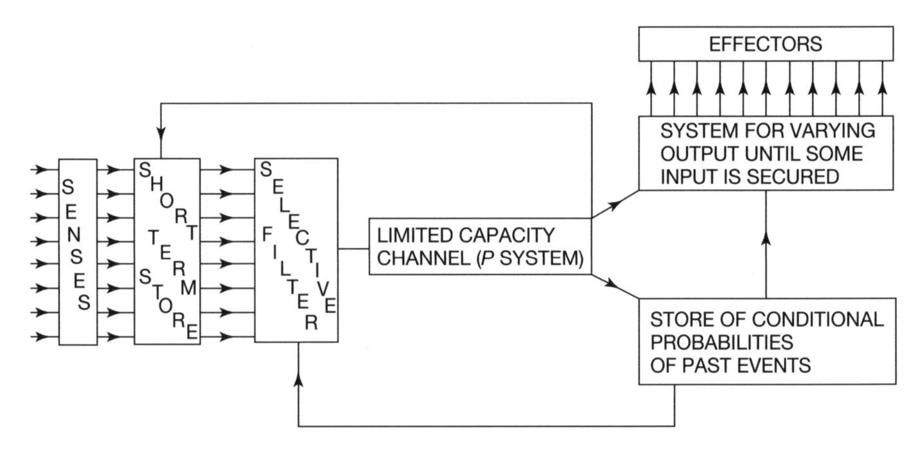

FIG. 6.2. Broadbent's "tentative information-flow diagram for the organism". Redrawn with permission from Broadbent (1958: 299)

recognized if it were possible to observe them directly: a filter or a short-term store might take different physiological forms, but it could be decided with reasonable ease whether any particular physiological structure was or was not describable by these terms. (1958: 302–3)

The physiology must always be borne in mind, and the psychological theory eventually explained by it. For instance, Broadbent cited the recent discovery that cells in a cat's auditory cortex are 'distracted' from responding to meaningless clicks if a mouse enters the cat's visual field (Galambos *et al.* 1956). But he pointed out—what will be emphasized throughout Chapter 14—that "it may often be preferable to explain a *physiological* fact by reference to its role in a well-understood *psychological* function" (p. 305; italics added).

Further, he said, there's room for purely psychological theorizing while remaining silent on the physiology. (The Gestaltists hadn't realized this: in the event, their mistaken ideas about the brain led people to downgrade their psychological work.)

In sum: "information theory is desirable as allowing future contact with physiology but never assuming physiological detail" (p. 306). Or as Jerry (Jerrold) Fodor (1968) later put it, psychology is an autonomous ("special") science, not reducible to neurophysiology even though mental processes are implemented in the brain (see Chapter 16.iii.b).

d. Information and computation

During his lifetime, Broadbent contrasted his information-theoretic approach with three alternative views: behaviourist, humanist, and computational psychology. He had only limited respect for the first, and none at all for the second. However, he was sympathetic to the third. In his eyes, it wasn't wrong-headed so much as premature.

In the late 1950s, he was defending himself against attack by the behaviourists, as we've seen. Twenty years later, he would defend himself against a very different enemy: humanism.

Humanism, in this sense, was the neo-Kantian view that psychology is—or should be—an interpretative discipline, forever trapped within the hermeneutic circle (Chapters 2.vi and 16.vi–viii). Moreover, since interpretation focuses on single texts, not on regularities, the subject matter of psychology was seen by the neo-Kantians as a patchwork of individual phenomena, or "stories". Comparisons could doubtless be made, much as *Hamlet* can be compared with *The Tempest*. But the 'Newtonian' assumption that psychology should look for general principles of explanation was roundly rejected.

This view hadn't imbued the Anglo-American Zeitgeist of the 1950s. In the late 1960s, however, it became fashionable, forming part of the general reaction against modernism (Roszak 1969; Toulmin 1999).

Many disciplines were affected. Terry Winograd, an AI researcher whose work had interested many hard-headed psychologists, was just one of those who turned away from empiricism (see 11.ii.g). Social psychology, in certain quarters, was renamed ethnomethodology—and concerned itself with highly detailed, non-generalizable, accounts of everyday behaviour (Garfinkel 1967). Anthropology took an overwhelmingly "narrative" turn, and literary/historical studies were in turmoil too (8.ii.a–c). And the turmoil often involved bad-tempered invective from both sides of the fence.

Within professional psychology, this counter-cultural movement spawned many—often vitriolic—attacks on orthodox scientific approaches. Liam Hudson, for instance, attacked the so-called "cult of the fact" (1972), and described psychologists as "prime sufferers of the infatuation with scientism" (1975: 17). "Evidence", he said, should give way to "interpretation", and we should accept "the daring assumption that psychologists should study the people around them".

Hudson was by no means the only one to rebel against the experimental orthodoxy. Even Boston University's Sigmund Koch, for ten years "a dauntless and virile rat-runner", joined in on the enemy's side (S. Koch 1961; 1974: 3). The "enemy" he was referring to was "humanistic" psychology in general, including those new forms of social psychology which saw it as descriptive/historical rather than scientific (e.g. Gergen 1973). The *Journal of Humanistic Psychology* was founded in the late 1960s to counter this approach (and behaviourism and Freudian psychoanalysis too), and many papers and fast-selling books of the 1970s opposed it.

Psychologists whose views gained some notoriety—and influence—in this period included John Shotter (1970), Amedeo Giorgi (1970), Robert Joynson (1970, 1974), Rom Harré and Paul Secord (1972), Kenneth Gergen (1973), and Alan Gauld (Gauld and Shotter 1977). And one psychologist who was hugely famous already, namely Jerome Bruner, suffered a humiliating ordeal at the hands of his Oxford colleagues when—at the height of this controversy—he gave a public lecture criticizing their more orthodox scientific approach (see Section ii.c, below).

If the notoriety has now faded, the conviction usually hasn't. For recent examples, see Harré's unorthodox account of what cognitive science should be like (1994, 2002), Shotter's Derridean views on social psychology (Parker and Shotter 1990), and Bruner's characterization of psychology as "narrative" (2002). An exceptionally clear statement of the core claim is Giorgi's unapologetic insistence that

> a complete break from the natural science conception of psychology would be profitable at this time, and only after psychology as a human science has had significant development, should the dialogue with psychology as a natural science be pursued. (Giorgi 2000)

Today, the humanists and "scientific" psychologists quietly follow their own paths, rarely asking how they might fruitfully meet (Giorgi 2005). But in the 1970s humanist psychology was perceived as an aspect of the counter-culture, and fervently favoured by many people for political as well as purely theoretical reasons (see Chapter 1.iii.c). As such, it represented a real threat to the more scientific forms of psychology.

Whereas the American Anthropological Association (as we'll see in Chapter 8.ii.b) was happy to embrace the counter-culture, the British Psychological Society (BPS) wasn't. Broadbent himself was an influential BPS figure, of course, but there were many other experimental psychologists with scant sympathy for the new movement. The BPS became so concerned by this division into warring camps that it convened a special conference in 1979 to discuss it (A. J. Chapman and Jones 1980).

Broadbent was unmoved. He refused to take refuge in the easy excuse that social psychology, from which most of these attacks had come, is humanist whereas cognitive psychology is scientific. ('Humanist' is a useful shorthand here: many in the opposing camp, namely those with postmodernist leanings, were almost as critical of classical humanism as of science.) In a book-length "defence of empirical psychology" written soon after these attacks had surfaced, he declared: "We are, in fact, studying one subject, not many" (1973: 141). (Whether that's really so is still controversial: see Chapters 14.x–xi and 16.)

Citing the raging language wars of post-Chomskyan linguistics (9.ix.a), Broadbent derided any approach to psychology "through the armchair, by the exercise of fallible human reason, intuition and imagination" (1970*c*: 96). Empirical, not interpretative, methods were required:

> [In] dealing with human beings, a proper sensitivity to other men demands that we should take an interest in what they actually do rather than what we think they do. The empirical method is a way of reconciling differences. If one rejects it, the only way of dealing with a disagreement is by emotional polemic... If we refuse to use experiment and observation on other human beings, we start to regard them as wicked or foolish. I think this is a serious danger, and I have no doubt whatever that the methods of empirical psychology are socially more hygienic, or to use the older and more robust phrase, morally better. (Broadbent 1970*c*: 95–6)

This wasn't mere rhetoric. In a book already in press, Broadbent (1971) had shown why bad decisions often result from stress, in the form of noise, time-pressures, and/or over-abundant input information (Broadbent 1971). To regard them as bad (mistaken) decisions is justifiable, but to damn them as "foolish" is not. Even 'pilot error' is tendentious, since it implies that full responsibility for the pilot's mistake rests with the pilot, not with (for instance) the designer of the dashboard in the cockpit.

At the same BPS conference, Broadbent expressed his sympathy for a third way of doing psychology: computer modelling. He was closer to this approach than to humanistic psychology or behaviourism, but remained sceptical nonetheless:

> The great merit of models which can be implemented on a computer... is that they avoid many... ambiguities. I would firmly believe that in the long run any adequate account of human beings will have to be capable of computer implementation. The problem at the moment is rather that such models are not adequate empirically; none of them behaves quite like a person behaves, and this tends to be obscured when one discusses this without looking at concrete data. (1980: 114)

> The models advanced in artificial intelligence... are not satisfactory in detail; but they have made the important contribution of showing that in principle a [formal–computational] system could show the interesting, creative, and purposive features of human behaviour without resorting to a little man at the centre of the model. (1980: 126)

In other words, the Church–Turing thesis (Chapter 4.i.c) was being tacitly granted, since "in principle" an AI system could model even creativity and purpose. The problem, rather, lay in the practicalities. Computational psychology *understood as computer modelling* was just too hard. The informational approach was—as a matter of fact, not principle—more effective.

Shortly before his death, Broadbent organized a series of eight public lectures on 'The Simulation of Human Intelligence'. His own lecture declared that AI models had much to offer to psychology (Broadbent 1993). But he still wasn't a computer modeller, and nor did he treat programs as a rich source of theoretical concepts. What 'flowed' between his diagram boxes was information, not control.

The switch from information to control would be made by Miller, soon after Broadbent's seminal book appeared (see Section iv.c). This switch was not a schism.

It was seen by many, including Miller himself, as an *improvement* rather than an *alternative*.

Walter Reitman, for example, who wrote one of the more interesting early GOFAI simulations (7.i.b), refused to distinguish between 'information-processing' and 'computer-modelling' theories and concepts:

> The term [information processing] is not meant to be synonymous with "data processing" or "computer". It is a label for a general approach to the study of psychological activity... [So] information-processing theories examine the representations and processes involved in cognitive activity. They emphasize the functional properties of thought and the things it achieves. (Reitman 1965: 1)

The information-processing approach, he said, encourages us—and enables us—to focus on the details of cognitive structures and processes, and on the higher-order strategies involved in thinking. Computer models are especially helpful, precisely *because* a programmed computer can do only what we've ordered it to do (1965: 15). But the basic theoretical advance is to theorize in terms of information-processing (including computational) concepts.

Despite such commonalities, the *methodology* of the informational and computational psychologists differed. And the theoretical emphasis was different too, on data and control respectively. These were two intellectual communities, not one.

Nevertheless, the sympathies and assumptions were very close. Broadbent's interest in computational psychology was always evident, in his private conversation as in many of his writings. Some of his disciples were less broad-minded, more concerned to protect their patch against rivals. In retrospect, however, it's clear that the two streams were part of a common intellectual enterprise, springing from cybernetics (broadly defined)—and with behaviourism and, of course, hermeneutic "humanism" as the common enemies.

e. Chomsky comes on the scene

Chomsky's first—and still his most well-established—theoretical contributions concerned different types of *computation* (see Chapter 9.vi.a). In 1956 he published a classification of computer languages that's still central to computer science today; and 1958 saw a paper on finite state languages co-authored by him and Miller. Indeed, it was largely because of Chomsky that Miller replaced 'information' by 'computation' as the core concept of psychology.

Very few of Miller's psychological colleagues were enthused by Chomsky's 1956 paper. Their attention was captured, rather, by his little book *Syntactic Structures*. When this hit the scene in 1957, it wasn't only linguists who were interested. Psychologists of language were interested too—though *not*, yet, psychologists in general. Chomsky had deliberately 'censored' his views on innate knowledge of language: to mention that would have been far too provocative (9.vi.e). Nor had he mentioned behaviourism *as such*, although he had rejected statistical/Markovian theories of language. Most psychologists would be introduced to Chomsky, whether to agree or disagree, by his assault on Burrhus Skinner two years later (9.vii).

But if the fourth tenet of behaviourism remained unchallenged, the first, second, and sixth did not (see 5.i.a). In rejecting them, Chomsky's work made psycholinguistics possible. Or rather, it made *what we now think of as* psycholinguistics possible. Quite a few people besides Skinner were already asking questions about the psychology of language (aka verbal behaviour). But they were doing so in the context of Bloomfieldian theory (9.v.b).

For instance, James Jenkins (who, unlike Skinner, was happy to posit intervening variables) had been seeking experimental evidence for structuralist linguistics. He first learnt of Chomsky's work from Miller, when they were both visiting the Stanford Center in 1958–9. Six years later, again at Stanford, he attended a sentence-by-sentence tutorial on *Syntactic Structures* given by Fodor and Sol Saporta, and co-published with them as a result (Fodor *et al.* 1967).

It wasn't easy for him to make the switch. In 1964 (a full seven years after *Syntactic Structures*) Jenkins recommended Chomsky to his verbal-behaviour colleagues, telling them: "You know, we've got to work on generative behavior . . . rule-governed behavior" (Jenkins 1986: 247). And their response?—"Everybody in the meeting jumped on me."

Besides being "jumped on" by his behaviourist colleagues, he was soon dismayed by the master himself. For Chomsky, in his second book, moved the goalposts:

> Then Chomsky pulled the rug out from under us . . . [We] were very busy trying to provide the apparatus for a theory of linguistics that at that moment was being discredited. It's a very disappointing position to be in . . . [By] the time we could supply the right kind of theory, the nature of what language was believed to be had changed. The whole theory was no longer appropriate. Very grim, very grim. (Jenkins 1986: 243)

What had excited Jenkins (and other refugees from behaviourism), and what persisted even after the "grim" rug-pulling, boiled down to three things. These were (1) Chomsky's view that language—or anyway, syntax—could be formally described as a generative system; (2) his claim that sentences must be represented on more than one level; and (3) his talk of grammatical transformations.

"Generative", "representation", and "transformation" can all be understood mathematically and/or psychologically. Chomsky had intended them to denote timeless relations between abstractly defined linguistic structures (see 9.vi.a–d). But psychologists naturally interpreted them as hypotheses about actual mental processes. Indeed, Miller's cognitive science 'manifesto' in 1960 encouraged them to do just that (Section iv.c, below).

An explosion of experimental work ensued, investigating the psychological reality of Chomsky's ideas. For example, researchers asked whether sentences are remembered more easily if fewer transformations are involved. They tried to find out whether listeners assign syntactic structure to a sentence word by word, or only after the whole sentence has been heard. And they asked whether people can understand sentences with many levels of grammatical nesting, or recursion.

Very early in the game, Miller—with Chomsky himself—formulated the derivational theory of complexity (G. A. Miller and Chomsky 1963). This proposed that people actually perform (unconscious) transformations when they produce, understand, or remember sentences. For a while, the experimental evidence seemed to support them. But trouble lay ahead. It soon became clear that derivational complexity *didn't* necessarily predict difficulty in understanding or remembering sentences (Fodor and

Garrett 1966). By the early 1970s this theory had bitten the dust, its obituary written by Chomsky's ardent champion (Fodor *et al.* 1974).

Chomsky had changed his mind about transformations by then (9.viii.b). So Miller reacted in much the same way as Jenkins. Remembering the early days, he wryly recalls: "[We] were working before Chomsky's 1965 book; so, as we were trying to test the psychological reality of syntax, the syntactic theory we were testing was moving out from under us." As a result, he said, "About '65–'66 I had given up on syntax" (G. A. Miller 1986: 209, 216). (However, he hadn't given up on language: see 7.ii.a and d.)

What matters, here, isn't whether Chomsky's œuvre was consistent, or whether any version was correct (considered either as linguistics or as psychology). These questions will be discussed in Chapter 9. The important point is that his early approach, part-mediated by Miller, encouraged psychologists to think about language in computational terms.

Even though most of them didn't turn to computer modelling, they began to think of language as a precisely describable generative system. Chomsky didn't try to measure information, nor to locate it on a flow chart, nor to incorporate it in a program. But he asked how it was *structured*, how it could be *generated* and *transformed*, and what (very abstractly defined) types of *computational* system would be able to represent this. That's why he was an important voice in the rise of computational psychology.

6.ii. The New Look

The New Look in haute couture was a heady liberation. The austere fashions of wartime gave way in 1947 to the glorious full, long, skirts of Christian Dior's A-line. And, with fabric rationing ended, the new dresses soon migrated joyfully from catwalk to high street.

Much the same was true of the 'New Look' in perception, launched around the same time. The major couturier was Bruner (1915–). The British stylist Richard Gregory (1923–) was important too, doing even more than Bruner to make *mind-as-machine* fashionable on the high street. But his New Look research was begun after Bruner's, and was narrower in scope, as we'll see. It was largely thanks to Bruner's imaginative cutting and stitching that American psychologists by 1960 were free at last to study full-skirted perception (instead of the narrower 'sensory discrimination'), to recognize joy (or anyway, values), and even to champion models—not on the catwalk, but in the mind/brain.

New Lookers reacted against the 'passive reception' view of perception that had been inherited from classical empiricism (Chapter 2.x.a). They saw the human mind as "proactive" rather than "reactive"—a position stressed also, in relation to personality theory, by Bruner's colleague Gordon Allport (1897–1967): see 5.ii.a and G. W. Allport (1960). Accordingly, they focused on the selective and constructive aspects of perception. They were interested less in *how much* input information can pass into/through the mind than in *how it's selected and represented* by the person (*sic*) concerned. They saw perception (and concept learning) as being guided by conscious and unconscious *expectancies*, *hypotheses*, or *models*, and *tested* by more or less reliable *cues*.

In short, everyday cognition was thought of as an informal version of scientific inference or reasoning, though with social/personal factors playing a larger role than they do in science. This 'reasoning' can trigger predispositions to act in specific ways, so our perceptual expectancies, said Bruner (1961), are similar to what the ethologists had called IRMs (5.ii.c). But they're mostly learnt, not innate. And, as we'll see, they depend on the person's values as well as their beliefs.

These psychologists didn't do computer modelling, but their theoretical approach became increasingly computational over the years. This reflected the influence not only of Miller but also of Broadbent, who (with Gregory) was one of Bruner's closest friends during his visit to England in 1955–6. In its fully developed form, the New Look wasn't a mere forerunner of computational psychology but an early version of it.

Moreover, it attracted the attention of the very early GOFAI researchers. Marvin Minsky's seminal 'Steps Toward Artificial Intelligence', for instance, mentioned Bruner's work as highly relevant to the programme of AI (Minsky 1961*b*: 450; cf. 10.i.g). That's partly why a publication by Bruner was one of the six events which defined the *annus mirabilis* of cognitive science—namely, 1956 (see Section iv.a, below).

a. Coins and cards

The first liberating New Look challenge to the austerities of behaviourism was the decision to study perception properly so-called (i.e. not just sensory discrimination), or what the world actually *looks* like. The learning theorists at the time, and the sensory psychophysicists too, regarded this question as "mentalistic, phenomenological, essentially European" (Bruner 1983: 67). A main achievement of the New Look was to make perception respectable (in America) again.

The second liberating challenge arose as what people *wanted*—and what they wanted to avoid—came to the fore. This included not only what they wanted (what they were trying to achieve) at a particular time, but also what they were generally disposed to want (or to fear), given their individual personality-structure:

> [Given] the interdependence of all aspects of personality . . . [within] the dynamical system that constitutes the person, [the problem is] to understand how the process of perception is affected by other concurrent mental functions and how these functions, in their turn, are affected by the operation of perceptual processes. (Bruner and Goodman 1947: 33)

Sigmund Freud (1925), in his essay on the "mystic writing pad", had described perception as selective, balanced between motivated hallucination and realistic ("ego-driven") response. But he'd cited anecdotes rather than experiments.

Immediately after the war, Bruner—with his "inseparable" friend Leo Postman (Bruner 1983: 75) and his undergraduate student Cecile Goodman—did a number of experiments exploring this general theme (e.g. Postman and Bruner 1946; Bruner and Postman 1947*a*,*b*; Bruner and Goodman 1947). Even they, however, had been impelled to do so by an anecdote:

> J.B.: [One needs] crazy ideas that connect things up. I think it was in Faulkner's *Intruder in the dust*, or one of the Faulkner novels, where the kid named Benji or something reaches his hand in his pocket and finds a half-dollar. And he says that it felt as though it could fill his whole palm.

B.S.: Is that where the idea for your little experiment came from?

J.B.: That's where the idea of the experiment came from . . . And I even remember that I happened to be having dinner that night with [my close friend the Gide scholar Albert Gérard]. Albert said, "Oh. That's fantastic! That's a wonderful passage." And he starts giving me other examples where people have come to him with the same sort of thing. Something looking brighter or bigger because it had gotten important. So I thought, "Marvelous experiment."

(Shore 2004: 15)

They found that perception can indeed be influenced by emotion and motivation. It wasn't only basic drives such as hunger which could do this (R. Levine *et al.* 1942), but culturally acquired needs too. In fact, one effect of the New Look was to bring social and cognitive psychologists closer together. This happened largely because Miller and Bruner were close personal friends. Despite being located in different Harvard departments, due to an administrative split in 1946 (Bruner 1983: 63), they were highly cooperative colleagues (see Section iv.d).

It seemed that the greater the social value of an object, and/or the greater the individual need for it, the more susceptible it was to perceptual biases from 'dynamic' factors. For example, Bruner and Goodman reported that children's perception of such an apparently objective matter as the size of a coin was affected both by their knowledge of the coin's face value and by their need/desire for money. The size of high-value coins was overestimated, and the poorer children overestimated more than the richer ones did. (The effect didn't work for half-dollars: the experimenters suggested that the 10-year-olds saw a half-dollar as less "real" than the more familiar quarter.) Bruner and Goodman posited a range of unconscious perceptual "hypotheses" to explain the bias, or selectivity, they'd observed.

These results were an eye-opener because previous experiments on the effect of social influence had dealt only with highly ambiguous stimuli, where objective information wasn't available. These concerned the 'autokinetic' effect, in which a spot of light in a darkened room is mistakenly seen as *moving.*

Muzafer Sherif (1935: 17–22) had shown that the extent to which the spot apparently moves can be altered by the statements of other people in the room. These "other people" were in fact the experimenter's stooges, instructed to describe the spot to the victim in a particular way. But the victim's perception might be explained as a case of rationality: if everyone else says they see the spot as moving *thus-and-so*, it's only reasonable to think/believe that it's moving in that way. And if perception involves judgement, as René Descartes had argued centuries before (Chapter 2.iii.d), then it's understandable that the subject should *perceive* it in that way too. This is a far cry from perceiving some 'objective' feature in a subjectively biased manner.

The coins paper, Bruner recalled later, "turned out to be the catalyst, the cloud seeder. It produced the New Look, and before that was done, it rained about a thousand articles and books" (1983: 69). An APA citation study done three years after the coins publication showed Bruner and Postman as second only to Freud in the number of bibliographical references made to their work (Bruner 1980: 107).

For instance, Solomon Asch (1956) caused a sensation by claiming that people's perception of relative line lengths (an objective matter, unlike the illusory movement of the autokinetic effect) could be affected by social pressures. He'd found—on debriefing—that although some subjects had merely conformed verbally (i.e. they'd said the same thing about the line as the other subjects were saying), others insisted that they *really had seen* the shorter line as longer. However, this was tricky. Sceptics objected that Asch's key subjects might be lying during the debriefing, not wanting to admit that they'd said something they didn't believe merely because everyone else was saying it. (Follow-up studies showed that people were less likely to conform if the judgement involved politics—presumably, because disagreement is common there.) Various criticisms were made of the methodology of the original coin studies too. In short, enthusiasm and critique both flourished.

Soon after the coin experiments, Bruner (1951) discussed a range of contemporary research indicating that different personality types show systematic biases in perception. For example, some people are biased towards tolerance or intolerance of ambiguity, others towards optimism or pessimism. (For the distinction between optimism as perceptual/affective bias and optimism as reasoned belief, see Boden 1966.)

That research included post-war studies of "the authoritarian personality", undertaken in an effort to explain the popular acceptance of fascism in Nazi Germany (Frenkel-Brunswik 1948, 1949; Adorno *et al.* 1950). Both Bruner and these researchers (whose methodology, for the record, was gravely flawed: Christie and Jahoda 1954; McKinney 1973) saw Freudian repression as a determinant of perception, as well as belief.

This conviction underlay Bruner's claim that "intrapunitive" individuals have a disposition (or "set") to evaluate ambiguous information as confirming their own guilt:

> The more marked the degree of intrapunitiveness, the less the appropriate information necessary to confirm self-guilt. As the [often unconscious] hypothesis attains greater and greater strength, intrapunitiveness attains neurotic proportions, which is to say that self-guilt hypotheses are confirmed by information judged by society to be grossly inappropriate or ambiguous. (Bruner 1951: 104)

Experimental studies of such matters, however, had been very thin on the ground, "in spite of the fact that Freud early referred to one aspect of the ego as 'perceptual consciousness' and despite the title of [Anna Freud's] first chapter ('The Ego as the Seat of Observation')" (Bruner 1951: 108). Bruner decided to provide some.

As an ex-student of McDougall (5.ii.a), his interests in personality theory and social psychology—and in their close relations with cognition—were only to be expected (cf. Bruner *et al.* 1956: 16; Bruner 1983: 59). Indeed, before the 1946/1947 papers (co-authored with Postman and Goodman) cited above, he'd already published some twenty articles on social and personal psychology. Several of these had had a 'cognitive' air, for they showed that people's preconceived political opinions acted as (often unconscious) "filters" of incoming news-items and propaganda (e.g. Bruner 1944).

By the late 1940s Bruner was already referring (as in the quotation, above) to perceptual selectivity as the result of "hypotheses", and to motivation as "strengthening hypotheses". In his defence, he quoted Bartlett's opening remark in a lecture he'd given in 1951:

> Whenever anybody interprets evidence from any source, and his interpretation contains *characteristics that cannot be referred wholly to direct sensory observation or perception*, this person thinks. The bother is that nobody has ever been able to find any case of the human use of evidence

which does not include characters that run beyond what is directly observed by the senses. So, according to this, people think *whenever they do anything at all with evidence.* (Bartlett, quoted in Bruner 1957*a*: 219; italics added)

To posit internal hypotheses was highly non-Newtonian, of course. But to some extent Bruner was protected from his fellow professionals' scorn by the fact that "hypothesis" was a term derived from the language of science. (He now sees it as equivalent to "intentionality", a philosophers' term that was "unheard of" in those behaviourist-dominated days—interviewed in Shore 2004: 180.)

Where Bartlett had spoken of schemas (Chapter 5.ii.b), Bruner spoke of hypotheses. These could be made stronger or weaker (i.e. more or less salient, or "accessible") by motivational factors. But they arose, directly or indirectly, from past experience: the more probable a cue category was found to be, the more likely that it would be applied. (By the early 1950s, he was expressing this in terms of the recognition of informational redundancy: G. A. Miller *et al.* 1954.)

Bruner was especially interested in Craik's question of how people use concepts to predict and anticipate the future (Bruner *et al.* 1956: 14; Bruner 1957*a*). In addition, he wanted to know how they test, or validate, their perceptual/conceptual hunches.

This assimilation of perception to scientific reasoning, in the active construction of percepts, became the third criterion of the New Look. As Bruner put it, "A theory of perception . . . needs a mechanism capable of inference and categorizing as much as one is needed in a theory of cognition" (1957*b*: 123). By "mechanism", here, he meant a system defined in psychological terms. But he didn't forget that there must be some (broadly Hebbian) neurophysiological mechanism too: see, for instance, the section on 'Mechanisms Mediating Perceptual Readiness' in Bruner (1957*b*).

Bruner used anomalous or blurred stimuli to show how stubbornly we protect our perceptual hypotheses despite conflicting evidence (Bruner and Postman 1949*a*; Bruner *et al.* 1951; Bruner and Potter 1964). For instance, he reported that people often see (briefly presented) anomalous playing-cards, bearing black hearts or red spades, as normal ones.

The delay in recognizing the colour red could be a whole order of magnitude greater than when the redness was expected, on the basis of the heart/diamond shape. (So much for the associationists' passively received red sense datum.) Sometimes, there was a perceptual 'compromise'—as when red spades are seen as "brownish" or "purple". Only very rarely was the strange playing-card immediately seen as it really is.

This experiment had required a certain amount of do-it-yourself ingenuity on Bruner's part. And the reason why that was so was a further confirmation of the psychological truth the experiments were taken to show:

J.B.: I tried to get an American playing card company to manufacture special cards which altered the colors for the suits. I had written on Harvard stationery so they wouldn't tell me "I'm going to get the cops on you. What kind of scam are you trying to pull off?" [But they wouldn't make them for me, either.]

B.S.: This resistance that they had to doing that was just as strong an evidence of the importance of category classification as the experiment was.

J.B.: Exactly. I only realized that years later. I mostly thought they were being a pain in the ass. And they were. So I went into an art shop on Beacon street with Mrs. Ware Eliot, T. S. Eliot's sister-in-law, with whom I had taken some drawing lessons. We brought in some playing cards and tried out different things until we found the right thing to paint them with.

(Shore 2004: 92–3)

In the case of playing cards, we're confident that we know the card "as it really is". Even if it's 'really' a black card, painted red with Mrs Eliot's help, we're confident that—now—it's *really* red. This is largely because of intersubjectivity: in normal conditions, we reach agreement with ease. (That's why Bruner and Postman had to use a tachistoscope, to show the cards only very briefly.) In the case of describing personalities, we aren't so sure.

People are always complex, and they sometimes dissemble. Any action, considered on its own, can be given many different interpretations (the hermeneutic circle, again: Chapter 5.iii.b). It's hardly surprising, then, that 'first impressions' are so important. The New Look approach implied that they should be both influential (in biasing later action interpretations) and difficult to change. For the more disputable each individual cue, the less chance of the original hypothesis being abandoned—and personality categories, unlike red/black, involve highly disputable cues.

This made sense of some intriguing mid-century experiments, done by Hull's student Carl Hovland (1912–61) and others. These showed that once a person had been categorized as (for instance) honest/dishonest, their later behaviour would be seen as reticence/evasiveness, and integrity/trickiness, accordingly (Haire and Grunes 1950; Hovland 1957).

The playing-card experiments (Bruner and Postman 1949*a*) interested not only psychologists but philosophers too. Thomas Kuhn, for instance, cited them in his account of "normal" science, much as he drew on Gestalt psychology in describing "revolutionary" science (Kuhn 1962: 63–4, 110–14). Nearly twenty years later, Paul Churchland (1979: 1–45) would draw heavily on the by-then-familiar fact of concept-based perceptual plasticity in his first statement of eliminative materialism (Chapter 16.iv.e).

In short, the New Look (and subsequent work) had convinced psychologists that our perception is deeply imbued with concepts (Bruner 1957*b*). Many philosophers, from Descartes to Karl Popper (1935), had already said this, of course. Now, there was added empirical support.

One shouldn't assume, however, that *all* philosophers were favourably impressed by the New Look. Wittgensteinians later objected strongly to its talk of non-conscious hypotheses and inferences. They didn't contest the experimental data. But they argued that, when seeking to explain perception, one shouldn't adopt—nor even adapt—language normally used to describe deliberate conscious thought. Cognitive scientists in general, and Gregory in particular, were rapped over the knuckles by two of the master's closest disciples: Norman Malcolm (1971) and Elizabeth Anscombe (1974). (Gregory defended himself with some spirit. In my view, he won that battle hands down: see Chapter 16.v.f.)

Nor should one assume that all *psychologists* welcomed the New Look. The sensory psychophysiologists caricatured the New Looker as someone "who tends to use a stimulus input of a relatively complex kind, without being able to specify very rigorously its nature, and then proceeds to vary a great many other conditions that have little directly to do with conventional descriptions of perception" (Bruner and Klein

1960: 121). And the New Look in general was felt by some critics to be unpersuasive, because of its reliance on the suspect concepts of *unconscious bias* and neo-Freudian *defence.*

Bruner and Postman had posited "perceptual defense" in 'normal' subjects as well as neurotics, to explain why people often fail to perceive emotionally threatening stimuli (Bruner and Postman 1949*b*; Bruner 1957*b*). Their subjects might *register* it at some unconscious level (measured by galvanic skin response, for example), but conscious *perception* was delayed or even prevented. Such data were new, and undeniably interesting (and prompted an explosion of work on 'subliminal' perception). But Floyd Allport (Gordon's brother), for one, was unconvinced by explanations in terms of defence or unconscious perceptual expectancies (1955, chs. 13–15). Quite apart from their air of paradox, he said, these didn't allow precise predictions. Bruner had an answer:

> To ask [the New Looker] to rule [defensive processes and other personality variables] out of his research in the interest of cleaning up his experiments is as silly as asking the psychophysicist at every turn to replicate his experiments under conditions of inattention or imperious need states, given that his interest is in sensory receptivity under optimal conditions.
>
> In the end, what we are saying is that full explanation of any phenomenon—be it perception or anything else—requires both a close study of the context in which the phenomenon occurs, and also of the intrinsic nature of the phenomenon itself under idealized conditions. The New Lookers have tended to do the former, the researcher raised in psychophysics and sensory physiology the latter. (Bruner and Klein 1960: 122)

He admitted, in retrospect, that the New Look in its early days had been "naive and inept and confusing", and like a "noisy and brawling adolescent... [had] often proceeded without enough attention to the lessons learned by its [elders]" (Bruner and Klein 1960: 119). However, the critics wouldn't have bothered if the ideas hadn't been so provocative. Bruner again:

> [The] New look has had an activating effect, a disturbing effect; it has created some useful models; it has got part way through some research that shows signs of being better done; and it has been bold enough to look at problems... [It] has at least had the virtue of not taking much for granted... (Bruner and Klein 1960: 119)

b. A study of thinking

The New Look culminated in *A Study of Thinking* (1956), by Bruner, Jacqueline Goodnow, and George Austin (BGA). This was both the last stage of the New Look and a first stage of computational psychology.

Bearing in mind the many criticisms of the "noisy and brawling adolescent", this mature version of the New Look involved painstaking and systematic experiments—and said almost nothing about issues of emotion and personality (but see pp. 79, 228). Less 'sexy' than the need-biased coins and the dodgy playing-cards, it wouldn't reach the newspapers. In theoretical psychology, however, it made even more waves than they had done.

The title itself was a challenge. No one 'respectable' (in S–R terms) was studying *thinking*—or if they were, they called it 'learning'. Only ten years earlier, Wolfgang

Kohler had lamented: "It is to be regretted that few psychologists take an active part in the investigation of thinking" (1945, p. iii). Even Hebb, with his unorthodox talk of "concepts", had merely outlined a theoretical approach to thinking (Chapter 5.iv.c).

BGA not only tackled this unfashionable topic, but disarmingly declared: "To the reader conversant with contemporary American psychology, [our] book will appear *singularly lacking in the more familiar forms of theoretical discourse*" (p. 23; italics added). That was true—and the book caused a revolution. (One textbook later described it as "the" classic work on adult intellectual behaviour—P. H. Lindsay and Norman 1972: 498.) It led to a lasting revival of interest in *cognitive* psychology, and helped lay the foundations for *computational* psychology—and for cognitive science as such.

The central topic was concept learning—or, as Pitts and McCulloch (1947) had put it, 'How We Know Universals' (see Chapter 12.i.c). But where Pitts and McCulloch had concentrated on the sensory aspects, BGA—in typical New Look style—assimilated perception to scientific reasoning. Moreover, they pointed out at the start (p. 10) that concept learning has a very wide scope. Besides being important in its own right, it *includes* crucial aspects of perception and *is included in* reasoning, memory, and creativity. In effect, then, it's the pathway that leads into cognitive psychology in general: research on what used to be called the "higher mental processes".

Some of the behaviourists had studied concept learning: Hull (1920), for example. But their subjects had been asked to learn the names of certain patterns, not to discover their defining features. A New Look psychologist had recently discovered that if specifically bidden to do the latter, people's success in identifying those features rose hugely (H. B. Reed 1946). Moreover, the informational aspects of the stimuli hadn't been considered or controlled. If a stimulus part was there, it was there: *how many alternatives* were also present wasn't considered relevant. And results that depended on the order of presentation weren't interpreted by Hull in terms of *how the subject had decided to tackle the task.*

BGA paid homage to Tolman, rather than Hull (5.iii.b–c). They described their work as a study of Tolman's intervening variables, or cognitive maps (p. vii). For they reported systematic experiments on the (largely unconscious) "strategies" that people use in learning a new concept. And this term, they said, wasn't meant metaphorically:

> A strategy refers to a pattern of decisions in *the acquisition, retention, and utilization of information* that serves to meet certain objectives ... [including, though not limited to,] the following:
>
> a. To insure that the concept will be attained after the minimum number of encounters with relevant instances.
>
> b. To insure that a concept will be attained with certainty, regardless of the number of instances one must test *en route* to attainment.
>
> c. To minimize the amount of strain on inference and memory capacity while at the same time insuring that a concept will be attained.
>
> d. To minimize the number of wrong categorizations prior to attaining a concept. (Bruner *et al.* 1956: 54; first italics added)

Instead of everyday stimuli such as coins, BGA used artificial ones. This was partly to minimize social/personal factors, but mostly to provide neatly defined concepts (more tractable than real concepts are: see 5.iv.c, 8.i.b, 9.x.a, and 12.x.b).

So they showed their subjects cards bearing crosses, squares, and circles in various numbers and colours, and with or without borders. In terms of these dimensions, they defined concepts of several kinds:

* conjunctive (A + B + . . .),
* disjunctive (A or B or . . .),
* relational (e.g. having more As than Bs), and
* probabilistic (if such-and-such conditions are satisfied, then it's probably an X).

And instead of focusing on isolated responses, like someone's estimation of the size of a coin, they looked at the behavioural patterns within *sequences* of responses.

Their experimental rationale was explained in terms of examples of (neuro)scientific reasoning (pp. 81, 246). So, they said, much as a neuroscientist investigating the role of six brain regions can choose the order, and combinations, in which to ablate them, so BGA's subjects could normally choose which card to ask about at any given time. (Occasionally, the cards were presented in an order chosen by BGA.) In each case, they'd be told whether it was or wasn't an example of the concept being learnt. In the vocabulary of Chapter 12.ii.b, this was supervised learning; in BGA's vocabulary, it was concept attainment, as opposed to concept formation (pp. 232–3).

The information (logical implications) available from *this* card, in *this* position in the sequence, for *this* type of concept, was known to the experimenters. (Such matters had recently been analysed by Yale's Hovland: 1952.) They defined four ideal strategies—"ideal" meaning logically pure, not best: simultaneous scanning, successive scanning, conservative focusing, and focus gambling. These had complementary advantages and disadvantages in terms of ease of remembering, informational content, and cognitive risk.

The overall question was whether people would follow identifiable search plans—and if so, whether these would match/approximate the ideal ones. (Accordingly, BGA saw their work as description, not explanation; in particular, they were agnostic on how strategies come about: pp. 241–2.)

BGA's subjects could distinguish exemplars from non-exemplars before being able to define the features on which their judgements were actually based. That wasn't new: Hull's subjects, nearly forty years before, had done so too. But BGA (unlike Hull) asked how people actively go about learning new categories.

They observed several different strategies, which varied according to the informational demands of the task (including the time allowed). For instance, some subjects started by using random sampling—an inefficient plan normally soon abandoned for more structured sequences of choice. Others waited till they found a positive instance of the concept, and then searched for a stimulus like it in all ways but one. If this was *not* an example, the 'missing' feature must be a relevant one. The procedure would then be repeated for each of the other features. As BGA pointed out, this strategy puts little strain on the memory but is time-consuming. When subjects were hurried, they abandoned it for more chancy strategies. These often failed, as people lost track of the implications of what they'd learnt. In that case (sometimes, even from the start), they'd use a more risky method, gambling that *several* dimensions were relevant and trying to test them all at once. With good luck, this could yield success quickly. With bad luck, the person was left floundering.

By and large, and especially in relaxed conditions (with no time pressure), people chose a sensible strategy—if not always the best. They tended to follow each strategy faithfully, though they sometimes switched from one to another. The experimenters concluded that they were constructing internal hypotheses, or representations, which guided their choices and were continually modified by the results.

Moreover, it appeared that certain sorts of hypothesis were difficult to represent and/or to modify. Conjunctive concepts were learnt more easily than disjunctive or relational ones. Probabilistic concepts caused difficulty. And the information available from positive instances (*this is an X*) was used more efficiently than that from negatives (*this is not an X*).

In addition, there was an important distinction between "defining" and "criterial" attributes. The former is a logically necessary condition, whereas the latter is a pragmatically useful feature (pp. 32–41). BGA gave the example of psychiatrists screening conscripts during the war: "Did you wet your bed as a child?" was quickly asked and answered, and was assumed to be a useful—but not a defining—indicator of maladjustment. Bruner's team drew here on the work of Tolman and Brunswik (1935), whose concept of "the causal structure of the environment" they interpreted in terms of informational redundancy and contingent probability.

This reference to real-world probabilities shouldn't be taken too literally. BGA themselves pointed out (p. 204) that easily available cues may be relied on *even when they're not realistic.* For instance, Bruner's Harvard colleague Gordon Allport had done important work on prejudice, showing that racial (and other) stereotyping depends on quick judgements based on easily available cues (G. W. Allport 1954). The general conservatism of perceptual hypotheses mentioned in subsection a (above) then protected the prejudiced first impression from change.

With hindsight, BGA's discussion of one experimental result was especially interesting. In situations of stress and/or cognitive overload (caused in various ways: p. 235), their subjects relied more than usual on criterial attributes. Easily available cues were used in preference to logically reliable ones.

What's significant about that is that BGA didn't reject this strategy as an error, nor as a lazy prejudice that should be avoided. Rather, they valued it as rational—or anyway, sensible (238 ff.). In other words, this was an early example wherein 'satisficing' was valued above logic. Later work in cognitive science would stress the adaptiveness of such thinking—and, in so doing, would challenge the philosophers' notion of rationality itself (see Chapter 7.iv).

c. Computational couture

Bruner's post-war tailoring hadn't been cut from computational cloth. In the mid-1940s, personality theory had contributed more threads than information theory. As for computers, these were still a dream—except in top-secret Bletchley Park (Chapter 3.v.d). By the early 1950s, however, things had changed. To be sure, Bruner didn't start writing programs. Nevertheless, the New Look had now acquired a clearly computational style.

For instance, the notion that perceptual input would be tested against internal models, or hypotheses, owed something to Craik as well as to Bartlett. More to the point, the guiding questions in BGA's 1956 book couldn't have been asked without the spur of information theory.

In fact, they'd been inspired by John von Neumann himself. In a group meeting at Princeton's Institute of Advanced Study, during Bruner's stay there in 1951, von Neumann had argued that any efficient information-seeking system would have a "strategy" that specified selectivity *not only of what information would be taken up, but also of how that information would be searched for*. This, Bruner said many years later, was "the germ of the idea that started us off on the experiments that went into *A Study of Thinking*" (1980: 112).

As for the New Look in general, Bruner described its main contribution as relating perception to various other forms of mental processing, including psychodynamics and personality traits, "by which information [*sic*] is acquired, retained, and transformed for future use". This non-homuncular language drew the problematic sting from talk of perceptual (or even neurotic) defence (Erdelyi 1974). Twenty years later, many judged the focus on information processing, not the attention to psychodynamics, to be the New Look's most important contribution (Greenwald 1992; Bruner 1992). (There are still some, however, who unfashionably resist *all* talk of unconscious perceptions: Holender and Duscherer 2004.)

The computing metaphor had been important too, he said. He spoke of "metaphor" because virtually no one at the time was actually *modelling* mental processes. His 1957*b* paper mentioned several early machines in passing, due to Oliver Selfridge and Uttley (12.ii.c–d and 14.iii.b), and cited a model sketched by Donald MacKay (1956*b*). It also referred to a toy system that could discriminate a small number of phonemes (Fry and Denes 1953). But these were very simple, and didn't deal with high-level thought. (An interesting AI program for game playing ran in 1949, and an impressive theorem prover in the autumn of 1956: see 10.i.b and Section iv.b, below.)

However, even merely thinking "metaphorically" about the mind in terms of computers helped psychologists to think clearly. It made them consider, for every identifiable point in the reasoning/learning process, *just what* information is being used, and *how*. When Bruner spoke of an inferential "mechanism" underlying perception, he was thinking computationally if not *program*-atically. (Programs based on Bruner's work were very soon written by others, however: see Chapter 10.iii.d.)

In addition, the computer metaphor eliminated the mystery that had formerly attended theories of perceptual selectivity, or "filtering". Whether the filters were ascribed to learnt expectancies or to motivations, they'd had an unsettling air of paradox. For, it seemed, the stimulus had to be perceived before being not-perceived. Now, the paradox was dissolved:

> Many of the forms of filtering required by a selective system could now be reformulated as ordinary operations of ordinary computers—*and man surely was one of the most high-powered versions of a computer, whatever else he might also be*. The old objections against filtering in perception as implying a Judas Eye that first had to have "a tiny look" before deciding what to exclude were at last being undermined. (Bruner 1983: 277; italics added)

By the same token, studies of perceptual masking, such as Miller's (1947*a*,*b*) and Broadbent's (1952*a*,*b*), were now less mysterious. In cases of 'meta-contrast', for instance, one stimulus masks (blots out), or sometimes modifies, another one. So visual stimulus

A may make stimulus B invisible. What's so surprising, and had previously been deeply mysterious, is that stimulus A is presented *after* stimulus B. How can an input that's presented later prevent (or even modify) conscious awareness of an input that was presented earlier? Once one realizes that a percept isn't a passively triggered sense datum, but a mental structure that takes time—and several information-processing steps—to be actively constructed, one can see that such things are *possible*. (Explaining them in detail is another matter; interesting work on meta-contrast was done by Paul Kolers, a visitor at Bruner's Center for Cognitive Studies in 1962–4: Kolers 1968, 1972.)

Bruner first engaged with computational ideas in the early 1950s. He'd come across psychologically oriented pioneers on both sides of the Atlantic. Most were based on his patch: Harvard–MIT.

One local colleague (at MIT's Lincoln Laboratory) was Selfridge, whose work on Pandemonium Bruner later recalled as "one of the first instances in which the new computing metaphor began to affect my way of thinking" (1980: 109; cf. Chapter 12.ii.d). Another was Miller, whose research on linguistic redundancies chimed with Bruner's interest in "how people constructed 'theories' or 'models' ". Chomsky's attempt to formalize language, and to see it as a *generative* system, was relevant also (Section i.e above, and Chapter 9.vi). (The influence was mutual: Chomsky later told Bruner that the New Look had prompted his own notion of a "Language Acquisition Device" generating hypotheses about well-formed utterances—Bruner 1980: 81.)

Craik, too, "was beginning to make an impression" (Bruner 1980: 110). Bruner entered Craik's (posthumous) circle of influence when he spent 1955–6 in Cambridge, England. There, Bartlett's student Gregory became one of his two "closest companions"—the other being Broadbent (1980: 101). Gregory was a perceptual psychologist, a follower of Craik, and a fan of engineering/computational models (see subsection e below, and Chapter 14.iii.a). Such models were being hotly discussed, for instance by the early connectionist Uttley, whose 1958 meeting in London—at which Gregory gave a paper—caused much excitement (Section iv.b, below).

The New Lookers had started out with an anti-Newtonian aim: "to liberate psychology from the domination of sense-data theory, the notion that meaning is an overlay on a sensory core" (Bruner 1983: 103). But—in Bruner's view (and mine)—they merely "readied the ground" for new theories of perception.

The real change was effected by "the cognitive revolution . . . [and] particularly the respected metaphor of information-processing automata—computers" (p. 104). By exploiting that metaphor, the New Lookers' hopes could be fulfilled. Computational psychology could integrate perception with other aspects of cognition: not only concepts and inferences, but also needs, wishes, styles of mind, and patterns of personality. And this was so "thanks, ironically, to *the liberating effect of the computer on the psychologist's image of what is humanly possible*" (p. 104; italics added).

By 1960, then, Bruner had used these new ideas to rescue perception from the psychophysicists' grasp, and even from 'pure' cognitivism. And although he himself didn't do computer modelling, BGA had given those who did some reasonably rigorous ideas to play with.

Indeed, *A Study of Thinking* "served as the basis for a lot of Alan Kay's work with Apple Computer" (J. S. Bruner, interviewed in Shore 2004: 43; cf. Chapter 13.v.c–d). And Douglas Engelbart's pioneering vision of personal computing, though it didn't cite

BGA by name, had explicitly remarked that understanding how people process different types of concept structure would be essential for improving interface technology (1962: 85 ff.; cf. 10.i.h).

Bruner's interests, already broad, now broadened further. Having started with personality and motivation, he passed on (in 1957) to development and education. His developmental psychology stressed the growth of increasingly powerful *systems of representation* in the child's mind: enactive, iconic, and verbal–symbolic (e.g. Bruner 1964, 1966*a*,*b*). This work, also, was taken up in the design of the Apple computer. His ideas on education (which had a huge influence in the USA) drew on that theory, and his widely read book *The Process of Education* (1960)—which appeared in over twenty languages—stressed neo-Craikian "models in the head". This focus on how knowledge is actively structured by the mind was evident, too, in his intriguing discussions of creativity and myth (1962).

Bruner had a huge influence on education in the USA, and was invited to become the policy top-dog in the field—an invitation which was declined:

> Mac Bundy wanted to know whether I would be available [for the post of head of the US Office of Education]. And I thought about it real hard and I said no, because the fact of the matter is I really think I do better not upholding official policy. I do better in the role of the oppositional [or rather, dialectical] critic. (Shore 2004: 36)

His approach to education is interesting, here, not least because it put pedagogy into a comparative biological context. In his wide-ranging paper 'The Nature and Uses of Immaturity' (1970), he argued that play is adaptive because it allows the young child (or mammal) to experiment with actions whose consequences might be costly, even disastrous, if done 'for real'. Having introduced biological comparisons into his account of education, he soon moved on to quasi-ethological studies of mother–infant communication.

Based in Oxford in the 1970s (having sailed his small yacht across the Atlantic in 1972 to take up the new Watts Professorship of Psychology), he and his student Michael Scaife pioneered what was to become a hugely influential experimental method for studying very young babies (see 7.vi.h)—and, eventually, for modelling social interaction in humanoid robots (15.viii.a).

The eye movements of adults had already been used as indicators of their internal information processing, namely, their perceptual strategies (Yarbus 1967). Now, Scaife and Bruner (1975) worked out a way of using babies' eye movements to discover what they're interested in, and in particular of tracking the joint attention of mother and baby. Mothers, they discovered, are able to get their babies to look at something by looking at it themselves—and the baby's gaze will follow.

They did this because they thought that "very early on, starting at three or four months, there had to be some primitive form of intersubjectivity" (J. S. Bruner, in Shore 2004: 82). And that, indeed, they found to be the case. Since interest is correlated with novelty (see Chapter 8.iv.b), these experiments also showed what babies of various ages could and couldn't recognize, or discriminate.

The results were an eye-opener (no pun intended!). Babies can discriminate a good deal more than had previously been thought—and what they attend to is greatly influenced by the direction of the mother's gaze. In that sense, 'the baby's looking' is *a joint action* of mother and baby combined. Social cognition had been found to be a key aspect of cognitive development. Later work, including that done by Bruner's Oxford students Colwyn Trevarthen (1931–) and George Butterworth (1946–2000), reiterated the importance of social "scaffolding" in what had previously been assumed to be 'pure' cognitive development, or 'individual' learning (see C. B. Trevarthen 1979; Butterworth 1991; and Chapters 7.vi.h and 13.iii.e).

As well as his concern for the mother–baby social dyad, Bruner was interested in the psychological effects of culture as such. (He was one of those who had always seen anthropology as a crucial aspect of cognitive science: see 8.i.a and 8.ii.a.) Culture, he said, is crucial to cognition: to its *inner nature*, not just the particular subject matter it happens to be concerned with. For he argued that cultural artefacts, especially widely shared representational systems—such as drawing, language, and numerals—provide various "cognitive technologies" (his term). These shape, and empower, the mind itself (e.g. Greenfield and Bruner 1969; Cole and Bruner 1971).

Bruner's work on language as a cognitive technology owed a great deal to the Soviet psychologists Alexander Luria (1902–77) and, especially, Lev Vygotsky (1896–1934). Both men were often mentioned in his seminars at the Harvard Center in the early 1960s. Indeed, he was largely responsible for the first publication in English of Vygotsky's posthumous book *Thought and Language* (1962), for which he wrote an appreciative Introduction (pp. v–x). (He also wrote a preface for the 1987 edition of Luria's *The Mind of a Mnemonist*, first translated in 1968.) Although Vygotsky became *persona non grata* with the Soviet authorities, largely because of his interest in the non-Marxist topic of consciousness (Kozulin 1990), his developmental psychology reflected the Soviet respect for social, communal, influences. Where Jean Piaget saw the origin of language in the sensori-motor behaviour of the individual child, Vygotsky saw it as the gradual internalization of external signs learnt from the child's social environment (cf. Chapter 12.x.g).

During his seven-year stay in Oxford in the 1970s, Bruner did further important work on cognitive technologies. For instance, he and Michael Scaife did pioneering research on the development of cooperative attention and turn-taking in mother–infant communication, showing how these social activities "scaffold" the growth of language—the most important cultural tool of all (e.g. Scaife and Bruner 1975).

In short, he anticipated some of the insights of the anthropologist Edwin Hutchins and the philosopher Andy Clark, who would write about such matters in highly provocative terms in the late 1990s (Chapters 8.iii and 16.vii.d; see also 13.iii.d). And he influenced Kay yet again, by implying that successful human–computer interfaces would draw on *several* cognitive technologies, not just one (13.v.b and d).

d. Costume change

Eventually, Bruner moved on still further, to cultural matters in general—and to "interpretation" rather than "explanation". The latter move ousted him from his position as a universally respected guru of developmental psychology and cognitive science. For it was seen by some of his previous admirers as an intellectual betrayal.

One can compare his move to the apostasy of Hilary Putnam, who—at much the same time—abandoned functionalism for a neo-Kantian philosophy of mind

(Chapter 16.vi). But in the eyes of Bruner's bitterly disappointed followers, it wasn't just comparable: it was worse. Because of the nature of philosophy, philosophers are relatively tolerant of people changing, even fundamentally overturning, their previous views. But experimental scientists, besides having intellectual convictions as deep as anyone's, have hard-learnt technical skills and expensive laboratory equipment, neither of which can be changed overnight. So the suggestion that their entire project has been fundamentally wrongheaded is experienced as even more of a threat.

That helps to explain why Bruner's sojourn in Oxford was brought to a premature end in the mid-1970s by opposition from his colleagues in the Experimental Psychology Department. They felt that in his public pronouncements, if not in his experimental work, he was bringing psychology into disrepute. In particular, his high-profile Herbert Spencer Lecture in 1976, on 'Intentionality and the Image of Man', caused a scandal.

In it, Bruner criticized scientific (including computational) psychology for ignoring meaning, intention, and responsibility. In doing that, he said, psychology was inevitably distancing itself from the human and cultural sciences—such as his old love anthropology, and jurisprudence and sociology too. As he put it later: "a psychology that is negligent or inattentive to the other social sciences and to philosophy will inevitably be bland, particular *and even trivial*" (1983: 280; italics added).

Retribution was swift. It was the sole occasion, says Bruner, on which the overly compartmentalized department ("I have rarely seen an unhappier one"—1983: 264) pulled together. And it was brutal:

> [A] weekly departmental seminar was organized; my friends referred to it merrily as the "Bruner-bashing seminar". It was in the high English academic abrasive tradition. I never was much good at it. But at least the issues were aired. (Bruner 1983: 265)

Week by week, his views were not only rebutted, but ridiculed—by the very people who'd invited him to Oxford in the first place.

This wasn't a mere personal vendetta. Nor was it grounded merely in intellectual, albeit passionate, disagreement. For the experimentalists' hackles had previously been raised by the counter-cultural turn in social psychology, and the general public's sympathy for it (Section i.d, above). The popular, not to say populist, political/intellectual attacks on their way of doing psychology were at their height. Had that not been the case, the Herbert Spencer Lecture might have been seen as a mark of eccentricity (highly valued in England), not professional betrayal.

In short, Bruner's departmental colleagues were already under siege. So this, from one of their own, was a bitter blow. Punishment, indeed *public* punishment, was called for. In other words, this exercise in "the high English academic abrasive tradition" was in fact a "ritual degradation" (Shotter 2001). It even fitted the classic criteria of a status degradation ceremony, whereby someone perceived as a deviant is expelled from a social group (Garfinkel 1956). Far from being a purely personal matter, it was a skirmish in the (still-continuing) war between opposing philosophical conceptions of mind and psychology (see 16.vi–viii).

Bruner left Oxford soon afterwards. Much as he loved the place, and much as he savoured its many eccentricities, the atmosphere surrounding him at the Psychology Laboratory had become insufferable.

But he was (and still is) unstoppable. In his mid-seventies he became adjunct Professor of Law at the New York University Law School, because of his interest in legal reasoning—which combines "coherent system" with "messiness" (Shore 2004: 40). He sees law as largely grounded not in formal rules but in case-based reasoning, where the "cases" concerned are *stories*, whose narrative structure has to be mapped onto legal concepts such as "negligence" or "attractive nuisance" (Bruner 2002; see 13.ii.c).

Clearly, then, he wasn't taken over by computing and information theory, even though he'd been heavily—and fruitfully—influenced by them. Some forty years after his New Look phase, he confessed:

> I think I am suspicious of "formal" models of human behavior—theories couched exclusively in mathematical terms or in abstract "flow diagrams". I have always been sympathetic to the metaphors of computation and information processing, but resistant to getting trapped in their necessary measurement constraints. Perhaps I feel that such systems of measurement trap you on their flypaper while you are still wanting to fly...
>
> I did not, in consequence, get deeply drawn into the Harvard–MIT "cognitive sciences" network—the remarkable group of people who were pursuing formal ideas about information processing and computing and decision-making. George Miller and Oliver Selfridge kept me abreast enough, I thought. I regret now not getting more involved earlier. (1983: 99)

However, if in 1983 Bruner still "regretted" not having done more specifically computational research, he felt differently fourteen years later. By then, he believed that computational metaphors and methods *had* trapped psychologists on the flypaper. He was still "a well-wisher", he said. But he felt that "computational models are not, on the whole, much concerned with my principal interest... [namely] how human beings *achieve* meanings and do so in a fashion that makes *human culture* possible and effective" (1997: 281; last italics added).

As a description of what computationalists have actually done, this was fair. Whether they're *necessarily* trapped on the non-cultural flypaper, as Bruner suggested (1997: 281–9), will be discussed later (Chapters 8 and 16.vi–viii).

e. Will seeing machines have illusions?

If Bruner was the leading New Look psychologist in the USA, his counterpart in the UK was Gregory. He was more explicitly committed to 'mind-as-machine' than Bruner was, and even more important in spreading this idea far and wide.

His huge influence in the 1960s was partly due to his provocative critique of the way in which brain-ablation studies were commonly interpreted at the time (see Chapter 14.iii.a). As he said, an engineer who applied the same logic in diagnosing faults in a radio set would be laughed out of court. But his influence was mostly due to his ingenious experiments, described in his unusually accessible book *Eye and Brain*, first published in 1966 and now in its fifth edition (1998).

By means of this publication, Gregory did more than anyone else to introduce the general public, and many generations of psychology students too, to the idea that sight involves active interpretation by the brain. The book skilfully integrated work in computational psychology, neuroscience, and philosophy to tell a fascinating story about vision.

One of the stories within the story concerned Gregory's experience with a man who'd been blind from 10 months of age, whom he called S.B. He'd first heard of him in January 1959. At that time, neuroscientists were suggesting—to huge excitement—that some visual interpretation is carried out by dedicated mechanisms in visual cortex (Chapter 14.iv). But as Bruner's playing-cards had shown ten years earlier, the higher brain centres can play a role too. In addition, Hebb had argued that early visual experience would have a lasting effect on the cell assemblies involved, and Hebb-inspired experiments on animals reared in darkness had confirmed this (Riesen 1947, 1958; see also Chapter 14.vi.b). So when Gregory read in his daily newspaper that a blind man had "immediately recovered his sight" after corneal replacements, he smelt a rat (Gregory and Wallace 1963: 15).

He sought permission to work with the man, predicting (correctly) that he would *not*, at first, be able to see—unless, through touch, he already 'knew' what to expect and could somehow transfer information from one sense to the other.

The work with S.B. addressed the question famously posed to John Locke by his friend William Molyneux (Chapter 2.x.a). Indeed, Gregory had first encountered it in Locke's own writing. On leaving school at 16 he was too young to go to university, and joined the RAF's Voluntary Reserve instead. He was called into the RAF proper a couple of days after his seventeenth birthday, and worked on the development of airborne radar (R. L. Gregory, personal communication). But in 1947 he was free to go to Cambridge, where he studied Moral Sciences (alias philosophy)—with Bertrand Russell as one of his teachers, and Locke as one of the set texts.

In 1949 he turned to experimental psychology, moving a mile along the Cam to become one of Bartlett's last students. Coming from the same Cambridge stable as Craik, whose engineering ideas were "dominating" the department at the time (Gregory 2001: 382), Gregory was well versed in the notion of anticipatory internal models. He still regards Craik's concept of physical representations in the brain as "surely the most important single idea in cognitive psychology" (p. 383).

As a youngster, he saw it as an exciting new version of a key insight due to his intellectual hero, Hermann von Helmholtz: namely, that "unconscious inferences" are necessary for perception (2.vii.c). But inferences may be faulty. Even if a perceptual inference or hypothesis is reliable for 99.9 per cent of the time, there's the other 0.1 per cent too. Gregory took this fact to heart, in deciding to study visual illusions. That was a "deeply unfashionable" topic in the 1950s, although one that had interested Helmholtz, among others, 100 years earlier (Gregory 2001: 388). Illusions, then, fitted his vision of the eye–brain as a (biologically evolved) *machine*, whose 'pre-programmed' functions might sometimes lead us astray.

Much as Craik had argued for close relations between internal models and highly general features of the external world, so did Gregory. Many illusions, for example, depend on shadows (1966: 184–7). That's to say, they depend on our expectations, whether learnt or inbuilt, that there's a single sun located above us (pp. 236–7).

When such fundamental expectations are false, as may happen in space travel, unprecedented errors of judgement result (pp. 229–37). In the early 1960s, Gregory (personal communication) was invited to give two lectures at NASA's Houston Space Centre, after the 'inexplicable' failure (twice) of their astronauts to dock in space—a simple enough manoeuvre, they'd thought. The problem, Gregory told them, was the highly unusual visual environment. Shadows weren't helpful, because of the anomalous position of the sun; and familiar distance cues were missing, because there were no physical objects to be seen, apart from the space dock itself. (The stars weren't visible as three-dimensional objects, but as mere spots of light.) After further training with these psychological matters in mind, the astronauts on the next voyage managed to dock successfully.

Largely because of his wartime engineering background, the mind-as-machine approach was more evident in Gregory's work than in his friend Bruner's. For example, he soon argued (1967) that *any* effective visual system, including a seeing *machine*, will be prone to illusions just as we are. Illusions aren't mere human fallibility. Rather, they're an inevitable result of a system's using internal representations to go beyond the information given (one of Bruner's favourite phrases: 1957*a*). By the same token, they're a promising way of finding out what those representations (hypotheses) are. Gregory's hugely ingenious demonstrations of newly discovered visual illusions became justly famous. But the point, of course, was to understand *normal* vision.

Gregory and Bruner communicated often during Bruner's visit to Cambridge in 1956. But Gregory didn't share his American colleague's interest in the perceptual effects of psychodynamics. He was much closer to psychophysics and neurology (Gregory 1966). (Which isn't to say that Bruner totally ignored neurology: see Bruner and Klein 1960.)

He was closer to AI and computational modelling, too. Although most of his research focused on human subjects, Gregory did some work in robot vision, while at Edinburgh's pioneering Department of Machine Intelligence and Perception—which he'd helped to found in 1967 (Section iv.e, below). The first edition of his hugely popular *Eye and Brain* ended thus:

> We think of the brain as a computer, and we believe that perceiving the world involves a series of computer-like tricks, which we should be able to duplicate, but some of the tricks remain to be discovered and, until they are, we cannot build a machine that will see or fully understand our own eyes and brains. (1966: 237)

Nine years later, he put it in a nutshell: "As man teaches machines to see, machines teach us *what it is to be able to see*" (1975: 627; italics added).

In the third edition of his best-seller, Gregory added a final chapter on 'Eye and Brain Machines'. And there, he expressed this Craikian view:

> The problem is to build in knowledge and allow machines to learn about the structure of the world, so that they can develop internal representations adequate for finding solutions within themselves. When machines have knowledge they can be imaginative. When they have imagination they, like us, may live with reality by acting on their postulated alternatives rather than as slaves to signalled events. (1977: 236)

He'd already built a machine with visual hypotheses, if not with "imagination", in the 1960s. He'd done so with his then assistant Stephen Salter (later, the inventor of the wave-power 'duck'), whose technical ingenuity was to be invaluable to him for many years (see Chapter 1.iii.f). First outlined in *Nature* in 1964, this early hypothesizing machine wasn't a digital computer. Nor was it programmed, as the

Edinburgh robot was. Rather, it was a specially adapted camera (Gregory 1964/1974; 1970: 170 ff.).

The camera was designed for use with earth-based telescopes, since it could compensate for potentially confusing effects in the image caused by atmospheric disturbances. It worked on broadly Craikian principles, in that it first built up an *average* image of the relevant portion of the night sky and then used this as a model against which future inputs were judged. When the difference between the input and the model was at its lowest (measured automatically, as electric currents), a high-speed photograph would be taken. Several such high-speed images would be superimposed. The result was always much clearer than the model itself, which was very blurred because it had been constructed by taking a long-exposure photograph.

Gregory didn't claim that this model was similar to the visual models in our brains. But he did see it as an example of a general (Craikian) psychological principle, that "sensory information is used to build up symbolic models of the world in the brain" (1970: 170). His extensive research on visual illusions aimed to identify some of those "symbolic models", and to distinguish those due to individual experience from those due (thanks to evolution and/or post-natal development) to the physiology of the eye–brain.

One familiar example is the Müller-Lyer illusion, in which lines of equal length *look* as though they're different (see Figure 6.3). Gregory reported that all human subjects so far studied—and even pigeons and fish—suffer this illusion to some degree (1966: 160 ff.). Nevertheless, people in whose environments there's not a straight line to be seen are much less susceptible to it. So rural Zulus—who live in round houses with round doors, keep their food in round pots, and plough their furrows in curves—are only slightly affected. (Later, it was found that Zulu children who visit the town for their schooling are more susceptible than their rural parents, but less than city-living Zulus and Westerners.) Evidently, said Gregory, this illusion is largely due to learnt expectations, being strongest in people surrounded by straight-sided artefacts such as boxes and buildings.

He didn't conclude that all illusions depend on learning. Although some are entirely absent in non-Western environments, others affect every cultural group so far studied—and some non-human species too. Presumably, the latter involve 'expectations',

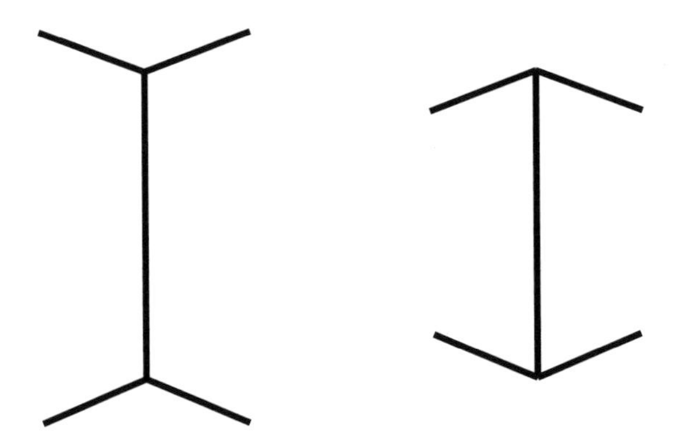

Fig. 6.3. The Müller-Lyer illusion. Adapted with permission from Gregory (1966: 136)

or 'hypotheses', built into the biology of the visual system as such—including visual cortex, but not the higher centres of the brain. But just what are they? In this early book and in much of his later research, Gregory suggested/rejected a host of detailed processing explanations for the phenomenological facts.

Gregory had encountered anti-Craikian views as a student, when he was "saturated in [James Gibson's] direct theory of vision" (see Chapters 5.ii.c and 7.v.e–f). But, thanks to his grounding in Helmholtz, he "never believed it" (Gregory 2001: 387–8).

His disbelief was buttressed by some early experiments he did at Farnborough air base, using their human centrifuge (Gregory 1974*a*: 295). These showed that the change of size in after-images, when the subject moves forwards and backwards (a phenomenon that was already familiar) may sometimes be set by the subject's *hypothesis* of distance, and of how he is moving through space. "It was these observations", Gregory recalled later, "which gradually made my thinking deviate sharply from the position held, and argued so well, by [James and Eleanor Gibson]" (Gregory *et al.* 1959/1974: 296).

In sum, the mind-as-machine hypothesis spread far and wide in the mid-1960s, partly as a result of Gregory's research and writings. Like Bruner, he'd warmed to the notion that perception involves predictive hypotheses partly because "it suggests links between [perception and] the methods whereby science gains knowledge" (2001: 388). In other words, as his friend Horace Barlow was then saying in a different context (14.iii.b), the very same psychological mechanisms probably underlie the heights of human thought and the perceptual capacities we share with many other animals.

6.iii. From Heuristics to Computers

The mid-century encounter between Newell (1927–92) and Simon (1916–2001) led to the most fruitful partnership in cognitive science. They caused a sensation at Dartmouth College (New Hampshire) in 1956, where they reported almost the first functioning program in psychological AI (Chapter 10.i.b). Indeed, it's thanks largely to their contribution that the Dartmouth meeting was one of the formative events of cognitive science (see Section iv.a, below).

After their Dartmouth bombshell, they spent the rest of their lives developing an increasingly powerful intellectual artillery, their guns finally silenced only by death. But the reverberations still persist—in AI, in certain areas of A-Life, in psychology, and in philosophy. In short, their echoes are everywhere.

Their mature psychological work, done at Carnegie Mellon, will be outlined in Chapter 7.iv.b. Here, the focus is on their role in getting cognitive science off the ground.

A warning, before we begin: They—and everyone else—naturally spoke of AI systems as searching, knowing, recognizing, trying, remembering, choosing, and the like. Such psychological language is a practical necessity, if one wants to describe what programs do. Even John Searle agrees with that (Pagels *et al.* 1984: 160). Whether it can properly be interpreted *literally* (which Searle robustly denies) is a deep and difficult philosophical question, to be discussed throughout Chapter 16. Newell and Simon eventually argued that it can (16.ix.b).

Many cognitive scientists are content to leave that issue open. But all are committed, at least, to "weak AI" (16.v.c). That is, they hold that computer programs can help in clarifying, testing, and developing psychological theories. This section shows how Newell and Simon's early work helped that view to arise.

a. The economics of thought

The two men had very different intellectual backgrounds: physics and mathematics for Newell, economics and political science for Simon. But both were interested from the start in *human behaviour.*

In Simon's case, the behaviour in question was economic and administrative decision making. His painstaking empirical research as a graduate student and tyro-academic was sociological rather than psychological. But it was underlain by intuitive psychological insights, which would eventually seed his work in cognitive science. Years later, he declared in his autobiography:

> I would not object to having my whole scientific output described as a gloss—a rather elaborate gloss, to be sure—on the pages of *Administrative Behavior* [his Ph.D. thesis, Simon 1947*a*]. (Simon 1991: 88)

That book, *Administrative Behavior*, was—and still is—widely read for its own sake by students of management, business economics, and administration. Besides its discussions of decision making, it devoted several chapters to social-psychological topics such as incentives, organizational loyalty, status, and authority. It has remained in print ever since 1947, the fourth edition appearing a few years before he died; and it has been translated eight or more times. Indeed, it was specifically mentioned in the official notification of his Nobel Prize in 1978. Our interest in it here, however, is as the forerunner of his research in (cognitive) psychology.

To begin with, Simon was more concerned with what goes on inside an organization than with what goes on inside a mind. For instance, in his first book he pointed out that *administrative organization* influences decision making. ("Decision making", not "problem solving" or "information processing": that terminology came later, in 1950 and 1952 respectively.)

The different divisions in a company start from different premisses. They employ different heuristics, or intuitive rules of thumb. (The word "heuristic" has the same root as Archimedes' famed "Eureka!"; Simon focused on problem-solving rules of thumb, but the term also covers default assumptions used in reasoning.) And they follow different goals—for the company as a whole, *sub*-goals. Indeed, the concepts of purpose, goals, and goal hierarchies were set out at the opening of the book as the root of everything in the following 250 pages (1947*a*: 4–8); and means–end analysis was highlighted too (pp. 62–6).

It follows, said Simon, that the various divisions make decisions—about budgets, for example—in different ways. There's no magical guarantee that their decisions will mesh sensibly. For instance, the marketing department may strive to produce highly creative advertising, largely irrespective of its effectiveness or expense. The management's task, then, is to ensure that the divisions are not only individually efficient, but mutually supportive.

Moreover, said Simon, individual employees have *bounded rationality*: they have incomplete knowledge and limited reasoning ability. The term "bounded rationality" didn't appear in the first edition, where Simon put the core idea like this:

> Rationality, then, does not determine behavior.... Instead, behavior is determined by the irrational and non-rational elements that *bound* the area of rationality. (Simon 1947*a*: 241; italics added)

So companies and administrative units should aim for satisfactory solutions, not perfect ones. Likewise, economists should drop their unrealistic vision of "a preposterously omniscient rationality" (1976, p. xxvi), and focus on *satisficing*, not optimizing (1955).

Much later, some of the "bounds" would be identified. For instance, it was found that people play competitive games at only one or two levels of strategic depth (Colman 2003), much as they employ only one or two levels of grammatical embedding (see 7.ii.b). In the 1940s, however, the concepts needed to consider such matters weren't available.

Nevertheless, Simon was already trying to find formal methods for talking about decisions.

* As a graduate student, for instance, he'd attended "with more diligence than usual" Rudolf Carnap's courses on logic and the axiomatic philosophy of science, and found them "particularly important" (Simon 1991: 53; cf. Chapter 9.v.a).
* In the late 1930s he'd co-authored some twenty papers on how to *measure* success in city administration.
* He'd been intrigued by the theory of games even before the key book (von Neumann and Morgenstern 1944) was published, and devoted Christmas 1944 to reading it night and day so as to write "the very first review"—for the *American Journal of Sociology* (1991: 108). Indeed, on glimpsing the title *Theory of Games and Economic Behavior* in a journal advertisement under someone's arm in a concert queue, he'd felt "a flush of envy so great he could remember it vividly thirty years later" (McCorduck 1979: 147).
* He'd described workers in organizations as, in effect, machines for doing logic: given *this* set of premisses and goals, *those* decisions would result. (He himself has identified this as a notion which would later "transform itself" into the mind-as-program view—McCorduck 1979: 127.)
* And when teaching law to engineers in the 1940s, he'd sketched Supreme Court decisions as wiring diagrams for electric circuits, the switches representing the yes-or-no choices of the court (Simon 1991: 96).

He'd even used "prehistoric" (punched-card) machines in the late 1930s—but for doing statistics, not modelling. Although they'd intrigued him greatly, he hadn't realized that similar machines might be used to study thinking:

> Of course, what the calculating punch could do was very limited... But the seed of an idea had been planted in my mind, and from that time on I was alert to any scraps of news I encountered about the progress of calculating machines. *I had no idea that I would find a use for them*; they simply fascinated me. (1991: 70; italics added)

By the late 1940s he was adept at wiring computer boards: hence his wired-justice law lectures. But those lectures formalized the *results* of the Supreme Court's ongoing decisions, not the mental processes involved in generating them.

It's all of a piece, then, that when he helped to found Carnegie Mellon's Graduate School of Industrial Administration (GSIA) in 1949, neither he nor anyone else expected to become involved in the computer simulation of psychology. He did expect that "we were going to experiment and see where these new ideas—operations research, or management science, as we preferred to call it, and organization theory—where they led" (interview in McCorduck 1979: 116). But no simulations were foreseen: they "came into GSIA by accident, so to speak".

Around 1950 Simon wrote several papers inspired by cybernetics and automata theory. One of these was a review of Nicholas Rashevsky's *The Mathematical Biology of Social Behavior* (Simon 1951). Rashevsky's seminars on mathematical biophysics had inspired McCulloch (Chapter 4.iii), and they worked their formalist magic on Simon too. But his enchantment wasn't unmixed:

> Rashevsky had a marvellous talent for building simple assumptions into models of biological systems... [But in] spite of his skills in building mathematical models, *Rashevsky was rather cavalier in his attitude toward data.* He never acquired a deep command of the biological phenomena he was treating and, as a consequence, was generally ignored by biologists. They could have learned a great deal from him, and a few of them [such as McCulloch] did, but the mutual respect was largely lacking on which effective interdisciplinary communication must build. (Simon 1991: 51–2; italics added)

He himself avoided a "cavalier" attitude to data, as we'll see. But he still couldn't express his theory of bounded rationality in formal terms.

Nor could he—or anyone else in the 1940s—conceptualize the *details* of decision making. Stafford Beer, for example, was embarking on a lifetime's work on management efficiency—but he thought about it in cybernetic terms, in which 'internal' decision making didn't feature (see Chapter 4.v.e). So although Simon indicated some of the heuristics used by administrators to compensate for bounded rationality, he did so at a very coarse-grained level. He didn't ask *just why* and *just where* rationality is bounded, merely pointing out that human minds are finite. Similarly, he took it for granted that people can apply heuristics, identify sub-goals, compare alternatives, infer from premisses, and solve problems satisfactorily. *Just how* they manage to do this wasn't a prime concern.

That all changed when he encountered the 24-year-old Newell, early in 1952. This was the "accident" which eventually led to GSIA becoming a hub for computational psychology.

b. A meeting of minds

Why Newell? After all, psychologists had been studying problem solving for decades (5.ii.b). Simon's attention was captured by Newell because the younger man—since 1950 a full-time researcher at the RAND corporation, Santa Monica—was describing human problem solving in *formal* terms.

Like Simon, he had a thorough grounding in game theory. Indeed, he'd worked as a research associate for Oskar Morgenstern for a while, as a student at Princeton. But he wasn't as enthusiastic about it as Simon had been. Although he'd initially applied to RAND because he'd heard that game theory was being studied there (McCorduck 1979: 118), he'd soon got involved in a very different approach. By the time he and Simon met, he was already speaking in terms of "information processing", and thinking about how to program computers to play chess (Newell 1955).

(He *wasn't*, however, thinking about simulating human thought on a computer. The idea of doing that would come to both men together in summer 1954, in a conversation started in a parking lot—McCorduck 1979: 132.)

Moreover, Newell was familiar with the notion of heuristics. He'd taken undergraduate courses with the mathematician George Polya (1887–1985), whose imaginative 'tricks' for arriving at proofs would eventually influence early AI (Polya 1945). (That had been at Stanford; before then, still in Europe at Zurich's Institute of Technology, Polya had also taught von Neumann.) Although Polya wasn't explicitly cited in Newell's early work (on the Logic Theory Machine and General Problem Solver, or GPS), he later acknowledged that Polya had implicitly affected his thinking for many years to come (Laird and Rosenbloom 1992: 20).

RAND saw Newell's work as technology: trying to make computers solve "complex tasks". The psychology was added by Newell, largely inspired by Selfridge's work on pattern recognition, described at a talk at RAND late in 1954 (see below). The crucial insight was that computers could become "truly non-numerical processors" (Newell and Simon 1972: 882). Even before Selfridge's talk Newell was already using 'numerical' machines (card-programmed calculators) to model non-numerical matters, namely, the blips on a radar screen. If pattern recognition, why not chess—or (less difficult) logic? And why not the *psychology* of chess or logic?

Predictably, Simon was sympathetic. For despite his disdain for "excessive formalism and shallow mathematical pyrotechnics" (1991: 249), he'd long sought formal theories of behaviour. He'd been happy to represent Supreme Court decisions as electrical circuits. And he'd very recently started speaking of "problem solving" as well as "decision making" (1991: 163). If he wasn't yet using computers to explore psychological questions, he was on the verge of doing so.

Newell enabled him to take the crucial step. Indeed, Simon soon showed the enthusiasm typical of new converts. In 1952, when he (and Newell) attended von Neumann's RAND seminar on the obstacles facing chess programmers, he felt that the master had "overestimated the difficulties substantially" (1991: 166). (Five years later, he made his notorious prediction that a computer would be the world champion in chess within another ten years: Chapter 10.i.f.)

To cap it all, Simon's "emphasis was on human simulation rather than chess prowess". (He would soon publish formal theories of "rational choice" in leading journals of economics *and psychology*: Simon 1955, 1956). In late summer 1954, they hatched the notion, together, of using computers to simulate human problem solving (McCorduck 1979: 132).—But how? The clue to that came in November, when Selfridge visited RAND to talk about his work (with Gerald Dinneen) on the infant Pandemonium (Selfridge 1955; Dinneen 1955; cf. Selfridge 1959).

As Newell recalled later (and as described in Chapter 12.ii.d), this system didn't merely recognize patterns, but "actually carried out transformations and had several levels of logic to it". For him, it was an epiphany:

> I didn't know Oliver at the time at all. We just sat in an office with five or six other people while he talked about this system they were programming, just in order to keep us people at RAND up on it. *And that just fell on completely fertile ground. I hate to use the phrase, but it really was a case of the prepared mind.* It made such an impact on me that I walked out after a couple of hours and walked into somebody's office—I don't remember whose—and gave them an hour's lecture on this thing. And then I went home that night and designed another system like it, for working on the air-defense center. (Newell, interviewed in McCorduck 1979: 133; italics added)

> I can remember sort of thinking to myself, you know, we're there. Those guys, Oliver and Jerry, had developed a mechanism that was so much richer than any other mechanism that I'd been exposed to that we'd entered another world as far as our ability to conceptualize. And that turned my life. I mean that was a point at which I started working on artificial intelligence. Very clear—it all happened in one afternoon. (p. 134)

Newell soon subjected Simon to more than an hour's lecture on the thing. The older man got the point immediately.

It had been Simon's prepared mind (i.e. formalist and computer-fascinated) which had enabled him to appreciate what Newell was trying to do even at their first meeting. The intellectual attraction was mutual:

> In our first five minutes of conversation, Al and I discovered our intellectual affinity. We launched at once into an animated discussion, recognizing that though our vocabularies were different, we both viewed the human mind as a symbol-manipulating (my term) or information-processing (his term) system. (Simon 1991: 168)

Newell soon enrolled for a Ph.D. under Simon in Carnegie Mellon's Business School (GSIA), and in 1961 joined him as a Professor of Psychology. Simon was already energetically building CMU's Computer Science department, to which—in his eyes—psychology was crucial (later, he was named Professor of *Computer Science and Psychology*). He took pains from the start to insist that they were equal partners, despite the difference in age and status. For instance, their joint publications listed their names alphabetically—although, due to Simon's fame (capped in 1978 by the Nobel Prize for Economics), people often mistakenly reversed the order. In addition, they alternated in accepting talk invitations (McCorduck 1979: 136).

Why they met in the first place was that both were involved in RAND's study of the behaviour in an air-defence facility in Tacoma, Washington. (As explained in Chapter 11.i.a, RAND was a military organization, set up to study techniques of air warfare.) The group was noting the detailed communications between radar operators and air controllers, to discover how these operatives made their decisions and how the stress involved could be reduced. This involved tape-recording and analysing all the phone calls between crew members: an early example of what Newell and Simon would later term "protocol analysis".

In practical terms, they succeeded—so much so that the US Air Force asked RAND to train its air-defence personnel. But the theoretical dividends were initially sparse. Whereas Broadbent, across the Atlantic, was investigating similar matters by considering information flow (Section i.c–d, above), Newell and Simon wanted to get inside the heads of the decision-makers concerned, to discover *what they were doing* with the information, and *how*.

At first, they had no luck. They toyed with geometry, in which the nature of the task seemed more clear. But they soon discovered that they couldn't reliably represent diagrams in the computer (McCorduck 1979: 138).

Then they decided to tackle logical problem solving. Logic was admittedly difficult for their human subjects, but that was deliberate. For Newell, on first arriving at RAND, had done some experiments in which he'd discovered that intelligent people solve simple problems by thinking about them silently and then announcing the answer: good for the subjects' egos, perhaps, but frustrating for the experimenter (McCorduck 1979: 126–7). With the first inklings of the Logic Theory Machine (see below), they felt they were beginning to get somewhere:

> [Despite the profusion of empirical data] Al and I suffered from continuing frustration in trying to write formal descriptions of the process. Somehow, *we lacked the necessary language and technology* to describe thinking people as information processors . . .
>
> [This] frustration . . . had major consequences, the first of which [was the foundation of AI]. In simplest terms, *it determined the rest of my life*. It put me in a maze I have never escaped from—or wanted to. (Simon 1991: 168; italics added)

Their mid-1950s work, which electrified many in the Dartmouth audience, had set the mould for all their later research. In the Logic Theory Machine (also called the Logic Theorist), computers and psychology were inextricably mixed: buffs on both sides were equally excited. This interdisciplinary mix endured. As they put it, some twenty years later:

> [Newell's] initial approach also established the precedents, followed in all of the subsequent work of our project, that *artificial intelligence was to borrow ideas from psychology and psychology from artificial intelligence* . . . (Newell and Simon 1972: 883; italics added)

c. A new dawn

The Logic Theory Machine (LT) was a new dawn in the psychology of reasoning. It solved problems by structuring them in terms of goal/sub-goals, and then searching the space of possibilities in sensible, though fallible, ways (Newell and Simon 1956*b*; Newell *et al.* 1957). The same was true of its more powerful sibling GPS, briefly reported in *Computers and Automation* in 1959, and described more fully two years later (Newell *et al.* 1959; Newell and Simon 1961; see also Ernst and Newell 1967).

If this sounds familiar, that's no surprise. For these two programs—which are described *as programs* in Chapter 10.i.d and v.b—drew on Simon's views on administrative decision making. They involved bounded rationality, satisficing, heuristics, the definition and achievement of distinct sub-goals, and efficient internal ordering and the identification and reduction of 'goal differences', an idea grounded in cybernetics (Chapter 4.vii.a). In other words, unlike McCulloch and Pitts' a priori mapping of propositional logic onto Turing machines (4.iii.e–f), they reflected problem solving *in practice*.

For the psychologists at Dartmouth, the Logic Theorist's triumphant debut was doubly exciting. Not only was it shockingly anti-behaviourist (transgressing all but the third and fourth of the six tenets: see 5.i.a), but it promised new experimental vistas. For instead of rat mazes and rote memories, its topic was logical reasoning—and of a significant kind, namely, proving the first fifty-two theorems of *Principia Mathematica* (Russell and Whitehead 1910).

Logic was chosen as the theme because of the research being directed at Pittsburgh by O. K. Moore (his initials represented the Persian poet Omar Khayyam), who'd asked his experimental subjects to think aloud while solving their problems (Moore and Anderson 1954*a*,*b*). That thinking aloud was the seed of Newell and Simon's lifelong methodology of "protocol analysis". As for the Russell–Whitehead book, this was chosen because it was already in Simon's personal library.

Some ten years earlier, Alan Turing had suggested using programmed search to prove *Principia* theorems. However, he didn't write such a program; and his suggestion remained unpublished outside the National Physical Laboratory (NPL) until 1969 (A. M. Turing 1947*b*; see 4.ii.b), although a briefer version had appeared in *Mind* (Turing 1950; see 16.ii.a). In fact, prior to the reprinting of Turing's 1950 essay in the *Computers and Thought* collection, his work had very little direct influence on NewFAI. It had enthused McCulloch and Pitts, to be sure. But Simon later recalled that Turing had "no particular influence on his work", and Minsky said much the same thing (McCorduck 1979: 95 n.).

Starting from the five axioms and three rules of inference given in *Principia*, Newell and Simon's Logic Theorist managed to prove thirty-eight of the Russell–Whitehead theorems for the propositional calculus. It even found a more elegant proof of one of them (no. 2.85) than that given by the original authors. Bertrand Russell was delighted when he was told this. But the *Journal of Symbolic Logic* refused to publish a paper with the Logic Theorist listed as an author (McCorduck 1979: 142). The editor's attitude was that the *Principia* theorems were old hat—and his readers were interested only in theorems, not in who/what had proved them.

Despite the LT team's unfashionable focus on internal processes, they couldn't be accused of woolly-mindedness, nor—despite the debt to Gestalt psychology—of "having religion up their sleeve" (Chapter 5.ii.b). For after all, it worked. And so did GPS, three years later. GPS could even solve the tricky missionaries-and-cannibals puzzle, which requires one to go backwards in order to go forwards. (Three missionaries and three cannibals on one side of a river; a boat big enough for two people; how can everyone cross the river, without cannibals ever outnumbering missionaries?) Purpose, thinking, and mental representations: all these seemed within reach at last.

It's little wonder, then, that Miller was energized to write a manifesto promising myriad successes for cognitive science (Section iv.b, below). There wasn't an equivalently broad-ranging manifesto for AI as such. Minsky's paper 'Steps Toward Artificial Intelligence', written in 1956, wasn't aimed at a wide audience—and it had been available only as an MIT Technical Report for five years, until it appeared in a journal in 1961. As for what *could* have been the first AI manifesto, this was still languishing unpublished at London's National Physical Laboratory (A. M. Turing 1947*b*; see 4.ii.b)—although it had been briefly outlined in *Mind* (A. M. Turing 1950; see 16.ii.a).

In one sense, of course, these "new vistas" weren't new at all. The Gestaltists, too, had studied structure, directedness, and heuristics in problem solving. Indeed, Simon had read their work with great appreciation. He'd read Otto Selz (1881–1943) with especial interest, for Selz had focused less on the structure of thoughts than on the process of thinking. Later, after defining *production systems* (Chapter 7.iv.b), Simon even credited Selz with "the idea of condition–action pairs", quoting Selz's remark:

> Intellectual processes ... [like bodily reflexes] are a system of specific reactions in which there is as a rule an unambiguous relation between specific conditions of elicitation and both general and special intellectual operations. (Selz, quoted by Simon 1991: 226–7)

But it was one thing to make such a remark, quite another to spell it out. That, Selz hadn't done. Nor could Karl Duncker and Max Wertheimer provide chapter and verse for their ideas about problem solving (Chapter 5.ii.b). With LT and GPS, Simon and Newell had provided a methodology for explaining *how* the Gestaltists' vague and (then) suspiciously mentalistic processes might be formalized and tested.

This "testing" took two forms, technological and psychological. On the one hand, the program must work. On the other hand, it must work in a human-like way, since (for Newell and Simon) it was equivalent to a psychological theory. Just what counts as 'human-like' and 'equivalent' were intuitively understood: these questions would be explored at length later (see 7.iii and 16.ix.b). Meanwhile, Newell and Simon (1961: 121 ff.) listed some of the relevant dimensions, and compared people with programs as closely as was then feasible.

Such comparisons required psychological knowledge, preferably based on experimental work as well as everyday introspection and common sense. Here, Simon's healthy respect for empirical data, evident ever since his earliest work on municipal administration, paid dividends. Having studied the detailed experimental reports of Wertheimer and Duncker, he and Newell had already begun their own experimental programme. The careful observation of human subjects, alongside theoretical interpretation and/or computer modelling, would become a lifelong project for each of them.

Their seminal *Psychological Review* paper was co-authored with the computer scientist J. Clifford Shaw (1922–91), a colleague of Newell's at RAND who did most of the initial programming. (He'd already helped Newell develop the radar simulator, and had written the assembly language for the JOHNNIAC—the machine on which the Logic Theorist was first run.) The paper didn't attempt detailed mind–machine parallels, although it compared the Logic Theory Machine with human thinking in general terms (Newell *et al.* 1958*a*). Simon describes this paper as "the first explicit and deliberate exposition of information-processing psychology, but without using that or any other trademark name" (1991: 221). ("Information psychology", of course, had already started, and Broadbent's key book was published in the very same year; but that was concerned more with information flow than with information processing: see Sections i.a–c, above.)

In 1958, however, GPS was already in preparation, explicitly intended as a psychological theory. Over two-thirds of the first GPS paper was devoted to human/program comparisons (Newell and Simon 1961). Besides, Newell and Simon were already running their own experiments, in which they asked people to talk aloud while solving a problem and then compared the results with the program trace.

At first, they used the type of logic problem that had been tackled by their Dartmouth program (the missionaries and cannibals came later). For instance, they gave an engineering student, ignorant of symbolic logic, twelve rules for transforming one logical expression into another. (These were 'rewrite' rules, as Chomsky's $S \rightarrow \mathrm{NP} + \mathrm{VP}$ and $\mathrm{NP} \rightarrow \mathrm{det} + \mathrm{N}$ were too.) Rule 1, for example, stated that A.B can be transformed into B.A, and that $A \vee B$ can be replaced by $B \vee A$. They told the subject that the logical symbols ⊃ (horseshoe), ∨ (wedge), ~ (tilde), and . (dot) stand respectively for "implies", "or", "not", and "and". Then they asked him to find a way of going from the left-hand expression, below, to the right-hand one:

(R ⊃ ~P).(~R ⊃ Q)	~ (~Q . P)

(The wedge didn't appear in the statement of this problem, but featured in some of the rules used in solving it.)

The young man's protocols included these four (taken from Newell and Simon 1961: 282):

> Well, looking at the left-hand side of the equation, first we want to eliminate one of the sides by using rule 8.
>
> Now—no,—no, I can't do that because I will be eliminating either the Q or the P in that total expression. I won't do that at first.
>
> I can almost apply rule 7, but one R needs a tilde. [S]o I'll have to look for another rule. I'm going to see if I can change that R to a tilde R.
>
> As a matter of fact, I should have used rule 6 on only the left-hand side of the equation. So use 6, but only on the left-hand side.

Whereas the first three of these protocols matched 'equivalent' actions of GPS, the fourth didn't. Nothing in the program trace corresponded to this self-correction, which involved the realization that applying rule 6 at one point would *undo* its application at a previous point. It would be another ten years before AI would be able to implement such self-criticism (Chapter 10.iii.d).

That's not to say that GPS couldn't provide *any* self-corrections, or changes of mind. For example, Newell and Simon saw a "good agreement" between its performance and this human report:

> Now I'll apply rule 7 as it is expressed. Both—excuse me, excuse me, it can't be done because of the horseshoe. So—now I'm looking—scanning the rules here for a second, and seeing if I can change the R to ~R in the second equation, but I don't see any way of doing it. (Sigh.) I'm just sort of lost for a second. (Newell and Simon 1961: 282)

Where others would later model program redirections on people's emotional hang-ups (7.ii.c), the GPS designers modelled them on the inner processes of rational thought. And the details of the subject's imperfect behaviour were taken seriously.

The authors admitted various instances where trace and protocol diverged. But they didn't despair of AI as a result. To the contrary, they tried to explain these divergences in terms of general features of problem solving (such as self-correction due to hindsight: see above) which GPS lacked, but which some future program might possess. That is, they showed—by Popperian "conjectures and refutations"—how an AI program can be used not only to *model* a psychological theory (so testing its coherence), but also to *criticize* it on empirical grounds, and to indicate ways in which it might be *improved*. The Gestaltists had been vindicated, but the vindicators had gone way beyond them.

That's not to say that behaviourism had been rejected entirely. Besides claiming descent from *both* Gestalt and behaviourist psychology (see Chapter 7.iv.a), Simon did some early modelling work that was closer to the latter. With Edward Feigenbaum, about to become famous as co-editor of *Computers and Thought* (and later more famous still, as the populist champion of expert systems: Chapter 11.v.a–c), he developed a simulation of rote learning (Feigenbaum 1961; Feigenbaum and Simon 1962; Simon and Feigenbaum 1964).

This skill had been studied fifty years before by Hermann Ebbinghaus (2.x.a), whose atomistic, objectivist, and meaning-less approach was highly congenial to the behaviourists. Indeed, they'd adapted his nineteenth-century methodology for their own experiments on "verbal learning". Simon and Feigenbaum's EPAM model (Elementary Perceiver And Memorizer) aimed to specify the information processes that make rote learning possible, and to explain some of the positional effects first reported by Ebbinghaus. Indeed, EPAM aimed—with some success—to model people's mistakes in memory tasks as well as their achievements (see 10.iii.d).

The moral was said to be that behaviourism was inadequate. Information processing must provide the theoretical terms for psychology:

> It is asserted that there are certain elementary information processes which an individual must perform if he is to discriminate, memorize and associate verbal items, and that these information processes participate in the cognitive activity of all individuals. (Feigenbaum 1961: 122)

Despite its 'mentalism', however, EPAM could be read as buttressing behaviourism, not rejecting it. For its authors apparently assumed that mastering lists of nonsense syllables is appropriately described as *verbal* learning, and even as the learning of *language*. Some years before EPAM was fully implemented, both Chomsky (1957, 1959*b*) and the manifesto authors (see iv.b, below) had declared that view a fundamental mistake.

The Logic Theory Machine was a new dawn in *philosophy* too, for it offered new philosophical vistas as much as psychological ones. Simon's interest in philosophy, expressed by his fascination with Carnap's lectures and by early papers on the philosophy of science (Simon 1947*b*, 1952), had been channelled into a 'functionalist' view of mind. Indeed, he himself saw the LT less as a new dawn in philosophy than as a final closing of the philosophical curtains:

> [We] invented a computer program capable of thinking non-numerically, *and thereby solved the venerable mind/body problem*, explaining how a system composed of matter can have the properties of mind. (Simon 1991: 190; italics added)

However, as we'll see in Chapter 16, philosophical problems aren't solved so quickly. Over the next three decades, both Newell and Simon would devote a good deal of time to justifying that claim—for instance, by articulating the theory of "Physical Symbol Systems" (16.ix.b). Many, including John Searle (1980), would remain unconvinced (16.v.c). In brief, it's still highly controversial whether they'd solved the mind–body problem—and even whether the 'problem' was a real one in the first place (16.vi–viii).

6.iv. The Early Church

A new church needs more besides a handful of theological leaders. It also needs consciousness-raising meetings, perhaps a sacred text, some mission stations, and working missionaries. Much the same is true of a new field of scientific research. Miller, Chomsky, Bruner, Selfridge, Simon, and Newell: all these high priests were crucial. But without enthusiastic disciples and organized church activities, their message wouldn't have spread as it did.

The first mission statement of cognitive science—hardly a sacred text, but hugely important as a manifesto nonetheless—appeared in 1960. It would have been less sympathetically received had it not been for the interest already aroused by three inspirational interdisciplinary meetings of the late 1950s.

The first mission station was established in New England, at much the same time—although an important precursor had existed in the same place for ten years. Before another ten years had passed, other early missions were starting to form on both sides of the Atlantic.

a. Consciousness raising

It's no accident that people still remember the three mid-1950s meetings that launched cognitive science—and also those which launched PDP connectionism in 1979, and A-Life ten years after that (Chapters 12.v.b and 15.x.a). For each of these occasions involved consciousness raising of a high order.

Consciousness raising isn't the same as making new converts. The former typically involves people who are *already* sympathetic to some challenging new idea, or anyway are on the verge of accepting it. The latter spreads the idea to people who previously had no inkling of it. So although consciousness raising may produce enormous, even infectious, enthusiasm, those who are infected are members/hangers-on of the same club. Outsiders aren't involved, or may remain immune. As we'll see, this applies to the three meetings considered here. Even the most famous, which took place at Dartmouth College, made very few new disciples: Simon later recalled that "Dartmouth got only half a dozen people active that weren't before" (quoted in Crevier 1993: 49).

The more interdisciplinary a new field is, the more important face-to-face gatherings are likely to be. The different specialists need to enthuse each other, to convince each other that their particular field *is* potentially relevant even though this may not be immediately obvious. Only then can substantive interdisciplinary research be done.

Conversations over the coffee cups are more helpful, here, than podium talks. Few professional academics have the confidence, or indeed the desire, to speak informally from the platform—excepting 'Panel' sessions. In any event, they may not know what interests, or bemuses, the other-disciplinary members of their audience. Far better to face them in person, where both individuals can speak in words of one syllable (unfamiliar jargon being explicitly questioned), and where one person's enthusiasm can excite the other's.

When this works well, the interchange quickly ratchets up to an intellectually provocative, and mutually enjoyable, level. Moreover, in small 'invitees-only' groups people accept that every other member is probably worth listening to, even if they've never even heard of them before. They're probably imaginative too, or they wouldn't have been invited in the first place. The social costs of saying something ignorant, or apparently foolish, are vastly reduced.

McCulloch had organized the interdisciplinary Macy seminars on this principle (4.v.b). There were fewer than thirty regular participants, and never more than five invited guests. The meetings were highly informal, and recorded by minuted notes of discussion rather than prose texts. No official *Proceedings* appeared until six years after the first meeting (held in 1944), and even these offered verbatim transcripts and discussion notes rather than polished essays.

Turning intellectual provocation into satisfaction, of course, requires not coffee cups but test tubes, or the equivalent. In other words, meetings and manifestos are all very well, but hard work is needed too. One of the Macy participants, the neurophysiologist Ralph Gerard, described both the excitement and the danger of coffee conferencing:

> It seems to me ... that we started our discussions and sessions in the "as if" spirit. Everyone was delighted to express any idea that came into his mind, whether it seemed silly or certain or merely a stimulating guess that would affect someone else. We explored possibilities for all sorts of "ifs". Then, rather sharply it seemed to me, we began to talk in an "is" idiom. We were saying much the same things, but now saying them as if they were so. (von Foerster 1950–5, seventh conference, p. 11)

The Macy meetings had caused much excitement among physiologists and engineers. But they'd been less provocative to psychologists. Admittedly, mathematical psychology arose, thanks to information theory. But the contents of the mind were all but ignored: *meaning* was hardly mentioned. Only a few mavericks, such as Gordon Pask (a 'lodger' at the interdisciplinary Cambridge Language Research Unit, or CLRU), used cybernetic ideas to discuss specific concepts or beliefs (Chapter 4.v.d–e). Even machine translation (MT), which had started in a number of places by the late 1940s (Chapter 9.x), wasn't being done in a psychological spirit. Nor did MT take *sentences* seriously: Lashley was still almost alone in focusing on hierarchical structure in behaviour.

Computational psychology would need more than cybernetics and MT, more than Macy and the Hixon symposium, and even more than the meaning-rich New Look, to get off the ground.

Some might argue that it got off the ground in 1943. For that year saw the introduction of the McCulloch–Pitts neurone, Craik's book on cerebral models, and the first publication of Arturo Rosenblueth's (and Norbert Wiener's) ideas on purpose. However, the scientific community, albeit intrigued by the last, wasn't instantly enthused by any of them (see Chapter 4.iii.f and vi.d). A few insiders got the point, but it's only with hindsight that this year seems so significant.

It wasn't until 1956 that the fog obscuring mind-as-machine from wider view began to clear. One year later, Chomsky's *Syntactic Structures* (1957) brought language—that is, sentences—publicly into the formalist fold. And one year after that, seminal ideas in connectionism and in computational neuroscience suddenly became a talking point. In short, it was only in the mid- to late 1950s that the new—and shocking—approach to mind captured the imagination of anyone but a few cognoscenti.

Asked to name the *annus mirabilis* for cognitive science, I'd choose 1956. That year saw a sudden growth in the invisible college, as six events made people aware that something new, and exciting, was happening. Four were publications, and two were interdisciplinary meetings. A third such meeting, held two years later, would turn out to be equally important (see below).

The publications were BGA's *A Study of Thinking*, Miller's 'Magical Number Seven', Nathaniel Rochester's paper on Hebbian theory (Chapter 5.iv.f), and—across the seas—Ullin Place's 'Is Consciousness a Brain Process?' Place's paper is the outlier here, for his mind–brain identity theory wasn't a contribution to cognitive science *as such*: it said nothing about mind as machine. Nevertheless, it was eagerly welcomed by scientifically minded readers. Moreover, its materialist spirit—though not its reductionist letter—was retained when mind-as-machine "functionalism" replaced it four years later (Chapter 16.i.d and iii).

The first—and longer—of the two 1956 meetings was the Summer Research Project on AI, held on the beautiful campus of Dartmouth College, New Hampshire. The second was the IEEE's three-day Symposium on Information Theory, convened at MIT in mid-September. (The "EEE" stands for Electronics and Electrical Engineers.) The 1958 meeting was held at the NPL in London—a resonant venue, given its post-war connection with Turing (3.v.c).

This was the mid-1950s, and the intellectual schism described in Chapter 4.ix hadn't yet happened. Each of the three gatherings featured people on both sides of the future divide, and welcomed both styles of contribution. For example, the MIT symposium saw signal detection theory being applied to psychophysics (Tanner and Swets 1954; D. M. Green and Swets 1966) and GOFAI to human problem solving.

The Dartmouth forum is widely regarded as the kick-off point for AI—considered as a communal endeavour, not just a few lone researchers (or BBC talks: Copeland 1999). Indeed, a "Fiftieth Anniversary of AI" meeting is now being planned for 2006 by the AAAI (American Association for AI), perhaps to be followed by a celebratory jaunt to Dartmouth.

The 1956 meeting at MIT, despite its being sponsored by the IEEE, was more significant for psychologists and linguists. But, occurring within such a short time-span, they had a joint effect in launching cognitive science. Hard on their heels in 1958, the London meeting combined AI and psychology with neurophysiology. At that point, cognitive science was truly on its way.

b. A trio of meetings

The beauty of the campus wasn't the prime reason why Dartmouth College was a fitting venue for AI's first meeting. The chairman of the Mathematics Department, from 1955, was the computer pioneer John Kemeny (1915–2001), co-inventor of the programming language BASIC. It was thanks to him that Dartmouth would soon be one of the first places in which students' computer literacy was taken almost as seriously as their 'real' literacy. Kemeny knew intellectual quality when he saw it, having worked on the

Manhattan Project under Richard Feynman, and having been employed for a year while still a graduate student to help Albert Einstein with his maths. (Yes, truly!) Accordingly, it was thanks to him too that, as soon as he had the power to make appointments, the young John McCarthy (1927–) was brought in to teach mathematics there.

And thereby hangs the tale, for the memorable Dartmouth event was initiated by McCarthy—with his ex-Princeton friend Minsky (1927–). The planning was done by them together with Rochester and Shannon. With hindsight (again!), this foursome is impressive.

Shannon was hugely famous already. The young Minsky and McCarthy knew him personally, for he'd given them summer jobs at Bell Labs in 1953. That's why they dared approach him to give respectability to their application for $13,500 from the Rockefeller Foundation. (Soon, all three would be colleagues at MIT: Shannon moved there from Bell Labs in 1958, McCarthy from Dartmouth in 1957, and Minsky from MIT's offshoot Lincoln Laboratory in 1958.)

Shannon's fame was due to his development of information theory, dating from the late 1930s (Chapter 4.v.d). But his role in the Dartmouth planning was grounded in his work on chess and maze-running, and on his insight—which he'd already made explicit (Shannon 1950*b*, 1953)—that digital computers could be used *non*-numerically, as general symbol processors. Despite Shannon's imprimatur, that idea still hadn't filtered through to most people, even those professionally concerned with computing. (Fully thirteen years later, Minsky would still feel it necessary to give a book on AI the title '*Semantic* Information Processing': Chapter 10.iii.a.)

As for Rochester, who'd also employed McCarthy in a summer job, he was familiar to computer scientists for co-designing the first mass-produced general-purpose computer (the IBM-701). He was about to become known to psychologists too, for his model of Hebbian learning—described at the Summer Project as well as in print (see Chapter 5.iv.f). His own particular hope for AI, as expressed in the proposal for funding, was to study "the process of invention or discovery" by asking "How can I make a machine which will exhibit originality in its solution of problems?" (McCarthy *et al.* 1955: 49). (His wide interests endured: much later, he'd encourage IBM to engage in the Media Lab at MIT—Brand 1988: 6; see 13.v.a.)

The two younger men, Minsky and McCarthy, would later be great names in cognitive science, and especially in AI (Chapters 7.i.e, 10, 12.iii). Indeed, McCarthy was already developing the ideas for his seminal paper on 'Programs with Common Sense' (1959), which he'd deliver in London two years later (see below, and 10.i.f). And Minsky distributed drafts of his influential 'Steps Toward Artificial Intelligence' to various Dartmouth visitors (Minsky 1961*b*: 10.i.g).

Dartmouth turned out to be the naming-party for AI. However, in contrast with Christopher Langton's 1987 party for A-Life (15.x.a), the new name wasn't welcomed by every guest there. The event had been announced as "The Dartmouth Summer Research Project on Artificial Intelligence". But there was no little disagreement at the meeting about which term should be adopted for the field: *cybernetics*, *automata studies*, *complex information processing* (the Newell–Simon favourite), *machine intelligence*, or (McCarthy's suggestion) *artificial intelligence*. (The phrase *knowledge-based systems* was dreamt up a quarter-century later, in the mistaken belief that this raises fewer philosophical problems: 11.v.c.)

McCarthy had insisted on using "artificial intelligence" in the meeting's official label partly because the book he'd just co-edited with Shannon, and which the senior author had insisted (against McCarthy's pleading) be called *Automata Studies* (1956), had attracted only highly mathematical papers. There was little or no mention of intelligence, language, game playing, or other psychological issues (McCorduck 1979: 96). His ploy was only partly successful.

On the one hand, the "AI" label did encourage people with interests in computing to think about computational *psychology*—whether human/animal intelligence or (McCarthy's concern) intelligence in general. On the other hand, it raised hackles unnecessarily, implying that AI *must* be seen as what Searle would later call "strong" AI (16.v.c). The seeming philosophical absurdity (to many others besides Searle) of that position deflected attention from the empirical question of just what tasks could, or couldn't, be achieved by computers, and how. So Minsky, asked in 1979 whether he'd had a sense of being at a historical gathering, replied:

> Well, yes and no. There was a false sense [at the Dartmouth meeting] that people were beginning to understand theories of symbolic manipulation and theories of cybernetics which dealt with concepts rather than simple feedback, and that things were going to be understood around the world on a wide scale. I think we had the feeling that these ideas were beginning to become popular, and maybe that's a historic event. It wasn't really true. *It took another ten years* before people could tolerate the idea of AI without thinking that it was funny and impossible. (interview in McCorduck 1979: 98; italics added)

Even then, of course, not everyone could tolerate it: almost exactly ten years after Dartmouth, Hubert Dreyfus (1965) published his first squib attacking the very idea of artificial intelligence (11.ii.a–b and 16.vii.a).

Unlike most scientific meetings, the Dartmouth affair was a long-extended conversation: a two-month Summer School. McCarthy (1989) still prefers to call it a Summer Project, since there was no separation into "lecturers" and "students", and no defined courses. It's often called a Summer School nonetheless, not least because most of those who turned up went there in order to learn something.

No invitation was needed, and many people dropped in and out. Indeed, McCarthy later expressed "great disappointment" that, partly because of this dropping-in-and-out, there was not "as far as I could see, any real exchange of ideas" (McCorduck 1979: 95). To the contrary, he said, "Anybody who was there was pretty stubborn about pursuing the ideas that he had before he came." Consequently, "[The] distance between what I had hoped to accomplish and what we did accomplish . . . was pretty large" (p. 99).

That "anybody" was too strong. Others, such as Bernard Widrow (1929–), remember Dartmouth as a major turning point in their lives (see below). But certainly, the ten core researchers, who were funded to be present throughout (though two played hookey), spent their time describing their own work—and their hopes for its future development.

Two of those core researchers were Newell and Simon, then at RAND Santa Monica and Carnegie Institute of Technology, respectively. Their names hadn't been put on the list by the local organizer, McCarthy (then teaching mathematics at Dartmouth): "I'd never heard of them before. I had no idea that anyone was doing logic with computers" (personal communication). McCarthy's ignorance was shared by others supposedly in the know. McCulloch, for instance, had recently summarized the "Points of Agreement" at the 1955 Macy conference without even mentioning digital computers (von Foerster 1950–5). Although Newell and Simon weren't named in the original "Proposal", they were featured in an Appendix written a few months later, early in 1956 (J. McCarthy, personal communication).

In the event, the two mould-breakers played truant for all but one week of the Dartmouth meeting. They weren't going fishing, but desperately trying to finish programming their Logic Theory Machine. Or rather, they were trying to convert the outline program (verbal instructions written on index cards), which had already been 'acted out' by friends and family members for about a year (Newquist 1994: 53; Simon 1991: 206–7), into a form that would actually run on a computer.

But the truancy paid off. The Logic Theorist, completed at the last minute, provided the only printout of a functioning AI program to be exhibited at Dartmouth (Newell and Simon 1956*b*; Newell *et al.* 1957). Compared with human thought, it was limited in many ways. But it *worked.* (Whether it was, as is often claimed, the *first* working AI program is another matter: see 10.i.b. And whether it had, as its programmers proudly proclaimed, "solved the venerable mind/body problem" is yet another—Simon 1991: 190; see Chapter 16.)

Other core members at Dartmouth included MIT's Ray Solomonoff (9.x.b, 10.i.g), and the AI pioneers Arthur Samuel (10.i.e), Alex Bernstein, and Selfridge—who was already a major influence on Bruner and on Newell and Simon, and also on the neurophysiologist Jerome Lettvin (see Bruner 1983: 99, Section iii.b above, and 12.ii.d). IBM staff were prominent: besides Rochester himself, Samuel and Bernstein were both employed by IBM. Among the visitors were Belmont Farley (12.i.b) and Widrow, each of whom was enthused by the general air of optimism. Widrow resolved on the spot "to dedicate the rest of my life" to AI (12.v.a).

Having been asked for $13,500, the Rockefeller Foundation provided $7,500 (McCarthy 1989). But it was money well spent. Besides the general consciousness raising, specific proposals emerged too.

For instance, Minsky—having seen the Logic Theorist—outlined a possible "geometry machine" (Minsky 1956*b*). While still at the meeting, he got Rochester to persuade an ex-physicist colleague, Herbert Gelernter, to undertake it (J. McCarthy, personal communication). With the advice—and programming skills—of several IBM staff, plus the technical resources of their Poughkeepsie office, it took him three years (Gelernter 1959). In part, this was because he had to design a new programming language, FLPL, in order to do so (10.i.c). This program is seen today as a seminal move in AI problem solving (13.iii.a).

The consciousness raising worked. Only five years after Dartmouth, Minsky would publish a "selected" bibliography on AI covering sixteen large pages—seventy, in the widely read reprint that appeared two years later (Minsky 1961*a*). Admittedly, many items dated from the early cybernetic era. But others were more recent. In short, the effect of this meeting on the budding field of AI, despite McCarthy's "great disappointment", was highly positive.

More to the point, here, Dartmouth also implied the possibility of a computational *psychology*. All four men on the planning panel were mathematicians and/or computer scientists, although Minsky's Junior Fellowship at Harvard was in "Mathematics *and Neurology*". However, they'd already embarked on some psychologically relevant projects (Minsky's is described in Chapter 12.ii.a), and Shannon's work had influenced psychology in various ways, as we've seen. Moreover, their overall agenda, outlined in their application for funding, had a markedly psychological air. They proposed research on machine models of language; goal seeking; intelligence; adaptation to the environment; problem solving; musical composition (eventually); sensori-motor control; learning; imaginativeness and originality; and neural processes (McCarthy *et al.* 1955).

Naturally, psychologists would prick up their ears at this. To be sure, their interest was rather different. Where the mathematicians wanted to show how phenomena *such as* human thought are possible, the psychologists wanted to understand human thought *as such*. In other words, the AI people didn't care if their machines didn't match the human details, whereas the psychologists did. Newell and Simon straddled this fence: besides making a huge contribution to technological AI, they did care about the psychological details. Even in the 1950s they were outlining a theory of *human* problem solving. (Later, they added multifarious empirical data: see Chapter 7.iv.b.)

The difference of emphasis, however, didn't prevent eager communication. At that time, even broad similarities between mind and machine were novel enough to be exciting. Faithful matching could wait. (And *completely* faithful matching is impossible anyway: 7.iii.d.) So Dartmouth, and tales thereof, excited psychologists too.

For psychology, however, the MIT symposium was even more fruitful than Dartmouth. Miller delivered his 'Magical Number Seven' paper there (for the second time: the first had been in April 1955, to the Eastern Psychological Association—Hirst 1988*a*: 71). And he experienced an intellectual epiphany that moved him from mathematical to computational psychology. He soon initiated a spirited declaration of intent for what's now called cognitive science (see subsection c).

In an autobiographical talk given more than twenty years later, he named the second day of the MIT meeting as the time when—for him—things suddenly came together. On that day, talks were given on the Logic Theorist by Newell and Simon and on formal grammars by Chomsky. The first, delivered at Dartmouth only shortly before, showed that a computer can prove theorems in logic. The second showed that language—considered as structured sentences, not just word strings—can be formally described. Miller instantly put the two together:

> I went away from the Symposium with a strong conviction, more intuitive than rational, that human experimental psychology, theoretical linguistics, and computer simulation of cognitive processes were all pieces of a larger whole, and that the future would see progressive elaboration and coordination of their shared concerns. (H. Gardner 1985: 29)

This larger whole, he continued, was cognitive science—in which he'd begun work "before I knew what to call it".

No one else knew what to call it either. At that time, what's now referred to as cognitive science was more likely to be dubbed "the simulation of cognitive processes" (as in RAND's summer 1958 "Research Institute", where Miller, Minsky, Newell, and

Simon all worked together) or "the mechanization of thought processes". Indeed, the latter phrase was chosen as the title of another hugely influential meeting, held two years later.

In November 1958 Uttley (1906–85) organized a select four-day seminar at London's NPL. He himself had worked for some years at the UK's Radar Research Establishment, but his interests went way beyond the technicalities of radar. He'd done pioneering work in symbolic AI (a logic program, for instance: Uttley 1951), in information-theoretic neurology (1954), and in connectionist modelling (1956). So he was eminently well suited to host a seminar that promised to set many intellectual hares running.

And that, it certainly did. What the participants lacked in quantity, they more than made up in quality. The event turned out to be important for both connectionist and symbolic AI, and for computational neuroscience too. Indeed, the list of papers given at this small meeting reads almost like a core course on cognitive science. For among NPL's many memorable moments were these:

* Selfridge first presented Pandemonium (Chapter 12.ii.d).
* Frank Rosenblatt introduced perceptrons (12.ii.e).
* Barlow announced his 'coding' theory of perception (14.iii.b).
* Gregory initiated his cat-among-pigeons critique of brain-ablation studies (14.iii.a).
* MacKay made public his ideas about building 'hybrid' (analogue–digital) machines, previously discussed only privately—with the Ratio members, for instance (4.v.b and 12.ix.b).
* McCarthy proposed that programs could be given "common sense" by expressing everyday knowledge in Russell's predicate calculus, using a programming language later developed as LISP (10.i.f, iii.e, and v.c).
* Minsky talked about heuristic programming, and told the soon-to-be-garbled story of the 'computer-generated' geometry proof (10.i.c). This was the first officially published version of core ideas in his 'Steps' paper (10.i.g), already handed out in draft at Dartmouth.
* Pask described his electrochemical model of a dynamically developing concept (4.v.e).
* And, as a dissenting voice, Yehoshua Bar-Hillel argued that McCarthy's ideas were "half-baked" and philosophically jejune, and subject to what was later called the frame problem (10.i.f and iv.iii).

Not surprisingly, then, this get-together was immediately recognized as an important step in developing theories of mind-as-machine. Instead of falling into a black hole, as so many conference Proceedings do, the two volumes of record—verbatim discussions included—were made generally available very soon (Blake and Uttley 1959).

The atmosphere at the meeting itself was electric (R. L. Gregory, personal communication). It was clear that something exciting was happening. McCulloch, in particular, made a great impression on the meeting as a whole. And Bartlett took a very active part. Colin Cherry "as usual, spoke at literally twice any normal speed, and managed to say practically twice as much as would normally be possible". The Pandemonium paper "created a visible stir". J. Z. Young was "very interesting", and McCarthy added colour to the event, wearing "hyper-Californian clothes with a necklace of big beads". (The young McCarthy was a member of the counter-culture, not least because his father was a Marxist union leader and both he and his brother were communists—Crevier 1993: 37.) As Gregory remembers it, no one realized just how difficult McCarthy's plan to give computers common sense would turn out to be.

Selfridge, too, recalls the NPL meeting as an event where "the general mood was excitement". He says this led him, and no doubt others, to make "preposterous predictions" (personal communication).

The small band of participants included no fewer than twenty-two people featured in various chapters of this book. Besides the thirteen just mentioned, were Ross Ashby, William Grey Walter, Wilfred Taylor, Stafford Beer, Richard Richens, Grace Hopper, N. Stuart Sutherland, Lucien Mehl, and A. J. Angyan.

Several were members of the influential Ratio Club—hardly surprising, since Uttley himself had been a founder member (Chapter 4.viii). Others—such as Young—had spoken there. Young was the first general anatomist to adopt a computational approach (which he would defend, years later, in the Gifford Lectures on natural religion: J. Z. Young 1964, 1978). Indeed, he'd already taken part in discussions with Turing on mind and computers (see 16.ii.a). And Gregory, for whom this was his "first significant meeting", only narrowly missed becoming a member just before it closed down (4.vi.c).

In short, this 1958 meeting brought together many of the older luminaries of the field, and several youngsters destined to become luminaries in their turn. Like its predecessors at Dartmouth and MIT, it was a memorable catalyst for the growth of an intellectual community.

c. The manifesto

If Miller didn't know what to call the new field in the late 1950s, he knew how to envision it. He did so in a hugely influential book, a rousing declaration of intent, or mission statement, for cognitive science: *Plans and the Structure of Behavior* (Miller *et al.* 1960). A good way of judging, today, how far cognitive science has—or hasn't—succeeded is to compare current theories with the hopes expressed there.

The book was written with Eugene Galanter (1924–) and Karl Pribram (1919–), after the three men (MGP for short) had spent 1959–60 together at Stanford's Center for Advanced Study in the Behavioral Sciences. They already knew each other before the trip to Palo Alto, for all three had worked at Stevens's Psycho-Acoustic Laboratory—as had Postman, Garner, and Licklider too. (Pribram had worked also with Lashley.)

Together with another half-dozen visiting psychologists, and faculty dropping in from the Stanford campus, MGP's general remit had been to develop "an experimental and theoretical capability to study natural language and communication" (Galanter 1988: 37). But mind-as-machine was a constant dimension. In the summer months before his arrival, Miller had been at RAND, working on what was then called "cognitive simulation" alongside Minsky, Newell, and Simon. Indeed, when the book appeared its declared goal was to discover "whether the cybernetic ideas have any relevance for psychology" (p. 3).

The idea of writing such a book arose at a conference organized by Miller at the Stanford Center. This occasion has been vividly recalled by Galanter:

> It was a wild affair. The people who participated were mathematical psychologists and empirically oriented experimental psychologists who tore into each other with a viciousness I had rarely witnessed. (1988: 38)

The experimentalists accused the mathematical psychologists of producing mere descriptive statistics, empty trivia seducing bright young graduate students away from real behaviour. The mathematicians, besides saying "this is only the beginning; tomorrow our theories, like Heinz catsup, will be thick and rich", suggested that even human behaviour isn't nearly as complex as it seems: what looked like complexity was merely randomness.

Galanter's—and Miller's—sympathies lay mid-stream. Yes, behavioural complexity was real, and not captured by mathematical psychology. But *theory*, not yet-more-data, was needed. What the theorists there were failing to see was that "there might be other logical and internally consistent modes of theory construction besides the analytical and mathematical" (Galanter 1988: 39)—specifically, *computational* theories. The point of writing *Plans*, then, was to enlarge the theorists' conceptual armoury while doing justice to the richness of behaviour.

Galanter had recently co-published with Stevens on scales of measurement (Stevens and Galanter 1957), and was co-editing the huge *Handbook of Mathematical Psychology* (Luce *et al.* 1963–5). But he wasn't interested only in numbers. Familiar with Turing machines and (tutored by Nelson Goodman) with computational logic, he already suspected that symbolic computation might be a way of describing the structure of behaviour. In addition, he'd recently published a plea for a "synthesis" of S–R and model-based (cognitive) theories in psychology (Galanter and Gerstenhaber 1956).

As for Pribram, he was a controversial neuroscientist who'd recently suggested (see 14.ii.b) that vision involved dedicated brain areas higher than striate cortex (and who would later offer a holographic account of memory: 12.v.c). He'd argued in 1957 that extreme behaviourism needed leavening by concepts derived from introspection (G. A. Miller *et al.* 1960: 212).

So both those men had something to contribute. But Miller was given most of the credit. I remember a quip going the rounds of Harvard's Emerson Hall in 1962: "Miller thought of it, Galanter wrote it, and Pribram believed it."

This was the Chinese-whispered version of George Mandler's remark "Miller wrote it, Galanter takes credit for it, and Pribram believes it." ("Others", says Mandler, "have claimed this invention as their own and I want, once and for all, to establish copyright and priority rights"—Mandler 2002*c*: 170.) Both versions, you'll have noticed, had Pribram believing it. This was amusing partly because Pribram held highly maverick views on other matters, and had attracted mockery accordingly (see 12.v.c and 14.ii.b, and cf. 16.x.a). Indeed, he was described as "crazy Karl Pribram, the *prima donna* of physiological psychology" (Walter Weimer, interview in Baars 1986: 309). Mostly, however, it was amusing because *no one* could seriously believe it.

The book was vague, simplistic, often careless. It had to be taken not with a pinch of salt, but barrelfuls. It presented the mind, *all* aspects of *every* mind, as a tote bag of TOTE units (see below). Psychology was assimilated to GOFAI, with parallelism acknowledged in two footnotes (pp. 50 and 198). And it was hopefully—or hopelessly?—ambitious. Animals and humans; instinct and learning; language and memory; habit and motor skill; chess and choice; values and facts; self-image and social role; knowledge and affect; intention and desire; hope and morality; personality and hypnosis; normal life and psychopathology (Miller's initial interest as a student) . . . *everything* was included. Even during my exhilarating ten-minute introduction to it (see Preface, ii), I was almost deafened by the sound of hand-waving.

The draft manuscript had been even more provocative:

> We sent our first draft, complete with no citations, to our friends and colleagues. They hated it. Where were the footnotes? What were the references to X's (for X read "my") papers? Where were the data in support of such dream work? It was depressing as, each day, a new fusillade of criticism and derision arrived. Karl and I were for damning the torpedoes and publishing at once. George (thankfully) displayed a cooler hand.
>
> "We will get the footnotes", he said.
>
> We moaned and went about the task. This period of grace also allowed us to clean up some[!] of the more egregious proposals and to lay in some more experimental-sounding text . . . [It was a] dark journey of the soul. (Galanter 1988: 42)

As remarked in Chapter 1.iii.g, the "X" above includes Newell and Simon. Alongside Hovland, Roger Shepard, and Edward Feigenbaum, they'd spent much of the summer of 1958 doing computer simulation with Miller, at RAND. But they felt that they'd taught him more than they really had. Miller recalls the unpleasantness thus:

> We showed [our draft] to Newell and Simon, who hated it. So I rewrote it, toned it down, and put some scholarship into it . . .
>
> Newell and Simon felt that we had stolen their ideas and not gotten them right. It was a very emotional thing . . . I had to put the scholarship into the book, so they would no longer claim that those were their ideas. As far as I was concerned they were old familiar ideas; the fact that they had thought of it for themselves didn't mean that nobody ever thought of it before. (G. A. Miller 1986: 213)

This fits in, of course, with the suggestion in Chapter 5.i.b, that the cognitive revolution was in fact a counter-revolution. The ideas beneath the mantle were being resuscitated.

Despite all that, the book was a revelation. It was unremittingly non-behaviourist. Its opposition was up-front and explicit:

> As for the stimulus–response business, it seemed the right approach for quite a long time. The limits were not really explicit until you got to Miller, Galanter, and Pribram. They made it explicit that stimulus–response really wasn't the right unit. (Ernest Hilgard, interviewed in Baars 1986: 298)

In addition, *Plans* brought *purpose* and *meaning* into the heart of psychology, in a way that promised (*sic*) precise formulation and rigorous testing.

Where S–R psychology had focused on the S and the R, *Plans* focused rather on the hyphen. It took Dewey, Tolman, Bartlett, Craik, and Lashley very seriously. Freud, the Gestaltists, and the ethologists were acknowledged too. It drew from Turing, Shannon, and McCulloch and Pitts—saluting the schematic sowbug and its mechanical cousins along the way (5.iii.c). Among MGP's contemporaries, it highlighted Bruner, Newell and Simon, and Chomsky (both his generative grammar and his critique of Skinner). And it offered a full chapter of neurophysiological speculation.

The core ideas—feedback, internal model, programs, hierarchy, and the flow of control—were drawn from both sides of the emerging cybernetic/GOFAI divide. A few brief quotes will convey the flavour of MGP's approach:

A human being—and probably other animals as well—builds up an internal representation, a model of the universe, a schema, a simulacrum, a cognitive map, an image. (p. 7)

[Our book tries] to describe how actions are controlled by an organism's internal representation of its universe. (p. 12)

The Image is all the accumulated, organized knowledge that the organism has about itself and its world. (p. 17)

A Plan is any hierarchical process in the organism that can control the order in which a sequence of operations is to be performed.

A plan is, for an organism, essentially the same as a program for a computer, especially if the program has [a] hierarchical character . . . (p. 16).

The organizing, or planning, operations in memorization are perhaps easier to see and recognize when the material to be memorized consists of *meaningful* discourse, rather than nonsense syllables, mazes, etc. (p. 133; italics added)

A fact not much recognized at the time—nor later, for that matter (13.iii.b)—was that MGP allowed that one might sometimes want to think of the Plan as partly contained in the environment, rather than residing entirely within the individual organism. That would be so when the observed behaviour seems to be integrated by an appropriate succession of environmental stimuli. An excellent example, though not one quoted by them, had recently been discovered—namely the courtship and nesting behaviour of ring doves (Lehrman 1955, 1958*a*,*b*). Examples in human beings included behaviour prompted by semi-permanent cultural signs (such as traffic lights: see Chapter 8.i.a) or transient jottings on pieces of paper.

But although they acknowledged that this approach wasn't unreasonable, MGP chose not to adopt it:

It is almost as if the Plan were not in the organism alone, *but in the total constellation of organism and environment together.* How far one is willing to extend the concept of a Plan beyond the boundaries of an organism seems to be *a matter of metaphysical predilections.* We shall try to confine our use of the term to Plans that either are, have been, or could be *known to the organism*, so that we shall not speak of [environmentally] concatenated behavior as part of a Plan even when it is highly adaptive. (p. 78; italics added)

One way in which the book was "careless" was that this metaphysical predilection seemed to confine Plans to *Homo sapiens*. Yet MGP had also applied it, helpfully, to instinctive behaviour in animals—speaking of inborn Plans, such as the IRMs and FAPs of the ethologists (5.ii.c).

(If they'd chosen the alternative meaning given above, the history of cognitive science might have been very different. For late-century work on situationism, and on cognitive technologies, would make a point of rejecting MGP-friendly psychology and philosophy as overly rationalist, individualist, and internalist: see Chapters 7.iv.a and g, 13.iii.b–e, and 16.vii.d.)

Another example of MGP's carelessness was that the core concept of Image, introduced (with its dignifying capital letter) in the opening sentence, was largely ignored.

Internal models, or representations, yes—but not the Image as a whole. It was declared to be 75 per cent of the human mind (the other 25 per cent being Plans), and to include not only our background knowledge but our values, so all our hopes and fears. Yet, in MGP's pages, values (motives, affect, and culture) were taken for granted: how they arose, or how they might be changed, wasn't considered. As Miller admitted:

> [The Image concept] was a sort of intellectual wastebasket for us into which (it seemed to me) we tossed all the problems that did not fit easily into the hierarchical plan structure . . . But as soon as you want to *do* anything with our ideas, of course, you must decide what a "test" is, and that can only be done if you know what part of the Image is to be tested against this input. Two ideas about it that I wish we had made more explicit in the book (but perhaps we didn't understand them then!) are: (1) some "operate" units serve merely to modify the Image and so to change what the "test" will find acceptable, and (2) it must be possible to *generate* parts of the Image as needed, since it is too detailed to have been all stored away in memory in advance. This opens up a huge spectrum of speculations, of course. (letter from G.A.M. to M.A.B., 27 April 1964)

Memory received a whole chapter in *Plans*. MGP regarded Ebbinghaus's deliberately *meaning-less* experimental paradigm of nonsense syllables (Chapter 2.x.a) as irrelevant to memory in real life. Bartlett had done so too, but he'd said very little about *how* memory schemas are formed. MGP discussed subjects' spontaneous use of mnemonics in terms of Miller's "Magical Number Seven". Such memory aids, they said, effectively shorten the length of the list to be memorized, by converting it into a hierarchical structure having only a few information-rich chunks on each level (pp. 130–6). Short-term memory was compared to a computer's input buffer, and retrieval from long-term memory was described as heavily dependent on language. And the notion of "working memory" was highlighted for the first time, implying a distinction between long-term *storage* and short-term *processing*.

Language too, "the very lifeblood of the thought processes" (Section i.b, above), was discussed at length. The central topic wasn't vocabulary, nor even MT, but Chomsky's grammar—here introduced to most of MGP's readers for the first time.

As for *the* central theoretical concept, the TOTE unit, this had been provided by Miller (Galanter 1988: 40). It was similar to Broadbent's flow diagrams, except that what 'flowed' between the boxes was not information but control. Miller explicitly intended the TOTE unit as a replacement for the S–R reflex:

> [The] fundamental building block of the nervous system is the feedback loop. (pp. 26–7)
>
> Obviously, the reflex is not the unit we should use as the element of behavior: the unit should be the feedback loop itself. [This can be defined as] the Test-Operate-Test-Exit unit—for convenience, we shall call it a TOTE unit . . . (p. 27)
>
> The TOTE represents the basic pattern in which our Plans are cast, the test phase of the TOTE involves the specification of whatever knowledge is necessary for the comparison that is to be made, and the operational phase represents what the organism does about it—and what the organism does may often [though it need not] involve overt, observable actions. (p. 31)

The most important difference between TOTEs and S–R reflexes was that TOTEs can be hierarchically structured:

> A central notion [of our book] is that the operational components of TOTE units may themselves be TOTE units . . . Thus the operational phase of a higher-order TOTE might itself consist of a

string of other TOTE units, and each of these, in turn, may contain still other strings of TOTEs, and so on. (p. 32)

MGP gave the example of hammering a nail: a *sequence* of observable actions, whose inner structure is defined—and controlled—by a holistic *hierarchical* Plan. This involves Tests such as looking at (or feeling) the nail, to see whether it's flush with the wood, and Operations such as raising and lowering the hammer (see Figure 6.4).

How we hammer nails, of course, isn't an enthralling topic. But MGP generalized the many-levelled hierarchy of TOTEs to *all* thinking and behaviour. Although not all behaviour is (as we say) planned, it *is* (as they said) all Planned. Even a 'reflex' knee jerk is generated and controlled by a (very simple) Plan.

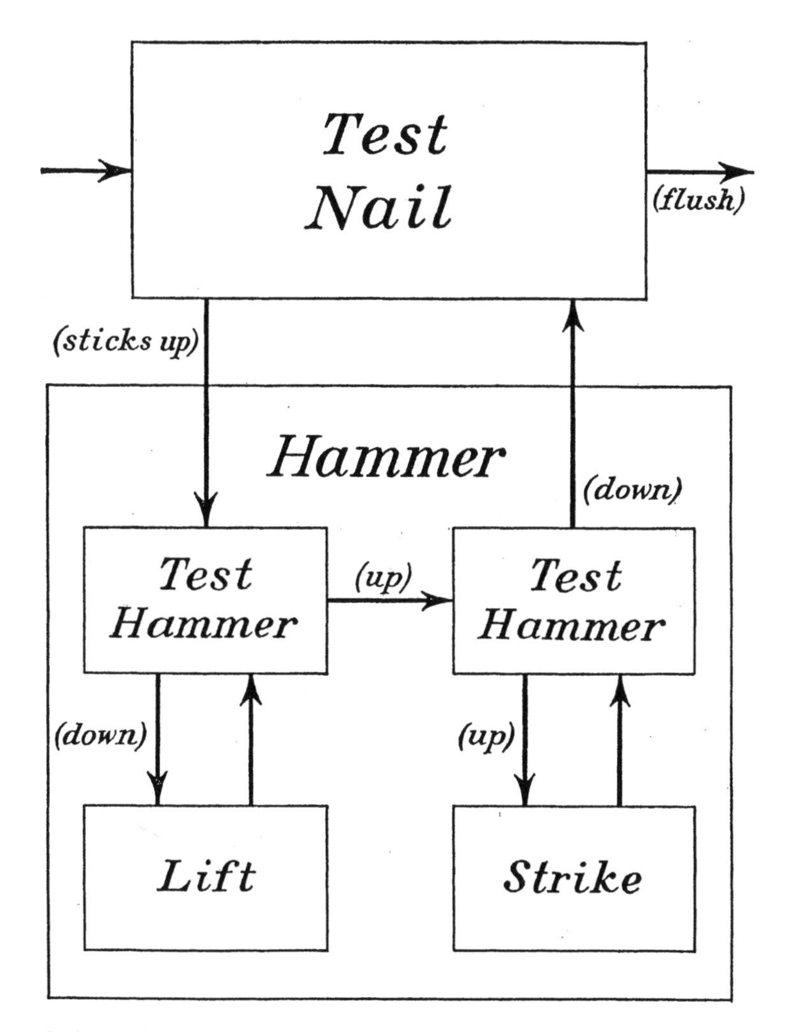

Fig. 6.4. The hierarchical Plan, or TOTE unit, for hammering nails. Reprinted with permission from G. A. Miller *et al.* (1960: 36)

Hypnosis, they suggested, involves "relinquishing" one's own Plans and adopting those of someone else (1960, ch. 8). Many personal idiosyncrasies depend not on different knowledge or motives, but on differences in the way individuals organize and manage their Plans (ch. 9). Various sorts of psychopathology, including neurosis and depression, result from a faulty integration of one person's Plans, whereas cooperation requires the socially integrated sharing of Plans (chs. 7–8). Memory and language involve hierarchical schemas, or rules, and so does problem solving (chs. 10 and 12–15). And whereas some Plans are inherited (instincts), others—such as habits, motor skills, and social roles—are learnt (chs. 5–6, 4–15).

All of this, MGP insisted, applies to human beings in general. Even very basic cultural differences, such as are sometimes described by anthropologists, didn't undermine their theory:

> Planning is a function we have seen span whatever boundaries separate Americans from the digger wasp and the computing machine. It is not the fact that Americans plan that makes them different from their fellow men. It is the way Americans plan . . . that distinguish[es] them from men who live in other cultures. *In its broader aspects, we would argue, planning is not an American idiosyncrasy. It is an indispensable aspect of the human mind.* (102; italics added)

Plans aroused great interest, earning praise from Hebb in his high-visibility address as President of the APA: "[MGP have provided] a fundamentally important line of analysis . . . [and] promise a new approach to some major problems of human behavior" (Hebb 1960: 744). It helped, of course, that Miller was already highly respected. The jokey quips withal, his book couldn't simply be shrugged off.

Or rather, it couldn't be shrugged off by anyone who understood what it was getting at. In 1960 not everyone did. For instance, the ecological psychologist/philosopher Walter Weimer, despite his growing scepticism about behaviourism, dismissed it as unintelligible:

> If you go back to 1960 there's the infamous Miller, Galanter, and Pribram book, *Plans and the Structure of Behavior*. A funny book. A lot of us bought it, looked at it, and said, What's this? We couldn't make head or tail of it. So we threw it on the shelf and never came back to it until *after* the revolution. (interview in Baars 1986: 309)

It didn't help matters that, as Weimer records, all three authors were seen as "oddballs" in psychology. Even Miller, well-respected because of 'The Magical Number Seven', was "a little suspect among the experimentalists. . . . basically a good guy, but sort of way out" (ibid.).

Another reason for being interested was that various thinkers from the past, even including Dewey, had been resurrected. Yet another, that many readers were intrigued by the new research cited by MGP: Chomsky's grammar and Newell and Simon's GOFAI planning. Further, people already sympathetic to cybernetics and/or GOFAI (remember: this was pre-schism) would at least be interested in, even if they weren't fully convinced by, any serious attempt to apply feedback and programs to *the entire range* of psychology.

Despite their talk of programs, however, MGP didn't realize just how useful computer modelling could be for psychology. Galanter, again:

> What we clearly failed to recognize at the time was the over-riding importance of the computer as an experimental device. We did not catch its signal potential contribution to cognitive science. This machine, with its ability to control complicated input–output relations, would open the way to new and complex kinds of behavioral studies. (1988: 42)

In 1960, of course, "complicated input–output relations" couldn't yet be modelled. There were fewer than a dozen AI programs of specifically *psychological* interest. Although MGP's visionary hand-waving could have been more careful, it couldn't—at that time—have been much more precise. In sum: for all its faults, *Plans* was the right book at the right time.

d. The first mission station

Simultaneously with the appearance of *Plans*, cognitive science got its first dedicated research centre, at Harvard. (It grew out of an enterprise that had already been bustling along for ten years, as we'll see.)

Granted, the first AI labs had been started a few years earlier at Carnegie Mellon and MIT (see Chapter 10.ii.a). And the CMU version, thanks to Newell and Simon, was markedly psychological in spirit. But the potential scope of cognitive science as a whole was better represented at Harvard. Most unusually, the invisible Chinese walls that normally separate academic disciplines were being deliberately ignored.

One of the two people responsible was the senior partner in the MGP trio. The other was Bruner. On Miller's return to Harvard from the Stanford think-tank, he and Bruner applied to the Carnegie Foundation for ten-year funding to set up the Center for Cognitive Studies.

Carnegie obliged, and the Harvard centre was established in 1960. A separate grant proposal soon earned them an in-house PDP-4 computer, perhaps only the second computer ever to be set aside for psychologists' use (D. R. Norman and Levelt 1988: 104). Intriguingly, the house they were assigned was one that had been lived in by the family of William James (G. A. Miller 1986: 214): whether this was more inspiring than inhibiting, who knows? In retrospect, Miller wrote:

> I think Jerry and I made a good pair . . . We shared a vision of cognitive psychology, but our intellectual playmates were very different. I gave him access to ideas growing out of communication theory, computation, and linguistics, whereas he gave me access to ideas from social psychology, developmental psychology, and anthropology. He broadened my view of cognition; I hope I helped to sharpen his. (quoted in Bruner 1983: 123)

The Center was a Solomon's House (2.ii.b) whose inhabitants and visitors were drawn from many fields. These included various areas of psychology, from 'respectable' psychophysics to developmental psychology—then still far from fashionable. Bruner himself had been deeply influenced by Piaget, who featured prominently on his reading lists—and in his own developmental writings, which helped to 'cognitivize' developmental psychology around the world (Bruner 1964, 1966*a*; cf. Chapter 5.ii.c).

The other fields were primarily linguistics and anthropology, though AI and philosophy were included too. History, art history, and sociology could be found in the visitor fringe. The common interest was in "the distinctively human forms of gaining, storing, transforming, and using knowledge of all sorts—what makes humans human" (Bruner 1983: 122).

These "cognitive processes", Bruner recalled later, "were certainly being neglected—particularly at Harvard". Quite a few psychologists felt thoroughly uneasy with the terminology. In fact, cognition—so Mandler remembers—was "a dirty word" among his own American colleagues. Cognitive psychologists were seen as "fuzzy, hand-waving, imprecise people who really never did anything that was testable" (G. Mandler 1986: 254).

The key principle in inviting visitors was interdisciplinarity:

> What [George Miller and I] had decided was that *psychology was too complicated a field to be left to the psychologists*, and that what we needed was an alliance with colleagues in other disciplines who were, each in his or her own context, concerned about how humans used and acquired knowledge. (Bruner 1988: 92; italics added)

The visitors invited to the Center were imaginatively chosen. For example, those I worked with while I was a graduate student there included Kolers, Goodnow, Roger Brown, Brendan Maher, Barbel Inhelder, Margaret Donaldson, Philip Stone, Ernst Gombrich, and Eric Hobsbawm—the last two, prominent scholars in art history and social–political history, respectively.

Others who were sometimes around (though I didn't have the opportunity to work with them) included the philosopher Nelson Goodman (1906–98). He was a research fellow at the Center in 1962–3, preparing the published version of his 1962 John Locke Lectures on cognition and aesthetics (Goodman 1968). Goodman's presence exemplified the reciprocal successes of interdisciplinarity that Bruner and Miller were trying to foster. For his influence on Bruner's thinking was considerable—and, apparently, vice versa:

> J.B.: *[One of the things] I'd like to be remembered for* is as another one of the guys among a long series of them [notably Bartlett: see 5.ii.b] who made an effort to reintroduce meaning into psychology. I intend meaning-making as a process. [I called my recent] book *Acts of Meaning*, because it made meaning into something that you have to get. This is something I found very sympathetically expressed first by Nelson Goodman.
>
> B.S.: What's your genealogy in terms of this focus on meaning-making. Where did it come from?
>
> J.B.: I think from Nelson. And oddly enough, when he was interviewed about his influences recently [i.e. in the 1990s], Nelson Goodman gave me as one of the important factors that led him along this line. (J. S. Bruner, in Shore 2004: 100; italics added)

Even Bruner and Miller themselves were often seen as coming from different disciplines. For they were located in two recently separated departments, with little love lost between them: Social Relations and Experimental Psychology. The former covered social and clinical psychology, anthropology, and sociology; the latter, psychophysics and behaviourism.

Indeed, the interdisciplinarity went even further than that. For many 'self-invited' visitors turned up at the Center's weekly colloquium, including physicists, historians (such as Stuart Hughes and John Fairbank), and lawyers (such as Paul Freund). Bruner saw this as hugely important for psychology as a discipline, not just at Harvard but in the intellectual world as such:

> Up to that point (after the James days) psychology had been (and [in 2004] it still is in many places) an oddly isolated field. I think the Center (with its emphasis of "knowledge acquisition and organization") had the effect of remaking psychology into a part of the university at large—and of the American university world at large. (J. S. Bruner, personal communication)

Bearing in mind the non-humanism, not to say anti-humanism, of the behaviourists who dominated academic psychology at the time, this isn't too surprising. However, as we saw in Section i.d, intellectual counter-currents would soon form which would undermine attempts to foster an experimentally based humanism.

One might say that the Center was born already 8 years old. For it had been anticipated by Bruner's Cognition Project, founded in 1952. The Project had set its stamp on the nascent cognitive science: not just psychology, but linguistics and philosophy too. Chomsky had been a visitor there for a year, and Jerry Fodor had attended many seminars—indeed, both were deeply influenced by it. Chomsky would soon borrow Bruner's ideas about perceptual/conceptual hypotheses, in his account of language acquisition (Chapter 9.vii.d). As for Fodor, he would later defend non-behaviourist psychology, and draw on Bruner's study of concept learning in his notorious *Language of Thought* (Fodor 1968, 1975; see Chapter 16.iii.b).

Alongside the Project, considered as a research centre, Bruner and Miller had been doing subversive undergraduate teaching for several years:

> [We] collaborated in teaching the (in)famous Psychology 148—The Cognitive Processes. It lured an astonishing number of bright kids into the field. (J. S. Bruner, personal communication)

Their collaboration didn't stop there. Bruner said later, "I have no doubt (1) that there would have been no Center without George Miller" (1988: 90).

However, he immediately added his conviction "(2) that George and the other actors in the events surrounding its establishment were corks on the waves of history". And the most powerful roller of them all, according to Donald Norman (an early visitor), was anti-Newtonianism:

> [The] Center at Harvard was not set up to be *for* anything in particular; it was set up to be *against* things. What was important was what it was not: not psychophysics (at the time, a major, mainstream activity in psychology), not animal studies, certainly not Skinner's operant psychology (whose world headquarters were just down the street). Basically, not contemporary American 1950s psychology. Late 1800s or early 1900s psychology, perhaps, but certainly not the contemporary American psychology of the 1950s. The enemy was the present. (Norman and Levelt 1988: 101)

In Europe, "the present" was of course very different. (We've already noted Mandler's remark, that behaviourism was "a *very* parochial event": 5.i.b.) Willem Levelt, on his arrival at the Center from the Netherlands (and studying with Albert Michotte), was familiar with the sitting tenants discussed in Chapter 5.ii—which his American peers were not. But, he recalled later, "behaviorism had been so completely absent from my horizon that I didn't even know the difference between classical and operant conditioning" (Norman and Levelt 1988: 102).

Similarly in Britain. When Miller gave his anti-behaviourist lecture on a visit to the UK in 1963, his host said, "What was all that about behaviorism? You know, there are only three behaviorists in England, and none of them were here today!" (G. A. Miller 1986: 212). That was a joke, of course. But it had a firm grounding:

> The one road which [when behaviourist ideas were overwhelming psychology in the USA] no British psychologist of standing took was the road of behaviourism . . . Most British philosophers and psychologists between the wars would have agreed with C. D. Broad's judgment that behaviourism was simply a "silly" theory with no claim to be taken seriously. (Hearnshaw 1964: 210)
>
> [Above] all British psychology was largely shaped by McDougall . . . (p. 212)
>
> From among contemporary foreign schools of thought a good deal was borrowed from psychoanalysis and something from Gestalt psychology. . . . The more extreme movements, American behaviourism on the one hand and German phenomenological schools on the other, made little headway in Great Britain. (p. 213)
>
> By 1940 [there was] a wide divergence between British and American psychology. (p. 215)
>
> In America the importance of the conditioned reflex was widely appreciated, and it became one of the main planks of behaviourism; in Britain it was not regarded as much more than a minor curiosity. (p. 216)

To be sure, by the late 1950s the conditioned reflex was no longer "a minor curiosity" in Cambridge: it was sufficiently prominent to dissuade me from my plan to study psychology there (see Preface, ii). Even so, the new Center was more shocking—and more courageous—in Cambridge Mass. than it would have been in Cambridge England.

That "would have been" is important. Although many, even most, European psychologists were *already* cognitivists, the Bruner–Miller enterprise had no 'twin' outside the USA. The nearest equivalent was the apple-orchard CLRU, the most visible the MRC's Applied Psychology Unit—both in Cambridge (Preface, ii, and Section ii, above).

What was distinctive about the Center wasn't only that it was cognitivist, but that it was deliberately interdisciplinary. As a result of their experiences during the war, many denizens of the CLRU and the APU were accustomed to interdisciplinarity. But that was primarily for practical reasons: if one was interested in how pilots or gunners use their high-tech instruments, one had better consult some communications engineers as well as psychologists. The Harvard Center, by contrast, valued interdisciplinarity in (all) the human sciences for its own sake. For without it, the human mind would never be properly understood.

The attack on the behaviourist "enemy" was reflected in the Center's name. Miller has remarked:

> To me, even as late as 1960, using "cognitive" was an act of defiance. It was less outrageous for Jerry, of course; social psychologists were never swept away by behaviorism the way experimental psychologists had been. But for someone raised to respect reductionist science, "cognitive psychology" made a definite statement. It meant that I was interested in the mind—I came out of the closet. (quoted in Hirst 1988*a*: 89)

So the "cognitive" was a defiant rejection of behaviourism. Very soon, Miller's introductory book *Psychology* (1962) barred the door against the behaviourists even more firmly. It was subtitled *The Science of Mental Life* (a phrase taken from the first sentence of William James 1890), and—shock, horror!—it paid homage to several of the sitting tenants mentioned in Chapter 5.ii.

Almost twenty years later, Miller would found a second interdisciplinary centre, the Cognitive Neuroscience Institute at Rockefeller University. Although he'd been interested in neuroscience in the 1950s (hence the P in MGP), he'd had little knowledge of clinical neurology. It wasn't until the late 1970s that he realized the importance of detailed clinical data for psychological theory in general (14.x.b).

By that time, the Harvard Center no longer existed. Not because its task was completed, or unimportant, but because of a lack of imagination and an excess of red tape. In Bruner's words:

> In the end, we closed shop. The deans had never been very keen about a Center after McGeorge Bundy [who'd enabled us to found the Center] left. We were self-supporting, *but a center such as ours is messy to a good administrator.* It duplicates the power of departments. Bundy, like John Gardner, our first Medici, had been out to reduce the power of departments. After we [presumably Bundy and Bruner, though maybe Miller and Bruner] left, the Center ceased to exist. Nobody ever asked whether that was a good idea. It just happened. The dean said it should. (Bruner 1983: 126; italics added)

The Chinese walls had got their revenge. (A quarter-century later, much the same happened, for much the same reasons, to the School of Cognitive Sciences at Sussex: see below.)

If the Center hadn't closed down, its emphasis might have changed. For both founders eventually altered their intellectual priorities. Miller turned to neuroscience, founding the first institute of cognitive neuroscience in the late 1970s (Chapter 14.x.b). And Bruner, as we've seen, moved away from computation to interpretation, and from pure psychology to psychological aspects of law and anthropology.

Nevertheless, the original vision hadn't been abandoned by either of them. Miller still acknowledged his "enormous debt to people like Simon, Newell, and Minsky" (1983*a*: 57). As for Bruner, he still allows that "the study of man" requires psychologists, anthropologists, neurophysiologists, and "computational scientists" too (1996, p. xiii). And he still sees Miller, whose current interests are very different from his own, as a hugely important colleague:

> George Miller is [somebody who's close by my desk, in that I always ask myself, "What would he think about that?"]. George is particularly valuable because we know that I know that he knows that I know that we disagree very deeply, but if I can make my point to him and he can say, "I understand what you are saying but I disagree", then that's okay. (interviewed in Shore 2004: 6)

Moreover, he sees an essential continuity between his late 1950s formal (computational) work and his current emphasis on narrative, as the core of humanistic interpretation:

> B.S.: The old classical theory of categories [assumed that] distinctive features converge into a well-formed category based on a kind of magnetic attraction.
>
> J.B.: Yes. That's the old Bruner–Goodnow–Austin view in *A Study of Thinking*. At least it characterizes the first half of the book, but the second half had to do with what we called thematic categories. You should see the amount of waffling we did. I mean, it is shameful. I re-read that section twenty-five years later and realized the extent to which I simply didn't pay attention or have the capacity or the courage to pursue what it meant that categories formed

thematically. Out of what kind of themes, Bruner? So I come back twenty-five years later and I re-discover the obvious, that most people think in terms of stories about what happened and blah blah blah. And so they form their categories. (Shore 2004: 147)

(Part of that "blah blah blah" is carried by the "themes" and "scripts", and the insights into the general structure of motivation, intention, and cooperation, that were analysed formally—though crudely—by Robert Abelson in the late 1960s and early 1970s: see Chapter 7.i.c.) In light of the charge of intellectual "betrayal" mentioned in Section ii.c (above), this insistence on continuity is interesting.

e. Missionary outposts

On the other side of the Atlantic, the pioneering CLRU still continued. This had been founded in the mid-1950s, but it wasn't *psychologically* oriented (see Preface, ii, and Chapter 9.x.a). Except for Gordon Pask, its 'take' was linguistic and/or technological.

A halfway-house between CLRU and Bruner's Center was formed in the 1960s at the University of Edinburgh. Whereas the Harvard group, its precious PDP-4 notwithstanding, was focusing above all on psychology, the group in Scotland was at least equally concerned with AI.

Donald Michie (1923–), then Reader in Surgical Science, had been speculating about what would later be called AI since the early 1940s. As a cryptographer at Bletchley during the war (see 3.v.d), he—with fellow code-breakers Turing and Jack Good—had formed "a sort of discussion club focused around Turing's astonishing 'child machine' concept", that is, "a teachable intelligent machine" (Michie 2002). "It gripped me," he now recalls, and "I resolved to make artificial intelligence my life as soon as it became feasible."

Feasibility, of course, was some years in the making. In 1948, still in close contact with Turing, he was writing a paper-and-pencil chess program in his spare time, while spending most of his time on genetics. These split intellectual loyalties continued through the 1950s.

Michie's hopes for AI were well known by his Edinburgh colleagues, most of whom were pretty sceptical. In 1959, when word was spreading about Uttley's epochal NPL meeting, one of them told Michie that learning machines were "impossible"—and challenged him to prove him wrong. That was a red rag to a bull. Since Edinburgh, to Michie's disgust, still lacked an equivalent of ACE or MADM or EDSAC (3.v.b), he built "a contraption of matchboxes and glass beads" called MENACE (the Matchbox Educable Noughts And Crosses Engine). MENACE not only won the bet, but also won him an invitation to Stanford—where he wrote a reinforcement-learning program based on a task that Widrow (12.ii.g) had set to one of his Adaline students (Michie and Chambers 1968).

On his return from the USA, he "lobbied everyone in sight" to overcome "the national computer-blindness" (2002: 3). Even the Minister of Science and Education knew nothing of the Enigma-busting triumphs of Colossus (3.v.d), and thought that "computing" meant "desk calculators". Michie, of course, couldn't enlighten him about the still-secret Colossus. But, as he often admitted, he still carried the Bletchley spirit of *Get the job done, and damn the bureaucrats!* So ignorance, even opposition, from those on high was less of an obstacle to him than one might imagine. During this relentless lobbying, Michie persuaded the Royal Society to provide "a few hundred pounds" to enable him (with Bernard Meltzer)—to set up a small research group in 1963.

This was made official as Edinburgh's Experimental Programming Unit in 1965. In 1966 it became the Department of Machine Intelligence and Perception, whose intellectual focus was much wider. It was devoted to research on "integrated cognitive systems", whose "intellectual" attributes were grounded in "sensorimotor and reflex capabilities" (Michie 1970: 75). (Meltzer now headed the Metamathematics Unit, specializing in the automation of maths—whose students would include Patrick Hayes and Robert Kowalski.)

The "and Perception" in the Department's title was due to Gregory, who left Cambridge to be one of Michie's leading triumvirate at Edinburgh. He headed their new Bionics Group, where he worked on machine vision for robotics as well as on visual illusions (Section ii.d). The robotics caused a hiccup in the Department's search for accommodation:

> Initially we were offered a deconsecrated Church of Scotland church for our laboratory and offices; but when they heard we were going to build a robot, it was withdrawn! (Gregory 2001: 389)

(This experience put them in the same honourable league as Jacques de Vaucanson, thrown out of his seminary for building automatic flying angels: Chapter 2.iv.a.)

The third member of the triumvirate was H. Christopher Longuet-Higgins (1923–2004). He was a highly distinguished theoretical chemist (many colleagues feel that he was unlucky not to have got the Nobel Prize) who had recently turned to psychology. (He ended up being almost as distinguished there: see Chapters 2.iv.c, 7.v.d, and 12.v.c.) He, too, had left Cambridge for Edinburgh.

At one level, Michie's Department was a resounding success:

* It pioneered symbolic AI in the UK (connectionist AI had started even earlier: see Chapter 12.ii).
* Robin Popplestone, who became the fourth staff member in 1965, wrote a new-style programming language (POP2) that was hugely important for British AI research (10.v.c).
* The Edinburgh hand–eye robot FREDDY, designed to assemble a toy boat and car from a few pieces dumped together on a table, raised some important issues with regard to vision and robotics in general (10.ii.a, 11.iv.a).
* And Michie's intellectual breadth encouraged AI ventures outside the usual limits, such as James Doran's pioneering work on AI agents (Doran 1968*a*) and his applications of AI to archaeology (Doran 1968*b*, 1970, 1977; Doran and Hodson 1975; cf. Doran and Palmer 1995).
* Psychologically relevant early research included Longuet-Higgins's analyses of associative memories (12.v.c)
* and musical perception (Longuet-Higgins and Steedman 1971);
* Gregory's (1967) argument that seeing-machines would necessarily suffer illusions (Section ii.d, above);
* and Sylvia Weir's neo-Gestaltist study of the perception of 'meaningless' geometrical figures in social/personal terms (Weir 1974; Weir *et al.* 1975).

More generally, Michie's group raised the visibility of computer modelling. On the one hand, he founded the influential—and still continuing—Machine Intelligence Workshops in 1965. These drew participants from across the world. On the other hand, his Department trained many year-long visitors in AI techniques. I was invited to go there in the late 1960s for that purpose, but a new baby took priority. A few years later, my Sussex-philosophy colleague Aaron Sloman did so, and soon became a highly imaginative leader of AI in the UK (see Chapters 7.i.f, 10.iv.b, 12.v.h, and 16.ix.c). Other people who cut their teeth at Edinburgh in the early days included Longuet-Higgins's student Geoffrey Hinton (12.v.h).

Without Michie's energy and commitment, none of that would have happened. However, he wasn't an easy man to work with. The triumvirate soon splintered, as it became clear that three vibrant egos couldn't exist in the same small unit. In 1970 Gregory moved to Bristol, where he founded the Brain and Perception Laboratory, and in 1974 Longuet-Higgins (soon followed by Hinton) went to Sussex as a Royal Society Research Professor, working alongside Sutherland and Uttley. Arguably, this split—although painful at the time—helped the development of British cognitive science. For it dispersed a combination of psychological interest and AI expertise around the country.

The effect in Edinburgh, however, was to reduce the prominence of psychological AI. Increasingly, Michie's group—administratively restructured several times—moved away from psychology towards technology. (Its work on speech perception was an exception: although the prime aim was to produce useful tools, this required extensive attention to *real* speech.)

Ironically, Michie—having founded AI research in the UK—eventually led to the worst assault it has suffered over the years (Chapter 11.iv and v.c). His over-optimistic predictions about AI applications, and his often abrasive personality, irritated many of his scientific colleagues. In 1971 his main sponsors, hearing of "the high level of discord" at Edinburgh (Howe 1994), commissioned Sir James Lighthill to investigate AI in general and the Edinburgh group in particular. The result was a disaster, and led to the UK's "AI Winter"—which pre-dated the American one, and lasted for over ten years.

The mid-1960s also saw the beginnings of cognitive science at the recently founded University of Sussex. Sutherland, like Michie, had been enthused by visits to MIT and CMU in the early 1960s. He lectured on AI within the Experimental Psychology course from its inception, and invited various AI experts from outside to give additional, more technical, lectures. Indeed, he was already negotiating with potential funders to establish a (psychologically oriented) AI Department in 1966, when Michie was doing the same—with more immediate success.

The earliest researchers at Sussex were various members of Sutherland's group (notably Uttley and Longuet-Higgins), and myself (Preface, ii). After the arrival of the AI vision pioneer Max Clowes, and Sloman's visit to Michie's AI empire, Sussex established the Cognitive Studies Programme (COGS) in 1975 to offer interdisciplinary degrees in cognitive science. Students took some combination of philosophy, psychology, linguistics, and AI. (The AI courses were offered also to anthropologists, but the students' anthropology tutors soon started advising them against it: see Chapter 8.ii.c.) An AI subject-group was born within the Programme a few years later. In 1987 the

Programme metamorphosed into a School, and in 2003 into the Centre for Cognitive Science (closely associated with the new Department of Informatics).

In the late 1960s, the San Diego psychologists—also newly established, with Mandler as the first Chairman—were building a lively research group. This was a West Coast version of the New England mission station. Indeed, three of the founding members, Mandler himself (in the late 1950s) and his appointees Norman and William McGill, had been close to Miller and/or Bruner at Harvard—and Mandler was about to accept an invitation to Bruner's Center when San Diego approached him.

McGill, a psycho-acoustician trained by Miller, had the least influence on the group: he stayed at San Diego for only three years. But Norman and Mandler remained active there for much longer. Norman's background was in electrical and computer engineering, so he was well placed to combine computation with psychology. Mandler was less explicitly computational, but he was a committed cognitivist. As a refugee from Nazi Germany, he brought 'non-Newtonian' assumptions from the European tradition—"to some extent German and French but to a large extent British" (2002*c*: 175). So the ideas of Craik and Bartlett, one long dead and the other a nonagenarian half a world away, were now working their magic in California.

The San Diego group, and its offshoots such as CHIP (the Center for Human Information Processing), would soon have a major influence on the field. Its members wrote the first widely used textbook of cognitive science (Section v.a, below), and an early computational text on associative memory (John R. Anderson and Bower 1973). They helped found the Cognitive Science Society (v.c, below). And they were later crucial in the renaissance of connectionism (Chapter 12.vi–vii).

In addition, anthropology was represented at San Diego by Roy D'Andrade, a student of Miller and Bruner at Harvard in the mid-1950s. And, last but not least, San Diego philosophers would provide a counter-intuitive philosophy of mind. This was based in part on the insights of New Look psychology, and (after their arrival in San Diego) it was integrated with connectionist ideas (16.iv.e). Eventually, San Diego neuroscientists would join in too (14.ii).

So, by 1970, mission stations had been established on both sides of the pond. Although the Harvard Center finally succumbed to strangulation by red tape, its mission lived on.

f. The sine qua non

Every church needs its donation boxes, and Descartes's wealthy philanthropists were as crucial as ever. They were available in more abundance than usual in late 1950s and 1960s USA: "We were lucky historically. Those were the days of post-Sputnik academic expansion" (Bruner 1983: 124).

Sputnik, the first man-made satellite, was put into orbit by the USSR in October 1957. Only a few weeks later, Sputnik-II was launched. It was not only heavier, but carried a live dog, Laika. (She died from heat exhaustion, having spent two days on the launch-pad and five hours in space.) The USA, too, had been working for some years on rocket technology. Indeed, in the final days of the Second World War the US Army had recruited the Nazi scientist Werner von Braun, designer of the V2 rockets that devastated London in 1945. For many months before the appearance of Sputnik, he had been trying—without success—to persuade the army to think beyond earthbound rockets, and to fund a test launch of his planned satellite. The worldwide sensation caused by the two Sputniks was a bitter pill to swallow, not just for him but for the US government. The money-taps for space technology were suddenly turned on, enabling von Braun's device (Explorer-I) to be sent into orbit some four months after Sputnik. What's relevant here, however, is that the money-taps were also turned on for science, and scientific education, in general.

That political background helped raise the funding made available through the USA's national research councils. That was partly because many cognitive science projects arguably had some military potential. Indeed, the US Office of Naval Research was already funding MT in the USA and UK, and the USA's Department of Defense would be crucial for the development of AI and some types of cognitive psychology (see Chapters 9.x.a. and 11.i).

Mandler remembers, for instance, that as the Chairman of the San Diego psychology group in 1966, he was "in the midst of raising federal money for our building and thanks to Sputnik [launched by the USSR in October 1957], there was lots of it around" (2002*c*: 192). "Under the aegis of national defense", money was poured into higher education, so that he was granted $1.3 million only a year after his new group had been founded.

We've seen that Bruner's Center for Cognitive Studies was supported by the Carnegie Foundation. One of the "luxuries" of the place, he recalls, was that the Carnegie grant (and its runner-up of $250,000 five years later) was unrestricted money:

> Even *small* unrestricted grants help . . . Every cent of unticketed money is a fortune . . . Do what you will with it but make it interesting. (Bruner 1983: 125)

"Every cent" is correct: the size of the grant isn't all-important. The Dartmouth meeting, for instance, showed that even so small a sum as $7,500 can be very effective if it's used imaginatively.

The sponsorship of cognitive science by the Alfred P. Sloan Foundation was especially generous. In the late 1970s, they promised $15 million (later rising to $20 million) for a "particular program in cognitive science". Admittedly, even this munificent gift horse was looked in the mouth. The guiding Report aroused such "virulent opposition" from people disagreeing on the definition of *cognitive science* that it wasn't published until five years later (H. Gardner 1985: 35–8; State of the Art Committee 1978).

Besides huge donations to several US universities, this programme included money for numerous conferences, workshops, and foreign visitors—and even for the first history of the field (H. Gardner 1985, p. xiii). (I was one of the many non-US citizens who benefited, being invited by Yale's AI group as a "Sloan visiting scientist" for June 1979.) The sponsorship was intended to support six different disciplines (State of the Art Committee 1978). However, we'll see in Chapter 8 that one of the six virtually disappeared from view.

The highly interdisciplinary Media Lab at MIT (13.v.a), founded in 1986, received its funding from a remarkable diversity of sources (Brand 1988: 12–13). Besides the usual suspects (governmental and philanthropic bodies, plus computer manufacturers), these included several Hollywood film companies, half a dozen TV channels, the LEGO toymakers, Polaroid, and Nippon Telephone and Telegraph. Clearly, these benefactors were driven by the practical/commercial potential of cognitive science, not its theoretical/scientific interest.

In the mid-1980s, AI research in the USA and UK was boosted by extra governmental funds made available because of the Fifth Generation scare—and the "Star Wars" scare too (11.i.c and v.b). Although most of this work was technological, some was psychological.

That's not to say that governmental funds were always easy to access. The Lighthill Report (SRC 1973) decimated AI funding in the UK, and US funding had been hit a couple of years earlier. In 1970 the Mansfield Amendment of the Defense Appropriations Bill required that all ARPA research (henceforth, DARPA: D for Defense) have some direct military application. The money didn't dry up. But blue-sky projects, or even perfectly feasible projects which looked non-military, were sidelined.

To be sure, "military" didn't have to mean missiles. For instance, AI's early work on virtual reality (VR), involving speech and telepresence, was funded by DARPA in the 1970s and 1980s (Brand 1988: 163). But others, besides DARPA committee members, had their own views about what research it was reasonable to fund with taxpayers' money.

In 1975 Senator William Proxmire of Wisconsin made himself a household name in the USA by announcing (in a newsletter sent to over 100,000 people) that he'd instituted the Golden Fleece awards for "unnecessary expenditures in the Federal Government". (Had he read *Gulliver's Travels*, he might have called them the "Lagado Academy" awards instead: Chapter 2.i.b.)

He started off worrying, as well he might, about the US Air Force spending $2,000 for each toilet seat put into a bomber. But he soon turned his attention to the research grants provided by the NSF (National Science Foundation) and NIH (National Institutes of Health). And there, his bluff common sense often let him down.

He had a lot of fun, for instance, mocking a grant project entitled "The Sexual Behavior of the Screw-Worm Fly". Unfortunately, he rarely got beyond the titles. His judgements (or rather, his ill-educated guesses) about the scientific, and even the practical, potential of candidate projects was so unreliable that the scientists he named as Golden Fleece recipients took to boasting about it on the Johnny Carson TV show. One NIH project scorned by Proxmire, because it involved a study of "Polish pigs", helped lead to a blood-pressure medicine used by millions of his compatriots. And the Media Lab's Aspen Movie-Map, built with $300,000 from DARPA and also castigated by Proxmire, led on to today's VR (Chapter 13.v.a and vi). Only rarely did he hit the nail on the head, as when he suggested that the money spent by NASA on SETI (the Search for Extra-Terrestrial Intelligence) might be better spent on searching for intelligent life in Washington DC.

With hindsight, science-policy researchers see the Senator as having done significant damage to the advancement of science in the USA (Guston 2000). Fortunately, the damage was temporary. Proxmire eventually moderated his views to some extent (R. C. Atkinson 1999: 416). Between 1976 and 1980, no NSF grant received a Golden Fleece award. Indeed, he even admitted, at a 1980 seminar on biological methods of pest control, that the work on the screw-worm fly had actually been important. Nevertheless, for a while (until the awards ceased in 1988) he did have significant influence. For example, he managed—by giving the dreaded Golden Fleece—to block funding of a

SETL project (Search for Extra-Terrestrial Life) proposed by NASA that had already been endorsed by a Congressional subcommittee (Dick 1993).

As for "artificial intelligence", that was a red rag to a bull. AI was declared a "racket" by Proxmire, on national TV (Raskin 1996: 4; V. Raskin, personal communication).

All the more reason, then, to agree with Bruner's observation (above) that even small amounts of money given with no strings are hugely valuable. There's no need to tell the funders that you plan to study the amatory antics of the screw-worm fly, still less to try to justify it.

6.v. Spreading the Word

A new church needs to spread the word, largely by epistles sent to the four corners of the earth. Similarly, science couldn't progress if scientists hid their lights under bushels. We saw in Chapter 2.ii.c how important publications are, both in forming the invisible college and in turning it into something more established. The history of cognitive science therefore involves its formative textbooks and journals, and the setting-up of relevant societies and conference series.

At the same time, the initiates must learn the church's rituals. For these not only help to make the new ideas familiar, but also bring them alive. In the case of cognitive science, this meant introducing people to computer programming *in practice*.

a. Training sessions

That last remark doesn't merely mean that if someone wants to build a computer simulation they need a computer to build it on. (Although that's true: the early history of cognitive science was hugely affected by who got access to computers. Rosenblatt, for example, was able to develop the perceptron, at a time when Cornell's Psychology Department had no suitable machine, only because he persuaded his aeronautical neighbours to let him work on theirs: 12.ii.e.) It also means that *using* computers gives one an added dimension of understanding, which can then naturally flow into one's thinking about other things—such as psychology.

This is a special case of what Bruner called "cognitive technology", and of what Gerd Gigerenzer (1991*b*, 1994) calls the "tools to theories" heuristic in scientific creativity. As one grows skilled in using a tool, it's experienced as available, or ready to hand, in a way it wasn't before. One aspect of that is that its use can now be attempted outside the context in which it was first encountered. Gigerenzer originally described this heuristic by reference to statistical tools, newly used in the context of experimental (including cognitive) psychology. But it applies also to cognitive science. The importation of heavy-duty mathematical statistics into connectionism, for example, wouldn't have surprised him (see 12.vi.f). And, what's relevant here, the heuristic applies to computational tools as well as statistical ones.

If the new gospel was to spread, therefore, people needed to get their hands on a computer—not just to read about them. Accordingly, in the summer of 1958 Newell and Simon ran a Research Training Institute on the Simulation of Cognitive Processes at RAND. As well as lectures and seminars, and demonstrations of three working simulations (the Logic Theorist, GPS, and EPAM), the Institute gave participants a chance to do some programming on the RAND machine.

They did this in IPL-IV, a programming language in which it's almost impossible for a novice—or anyone else, for that matter—to do anything of interest (see 10.v.b). But that didn't matter. Nor was it a disaster, although it was a drawback, that the eager novitiates might not be able to access a machine after returning home. For they would have learnt a hugely important lesson at RAND, one which could have been learnt in no other way.

Even the minimal programming that my fellow graduate students and I did at Harvard–MIT in 1962 (Preface, ii) awakened us to the clarity of thought, and the unforgiving precision, that's needed to make a program do what one expects. In that sense, there's more difference between someone who's never programmed at all and someone who's done only Mickey Mouse programming, than there is between the latter person and a world-class AI wizard.

So if the Institute did nothing else, it provided attendees with a wholly new vision of theoretical rigour. Hand-waving was still often unavoidable in psychology (as it is today). But it could no longer be complacently mistaken for accuracy.

However, the Institute did do something else: it gave people hands-on experience of specific programming/theoretical concepts, such as *instruction*, *subroutine*, *goal–subgoal hierarchy*, *list*, *search-space*, *search*, *iteration*, *transfer of control*, and so on. Under Newell and Simon's expert guidance, people learned to use these new tools and in so doing sensed their potential power. Even though what they could do on the RAND computer was hugely limited, the relevant concepts had got inside their heads and could then be used (*sic*) to think about psychological processes in newly creative ways.

Among those attending this pioneering summer Institute were Miller himself, the social psychologist Robert Abelson, the cognitive psychologist Roger Shepard, and the NewFAI researcher Bert Green (Gigerenzer and Sturm 2004: 25). Abelson would soon embark on highly influential work on the representation of plans, intentions, social roles, and emotional relationships (see 7.i.c). Shepard would later revolutionize the study of mental imagery (7.v.a). And Green, with Carol Chomsky and others, would soon add impetus to natural language processing by writing the BASEBALL program (9.xi.b and 10.iii.a). Evidently, the summer was well spent.

However, summer schools weren't enough. The new gospel couldn't spread really widely until most research universities had their own computer, and encouraged their psychologists—not just their aeronautical engineers—to use it. This would take many years (Aaronson *et al.* 1976).

The Harvard Center itself, as we've seen, didn't actually *do* much computer modelling—even though it possessed a computer. The reason wasn't just that Bruner wasn't computer-minded. After all, his co-founder Miller was enthusiastic about programming, and had enrolled for the RAND summer school. The major problem was the difficulty of getting the computer to work. On an average week in 1965–6, the Center's machine (a PDP-4C) saw eighty-four hours of use—but fifty-six of these were spent on debugging and maintenance. The Annual Reports for 1963–9 contained "several remarks of the type 'It is difficult to program computers.... Getting a program to work may take months' " (Gigerenzer and Sturm 2004: 26). The title of a 1966 technical report issued by the Center said it all: *Programship, or How to Be One-Up on a Computer Without Actually Ripping Out its Wires.* (This mid-1960s title owed its resonance to

Stephen Potter's best-selling, and hilarious, manuals on the rituals of *One-Upmanship* and *The Theory and Practice of Gamesmanship, or The Art of Winning Games Without Actually Cheating.*)

Given such problems, people who couldn't yet do anything else—such as graduate students—might be willing to persevere in learning the new technique. But established professional psychologists, with other (professionally respected) investigative skills at their command, would be tempted to give up—or not to bother even to try. So hearing about the computer modelling approach, and even playing around with it for a time on a (necessarily) primitive machine, didn't always lead to a conversion experience.

For many years, then, the take-up of computing by psychologists was patchy. (Some men of my acquaintance refused even to try, because they couldn't type—having always relied on their female secretaries to do that job for them.) In 1972, fourteen years after the summer Institute, Newell and Simon's epochal book *Human Problem Solving* appeared (7.iv.b). It could function both as a sermon, or pep talk, and as a training manual. But even then, the majority of researchers *in their own Department* remained reluctant to use the new tool (Gigerenzer and Sturm 2004: 27).

If this was true of CMU, it was inevitably true elsewhere. Across the USA and UK, the social scientists' take-up of the computational approach was slow. For example, when Feigenbaum and Julian Feldman tried to set up an AI group at the Berkeley Business School in the early 1960s they got little support from their fellow faculty members—despite having just published their hugely exciting *Computers and Thought* (see below). That's why Feigenbaum left Berkeley for the more welcoming Stanford in 1965 (Crevier 1993: 148). Psychologists, too, were slow to respond to the clarion call. It's true that in 1969, even before the publication of the Newell-and-Simon tome, the APA gave Simon their award for a Distinguished Scientific Contribution: clearly, then, some professionally influential psychologists valued this new approach. Nevertheless, it was as late as 1983 when a cognitive psychologist told me that computational psychology was still "too specialized" to be discussed by the newly founded BPS sub-group on the History and Philosophy of Psychology (see 7.vii.d).

A quarter-century after the brief training session at RAND, another seminal tutorial was run at CMU (see 12.vi.a). The motives were the same, but the tool-for-training was different. The 1986 summer school was focused on the relatively novel form of computer modelling called PDP (parallel distributed processing) connectionism. Well aware of the endemic professional inertia remarked above, the organizers were happy to welcome interested colleagues but (like the Jesuits?) even happier to welcome the youngsters. Indeed, we'll see that they went to great lengths to make their training manual affordable to impecunious graduate students.

By that time, the need for training sessions had long been recognized. RAND was the first of several devoted to NewFAI methods—from the early 1970s, including production systems (10.v.e and 7.iv.b). Little by little, the rituals had been spreading.

b. Library tickets

Early computational work was scattered over many different journals, of which only a few were psychological (Simmons and Simmons 1962). If psychologists were to be made aware of the new approach, *collections* of papers would be needed. And if students were to be attracted, accessible textbooks would eventually be required.

A slim volume on *Simulation in Social Science* appeared in 1962 (Guetzkow 1962). This contained contributions from Newell, Hovland, and Robert Abelson—but most of the topics were drawn from sociology, economics, or management studies. The same year saw Harold Borko's collection *Computer Applications in the Behavioral Sciences*, a textbook based on a course he'd been teaching at the University of Southern California (Borko 1962, preface). However, only a few chapters dealt with psychological themes (binary choice, language, and diplomacy). Borko's book wasn't going to set the world alight.

What did set the world alight, just one year later, was the ground-breaking collection *Computers and Thought* (Feigenbaum and Feldman 1963). This had been put together using some of the money from a $70,000 grant from the Carnegie Corporation, awarded to the two young editors to look into "the potential of artificial intelligence" (Newquist 1994: 176).

The editors, then at Berkeley, were both ex-students from CMU. Feigenbaum was the one who'd raised the Carnegie money. He was a psychologist in the process of moving over to computer science, and had already written a pioneering program under Simon's tutelage: a model of rote memory (Feigenbaum 1961). But it was the book which thrust him onto the international stage. It sold so well that some of the royalties were eventually set aside to fund a biennial Computers and Thought Award. The first recipient was Winograd in 1971, at the second meeting of IJCAI (the International Joint Conference on Artificial Intelligence), in London (10.iv.a).

The psychological models described in the collection included five that would be repeatedly cited later. These were Newell and Simon's Logic Theorist and General Problem Solver; Selfridge's Pandemonium and a 'learning' version thereof (Chapter 12.ii.d); Earl Hunt and Hovland's model of concept learning (1961; cf. Hunt 1962; and see 13.iii.f); and Feigenbaum's EPAM (1961).

The next important collection was ten years later: *Computer Models of Thought and Language* (Schank and Colby 1973). This provided several new papers by leading researchers, including one by the wunderkind Winograd (9.xi.b). In the opening chapter, Newell (1973*b*: 25) boldly declared that AI might as well be called "theoretical psychology". (He'd already said as much in the 920-page Newell-and-Simon book on problem solving, but that had a more restricted audience.)

In 1975 the label "cognitive science" was used in two more collections: the San Diego psychologists' *Explorations in Cognition* (D. A. Norman and Rumelhart 1975) and the AI-based *Representation and Understanding* (Bobrow and Collins 1975). Both referred to their field of interest as *cognitive science*. The latter did so up front: in its subtitle and in the Preface, which referred to "a new field we call *cognitive science*" (p. ix). The former saved this unfamiliar term for the very last page (and avoided using it on the publisher's mail-order leaflet, which I still have). On that final page, the San Diego pair declared: "The concerted efforts of a number of people from the related disciplines of linguistics, artificial intelligence, and psychology may be creating a new field: *cognitive science*" (p. 409). These were the first *books* to identify their theme in that way. However, the plural version—cognitive *sciences*—had been used for some time by Michie's group, and had appeared in print in Longuet-Higgins's response to the infamous Lighthill Report (SRC 1973: 37; see Chapter 11.iv.b).

"Cognitive psychology" had been labelled some years earlier. In 1967 Neisser (1928–) had published the first monograph explicitly devoted to this area. "I got letters", he recalled later, "from people saying that they were glad that I had given it a name, because they were interested in all the topics I considered, but the area had not had a theoretical identity" (1986: 278).

The Gestaltists—who, in the person of Kohler, had influenced Neisser during his postgraduate course at Swarthmore—were cognitive psychologists too, though not under that label. In other words, not all cognitive psychology is part of *cognitive science*. But these terms are often assumed to be equivalent, partly because Neisser, when he defined the new term, was a computationalist.

An ex-student of Miller's, he had now—like Miller himself—passed beyond " 'bits' and 'phonemes' and the like" (see i.b, above). He'd already declared an interest in computer modelling some years before—together with a healthy scepticism about the extent to which (then current) simulations could match human thinking (Neisser 1963). For instance, he'd complained that "When a program is purposive, it is too purposive," and that programs aren't emotional—because they don't "get tangled up in conflicting motives", as humans often do. Nevertheless, his book followed MGP in declaring that "Although information *measurement* may be of little value to the cognitive psychologist, another branch of the information sciences, computer *programming*, has much more to offer" (1967: 7–8; first italics added). This meant GOFAI—but GOFAI, of course, included the parallelist Pandemonium. Indeed, Neisser was greatly impressed by Pandemonium, and had already written two commentaries on it (Neisser 1959; Selfridge and Neisser 1960).

Compared with MGP's catholic manifesto, however, Neisser's book was narrow in compass. It focused on experimental research, and especially on the problems of coding and representation involved in cognition. Bartlett, the New Look, and Chomsky all loomed large. As a result, it was much more influential than Reitman's earlier *Cognition and Thought* (1965) which, although largely 'methodological', *did* tackle motivational conflict and emotion (see Chapter 7.i.b)—and creativity as well.

Admittedly, cognition—the term used with such defiance by Miller and Bruner (and Reitman too) only a few years earlier—was said to be "involved in *everything a human being might possibly do*", so that "*every* psychological phenomenon is a cognitive phenomenon" (1967: 4; italics added). This followed from Neisser's definition of it as "all the processes by which the sensory input is transformed, reduced, elaborated, stored, recovered, and used". Dreams, schizophrenic hallucinations, and hypnosis were briefly mentioned in the chapter on "visual memory". Freud's primary process thinking was described, following Pandemonium, as "a shouting horde of demons"; and Freudian slips were attributed to the parallel processing of learnt associations (pp. 298–9).

These personal matters were included, if only fleetingly, in part because Neisser had been influenced "quite a bit" by the Third Force psychologist Abraham Maslow (5.ii.a). Maslow had been his chairman during his first job, at Brandeis (Neisser 1986: 275). Indeed, he was "pleased with the book" because it had "a systematic scientific feel to it . . . [yet] also had a definite humanistic quality" (p. 279). And both "clinical" and "humanistic" colleagues approved of it (p. 280).

Nevertheless, his closing words admitted that he'd ignored purpose and motives, even though cognition and motivation—and personality, too—were probably "inseparable" (p. 304). His final sentence ran: "The study of cognition is only one fraction of psychology, and it cannot stand alone" (p. 305).

Clearly, then, library tickets would eventually be needed for more wide-ranging volumes, reaching as far as personality and hypnosis. As it turned out, those tickets wouldn't be usable for many years to come. (More accurately, they were usable already—but only for a handful of relatively unilluminating texts: see Chapter 7.i.a–c.) Meanwhile, the developing cognitive science was more cognitive than either Neisser or MGP had intended.

Neisser's text was computational through and through. The S–R psychologists, he said, were loath to make hypotheses about mental processes because they were "afraid that a separate executive would return psychology to the soul, the will, and the *homunculus*; it would be equivalent to explaining behavior in terms of a 'little man in the head' " (p. 295). But, recently, things had changed:

> It now seems possible that there is an escape from the regress that formerly seemed infinite. As recently as a generation ago, processes of control had to be thought of as *homunculi*, because man was the only known model of an executive agent. Today, the stored-program computer has provided us with an alternative possibility, in the form of the *executive routine*. This is a concept which may be of considerable use to psychology. (Neisser 1967: 295)

> A program is not a device for measuring information, but a recipe for selecting, storing, recovering, combining, outputting, and generally manipulating it . . . [This] means that programs have much in common with theories of cognition. Both are descriptions of the vicissitudes of input information. (Neisser 1967: 8)

In other words, the sixth tenet of behaviourism could be dropped.

It didn't follow, said Neisser, that computational psychology implies "a commitment to computer 'simulation' of psychological processes" (p. 9). Computer models couldn't do "even remote justice to the complexity of human [minds]". But this didn't matter. The crucial point was to realize that psychology is as much about "processes" as "contents" (the distinction that had eluded me in the Cambridge apple orchard, but which was made crystal clear by MGP: Preface, ii).

That done, questions could be asked in a new way. For example, Neisser's "analysis by synthesis" theory of speech perception suggested that the hearer actively constructs an internal model that matches the attended speech signal, whereas unattended signals are perceived more passively (1967: 193–8). (In the 1970s, Neisser's analysis by synthesis would influence some of the AI vision work done at the University of Edinburgh: A. Sloman, personal communication.)

For the record, Neisser's commitment to computationalism didn't last. He soon turned to ecological psychology, dedicating his next book (1976) to James and Eleanor Gibson, whom he'd encountered on moving to Cornell in the late 1960s. Quite apart from the Gibsonians' general suspicion of 'computation' (see 7.v.e–f), real-life situations are even less amenable to computer modelling than laboratory tasks are. Neisser now accused cognitive psychology of being "indifferent to culture", and feared that it "could become a narrow and uninteresting specialist field" (1976: 7; see Chapter 8). In particular, he said:

> The villains of the piece are the mechanistic information-processing models, which treat the mind as a fixed-capacity device for converting discrete and meaningless inputs into conscious percepts. (Neisser 1976: 10)

He didn't change his mind back again. In his late-century retrospective (Neisser 1997), he gave computational theories much less credit than he'd given them thirty years before.

Neisser's 1967 book convinced many professional psychologists that computational theories were needed. But it didn't score highly with first-year undergraduates seeking a user-friendly introduction. Gregory's 'New Look' *Eye and Brain* (1966) had already done that, as we've seen. But it had concentrated on vision, not (like Neisser's volume) on cognitive psychology as a whole. The ice-breaker appeared in 1972: Peter Lindsay and Donald Norman's *Human Information Processing.*

This was the first of several hugely influential textbooks to come from the San Diego stable, the most famous being the PDP 'bible' of 1986 (12.vi.a). The 1972 volume had been six years in the making, dating from when Norman (fresh from Bruner's Harvard Center) and Lindsay helped set up UCSD's first degree programme in psychology. After some teething troubles, "The course took hold, the enrollment climbed. But we were hampered by the lack of a suitable text. The only way to get one seemed to be by writing it ourselves" (Lindsay and Norman 1972, p. xvii).

Lindsay and Norman boldly subtitled their book *An Introduction to Psychology*: not cognitive psychology, but psychology *tout court.* Admittedly, fifteen of the seventeen chapters dealt with perception, memory, language, learning, problem solving, and choice. And although amnesia was considered (and aphasia mentioned in passing), there was no discussion of personality, whether normal or abnormal. Nevertheless, social psychology was given one full chapter, and motivation (including stress, conflict, and emotions) another.

Among the book's most memorable features were its pictures, especially Leanne Hinton's impressions of Pandemonium. Several chapters included demonic diagrams as entertaining as they were educational (see Figures 6.5 and 6.6). And the paperback cover bore thumbnail sketches of the two authors, clean-shaven and bearded respectively: each had horns, cloven feet, and a demon's tail—no favouritism here. It also carried a cogs-and-pulleys depiction of illumination—the pun was clearly intended. There was even a private joke for the readers of the MGP manifesto: a lone demon, carefully hammering a nail into the "H" of the title (Figure 6.7).

These engaging illustrations reminded readers that Norman and Lindsay's approach was *computational,* even though they didn't claim that *computer modelling* was necessary. Their amusingly demon-ridden volume soon made computational psychology into a familiar undergraduate experience. The pages bristled with introductory bibliographies and other helpful advice. It's a telling comment on how much things had changed since the 1950s that their advice included this:

> Do not shun the older issues [of the journals]. Because of the peculiar history of psychology, the most fascinating papers seem to have been published in the years around 1890 through 1910. (Lindsay and Norman 1972: 686)

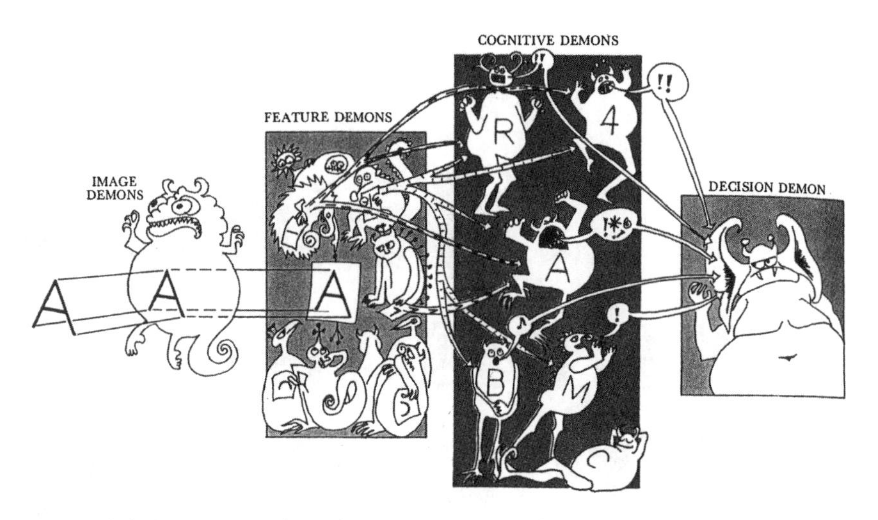

Fɪɢ. 6.5. Artist's impression of Pandemonium. Reprinted with permission from Lindsay and Norman (1972: 116)

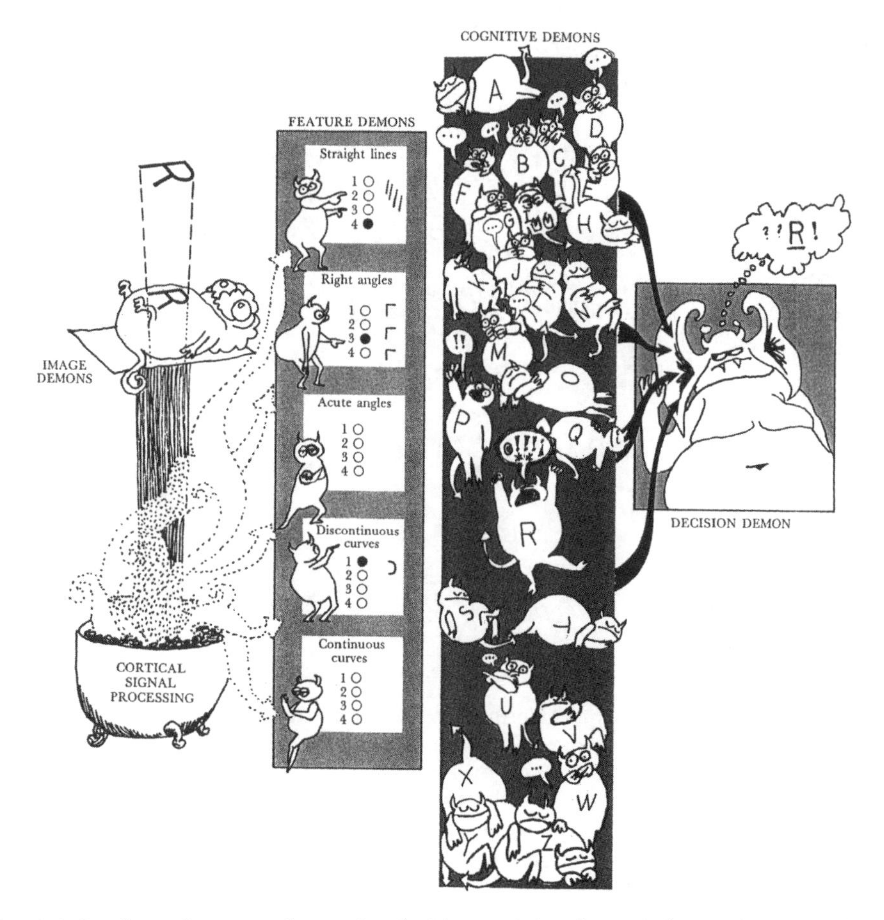

Fɪɢ. 6.6. Pandemonium at work. Reprinted with permission from Lindsay and Norman (1972: 125)

FIG. 6.7. Front cover of Lindsay and Norman (1972). Reprinted with permission

In other words, the elderly sitting tenants (Chapter 5.ii) should be invited to tea after all. (I can't resist backing up their claim by recommending George Stratton's paper of 1896: you'll rarely find more "fascinating" reading matter.)

Lindsay and Norman couldn't mention many programs other than Pandemonium, for only a few had then been written. (One of these was their colleague John R.

Anderson's HAM: see Chapter 7.iv.c.) By 1977, when the UK's Open University launched its first course in Cognitive Psychology, there were many more. One of the two set books chosen for the OU's course was my *Artificial Intelligence and Natural Man* (Boden 1977), soon adopted for other psychology (and AI) courses in the UK and USA. "Natural Man" was what the book was really about: it explored a wide range of psychologically relevant AI programs, detailing not only their successes but also their many failures, and asking how these might be overcome. The philosophical (and social) implications of AI and computational psychology were discussed too. In short, it was an exercise in *interdisciplinary* cognitive science.

Another disciplinary mix appeared a few years afterwards, a collection on *Mind Design* edited by the philosopher John Haugeland (1981*a*). His volume included papers by AI researchers, computational psychologists, and philosophers—and it forefronted some of the radical philosophical criticisms that were being made of the field. As a way of introducing cognitive science to outsiders it was highly successful, and was reprinted several times. (A revised edition appeared recently, now including papers on connectionism and dynamical systems theory: Haugeland 1997.)

In 1982 "cognitive science" featured in the final volume of *The Handbook of Artificial Intelligence*. Chapter 11 focused on GPS, EPAM, and semantic networks (10.iii.a); on a model of humans' "opportunistic" planning (B. Hayes-Roth 1980; B. Hayes-Roth and Hayes-Roth 1978); and on the psychologists' studies of belief systems and reasoning to be discussed in Chapter 7.i.a, i.c, and iv.c. The chapter was specifically described as "an introduction to cognitive science" (P. R. Cohen and Feigenbaum 1982, pp. xiv, 4).

Yet another interdisciplinary volume was supported by the Sloan Foundation itself. They sponsored Howard Gardner's history, *The Mind's New Science* (1985). This accessibly written book raised the profile of cognitive science, and doubtless drew many people into it as a result. Just as important, it helped those *within* the field to appreciate the links between the various disciplines.

By the mid-1980s, then, the epistles had multiplied. Their enthusiasm and interdisciplinarity had spread the gospel far and wide. Cognitive science might not be to everyone's taste. But at least it was visible.

c. Journal-ism

The eponymous journal *Cognitive Science* was launched in 1977 by three American researchers in "psychological" AI: Roger Schank, Eugene Charniak, and Allan Collins. This soon became the official journal of the Cognitive Science Society, established in 1979. The founding conference was in San Diego—in the same place and year, and thanks to some of the very same people, as the seminal interdisciplinary meeting on PDP connectionism (Chapter 12.v.b). The invited papers appeared in the journal, and also as a stand-alone volume (Norman 1981).

Even before that, the quarterly *Cognitive Psychology* had been founded in 1970, by Reitman in the USA. Noam Chomsky was on the Editorial Board, and so was Hunt, who later edited it for twelve years and is still on the Board as I write. And in 1977 Martin Ringle (at Vassar) and Michael Arbib (at UMass, Amherst) had started *Cognition and Brain Theory*, covering "philosophy, psychology, linguistics, AI, and neuroscience". (In 1982 this swallowed up Arbib's *Brain Theory Newsletter*, which had failed owing to neuroscientists' suspicion—in those days—of computational modelling: Chapter 14.vi.c.)

One journal which looked as though it was new wasn't new at all, except in its approach. The *Journal of Verbal Learning and Verbal Behavior* had been started in 1962, when stringent criticisms of S–R accounts of language were already being heard—from psychologists as well as Chomsky. From 1985 (volume 24), however, it appeared as the *Journal of Memory and Language*. The old title was too reminiscent of the behaviourist approach, which had still infused its pages in the early days but was now an embarrassment.

In mid-1960s England, the British Computer Society established a Study Group on Artificial Intelligence and the Simulation of Behaviour (AISB). Started by Clowes in 1964, within a decade this had become the AISB Society, under the interdisciplinary leadership of researchers at the universities of Edinburgh, Sussex, and Essex. The name was a psychological pun on Clowes's part: "A is B" captured the New Look ideas of visual interpretation and ambiguity, the focus of Sussex's AI research at that time (see Chapters 10.iv.b and 12.v.h).

The informal *AISB Newsletter*, now called the *AISB Quarterly* (and recently joined by the peer-reviewed *AISB Journal*), was started in the late 1960s. This low-profile serial carried psychological as well as technological items, and some telling spoofs too. One of these, by "Sir Grogram Darkvale FRS", featured the Lighthill Report (and appeared alongside a more serious critique: P. J. Hayes 1973; see 11.iv.b). Many others were credited to the pseudonymous Father Hacker, whose real identity was a closely guarded secret (*Now it can be told!*: it was Clowes). Father Hacker remains alive and well today, penned by various authors (his tart 'Guide for the Young AI-Researcher' is quoted in Chapter 13.vii.a). In other ways, however, AISB has changed. Over the years, the "AI" gradually swamped the "SB", so AISB's publications are now of more interest for AI scientists than for computational psychologists.

Across the English Channel, the high-profile journal *Cognition* was founded by Bruner's associate Jacques Mehler in 1971. It joined *Communication and Cognition*, founded in 1967 (with Piaget and Leo Apostel, among others, as advisory editors) by a research group at the University of Ghent. Both of these publications welcomed theoretical and methodological papers as well as experimental ones.

Perhaps the most important journal of all was Stevan Harnad's (1945–) interdisciplinary *Behavioral and Brain Sciences*, or *BBS*. And perhaps there's no "perhaps" about it: a quarter-century later, in 2004, *BBS* was the leader of *all* the behavioural science journals, according to the ISI citation index.

Harnad was a philosopher and psychologist, with strong interests in both computing and neuroscience. So he was exceptionally well suited to edit a wide-ranging journal. AI and computational psychology were featured in *BBS* right from the start. The first number carried Zenon Pylyshyn's (1978) paper on 'Computational Models and Empirical Constraints', and the first volume also included Haugeland's (1978) critique of "cognitivism". Volume 3 devoted a whole number to *The Foundations of Cognitive Science*, comprised of three hugely influential papers (Fodor 1980*a*; Pylyshyn 1980; Searle 1980).

One might even say that "Pylyshyn's paper" had *over thirty* authors, nearly all leading names in cognitive science (twenty-one are mentioned in this book). And much the same could be said about each of the other papers just cited. For *BBS* was a journal like no other—except *Current Anthropology*, which had pioneered the genre. It published up to four dozen peer reviews (plus the author's Reply) alongside each target article. That, itself, would have been chosen by a "substantial" number of referees as presenting "a controversial viewpoint worthy of argument and discussion from various subdiscipline perspectives" (Harnad 1978: 1). So publication in *BBS* not only started a debate. It provided commentary drawn widely from cognitive science as a whole.

Much later, Harnad would seek an even wider audience on the World Wide Web. He was one of the first scholars/editors to champion electronic publishing (Harnad 1995*a*,*b*; Fuller 1995). Indeed, he was the founder editor in 1990 of *Psycoloquy*, an international and peer-reviewed publication sponsored by the APA, which—like *BBS* before it—sends out widespread calls for comments on target papers. Thanks to the Web, interdisciplinary discussion is easier now than it used to be. But the *trust* that's needed for one scientist to build on another's work (2.ii.b) tends (in 2005) to accrue more strongly to the printed word, even if the electronic journal is peer-reviewed too. So *BBS* continues, albeit in both printed and electronic forms, and still plays a pivotal role.

Later, of course, further relevant journals would appear. Many were forbiddingly specialist. Others applied cognitive science to practical fields, such as law, medicine, or education. And some reflected unforeseen theoretical changes within cognitive science as such. For instance, one post-millennial arrival was *Phenomenology and the Cognitive Sciences*, a title that would have been unthinkable in the early days. But those I've mentioned here were the first to place cognitive science on librarians' shelves—and to help the invisible college become visible *to itself*.

7

THE RISE OF COMPUTATIONAL PSYCHOLOGY

Although the computational psychology of the late 1950s filled its few proponents with a messianic zeal (Chapter 6.iv), it wasn't yet an identifiable *movement*. The gospel had hardly begun to spread (6.v). So it wasn't mentioned by the President of the American Psychological Association (APA) when, in 1957, he looked down from a great height and identified "the two disciplines of scientific psychology" (Cronbach 1957).

These, Lee Cronbach (1916–2001) said, were the *correlational* and the *experimental*. The first focused on individual differences. It was exemplified by Sir Cyril Burt's work on IQ, and by comparisons between different ages and species, or different classes and cultures. The second, aiming at general laws/theories, included Sir Frederic Bartlett's studies of memory and the behaviourists' of learning. But for Cronbach it was methods, not topics, which defined a discipline:

> A discipline is a method of asking questions and of testing answers to determine whether they are sound... [It is our methods of enquiry] which qualify us as scientists rather than philosophers or artists. (Cronbach 1957: 671)

The "philosophers or artists" he had in mind were *other* professional psychologists. Cronbach bemoaned the fragmentation of psychology, but he was uninterested in many of the fragments. So the Freudians and Third Force movement, for instance, weren't mentioned (5.ii.i). They were tacitly included among "the agile paper-readers swinging high above us [like circus acrobats on a trapeze] in the theoretical blue, saved from disaster by only a few gossamer threads of fact" (p. 671). (Little did he know that those gossamer threads would cause a major scandal some fifteen years later: see 6.i.d and ii.d.) It was *scientific* psychology which concerned him.

Even those psychologists who shared a commitment to science, Cronbach complained, had for many years had little or nothing to say to each other:

> The personality, social, and child psychologists went one way, the perception and learning psychologists went the other; and *the country between turned into desert*. (Cronbach 1957: 673; italics added)

That was true. Indeed, any glimpse of an oasis was dismissively declared a mirage. For the two sides offered differing, often mutually contemptuous, accounts of psychological *explanation*: "nomothetic" for the generalizers, "idiographic" for the individual-oriented psychologists (see iii, preamble, below). Indeed, the invective sometimes reached fever pitch. Cronbach was hopeful, however: people were starting to see that one and the same topic—ego involvement, for instance—could be addressed from both sides (p. 682).

At that very moment, a third scientific discipline—namely, computational psychology—was emerging. For Cronbach's Presidential Address came halfway between the hugely exciting meetings of 1956 and 1958 (Chapter 6.iv.b). Methods based on information theory and computer modelling were being applied to perception, language, and memory by George Miller, Donald Broadbent, Jerome Bruner, Richard Gregory, Oliver Selfridge, and Frank Rosenblatt, and to problem solving by Allen Newell and Herbert Simon. And these researchers were now seeing themselves as joint pioneers in a new scientific project—with a new notion of psychological explanation.

It was by no means accepted by psychologists in general, however, that this approach might be fruitful in answering their questions. Indeed, over twenty years later a leading journal in cognitive psychology published a paper defending the—clearly, still heretical—view that AI concepts and methods offered "a new theoretical psychology" (Longuet-Higgins 1981).

By that time (the early 1980s), there were already more heretics around. For the first quarter-century of computational research in psychology had raised many new questions, and found some new answers. The second quarter-century raised the stakes still further. In those fifty years, the new approach was applied to countless different topics. (Even ego involvement: Section i, below.)

That being so, the story of how computational psychology matured can't be intelligibly condensed into a single narrative. Instead of trying to do that, I've selected six topics and followed each one chronologically. That is, the time line starts anew (from about 1960) with each of the relevant sections.

The first topic is personal psychology, including emotion. The next is language. Changing views on the nature of psychological explanation are featured in Section iii. Sections iv and v deal with reasoning and vision, respectively. And Section vi recounts the changing status—and interpretation—of nativism (the claim that some psychological powers are inborn). Finally, Section vii offers an overview of the development of computational psychology considered *as a whole.* It ends by comparing the state of psychology today with what Cronbach was describing in 1957.

A warning, before we begin: several important psychological topics have already been dropped like hot potatoes, and won't be picked up yet:

* I'll say almost nothing about natural-language processing (NLP) or neuroscience; these are tackled in Chapters 9.x–xi and 14, respectively.
* Similarly, learning is hardly mentioned: it was introduced in Chapter 5.iv.b–f, and is considered at length in Chapters 10.iii.d, 12, 13.iii.e, and 14.v–vi and ix.
* The nature of concepts, already raised in Chapters 5.iv.c and 6.ii.a–c, is merely touched on below (in Section ii.d), but is taken up again in Chapters 8.i.b and 12.x.
* Knowledge representation *as such* (e.g. logic, semantic nets, neural networks) is discussed at length in the AI-related chapters (10.iii.a and e, 12.iii–x, and 13.i); much of the research discussed in this chapter was deeply influenced by that work.

* As for memory, this is closely related to language use, learning, and knowledge representation; so it's touched on in Sections ii.d and iv.b–e, and in several other chapters too (2.x.a, 5.ii.b and iv.c, 6.i–ii, 8.i.b, 10.iii.a, and 12—especially 12.v.c–f, vi, and viii–x).
* Creativity isn't discussed here, but in Chapters 9.iv.f and 13.iv.
* I say very little (except in Section v.b–d) about connectionism: not only were most early (and some recent) models based on GOFAI, but connectionism will be discussed at length in Chapter 12. Many psychological topics are included.
* And evolutionary psychology, touched on in Sections iv.g and vi.d–f, is discussed at greater length in Chapter 8.iv–v.

As that list shows, psychological topics are discussed in virtually every chapter. (So Section vii's "overview" also contains multiple cross-references, and will be fully intelligible only after reading the whole book.) That's inevitable. It follows from the definition of cognitive science (1.ii.a) that although AI, in its various forms, is the *theoretical* heart of the field, computational psychology (and neuroscience) is its *thematic* heart.

7.i. The Personal Touch

Meetings and manifestos are all very well, but—as remarked in Chapter 6.iv.a—hard work is needed too. From the 1960s on, there was a lot of that. However, the *problems* were hard as well. Indeed, some pioneering efforts in computational psychology were hugely premature—namely, those aimed at the computer modelling of personality and emotion.

These themes had been of interest to several cyberneticists, including psychiatrists Warren McCulloch and Lawrence Kubie—who'd almost come to blows over the value of Freudian psychoanalysis (Chapter 4.iii.f). But because cybernetics had had little room for meaning, the *specific content* of anxiety neuroses and other personal phenomena was usually ignored. Even when it was addressed, by Gregory Bateson for instance, it was still considered in relatively general terms (4.v.d–e).

Symbolic computing seemed to offer a solution. Accordingly, GOFAI programs focused on personality, and personal belief systems appeared as early as the late 1950s. In 1962 they even enjoyed a dedicated conference at Princeton (Tomkins and Messick 1963). The main organizer was the psychiatrist Silvan Tomkins (1911–91), whose theory of "positive" and "negative" emotions—despite its subtle descriptions of phenomenology—owed as much to cybernetics as to Sigmund Freud (Tomkins 1963; cf. Sedgwick and Frank 1995).

The early birds, however, didn't catch the worm. Even considered as deliberate idealizations (inclined planes for psychologists), these systems were grossly oversimplified—although they did indicate some of the issues that would eventually need to be faced. By the time that Ulric Neisser defined cognitive psychology, albeit with "a definite humanistic quality", the initial interest in modelling personal matters had abated. With only a handful of exceptions, the practice was put on ice for a quarter-century. (However, a *cognitive*, though not explicitly computational, approach to psychoses and anxiety disorders would soon arise, and become increasingly influential: e.g. Beck 1976; Morrison 2002; D. M. Clark 1996.)

By the millennium, the ice had thawed. The computer modelling of emotions was now hotly debated even in the media. This was due in part to the 2001 Spielberg–Kubrick film *AI*, whose child-android David was an (unworthy) successor to HAL of *2001*. More importantly, it was due to popular writings by neurologists—whose general moral was that the human personality *can't* be compared with computers in any significant way. But that moral was disputed: the turn of the century saw computational theories of grief and mourning, and of hypnosis and clinical psychopathologies too. Cognitive science had come full circle, returning to topics discussed in its earliest days.

In brief, this section justifies the claim made in Chapter 1.ii: that "cognitive science" isn't merely "the science of cognition". And in comparing the early examples (described in subsections a–c below) with today's efforts (subsections d–i) it shows how far computational theorizing about personal matters has advanced in fifty years.

a. The return of the repressed

Psychiatry at mid-century relied heavily on the consulting-room couch, not only for diagnosis but also for therapy. Kenneth Colby (1920–2001), a psychiatrist and Freudian analyst as well as a GOFAI pioneer, tried to move the action from couch to computer.

In the late 1950s, at Stanford, he started work on a 'neurotic' program, which he improved over almost ten years (K. M. Colby 1963, 1964, 1967; Colby and Gilbert 1964). Unlike Ulrich Moser's group in Zurich, whose Freudian simulations would focus on 'energy-flow' (Moser *et al.* 1968, 1969), Colby tried to tackle *meanings*.

The program's "beliefs" were based on one of Colby's long-term psychoanalytic patients: a woman unable to admit to her unconscious hatred of her father, whom she believed to have abandoned her. Aiming to clarify Freudian ideas about repression, he modelled eight of the defence mechanisms listed in Chapter 5.ii.a. (Rationalization wasn't one of them, for reasons explained in subsection c, below.) Each simulated belief carried a number, reflecting its emotional importance (cathexis). Both *I must love father* (a prime diktat of the superego) and *I hate father*, for instance, had high emotional import. Each defence mechanism was selected in a particular class of anxiety-arousing circumstances, and each transformed the troubling belief in a different way. For example, applying Projection to *I hate father*, required merely that subject (*Self*) and object be switched, giving *Father hates me*. For Displacement, the program had to find an analogue of the father to be the new Object, and also weaken the verb (perhaps giving *I see faults in Raymond*).

Many psychological questions were raised, and some clarified, by Colby's early work (for details, see Boden 1977: 22–63). But few, if any, were answered. Even the best-developed version of the neurotic program, despite being complex for its time, was far too simple to advance understanding much.

Its major drawbacks were its crudeness in modelling anxiety and analogy. Anxiety can't be captured by semantic-clash-plus-numbers. It arises from a complex computational architecture that wasn't even minimally modelled until the end of the century (see subsection f). As for analogy, Colby's program—by means of the FINDANALOG procedure—treated this simply as *sharing x properties*, where *x* is a number that varied with circumstances. But even to explain intuitively obvious instances of neurotic displacement would require a much better theory of analogy than this. And psychoanalysts are able to see analogies where most of us cannot (even if they sometimes overdo it).

In the late 1960s, Colby switched from neurosis to paranoia. Although there was 85 per cent agreement between psychiatrists in diagnosing paranoia, there was less unanimity on its explanation. Colby drew partly on Freudian theory, but even more heavily on Tomkins's (1963) analysis of the emotions. The central idea was that paranoia is rooted in defence mechanisms whose goals are to protect the self against shame, the core negative emotion. This (so he said) accounted for common phenomena that otherwise seem anomalous, such as the occurrence of paranoid reactions after false arrest or on the birth of a deformed child (K. M. Colby 1977).

Colby saw his theory of paranoia as a significant "reconceptualization" of the syndrome. He viewed it "as a mode of processing symbols", where the patient's remarks "are produced by an underlying organized structure of rules and not by a variety of random and unconnected mechanical failures". That underlying structure consists of "an algorithm, an organization of symbol-processing strategies or procedures". In order to change it (to 'cure' the paranoia), "its procedures must be accessible to reprogramming in the higher-level language of the algorithm". In general, he said, "other types of psychopathologies might be viewed from a symbol-processing standpoint" (K. M. Colby 1975: 99–100). So paranoia was just an example: mental illness as such was the ultimate explanatory target.

To illustrate his theory, he wrote a program called PARRY (K. M. Colby *et al.* 1971; Colby 1975, 1981). This was a language-using system describable as ELIZA-with-attitude (10.iii.a), where the attitude was systematically grounded in its "anxieties" and "beliefs". It responded in a paranoid fashion to various keywords related to its particular danger themes. Again, these recalled one of Colby's patients: a man whose delusions featured dishonest bookmakers who claimed that he owed them money, and set the Mafia on him when he refused to pay.

PARRY soon became notorious for "passing the Turing test". Strictly speaking, this was inaccurate—as Colby realized (see 16.ii.c). But it's true that PARRY's responses were often diagnosed as "paranoid" by doctors unaware that they were dealing with a program.

Colby's aim in writing PARRY had been practical as well as theoretical. He hoped it might help in making therapeutic decisions, and also in training student therapists before they were let loose on real patients (K. M. Colby 1976). In effect, then, he was designing a virtual-reality system for training in psychotherapy, much as today's VR engineers are designing simulated 3D brains for training brain surgeons (see 13.vi.b). He'd had similar practically oriented hopes for PARRY's neurotic ancestor (K. M. Colby *et al.* 1966).

After moving to UCLA in 1974, he started using an ELIZA-like program to conduct initial diagnostic interviews at the Los Angeles Veterans' Hospital. The transcript would be discussed by the patient with Colby himself. Many people, he discovered (personal communication), found it easier to express their anxieties to a non-judgemental computer program than to an unknown psychiatrist (these were *initial* interviews).

In 1989 he retired to found a company marketing therapeutic software. One program was designed to help people decide whether they might be clinically depressed, and to advise them on what to do (e.g. to go to see a human doctor). Its seven "Lessons" dealt, for instance, with negative self-comparisons; mood and value; ideal standards; reprogramming oneself; and suicide and antidepressant drugs (K. M. Colby and Colby 1990).

The Manual opened with these three questions: "Are you sad for days or weeks at a time? Do you have a low opinion of yourself? Do you feel hopeless and helpless?" This was casting a wide net (fully 25 per cent of the US population, so readers were informed, are depressed at some time in their lives). And the Introduction closed with this seductive reassurance:

> The dialogue mode offers the world's first-ever computer therapeutic learning program for depression which allows you to express yourself freely in your own words *and which responds accordingly in natural language.* The program's dialogue responses are designed... to facilitate therapeutic learning through the emotional arousal stimulated by *real-life conversation.* (K. M. Colby and Colby 1990: 2; italics added)

If you know anything about ELIZA (10.iii.a), you'll know that this was a highly tendentious, not to say deceptive, way of describing the matter. True, the users could express themselves freely. And true, doing that—and reflecting on it (partly due to the promptings of the program itself)—might help to increase their self-knowledge and/or self-confidence. But a "real-life conversation" this was not. The program's responses were all essentially empty—and not just because *all* NLP responses are essentially empty (see 16.v.c). They were empty *even compared with other NLP systems,* for by the time this product was marketed there were many programs whose interlinking of verbal concepts was very much richer (see 9.x–xi).

The OVERCOMING DEPRESSION program was used on the mainland and in Hawaii, by the US Navy and the Department of Veterans Affairs (Chang 1993). In addition, from 1992 it was sold on the High Street (as "DEPRESSION 2.0")—to individuals who would be using it (on an IBM PC) without supervision from a psychiatrist. As one would imagine, the press had a field-day. Challenged by one journalist, Colby retorted provocatively that programs could be better than human therapists because "After all, the computer doesn't burn out, look down on you, or try to have sex with you" (quoted in Turkle 1995: 115).

This was just the sort of thing which the computer scientist Joseph Weizenbaum, the author of ELIZA, had denounced as "obscene" (see 11.ii.e). There was little love lost between him and Colby. Indeed, I witnessed a heated public interchange between them at the third international AI conference, held at Stanford in 1973. I also had many private conversations with Colby (and one with Weizenbaum) in 1972 and 1973. Each was bitter in his comments about the other.

This antagonism was partly grounded in a priority dispute concerning ELIZA. In the early 1960s Colby had written a similar program called DOCTOR. (DOCTOR was soon supplanted by a more interesting "dialogue" program, reported by Colby at the very first international conference on AI: K. M. Colby and Smith 1969. That system was less empty-headed than either DOCTOR or ELIZA: like PARRY, it had an artificial belief system on which it drew in responding to the human's remarks.) The brief publication that introduced DOCTOR to the world appeared shortly *before* Weizenbaum published on ELIZA (K. M. Colby *et al.* 1966; Weizenbaum 1966). It's true that they'd discussed these matters together before either of those publications appeared. But Colby certainly felt that the priority was his (personal communication). However, it was ELIZA who/which became famous.

Mostly, however, Weizenbaum's antipathy was due to his worries about "dehumanization". When he learnt of Colby's plans to develop therapeutic programs, he was appalled. Such uses, he said, would offer patients a "profoundly humiliating" self-image (1972; 1976, esp. 268–70).

Colby's response was that it's "dehumanizing" to herd thousands of patients into understaffed mental hospitals where they will hardly ever see a doctor (K. M. Colby 1967: 253). For him, the proof of the pudding was in the eating. By the 1980s, when computers had actually been used in psychiatric contexts, he could look at the clinical facts. Investigating computer-assisted therapy for stress, he found clear advantages and no obvious bad side effects (K. M. Colby *et al.* 1989).

Whether there were *non-obvious* bad side effects is another matter. The Third Force psychiatrist Rollo May had already reported that his patients' ability for self-control had been damaged by science's tendency "to make man over into the image of the machine" (May 1961: 20). Science in general, and behaviourist psychology in particular, had no room for concepts such as freedom, deliberation, purpose, and choice. Given the high cultural status of science (which would soon be resisted by the counter-culture: 1.iii.c–d), the result was that these concepts were insidiously downgraded—not to say denied. And, said May, if people didn't believe themselves to be capable of autonomous choice then they were unlikely to try to engage in it—with the result that they fatalistically accepted their current life situation instead of trying to change it. This "sapping of willing and decision" was bad news, since—like all the Third Force psychologists—he took people's conscious decisions to be crucial if therapy was to succeed (see 5.ii.a).

Colby's AI clinician may have had a similar effect. Some unsupervised users, perhaps with a tenuous grip on reality in the first place, may have imagined that there was genuine understanding there. And/or they may have tacitly inferred that they themselves were "no more than a machine"—where their concept of "machine" *did not* include the subtleties of cybernetics or computation. If so, their self-image and sense of personal responsibility could have been undermined. (This is true even though many cognitive scientists see their approach as *supporting* ascriptions of "freedom" to human beings: see subsection g, below.)

Whether this would actually happen or not would depend to some degree, of course, on the prevailing Zeitgeist. In the early 1970s, when Colby's interview program was first used in a real-life clinical setting, computers were still hugely unfamiliar and the counter-culture was at its height. But the Zeitgeist can shift. Sherry Turkle (1948–), another psychoanalyst (and a colleague of Weizenbaum's at MIT), noticed over fifteen years that there was a change in the attitudes of MIT students invited to discuss computer psychotherapy.

At first, there was heated disagreement with the project: almost all the students were highly sceptical. By 1984, the Colby fans were still a minority, although there were twice as many. But by 1990 the youngsters "saw nothing to debate" (Turkle 1995: 115). They now regarded therapy programs as like self-help books (some of which are actually helpful, some not), but better—and very much cheaper, and less threatening—than a human therapist. Turkle reports that this was part of a widespread disillusion with psychotherapy in general and Freudian theory in particular: "In the main, [MIT] students in the 1990s did not consider a relationship with a psychotherapist a key element in getting help" (p. 116).

MIT students, of course, aren't a representative cross-section of society. It would be interesting to know what the general public, today, would think of Colby's clinical aims. (Or, for that matter, how they would react to current attempts to use "social" robots to help autistic children, not least to mediate communication between the child and other human beings: see Chapter 13.vi.d.) As for Colby's theoretical aims, they were greatly premature—but, as we'll see, not intrinsically absurd.

b. Argus with 100 eyes

Walter Reitman (1932–) started modelling the mind at much the same time as Colby, but his approach was very different (Reitman 1963; 1965, esp. chs. 8–9; Reitman *et al.* 1964). He saw personality as "a problem-solving coalition" (1963: 81–6), incorporating multiple "distinct and perhaps conflicting needs and motives". And instead of relying on pure GOFAI sequentialism, he constructed a simple version of what would now be called a hybrid system (Chapter 12.ix.b).

His Argus program addressed Neisser's (1963) complaint that current problem-solving computers were too single-minded, non-distractable, and unemotional. This complaint, said Reitman, is relevant because "most theories of personality... allow for the possibility of several independently originated activities simultaneously under way" (1963: 77). The system as a whole must organize those simultaneous motives and activities sensibly. If it can't, various types of psychopathology will ensue.

The task for the simulator, then, is to describe the "intrapsychic communication" involved:

> [The theorist must specify on his code sheets] the manner and form in which information, commands, and requests at one level in the system are transmitted elsewhere... [His difficulties are increased if he has to consider] a system in which sub-systems are able to do such things as induce concealment or refuse access to information which other systems require to achieve their aims. (Reitman 1963: 73, 85)

These things must be specified because the computer modeller can't assume that the right hand knows what the left hand is doing: the right hand must be told. Today, this is a banality (though no less important for that). In Reitman's time, it wasn't.

Reitman, in the early 1960s, was a colleague of Newell and Simon at the Carnegie Institute of Technology. He was greatly influenced by them, not least by their interest in trying to match the tiniest details of behaviour (1965: 32–3). But Argus, despite being written in their new language IPL-V (10.v.b), wasn't another GPS.

Named in honour of the mythical Argus with 100 eyes, this GOFAI program (like Pandemonium) was conceptualized as a parallel system involving many simultaneously active elements. Reitman compared these to Donald Hebb's cell assemblies, but confessed that he hadn't been able "to imagine any way in which a system consisting entirely of Hebbian cell-assemblies might be made to [account for goal-directed thinking]" (1965: 208). (A similar failure of imagination still besets us today: Chapter 12.viii–ix.) So he implemented a "mixed" system, "linking a limited sequential control to an underlying structure of active [Hebbian] elements" (p. 209).

Moreover, Argus wasn't sequentially single-minded, like GPS. While it was following goal A, it could be distracted onto goal B by some content (idea) that had just arisen in its processing. That distraction could be temporary or permanent. At each point, it had to decide which of the competing alternatives it would follow.

Reitman's aim was to model how people recover from interruptions of various kinds. For instance, someone who's been distracted by an anxiety-ridden idea can usually pick up where they left off. Sometimes, the idea can be ignored altogether—but not if it's urgent and/or important (see subsection f below), nor if one's already in a highly emotional state. Again, something being taken for granted may turn out be false, so that one must reorganize one's thoughts about how to reach the goal (see 13.i.a). These self-organizing recoveries, Reitman argued, require top-down influences: what's needed is "a sequential executive able to some extent to modulate and focus the changes taking place in the cognitive structure" (1965: 203).

Argus illustrated what he meant. One of its first tasks was to complete very simple analogies. For instance, given *hot:cold::tall:(wall, short, wet, hold)*, the program would choose *short*. Its basic strategy is shown in Figure 7.1.

As the caption indicates, the activation states of the "Hebbian" elements (implemented as list structures) varied throughout the run (1965: 212 ff.). The numbers representing activation and inhibition weren't summed: a single element, or concept, could be represented as high on *both*. The executive would recognize this as a conflict, and decide (according to the current strategy and activation patterns) whether it should be resolved immediately, postponed, or tolerated. The elements could be affected by non-firing too: long-quiescent concepts wouldn't fire spontaneously, although they could be activated directly by some other element. These continuous state changes meant that a problem might be addressed differently at different times.

Many analogy problems, however, are more difficult than this. Reitman pointed out that, faced with *Sampson[sic]:hair::Achilles:(strength, shield, heel, tent)*, someone could get the right answer in several different ways (1965: 227–8). These include reliance on word association (Achilles' heel), superordinate category (hair and heel as body parts), and the concept of vulnerability (not straightforward: Samson's hair was his strength, Achilles' heel his weakness). In other words, several strategies are available for completing this analogy correctly. Which one will be chosen, and why? And will it be followed through to the bitter end, or abandoned (temporarily/permanently) in mid-stream?

Of course, Argus could have been set up beforehand to solve this problem just as it solved the *hot:cold::tall:?* puzzle. But Reitman's point was that, at least for the third strategy, humans have to think about it. To enable Argus to do that would require (among other things) giving it a mass of general knowledge that's not tagged for any particular goal, and wide-ranging ways of recognizing the relevance of specific items in it. Accordingly, he hoped to develop Argus so as to study cognitive dissonance and language comprehension (1965: 230–53), and to explore individual differences—between people, and between one person's thinking at different times.

Argus didn't go unnoticed. For instance, Neisser (1967: 299–300) praised it as a rare attempt to combine parallel and sequential processing (see Chapter 12.ix.b). And I referred to it in outlining a computational version of William McDougall's personality theory, which was broadly consonant with Reitman's (Boden 1965: 14, 16; 1972: 227).

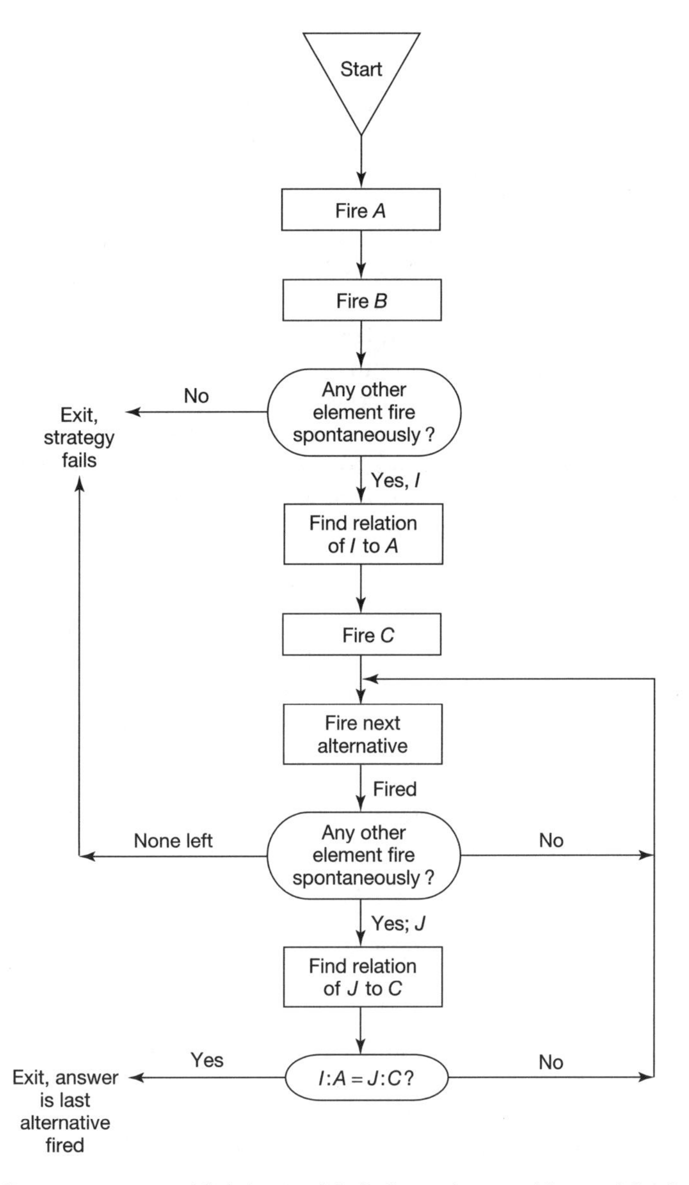

Fig. 7.1. Rudimentary strategy (slightly simplified) for analogy problems of the form *A*:*B*::*C*: (W, X, Y, Z). When an element is *fired*, activation and inhibition increments are added to related elements, altering their states. Redrawn with permission from Reitman (1965: 211)

However, the program was soon near-forgotten. It was ignored in the ice-breaking textbook of cognitive science (Lindsay and Norman 1972), despite the authors' enthusiasm for the parallelist Pandemonium. This was partly because it wasn't tied closely to experimental evidence. More to the point, Reitman's hopes for future versions of his program hadn't been fulfilled—or even approached (1965, ch. 9). The problems were far too difficult.

c. From scripts to scripts

Yet another early bird was Yale's Robert Abelson (1928–). He did more than anyone else to bring (not just personal but) interpersonal issues into computational theory.

A few others, to be sure, were modelling interpersonal issues. In anthropology, for instance, Anthony Wallace (1965) was applying TOTE units to socially organized behaviours such as driving to work (Chapter 8.i.a). In psychology, the best-known were John and Jeanne Gullahorn (at Berkeley's School of Business Administration), whose HOMUNCULUS program was featured in the widely read *Computers and Thought* (Gullahorn and Gullahorn 1963, 1964, 1965).

Their approach was very different from Abelson's, being based on George Homans's positivistic social psychology. Homans relied on five "axioms", including (2) *The more often within a given period of time a man's activity rewards the activity of another, the more often the other will emit the activity*, and (5) *The more to a man's disadvantage the rule of distributive justice fails of realization, the more likely he is to display the emotional behavior we call anger* (Homans 1961: 64, 75). No mental causation, no psycho-logic... just behaviourism, pure if not simple. In short, the Gullahorns offered a computer model of psychology but *not* an example of computational psychology.

Abelson himself was a convert to computationalism. But this wasn't a Damascene conversion. It happened gradually, as he became increasingly dissatisfied by the constraints of "Newtonian" theories (Chapter 5.i.a).

By the late 1950s, he was already acclaimed as a leader of operationalist social psychology. With Carl Hovland and others, he'd been studying the *coherence* of beliefs—assessed not by logic but by "psycho-logic", in which values and emotions play a major role (Rosenberg *et al.* 1960). So *anger*, for instance, might render a certain interpersonal response psycho-logical, if not strictly logical.

Psycho-logic was also central to Abelson's then popular theory of "cognitive balance". Inspired by the Gestaltist Fritz Heider (see 5.ii.b), this concerned how attitudes are changed by value clashes (Heider 1958; R. P. Abelson and Rosenberg 1958). Suppose a highly valued person says something you disagree with: how does this alter your attitude to them and/or to the proposition they've asserted? To cut a long story short, Abelson held that the high value is lowered, and the low value raised. *Just how much* depended on numerical calculations of the "balance" between positive and negative (lots of positivistic brownie points there!).

Attitude change—and resistance to it—was a key topic for 1950s social psychology. The central discussion was Leo Festinger's theory of cognitive dissonance (Festinger 1957; Brehm and Cohen 1962). This taught, for instance, that someone who has bought an expensive item will actively seek information that confirms its quality, while avoiding comparisons that favour alternatives. By the mid-1960s, Festinger's ideas were being described as "probably the most influential ideas in social psychology" (R. Brown 1965: 584).

Dissonance theory had swept the popular press too. For besides illuminating the power of advertising, it seemed to explain certain aspects of the strange behaviour of members of eschatological cults—and Christianity.

The newspapers had reported that a cult leader in the USA's mid-West was prophesying her hometown's destruction by flood. Her followers were abandoning their jobs, to enjoy life while they could. Festinger predicted that when the prophecy failed, the group's commitment would be *increased*, and 'reasons' found to 'explain' the unexpected non-event.

He was right. Having infiltrated the cult with two of his colleagues, he was present when the leader reported "a message from the Guardians on the planet Clarion" promising a midnight rescue by flying saucer some days later. The saucer's non-arrival prompted frozen shock (and frantic time checking), followed by fearful disappointment. But this didn't last. Before dawn, another "message" was reported (couched in suitably portentous quasi-biblical language—Festinger *et al.* 1956: 168 ff.). The flood, it announced, had been averted by the faith of the cult members. Immediately, they started spreading the happy news—even to the media, which initially they'd shunned.

For Abelson, however, even such telling anecdotes as this one weren't enough. Other psychologists had raised various 'factual' problems facing Festinger's theory, noting that we sometimes *favour* (a low level of) dissonance: think of Robert Herrick's poem 'A sweet disorder in the dress'... But Abelson was more interested in a theoretical problem, namely, that Festinger's approach predicted behaviour without really explaining it. (Similarly, Homans had predicted the "behavior we call anger" without really explaining it.) *How* someone comes up with this reason, or that one, wasn't considered. (Why say that faith had prevented the flood? Why not say that the Guardians had been mistaken? Or teasing?) The same was true of Abelson's own brainchild, balance theory. This said nothing about *how* evaluations can be changed, or *how* this can trigger a host of implications in the person's mind.

In the early 1960s, then, Abelson—shockingly—abandoned operationalism for mental processes. The focus of his address at the Princeton meeting was on inner information processing, aka thought. He focused on everyday cases of "hot" (value-laden) cognition whose emotional temperature is lower than it is in neurosis. And he described a program that applied *rationalization* and *denial* to input beliefs whose valuation clashed with pre-existing attitudes (R. P. Abelson 1963; Abelson and Carroll 1965).

Colby's neurotic program had modelled denial, but not rationalization—and that's no accident. For rationalization is more taxing. One must generate a belief with a different subject matter, not merely a logical/grammatical transform of the original. And the beliefs must be linked intelligibly. Abelson's "hot cognizer" selected one of three rationalizing subroutines: REINTERPRET FINAL GOAL, ACCIDENTAL BY-PRODUCT, and FIND THE PRIME MOVER. As these labels suggest (though over-optimistically: see 11.iii.a), this involved reasoning about cause and personal responsibility. His concern with these everyday linkages between beliefs would inform all of Abelson's later work.

Soon after Princeton, he developed an example he'd briefly mentioned there: the effect of different political attitudes on the acceptance of new information. Besides being a committed political campaigner for the Democrats (personal communication), Abelson was professionally interested in how politics works. For instance, he'd helped build a computer simulation of the strategies informing the recent US presidential campaigns (Da Sola Pool *et al.* 1965). Now, his IDEOLOGY MACHINE modelled the internal coherence of the thought processes of the Democrats' arch-opponent, Senator Barry Goldwater (R. P. Abelson and Reich 1969; Abelson 1973: 287–93).

Putting this in the terms that had recently been used in Bruner's *A Study of Thinking*, Abelson was investigating the categories, heuristics, and mental strategies employed by people with differing political viewpoints. Putting it (anachronistically) in the terms that would later be used by Philip Johnson-Laird, he was trying to define distinct mental models (iv.d–e, below).

What's especially relevant here is that he was beginning to see these categories/heuristics/models as *sharing* some underlying computational structure. Goldwater and his Democrat opponents favoured very different political dress. But according to Abelson, the contrasting fabrics were cut with similar computational patterns. Much as an anthropologist may seek universal structures and processes underlying a profusion of cross-cultural particularities (see 8.vi), so Abelson was doing this within a small part of his own culture.

When Goldwater ran for President, Democrat car stickers proclaimed, "In your guts you know he's nuts." (This was a riposte to the Republicans' "In your heart, you know he's right.") However, Abelson wasn't trying to ridicule him. On the contrary:

> The tendency to caricature and trivialize the motives and character of the enemy and to glorify—but also trivialize—the motives and character of one's own side has often been remarked by students of the human condition. My purpose is to try to anchor the phenomenon of ideological oversimplification in general psychological theory and data. (1963: 287)

Abelson's new model of attitude change was designed to generate politically plausible answers to these questions: *Is the news-report of event E credible? If and when E, what will happen? And what should political actor A do? When E happened, what should A have done? How come E? (That is, what caused E, or what is E meant to accomplish?) Sir, would you please comment on E?* (R. P. Abelson 1973: 290). For instance, how would Goldwater assimilate a news item such as "The East Germans have erected the Berlin wall" or "The USA has attacked neutral Vietnam"?—And how would the Communist leader Nikita Khrushchev respond to the very same items?

The program generated its answers by means of an information-processing mechanism called a *master script*. Abelson likened this to the Gestalt ideas of "good form" or "cognitive balance" (p. 289). The Goldwater version contained an internally coherent master script saying (among other things) that *if the Free world really uses its power then Communist schemes will surely fail* (p. 291). The seven core concepts of this master script—and, on Abelson's view, of Goldwater's political thinking—were: *Fuzzy liberal thinking*; *Our call to action*; *Free World paralysis*; *Use of Free World Power*; *Communist schemes*; *Free World Victory; Communist Victory*. Khrushchev's master script would contain partly different concepts, linked in somewhat different ways. Both statesmen, however, would accept the schematic belief that *The enemy attacks and subverts, and we defend, but never the other way round* (p. 290). Hence the problem, for Goldwater though not for Khrushchev, in assimilating and explaining the news item about the US attack on Vietnam.

To make this model work, Abelson had had to consider 'cold' cognition. He'd been forced to ask (again) how a belief can imply attributions of personal responsibility, or predictions of causal consequences. In 1970, then, his research shifted to embrace belief systems *in general*. He studied their internal structure, and how to integrate new information without compromising the system's core.

(Abelson had already dipped his toe into these waters, in his research on "implicational molecules": R. P. Abelson and Reich 1969. This work followed Heider's studies of 'personal' perception, mentioned in Chapter 5.ii.b. It was an early example of attribution theory, which became a prime topic in 1970s social psychology: e.g. Jones *et al.* 1972; Schmidt 1976.)

Very soon, he started to consider the internal structure of *intentions*, *plans*, and *roles* (1973). For he'd found that when his IDEOLOGY MACHINE generated implausible responses this was often due not to its factual ignorance but to its unstructured representation of these basic concepts. Much influenced by the cognitive science manifesto (6.iv.c), he outlined the ways in which two actors, each with their own agenda (plan), can intervene in the plans of others, whether to assist or to obstruct them. In other words, he explored the psychological structure of cooperation (and sabotage).

But he gave up, for a while, on computer programming. Like the manifesto authors, he now offered a schematic computational approach rather than a functioning model. That was no accident: the problems involved were way beyond the state of the art. It wasn't until the 1980s–1990s that AI researchers were able to model cooperation (and the lack of it) between autonomous mindlike agents (13.iii.d). When they did, they had to consider the problems of mutual plan-alignment that Abelson had sketched in this early essay.

Based on his analysis of plan-structures, Abelson defined a wide range of interpersonal concepts called *themes*. Examples included cooperation, betrayal, humiliation, victory, love (of which, more in subsection f), and many more (1973: 315–31). These weren't defined by dictionary entries, nor even by conceptual analysis. Rather, they were defined systematically, by placing what seemed to be the most appropriate ordinary-language words in the various locations available within a formal matrix. The dimensions of the matrix concerned actors' interests in, and abilities to influence, other actors' plans (see Figure 7.2).

Sentiments Toward Other \ Influence of Actors	Neither Influences Other	One Influences Other	Both Influence Other
Some Positive, No Negative	Admiration	(T_1) Devotion (T_2) Appreciation	(T_3) Cooperation (T_4) Love
One Actor Negative	(T_5) Alienation (also, Freedom)	(T_6) Betrayal (T_7) Victory (also, Humilation) (T_8) Dominance	(T_9) Rebellion
Both Actors Negative	(T_{10}) Mutual Antagonism	(T_{11}) Oppression (also, Law and Order)	(T_{12}) Conflict

FIG. 7.2. A taxonomy of themes. Adapted with permission from R. P. Abelson (1973: 320)

In other words, he was exploring *a structured space of computational possibilities.* These interpersonal themes, and the "scripts" defined in terms of them (see below), identified crucial aspects of a wide range of intelligible/plausible human stories. (That's why I said, at the close of Chapter 6.iv.d, that what Bruner terms the "blah blah blah" of human narratives includes the matters addressed by Abelson.)

For example, the box labelled *betrayal* defined a theme in which "actor F, having apparently agreed to serve as E's agent for action A_j, is for some reason so negatively disposed toward that role that he undertakes instead to subvert the action, preventing E from attaining his purpose" (p. 323). This abstract schema covered many varieties of betrayal, definable by varying the relative power of F and E and/or the importance (to E) of the thwarted plans. So abandonment is a specially nasty form of betrayal, since only the strong can abandon and only the weak can be abandoned; by contrast, anyone can let down, or be let down (Boden 1977: 80–6).

Abelson's 1973 paper ended by defining a number of empirically familiar sequences-of-themes, called *scripts*. These were less specific than the "master scripts" used by the IDEOLOGY MACHINE. Instead of fears about Communist domination, they included everyday notions such as *rescue, blossoming, turncoat, end of the honeymoon, revolution, romantic triangle, alliance*, and *the worm turns*. The relevance to personal (and even political) matters is intuitively obvious. But Abelson defined these scripts as coherent successions of formally specified themes. One theme leads "naturally" to the next as the relations (of power, success, and/or sympathy) between the two actors change. He suggested that whereas themes are probably human universals, anthropology might show some scripts to be culturally specific (see Chapter 8).

From the mid-1970s on, Abelson's work became increasingly influential in computational psychology and GOFAI. I've found, however, that his name is less well known than his ideas, having been largely eclipsed by that of Roger Schank (1946–). Abelson had used Schank's Conceptual Dependency (CD) theory in his paper on belief systems, to formalize the notion of intention and to express various causal/teleological implications (1973: 293–310: see Schank 1973). Soon afterwards, when Schank moved from Stanford to Yale, the two men began a long-lasting collaboration.

They soon published a highly influential book, *Scripts, Plans, Goals, and Understanding* (Schank and Abelson 1977). These "scripts" weren't the same as the scripts of Abelson's 1973 paper (rescue, and the like), or as the master scripts of the IDEOLOGY MACHINE. Rather, they were defined as *stereotyped ways of behaving*—in a hamburger restaurant, for example. That is, they were closer to the concept of *role*.

The social psychologist Roger Barker (1963, 1968) had recently rejected roles *considered as internal mechanisms generating an agent's behaviour* in favour of automatic triggering by "situational" cues. So he spoke of the behaviour-triggering power of clothes, furnishings, buildings, and ritualistic behaviour—such as greeting. Simon himself had taken a similar position, in his 1969 fable of the ant (Section iv.a, below). However, most computational psychologists at the time ignored Simon's ant. (They ignored Barker too, because as a *non*-computationalist he took it for granted that behavioural triggers can function, without asking precisely how.) Even those who did take Simon's fable seriously tried to find 'antlike' ways of accommodating scripts (see Section iv.b–c).

Both psycholinguistics and NLP were hugely influenced by Schank and Abelson, despite CD theory's many shortcomings (Chapter 9.xi.d–e). Their ideas helped fuel experiments on psychosemantics, asking what implications people spontaneously draw on hearing a sentence (R. P. Abelson 1976, 1981*a*; see also Schank and Langer 1994). In a roundabout way, philosophy was lastingly affected too: it was the script-driven Yale programs which were mocked by John Searle in his notorious Chinese Room argument (16.v.c).

Within professional social psychology, where Abelson had already been a 'name' in the late 1950s, recognition was more patchy. He received Distinguished Scientist awards for Experimental (*sic*) Social Psychology in 1990 and for Political Psychology in 1996. But social psychologists in general were less ready to accept computational ideas than their 'cognitive' peers. (Even Bruner reneged eventually: see 6.ii.d and 8.ii.a.) Moreover, we saw in Chapter 6.i.d that many in the 1970s argued that their field lay within history or hermeneutics, not science. It was at that very time that Abelson published his accounts of scripts (in all three senses). In brief, he was following an intellectual stream very different from either of the then fashionable approaches to social psychology—namely, operationalism and hermeneutics.

Nevertheless, it was largely thanks to his own early influence that he was able to say, in 1973:

> The modern trend in both experimental and social psychology is away from a behavioristic emphasis upon stimuli and responses toward a Gestaltist focus on cognitive capacities and performances, with experimental psychologists talking of information processing and social psychologists of cognitive consistency and causal attribution processes. (R. P. Abelson 1973: 287–8)

The Newtonian ramparts listed in Chapter 5.i.a were rapidly crumbling.

d. Emotional intelligence

One of the high priests of GOFAI, namely Simon (1967), acknowledged early in the game that emotion is integral to intelligence. (He was responding to Neisser's 1963 criticisms of GPS: Chapter 6.v.b.) It's needed, he said, to set our priorities when choosing which goals to follow. Purely 'intellectual' reasoning tells us what can be done, and how, and what might happen next . . . but not *what to do*. (Compare David Hume's " 'Tis not contrary to reason to prefer the destruction of the whole world to the scratching of my finger": Chapter 2.x.a.)

However, Simon didn't follow this up. In both GPS and production systems (Section iv.b below), top-level goals were taken for granted and sub-goals chosen by reference to them. Emotion wasn't featured. In his massive book on problem solving, it was mentioned only four times—one of which was a historical reference to Abelson (Newell and Simon 1972: 887).

This wasn't a mere oversight, but a deliberate strategy. Since emotion operates "through the lens of the cognitive system", he said, "a plausible scientific strategy is to put our cognitive models in order before moving to the other phenomena" (Newell and Simon 1972: 8). A theory of "normal" thinking can ignore emotion, to a first approximation (p. 866). Indeed, *only one* of the 321 protocols observed in his study of mental arithmetic was a "clear example of the injection of emotionally toned behavior" (p. 206; cf. p. 166). Even his later SOAR models didn't prioritize emotions, although he still recognized their importance (Simon 1994*a*).

This attitude was typical. Emotion, having been fairly prominent in 1960s cognitive science, had fallen out of sight. (Within psychology in general, of course, it was still visible. It was being studied in cognitive as well as psychodynamic contexts, and in tiny babies as well as in adult subjects: see Section v.e, below.) It was superficially represented in Abelson-influenced NLP programs (e.g. Dyer 1983), because most stories contain emotion words. But (with a few exceptions, discussed below) the role of emotions in mental processing *in general* wasn't addressed.

Some philosophers saw this as no accident, believing that GOFAI-based cognitive science is *in principle* unable to deal with all aspects of emotion. One of the first to say this was Keith Gunderson (1935–). Gunderson wasn't opposed to GOFAI-based cognitive science, nor even to computational models of emotion, in so far as these dealt with aspects *other than* the conscious feelings involved. But he drew a distinction between "program receptive" and "program resistant" features (1963, 1968, 1971). Problem solving was said to be an example of the former, emotion of the latter.

The program-resistant features of the mind were its *qualitative* aspects, of which consciously experienced feelings of emotion were a special case. Gunderson saw these as lying outside the type of computational psychology favoured by Simon, because—so he said—they are features of the underlying neural mechanism in which our mental functions are implemented. Possibly, he believed, such features might be *simulated* in computers. However, the simulation problem wasn't a matter of writing programmed routines, but of "somehow imparting to the machine analogues of basic capacities *presupposed* by possible software strategies" (1981: 538; italics added). This, he said, might be in principle impossible, because of subjectivity (see Chapters 14.xi and 16.iv–v).

As for Colby's "conceptual" representation of feelings of shame in PARRY, this—said Gunderson—was no better than

> drawing a pineal gland, making a dot in it, and saying "This," pointing to the dot, "represents the human soul as it is joined to the human body". Well, of course it does, but that tells us nothing illuminating about the human mind, but rather something about how easygoing and unilluminating certain forms of representation can be. (Gunderson 1981: 538)

He added, however, that Colby might have made an important contribution to the "taxonomy" of paranoia, for one doesn't need to solve the mind/brain problem to do that. (Whether functionalism in general is defeated by conscious experience will be discussed in Chapters 14.xi and 16.iv–v.)

The philosopher Hubert Dreyfus went much further: even problem solving, on his view, couldn't be understood in AI terms (see Chapter 11.ii.a–d). He said hardly anything about emotions as such. But he argued that human *needs* were being treated by AI researchers as pragmatic "values", conceptualized as *additional properties* to be computed alongside others (Dreyfus 1972: 184–90). Rather, he said, our needs belong "to the structure of the *field* of experience, not the objects in it" (p. 186).

It followed (although Dreyfus didn't explicitly say so) that one should reject any account of emotion, such as Simon's, which saw its role as prioritizing values during planning. In addition, Dreyfus stressed the role of the *body* in our psychology, and complained that GOFAI ignored this (1967; 1972, ch. 7). Again, it followed (though he didn't say so) that if emotions were thought of—as they generally were—as being somehow closer to the body than rational thinking is, they would inevitably be played down by a GOFAI-inspired cognitive science. Dreyfus's critique is considered in Chapters 11.ii.a–d and 16.vii.a. Here, the point to note is that he *wasn't* the leading philosophical critic of *computational accounts of emotion*, because he said so little about it. Rather, the main critics were Gunderson and (especially) John Haugeland (1945–).

Haugeland's doubts about cognitive science's ability to model affect drew more attention from the field than Gunderson's earlier criticisms had done. That was partly because his critique appeared as a target article in the new—and ground-breaking—journal *Behavioral and Brain Sciences,* or *BBS* (see 6.v.c).

In the 1970s Haugeland was excited by the promise of "cognitivism", by which he meant "roughly the position that intelligent behavior can (only) be explained by appeal to internal 'cognitive processes,' that is, rational thought in a very broad sense" (1978: 215). Later, he would cross the philosophical divide and opt for neo-Kantianism, becoming an enthusiastic advocate for Dreyfus (Chapter 16.vii.c). In fact, the acknowledgements of his *BBS* paper already mentioned "more of a debt than I can properly express to the inspiration and constant guiding criticism of H. L. Dreyfus". But even when he was still prepared to give cognitive science house room, he had misgivings about emotions and moods (1978, sect. 7). "I cannot prove that Cognitivist accounts of these phenomena are impossible," he said, but he thought their possibility was "dubious".

For Haugeland, cognitivist accounts were even less likely to succeed for moods than for emotions. He argued that emotions might be no more problematic for cognitivism than the "pre-cognitive" senses are. He was happy to allow that the bodily basis of conscious feelings and sensory images might be delivered as representational "input" to the information-processing mind, which would then focus on the cognitive content of the emotion (gratitude, for instance). But moods are different, he said.

On the one hand, they're "pervasive and all-encompassing" in a way that feelings aren't:

> The change from being cheerful to being melancholy is much more thorough and far-reaching than that from having a painless foot to having a foot that hurts. Not only does your foot seem different, but everything you encounter seems different . . . greyer, duller, less livable. (Haugeland 1978: 223)

On the other hand, moods are neither quasi-linguistic (representational) nor rational. This was a puzzle, even a paradox. Since they "permeate and affect all kinds of cognitive states and processes", they can't be theoretically segregated from cognition, as conscious experiences (arguably) can. Nevertheless, "they don't seem at all cognitive themselves".

One didn't have to be a professional philosopher to believe that affect and computers don't mix. Young children in the late 1970s who'd had access to MIT's computers for some years, in school and/or at home, had no qualms about saying that computers are intelligent, and can think. But they insisted that they don't have feelings, that they don't "care"—in short, that they don't have emotions (see 16.ii.c).

However, they weren't saying that computers *aren't as intelligent as they might be*, because they lack emotions. Nor was Haugeland saying that a computer's intelligence is compromised by its lack of moods. Affect was one thing, intelligence another. They were even opposed, in the sense that emotions were assumed to cause irrationality: love, after all, is blind. In short, "emotional intelligence" was seemingly a contradiction in terms.

By the late 1990s, that had changed: emotional intelligence had become a focus of research across the whole of cognitive science. (George Mandler sees the 1990s "tidal wave" of emotion studies as having begun with a "watershed conference" in Stockholm in 1972, at which the link with *cognition* was given due prominence for the first time: 2002*c*: 231–2.)

The new century saw several interdisciplinary conferences on it, including one organized in 2001 by the Royal Institute of Philosophy (D. Evans and Cruse 2004). For the philosophers, having largely ignored emotion for years, now started paying attention (e.g. D. Evans 2001; Prinz 2004*a*,*b*). Many psychologists got involved. For example, Paul Ekman linked the facial expression of emotions, which he'd been studying since the early 1970s (e.g. Ekman 1979), with the cognitive phenomenon of lying (Ekman 1985/2001, 1992, 1998; Ekman and Davidson 1994). AI workers got in on the act too, as we'll see (e.g. Picard 1997; cf. Sloman 1999; Picard 1999). Last but not least, the neurologists had plenty to say.

The topic even became trendy with the public at large, because of several trade books written by neurologists. Antonio Damasio's account of *Descartes' Error* (1994), for example, was so popular that it had appeared in twenty-three languages by 2002, and his 3-year-old book on feeling (1999) had already reached eighteen. Other widely read neurological accounts of emotion and/or the self came from Joseph LeDoux (1996) and Gerald Edelman (1992, chs. 17–18).

This neuroscientific research fed on, and fed, the 1990s upsurge of interest in the tricky topic of consciousness (discussed in Chapter 14.x–xi). But the psychological *function* of emotion was also at stake, and that's our concern here. "Descartes' error", said Damasio, had been to ignore the role that emotions (and the body) play in reason.

One case study was much cited: Damasio's patient "Elliot". Prefrontal lobe damage had left this man's intellectual intelligence intact. He had the normal stock of knowledge, including the social/personal concepts and scripts described by Abelson, and was aware of moral principles too. Moreover, he could compute sub-goals, compare possible consequences, construct contingency plans, and even perform individual sub-tasks with no trouble at all—often, better than most of us. What he couldn't do was choose sensibly between alternative goals, stick with a plan once he'd chosen it, or assess other people's motives and personality effectively. As a result, his personal and professional life, and his financial affairs, disintegrated in a series of bizarre disasters.

As Damasio put it:

> The tragedy... was that [Elliot] was neither stupid nor ignorant, and yet he acted often as if he were. The machinery for his decision making was so flawed that he could no longer be an effective social being. (1994: 38)

> After [many] tests, Elliot emerged as a man with a normal intellect who was unable to decide properly, especially when the decision involved *personal or social* matters. (p. 43; italics added)

Elliot was unable to choose effectively, or he might not choose at all, or choose badly . . . As we are confronted by a task, a number of options open themselves in front of us and we must select our path correctly, time after time, if we are to keep on target. Elliot could no longer select that path. Why he could not is what we needed to discover. (p. 50)

The germ of the answer was already available:

Elliot was able to recount the tragedy of his life with a detachment that was out of step with the magnitude of the events . . . Nowhere was there a sense of his own suffering . . . He was calm. He was relaxed . . . He was not inhibiting the expression of internal emotional resonance or hushing inner turmoil. He simply did not have any turmoil to hush. (p. 44)

I never saw a tinge of emotion in my many hours of conversation with him: no sadness, no impatience, no frustration . . .

Later, and quite spontaneously, I would obtain directly from him the evidence I needed. [On being shown pictures of ghastly events] he told me without equivocation that his own feelings had changed from before his illness. He could sense how topics that once had evoked a strong emotion no longer caused any reaction, positive or negative. (p. 45)

In short, Elliot's disabling predicament was "*to know but not to feel*" (p. 45).

The case histories of twelve other brain-damaged patients, said Damasio (p. 54), provided further evidence for (or illustration of: see below) the fact that intellect isn't rationality. As Abelson might have put it, all-beliefs-and-no-attitudes doesn't make a rational system. Our powers of reason depend on emotional valuation to set (and to retain) overall priorities, and to guide decisions at choice points in the plans that structure our behaviour. In short, MGP shouldn't have neglected the *Image* (Chapter 6.iv.c).

Damasio suggested that various regions of the brain underlie some or all emotions. And he outlined a theory of "somatic-markers": complex physiological responses that act as clues to the likely outcome—advantageous or not—of specific actions (1994, ch. 8). In short, "high reason" (which René Descartes had divorced from the body) was replaced by body-based judgement.

Because of its stress on emotion and the body, Damasio's work was widely regarded—not least by the general public, always ready to hear that the mind/brain *is not* like a computer—as a knock-out blow to computational theories of mind. Simon's long-standing argument that emotion is needed for goal setting was forgotten.

That was understandable, since Simon had seemingly forgotten it (or anyway, ignored it) too. But whether Damasio's—or anyone's—recent neurological research really was a knock-out blow is another matter. When his readers dismissed computational psychology, with a sigh of triumph or relief, they did so without knowing that important relevant work *had already been done.*

e. Architect-in-waiting

During the last quarter of the century, and despite the widespread neglect of emotion in cognitive science at that time, a few computationalists were working steadily on emotion and personal phenomena. They were doing so in the context of *mental architecture* in general.

Mental architecture is the overall structure of the mind (and/or possible minds), considered as a *virtual* machine—including the potential for various types of processing within and between the component parts. These theorists painted the mind with a broad brush, for claims about mental architecture need not involve hypotheses about specific processing details (although some do, such as Newell's SOAR: see Section iv.b, below). So they asked, for example, what general types of computation are involved in emotion, and how emotions fit into the overall computational system. As we'll see, however, their work wasn't being taken seriously by (most of) their peers.

The foremost examples of architectural theorists focusing on emotion and personality were Marvin Minsky (1927–) at MIT, and Aaron Sloman (1936–) at Sussex—and after 1991 at Birmingham. (Minsky was still communicating with Seymour Papert, but Papert's own research was now even more focused on educational matters: see 10.vi.) Their work is discussed, respectively, in this subsection and the next.

In addition, Daniel Dennett, who was much influenced by Minsky, sketched a computational account of the construction of the self, seen as "a center of narrative gravity" that unifies and directs thought and action (1991, ch. 13; and see subsection g, below). And the neuroscientist Michael Arbib, who described mental architecture in terms of *schemas* (Chapter 14.vii.c), claimed to have shown "how the hundreds of thousands of schemas in a single brain may cohere to constitute a single person" (1985, p. viii; cf. Arbib and Hesse 1986, ch. 7). But he did this only in the sketchiest terms. Minsky and Sloman, by contrast, tried to specify some relevant schemas and their interactions.

Minsky was already considering architectural questions by the late 1950s. He argued, for instance, that a creature capable of answering a question about a hypothetical experiment without actually performing it must, as Kenneth Craik had said (Chapter 4.vi.b), possess knowledge in the form of symbolic models of the world (Minsky 1965, sect. 2). Moreover, any creature capable of answering questions about itself *considered as an intelligent system* (so excluding questions like *How tall am I?* but including *What are my current goals and resources?*) would need to construct a model of itself—again, considered as an intelligent system, or mind. This self-model would omit some of the detail present at the level being modelled (hence giving rise to the empty illusion of 'free will': sect. 8), and would distinguish to some extent between the creature's physical body and its modelling activities (hence encouraging mind–body dualism: sect. 5).

The same—he said—would apply to successful AI systems, when they were eventually constructed:

> We should not be surprised to find [intelligent machines] as confused and as stubborn as men in their convictions about mind–matter, consciousness, free will, and the like. For all such questions are pointed at explaining *the [invisible] complicated interactions between parts of the self-model.* A man's or a machine's strength of conviction about such things tells us nothing about the man or about the machine except what it tells us about his model of himself. (Minsky 1965, sect. 9; italics added)

As for what those "complicated interactions" might be, Minsky couldn't say. Thankfully, they're not introspectible.—"Thankfully", because (as Patrick Hayes later put it) if they were, we'd all be "cast in the roles of something like servants of our former selves, running around inside our own heads attending to the mental machinery which currently is so conveniently hidden from our view, leaving us time to attend to more important matters". As Minsky remarked in response to Hayes, consciousness is marvellous "not because it tells us so much [about how we manage to think of a name, for instance], but for protecting us from such tedious stuff!" (Minsky in preparation, sect. 4-3).

The downside is that *if* we want to know how we manage to think of a name, hammer a nail, or write a love letter, it's very difficult to find out. In the mid-1960s, it seemed to Minsky that a creature's knowledge is recursively contained in its model-of-its-knowledge (if it has one)—and vice versa. This suggested:

> first, that the notion "contained in" is not sufficiently sophisticated to describe the kinds of relations between parts of programlike processes and second that the intuitive notion of "model" used herein is likewise too unsophisticated to support developing the theory in technical detail... *An adequate analysis will need much more advanced ideas about symbolic representation of information-processing structures.* (Minsky 1965, sect. 3; italics added)

Much of his later work would be an attempt to develop that "adequate analysis", using approaches from both connectionism and GOFAI (see 12.iii).

By the late 1970s, he was developing an architectural theory of the human mind as a whole. Emotions were briefly discussed in his trade book *The Society of Mind* (Minsky 1985), and at the turn of the century became the prime focus of *The Emotion Machine.* (As we'll see in Chapter 12.iii.d, the draft of *The Emotion Machine* has been in continuous change for nearly ten years; the quotations given in this section date from April 2002.)

In both books, Minsky paid homage to Freud. In 2002 he said: "few researchers in 'cognitive science' yet appreciate Freud's idea: that 'thinking' is a collection of schemes for appeasing both instincts and ideals" (in preparation, sect. 1-5). In Minsky's terminology, he assimilated the superego to "learned values, censors, and self-ideals", the ego to "conflict-resolving processes", and the id to "innate, instinctive wishes and drives". And this terminology, in turn, was interpreted by reference to a wide range of specific computational mechanisms, drawn from nearly fifty years of AI research. Among other things, these helped to make explicit the complex interactions within and between the various 'levels' of the mind.

Minsky was (is) nothing if not ambitious: in April 2002, the opening section of *The Emotion Machine* was called 'Falling in Love'. He didn't attempt a conceptual analysis of *love* (cf. Fisher 1990, and subsection f, below). But he did point out that personal love (of a parent for a child, or a child for its parent, between friends or lifelong companions, or for a group and/or its leader...) differs from other things called love (a patriot's allegiance to country, a convert's adherence to doctrine, a scientist's passion for finding new truth...) (Minsky in preparation, sect. 1-2). And he followed Sloman (see below) in listing several—very different—psychological features normally associated with personal love, including: grieving for a lost child, excited anticipation of a loved one's arrival, and jealousy of someone favoured by the person you adore.

Minsky didn't offer a conceptual analysis of *emotion*, either. But he pointed out that dictionary definitions differ, and that English has hundreds of 'emotion' words naming a wide variety of phenomena—from admiration, agony, and alarm; through curiosity, dismay, and disgust; to hope, impatience, and infatuation (sect. 1-8). Alarm and hope are intuitively (and computationally) very different, as are agony and infatuation—yet these two are also linked. Pride and shame are 'opposites'—yet both can help us learn to evaluate our goals in a new way and to generate new goals appropriately (sect. 2-1).

We can't go into the (fascinating) details here. The crucial point is that Minsky's aim was to outline the computational architecture—an integration of detailed aspects of cognition, affect, and motivation—that *generates* emotions, and that *enables them to perform their psychological functions.* Some of these functions involve self-models (e.g. those associated with pride and shame). Others don't (e.g. fear).

There are two reasons why one might call Minsky an architect-in-waiting. The first is that, as he'd readily admit, he hasn't finished the job. Indeed, "finishing" the job may be too much to ask, even by the *next* millennium. It would be enough (for example) to have outlined the computational structures and processes enabling love to be expressed in the varied ways familiar to novelists and psychologists alike (even within a single culture). Minsky has made a start, but there's a long way to go.

The second reason is that Minsky's ideas on personality still (in 2003) aren't widely known. Why is this? With respect to *The Emotion Machine*, the answer's easy: it isn't officially published. But why was *Society of Mind* largely ignored? After all, Minsky was already hugely famous, and the book had been eagerly awaited for many years. The prime obstacles were its (deceptively simple) rhetorical style, and the lack of programming (see Chapter 12.iii.d).

Few of Minsky's AI colleagues were impressed, and some were highly scornful (12.iii.d). Most psychologists weren't impressed either. Minsky's text alluded to many AI results without identifying them, so people lacking a wide knowledge of AI didn't find it easy to understand. Moreover, its message was largely tacit. Readers were left to make important connections for themselves—and many couldn't. Some psychologists, no doubt, were put off also by the lack of experimental evidence, quantitative measurement, and testable claims. (They may have been wrong to dismiss the book on these grounds: see subsection f, below.) So Minsky is an architect-in-waiting: waiting for his contribution to be given its due.

Probably more people became (superficially) aware of Minsky's ideas by reading Dennett than by reading Minsky himself. For in his runaway best-seller *Consciousness Explained* (1991), Dennett included a chapter on the self that was deeply influenced by Minsky. This was welcomed, among others, by postmodernists who'd already rejected the unity/stability of the self (e.g. Turkle 1995: 261; see 13.vi.e). But it didn't produce a visible upsurge of interest in Minsky's own writings.

Dennett's discussion had appeared in a volume devoted to consciousness. Minsky, by contrast, was trying to work out how the various psychological functions (of which consciousness and self-consciousness are but examples) cooperate within the mind as a whole. For most of Dennett's readers, that seemed less interesting. They didn't realize that the various meanings of the word "consciousness" can't be properly understood without using *architectural* ideas. (Indeed, this is still widely doubted: 14.x–xi.)

f. Of nursemaids and grief

Sloman, too, focused on the computational architecture of the mind as a whole. His core interest, ever since the mid-1970s, has been in what he calls "the space of possible minds", of which existing (animal and human) minds are just examples (Sloman

1978, ch. 6; cf. Dennett 1996). Even his early research on the POPEYE vision program discussed visual psychology in light of architectural principles (and, in so doing, clarified some aspects of familiar concepts such as *interest*, *conscious*, and *experience*). In that sense, it was way ahead of its time (see Chapters 10.iv.b and 12.v.h).

His mature work on mental architecture is broadly like Minsky's, but instead of Minsky's "society of mind" he speaks of the "ecosystem of the mind" (Sloman 2003*a*). Mental mechanisms, he argues, are even more closely interdependent than the members of a society, as they're co-evolved sub-organisms. It's no wonder, then, that vision is intimately involved in reasoning (see Section iv.d, below), and that rational and emotional processes are intertwined too.

Where personal psychology is concerned, Sloman was willing to take the bull by the horns. One of his papers was subtitled 'What Sorts of Machine Can Love?' (That paper, Sloman 2000, is the most accessible statement of his mature approach. The best technical introduction is Sloman 2001; the latest developments can be found on his web site: Sloman n.d.)

One doesn't have to be a computationalist to give a broadly functionalist analysis of love. Mark Fisher (1990) has shown that the essence of personal love is a deep commitment to the goals and motives of the loved one, even to the extent of preferring them over one's own. Such a commitment involves many different cognitive and emotional phenomena, working together to discover and advance the interests of the other person. As a 'pure' philosopher, Fisher was content to take the *possibility* of these phenomena for granted. Sloman, by contrast, aimed also to outline the computational resources required to generate them. (Both Fisher and Sloman were considering "love" as it's conceptualized in Western traditions. Near-equivalents in other societies, such as the Samoan *alofa*, would have to be analysed somewhat differently: see Chapter 8.i.d.)

Like Fisher, Sloman was an accomplished analytic philosopher (though originally trained in maths and physics). He was greatly influenced by Gilbert Ryle and John Austin, who showed that a structured folk psychology is tacit in our everyday language (e.g. Sloman 1978, ch. 4).

Austin, the quintessential 'ordinary-language philosopher', had argued that although ordinary language isn't the last word for theoretical psychology, it is a helpful first word. Its many subtle psychological distinctions have been "honed for centuries against the intricacies of real life under pressure of real needs and therefore give deep hints about the human mind" (Sloman 1987*a*: 217). But they're no more than hints, for everyday language is often ambiguous—or worse. For example, the concept of *pain*, commonly assumed to be the epitome of clarity/intelligibility, is in fact hugely confused and even contradictory (Dennett 1978*a*).

Moreover, the grammar of emotions isn't easily uncovered. As Sloman put it, "our ability to articulate the [psychological] distinctions we use and grasp intuitively is as limited as our ability to recite rules of English syntax" (1987*a*: 217). When we try to articulate these distinctions in expressing some psychological (or philosophical) theory, we typically lose sight of the fine print.

A telling example of this was described long ago by Austin (1957). He showed that the language used by an uneducated hospital assistant giving evidence in a law court may make clear just what went on in his mind (though not, of course, *how*). One can see just what it was which led to the unfortunate event in question (the severe scalding of a patient being given a bath). By contrast, said Austin, the legal terminology used by the judge in summarizing it may obscure many crucial details. And the same is true, he said, of philosophers' jargon and theories, such as their talk of "voluntary/involuntary" actions, or of "free will" (see subsection g, below).

Austin's strictures can be applied also to psychological theories of personal matters, including most computational theories of emotion. The psychologist should aim to recognize, and to explain, *at least* those distinctions which are intuitively drawn in everyday language. These will certainly have to be supplemented, and often even corrected. But they shouldn't be ignored, nor shoehorned into ludicrously over-simple theoretical categories. Sloman's work is an attempt to avoid such shoehorning.

Sloman defined emotions as datable *episodes*, closely related to attitudes—which are what Ryle (1949) called *dispositions* (see 16.i.c). (Thanks to his study of grief, mentioned below, he now defines them as "dispositions which can trigger episodes": personal communication) This wasn't a report of, nor a recommendation for, everyday usage. Rather, it was a theoretical clarification enabling the psychologist to distinguish emotions from other mental phenomena—which may or may not be described as emotions in ordinary conversation.

So Sloman argues (personal communication) that Damasio's case studies don't provide *empirical evidence* for a link between emotion and intelligence, as Damasio claims. Rather, they *confirm*—in a startling and memorable fashion—the relation that's already implicit in these familiar concepts. (Hence my suggestions, above, that Damasio's work offers illustration rather than evidence, and that asking "how one could test" Minsky's society theory may not be appropriate.)

What the computational psychologist should do, then, is (first) make explicit the distinctions tacit in ordinary-language terms as carefully as possible, and (second) "extend colloquial language with theoretically [i.e. computationally] grounded terminology that can be used to mark distinctions and describe possibilities not normally discerned by the populace" (Sloman 1987*a*: 217).

Accordingly, Sloman didn't merely unpack the implications of ordinary language, as Fisher did. He also exploited his experience of software design, and his deep understanding of various types of computation (Chapter 16.ix.c). He drew on these skills in discussing a wide range of *possible* mental architectures, with special emphasis on the generation and control of human motives, emotions, and attitudes.

To do that, he didn't need to do any programming, nor even decide just what types of virtual machine—e.g. GOFAI, localist-net, PDP, GasNet . . . or some form of computation yet to be discovered—might be most appropriate for particular functions. Similarly, Minsky didn't need to make those decisions when asking broad architectural questions. The price they both paid was to lay themselves open to the implementation hunger of their AI colleagues, many of whom ignored—or even scorned—their work for years because it wasn't immediately implementable (see 12.iii.d).

But times change. In the late 1980s a publisher's reviewer advised me to drop Sloman's (1987*a*) paper from a collection I was preparing on the philosophy of AI (Boden 1990*b*). He/she said it was unrepresentative (true) and irrelevant (words fail me!). I insisted that it stay in. Today, there wouldn't be a peep of protest.

Largely because of the revival of interest in emotional intelligence, Sloman's long-standing work on mental architecture has now achieved international recognition.

Indeed, in 2002 DARPA invited him to participate in a small workshop related to their new cognitive systems initiative, where Rodney Brooks (15.viii.a) discussed Sloman's work in one of the introductory papers. This was more than a straw in the wind: the significance of DARPA's support within cognitive science is illustrated in Chapter 12.iii.e and vii.b. Two years later, the AAAI held a cross-disciplinary symposium on 'Architectures for Modelling Emotion'.

Some of Sloman's current recognition is undeserved, however, as he's the first to admit:

> I've even had someone from a US government-funded research centre in California phone me a couple of months ago [i.e. mid-2002] about the possibility of modelling emotional processes in terrorists. I told him it was beyond the state of the art.
>
> He told me I was the first person to say that: everyone else he contacted claimed to know how to do it (presumably hoping to attract research contracts). (Sloman, personal communication)

It's difficult enough to limn the mind of a cricket (15.vii.c). To depict the complex architecture of the human personality, whether terrorists or just plain folk, is orders of magnitude harder.

That's why, nearly forty years ago, Reitman had to give up on Argus. He'd have been well aware, for instance, that a nursemaid dealing with several hungry, active, and attention-seeking babies (with an open door leading onto a water-filled ditch) is subject to countless perceptual and emotional interrupts, and consequent changes of plan. She has to distinguish between important and trivial goals, and decide on urgency and postponement. (Feeding a baby is important but not highly urgent, whereas preventing it from falling into a ditch is both.) She'd welcome Argus' 100 eyes, and 100 hands besides. But she has only two of each, so must schedule her limited resources effectively—which is what emotion, on Sloman's view, is basically about. Reitman knew this. But he couldn't express it, still less model it, in computational terms.

Thanks to thirty years' progress in AI, Sloman's architectural account of such matters is a huge advance on Reitman's. What's more, a small part of his theory was implemented in the 1990s, in the nursemaid-and-babies model mentioned in Chapter 1.i.b (Wright and Sloman 1997; see also Beaudoin 1994; Wright 1997). The model is still being continually improved. (My account here ignores many details, including the differences between successive versions.)

Sloman's simulated nursemaid coped with many fewer problems than a real nursemaid does. "She" merely had to feed the "babies" (i.e. recharge them), try to prevent them from falling into ditches, and—if they'd already fallen in—deliver them to a first-aid station. (That's a euphemism: a baby in a ditch was in effect dead, and had to be delivered to the exit for dead babies.) She didn't have to worry about cuddling them, bathing them, changing them, singing to them . . . and there were no live electric plugs to be avoided. And she herself was a pretty simple creature, for she had only seven motives to follow:

* feeding a baby (if she believed it to be hungry);
* putting a baby behind a protective fence (if it's already been built);
* moving a baby (if it's been damaged by falling into a ditch) to a first-aid station;
* patrolling the ditch, to see if any babies have fallen in;
* building a protective fence;

* moving a baby to a safe distance from the ditch;
* and wandering around the nursery (if no other motive is currently activated).

Even these few motives, however, could conflict. (What if she was feeding a very hungry baby, and another crawled near the ditch?) Further, they could arise unexpectedly as the result of environmental contingencies. For each baby was an autonomous agent, whose crawling and hunger crying was independent of other babies, and of the nursemaid's actions and motives.

By and large, Sloman's program enabled the carer to act with (minimal) intelligence in the circumstances. If she was focusing on feeding a baby, she'd instantly put it down if she realized that another baby was approaching the ditch. But she'd remember that the first baby still needed feeding, and would do so as soon as possible. Moreover, the hungrier babies would get fed first. If there were *two* babies approaching the ditch, the nursemaid would judge which was the nearer so as to prioritize her currently active motives: *feed baby a, rescue baby b, rescue baby c.* (Given baby a ... to baby z, however, there'd be trouble: the virtual nursemaid's performance deteriorated rapidly as the number of babies increased—as a real nurse's does, too.) Further flexibility sprang from the program's ability to adjust the filter that allowed a pre-attentive motive to enter attention.

There were many obvious weaknesses. For example, the system didn't implement visual processing, but merely simulated it. The carer's vision was assumed to be perfect, up to a certain distance—but human nursemaids may have to cope with dusk, or with vision blurred by tears or smoke. And babies behind obstacles were fully "visible" (i.e. their location was known by the nursemaid), because occlusion wasn't simulated.

Other weaknesses were less obvious, unless one's aware of the many psychological distinctions within Sloman's theory. He listed several important limitations, and admitted that the simulated anxieties ("proto-perturbances") are but a pale shadow of a real nursemaid's perturbances (Wright and Sloman 1997, sects. 3.7.2, 4.3). He also pointed out that "perturbances" are only one of several architecturally distinguishable forms of anxiety. In short, *anxiety* is really *anxieties*—but Sloman had provided clear computational distinctions between them.

However, the weaknesses weren't terminal. For this wasn't a one-off tinkered program, but the first in a theory-based series. Sloman had long taken on board Drew McDermott's advice on how AI should be pursued (see 11.iii.a). That's why his computational "toolkit", developed for designing autonomous agents in general, was made available on the Internet for others to experiment with (Sloman and Poli 1995; Sloman 1995). The 1997 version of the nursemaid program was called MINDER1. Eventually, MINDER2 and MINDER3 ... will follow, as Sloman's implementations approach his architectural theory more closely.

Sloman's theory of emotions—like Simon's and Minsky's too—isn't simply a phenomenological (descriptive) classification, as Tomkins's was. It provides an architecture-based *explanation* of the differences between what Damasio terms "primary" and "secondary" emotions (Sloman 1999). And it makes many fruitful distinctions.

For instance, emotions differ in terms of three main architectural "levels", which involve reactive, deliberative, and meta-management mechanisms. A cricket's mind is mostly reactive, depending on learnt or innate reflexes (Chapter 15.vii.c). As such, it's capable only of "proto-emotions": inflexible reactions that have much the same adaptive function as (for example) fear in higher animals. A chimpanzee's mind is largely deliberative, capable of representing and comparing past (and possible future) actions or events. So backward-looking and forward-looking emotions can now arise: non-linguistic versions of anxiety and hope, for instance.

In human adults, the deliberations can include conscious planning and reasoning, generating more precisely directed anxieties accordingly. (Remember Colby's neurotic and paranoid patients.) In general, language makes possible emotions with propositional content—unknown to crickets, or even to chimps. An adult human mind also has a rich store of reflexive meta-management mechanisms, which monitor and guide behaviour. (McDougall's "master sentiment of self-regard" and Dennett's "center of narrative gravity" are examples, though much less well specified.) So emotions centred on the self—such as vainglory and embarrassment—are now possible.

One might call this theory a generative grammar of emotions (cf. Sloman 1982), provided one remembers that Sloman (unlike Noam Chomsky) was interested in the underlying processes, not just the abstract structure of the phenomenology (see Section iii.a, below). Better, it's a generative grammar of *the mind as a whole*. For Sloman—like Minsky—offered many promising ideas about how the various types of emotion are functionally related to other mental features, including the processes involved in intelligent action.

For example, his computational analysis of grief (on the death of someone truly loved) shows why it's inevitable, why it lasts a long time, why it eventually disappears—replaced, perhaps, by an enduring sadness—and why it's *cognitively* debilitating in many specific ways (Wright *et al.* 1996).

In a word, to mourn is to dismantle. What's dismantled is much of the cognitive–motivational core of the personality concerned. In Fisher's terms, this is a complex self-denying (yet also self-enhancing) commitment to another, now departed; in McDougall's, it's a powerful sentiment of love, focused on the deceased person and closely integrated with the master sentiment of self-regard. Dismantling that can be neither easy nor pleasant. It involves both cognitive and emotional effects, some—but not all—of which are introspectible/observable episodes, as opposed to dormant dispositions. And for sure, it's not the work of a moment (see 1.i.b).

These effects will differ across cultures, because concepts of love, personal obligation, and so on vary. For instance, perhaps I should have said (above), "why it's inevitable *in privileged Western cultures*". For the medical anthropologist Nancy Scheper-Hughes has shown in distressing detail how some cultures protect themselves from what would otherwise be an intolerable excess of grief. Her book on Brazil has a chapter whose subtitle is 'The Social Production of Indifference to Child Death'—1992: 268–339. But that "indifference" is rooted in pre-existing schemata of parental love that are different from ours in culturally relevant ways. It's those schemata which mould whatever grieving goes on.

According to another anthropologist, Renato Rosaldo, headhunters often decapitate their fellow men out of rage, born of grief (see Chapter 8.ii.b). Whether their grief is—as Rosaldo (1989) claims—closely comparable to his own, on mourning his wife, is quite another matter. Sloman's analysis would imply that, beyond a common human core, it is very different. (However, that common core may be what leads virtually all religions to pay ritual attention to the bodies of the dead, especially of people whom the mourners had loved: 8.vi.d.)

Lasting personal attachments falling short of personal/romantic love also require computational dismantling to some degree. A funeral service for a friend, or a memorial service for a much-admired person, are each distressing in their own way. But the grief involved is different, because the mourner's cognitive/emotional schemas linking them to the deceased are different.

A final word: Sloman's theory of grief could doubtless be improved in detail, or perhaps replaced by an alternative. But anyone who's sceptical about *the very idea* of giving a computational analysis of grief should know that the journal editor who published it is a practising psychiatrist (K. William Fulford, of Oxford's Radcliffe Infirmary). As such, he's more familiar than most of us with the multifaceted cognitive–emotional ravages of mourning. So one should beware of dismissing computational treatments of grief simply because of prejudices about the emotional irrelevance of computers.

In particular, one should set aside the Spielberg–Kubrick puzzle over whether computers could "really" experience emotions (see Chapters 14.xi and 16.v.b). Cognitive science aims to help us understand *biological* minds, by showing (for instance) how such humanly significant phenomena as personal love, grief, mourning, and funeral rituals are *possible* (cf. Section iii.d, below). It may also aim to help us understand the range of *all possible* minds. Even so, if its philosophical implications about the "reality" of the mental lives of robots are unclear, that needn't compromise its power as a theory of human and animal psychology.

g. Free to be free

One of the ways in which the Third Force personality theorists differed from the behaviourists was their emphasis on freedom (see 5.ii.a). Whereas Skinner (1971) famously scorned this notion, as not only illusory but socially pernicious, the Third Force saw freedom of choice as the crucial aspect of human minds. Most 'laymen' agree with them.

It's certainly true, as remarked in Chapter 1.i.b, that an adult human being is free in a way in which a dog, or a cricket, is not. In other words, the range and flexibility of their behaviour is markedly different. So much is undeniable. What's controversial is the *philosophical analysis* and/or *psychological explanation* of that difference.

Before the rise of cognitive science, this issue was normally put in terms of determinism versus indeterminism. So the neuroscientist John Eccles, for example, posited mysterious quantum phenomena acting on "critically poised neurones" in the brain (see Chapters 2.viii.f and 16.i.a). Why such phenomena are absent from the brains of dogs and crickets—or, for that matter, of human babies (who *aren't* yet free agents: see Section vi.f below)—wasn't discussed. (Nor is this discussed when similar arguments are put forward today: see 14.x.d.)

Eccles's neurological hypothesis found few followers. But even if they found his "explanation" implausible, many people agreed with him that it's obvious that free choice is incompatible with determinism. The trouble was that the opposite seemed just as "obvious" to others. Compatibilists argued persuasively that indeterminism at the origin of action would *destroy* human freedom, because it would make it impossible to ascribe moral responsibility to people (e.g. Hobart 1934).

Cognitive science showed a way out of the impasse, by sidelining the determinism/indeterminism issue and focusing instead on the huge cognitive and motivational differences between humans, dogs, and crickets.

It's the computational *architecture* of the human mind/brain (including, crucially, its ability to support language: see 12.x.g) which generates deliberation, self-reflection, and multifaceted choice—wherein the values inherent in MGP's Image are used in comparing the many possible Plans. And it's our (largely tacit) understanding of those features which underlies our ascriptions of freedom, and of blame, to everyday actions (cf. Austin 1957). The subtle distinctions between different actions that are so often marked in gossip are *real* distinctions, made possible by the psychological abilities possessed by persons.

This view was hinted at in almost the first paper on GPS:

> [The key feature of sophisticated problem-solving programs] is . . . that they finally reveal with great clarity that the *free* behaviour of a reasonably intelligent human can be understood as the product of a complex but finite and determinate set of laws. (Newell and Simon 1961: 124; italics added)

Early versions of it were argued at greater length by a number of writers, including Minsky (1965), Dennett (1971), Sloman (1974), and myself (Boden 1972: 327–33; 1978, 1981*c*). Later, Dennett developed it in his essay 'On Giving Libertarians What They Say They Want' (1978*b*: 286–99), and especially in his book *Elbow Room: The Varieties of Free Will Worth Wanting* (1984*a*).

This book was a philosophical analysis, deepened by a cognitive scientist's appreciation of computational architecture. (Many years later, Dennett would return to the topic—now, with a focus on the *evolution* of the relevant architecture: Dennett 2003*a*.) It provided a host of examples illustrating the psychological processes that underlie our freedom of action, and various (normal or pathological) restrictions on it. It made clear that only creatures with a certain type of cognitive–motivational complexity are capable of choosing freely. Or, rather, to choose freely *just is* to exercise those architectural features.

In other words, determinism/indeterminism is largely a red herring. Although (Dennett argued) there must be *some* element of indeterminism, this can't occur at the point of decision—for that would give us a type of "free will" that's *not* one worth wanting. The indeterminism, then, affects the considerations that arise during deliberation. The person may or may not think of *x*, or be reminded (perhaps by some environmental contingency) of *y*—where *x* and *y* include both facts and values. The deliberation itself, by contrast, involves deterministic processes of selection, weighting, and choice. (Deterministic, but not universalist: moral principles, cultural conventions, and individual preferences are all involved.)

Elbow Room, in my opinion, didn't get the recognition it deserved. Analytic philosophers mostly dismissed it, despite Dennett's already high reputation and the book's origin as the 1983 John Locke Lectures in Oxford (see 16.iv.b). Besides being hung-up on the determinism/indeterminism issue, they saw Dennett's approach as "psychologism"—defined by Gottlob Frege as philosophy's unforgivable sin (2.ix.b). And the neo-Kantian philosophers, of course, would reject *any* naturalistic, science-based, account of freedom (16.vi–viii).

Some neuroscientists, too, analysed freedom in architectural terms. I don't mean Benjamin Libet, who showed in the early 1980s that the "intention" appears in consciousness *after* the relevant messages have been sent to the muscles—and who was highlighted in Dennett's 1991 book on consciousness, accordingly (14.x.b). Rather, I'm thinking of the schema theorist Arbib (14.vii.c), and of the clinical neurologist Timothy Shallice (1940–).

In his Gifford Lectures, written with the philosopher Mary Hesse, Arbib applied his neuropsychological theory of schemas (Chapter 14.vii.c) to cultural attitudes and personality—and devoted an entire chapter to freedom. Human behavioural flexibility was explained in architectural (schematic) terms, much as it was by Minsky, Sloman, and Dennett. Indeed, Dennett was praised for having made the best sense of the free-will supporters' position (Arbib and Hesse 1986: 96). But given the religious context of the Gifford Lectures (2.vii.b), Dennett's analysis was considered in light of its implications for theology as well as psychology.

In those terms, Dennett's arguments were said to "still leave everything to play for" (p. 98). Indeed, that applied to *any* scientific explanation of freedom. Whether humans (or robots) really have freedom is at base an "evaluative" question, not a scientific one (p. 99). That is, it's a question about *human relations*—and, for the theist, also about God–human relations. To have personal freedom, on this view, is to possess the type of mental architecture described by cognitive science *and* to participate in a community whose communications are imbued with trust, consideration, and respect.

That's not to say that Arbib's neuropsychology was essentially religious, for it wasn't. On the one hand, he accepted the *fact/value* distinction, as we've seen. So no scientific psychology can entail (or forbid) religious commitment. On the other hand, his theory applied to theists and atheists alike (Arbib and Hesse 1986, chs. 10–12). Theists believe in (possess central schemas positing) a God–human relation. Secularists don't. As educated persons, they possess the God schema; but they use it only in scare quotes. (In McDougall's terminology, it's not a *sentiment*—or if it is, it has a very different, perhaps even negative, emotional contribution.)

As for Shallice, his clinical work on the psychopathology of action made him all too familiar with how brain damage can constrain our freedom. That is, how brain damage can constrain the subtle complexities of our cognitive and emotional processing. Countless examples are described in his book (1988*a*), and in his research (with Donald Norman) on the classification of errors (see Chapter 12.ix.b).

With respect to deliberate conscious control, including what William James had called "the slow dead heave of the will" (1890: ii. 534), Shallice outlined how a "supervisory attentional system" can control action, given certain types of information about the various action schemas potentially involved (Norman and Shallice 1980/1986). He showed that the normal exercise of "the will" doesn't merely trigger action, but involves ongoing error correction too—not to mention the intelligent anticipation of consequences. His theory of control distinguished between schema-driven *contention scheduling*, which covers habitual actions done unthinkingly (and clinical cases of "automatism": see Preface, ii), and *executive control* in which conscious monitoring is involved (see 12.ix.b). In short: "The phenomenology of attention can be understood through a theory of mechanism" (Shallice 1988*a*: 388).

To "understand through a theory of mechanism", of course, needn't mean to do computer modelling. What's more, computer modelling needn't be seen as instantiation. So whether a computer could "really" have free choice is irrelevant here, just as the Spielberg–Kubrick question is irrelevant when evaluating computational theories of emotion. The question, rather, is *whether we can use computational concepts to help us understand human personality and freedom.*

(A caveat: I said, above, that cognitive science "showed a way out of the impasse". But not everyone agrees. In fact, the analysis sketched above is still a minority view. For instance, several people used quantum physics to bring indeterminism in again: R. Penrose 1994*a*; D. Hodgson 1991, 2005. Their arguments, in turn, were opposed: see the peer commentaries to D. Hodgson 2005. Like many other philosophical questions, this one will doubtless run and run. But, for what it's worth, my own view is that it has been answered—unlike the questions of conscious experience or realism/anti-realism: see 14.xi and 16.vi–viii.)

h. Some hypnotic suggestions

If cognitive science really can explain the possibility of freedom, it ought also to be able to explain the seeming *loss*, *surrender*, or *compromise* of freedom that's seen in hypnosis. Loss, because the hypnotized person may feel impelled to do something even though they don't consciously want to do it. Surrender, because they seem to be driven by the hypnotist's intentions rather than their own. And compromise because, even though they're able to perform complicated bodily actions just as effectively as in normal voluntary behaviour, they don't consciously want to do so.

Notoriously, someone under hypnosis may not only do strange things, such as bursting into song whenever the hypnotist coughs, but may see or not-see strange things too: even counting books wrongly, having been told by the hypnotist—falsely—that there are no books by Jane Austen on that shelf. Notoriously too, some people are easier to hypnotize than others. All that is common knowledge—not least, to stage hypnotists. It's also common knowledge that surgical anaesthesia can sometimes be effected by hypnosis rather than drugs, the pain either disappearing or not being experienced as "real" pain. Even if the concept of *pain* is by no means as clear, or coherent, as we normally think it is (Dennett 1978*a*), that fact can't be denied.

Experimentalists have studied these familiar, but highly mysterious, phenomena in some detail. So they've asked (for example) what sorts of task can/can't be carried out by hypnotized subjects; how individuals differ in their susceptibility to hypnosis; and what mental or environmental conditions raise/lower the likelihood of the hypnotist's success. And they've discovered countless intriguing data.

For instance, some sorts of hypnotic suggestion are easier to communicate, or instil, than others (Perry *et al.* 1992). The easiest are "motor" examples, such as *Your arm is becoming so light it is rising in the air.* Next come "challenge" examples, such as *Your arm is rigid* (from which it follows, in successful cases, that the person won't be able to bend their arm when asked to do so by the hypnotist). Negative cognitive demands are more difficult still: *Whenever you count, you'll forget the number four.* And the most difficult of all are positive hallucinations. But even here, experiments show systematic differences. It's easier to make someone hallucinate as required to the suggestions *You can taste something sweet* or *You can hear/feel a mosquito* (50 per cent of people are fooled) than to *You can hear a voice speaking* (only 10 per cent).

In short, a huge amount is now known about the behaviour, and the accompanying phenomenology (or rather, the lack of it), characteristic of hypnosis. What's been very much less clear is the explanation for these things.

As MGP put it in 1960, confessing to "an embarrassing lack of firsthand experience" due to the general mistrust of anyone who'd worked with hypnosis:

> One of the seven wonders of psychology is that so striking a phenomenon as hypnosis has been neglected. Some psychologists literally do not believe in it, but consider it a hoax, an act put on by a cooperative stooge ...
>
> [...Our] conception of hypnosis is quite simple. It is based on a naive faith that the subject means what he says when he tells us that he surrenders his will to the hypnotist. Is there any reason to doubt him? It is just as good a theory as any of the others that have been proposed.
>
> The trouble with such a theory, of course, is that *no one knows what the will is*, so we are scarcely any better off than before. (G. A. Miller *et al.* 1960: 103–4; italics added)

Their remark that "some psychologists literally do not believe in it" presumably referred, though they didn't say so, to Theodore Sarbin and/or to Martin Orne. Sarbin (1950) had argued that hypnosis is a matter of two people playing out familiar *roles.* A role, as Abelson would make clear (subsection b, above), involves two (or more) people with interlocking goals, intentions, and choice points. Sarbin's claim was that the role of hypnotic subject is part of people's general knowledge, so it can be played out when there's someone available (the hypnotist) to take the complementary role. In short, hypnosis is more a matter for the social psychologist than for the clinician.

When MGP were writing their book, Orne's views were even more in the air than Sarbin's. His notorious claim—which I remember filling the newspapers at the time—was that some cases of hypnosis were actually a species of suggestion, and that hypnotists themselves were often deceived accordingly (Orne 1959; cf. O'Connell *et al.* 1970). Although (unlike Sarbin) Orne allowed that high-hypnotizable people do pass into a genuine hypnotic state, he argued that low-hypnotizables could *simulate* compliance with the hypnotist's expectations (remember MGP's "hoax" and "cooperative stooge"). He also suggested ways of identifying the deceivers, although experiments showed this to be more difficult than he'd expected. (Many years later, some would claim that *all*, or virtually all, cases of hypnosis are mere social "compliance", as opposed to a special mental state: Wagstaff 1981.)

Orne implied (what was later confirmed) that easily hypnotizable subjects are, in general, more suggestible in other ways. As he put it, they're more likely to accept, and comply with, the "demand characteristics" of the situation laid down by the hypnotist. But this didn't solve the mystery. His theory allowed that some cases of hypnosis were genuine, without explaining *how it's possible* for someone to accept the suggestion to do X, but to do it without any awareness of the delicate monitoring that seems to be necessary when X is done of one's own free will (*sic*).

MGP described hypnosis in terms of someone's accepting the hypnotist's "Plans" as their own. They acknowledged, however, that

> Obviously, a hypnotized subject does not relinquish all his capacity for planning. It is still necessary for him to supply *tactics for executing* the strategy laid down by the hypnotist. If

his cessation of planning were to extend to the point where he could not even work out the step-by-step details of the hypnotist's Plans, then we would expect to find him in a state of stupor . . . (p. 108; italics added)

Discussing the various levels of hypnosis, they spoke of "the kind of planning by inner speech that sets the voluntary strategy for so much that we do". But even if that level of planning is "captured" by the hypnotist, "there must be still other planning functions that are performed more or less automatically as a kind of 'housekeeping' ".

In sum, MGP had many insightful things to say about hypnosis. These included their distinction between "thoroughly habitual" behaviour, such as speaking grammatically, and the "spontaneous inner speech" that guides normal voluntary behaviour (p. 111). Even so, their suggestive remarks didn't amount to an adequate theory.

Talk of "dissociation" soon seemed more promising—and in my opinion it still does (for a critical review, see Kirsch and Lynn 1998). This old idea (Chapter 5.ii.a) was revived in the 1970s by Stanford's one-time behaviourist Ernest Hilgard (1904–2001).

His pioneering experiments on hypnotic susceptibility in the early 1960s had already provided a plethora of new data—as well as a standardized measure of susceptibility to hypnosis (Hilgard 1965). (Hypnosis was so unfashionable in those behaviourist times that Hilgard had been the only person to apply for money set aside by the Ford Foundation for research in this area—Baars 1986: 296.) Now, he turned to dissociation as an explanatory construct (1973, 1977/1986).

Hilgard wasn't a card-carrying computationalist, although in the 1950s he'd been fairly close to Bruner (Hilgard 1974). It was at Bruner's invitation that he'd visited Cambridge, England, to attend a small meeting on cognitive processes—a maverick topic at that time, at least in the USA (6.iv.d). Among the other mavericks he met there were Broadbent and two-thirds of MGP: namely, Miller and Karl Pribram. (He also took the opportunity to meet Anna Freud, at Freud's house in Hampstead.)

His theory of "neo-dissociation", which he was still developing in his eighties (Hilgard 1992), posited a hierarchy of relatively autonomous subsystems, headed by a central controller (the "executive ego") capable of planning and monitoring action. So much wasn't new (5.ii.a). What was new—besides the greater (post-cybernetic) understanding of the notion of *control*—was his attempt to use dissociation (i.e. varying degrees of breakdown in intra-systemic communication) to explain not only abnormal phenomena such as hypnosis but a host of everyday details concerning *attention*, and the lack of it.

Hilgard intended his theory of attention to explain the nature of consciousness, and of the ways in which it can be compromised while behaviour carries on nevertheless. One could also say that it was a study of *freedom*, and the ways in which it can be adulterated. (Predictably, many were sceptical. Neisser, for instance, argued that psychology wasn't yet "ready" for consciousness: Neisser 1979.)

Of the explanations of hypnosis offered by today's computational psychologists, perhaps the most persuasive is that of Zoltán Dienes and Josef Perner (forthcoming; cf. D. A. Oakley 1999). They see their theory as closely related to Hilgard's but, modulo certain ambiguities in Hilgard's writing, different in one important respect (see below). They outline computational mechanisms whereby hypnosis of varying types can occur, and explain the ranked difficulty-differences mentioned above.

For example, they argue that the more computational effort goes into performing a task the harder it is to suppress higher-order thoughts (HOTs) of intention. This may explain "why challenge suggestions are more difficult than simple motor suggestions" (p. 16). Similarly, positive hallucinations may be especially difficult to induce because they involve more cognitive effort, of a type that would normally involve HOTs (pp. 15–16). In each of these cases, and others too, they outline experiments that could distinguish between various computational possibilities. In addition, their theory generates predictions (as I write, not yet fully tested) about how "high-" and "low-hypnotizable subjects" will differ in several tasks where hypnosis *isn't* involved.

Dienes and Perner, a cognitive and a developmental psychologist respectively, base their account on six main roots:

* One is the Norman–Shallice theory mentioned in the previous subsection (and described more fully in Chapter 12.ix.b).
* The next two are Perner's own long-standing research on the nature of representations of various types/levels (Section vi.f below), and his more recent account of a hierarchy of voluntary action (Perner 2003).
* The fourth is Annette Karmiloff-Smith's developmental psychology, in particular her theory of representational redescription (Section vi.h, below; and see Dienes and Perner 1999: 748–9).
* Next, is their previous joint work distinguishing various levels of "explicitness" and "implicitness" in representations—a nineteen-page *BBS* paper that prompted thirty-five pages of peer commentary (Dienes and Perner 1999).
* And last, is the CUNY philosopher David Rosenthal's (1986, 2000, 2002) concept of Higher Order Thought—HOT, for short. (HOT is intended as a theoretical analysis of conscious awareness. It's primarily because Hilgard helped himself to the notion of conscious awareness without analysing it that his theory differs from this one.)

In a nutshell, Dienes and Perner explain hypnosis in terms of "cold control". By *control*, they mean what Norman and Shallice called executive control—as opposed to contention scheduling (which underlies what MGP termed "thoroughly habitual" behaviour). By *cold*, they mean *absence of HOT*. That is, there's no higher-order thought.

More precisely, there's no explicit (self-reflexive) representation of the fact that the hypnotized person has the intention to do X. (As MGP put it, there's no spontaneous inner speech expressing the Plan to do X.) Nevertheless, the person does have that intention, communicated to them by the hypnotist. Moreover, they can execute it flexibly by means of detailed perceptual monitoring and motor adjustment. For that to be possible, the intention needs to be (in a particular sense explained by the authors) *explicit*. That is, it indicates some feature or state of affairs "directly" (2004: 4). But it's *not* represented at the higher level of explicitness that's required for conscious thought and verbal report. In other words, the person doesn't *know* that he/she has the intention to do X.

Dienes and Perner added the "cold" to the "control" because they needed to be able to explain *unconscious but planned* behaviour. Executive control as described by Norman and Shallice is conscious by definition. It applies to paradigm cases of planned, monitored, voluntary action—as opposed to absent-mindedness or pathological error

(12.ix.b). Much hypnotic behaviour is also planned and carefully monitored, as MGP had pointed out—but it's not guided by conscious intent or deliberation.

As for when and why the relevant HOT gets turned off, or inhibited from forming in the first place, this depends on a wide range of factors discussed in some detail by Dienes and Perner. They include internal cognitive and motivational conditions, and external social conditions. Norman and Shallice's "supervisory attention system" enables some action schemata to be more heavily weighted than others (12.ix.b), and Dienes and Perner adapt this idea in discussing how attentional mechanisms can aid/inhibit someone's falling prey to the hypnotist's suggestions.

Their theory, which takes account of computational work in many different areas (perception, memory, sensori-motor control . . .), indicates what types of turn-off/inhibition are *possible.* But experimental data are needed to discover which possibilities are realized, and with what frequencies. Sometimes, the data (such as the differential frequencies mentioned above) can be explained by the two authors in terms of the computational architecture they posit.

Dienes and Perner relate their theory to the notion of "multiple selves" (24 ff., 31–2). By implication, it can help us to understand the nature, and to some extent the aetiology, of multiple personality disorder.

As remarked in Chapter 5.ii.a (and in Preface, ii), the very possibility of multiple personality is a puzzle. Postmodernists insist that each of us 'possesses' a diversity of selves (and they welcome virtual reality avatars, as a way of playing with them: see 13.vi.e). But they don't explain how such diversity, or its extreme pathological version, is possible in the first place. The computational approach in general can demystify this strange phenomenon (Humphrey and Dennett 1989; Boden 1994*d*; and cf. Flanagan 1994). And Dienes and Perner's specific version of it helps us to distinguish competing accounts of *just what* may be going on. (Brain-scanning studies, given a good psychological/computational theory to help us to interpret the data, may help too: Reinders *et al.* 2003.)

In addition, though they don't say so, certain aspects of schizophrenia can be glossed in these terms. Hallucinations of inner voices, or delusions of alien control of one's own body, apparently involve planned speech or action *without* the normal HOTs. (Similarly, neurosis involves non-HOT beliefs and intentions, which guide thought and behaviour unknown to the person concerned: see subsection a, above.)

What causes the absence of HOTs in schizophrenia is another matter. Some recent neuroscientific studies suggest that a prime cause may be delay to, or inhibition of, reafferent motor signals—so that the person *does not* attribute responsibility to the self (Cahill *et al.* 1996; Silbersweig *et al.* 1996; C. D. Frith *et al.* 2000; S.-J. Blakemore *et al.* 2003). As for *which* person, imaginary or otherwise, is deemed responsible, that will naturally depend on rationalizations, or confabulations, guided by the idiosyncratic belief structure and anxieties in the person's mind (subsection a, again).

Cold control theory, then, fulfils some of the more startling promises made by MGP nearly half a century ago (Chapter 6.iv.c). Strictly, "fulfils" isn't quite the right word. For the theory has gaps. It doesn't explain, for instance, why "only about a third of people pass amnesia suggestions compared to 50% of people who pass the mosquito hallucination" (p. 17). Nevertheless, it appears to be very much on the right path. (Dienes and Perner themselves regard their theory "more as a means of theoretically orienting in the right direction rather than as a final explanation of hypnosis": p. 33.)

i. An alien appendage

Given a cognitive system capable both of Shallice's "executive control" and of all the layers of representation specified by Dienes and Perner, the possibility naturally (*sic*) arises of the first being activated *without* any guiding representation at the very highest, self-reflexive, level. If this happens, the person naturally (*sic*) won't be able to report any such non-existent representation—whether to themselves in introspection, or to others.

I say "naturally" here, partly because this computational possibility is *inherent* in any such system (cf. Section iii.d, below). But in addition, it's a significant fact about *Homo sapiens* that this possibility is often realized—and that when it is, other members of the species sit up and take notice.

I'm not thinking primarily of ordinary hypnotism, nor of the always eager audiences for the stage variety. Rather, I'm thinking of religious experiences, and their effects on religious believers.

As James (1902) noted long ago, there's a wide variety of "religious" experiences, ranging from inchoate feelings of some numinous presence, through visions and voices, to states of "possession" and/or alien bodily control like those mentioned above (Bourguignon 1976; Ferrari 2002). Sometimes, these come unexpectedly. At other times, they're more or less deliberately sought after.

Typically, they're interpreted by the individual concerned (and often by their fellows, too) in terms of some alien power. This, in turn, is glossed in terms provided by the religious tradition of the person concerned. Perhaps a fire god or an ancestral soul, perhaps a saint, perhaps some utterly transcendent Being . . . or perhaps even a nine-day-wonder belief in the Guardians on the planet Clarion (subsection c, above).

Besides drawing on those traditions, these anomalous experiences strengthen them further. The near-universal forms of religious culture discussed in Chapter 8.vi are typically buttressed by the religious experiences of priests, shamans, or just plain folk—and, in some cases, of schizophrenics too. Indeed, we'll see that one common explanation for the origin of religion refers specifically to those experiences (item 2 on the list given in 8.vi.b).

Given that the core religious *concepts* arise naturally for other reasons (8.vi.d), they can be maintained—and the power of the priests protected—partly because of visions and states of possession (and partly by mechanisms avoiding cognitive dissonance: subsection c, above). These involuntary states are attributed to the influence of gods or spirits largely *because* their controlling causes aren't introspectively accessible to those undergoing them. (That is, there are no HOTs.)

From a computational point of view, it's no surprise that the specific *content* of religious experiences can be communicated, or "suggested", by the culture concerned—and especially by charismatic shamans. Much as Senator Goldwater's belief structure led him to react in predictable ways to news items about the Berlin Wall (subsection c, above), so someone committed to a particular structure of religious belief will react in broadly predictable ways to alien—that is, involuntary and otherwise inexplicable—experiences.

In many societies, religious rituals have developed which encourage the occurrence of such anomalous psychological states, and which lead the persons concerned to interpret them in culturally preordained terms. They can be induced not just by charismatic suggestion, but by anything from monotonous Gregorian chants to psychotropic drugs such as mescaline. Once induced, the ritual fixes their interpretation in a culturally appropriate manner.

Dienes and Perner themselves point out that "spirit possession" can be seen as a species of hypnosis (2004: 30). Indeed, this suggestion is a very old one. The novel insight, here, is that such phenomena *are only to be expected* in minds with a computational architecture of the type described above. The specific *content* of religious belief may be very surprising, especially if we focus on the many cultural variations rather than the near-universal aspects (Chapter 8.vi). But *that religious experiences occur at all* is not.

MGP had pointed out in their manifesto that the overall "Image", or value system, differs in different cultures (1960: 122–3). Indeed, a culture *just is* an Image that's widely shared within a particular community, with various ritual and spontaneous Plans springing from it. However, they said very little about the Image in general terms. And they said almost nothing about cultural differences, except to remark that human societies adopt differing views about the importance of time, and of attempts to anticipate the future (p. 123).

If they'd tried to say more, their hand-waving would have been even more evident than when they discussed hammering. When Wallace (1965) applied their ideas in anthropology, his hands didn't stay firmly in his pockets either (Chapter 8.i.a). Even today, forty years later, computational analyses of cultural matters are still largely speculative, and usually based on verbal theories rather than computer models. But, as the example of 'auto-hypnotic' possession indicates (other examples are given in Chapter 8), our improved grasp of the nature of human minds can help us come to grips with phenomena that were previously mysterious in the extreme.

In sum, cognitive scientists have always believed that their approach could help us understand personality, psychopathology, and even culture. MGP were explicit about that. But the early efforts were grossly over-ambitious, and had to be dropped (subsections a–c, above). It was many years before they could be taken up again. Now, it's clear that MGP were right.

For instance, Dennett's work, and Shallice's, helps us understand human freedom. Dienes and Perner's throws light not only on hypnosis, but to some extent on Colby's concerns (neurosis and paranoia) too. An account of "relevance", to be described in Section iii.d below, throws light on the communicative processes involved in what Abelson called attitude change, and on what he termed structured belief systems. And architectural theories such as Minsky's or Sloman's have put nourishing meat onto the bare bones of emotion and personality sketched by Reitman (and Colby and Abelson) in the 1960s.

All these areas of research need to be developed further. To be honest—as Sloman was, in responding to the ludicrous query about simulating terrorists—the computational birds *still* can't catch the personal worm. (Sloman intends to work on his approach "for the next 300 years or so"—2003*b*: 7.) But in the new millennium the worm is, at last, in sight (see Chapter 17.iv).

7.ii. The Spoken Word

If 1960s computational psychology wasn't ready for personality, it was ready for certain aspects of language. As outlined in Chapter 6.i.e, language—thanks to Chomsky—was now being seen in a new way. Specifically, it was now regarded as a generative system whose rules (conceptualized in GOFAI terms) play a role in psychological processing—and may be implemented in computer models. This approach prompted research into NLP (described at length in Chapter 9.x–xi). In addition, it spawned a new branch of experimental psychology.

Because of Chomsky's (1957) influence, most of the very earliest experimental work focused on syntax. Soon, however, psychologists also studied meaning—in vocabulary, sentences, and connected texts.

The most important computationalist influence here wasn't Chomsky, but M. Ross Quillian. His 1960s theory—and modelling method—of "semantic networks" (see 10.iii.a) was widely used in AI and computational psychology alike. That had been his intention all along:

> The purpose of [my research] is both to develop a theory of the structure of human long-term memory, and to embody this theory in a computer model such that the machine can utilize it to perform complex, memory-dependent tasks. (Quillian 1967: 410)

Besides his own experimental work on language and memory (e.g. A. M. Collins and Quillian 1972), many others used theories based on semantic networks to ask psycholinguistic questions. One such was Schank, whose account of conceptual dependencies will be featured in Chapter 9.xi.d. Some other examples of the (voluminous) research involved are described in this section.

Three areas of psycholinguistic research, however, will be postponed to later sections. The early work on "procedural semantics" is described in Section iv.d–e. And two hugely controversial topics will be explored in Section vi.

One of these is the question of non-human languages. Chomsky's *Syntactic Structures* had already implied that to attribute language to any species is to say that it has not only meaningful communication but also syntax; and after the publication of the nativist *Aspects* in 1965, Chomsky fans predicted that no non-human species would qualify. The other is how language develops in human children. Chomsky's emphasis on syntax had raised new questions about this ancient puzzle. It will be discussed later, in the context of nativism in general. So the relative brevity of this section on 'The Spoken Word' shouldn't mislead you. Psycholinguistics has been a thriving area of cognitive science ever since the 1960s.

If psycholinguistics eventually burgeoned beyond Chomsky, so did (some) linguistics as such (see Chapter 9.ix). Today, a new field has been named: *cognitive linguistics.* Two examples, briefly mentioned in Chapter 12.x.g, are situation semantics (Barsalou 1999*a*) and blending theory (Fauconnier and Turner 2002).

Cognitive linguistics has abandoned Chomsky's purist vision of language-in-the-abstract (and the logician's model-theoretic view of meaning: 9.ix.c). Instead, it focuses on the cognitive processes involved in language use. Its researchers may be less likely than psycholinguists to get involved in laboratory experimentation or neuroscientific studies, but they're just as ready to use the empirical results in their work. In short, the line between psycholinguistics and cognitive linguistics is a fine one, more a matter of background emphasis and preferred methodology than of theoretical principle.

a. Psychosyntax

We saw in Chapter 6.i.e that psycholinguistics was born around 1960, thanks to Chomsky—and to George Miller's championing of him. Their "derivational theory of complexity" prompted an explosion of studies on the psychological reality—or, as it turned out, unreality—of transformations.

This Chomsky-inspired experimentation included examples concerning the effect of grammatical knowledge on perception. That knowledge can affect perception top-down was by this time 'old hat', having been supported by Gestalt psychology and the New Look—and by early GOFAI too (see Chapters 6.ii and 10.iv.b). Now, psychologists asked whether knowledge *of syntactic structure* could do so.

Among the experiments designed to find out were some in which random clicks were conveyed to one ear, while spoken sentences were being fed into the other ear (Fodor and Bever 1965; Bever *et al.* 1969). To cut a long story short, the clicks were heard as being at, or near, the phrase boundaries. So *That he was [click] happy was evident from the way he smiled* might be heard as *That he was happy [click] was evident from the way he smiled.*

Even here, there were many complications. For instance, it seemed that it was the boundaries in the *deep* structure which were important. The subject's production of the sentence in speech or writing, when informing the experimenters of the position of the click, could have influenced the result. And semantic influences couldn't always be ruled out. Nevertheless, and unlike the contemporaneous work on the psychological reality of transformations, these findings did hold up over the years.

As for *why* the clicks were 'displaced' in this way, the reason was believed to be one which Donald Broadbent had shown to be generally important. Namely, it was assumed that the moments at which clicks are relatively difficult to perceive are moments when many other information-processing tasks (involved in parsing the sentence) are being tackled. But which tasks are these, exactly? Some psycholinguists used ATNs (augmented transition networks: Chapter 9.xi.b) in answering that question (e.g. Wanner and Maratsos 1978).

(These experiments, like other work on auditory and visual masking, implied that a conscious percept *takes time* to form. This fact would later be highlighted in various discussions of consciousness: 14.x.a–b.)

Once Chomsky started talking also about *semantics* (9.viii.c), experimenters asked—for instance—whether parsing is done independently of meaning (as it seems to be for *'Twas brillig, and the slithy toves | Did gyre and gimble in the wabe*), or whether syntax and semantics are assigned to words simultaneously. In addition, when he confessed to his long-standing nativism—positing "innate ideas" concerning grammar—developmental psycholinguistics suddenly became a hot topic (see Section vi.a). In brief, it's impossible to overestimate the role that Chomsky played in getting modern psycholinguistics off the ground.

However, to say that psycholinguists were hugely excited by Chomsky isn't to say that they took his writings as gospel. This was just as well, for he pulled the theoretical rug out from under them many times over the years (see 6.i.e and 9.viii.b–c). After the very earliest days, most psycholinguists discreetly ignored his current theory. (A few explicitly rejected it, offering their own alternative: 9.ix.b.)

In any event, they were interested in many questions of little concern to Chomsky—specifically, how people *actually* speak.

b. Up the garden path

One intriguing aspect of actual language use is the occurrence of spoken sentences that are unproblematic from the linguist's point of view, but which cause problems in practice.

This isn't a matter of length, for sentences of indefinitely many words may be readily understood. Think of a page-long sentence by Marcel Proust, or—better—of a seventy-four-word example that even young children can delight in: *This is the farmer sowing his corn, that owned the cock that crowed in the morn . . . that lay in the house that Jack built.* But multiply-embedded sentences, which are perfectly legal according to the rules of grammar, are well-nigh unintelligible. Consider, for instance, *The rat the cat the dog the man the woman the car injured married stroked bit chased squeaked.* The only way to get one's head round this is to use paper and pencil to link the right noun with the right verb. Indeed, Miller found that even shorter nested sentences than this one will elicit errors in comprehension and recall: three levels of embedding are enough to cause trouble (Miller and Isard 1964). In other words, grammar as such allows for indefinite complexity, but there's a limit to the grammatical complexity that people can easily handle.

Moreover, there's the strange phenomenon of "garden path" sentences, in which the final words require one to reparse the whole of the preceding string. Karl Lashley gave one of these in his 'Serial Order' paper, as we saw in Chapter 5.iv.a: "Rapid righting with his uninjured hand saved from loss the contents of the capsized canoe". This is normally heard as "Rapid writing with his uninjured hand . . . ", until the occurrence of "capsized canoe". Or consider a much shorter example: *The horse raced past the barn fell.*

Many people find this sentence baffling, even after thinking about it for a few moments. If you're one of them, compare it with *The horse stabled in the barn neighed.* But why should anyone need such a clue? How can seven little words cause such difficulty?

Most Chomskyans didn't care. (They'd retreated to the mathematics.) But one, MIT's Mitchell Marcus (1950–), did. In the late 1970s he wrote a program called PARSIFAL—because what it did was to *parse* incoming sentences (M. P. Marcus 1979, 1980). This was one of the very few computer models to be based on Chomsky's grammar.

(Chomsky's grammar, but not *transformational* grammar. Hardly any models had been based on this, because comparing two tree structures—namely, deep and surface forms—is computationally expensive. Many psychologists assumed that what's computationally expensive is psychologically implausible, so that realistic NLP systems should aim for computational economy. Some *linguists* assumed this too, eschewing transformations as a result: Chapter 9.ix.d–f. Marcus used a post-rug-pulling version known as "trace" theory. In this variant, transformations were fewer and less complex, and sentences were represented by *annotated surface structures*, in which there were *traces* explicitly reflecting the deep structure and derivation.)

Miller's 'Magical Number Seven' had implied that there's a limit to the short-term memory load one can carry when parsing a sentence, and Marcus wondered whether this sort of thing might account for the garden-path phenomenon. More generally, he wanted to discover what type of parsing procedure would necessarily result in language's having certain "universal" features posited by Chomsky (specifically, the "complex noun phrase", "subjacency", and "specified subject" constraints).

That is, he wanted to know how the processing properties of the *mind* might have affected the abstract nature of *language*. Other types of language might be theoretically conceivable, but if their sentences couldn't actually be parsed by human minds then they wouldn't exist. Chomsky himself had suggested that highly abstract features of syntax result from properties of the "mental organ" responsible for language. But, apart from vague references to "memory limitations" for instance, he hadn't said just what those properties might be. Marcus aimed to find out.

His key hypothesis was that the parser inside human heads is *deterministic*. That is, it allows only one choice at every point during the parsing process: namely (in nearly all cases), the right one. This would account, he said, for certain grammatical rules and "universals" that Chomsky had described but couldn't explain. In addition, and especially interesting to psychologists, it would explain the introspective ease and speed of speech understanding.

Here, Marcus was swimming against the tide. Other NLP programs at that time were *non-deterministic*. ATNs, for example, allowed many choices at a given node, and sometimes all but one had to be traversed—and their computations at least partly undone—before the correct parsing was found; the same applied to Terry Winograd's SHRDLU, too (Chapter 9.xi.b). This was computationally expensive, and therefore—so Marcus felt—psychologically implausible.

But deterministic parsing, it seemed, was implausible too. On the one hand, there were the garden-path sentences, where the *wrong* choice is made early on, and this isn't realized (if it's realized at all) until the end of the sentence. On the other hand, there were sentence pairs such as "Have the girls taken the exam today?" and "Have the girls take the exam today!" We can't know whether the initial word "Have" should be parsed as an interrogative (auxiliary) or an imperative (main) verb until we've encountered the fourth word—"taken" or "take", respectively. Yet these two sentences, even without intonation cues, cause us no difficulty. How, then, could parsing possibly be deterministic?

Marcus dealt with the 'take/taken' objection by saying that a deterministic parser can't rely on a simple left-to-right pass through the sentence. It requires some degree of look-ahead, enabling it to inspect the future context. Clearly, if the parser could always inspect the whole sentence, the determinism claim would be vacuous. It's a substantive (and testable) hypothesis only if the size of the look-ahead is specified, and strictly limited. As for the puzzling garden-path sentences, if determinism-with-look-ahead could explain and/or predict *them*, that would be further evidence in its favour.

The crucial questions, then, concerned the size of the look-ahead and how it worked. The answers lay in PARSIFAL's "constituent buffer", described by Marcus as the "heart" of the program. (More accurately, the answers lay in *just how* the buffer was used by the program; those details are ignored here, but see Boden 1988: 110–14.)

The constituent buffer was a left-to-right sequence of three 'boxes', each of which is either empty or holds only one item. Each item is a syntactic category seeking a higher-level attachment—where S is higher than NP, NP is higher then Det, and Det is higher than "the" (see 9.vi). So "item" here means *any grammatical constituent*, including not only words but also syntactic substructures that the parser has already built—such as the NP ⟨ the big black cat ⟩.

How do the boxes get filled/emptied? Items can enter the buffer either from the right (if they're words) or from the left (if they're partially parsed constituents). However, words aren't read in to fill all the empty buffer places. If they were, the buffer would soon overflow. A word can enter the buffer only if some parsing rule has asked to see the contents of a buffer place which happens to be empty. Once an item has entered the buffer, it can be pushed rightwards, if there's an empty box to receive it. But it can't be pushed out of the buffer at the rightmost end. An item can leave the buffer only at the left end, and only after it has been attached to some higher-level constituent.

Never mind the details. The core idea embodied in the buffer was that the parser can look at up to three items *but no more.* So Marcus was claiming that a three-item look-ahead will suffice to parse *all* the sentences, no matter how long they may be, which people can understand without conscious difficulty. The farmer sowing his corn can be easily linked with the house that Jack built, despite the many intermediate clauses (cock, priest, man, maiden . . .), by a succession of automatic bottom-up processes filling, emptying, and refilling . . . only three buffer boxes.

Garden-path sentences, however, do cause difficulty. The reason, said Marcus, is that they need *more* than three boxes. When he made up new sentences that would require four boxes, he found that twenty out of forty people did indeed find them difficult to understand.

What about the other twenty? Well, some of them may have picked the right choice initially by chance. Some may have picked it because of intonation cues. Some may have been less conscious of backtracking (to undo wrong choices) than others. And some—said Marcus—may have learnt to extend their buffer so as to allow four, or even five, boxes.

However, that last suggestion went against the spirit of his theory, which assumed *inbuilt* processing constraints. Possibly, some people may be genetically anomalous and have four boxes (as some have six fingers). But in that case, "learning" isn't in question. Or perhaps some people have learnt to use compensatory processing techniques, which make it appear as though there were a larger buffer when in fact there isn't. In other words, perhaps they've changed the virtual machine involved in parsing, even though the genetically grounded hardware remains the same.

Psychologists could try to address all of those questions. Moreover, they didn't have to accept Chomsky's trace theory in order to do so, even though Marcus himself had hoped to find support for it. Their interest in his work was that he'd given them a way of raising, and thinking about, detailed psychological questions about the effortlessness of everyday speech—and why it sometimes flounders.

c. You know, uh, well...

In speech, as opposed to writing, grammatical mistakes abound: the linguist's law-abiding NP–VP is an idealization. In the language of Section iii.a below, real speech is (flawed) performance whereas grammar is (ideal) competence. Some mistakes in performance involve phonetics, not syntax (e.g. spoonerisms such as "shoving leopard" instead of "loving shepherd"). Yet others involve an unhappy choice of vocabulary (think of Richard Sheridan's Mrs Malaprop). Psycholinguists interested in actual language use studied a wide variety of common speech errors, and explained many of them in cognitive terms (Kempen 1977; Cutler 1982).

However, Freud had pointed out in *The Psychopathology of Everyday Life* that some speech errors are due to affect—notably, to anxiety. This may be repressed, in the Freudian sense, or open to introspection. All-too-conscious worries probably underlay some of the grammatical infelicities in the Watergate tapes, for example. (And, of course, the "expletives deleted" too.)

John Clippinger, for his University of Pennsylvania Ph.D. in the early 1970s (Clippinger 1977), tried to show how such real-world details may be sculpted by processes integrating many different aspects of mind. His prime concern, here, wasn't speech as such. Rather, he saw anxiety-based speech errors as a window onto the computational architecture of the mind as a whole.

He'd been alerted to this topic by the anthropologist Claude Lévi-Strauss, and was influenced also by Freud and—especially—Bateson (Clippinger 1977, pp. xv, 158–70). Above all, he was inspired by Winograd's work (and his seminars) on NLP, which gave him "a handle on how to think about and represent cognitive and linguistic processes" (p. xv).

As data, Clippinger used the transcripts of four years of taped therapy sessions of a depressed woman undergoing psychoanalysis. He analysed her speech errors, and tried to explain them in terms of general principles of cognition and emotion.

Besides certain types of grammatical error (e.g. restarting a sentence, or rephrasing a clause in mid-sentence), he noted factors such as hesitation, grunting (*uh*), qualification, euphemism, and the use of contentless expressions such as *you know* or *well*. Each of these occurs in this fragment of the woman's speech:

> You know, for some reason, I, uh, just thought about, uh, about the bill and about payment again. That (pause 2 seconds), that, uh (pause 4 seconds), I was, uh, thinking that I—of asking you whether it wouldn't be all right for you (pause 2 seconds), you know, not to give me a bill. That is, uh—I would (hesitates), since I usually by—well, I immediately thought of the objections to this, but my idea was that I would simply count up the number of hours and give you a check at the end of the month. [And on she goes, and on...] But maybe because of it would—uh (pause 2 seconds), it would reduce some—I—the amount of, uh (pause 2 seconds), I—exchange or, uh (pause 2 seconds), I was going to say interchange, but that's not right. And then I thought of, uh (slight laugh) (hesitates), intercourse between us. (Clippinger 1977: 29–30)

Supported by an analysis of similar passages in the body of transcripts, Clippinger suggested that what underlay this whole speech episode was the patient's initially wanting to request intercourse with the therapist—meaning not simply "sex" (forbidden by the superego) but also/instead "a warm, affective, and intimate relationship" (p. 114).

She'd quickly recognized, however, that even this wasn't possible, because of the financial/contractual nature of the transaction.

In listening to such troubled speech, the psychoanalyst (and the sensitive layman, too) can often pick up what's bothering the speaker. But Clippinger wanted to know *just how* the mind's resources can be marshalled to modify the verbal expression of anxiety-ridden topics. His theory of mental processing was much more detailed than Freud's, because he had nearly fifteen years of cognitive science (including Winograd) to draw from. He expressed it as a computer program called ERMA, designed to generate something like the transcript quoted above.

Much as Colby had seen the level and semantic content of anxiety as selecting a particular defence mechanism, so Clippinger saw these factors as determining the types of speech error. ERMA monitored its verbal expressions while they were being generated, and continually adjusted its previous plans accordingly.

A top-level SPEAK-UP program was designed to initiate possible topics for expression. After a topic had been selected, it was passed on for processing organized in five sub-programs, each of which was a complex system in itself. One, named LEIBNITZ (*sic*), contained the system's general knowledge, and was accessed by all the others. Another (CALVIN) dealt with the initiation *and censoring* of the content of thoughts, and of verbal expressions. The third, MACHIAVELLI, found ways of structuring complex content that had already been approved by CALVIN, so that the fourth (CICERO) could then express these concepts in words. As for FREUD, this sub-program tried to work out *why* certain concepts had been thought or expressed. In certain circumstances, it could pass its findings on to MACHIAVELLI for eventual expression in speech (see Clippinger 1977, ch. 7).

These five aspects of processing were called "contexts", not stages. For ERMA wasn't conceptualized as a rigidly sequential program, even though it was implemented by GOFAI techniques. Continual monitoring led to continual interruptions, as the various sub-programs diverted the control of processing away from the current goal. (This was an example of what Winograd had called "heterarchical" processing: Chapter 10.iv.a.) The results of such interruptions were reflected, directly or indirectly, in ERMA's output.

Hand simulations of the program (in which the topic was provided and/or the rules partly applied by Clippinger himself, using paper and pencil) produced non-Chomskyan strings such as: "That, that I was thinking that I—of asking you whether it wouldn't be all right for you, you know, not to give me a bill" (p. 118) and "You know, I was just thinking about, uh . . . well, whatever it was isn't important" (p. 202). In the latter case, of course, the primary goal—to express a desire for "intercourse" with the therapist—hadn't been modified, but completely suppressed.

As for the wholly computer-generated output, this included the following extract (sentences that were realized by ERMA during processing, but didn't appear in the 'spoken' output, appear in parentheses):

> You know for some reason I just thought about the bill and about payment again. (You shouldn't give me a bill.) ⟨Uh⟩ I was thinking that I (shouldn't be given a bill) of asking you whether it wouldn't be all right for you not to give me a bill. That is, I usually by (the end of the month know the amount of the bill), well, I immediately thought of the objections to this, but my idea was that I would simply count up the number of hours and give you a check at the end of the month. (p. 146)

If you compare this with the woman's own words quoted above, you'll see a great similarity—marred, for instance, by her use of one more *you know* and several more *uhs*. Clippinger remarked (p. 146) that the extra *uhs* would be more difficult to model than the *you know*, because they come at points where the speaker interrupts her thoughts to evaluate the impact of what she's going to say. Additional procedures would be needed in both CALVIN and CICERO to make—and make use of—such evaluations. His remark indicates, even without going into the details of the program (which, for its time, were considerable), that ERMA wasn't a mere cheat. It wasn't simply ELIZA-with-knobs-on.

As regards the psychological plausibility of ERMA's output in general, quasi-Turing Tests similar to those carried out by Colby (16.ii.c) showed that psychotherapists found it all too familiar. Moreover, they usually ascribed the same anxiety topics to the 'patient' as had actually been used by the program (pp. 196–210).

Clippinger hoped that his theory might aid practising psychiatrists, in both diagnosis and therapy. The crucial concept here was debugging, since "debugging a computer program and 'curing' a patient are similar processes, both of which require teleological descriptions to proceed" (p. 191).

Expanding on Gerald Sussman's pioneering AI work on self-debugging (10.iii.c), Clippinger pointed out that such descriptions (bug identifications) are often difficult to achieve. That's especially true if the manifest problem is due to more than one bug. For instance, suppose that a patient diverts attention from the original goal by formulating one or more less anxiety-ridden substitute goals. It may be difficult for her to know what her original goal was—particularly if (as could happen in ERMA) the substitute goals weren't *sub-goals*, but were produced by independent computations. It's difficult for the therapist too, but experience (and disinterestedness) helps.

'Cognitive' behaviour—dream reports, free association, and discourse on the couch—can provide clues as to just which thought structures and frustrated goals are involved. And if the Third Force are to be believed, ego-driven discourse can help suggest which alternative structures and goals might be most fruitful. But only the superego (which was responsible for the patient's problems in the first place) can effect a cure, by constructing alternative processing mechanisms. If that's done, the anxiety-fuelled speech errors (and other behavioural problems) will disappear, because they'll no longer be generated.

The bad news is that this self-debugging—like the computational deconstruction/reconstruction involved in grief and mourning (Section i.f, above)—will inevitably take a long time. Clippinger suggested experiments to discover whether training patients to think in terms of simple forms of debugging would help them to resolve their life problems. (I've been unable to discover whether he ever followed up that suggestion. But "cognitive therapy" in general often involves analogous exercises: D. M. Clark and Fairburn 1996.)

Clippinger gave two talks at an invitation-only TINLAP workshop (Theoretical Issues in Natural Language Processing) in Cambridge, Massachusetts, in 1975. One was on ERMA, the other on impersonal aspects of NLP. Despite this airing, his work on pathological speech fell into a black hole, and his hopes (217–18) for continuing interdisciplinary research into the effects of affect on cognition weren't realized.

Partly, this was because he left academia to set up a commercial NLP company. As an outsider to the scientific research community (Chapter 2.ii.b–c), his work was hardly known. (I've cited it myself, and so has Colby—K. M. Colby and Stoller 1988: 44; but I've never come across another reference to it.) More to the point, it was too early: he'd followed this theoretical path as far as it was then possible to go.

Perhaps it's still too early, even now. Current work on user-friendly "soft-bots" or computer/VR agents, for example, includes attempts to give their speech an 'emotional' dimension (see 13.vi.b–d). These are painfully crude in comparison with Clippinger's model, constructed almost thirty years ago. However, the growing attention to theories of computational architecture may make it possible for his ideas to be developed.

d. Meaning matters

Although—because of Chomsky—psycholinguistics started with a focus on syntax, it soon moved on to meaning. For that, after all, is what syntax is *for*. Broadly speaking, questions could be asked about the meaning of vocabulary, or sentences, or whole texts.

Vocabulary meaning was the main focus of an intriguing book by Miller, co-authored by Johnson-Laird (1936–). Described as a study of "psycholexicology", it sought to explain the development and use of words by means of procedural semantics.

A procedural semantics represents the meaning of language in terms of how the mind (or AI system) actually processes it. Words and sentences are thought of as mini-programs, as sets of instructions to the hearer to search for or set up certain representations in their mind and to perform certain computational/inferential operations on them. Winograd (1972) was one of the first cognitive scientists to recommend this type of approach, but as explained in Chapter 9.xi.b he didn't claim that his program SHRDLU was psychologically realistic. It was Miller and Johnson-Laird who introduced procedural semantics into empirical psychology.

Their 750-page book *Language and Perception* (1976) aimed to show how we develop word meanings by means of our sensori-motor interactions with the world, and how we use them thereafter. It ranged over perceptual psychology and psycholinguistics; the psychology and neurophysiology of motor action; non-extensional and deontic logic; and philosophical research on (for example) definite descriptions, semantic primitives, and concepts in general. And it offered detailed procedural definitions of many words, including prepositions such as *on*, *in*, *under*, *above*, *left*, and *here*.

In justifying their fine-grained *and* interdisciplinary approach, they said:

> So detailed a discussion of "That is a book" runs the risk of making an almost reflex act [i.e. understanding that sentence] seem peculiarly complicated. The very simplicity of the example makes analysis verbose, but without it there is a danger of suggesting something more than, or something different from, what we have in mind. The meaning of "book" is not [1] the particular book that was designated, or [2] a perception of that book, or [3] the class of objects that "book" can refer to, or [4] a disposition to assent or dissent that some particular object is a book, or [5] the speaker's intention (whatever it may have been) that caused him to use this utterance, or [6] a mental image (if any) of some book or other, or [7] the set of other words associated with books, or [8] a dictionary definition of "book", or [9] the program of operations (whatever they are) that people have learned to perform in order to verify that some object is conventionally labelled a book. We will argue that the meaning of "book" depends on a general concept of books; *to know the meaning is to be able to construct routines that involve the concept in an appropriate way*, that is, routines that take advantage of the place "book" occupies in an organized system of concepts. (1976: 127–8; italics added)

As you'll have guessed, if you didn't know it already, every one of those nine rejected disjuncts was a semantic theory held by some psychologists and/or philosophers.

Miller and Johnson-Laird posited a set of semantic primitives (see Chapter 9.viii.c), but of a novel kind. They allowed that some words are undefinable by other *words*. But they insisted that they're analysable psychologically. The meaning, they said, is carried by underlying concepts ("ultra-primitives") that don't correspond exactly to any natural-language words. These are contained—and often combined—in interpretative procedures that may be recursive and/or hierarchical. In short, to learn even an apparently 'simple' word may be to acquire a fairly complex procedure for using it.

On this theory, the semantics of those words whose meaning depends on our material embodiment is biologically plugged in—much as the semantics of low-level descriptions in vision is plugged in (Section v.b–d, below). And that includes a lot of words: hierarchical class inclusion, for instance, depends—according to Miller and Johnson-Laird—on the more primitive concept of spatial inclusion. But it didn't follow that the meanings of lexically indefinable concepts are innate. On the contrary, the two authors used up many pages in describing how infants learn the meanings of words.

They argued, for example, that some meanings (some interpretative procedures) can't be acquired without movement in and action on the material world. They cited evidence showing that infants learn the word "in" before they learn "on", and both of these before "under" and "at". And they pointed out that understanding these words requires the availability of bodily action schemata. For if it's physically possible to put *x* in *y*, then a 20-month-old child asked to put *x* on *y* will instead put it inside *y*; and if it's possible to put *x* on *y*, then a child asked to put *x under y* will place *x* on top of *y* instead. Their explanation of why children learn the word "at" later still was that its meaning is more complex: it involves the abstract notion of *region* and (when fully developed) concepts of *size*, *salience*, and *mobility*.

The computer programs (not actually implemented) that Miller and Johnson-Laird provided for *on*, *in*, and *at*, and many other words besides, were written as ATNs (9.xi.b). They were designed to interpret various linguistic features, such as the tense and aspect of verbs, and the use of definite descriptions (*The* such-and-such . . .). The chosen features were very general—within one language, or even across many; they were clearly defined in syntactic terms; and their meanings (and truth conditions) were relatively clear.

Syntactic cues helped determine the sort of processing that was specified in their ATNs. So an interrogative sentence would immediately prompt a search for the information requested, whereas a declarative sentence would (indirectly) lead to a search only if this was necessary in order for the program to construct its semantic representation. Likewise, the active verb in *Did Mary meet John at two o'clock?* instructs the hearer to search their memory for events in which Mary meets someone, to locate any in which she meets John, and to see if one of those happened at two o'clock. By contrast, the passive verb in *Was John met by Mary at two o'clock?* instructs the hearer to locate John's meetings first, then the subset in which he meets Mary, and finally the one (if any) which occurred at two o'clock. This analysis, said Miller and Johnson-Laird, explains our intuitions that the two sentences *both do and don't* have the same meaning.

Despite having been Chomsky's earliest collaborator, Miller was now swimming against the Chomskyan tide. For he and Johnson-Laird opposed Chomsky on the "autonomy" of syntax (9.ix.c). So did the philosopher Richard Montague, who in 1971 had "died in his home in Los Angeles, at the hands of persons still unknown" (Furth *et al.* 1974). But his radical ideas wouldn't be picked up by them—more strictly, by Johnson-Laird—until later (see Section iv.d). Meanwhile, they argued that the psychological evidence didn't support Chomsky's claim that a syntactic representation is built before the semantic one. Rather, syntactic cues are used during interpretation (as in the Mary/John examples), to help construct a representation of the meaning.

Their book attracted some interest, although not as much as it deserved. George Mandler (personal communication) feels that "the title was all wrong—in 1976 people interested in language weren't interested in perception, and also *vice versa*". A quarter-century later, the citations totalled only about 700. Moreover, very few of the citations in the first fifteen years were in mainstream psychological journals, as opposed to journals of linguistics, AI, and developmental psychology (at that time, not a high-status field).

In retrospect, Johnson-Laird (personal communication) believes their research would have made more of an impact if they'd published the 200-page paper they'd written by 1972. But they decided to turn it into a book—which grew, and grew, and grew:

> There was a great exchange between George Miller and Phil Johnson-Laird when they were finishing that book on perception and language. They had gone through revision after revision after revision after revision and I can't remember whether it was Phil or George who said, "Well, we've got to decide whether it's going to be perfect or Thursday." So they decided it was going to be Thursday, to the great relief of Harvard University Press. (J. S. Bruner, on whether one can ever publish "the final account of anything", quoted in Shore 2004: 8)

Maybe there'd been too many Thursdays already. For few people read the book carefully from cover to cover. One critic, for instance, accused them of reducing language to perception, or perceptual primitives—which they'd specifically said isn't possible. Many people didn't bother to read it at all, restricting their view to the journals. And the two authors didn't publish any follow-up research. (Johnson-Laird turned to other topics, namely inference and problem solving by means of mental models: see Section iv.d–e, below.)

Together with Winograd's SHRDLU, *Language and Perception* did encourage its readers to aim for some kind of *procedural* semantics. But the details were largely ignored. Instead, most of the opposition raised broad issues in the philosophical psychology of language. For instance, Jerry Fodor (1978*a*, 1979) accused Miller and Johnson-Laird of being crypto-verificationists. (For a reply, see Johnson-Laird 1978.) That was fair enough: after all, they'd raised the philosophical issues in the first place. But they'd also offered detailed hypotheses about the meaning and acquisition of particular words, and those hypotheses weren't carefully addressed. Alternative procedural models of individual words, specified in comparable detail, were rare—not to say conspicuously lacking.

Given the unsuspected complexities of *at*, and even the hidden complexities of *on*, that's hardly surprising. But it was due also to the fact that, by the late 1970s, many psycholinguists were considering not single words, but sentences, conversations, or even stories.

They weren't (usually) writing computer programs. But they were doing research prompted by computational ideas—specifically, by recent AI research on NLP (9.xi.d).

The important influences were Schank and Abelson's theories of conceptual dependency (CD) and scripts, including Wendy Lehnert's work on question answering (soon to be mocked in the "Chinese Room" argument), and David Rumelhart's models of story grammars. These NLP ideas were followed up by experimentalists, many of their papers gracing the pages of the recently founded journals *Cognitive Psychology* and *Cognitive Science*.

For example, John Bransford and Marcia Johnson (1972, 1973) studied the spontaneous inferences—about cause, intention, or spatial location—that people make in interpreting single sentences or very brief texts. Gerard Kempen (1977) applied CD theory to explain common hesitations and speech errors, less florid than those studied by Clippinger. As for story grammars, which represented the plots of stories, these were studied in psychological terms—and related to computer models of text interpretation (van Dijk 1972; Rumelhart 1975; J. B. Black and Wilensky 1979; J. M. Mandler 1984). Many psychologists, including Bower and Jean Mandler, tried to find out whether people's understanding of and memory for stories are grounded in high-level concepts like the schemas, plans, goals, scripts, MOPs, TOPs, and TAUs emanating from Yale in the 1970s (J. M. Mandler and Johnson 1976; Thorndyke 1977; G. H. Bower 1978; Black and Bower 1980; Lichtenstein and Brewer 1980; R. P. Abelson 1981*a*; J. M. Mandler 1984). These concepts were grounded in Abelson's earlier work on plans and scripts (see Section i.c, above). So a MOP (memory organization packet) denoted the central features of a large number of episodes or scripts unified by a common theme, such as *requesting service from people whose profession is to provide that service* (Schank 1982). TOPs (thematic organization points), too, were high-level schemata that organized memories and generated predictions about events unified by a common goal-related theme (such as *unrequited love* and *revenge against teachers*); but, unlike MOPs, they stored detailed representations of the episodes concerned, rather than their thematic structure alone. And TAUs (thematic abstraction units) were defined as abstract patterns of planning and plan adjustment, such as *incompetent agent, hypocrisy*, and *a stitch in time saves nine*, that aid not only language understanding but analogy and reminding as well (Dyer 1983; and see Chapter 9.xi.d).

Just what these psychologists found out is a tricky question. Because the ideas about schemas and scripts were higher-level and less precise than ideas about syntax, or than Miller and Johnson-Laird's psycholexicology, their research was less detailed accordingly.

Bransford's experiments, for instance, couldn't show that CD theory captured the specifics of human psychology. At best, they showed that it reflected *something like* what goes on in human minds. Similarly, the discovery that people do indeed remember stories in terms of their thematic structure showed only that *something broadly like* the Yale group's TOPs and TAUs was at work. It would have been surprising if that weren't so, for these concepts were intuitively identified in the first place, and TAUs were even named by clichés (see Chapter 9.xi.d). Critics therefore objected that CD theory—whether programmed or not—was no more than common sense dressed up in novel jargon (Dresher and Hornstein 1976).

However, theories constructed with *no* thoughts about computational processes, such as Frederic Bartlett's pioneering version of schema psychology (which the NLP scientists had picked up), were even vaguer. Accordingly, they were even less likely to prompt detailed questions about what actually goes on inside our heads. Given what was going on in AI at the time, it's no accident that a host of experimental studies on language understanding by means of semantic schemas were done in the mid- to late 1970s.

7.iii. Explanation as the Holy Grail

Chomsky's work didn't turn psychologists' eyes towards only language: it turned them towards psychological explanation as such. Over the next quarter-century, *what counts* as an explanation was a controversial topic within cognitive science.

Today, it's not easy to appreciate how novel this was. Quite apart from *what* computational psychologists were saying about explanation, *that they were saying anything at all* outside the opening pages of an introductory textbook was surprising. Despite Craik's having devoted a whole book to the "philosophy" of psychological explanation in the early 1940s (Chapter 4.vi), this topic wasn't much debated. Worse, it was widely shunned. As late as the 1960s, student texts would still celebrate psychology's "liberation" from philosophy—only to follow with a tendentious statement of operationalist positivism, presented as though it were plain fact.

To be sure, the Third Force personality theorists had questioned the positivist party line. Perhaps their most philosophically sophisticated work was Gordon Allport's defence of "idiographic" (and proactive) explanations, as against "nomothetic" (and reactive) ones (G. W. Allport 1942, 1946, 1955). Very broadly speaking, these mapped onto Cronbach's "applied" and "academic" approaches. Idiographic explanations, Allport said, were based in human empathy, and they applied to *particularities*: what his colleague Bruner has recently called "human beings in specific situations" (Chapter 1.iii.f).

In response, the psychologist Paul Meehl (1920–2003) at the University of Minnesota, where philosophy of science was a special strength, had recently provided a thorough discussion of this issue (Meehl 1954). He championed the nomothetic approach. For Meehl, reliable personal prediction could be based only in some "statistical cookbook", whether represented in the tables and tick-boxes of the MMPI (a personality test developed at Minnesota) or tacitly in clinicians' heads. (Prediction as pattern recognition.)

In choosing to write at length on these issues, however, Meehl was an exception. Most experimentalists agreed with him, but didn't bother to say so. In short: for 'Newtonian' psychologists, philosophy was a dirty word. Hence the scandal when Chomsky not only started talking about explanation, but defiantly placed himself within a highly unfashionable—namely, rationalist—philosophical tradition (9.ii–iv).

By the early 1980s, things had changed. A hugely influential book on vision opened with a chapter on 'The Philosophy and the Approach' (D. C. Marr 1982: 8). One year later another influential book, on reasoning, started by defending a Craikian analysis of explanation—and a functionalist philosophy of mind (Section iv.d, below). *BBS* featured a lengthy debate on the philosophical relation between computation and cognition, soon followed by an even lengthier book from the main protagonist (Pylyshyn 1980, 1984). And when Colby (1981) described PARRY in the pages of *BBS*, he—and the peer commentators—devoted many *BBS* column inches to discussing *what it means* to say that a model is "equivalent" to the theory and/or phenomenon being modelled (Colby 1981: 532–4).

A host of other examples could be given. In a word, the philosophy of psychology had become respectable again. (It was always respectable for *philosophers*, of course: see Chapter 16.)

a. Competence and performance

I said, in Chapter 6.i.e, that Chomsky thought of transformations as timeless mathematical functions whereas psychologists thought of them as actual mental processes. That's true, but misleading. He was happy enough to interpret them *also* as psychological processes, when the evidence went that way—as it did, for a while, with respect to his and Miller's "derivational" theory. But when the evidence became problematic, he declared it irrelevant and retreated to mathematical linguistics (cf. Itkonen 1996: 493).

The reason why he "pulled the rug from under" the psychologists in the 1960s *wasn't* that experiments were casting doubt on the psychological reality of deep structure and transformations. It was that he believed he'd found a more formally elegant way of representing the abstract structure of language.

A 'pure' mathematical theory isn't under any threat from facts. But the claim that it describes something in the real world is. By the mid-1960s Chomsky had moved beyond pure mathematics, for he was now making claims about language (abstractly conceived) and *mind* (9.ii and vii, and Section vi.a below). He even confessed that "linguistic theory is mentalistic, since it is concerned with discovering a mental reality underlying actual behavior" (1965: 4). This meant that he couldn't justify his ignoring psycholinguistic experiments simply by saying, "I'm a mathematician: get off my back!" Rather, he justified it by drawing a psychological distinction between *competence* and *performance.*

Performance concerns what people actually do when they speak on particular occasions (English as she is spoke). It's complicated by many different factors, including "memory limitations, distractions, shifts of attention and interest, and errors (random or characteristic) in applying [one's] knowledge of the language in actual performance" (Chomsky 1965: 3). And it's what experimenters have to work with when they observe their speaking subjects.

Having defined performance, however, Chomsky said almost nothing about it. He'd have had no interest in Clippinger's work, for instance. So far as he was concerned, performance got in the way of the *really* interesting issues. It was the other concept, competence, which fascinated him.

For the record, Chomsky's neglect of performance eventually led to "a gulf between linguistics and the rest of cognitive science that has persisted until the present" (Jackendoff 2003: 652). This explains the "irony" mentioned in the preamble to Chapter 9, namely, that although theoretical linguistics was hugely important in the origins of cognitive science, it's now almost invisible. Various people, including two ex-pupils of Chomsky, have recently tried to effect a rapprochement between theoretical and empirical studies of language (e.g. Pinker 1994; Jackendoff 2002, 2003). In *Aspects of the Theory of Syntax*, however, Chomsky had erected a solid firewall between the two.

The notion of competence applied not to real-world speakers, with their *ums* and *ahs* and egregious grammatical errors, but to "the ideal speaker–listener" (1965: 3). Indeed, his second major book (*Aspects*) opened with the words "Generative Grammars as Theories of Linguistic Competence".

Competence, he said, is the underlying knowledge of grammar that's possessed by every native speaker—considered not as a list of facts but as a system of generative processes. A "fully adequate" grammar must assign a structural description to any sentence of the language, indicating "how this sentence is understood by the ideal speaker–hearer" (1965: 4–5).

Introspection is useless for discovering this: "A generative grammar attempts to specify what the speaker actually knows, not what he may report about his knowledge" (1965: 8). Since "any interesting generative grammar" will be dealing with "mental processes that are far beyond the level of actual or even potential consciousness", it follows that "a speaker's reports and viewpoints about his behavior and his competence may be in error" (p. 8).

The native speaker is needed, to judge word strings as acceptable or not. But beyond that, the study of competence is an abstract exercise—to be done by linguists, not experimentalists. (How Chomsky went about it is described in Chapter 9.)

His notion of competence underlay his view of *explanation*. To explain language understanding, he said, was to show that it exemplified the structural principles defined by his generative grammar. A theory may be merely "descriptively adequate", in that it reflects/predicts a relatively small set of data. But one which is "explanatorily adequate" identifies the underlying mechanisms, so can be extended to many other examples outside the initial data set. Chomsky's view was that in any fully "adequate" explanation, the abstract ideal must always come first.

This prioritizing of competence quickly became one of the most well-known aspects of his work. (Newell and Simon, in emphasizing "task analysis", were making a similar point at much the same time: see Section iv.b.) Clearly, Chomsky's sort of explanation was very different from operationalist laws linking dependent/independent variables, and negating some theoretically unmotivated null hypothesis.

Others soon followed him in asking what a psychological explanation *is*. And many focused on something he'd ignored: how *a functioning computer model* can be seen—and judged—as a psychological theory.

The first serious discussion of computational explanation in psychology was written around the time of *Aspects*, by Chomsky's disciple and colleague Fodor (1968; see 16.iv.c). Granted, Hilary Putnam (1960) had already argued that psychology should be computational (16.iii). But he'd been thinking of folk psychology rather than the professional variety—and he hadn't even mentioned computer modelling. (To be fair, there wasn't much to mention—but the Logic Theorist had already made its mark.) Fodor went further. Dismissing the reductionist assumption (held by the behaviourists and by Hebb) that any explanations of S–R correlations must be neurological, he saw the "special science" of psychology as offering computational explanations.

But even Fodor didn't discuss particular computer models in detail. As computer simulation became increasingly common, many people started asking *just how* a program could count as an explanation. Some were professional philosophers, such as

Gunderson (1968, 1971), then at UCLA (later, at Princeton). Others were psychologists, able to relate their philosophical arguments to specific data and theories in the field.

Foremost among these were Newell and Simon, whose "Physical Symbol System" (PSS) hypothesis was a general philosophy of mind *and of explanation*. The next subsection could have been devoted to them—but their PSS hypothesis will be discussed in Chapter 16.ix.b, instead. However, the vision researcher David Marr (1945–80) was important too.

b. Three levels, two types

Some ten years after *Aspects*, Marr developed a "three-level" account of psychological explanation which he explicitly likened to Chomsky's, and which received almost as much attention as Chomsky's had done.

Marr's first, theoretically most basic, level was the "computational" level. This identifies *the task* of the psychological domain concerned. Chomsky had shown, he said, that the core task of language is parsing. And the task of vision, he argued, is mapping from 2D information on the retina to 3D information about the scene. At the computational level, it's defined in terms of timeless mathematical functions, saying nothing about *processes*. (Compare: "competence" and the "ideal" speaker.)

Unlike Chomsky, however, Marr was a psychologist: he wanted to know *how things happen* in the mind/brain. So he posited two more levels: one for the mind, one for the brain.

The "algorithmic" level specifies a set of information processes capable of computing the abstract functions identified at the computational level. In principle, many different processes will be capable of doing this. The psychologist's job, then, is to propose candidate algorithms, to test their coherence/results in computer models, and to find out which ones are used in the real world. Finally, the third level deals with "implementation". It describes the material embodiment of the chosen algorithm, showing how *those* processes can be carried out by *this* stuff—namely, the neurones (or, in a computer model, the hardware).

It had been Minsky, not Chomsky, who'd first alerted Marr to the idea of the computational level. Minsky had remarked, at a talk on the cerebellum given by Marr in 1972, that to know *what questions we should be asking* about the cerebellum we must concentrate on "the problem of motor control", considered as a set of abstract constraints (see 14.v.f). For Marr, this was an epiphany. He generalized Minsky's remark to vision, and later (recognizing Chomsky as a soulmate) to psychology as a whole.

At a vision conference in late 1973, he spoke of two 'Levels of Understanding'. He developed this idea with his colleague Tomaso Poggio, who was thinking along similar lines (W. Reichardt and Poggio 1976; Marr and Poggio 1977). By 1976, the two levels had stretched to three. And in his book, the three-level theory took pride of place.

Previous research on vision—theories of stereopsis, for example—was dismissed by Marr as largely irrelevant. He criticized his predecessors less for getting their facts wrong than for having asked the wrong questions in the first place. And that, he said, was because they'd ignored the computational level. He even said the same thing about Newell and Simon's work on mental arithmetic (Section iv.b)—although he admitted that he couldn't say what basic "task" it was which they were ignoring (1982: 348).

In short, his three-level theory was supposed to cover *all* psychological explanation. Psychologists who didn't adopt this approach might provide intriguing *descriptions* of behaviour—but they were doing natural history, not science.

There were three important caveats, often raised over the following years. First, there's no such thing as "the" algorithmic level. There are usually many different layers of processing, many different virtual machines, involved in a 'single' psychological phenomenon. Marr himself posited several stages of visual processing (see Section v.b). And the architectural theories discussed in Section i.e–f are more complex yet—though Marr, of course, wouldn't have regarded them as *explanations.* (One philosopher proposed an intermediate explanatory level, "Level 1.5", which identifies the function being computed *and* the information used to do so, but without specifying an algorithm: Peacocke 1986; cf. Peacocke 1996.)

Second, there's no such thing as "the" implementation level. There are retinas, neural networks, neurones, dendrites, synapses, neurotransmitters, ions . . . all of which must figure in an adequate neuropsychology (and in persuasive computer models thereof: see 14.ii). And third, what's theoretically prior needn't be chronologically prior. One can usefully study the retina, for example, without first having an abstract definition of the task of vision (see 14.iv.a).

Marr had given the computational level priority: questions at the other levels must be posed in light of analyses at the first. And the theoretical priority, for him, went from second to third level too: knowing what the algorithm is (or may be) enables us to ask the right questions, to look for the right things, when we study the hardware. Chomsky had said the same, with respect to linguistics and the neurophysiology of language:

> [The] mentalistic studies [of language] will ultimately be of greatest value *for the investigation of neurophysiological mechanisms*, since they alone are concerned with determining abstractly *the properties that such mechanisms must exhibit and the functions they must perform.* (Chomsky 1965: 193; italics added)

In short, both Marr and Chomsky (like Broadbent before them) pointed out a main theme of Chapter 14: that a computational psychology can tell the neuroscientist what to look for.

In practice, however, there's a back-and-forth dialectic between the levels. Marr allowed that discoveries about neurophysiology (lateral inhibition or feature detectors, for instance: 14.iii.b and iv) can sometimes suggest hypotheses about the information processing involved. He even allowed that, in principle, processing constraints might affect the theory at the computational level:

> Finding algorithms by which Chomsky's theory may be implemented is a completely different endeavor from formulating the theory itself. In our terms, it is a study at a different level, and both tasks have to be done . . . [Nevertheless, it] even appears that the emerging "trace" theory of grammar (Chomsky and Lasnik 1977) may provide a way of synthesizing the two approaches—showing that, for example, some of the rather ad hoc restrictions that form part of the computational theory *may be consequences of weaknesses in the computational power that is available for implementing syntactical decoding.* (D. C. Marr 1982: 28–9)

So Marr's "first/second/third" labelling was what Karl Popper (1935) had called a "rational reconstruction" of scientific psychology, not a description of what actually goes on when psychologists do their work.

As well as defining three "levels" of explanation, Marr defined two "types" of theory. Type 1, he said, is based on an abstract identification of the task, as described above. Type 2 is very different—and, in practice, difficult or even impossible to achieve.

A Type 2 theory applies if and when the information processing is carried out by "the simultaneous action of a considerable number of processes, *whose interaction is its own simplest description*" (D. C. Marr 1977: 38). So a Type 2 explanation would be nothing less than a complete description of the processes responsible for the phenomenon concerned. What happens, happens: in explaining how it happens, there's no more—and, crucially, no less—to be said.

Marr wasn't the first to suggest that some psychological phenomena might be of this kind. In his Hixon Lecture in 1948 (Chapter 12.i.d), von Neumann had declared:

> [It] is futile to look for a precise logical concept, that is, for a precise verbal description, of "visual analogy". It is possible that *the connection pattern of the visual brain itself* is the simplest logical expression of this principle. (von Neumann 1951: 24; italics added)

But if that's so, it's not clear that we should speak of explanation (or theory) at all. Explanation is a psychological concept: it covers any account, scientific or otherwise, which makes something *more intelligible to someone* than it was before (Boden 1962). Sometimes, the neural "connection pattern" may be simple enough for us to understand how it produces the behaviour. This may apply to the female cricket's recognition of her mate's 'song' (15.vii.c). But mammalian nervous systems are far more complex. Certainly, one can (today) simulate a vast amount of psychological and neurological data in a single computer model. But one runs the risk of having an all-singing, all-dancing model that performs well (i.e. that matches the experimental data) but which no one can actually *understand* (see 14.vi.d).

c. The sweet smell of success

The pyrrhic victory promised by a complex Type 2 simulation prompts the question of what a "successful" computer model would be. What counts as a "good match" between the model and the *theory* it's supposed to be modelling? And what counts as a "good match" between model and *data*?

In the very early days of cognitive science, these questions weren't at the top of the list. Probably the people who took them most seriously were the AI pioneers Newell and Simon. It was difficult enough to get a computer program to do anything interesting, never mind worrying about degrees of match. All too often, the lack of match was painfully obvious—as in the quasi-personal models described in Section i.a–c.

In the late 1970s and 1980s, however, a great deal of ink was spilled on these philosophical/methodological questions. That was inevitable, for by that time many programs could at least *pass* as psychologically relevant. Short of raising one's hands in wonder (or horror), how could they be evaluated?

The most widely discussed penman was the Canadian psychologist Zenon Pylyshyn (1937–), already well known because of his counter-intuitive theory of imagery (Section v.a, below). The very first issue of *BBS* included a philosophical paper by him, and a similar piece joined it two years later (Pylyshyn 1978, 1980). A book chapter appeared in 1979, and was included in a popular anthology in 1981. And his book carried the argument further (Pylyshyn 1984). Thanks in large part to the interdisciplinary peer commentary in *BBS*, his ideas gained a wide audience within the field.

Pylyshyn championed a strong version of mind-as-machine. He had no time for "the computer *metaphor*". Computational psychologists should bite the philosophical bullet, and allow that a computer model might do *the very same things* as the mind does, *in the very same way*. In other words, mental processes *really are* computations (by which he meant formal transformations defined over representations: 16.iv.c).

A *successful* computer model, according to Pylyshyn, is one in which some (specifiable) aspects are "strongly equivalent" to the mental processes concerned. That is, both can be represented by the same program in some theoretically specified virtual machine. As for how psychologists can test for this, he said, they needn't depend only on direct evidence, such as 'matching' protocols in problem solving. That one system—mind or machine—takes computational steps which the other one doesn't, may show up (indirectly) in memory use, error patterns, and time factors.

Not all computationalists were entirely happy with Pylyshyn's account. Sloman (1978, ch. 5), for example, argued that whether a program did things "in the same way" as human minds was an ill-defined question, not least because which similarities/differences were relevant could depend on context. Nevertheless, the discussion spurred by Pylyshyn's work sharpened the issues significantly.

Besides helping cognitive scientists to decide which programs were genuinely interesting, Pylyshyn's work on explanation addressed a truism that was increasingly being used as an objection. Computational psychologists were often told: "Just because a computer does something in a certain way, it doesn't follow that the mind/brain does it in that way too."

This was (and still is) trotted out triumphantly as though it applied *only* to computer modelling. In fact, it was a special case of something much more general: what philosophers of science call "the underdetermination of theory by evidence" (e.g. Laudan 1998). In principle, *every* empirical data set can confirm (can be explained by) indefinitely many theories. (Compare: infinitely many curves can pass through a given set of points.) In short, chemists and computationalists are in the same philosophical boat—but most people notice the boat only in the latter case.

Usually, scientists would be lucky to have even two or three well-confirmed theories to choose from. As Chomsky had (in effect) said in his own defence, *My grammar fits the facts. If someone wants to oppose it, let them produce a rival that fits as well, or better*. (Eventually, they did: see 9.ix.d–f.) Pylyshyn's work suggested rational ways of choosing between superficially equivalent computer models.

Even more to the point, computational concepts and/or modelling had enabled psychologists to formulate rigorous explanatory hypotheses *for the first time*. As Chomsky, Newell and Simon, Marr, and Pylyshyn all pointed out, the vagueness quotient of previous theories was immeasurably higher than that of those couched in computational/informational terms.

d. Chasing a will-o'-the-wisp?

Could psychological explanation cover individual thoughts? Could it show precisely why *that* idea, rather than any other, came to mind? (Why *that* association, *that* slip of the tongue, *that* dream content?) And could it explain why *this* particular belief is accepted, rather than some other one?

Many people have answered "No". For Descartes, reasoning involved human "freedom": even in a deductive argument, he said, we can freely choose not to assent to the conclusion. Similar views are popular today, especially in the contexts of moral choice and creativity.

Eventually, cognitive science would address both of these (see Section i.g, above, and Chapter 13.iv, respectively). But its early focus was on more mundane types of thought. These included the New Look's studies of how concepts can affect perception; various forms of 'hot' cognition; and problem solving of diverse kinds (Sections iv.b–e, below). The hope—indeed, the assumption—was that conceptual thinking could be corralled by science.

This hope flourished for some twenty years. But in the early 1980s, Fodor caused a sensation (again!) by attacking it as a delusion. Having championed computational psychology since its inception, he now argued that it couldn't cope with the higher mental processes (Fodor 1983: see 16.iv.d). More specifically, it couldn't explain the origin of individual beliefs.

Explaining individual beliefs in scientific terms, he said, is impossible, because of the many degrees of freedom—i.e. the many generative possibilities—available in conceptual thought. Non-demonstrative inferences are less constrained than deductive ones are. Indeed, virtually any concept or belief can turn out, with sufficient subtlety and imagination, to be relevant to any other. For instance, skilled teachers or poets can draw previously unsuspected analogies to mind, so that their audience accept new beliefs and/or come to see things in a new way.

According to Fodor, this doesn't apply only to poetry, metaphor, or imaginative teaching. It applies also to what he called "the fixation of perceptual beliefs" (1983: 102–3). To believe (because we can see it) that there's a cup of coffee on the table involves not only vision, but maybe smell, and certainly memory—not least, for conceptual/linguistic knowledge about cups, coffee, tables, social rituals, and perhaps even table manners.

The integration of all these psychological sources can't be scientifically captured, said Fodor, because common sense (central processing) isn't *modular* (see Section vi.d–e below, and Chapter 16.iv.d). Modular computations are *encapsulated* and *domain-specific*: that is, they can't be influenced by each other, nor by high-level concepts or beliefs. Common-sense reasoning isn't encapsulated, since it consults many different kinds of information in computing answers to its questions. Nor is it domain-specific, since it can answer questions about many different things (1983: 104 ff.). (To do so, Fodor claimed, it doesn't use a proprietary form of representation but a common, domain-general, "Language of Thought": 16.iv.c.) In short, it's scientifically unmanageable.

We can have hunches, of course: everyday gossip abounds with them, and so does Freudian psychology. Sometimes, we can even find experimental support. (For instance, Maier's 1931 study of functional fixedness showed that the idea of tying a weight to a string, to turn it into a pendulum, can—note: *can*, not *must*—be prompted by seeing the experimenter brush against it 'accidentally', causing it to swing: 5.ii.b.) But it's impossible for a scientific theory to predict someone's detailed thoughts, or even to explain them *post hoc*. Even if we knew of every single idea in someone's mind, we couldn't be sure just which ones had led them to think *this* thought: the system is too complex.

This made sense of Allport's defence of idiographic explanation (see above). When we want to explain particular thoughts of particular people, we must rely largely on common sense, personal empathy, literary sensitivity, or a psychotherapist's 'intuition'. But whereas Allport was content to regard these as scientific "explanations", Fodor wasn't. He saw them merely as more or less well-informed, more or less plausible, hunches—or, in other words, hermeneutic interpretations (16.vi–viii).

For Fodor, then, individual thoughts are inexplicable. It followed (or so he claimed) that computational psychology—"the only scientific psychology we've got"—must be restricted to "modular" phenomena such as syntax and low-level vision. He regarded these as scientifically tractable, being innate competences unaffected by conceptual thought (see Section vi.d–e below, and 16.iv.d).

The implication was that computational psychologists should abandon the more 'sexy' topics: neurotic rationalization, political attitudes, creative analogy, and even rational problem solving. Fodor claimed that "it is becoming rather generally conceded" that AI work on problem solving, "despite the ingenuity and seriousness with which it has often been pursued", had produced "surprisingly little insight". In short, it was "a dead end" (1983: 126). To try to explain thinking is to chase a will-o'-the-wisp.

Many readers interpreted Fodor's 1983 book as the death knell of computational psychology, with respect to the higher mental processes. Those who'd been unsympathetic to cognitive science in the first place were especially easily persuaded. As they saw it, they'd been vindicated. Hermeneutics had won! Indeed, Fodor himself had almost said so:

> [In explaining central cognitive processes], cognitive science hasn't even *started*: we are literally no farther advanced than we were in the darkest days of behaviorism . . . If someone—a Dreyfus, for example—were to ask us why we should even suppose that the digital computer is a plausible mechanism for the simulation of global cognitive processes, the answering silence would be deafening. (1983: 129)

If Fodor, an arch-priest of cognitive science, was declaring much of it to have been a monumental waste of time, why would they want to argue with him?

However, this reaction was justified only if (1) Fodor was right in denying modularity to central processes, and only if (2) one *expects* a psychological explanation to cover the detailed origins of particular thoughts.

The first point drew two rather different replies. These were closely interrelated however, since the anthropologist Daniel Sperber (1942–) was a prime proponent of each of them. On the one hand, evolutionary psychologists would soon posit modules for many types of thought normally regarded as "central" (recognizing cheats, for instance: Cosmides and Tooby 1992, 1994)—and Sperber agreed (2002). (Fodor himself was unpersuaded, as we'll see: Section vi.e below.) On the other hand, Sperber challenged Fodor's claim that any belief can be inferentially linked to any other.

He did this, together with the linguist Deirdre Wilson (1941–) , by formulating a theory of communication and thought in which the philosopher H. Paul Grice's (1967/1975) advice on conversation—*Be cooperative—and in particular, be relevant!*—was taken seriously (Sperber and Wilson 1986, 1996). Indeed, "advice"—like Grice's terminology of "conventions", "norms", and "maxims"—was the wrong word. Their approach didn't merely go beyond Grice, but replaced his convention-based analysis with a new "principle of relevance"—*Every act of ostensive communication communicates the presumption of its own optimal relevance*—which was involuntary, exceptionless, and evolved (1986: 155–71).

Sperber and Wilson rejected Fodor's suggestion that laboured scientific inference is a good model for everyday, instantaneous, understanding (1986: 66–7). Similarly, they denied the GOFAI assumption that deliberate reasoning (needed by literary scholars and historians, puzzling over obscure texts) is required for spontaneous interpretation (p. 75). This, they said, isn't done by logical inference, whether deductive or inductive, but by non-demonstrative guessing. However, the guesses are constrained by what we take to be relevant—where *(pace* Fodor) some things are more relevant than others. Without such constraints, effective communication simply couldn't happen. (Similarly, effective problem solving couldn't happen if *just anything* might be useful; in other words, this was a verbal/conceptual version of the notorious frame problem: 10.iii.e.)

Sperber and Wilson defined "relevance" in terms of a cost–benefit analysis, weighing effort against effect. The more information-processing effort it would take to bear *x* in mind in the context of *y*, the more costly this would be: and high cost gives low relevance. The more implications (regarding things of interest to the individual concerned) that would follow from considering *x*, the more effective it would be: and high effectiveness gives high relevance.

They didn't suggest (paradoxically) that we pre-compute just what effort/effect would be involved in considering this concept/belief, or that one. But they did say that there must be psychological mechanisms having much the same result: quasi-modules, evolved for recognizing relevance in speaker's utterances—and in other problem situations too.

For example, our attention is naturally (*sic*) caught by movement, because moving things are often of interest. (Think tigers!) Similarly, even the newborn baby's attention is preferentially caught by human speech sounds (see Section vi.h, below). In general, current sensory input indicates relevance. So (the potentially ambiguous) "Put the blue pyramid on the block in the box" is assumed—*without* conscious inference—to apply to *the one and only blue pyramid already sitting on the block* if perception shows that such a thing exists; otherwise, the pyramid is put *onto* the block (see 9.xi.b).

Besides being built into our sensory systems, relevance recognition is built into our memories. It's no accident, Sperber and Wilson said, that similar and/or frequently co-occurring memories are easily accessible, being 'stored' together in scripts, schemas, and conceptual hierarchies.

Different individuals may adopt different cognitive strategies, which vary in the measure of cost or benefit they attach to a given conceptual 'distance' (see Chapter 4.v.e). Similarly, different rhetorical styles involve different levels of cost and/or different types of information processing in both speaker and hearer.

In 'literal' speech, potential ambiguities usually aren't noticed. Rhetorical styles such as hyperbole, irony, sarcasm, and metaphor are different. Here, utterances are like garden-path sentences (Section ii.b, above) in that the first interpretation is implausible: they must be recognized as non-literal before they can be properly understood.

Again, Grice had said something like this already. But Sperber and Wilson went into much greater depth. They pointed out, for example, that his definition of irony as *saying one thing but meaning the opposite* doesn't capture the "echoic" nature of irony, nor explain "why a speaker who could, by hypothesis, have expressed her intended message directly should decide instead to say the opposite of what she meant" (p. 240). Ironic utterances require second-order interpretation, and assumptions about the speaker's knowledge and intentions, for their actual implications to be worked out (pp. 237–43). In short, decisions are required about what is—what possibly could be—relevant.

With respect to Fodor's pessimism about understanding the higher mental processes, they didn't deny that one could, at a pinch, find a tortured 'relevance' in virtually anything. To that extent, he was right. But they did deny that this would be possible in general: life's too short. Even poets have to provide enough context to make their meaning communicable. And everyday speech, in general, has to be understood *immediately.*

This raises the second point distinguished above: Fodor's assumption that a scientific explanation of thought (if such were possible) would explain every individual belief in detail. That assumption was unreasonable.

Some years before the appearance of Fodor's book, Sloman had argued that scientific explanation *in general* identifies abstract structures generating distinct classes of possibilities (Sloman 1978, chs. 2–3). Correlational "laws" and event predictions are sometimes available, but they're a special case. Physics is thus an example of science, not the paradigm case. Its laws (including the mathematical patterns rediscovered by AI programs: Chapter 13.iv.c) define *structured sets of possibilities*, much as computational psychology does. In physics, the structures are often simple enough, and the initial conditions accessible enough, for us to deduce 'full' explanations (and even predictions) of specific events which interest us. (Hence our ability to predict the time and place of a space vehicle's landing on the moon.) In the psychology of human thought and personality, that's not so.

This is why computational psychologists tended to pay less attention to statistics than their orthodox peers—an attitude guaranteed to bemuse, even shock, many experimental psychologists. If one's question is "What do people, mostly, do?", then statistics (such as Meehl's "cookbook") are essential. But if one's question is "*How is it possible* for someone to do *this* (which may have been done only once)?", then statistics are irrelevant. What's needed instead is a generative theory showing how the episode concerned could arise. That counts as explanation, too.

Statistics are needed, of course, to show that information-processing mechanisms of various kinds are indeed present. In Chapter 8.i.b, for instance, we'll see that the discovery of conceptual "prototypes"—which replaced Bruner's feature-list picture of concepts—involved many careful statistical measures of behaviour. Similarly, psychologists' work on schemas, from Bartlett onwards, had used statistical methods to discover them. By contrast, AI theorists relied mainly on introspection and common sense in positing specific schemas and scripts. That didn't matter if the main problem was *how schemas in general might be structured and processed.* But it mattered more if the question was *which specific schemas people in culture x have in mind* (hence much of the suspicion of AI, on the part of empirically trained psychologists).

Sloman's account of scientific explanation was compatible with Fodor's claim that the human mind is too complex, and our knowledge of the contents of any specific mind too limited, for us to explain every individual belief. What a computational psychology of thinking could do, however, is to show—in general terms—*how a particular thought is possible* (and maybe even unsurprising) in the first place. To identify the underlying processes and structures—from specific schemas to personal architecture—that make individual thoughts possible is, in an important sense, to *explain* them.

(That is, I agree with Sloman here. Some psychologists wouldn't. Bruner, for instance, now explicitly contrasts an "explanatory", generalizing, psychology with an "interpretive" or "narrative" one, which deals with *particularities*—see Bruner 2002, and Shore 2004: 40–1, 157.)

A prime example of such 'possibilistic' explanation was already in the pipeline while Fodor was writing his 1983 book. Karmiloff-Smith's (1986, 1990*a*) theory of "representational redescription" explained how increasingly imaginative, or creative, thinking can develop from limited and uncomprehending skills. It didn't predict *just which* imaginative idea would arise. But it did say what *structural types* of idea could arise, and (up to a point) how. Eventually, Karmiloff-Smith (1992) would use that theory as part of a wide-ranging critique of Fodor's views on modularity (Section vi.h, below).

In sum: despite being right about the explanatory elusiveness of individual thoughts, Fodor's despairing obituary of non-modular psychology went too far. Fortunately, it didn't bring computational studies of thinking to a halt.

7.iv. Reasoning and Rationality

How people reason was a prime topic of cognitive science from the start, thanks to the AI thunderclap caused at Dartmouth by Newell and Simon (6.iv.b and 10.i.b). The electric charge was carried on Simon's early 1940s ideas of bounded rationality and heuristics. Over the next half-century, those ideas would increase in power. In the early years of the new millennium, the philosophers would hold meetings on "dual-process" theories of reasoning (at Fitzwilliam College, Cambridge, for example), and—shockingly—would invite empirical psychologists to speak there. Frege must have been turning in his grave (see Chapter 2.ix.b).

The psychologists' message, in a nutshell, was that rationality-in-practice can't be bound by logic. To be sure, Frege's distinction between logical norms and psychological processes was genuine. But it didn't follow that real thinking should always match the standards of logical rationality. For 'ought' implies 'can'—and, considered as real computational systems, we *can't* think in a purely logical way. There's not world enough, or time.

Newell and Simon themselves, and John R. Anderson (1947–) too, used a new method of programming from about 1970 to model bounded rationality. And in the 1970s–1980s Johnson-Laird argued that non-logical models *in the mind* are used to ground even 'logical' thought.

As for those cognitive psychologists who weren't computer modellers but who stressed people's reliance on heuristics, their research spoke with forked tongue. On the one hand, it confirmed that people typically don't employ so-called 'rational' methods, and often make *mistaken* judgements accordingly: the *bounds* were more prominent than the rationality. On the other, it showed that the unthinking use of very simple heuristics can be astonishingly *effective*: the bounds made rationality possible. The first point was due primarily to Daniel Kahneman and Amos Tversky (and to Peter Wason and Johnson-Laird) in the 1970s, the second to Gerd Gigerenzer in the 1990s. Gigerenzer's specific suggestions were new, and shocking. But as we'll see, his central insight had been expressed by Simon some twenty years earlier.

This work escaped from the laboratory, and excited the general public. The view that our everyday judgements are "irrational" was as threatening to our self-esteem as Freud's stress on unconscious urges. N. Stuart Sutherland's trade book *Irrationality: The Enemy Within* (1992) sold briskly, and the topic was widely featured in the media. Ten years later, the public (their appetite already whetted by Sutherland's book) were fascinated by Gigerenzer's suggestion that we often think *best* when we ignore maths and logic—as our evolutionary ancestors did.

Moreover, a committee in Stockholm pricked up their ears. The only Nobel Prizes yet awarded for Psychology—well, technically for Economics (there's no prize set aside for Psychology)—went to Simon in 1978 and to Kahneman in 2002. (Tversky had succumbed to melanoma in 1996, and Nobels aren't awarded posthumously.) Both were honoured for discovering that the realities of decision making are very different from the idealized 'rational decision theory' of neo-classical economics. Kahneman, for example, had found that people will take a special trip to buy a $15 calculator for $10, but won't do so to save 'the same' $5 on a $125 jacket. In purely economic terms, that made no sense. In other words, cognitive science had shown how very *inhuman* the dominant picture of 'rational man' had been.

A non-psychological coda: in fact, Simon won his Nobel not for humanizing economics but for attempting to do so. To his lasting regret, the Stockholm committee—and people concerned with practical business management (who'd appointed him Professor of Industrial Administration at Carnegie Mellon)—were more appreciative than the 'pure' economists:

> Economists did not flock to the banner of satisficing with its bounded rationality. These ideas still remain well outside the mainstream of economics. (1991: 364)

This wasn't self-regarding paranoia. For it's confirmed by a recent remark from a professional economist:

> It is striking that economists, *after neglecting Simon's ideas for decades*, are now close to [still only "close to"!] accepting them, *but still cite Simon only rarely*. For example, Stigler is widely cited as the economist who brought optimal search theory into economics in the early 1960s, whereas Simon had already brought both optimal and satisficing search into economics in the mid 1950s. (Conlisk 2004: 193; italics added)

The only economists who *have* been influenced by him for many years are the computational economists, or market modellers—including, now, evolutionary economics (see Chapter 15.ix.d and Mirowski 2002: 529–30).

Simon's explanation for this blindness on the part of economists was the need for "backbreaking empirical work", requiring "close, almost microscopic, study of how people actually behave". The economists, he said, had never been interested in that.

In his autobiography, he remembered "armoring myself against the aesthetic lures of neoclassical economics, so responsive to mathematical elegance and so indifferent to data" (1991: 53). He'd tried to persuade mathematical economists to take account of data, aka cognitive psychology (Simon 1978). But they were almost as dismissive of it as 'Fregean' philosophers would be.

a. Simon's ant

If Newell and Simon had moved far away from orthodox behaviourism by the mid-1950s, they hadn't dismissed it as worthless. On the contrary, they claimed to be "natural descendants" of both behaviourists and Gestaltists (Newell *et al.* 1958*a*), and said:

> Today [i.e. 1961] psychology lives in a state of relatively stable tension between the poles of Behaviorism and Gestalt psychology. *All of us have internalized the major lessons of both.* (Newell and Simon 1961: 110; italics added)

The main strengths of behaviourism were its detailed observation and its recognition of environmental influences. They valued both. And the second was epitomized by the most famous insect in cognitive science: Simon's ant.

In his (still) widely read Karl Taylor Compton Lectures on *The Sciences of the Artificial*, given at MIT in 1968, Simon described an ant walking on the ground (1969, ch. 3; cf. Simon 1962). Its behaviour appears highly complex, zigzagging around countless tiny pebbles and lumps of earth. But if the *behaviour* is complex, the mechanisms generating it aren't: "An ant, viewed as a behaving system, is quite simple." The creature can't plan its locomotion ahead of time: it doesn't have the brain. Instead, it responds to environmental cues, encountered from moment to moment. If it meets an obstacle, it just turns away and continues walking.

That is, it's a biological version of Grey Walter's ELSIE (Chapter 4.viii.a). (But much more interesting: recent work on insects has discovered many different navigational mechanisms—see 15.vii.) Moreover, it's potentially open to what A-Lifers would later study as "stigmergy": a 1950s biological term meaning *social integration brought about by individuals responding to environmental signs deposited by other individuals*, as in ant-trails caused by the laying-down of pheromones (see Bonabeau 1999, and 15.x.a). In short, said Simon, the complexity isn't in the ant, but in "the environment in which it finds itself".

So what?—Well (Simon continued), as with ants, so with people: "Human beings, viewed as behaving systems, are quite simple." For instance, someone doing arithmetic is often driven by environmental cues, such as the relative positions of the numbers written on the paper. In brief, the physical (and, for humans, cultural) world constitutes an "external memory" storing crucial information outside the organism. He even suggested that long-term memory itself should be viewed "less as part of the organism than as part of the environment to which it adapts".

Like most of Simon's psychological ideas, this one too was rooted in his economics. It had arisen in the 1940s, when he read the town planner Lewis Mumford's book on medieval cities. Their beauty wasn't planned, but grew "out of [today, many would say 'emerged' from] the interaction of many natural and social forces" (Simon 1991: 98). He'd realized then that market economics needed to allow for the influence of "externalities (for example, noxious odors wafted from the stockyards to surrounding neighborhoods)". In short, internal mechanisms don't suffice to explain the richness of what actually happens.

The phenomenologist Maurice Merleau-Ponty (1945/1962) had made a similar point, at greater length (16.vi–vii). Simon didn't cite him, and perhaps—despite his exceptionally wide reading—wasn't familiar with him. If that's so, he wasn't alone. As we saw in Chapter 2.vi, this branch of philosophy had diverged from the scientific/empiricist tradition long before. It was unusual enough for someone with Simon's intellectual background to have read the Gestaltists carefully; to have devoured Merleau-Ponty too would have been highly eccentric.

For some twenty years, Simon's ant was largely ignored by cognitive scientists, especially in AI. Some 1960s social psychologists had reached a similar position by a different path (Section i.c above), but they weren't much read by cognitivists. And James Gibson (1977, 1979) soon made an essentially similar point. But he was *persona non grata* because of his rejection of "representation" and "computation" (Section v.e–f, below).

In the late 1980s, however, the importance of automatic response to the environment would be rediscovered by cognitive psychologists (especially Gigerenzer, as we'll see), and by the situated roboticists—who mistook it for a revolutionary insight. In A-Life generally, much would be made of the fact that organisms are *situated* in their physical and/or cultural environment (15.vii–viii.a). And in A-Life-oriented philosophy, the long-neglected phenomenologists would become names to conjure with (16.vii).

Why the delay? After all, Simon was a towering presence in cognitive science. If he'd realized the importance of environmental cues in the 1940s, hinted at it in scattered 'asides' in the GPS papers, explicitly highlighted it in 1962 and 1969, and (as we'll see in the next subsection) implemented it in the late 1960s–1970s, why was it so often overlooked?

In a sense, he himself—with Newell—was to blame. First, the excitement caused by their top-down planning programs was so great that bottom-up 'ant insights' were ignored in the very early days of GOFAI. Second, even when they abandoned GPS planning for productions (to which the ant insight was crucial), their main emphasis, still, was on information processing *within* the mind. In short, despite their obeisance to external memory, they didn't prioritize *material embodiment in the physical world*, as the situated robotocists later would—and as the phenomenologists already had.

Nevertheless, Simon's 1940s notion of bounded rationality had always had a situated ant hidden inside it. For when asked what the difference was between *bounded* rationality and *irrationality*, he used the analogy of a pair of scissors: the mind is one blade, the structure of the environment is the other. To understand behaviour, one has to consider both—and, in particular, *how they fit*. Looking at one only (as cognitive psychologists sometimes tend to do) is about as fruitful as cutting paper with a single scissor blade.

b. Productions and SOAR

Simon himself made obeisance to the ant by means of a new form of programming he developed with Newell in the late 1960s: production systems. This approach revolutionized AI, and gave a very different flavour to psychological modelling.

Even while GPS was still being hailed as a breakthrough, they'd realized that it needed 'humanizing' much as neo-classical economics had done. GPS possessed full knowledge of strictly limited, and static, domains. But life's not like that. A theory of problem solving should explain how we manage to be rational in a largely uncertain, and continually changing, world. That, said Newell and Simon, was what production systems could do.

These programs were sets of IF–THEN rules, or "productions". Each rule specified (one or more) actions to be taken in response to (one or more) particular cues, or conditions. Some of the cues and actions arose from, or modified, the external environment. But most specified information processing within the system. In other words, these were 'ant programs'—with most of the action being triggered *inside* the ant.

The internal actions included the setting-up of a new goal, or sub-goal (implemented as a new condition); and one type of cue was a reminder of the current goal. If goal A was current, in *these circumstances* (also specified in the IF-side of the rule), then the system would perform *those actions*; if goal B was current, it would do something else. So hierarchical goal seeking was included in production systems, not by top-down planning as in GPS but by bottom-up procedures designed to work in an untidy world. Since a given condition might crop up at any time, behaviour could be interrupted by, and/or instantaneously adapted to, unpredicted events (including thoughts like *Oh, that reminds me . . .*).

They were intended to be biologically plausible in another way, too. Simon's paper on 'The Architecture of Complexity' (1962) had argued that, in evolution, individual mutations must give rise to *distinct* sub-procedures. Accordingly, each production was logically independent—so a new one could be added at any time. (Damaging interactions with pre-existing rules had to be avoided by foresight and/or debugging, since evolutionary programs weren't yet available: see 15.vi.)

Production systems (as described so far) were a universal programming language: see Chapter 10.v.e. It's no wonder, then, that AI technologists used them for many purposes. The first volume of the journal *Artificial Intelligence* described a program that *learnt* to play poker, written by one of Newell and Simon's students (Waterman 1970). The first 'practical' applications were DENDRAL and MYCIN (see 10.iv.c), and the HEARSAY speech system—which was part-planned by Newell himself (Newell *et al.* 1973; Reddy *et al.* 1973; Reddy and Newell 1974). Others soon followed, including the "expert systems" that galvanized AI funding around the world in the mid-1980s (11.v).

But Newell and Simon weren't primarily interested in technology. Their 900-page *Human Problem Solving* (1972), which introduced production systems to psychologists, was an exercise in *psychology*. Because of this, they deliberately constrained the power of their production systems, to match the bounded rationality of human beings.

For example, they took Miller's "magical number seven" seriously (Chapter 6.i.b). So the eight-point summary of their theory included these two claims:

(1) [The human information-processing system] is a *serial* system [i.e. only one rule fired at a time] consisting of an active processor, input (sensory) and output (motor) systems, an internal LTM and STM [long-term and short-term memory] and an EM [the "external memory", or perceptible field].

(3) *Its STM holds about five to seven symbols, but only about two can be retained for one task while another unrelated task is performed.* All the symbols in STM are available to the processes

(i.e., there is no accessing or search of STM) [in effect, it's a blackboard: see Chapter 10.v.e]. (1972: 808; italics added)

These were no mere metaphors, but literal hypotheses:

(7) *Its program is structured as a production system*, the conditions for evocation of a production being the presence of appropriate symbols in the STM augmented by the foveal EM. (p. 809; italics added)

In short, Newell and Simon were trying to describe the mind's *general architecture*, not merely how it functions in specific situations.

Several of the eight theory points referred to the *time* needed for the mind/brain to process information. "Elementary processes", it was said, take about fifty milliseconds, while "writing [into LTM] a new symbol structure that contains K familiar symbols takes about 5K to 10K seconds", and "accessing and reading a symbol out of LTM takes a few hundred milliseconds". Out of context, that might suggest that theirs was a *neurological* theory. But it wasn't. The symbols were presumably implemented in the brain by Hebbian cell assemblies, but Newell and Simon weren't concerned with that. They defined *symbols* in an abstract, computational, way (see Chapter 16.ix.b). As they put it, their information-processing theories represented "a specific layer of explanation lying between the behavior, on the one side, and neurology on the other" (p. 876).

Counting milliseconds wasn't the only evidence of attention to detail. The book reported levels of program/protocol matching, in domains such as chess and crypt-arithmetic, that far surpassed any previous computer models (for a fuller discussion, see Boden 1988: 154–70). As a result, their new methodology was adopted by many computational psychologists—and not just for problem solving. It was soon used to model child development (Richard M. Young 1974, 1976), and later applied to detailed motor skills such as typing (Card *et al.* 1983).

Their own production systems were both precise and economical. For example, consider this cryptarithmetic problem:

```
  DONALD   (D = 5)
 +GERALD
  ------
  ROBERT
```

As you'll discover if you try to solve it yourself, this isn't a straightforward exercise. Nevertheless, Newell and Simon explained their subject's behaviour by *only fourteen* IF–THEN rules (p. 192).

This was logician-as-ant: "Human beings, viewed as behaving systems, are quite simple." But the emergent complexities of human behaviour can be considerable. So some of the fourteen rules dealt with goal–sub-goal organization. Some directed the attention (e.g to a specific letter, or column, in the sum shown above). Some recalled previous steps in the process, so took account of intermediate results. Some handled the generation—and recognition—of false starts. And some enabled the backtracking needed to recover from them. In short, purposive behaviour was being modelled—but not by top-down planning.

Just as their earlier programs had been designed only after careful "task analysis"—their term for the identification of what needs to be done if the problem is to be solved—so were these (1972, ch. 3). The logical constraints of cryptarithmetic and chess were carefully thought out beforehand. Moreover, their eight-point architectural theory provided the general context for every individual program. So they didn't fall foul of Drew McDermott's (1976) complaint that early AI workers were often playing around, rather than theorizing in a systematic way (see Chapter 11.iii.a).

They did, however, eventually fall foul of Marr, for whom "computational" explanation referred to unconscious, automatic, modular processing (Section iii.b, above). Cryptarithmetic is far from automatic, as anyone who's torn their hair out over "SEND + MORE = MONEY" can testify. Marr had no quarrel with Newell and Simon's emphasis on task analysis, nor even with their analyses of their chosen tasks. But he rejected their choice of tasks to be analysed. As he put it: "I have no doubt that when we do mental arithmetic we are doing something well, but it is not arithmetic" (1982: 348). And he added that "we are very far from understanding even one component of what that something is". In other words, he saw their apparently human-like programs as superficial gimmickry, not explanation.

A more common critique was that Newell and Simon were behaviourists in disguise, since individual productions are essentially similar to S–R connections. As self-confessed "descendants" of both behaviourist and Gestalt psychologists, they accepted the parallel. But they pointed out that, unlike the behaviourists (even including Hull: 5.iii.b), most of their stimulus–response—i.e. condition–action—pairs concerned *internal* processes, described in *information-processing* terms.

Newell and Simon's remarks about the brain met with resistance, too. If they'd followed the doctrine of multiple realizability strictly, they'd have been immune to criticism on this count. As it was, their brain theorizing led some psychologists to ignore their approach. So Walter Weimer, for instance, said this:

> Everybody in psychology knows [Herb Simon] is great, but nobody's ever read him. I never have taken him seriously, and I can tell you why. When you have a man who sits there and looks you straight in the eye and says the brain is basically a very simple organ and we know all about it already, I no longer take him seriously. And Simon tells you that constantly in his books and lectures. ... [He's] only kidding himself when he tells us that we already know how the head works. That is not cognitive science, but abject cognitive scientism. ... [He] does not do cognitive psychology. He does something distantly related to it, but he doesn't do it. (interview in Baars 1986: 307–8)

In 1980 Newell—with John Laird (1954–) and Paul Rosenbloom (1954–)—started work on a new type of production system, to model the architecture of cognition *as a whole.* SOAR—the acronym stood for Success Oriented Achievement Realized—was implemented in 1983 (Laird *et al.* 1987). Reasoning in SOAR was a multidimensional matter, for the system integrated perception, attention, memory, association/inference, analogy, and learning.

SOAR differed in many ways from the 1972 production systems, which had modelled specific tasks (such as cryptarithmetic) rather than *general* intelligence. Its increased flexibility was due to several factors.

For example, different types of problem (some 'closed', others more open-ended) could all be handled within the same framework for defining problem spaces. Conflict resolution, needed when *several* rules have matching conditions, was handled differently—so that all the unfired (but potentially excited) rules remained visible to the system as a whole, instead of being repressed. Ant-like responses, or 'situated' behaviour, were combined with internal deliberation (in that sense, SOAR was a 'hybrid' system). Indeed, deliberation was often turned into reflex responses, so that a problem initially solved by the former was later dealt with by the latter. This involved "chunking", whereby sub-goal settings that had frequently been executed in sequence were rolled into one rule, which improved efficiency (and respected Miller's magical number seven).

Nevertheless, SOAR built on the insights pioneered in GPS and the 1972 programs. It treated problem solving as goal-directed movement through a problem space, and defined actions and operators suitable for various (specified) situations. In addition, it handled both procedural and declarative knowledge (as ACT had done earlier: see subsection c, below).

SOAR attracted huge attention across cognitive science. A variety of comments, including many by 'big names', plus a lengthy reply were published in *BBS* (Newell 1992), and another batch of comments/reply in *Artificial Intelligence* (Stefik and Smoliar 1993). The interest concerned not only its technicalities but also its general philosophy, elaborated in Newell's William James Lectures, given at Harvard in 1987 (Newell 1990, chs. 1–3, 8). (Because of the notice it received, he soon wrote several summaries: Newell 1992; Newell *et al.* 1993.) It consisted of two main claims.

One was already familiar, since the early days of GOFAI: a highly abstract account of mind-as-machine known as the Physical Symbol System (PSS) hypothesis (16.ix.b). This held that psychology concerns the generation and transformation of symbolic expressions, conceptualized at "the knowledge level". Like the IF–THEN rules for "DONALD + GERALD = ROBERT", those expressions could be (and sometimes are) verbalized by the subject. A paper on the SOAR updating of PSS was included in a 1990s collection on the foundations of AI (Rosenbloom *et al.* 1992). But by that time, the PSS debate—though still 'live'—was old news.

The second claim was more novel, and more interesting to model-builders. It was a methodological directive, underlain by the assumption that there are psychological mechanisms underlying *intelligence in general.* Instead of "microtheories" dealing only with specific tasks, Newell now called for "unified theories". These, he said, try to capture the general principles underlying all cognition: not just reasoning, but memory, language, perception, and attention too. In other words, the *architectural* aspect, which had been in the background of his 1970s research with Simon, was now brought to the forefront.

Simon himself didn't approve, as we'll see. Arbib, too, had reservations: "no single, central, logical representation of the world need link perception and action—the representation of the world is *the pattern of relationships between all its partial representations*" (1994: 29).

But many people were persuaded, and SOAR has been in continuous development ever since. Among the philosophers who thought well of it were Dennett (1993*a*) and, in particular, Richard Samuels (forthcoming). Today, it's widely used for both technology and psychology. A recent account runs to no fewer than 1,438 pages (Rosenbloom *et al.*

1993), and Soar Technology Inc. is one of the commercial offshoots. The practical tasks being handled run from medical diagnosis to factory scheduling.

Charles Babbage, given his own pioneering work on factory management, would have been impressed (3.i.a). But Simon wasn't. SOAR was Newell's baby, not his. They were still colleagues at Pittsburgh, but their thirty-year collaboration had ended. The 1960s and 1970s had seen a host of joint publications, but the 1980s saw only two: both historical commentaries on their earlier work (Simon and Newell 1986; Newell and Simon 1987). Although both men were still doing psychological AI, their interests had diverged.

Simon was dubious about the whole SOAR exercise. In his autobiography, he wrote:

> *In cognitive science there is currently [i.e. in 1990] a preoccupation with questions of general architecture, which I do not share.* There are great debates about whether the human mind is to be modelled by SOAR (Allen Newell), Act* (John Anderson) [see subsection c, below], connectionist nets (Jay McClelland) [see Chapter 12], or something else. I have been more interested in ... "theories of the middle range"—programs such as GPS, EPAM, ... and BACON, which simulate human behavior over a significant range of tasks but do not pretend to model the whole mind and its control structure.
>
> *It is not that I regard the broader architectural issues as unimportant; but, even when solved, they do not explain how very general schemes are adapted to perform particular classes of cognitive tasks. The architectures have almost more the flavor of programming languages than of programs.* (1991: 328; italics added)

While SOAR was occupying his one-time collaborator, Simon was investigating learning, creativity (13.iv.c), and mathematical education. He saw the last as especially important—indeed, it had been "a major hidden objective" in his research on learning. Worried about the prospects for democracy in high-tech societies full of "alienated" (non-scientific) intellectuals, he wanted to know how to overcome the common resistance to mathematics: "There is no question I would more like to answer than this one before my research career ends" (1991: 330).

The implication that he might outlive his research career wasn't fulfilled. He died in 2001, still at the height of his intellectual powers. (Newell had predeceased him, in 1992. For a discussion of Newell's AI obituary, see Chapter 10.v.b.)

c. The ACTs of Anderson

If Newell and Simon's 1972 theory was the first attempt to model the computational architecture of cognition as a whole, John R. Anderson's ACT (1976) was a close second. Indeed, early-ACT was even more clearly guided by this aim, which for Newell became paramount only with the birth of SOAR.

John R. Anderson (1947–) was a Canadian psychologist educated at Stanford, and employed at Michigan, Yale, and (from 1978) Carnegie Mellon. (He shouldn't be confused with the Brown/UCLA neurophysiologist James A. Anderson, who also worked on associative memory: Chapter 12.v.b and e.) His early work on human associative memory (acronym: HAM) included computer models co-published with Gordon Bower, whose prime research interest was language understanding (Anderson and Bower 1972, 1973; cf. G. H. Bower 1970, 1978).

The HAM theory (and mini-models) was based on a propositional network representing the structure of memory and linguistic meanings (Anderson 1976: 39–47). That is, it was more concerned with the *contents* of memory than with its *function*. Accordingly, it was soon superseded by ACT, a model (and an acronym) based on the adaptive control of thought. (Some of the differences between HAM and ACT were described in Anderson 1976: 270–90.) After an unproductive flirtation with ATNs, Anderson picked up Newell and Simon's new technique of production systems to focus on the *interactions* between memory and other cognitive processes.

Since a system as complex as ACT couldn't be built overnight, there were in fact several versions, named (when people could be bothered) with successive letters of the alphabet. The "ACT" described at great length in Anderson's 1976 book was in fact ACTE (Anderson 1983: 17). Soon afterwards, ACTE spawned ACTF, which modelled the *acquisition* of productions (Anderson *et al.* 1977/1980). So meaning, memory, problem solving, and learning were all grist to ACT's mill even in the early 1970s. The theory—and its implementation—was continuously developed over the years. The first comprehensive account appeared in 1976, and the learning version (ACT*) was described some six years later (Anderson 1982, 1983).

In 1980 Anderson published a textbook on cognitive psychology (now in its fifth edition) in which ACT as such was downplayed—but AI in general, and production systems in particular, were prominent. Moreover, the guiding aims underlying ACT were evident throughout the text. In the opening chapter, for example, he declared:

> Certain subdomains within the field—for example, perception, memory, problem solving, and language—are becoming well understood. Still, the form of a theory that would specify *how all the subfields in cognitive psychology interconnect* is still very unclear. Later chapters will present some of the theories that have been proposed to explain how these subareas are connected... (1980: 17–18; italics added)

And in Part II on 'The Representation of Knowledge', which began with an account of the striking early 1970s experiments on mental imagery, he supported "dual-code" theories, according to which visual images and verbal memories are stored in different ways (see Section v.a, below). However, he argued that (as in ACT*) *long-term* memories in either case are stored in some abstract "propositional" code that's neither visual nor verbal.

In comparison with other forms of computational psychology in the 1970s, ACT was unusual in three ways. First, it employed both declarative and procedural representations, of domain knowledge and practical skill respectively (see 10.iv.a). These resembled Gilbert Ryle's "knowledge that" and "knowledge how" (16.i.c). But Anderson's focus was on the fact that a person may know (for instance) that a certain Euclidean theorem is true, without being able to use it in a geometrical proof. In acquiring the skill of doing so, the learner gradually constructs a set—maybe, thousands—of production rules by trial-and-error application of the declarative proposition in many different circumstances.

Second, ACT combined GOFAI insights with (localist) connectionism, in the form of a network allowing spread-of-activation through its links. When the system was instigated, most people thought of connectionism as being very different from, or even opposed to, GOFAI (see 4.ix and 12.ii–iii). But there had been an earlier 'half-way-house': Quillian's (1968) semantic networks (9.xi.e and 10.iii.a). ACT could be seen as "a special case of the Quillian model—a special case sufficiently well specified to make predictions and be proven false" (Anderson 1976: 291).

Third, it was an *architectural* theory. That is, it was "a theory of the basic principles of operation built into the cognitive system" (1983, p. ix). Newell and Simon's 1970s work was informed by architectural assumptions too, as we've seen, but it didn't stress mental integration as heavily as Anderson did. It's no accident that his first book (1976) was called *Language, Memory, and Thought.*

So Anderson swam against the tide when, in the early 1980s, the mind was commonly viewed as a number of non-interacting modules (Section vi.d–e, below). Resisting this intellectual fashion, he directly contradicted a 'modular' quotation from Chomsky (1980*a*) on his opening page. He still stressed mental *integration*: "Memory, language, problem solving, imagery, deduction, and induction are different manifestations of the same underlying system" (1983: 1). Modularity, he said, applied only to "peripheral" processing. Whereas Fodor (1983) had recently ruled conceptual thinking out of court for a scientific psychology, Anderson's work was unabashedly aimed at the *higher* mental processes:

> *A major presupposition ... is that higher-level cognition constitutes a unitary human system. A central issue in higher-level cognition* is control—what gives thought its direction, and what controls the transition from thought to thought. Production systems are directed at this central issue ...
>
> It needs to be emphasized that production systems address the issue of control of cognition in a precise way that is relatively unusual in cognitive psychology. Other types of theoretical analyses may produce precise models of specific tasks, but *how the system sets itself to do a particular task in a particular way* is left to intuition. In a production system the choice of what to do next is made in the choice of what production to execute next. Central to this choice are the conflict resolution strategies ... Thus production systems *have finally succeeded in banishing the homunculus from psychology.* (1983, pp. ix–x; italics added)

(Here, both Fodor and Anderson could have been right. One may be able to discover the general principles underlying conceptual thought without being able to explain particular instances of it: see Section iii.d, above.)

Someone may have to construct "thousands" of production rules in acquiring a skill because the relevant declarative knowledge (e.g. a Euclidean theorem) can be used in many different ways. According to Anderson, the person has to learn what task goals (and sub-goals, and sub-sub-goals . . .) are relevant in which task circumstances, and what results a particular action will give in the various circumstances. This requires both teleological sensitivity and immediate feedback during problem solving. It also requires time:

> The acquisition of productions is unlike the acquisition of facts or cognitive units in the declarative component. It is not possible to simply add a production in the way it is possible to simply encode a cognitive unit. Rather, *procedural learning occurs only in executing a skill; one learns by doing.* This is one of the reasons why *procedural learning is a much more gradual process than declarative learning.* (1983: 215; italics added)

One shouldn't assume that Anderson's production rules were simple. Some were. But others, built up only after long practice, were not. An example of a production used in reading the word "EACH" contained no fewer than eighteen IF-conditions (some of which were themselves conjunctions), and five THEN-results (1983: 144).

According to Anderson, this reflects an important difference between how beginners and experts achieve a task (recognize a pattern, understand a sentence, solve a problem, manipulate a tool . . .). Experimental data, including some from Simon's research on scientific reasoning, suggest that where beginners use several steps, the expert uses only one. Simon, and Newell too, explained this in terms of declarative chunking, as we've seen. Anderson preferred a form of 'procedural chunking'.

As he put it, borrowing AI jargon, the expert has "compiled" knowledge of the task whereas the novice has "interpreted" knowledge (1983: 216, 255). So in ACT*, the 'composition learning operator' would convert a set of individual production rules that had often been carried out sequentially into a *single* production, that could be executed *without* having to access the declarative knowledge retrieved and interpreted by the beginner (1983: 235–41).

This throws light on an old puzzle in philosophy. The fourth "rule of method" identified by Descartes (1637), i.e. *recapitulation*, seems at first sight far too boring to constitute 25 per cent of a revolutionary 'Method of Rightly Conducting the Reason and Seeking for Truth in the Sciences'. But Descartes wasn't concerned here only with someone's remembering what they'd already read. He was interested also in their *coming to understand it better*. He said that someone who read his argument, or a complex geometrical proof, over and over again would eventually come to see the relation between the initial premisses and the final conclusion directly, or intuitively. Before that point, they'd have to rely on remembering that they'd arrived at the conclusion by going through many previous steps—and memory, he pointed out, is fallible. Introspectively, this seems right; and it fits the phenomenology of other skills besides reasoning, such as improvising on the piano (Sudnow 1978/2001). Anderson's theory suggests a way in which this mental progression could happen.

Anderson was eager to roam beyond the ivory tower. By the late 1980s, he was working on educational applications of his ideas about how people acquire new skills. So he designed 'intelligent tutors' that gave instruction in LISP, geometry, and algebra. These were used in the Pittsburgh Public School System, and elsewhere.

On the basis of this experience, he developed the ACT-R version and changed his approach to automated tutoring (Anderson 1993). Instead of trying to write programs that *emulate* the student, he now aimed at providing helpful learning environments, offering domain knowledge and feedback. The "general principles" he followed in designing computer tutors included communicating the goal–sub-goal structure that underlies learning, minimizing the load on working memory, providing immediate feedback, and adjusting the grain size of instructions as learning progressed (Anderson *et al.* 1987, 1990, 1992). The last of these reflected his theory of how novices differ from more advanced learners (see above).

ACT is still being improved, giving us ACT-R, ACT-RP, ACT-RPM . . . (Anderson 1993, 1995). Like Stephen Grossberg's ART family (14.vi.c–d), however, it risks becoming too complex to be readily intelligible. Moreover, not all computational psychologists are convinced that the underlying philosophy is sound. Newell's collaborators, for instance, regard SOAR as a superior hybrid architecture (they compared the two systems in Newell *et al.* 1989). And Simon, as we've seen, remained sceptical about *all* attempts to model the general principles of the mind.

d. Models in the mind

Where Newell and Simon, and Anderson too, had prioritized models *of* reasoning, Johnson-Laird—in the 1980s—focused rather on models *in* reasoning. Granted, he supported (and developed) his theories by doing computer simulation—and chided those colleagues who didn't: "Sadly, many experimental psychologists make no use of computer modelling" (p. xii). But his core hypothesis was self-confessedly Craikian (Chapter 4.vi): that "human beings understand the world by constructing working models of it in their minds" (Johnson-Laird 1983: 10 ff.).

Starting with seven talks given at Stanford early in 1980, followed by a 500-page book in 1983, he elaborated this idea as a novel computational theory of thought. It was a development of the procedural semantics that he and Miller had initiated at the end of the 1960s (Section ii.d, above). But it was even more closely integrated with work in the *philosophy* of meaning, as we'll see.

He'd already done some widely cited experiments on reasoning in the early 1970s, with his adviser Wason (1924–2003) of University College London (Johnson-Laird and Wason 1970; Wason and Johnson-Laird 1972; Johnson-Laird 1975). Using the "card selection" task originated by Wason some years earlier (1966), they'd shown that problems of identical logical form are much easier to solve when expressed as 'real' examples (e.g. involving the postage required for letters of different kinds) rather than as abstract *p*, *q*, and *r*. Indeed, this was true even of professors of formal logic.

The usual response had been to bemoan the "irrationality" of mere mortals, and pass on. For Johnson-Laird, however, the failures of the professors of logic sounded a warning bell. Apparently, something powerful was involved here, which enabled every Tom, Dick, or Harry to succeed in realistic cases but could desert even professional logicians when abstract examples were in play. He spent the next decade trying to discover what this "something" might be. And his answer was: mental models. (Later, two evolutionary psychologists would suggest a different "something", namely, a mechanism evolved to detect cheaters: see Section vi.d, below.)

His version of bounded rationality acknowledged the successes as well as the failures of human thought. But it didn't explain them in terms that Fregeans would respect. For Johnson-Laird, even logical reasoning doesn't depend on formal logic chopping, but on quasi-perceptual representations within the mind. As he put it:

> Mental models owe their origin to the evolution of perceptual ability in organisms with nervous systems. [David Marr has] outlined a computational theory of vision that largely accounts for the derivation of perceptually based models of the world [see Section v.b–d, below]...
>
> Mental models can take other forms and serve other purposes, and, in particular, they can be used in interpreting language and in making inferences. These roles are a natural extension of their perceptual function... Discourse, however, may be about fictitious or imaginary worlds, and hence our propensity to interpret it by building models of the states of affairs it describes frees us from the fetters of perceptual reality. (1983: 406–7)

Consider syllogisms, for instance. Johnson-Laird dismissed all previous psychological theories as—in Chomsky's terms (Section iii.a, above)—inadequate. They failed to explain why some syllogisms are intuitively easy and others hard, and couldn't be generalized to the development or teaching of syllogistic reasoning—nor to other types of inference (1983: 65–6).

Here's an example of an easy syllogism:

Some of the artists are beekeepers.

All of the beekeepers are chemists.

Nearly everyone (correctly) infers that some of the artists are chemists—and some people also (correctly) infer that some of the chemists are artists. "Nearly" everyone—but not quite. It's significant, said Johnson-Laird, that "the only person whom I have ever known to get the answer wrong is a distinguished philosopher who tried to exploit his logical expertise!"

I've seen the same thing happen, when playing around with Lewis Carroll's Sorites. Everyone in the room, including my 16-year-old daughter, got it right—except the friend with several logic publications under his belt. (A Sorites puzzle provides several premisses, and you must discover the one and only conclusion which follows from all of them taken together. In Carroll's hands there might be as few as three or as many as fifty premisses (Carroll 1977: 386–9), ranging from *No shrimp is remarkable for sagacity* (p. 407), through *No discontented judges are chickens* (p. 420), to *Any good-tempered man, who has lent me money and does not care for appearances, is willing to shake hands with me when I am in rags* (p. 387). Not all his premisses were as credible as these, so common-sense guessing couldn't help in dealing with them: e.g. *Brothers of the same height always differ in Politics* (p. 421) or *All spiders are healthy, except the green ones* (p. 406).)

By contrast, hardly anyone can deal with this:

All of the bankers are athletes.

None of the councillors are bankers.

(In case you're wondering, the only valid conclusion is *Some of the athletes are not councillors.*) "Few people", reported Johnson-Laird, "are able to cope correctly with [these] premises" (p. 67).

But why? And why is the "easy" syllogism more likely to yield *Some of the artists are chemists* than the equally valid *Some of the chemists are artists*? Indeed, why do these differences (and many others discovered by Johnson-Laird) persist when the argument is expressed informally in everyday conversation, instead of 'artificially' as a syllogism?

Johnson-Laird's view, in brief, was that the verbal formulation of the problem—*any* problem—prompts us to build an internal analogue of the state of affairs portrayed in it. This model is constructed in *understanding* the incoming sentence/s. Once built, it's then available for use in making inferences.

An initial model may be more or less easy to change, on receipt of further information of various kinds. And it may yield the answer to a specific 'question' (the search for a specific inference) more or less readily. Sometimes, mere (mental) inspection will suffice; at other times, further modification of the model will be required. This explains the ease/difficulty of problems of different forms, or of the same problem expressed in different ways.

For example, suppose someone is told the following:

A is on the right of B.

C is in front of B.

D is on the left of C.

Johnson-Laird claimed that they would construct an internal spatial representation, in which—in true Craikian style—the items standing for A, B, and C would bear relations to each other analogous to those between the real things (as described by the three sentences). With this representation in place, the person could then use it, for instance, to infer that A *is not* on the left of D.

His claim was perhaps false. But it wasn't hand-waving. For in this case, and many others, he provided a computer program detailing computations capable of constructing the model and of drawing inferences from it. His effective procedure for syllogistic inference (pp. 97–110), for example, had three main parts:

1. Construct a mental model of the first premise.
   [For instance,] the representation of a universal affirmative assertion has the following structure:

   All of the X are Y: x = y
   x = y
   (y)
   (y)

   where the number of tokens corresponding to x's and y's is arbitrary, and the items in parentheses represent the possible existence of y's that are not x's. [Comparable models were given for *Some of the X are Y*, *None of the X are Y*, and *Some of the X are not Y.*] (1983: 97–8)
2. Add the information in the second premise to the mental model of the first premise, taking into account the different ways in which this can be done. (p. 98)
3. Frame a conclusion to express the relation, if any, between the 'end' terms that holds in all the models of the premises. (p. 101)

The logical modalities were defined/explained accordingly. A conclusion was *possible* if it was true in at least one model, *probable* if true in most models, *necessary* if true in all models, and *impossible* if true in none.

Devotees of 'pure' rationality would baulk at this. For instead of a formal deductive proof, step 3 involves an *inspection* of all the models that happen to have been built. If the program (or person) had missed some out during construction (steps 1 and 2), an invalid conclusion could result. To be sure, Johnson-Laird's programs included procedures designed to avoid this (for instance, the inclusion of y's *in parentheses*, above). In the general case, however, complete modelling wasn't—and, of course, isn't—guaranteed.

In addition, the (various) "figural" effects on ease/difficulty of solution were modelled by computational rules based on the assumption that "working memory operates on a 'first in, first out' basis" (p. 105). It follows, for instance, that the "natural" order in which to state a conclusion is the order in which the terms were used to construct a mental model of the premisses. This explains the different probabilities of the two correct conclusions in the artists/beekeepers/chemists example.

(For the record, Johnson-Laird's views haven't changed radically since then. He says now that "the biggest development in my theory of reasoning since the '83 book is probably the discovery that people only represent what is true": personal communication, 2003. After detailing various common errors, including some which virtually *no one* can resist, he concludes: "Such illusions occur in many, many domains of reasoning. Their moral is that we normally think only about what is true, and not what is false.")

e. The marriage of Craik and Montague

Johnson-Laird's computer programs for building and manipulating mental models helped support his theory. But as he pointed out, full explanatory adequacy would require a principled account of *all possible* mental models. That he couldn't provide. However, he did offer an incomplete "typology" of models: six "physical" (including images) and four "conceptual". Moreover, he listed ten general constraints which all such models must satisfy (1983, ch. 15).

These constraints were inspired partly by considerations of computational feasibility (i.e. bounded rationality), and partly by Montague's model-theoretic semantics. This was one of the few semantic theories which Johnson-Laird *hadn't* discussed in his 1976 book—even though Montague had written (and died) some years earlier (Chapter 9.ix.c). Now, he made up for lost time.

Montague's philosophy, he said, laid out the structure of semantic interpretations "with a pure but almost unreal clarity" (p. 167). It specified "what is computed in understanding a sentence", whereas psychological semantics should specify "how it is computed". As he remarked, this was a version of the logic/psychology distinction formulated by Frege (2.ix.b).

The ten general principles of Johnson-Laird's theory were these:

1. The principle of computability: Mental models, and the machinery for constructing and interpreting them, must be computable (p. 398).
2. The principle of finitism: A mental model must be finite in size and cannot directly represent an infinite domain (p. 398).
3. The principle of constructivism: A mental model is constructed from tokens arranged in a particular structure to represent a state of affairs (p. 398). [It is "a *Craikian* automaton" (p. 403).]
4. The principle of economy in models: A description of a single state of affairs is represented by a single mental model even if the description is incomplete or indeterminate (p. 408).
5. Mental models can directly represent indeterminacies if and only if their use is not computationally intractable, i.e. there is not an exponential growth in complexity (p. 409).
6. The predicability principle: One predicate can apply to all the terms to which another applies, but they cannot have intersecting ranges of application (p. 411).
7. The innateness principle: All conceptual primitives are innate (p. 411). [He explicitly rejected Fodor's claim that all *concepts* are innate: see 16.iv.c, below.]
8. There is a finite set of conceptual primitives that give rise to a corresponding set of semantic fields, and there is a further finite set of concepts, or 'semantic operators', that occur in every semantic field serving to build up more complex concepts out of the underlying primitives (p. 413).
9. The principle of structural identity: The structures of mental models are identical to the structures of the states of affairs, whether perceived or conceived, that the models represent (p. 419).
10. The principle of set formation: If a set is to be formed from *sets*, then the members of those sets must first be specified (p. 429).

In illustrating what he meant by these ten constraints, he showed how mental models (in his sense) differed from other forms of internal representation that had been suggested by computational psychologists—such as schemata, prototypes (Chapter 8.i.b), images, and propositions.

Ambitiously wide-ranging as it was, Johnson-Laird's new book attracted considerable interest—and no little disagreement. One critic, a long-time researcher on syllogistic reasoning, complained of Johnson-Laird's 'Mental Muddles' (Rips 1986). But his erstwhile collaborator Miller, assessing the theory in the *London Review of Books*, described it as 'A Model Science' (G. A. Miller 1983*b*). While much remained to be done, he said, Johnson-Laird was on the right track.

(He stayed on that track. Recently, he's used mental-model theory to describe real-life choices, like those which had interested Simon as an economist. In particular, he's stressed the dynamic interplay between reasoning, judgement, and decision making: Johnson-Laird and Shafir 1994.)

Philosophers were interested too, not least because Johnson-Laird had trespassed on their territory. He used Montagovian semantics, for instance, in offering a *psychological* explanation of our grasp of "reference" and "truth". And in so doing, he made various assumptions about ontology, a classic problem of metaphysics (see constraint no. 9)—and of GOFAI (10.iii.e and 13.i).

The prime bone of contention was whether how we *actually* reason should be of any concern to philosophers. This bone had been nibbled already, in response to Johnson-Laird's (and Wason's) earlier work. For example, at a meeting held in 1978 the philosopher of science Henry Kyburg (1980) had declared:

> I lost track of the number of times the maxim "Ought implies can" was solemnly enunciated [in two papers on inductive reasoning written by Winograd and Boden]. Whatever its virtues in ethics, the maxim is false for logic. One ought to be consistent; one ought not to offer invalid deductive arguments. No one lives up to these norms; but they are valuable precisely because they can be approached; formal logic gives us standards by which we can measure our approach to them. (Kyburg 1980: 376)

The task of "inductive logic", he continued, is to develop comparable standards for inductive arguments: formal measures that tell us when someone is leaping to conclusions, or resisting overwhelming evidence. He granted that AI modelling might help. To that extent, he welcomed communication across the disciplinary boundary. But it was AI, not psychology, that was to be the partner discipline: "an inductive program should embody standards and norms reflecting the way people *ought* to argue inductively" (italics added).

That complaint was made two years before Johnson-Laird's Stanford lectures. Unknown to Kyburg, he was already planning to use the ten constraints to ground not only deductive reasoning (artists and beekeepers) but also the inductive standards (*sic*) that Kyburg had been asking for.

Whereas philosophers found the abstract constraints intriguing (whether satisfactory or not), most psychologists ignored them. They focused rather on the study of the *particular* mental models described by Johnson-Laird. One reason for this was the fearsome difficulty of Montague's work, which had to be 'translated' even for *linguists*, who were well accustomed to abstract formalisms and argument about semantics (see 9.ix.c). (That's probably why Johnson-Laird hadn't discussed it earlier: the first 'translation' appeared when *Language and Perception* was already in press: Partee 1975.) Psychologists were thankful to escape into their laboratory.

Many even forgot Johnson-Laird's insistence that mental models are Craikian, never mind Montagovian. So 'unprincipled' talk of *schemas*, for example, continued to flourish. In short, the search for Chomskyan explanatory adequacy was taken more seriously by Johnson-Laird than by his experimental followers.

f. Irrationality rules—or does it?

Where Johnson-Laird looked to Montague and even to ontology (cf. constraint no. 9, above), others used a more traditional approach: experimentation. The Israeli psychologists Kahneman (1934–) and Tversky (1937–96) showed in the 1970s that intuitive judgements aren't ruled by the standards of logic or probability theory.

Both one-time (mid-1960s) postdocs at Bruner's Center for Cognitive Studies, Kahneman and Tversky—K&T for short—started off from *A Study of Thinking* (Shore in 2004: 133). They'd accepted Bruner's view that perception and thought are guided by expectations, and by various heuristic strategies for working with them (see 6.ii.b–c). They wanted to discover the constant ('base rate') error that people make because of their expectations about what things are most likely. But if their approach was traditional, their findings weren't. Indeed, the editor of the *Psychological Review*, namely Mandler, was forced to reject their "path-breaking" work, having sent it to "an overrated elder statesman of psychology" for peer review (G. Mandler 2002*c*: 205).

They didn't appeal to computer models, nor even to specific AI ideas. But they did stress heuristics (they'd been influenced by Simon as well as by Bruner). They'd both been trained in philosophy as well as psychology, but instead of model-theoretic semantics their interest was in the philosophy of science and probability theory.

K&T focused on probabilistic thinking, which is endemic in everyday life (Kahneman and Tversky 1972, 1973; Tversky and Kahneman 1973, 1974, 1981, 1982). Like Babbage (Chapter 3.i.b), they were well aware that inductive thinking can let us down—but they wanted to know just how it works. And like Simon, they were more interested in practical decision making than in 'purely intellectual' judgements (Kahneman and Tversky 1979).

Simon himself was sympathetic to their views. Indeed, when Gigerenzer told him that he thought K&T's work to be inconsistent with Simon's notion of bounded rationality, he was "surprised". Apparently, however, Gigerenzer persuaded him. For later, when he asked Simon what he thought of K&T's followers describing their work as a study of bounded rationality, he replied:

> That's rhetoric. But Kahneman and Tversky have decisively disproved economists' rationality model. (quoted in Gigerenzer 2004*a*: 396)

The reason for Simon's word "rhetoric" was that K&T had forgotten the ant. The ant's behaviour is partly explained by *the detailed structure of the environment*, but K&T ignored this. Besides focusing on mental processes (cognitive heuristics and illusions), they considered 'the world' only in the guise of the laws of probability.

The heuristics they claimed to have identified—though *without* using computer models or computational concepts—were said to involve "natural assessments", of ease of recall, or similarity, or relation to some prior judgement. So K&T discovered that people tend to overestimate the probability of a state of affairs if actual instances of it are easily recalled ("availability"), or if it fits their pre-existing stereotypes ("representativeness"). In addition, they're reluctant to adjust their initial judgements in light of potentially relevant information that's provided later ("anchoring and adjustment")—an effect previously noted with respect to affectively laden decisions by the theory of cognitive dissonance (Section i.c, above).

K&T argued that these heuristics guide judgements as to whether (for instance) a brief personality sketch describes a lawyer or an engineer, or whether I'm likely to get cancer if I smoke thirty cigarettes a day. In general, they're reliable. But sometimes they let us down, even when the 'correcting' data are available.

Thanks ("thanks"?) to the representativeness heuristic, for instance, the information that the population from which the mystery personality was drawn consists of thirty engineers and seventy lawyers (or vice versa) has little or no effect on people's decisions to assign the person to one class or the other. It's their preconceived ideas about engineers and lawyers which make the difference, not the base rate in the population. In the laboratory, who cares? But it's another matter when we consider a jury having to decide, supposedly *on the basis of the evidence presented in court*, whether a librarian is guilty of murder. (This example may ring warning bells: see the discussion of "cognitive illusions" below.)

Some heuristic-grounded mistakes are highly predictable. One is the Monte Carlo fallacy, in which a roulette player wrongly believes that a preceding run of blacks increases the probability of the next throw's being red, or a coin-tosser imagines that twenty heads must be followed by a tail. Even if a clear warning-explanation has been given, only the most resolute among us can withstand this judgemental bias entirely.

Another is what K&T called the conjunction fallacy. When their experimental subjects were told that Linda is single, intelligent, a graduate in philosophy, and has demonstrated against discrimination and nuclear weapons, 85 per cent decided that she was "more likely" to be *a bank teller and active in the feminist movement* than to be *a bank teller* (Kahneman *et al.* 1982: 91 ff.). But a conjunction can't be more probable than one of its conjuncts. Clearly, K&T said, people were being influenced by the representativeness heuristic.

As the persistence of the Monte Carlo fallacy illustrates, it can be difficult to counter or prevent biased reasoning. Indeed, it's often deliberately encouraged. Effective rhetoric presents the evidence in such a way as to maximize the likelihood that the hearer will make *this* judgement rather than *that* one—which may or may not be the correct judgement. Often, such persuasive presentation is done unthinkingly. But K&T's work led to a flurry of research on how to do it in a relatively systematic way.

For instance, they discussed the rhetorical implications of the fact that people generally give more weight to potential losses (risks) than to potential gains (Kahneman and Tversky 1979). This fact had been 'discovered' nearly thirty years earlier by the economist Harry Markowitz, who eventually received the Nobel Prize for his work on risk minimizing in the stock market (Markowitz 1952; Rubinstein 2002). But, as K&T acknowledged, canny communicators had sensed it intuitively long before that. Indeed,

Tversky's *New York Times* obituary reported him as having said that he merely examined in a scientific way things about behaviour that were already known to "advertisers and used-car salesmen" (and, one may add, politicians).

K&T's work, and similar research directed by the social psychologist Richard Nisbett (Nisbett and Ross 1980), promoted considerable discussion in the 1980s about the extent to which people are rational, and also about *what it means* to say that a judgement is rational. K&T themselves believed they'd shown that we're largely *irrational*, declaring that humans are guided by "cognitive illusions". (Likewise, Nisbett highlighted our "shortcomings" in the title of his influential book.)

However, the Oxford philosopher L. J. Cohen (1981) used the pages of *BBS* to deny that they'd done any such thing—and even to question whether *any* experimental methodology could do so. He pointed out, for instance, that to start from a false premiss doesn't prevent one's arguing validly on the basis of it. Gamblers who commit the Monte Carlo fallacy won't win a fortune. But they're thinking rationally rather than irrationally, given their false belief about probabilities. Similarly, given their mistaken (non-Bayesian) assumptions about probability, their inference to the false premiss is itself rational. For sure, they aren't acting *non-rationally*, for their behaviour is determined by the semantic content of their beliefs and desires. That is, they're rightly viewed from the "intentional stance"—which presupposes rationality, so defined (Chapter 16.iv.b).

Even Cohen's defence of the rationality of everyday reasoning, however, was less startling than what happened next.

g. Evolved for success

In the late 1980s, Gigerenzer (1947–) came onto the stage—and rewrote the script. For if K&T had largely codified common sense, Gigerenzer came up with some highly counter-intuitive findings. (For recent compilations, see Gigerenzer and Todd 1999; Gigerenzer and Selten 2001.)

His 'negative' position was stated in the subtitle of one of his papers: 'How Intelligent Inferences [Only] Look Like Reasoning Errors' (Hertwig and Gigerenzer 1999). His 'positive' position harked back to Simon's ant. (Indeed, he took Simon's phrase "bounded rationality" as the title of one of his books: Gigerenzer and Selten 2001.) For his central claim was that intelligent animals—including *Homo sapiens*—have evolved decision-making mechanisms that let *the environment* do the work.

Certainly, he said, some highly educated humans can use formal logic or Bayesian probability theory. But extraordinarily simple—"fast and frugal"—heuristics can match, or sometimes even surpass, these 'rational' methods. With respect to reasoning, one might say, less is often more (compare the developmental advantages of "starting small": 12.viii.c.) "The rationality of heuristics", Gigerenzer said, "is not logical, but ecological."

Describing people as "intuitive statisticians" (Gigerenzer and Murray 1987), Gigerenzer used frequency-based statistical theory as a model of what goes on in the mind. He saw this suggestion as just one example of a common creative strategy in science, the "tools-to-theories" heuristic—*mind-as-computer* being another (Gigerenzer 1991*b*; Gigerenzer and Goldstein 1996*b*). That fitted his overall position, which was that the environment inspires most of our thinking. But whereas his theory had come from (environmentally triggered) creative thought, the intuitive statistics—he said—had come from biological evolution.

That didn't mean, of course, that they couldn't be encouraged and supported by education. At present, cultural numeracy is thought of in terms of precise (maximizing) arithmetic. Gigerenzer (2002*b*) semi-seriously foresees a time (fifty years hence) when the President of France and the President of the World Health Organization may share a flower-decked platform with historians of psychology and economics to celebrate the WHO's latest achievement: abolishing innumeracy in the developed world, much as illiteracy has been abolished already. But innumeracy will be defined by them—as it was by H. G. Wells, according to Gigerenzer (but see Tankard 1979)—as statistics, not arithmetic.

(That's *not* to say that Gigerenzer recommends the usual statistical practices of experimental psychology. On the contrary, he regards these as often "mindless" and irrelevant: "Statistical rituals largely eliminate statistical thinking"—Gigerenzer 2004*c*: 587.)

In a variety of experiments, Gigerenzer found that if problems are posed in terms of population frequencies instead of individual cases, people are much less likely to make a misjudgement. For instance, he provided the 'profile' which K&T had used to describe "Linda", but then told his subjects that "There are 100 people who fit the description above," and asked "How many of them are *bank tellers*, and how many are *bank tellers and active in the feminist movement?*" (Gigerenzer 1991*a*). The error rate, or conjunction fallacy, dropped sharply: from K&T's 85 per cent to between 10 and 20 per cent. (For a recent paper casting even more doubt on the prevalence of the conjunction fallacy, see Hertwig and Gigerenzer 1999).

Thus far, psychologists (and philosophers) could—and many did—insist that bounded rationality is inferior to 'real' rationality, as identified by logic or mathematics. Gigerenzer might have shown that our intuitive reasoning isn't quite so bounded as K&T had claimed—but bounded it still was. After all, the ideal error rate is zero. Now, Gigerenzer came up with another surprise, not to say a bombshell.

Using computer modelling, he and Daniel Goldstein ran a competition between optimal mathematical algorithms and his own suggested heuristics (Gigerenzer and Goldstein 1996*a*, 1999; Gigerenzer 2000, ch. 8). Amazingly, they found that simple intuitive heuristics can often do *just as well as*, or even *better than*, the supposedly optimal rational methods. (Remember: this was being done on computers, which have no problems with the maths.)

In one experiment, for example, the best mathematical methods for doing integration, such as multiple regression, were compared with a heuristic called "Take the Best (Ignore the Rest)"—TtB for short. What TtB tells the system to do is to base its inference *only* on the cue which in the past has discriminated best between alternatives. All other cues are ignored. In other words, a large amount of supposedly relevant information is discarded.

One can well believe that the one-item ("frugal") TtB is *useful*, in an unfriendly and fast-changing world. But surely, mathematically optimal methods would be better? Not so, apparently. In the computer models run by Gigerenzer and Goldstein, Ttb surpassed all the mathematical methods in speed, and matched or surpassed them all in the proportion of correct inferences. (Further computer simulations run by Nick Chater suggest that other simple heuristics are at least as plausible as TtB: Chater *et al.* 1997.)

Another set of counter-intuitive results involved Gigerenzer's "Recognition" heuristic. (This wasn't the same as K&T's "availability", which referred to ease of recall, not recognition.) Like TtB, it asked only one question: "If one of two objects is recognized and the other is not, then infer that the recognized object has the higher value with respect to the criterion." (Notice: *any* criterion.) This simple rule accounted for the success of 'pure' (i.e. uninformative) brand-name advertising. And it explained more bizarre phenomena, too.

For instance, Gigerenzer's German students were much better than US students at deciding whether San Diego or San Antonio has the larger population—even though most of them had never even heard of San Antonio. But that was the point. If one's heard of a foreign town, it's likely to be populous. There are exceptions: one can hear a visitor reminiscing about their home town, or see a foreign movie featuring a small village . . . But whenever our ignorance is systematic rather than random, so that recognition is strongly correlated with the criterion of interest, then the question "Have I encountered it?" is likely to be the best clue.

The Recognition heuristic evolved because, for species other than *Homo sapiens*, ignorance almost always is systematic. Ecologically important features tend to be noticed; and features that are noticed tend to be ecologically significant. It's only humans who can generate interest in largely irrelevant matters, like those featured in pub quizzes and Trivial Pursuit. Facing pointless questions such as these, the Recognition heuristic can let us down. But it can let us down in important decision making too (including decisions made in the law courts), where our culture has provided us with knowledge divorced from systematic significance. So although Gigerenzer (2004*b*: 80) doesn't say that ignorance is bliss, he does say that—often—"Less is more."

In short, Gigerenzer offered a radically new vision of "bounded rationality". It's not a regrettably necessary second-best—as Simon and everyone else had assumed. Rather, it can (often) give accuracy, not mere approximation. Admittedly, what's being approximated are the *norms* of rationality: they hadn't been forgotten. But they were being used by Gigerenzer as evaluative ideals, not as blueprints of 'the best' reasoning processes.

(Recently, he's raised the stakes still further. Now, he considers not just rationality but happiness, too:

> Satisficers are reported to be more optimistic and have higher self-esteem and life satisfaction, whereas maximizers excel in depression, perfectionism, regret, and self-blame. (Gigerenzer 2004*b*: 80; cf. B. Schwartz *et al.* 2002)

No wonder, then, that he follows this sentence with "Less can be more.")

To put it another way, Gigerenzer's claim was that *the environment itself* gives us the information we need to get the right answers: we don't need recourse to maths or logic. This ant insight about the importance—indeed, the sufficiency—of environmental cues was largely responsible for his quarrel with K&T about cognitive illusions (Gigerenzer 1991*a*, 1996; Kahneman and Tversky 1996).

He made two charges against K&T. First, that their one-word "heuristics" were so vague that they explained everything and nothing (as had been said of cognitive dissonance theory in the late 1950s). For instance, both the Monte Carlo fallacy and its opposite, expecting a long run of blacks to *continue*, were attributed to *representativeness* by K&T (Gilovich *et al.* 1985: 295; Tversky and Kahneman 1974: 1125). Second, that our minds are guided not by cognitive illusions (i.e. subjective probabilities) but by veridical perceptions of real-world patterns (i.e. frequencies). K&T countered that judgements of frequency are equally 'subjective'. Gigerenzer replied, in turn, that his computer models enabled him to say just when frequency judgements would be valid and when they would not.

K&T also declared that "representativeness (like similarity) can be assessed experimentally; hence it need not be defined a priori" (1996: 585). Gigerenzer (2004*b*) sees this as an egregious example of *Hypotheses non fingo* (Chapter 5.i.a). K&T, he says, had regressed to behaviourism. Indeed, decision-making research in general abounds with "surrogates for theories . . . from one-word explanations to mere redescription to ying–yang dichotomies". Yet testable computer models of heuristics have been feasible for many years. Evidently, then, even in the twenty-first century not all cognitive psychologists are cognitive scientists.

The general public, of course, couldn't have cared less about the arcane reaches of probability theory ("frequencies", and the like). Nor did they get excited about "theories" versus "surrogates for theories". But much as they'd previously been interested in popular accounts of K&T's work, and Nisbett's too, so they were now intrigued by Gigerenzer's findings. His own popular writings were widely read, not least because they didn't merely describe irrationality in important life decisions but also gave advice on how to avoid it.

The public's interest in A-Life and evolutionary psychology, which burgeoned in the early 1990s (Chapter 15.x.a and Section vi below), added power to his elbow. By this time, Gigerenzer was Director of the Centre for Adaptive Behaviour (*sic*) and Cognition at Berlin's Max Planck Institute. The ant insight, indeed, was flagged in the title of his millennial book, *Adaptive Thinking: Rationality in the Real World* (2000). Accordingly, the statistical heuristics he described were presented as *evolved adaptations*, essentially comparable with the behavioural reflexes of crickets (15.vii.c). Most evolutionary psychologists had concentrated on motivation (goals, sexual preferences, selfishness, altruism . . .) and perception (a species' sense organs will 'fit' its habitat and motor repertoire). Now, Gigerenzer had moved beyond these, into topic-neutral reasoning.

It's little wonder, then, that Steven Pinker (1954–)—author of *The Language Instinct* (1994) and scourge of "the blank slate" (2002)—approved. He even provided a puff to die for. The cover of Gigerenzer's trade book *Reckoning with Risk* (2002*a*) was emblazoned with Pinker's accolade: "Gigerenzer is brilliant and his topic is fabulous."

h. Give thanks for boundedness

Gigerenzer's "fabulous" topic received an intriguing added twist a couple of years later, when two of his colleagues reversed the causal direction assumed in his work.

He'd said, in effect, that simple heuristics evolved because of inescapable processing limits in the brains of cognitive creatures, *Homo sapiens* included. But Ralph Hertwig and Peter Todd now argued that those very processing limits have evolved because they're required by simple heuristics—which have been selected because of their (highly adaptive) speed and robustness (Hertwig and Todd 2003: 223–8).

To be sure, no brain can have absolutely unlimited processing capacity. But its specific limits, on this view, aren't inevitable—and, way back in our phylogenetic history, they didn't exist. They've been evolved, just as successful heuristics have been—with the latter driving the former, not the other way around. So Gigerenzer's "simple heuristics" aren't *compensations* for the boundedness of human rationality: in evolutionary terms, they're partly *responsible* for it.

Counter-intuitive though this may seem, it's actually a special case of a more general phenomenon. For there's persuasive evidence that processing limits can *help* one to learn what must be learnt (Chapter 12.viii.c–d). Research on "starting small" in the development of language (Elman 1993), and on "representational trajectories" in general (Clark and Thornton 1997), implies that bounds on rationality are a Good Thing, especially—but not only—in infancy.

So Simon's insistence on bounded rationality has been stood on its head. Limits on our processing power are to be welcomed, not merely acknowledged. Without them, we'd be overwhelmed by the richness of the environment. Our minds would be swamped by what James called the "big blooming buzzing confusion" which is our "immediate sensible life"—that is, perception shorn of all conceptual interpretation (James 1911: 50).

The particular ways in which our rationality happens to be bounded are neither inevitable nor, in some cases at least, accidental. Whether comparative psychology and/or computational studies will ever enable us to say just which limits have been specifically selected, and why, remains to be seen. What about the "magic number", for instance: why not ten, or four, rather than seven? (see 6.i.b). Such questions should ideally be considered by people thinking about the computational architecture of a wide range of *possible* minds (Sections i.e–f and iii.d, above).

In sum, Gigerenzer and his colleagues argued for an especially strong version of the thesis that, despite our many failings, we're actually pretty good at doing what we need to do.

A recent commentary on twentieth-century psychology, especially social psychology, has pointed out that the dominant strategy has been to identify, and attempt to explain and/or "fix", *deviations* from accepted norms—whether these be norms of rationality or of civilized interpersonal behaviour (J. I. Krueger and Funder 2004). Festinger's study of the cult of the Guardians on the planet Clarion (i.c, above) was one of many examples; Stanley Milgram's (1963, 1974) astonishing work on obedience was another; and—initially—the study of visual illusions yet another. One reason for the popularity of this strategy is that counter-intuitive findings, besides being interesting in themselves, have a better chance of escaping the scorn of Senator Proxmire and his ilk (6.iv.f; cf. Krueger and Funder 2004: 316). A more "balanced" psychology would concentrate also on our *strengths.*

Within cognitive psychology, that's started to happen. Work on visual illusions, for example, has increasingly seen these not as failings but as grounded in normally *successful* adaptive mechanisms, which don't happen to fit the unusual circumstances concerned (6.ii.e). Gigerenzer, too, saw not failings, but strengths.

It's not that he diverted his eyes from the failings, looking only at the strengths. Rather, he saw strength in a given aspect of behaviour where others saw weakness. Even Simon, the major champion of bounded rationality, generally implied that it's an unfortunate—though necessary—limitation. Gigerenzer, by contrast, would expect to find "simple heuristics that make us smart" even in an ideal world. (This switch from negative to positive hasn't yet spread across psychology as a whole: "so far [these cognitive psychologists'] influence on social psychology has been limited"—Krueger and Funder 2004: 311.)

7.v. Visions of Vision

Gregory and his New Look colleagues weren't the only ones to look anew at vision (Chapter 6.ii). In the early 1970s there was a sudden resurgence of experimental work on the forbidden topic of imagery. The core questions were computational, concerning the nature of 'imagistic' representations and the processes that could affect them.

The 1980s saw a sea change in theoretical styles. The key questions were still computational—if anything, more so than ever before. But the interest switched from top-down hypothesis-based interpretation to interpretation by means of bottom-up automatic processing.

In some circles, the New Look had never found favour. Wittgensteinian philosophers had no time for it (see 16.v.f). Nor did all psychologists. Even as it was being designed in the late 1940s, a distinctly *non*-identical twin was approaching the catwalk (J. J. Gibson 1950). In his theory of "direct" perception, Gibson resolutely avoided talk of models and interpretation. And when psychologists later spoke of computational processes in perception, Gibson was lying in wait to ambush them.

a. Icons of the eyes

The major steps in the renaissance of imagery as a fit topic for psychology were two publications in 1971. At first sight, they were very different: one ran to 600 pages (including over forty pages of bibliography), the other only to three.

The first was Allan Paivio's book on *Imagery and Verbal Processes*. This added chapter and verse to the core argument that he'd published two years earlier in the *Psychological Review* (Paivio 1969). The second was the paper on 'Mental Rotation of Three-Dimensional Objects' by Roger Shepard (1929–) and his student Jacqueline Metzler. These offerings, from the universities of Western Ontario and Harvard respectively, revivified a long-neglected dimension of cognitive psychology.

Paivio (1925–) reminded readers of a wide range of largely forgotten research. This included various counter-intuitive findings, such as a classic demonstration of confusion between visual imagination and perception (Perky 1910). If someone is looking at a screen, and is asked to imagine seeing a ship depicted on the screen, they can do so. However, if an image of a ship is secretly projected onto the screen, the person may not realize this. In Hume's terminology, they mistake an *impression* for an *idea*. (An anti-hallucination, one might say.) Paivio's work was startling primarily because it had until recently been forbidden, being incompatible with the behaviourist hegemony.

Shepard and Metzler were being daringly unfashionable too. But their main claim to fame was that they reported startling new results. They'd studied people's ability to imagine rotations of 3D objects (see Figure 7.3). In a nutshell, they'd found that

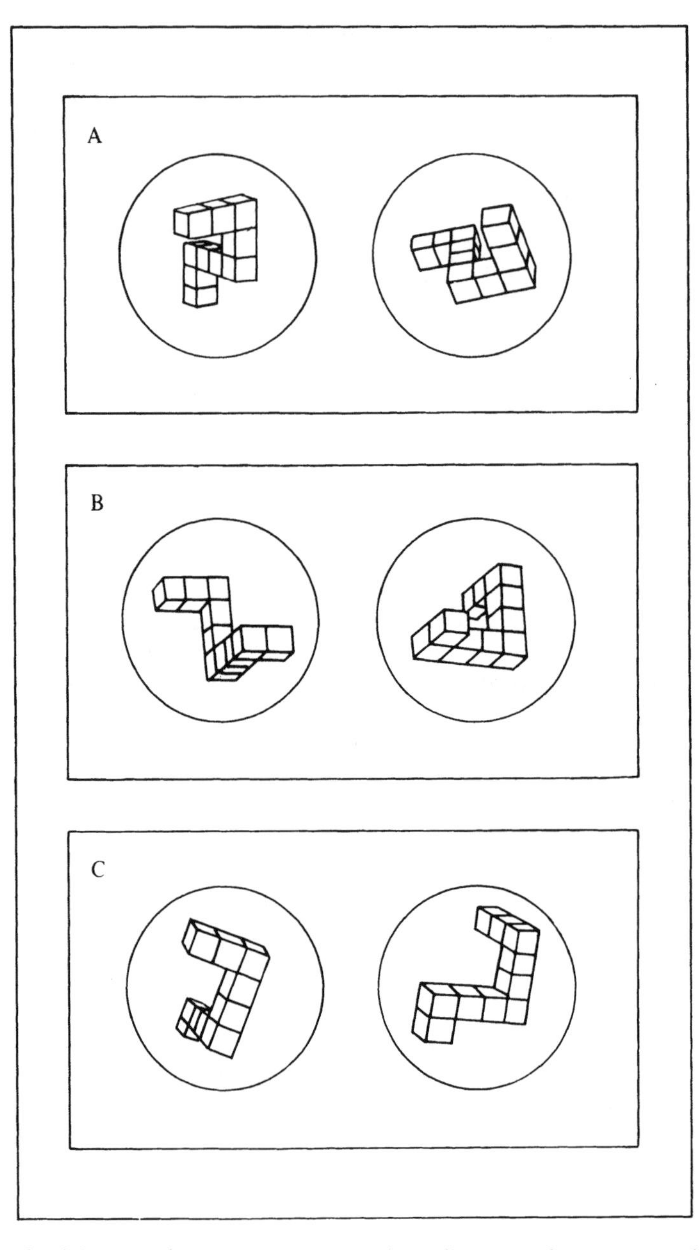

Fig. 7.3. Each of these six drawings represents a three-dimensional structure made up of ten cubes stuck together. People were asked to look at pairs of drawings and to decide whether the drawings showed one and the same object (rotated) or two different objects. Three of the 1,600 pairs used are shown here. The A-pair and the B-pair show rotations (in the plane of the page and perpendicular to the page, respectively), but the C-pair does not. Adapted with permission from Shepard and Metzler (1971: 702). Copyright 1971

the time taken for someone to decide whether two objects are identical, but shown in different orientations, is directly proportional to the degree of rotation involved.

In both cases, the researchers' main focus wasn't on the bizarre experimental data, but on their theoretical implications. Both Paivio and Shepard and Metzler argued that internal visual representations ("images") exist. Further, they both insisted that they're fundamentally distinct in nature from verbal representations—and, by implication, from the symbolic representations favoured by GOFAI.

Work on computer vision ("scene analysis") at that time represented visual/perspectival features by symbolic expressions like those used to represent verbal meanings (10.iv.b). But Paivio argued that "iconic" representations are very different. Similarly, Shepard and Metzler took their rotation experiments to show that some analogue (*sic*) of real 3D rotation was actually going on in the brain.

Immediately, experimental work on imagery blossomed. Besides a host of follow-up studies by Shepard and others (e.g. Shepard and Cooper 1982), the Harvard psychologist Stephen Kosslyn (1948–) initiated a rather different series of studies—which caused even more of a sensation than the rotating multi-cubes had done (Kosslyn 1973, 1975; Kosslyn and Pomerantz 1977).

For instance, he asked people to memorize a map of a fictitious island, showing the church, the lighthouse, the river and its bridge, and so on. Next, he asked them to visualize the map in their minds and to focus on a certain place: the lighthouse, say. Then, they were told to imagine a spot moving from the lighthouse to the bridge (or the church), and to press a button as soon as the spot reached the destination. Amazingly (or so it seemed to Kosslyn), the time taken to press the button varied according to the distance on the map between the positions of lighthouse, bridge, and church.

Again, the conclusion drawn by the experimenter was that there must be some property of the mental image which is analogous to real distance, and some process of traversing an image which is analogous to crossing real space. Kosslyn claimed, for example, that in image traversal as in actual traversal, all the intermediate points on some path between two points have to be visited in getting from one to the other. That's implied, he said, by there being some quasi-spatial form of representation, whose inherent properties allow transformations (such as imagined movements) isomorphic with spatial transformations. He used the tendentious term "image-scanning" to describe how this representation is supposedly used.

The Shepard–Kosslyn data caused a furore. Nonetheless, some people felt they were too good to be true.

Shepard and Metzler's subjects had remarked that they were "rotating images in their minds". But so what? Psychologists had known for sixty years that such remarks can't be taken at face value. For instance, people may be mistaken when they sincerely claim to be 'reading off' a visual image (of letters arrayed in rows and columns) much as they can read off a visual perception; for information (about diagonals, for example) that's available in the perceptual case isn't accessible here (Fernald 1912). In other words, the conscious phenomenology needn't match the underlying representation and/or the method of accessing it.

Even more to the point, some psychologists argued (*a*) that these new results weren't exciting at all, since they were only to be expected, and (*b*) that alternative explanations were possible. Indeed, (*a*) and (*b*) were closely linked.

The key critic here was Pylyshyn. In his 1973 paper 'What the Mind's Eye Tells the Mind's Brain' (a cheeky reminder of the classic 'What the Frog's Eye Tells the Frog's Brain': 14.iv.a), he argued that our power of imagery is deeply imbued with knowledge. As he put it, it's *cognitively penetrable.* It's not determined by the mind/brain's fixed functional architecture (FFA), but by the person's knowledge, or beliefs (see 16.iv.d). That knowledge, he said, is reflected when we're asked to imagine something.

We know, for instance, that it takes more time to walk a greater distance. So only an idiot would press Kosslyn's destination button immediately, having been asked to imagine a long traversal. By the same token, if Kosslyn had asked his subjects—as Pylyshyn now did—to imagine a magical transposition from the lighthouse to the church ("Beam me up, Scottie!"), the button would have been pressed instantly. (There were some puzzling exceptions: people *ignorant of psychophysics* experience imaginary after-images in complementary colours—Finke and Schmidt 1977; Pylyshyn 1984: 247 ff.)

The crux, for Pylyshyn, was that knowledge is stored propositionally, as descriptions. These were assumed to be computational *in the sense defined by Fodor* (1975: see Chapter 16.iv.c–d). And that, he said, gave us two ways of approaching the debate. On the one hand, *if imagery involves descriptions, then it will be cognitively penetrable.* On the other, *if images are quasi-spatial analogues, then the experimental results will not be cognitively penetrable.* The Shepard–Kosslyn data, he said, could be explained by cognitive penetrability. They therefore hadn't shown the need for any special, quasi-spatial, form of representation, provided by the FFA.

Various philosophers and psychologists, even including the Harvard behaviourist Richard Herrnstein, discussed imagery at length (e.g. Fodor 1975: 174–94; Sloman 1978, ch. 7; N. Block 1981, 1983; R. Brown and Herrnstein 1981).

Several of the philosophers warned against reification. To say "He has (I have) an image of a bridge" needn't imply that there's an object—namely, an image of a bridge—which the person has. Simply, the person *is imagining* a bridge—meaning that they're able to talk about an absent bridge in sensible ways. *Having an image* is a predicate (applying to the person) rather than a relation (between the person and a mental object). So the debate that was raging at that time was largely ill-posed. Instead of asking, "Are images descriptions?", people should have been asking, "In what ways are the representations underlying the power of imagery like, and unlike, descriptions?" They should have thought less about *images* and more about *the ability of imagination*: what we do when we're asked to imagine something—and how we manage to do it.

"How we manage to do it" was highlighted by Sloman (1971; 1975; 1978, ch. 7). Where Paivio had spoken of "iconic" and "verbal" representations, Sloman distinguished "analogical" and "Fregean" types.

In an analogical representation, he said, there's some interpretative mapping, or significant isomorphism, between the structure of the representation itself and the structure of the thing represented. To understand an analogical representation is to know how to interpret it by matching these two structures, and their associated inference procedures, in a systematic way. (In other words, similarity isn't enough: Chapter 4.vi.b.) To understand a Fregean representation, such as a sentence or a LISP expression, is to use some procedure essentially comparable to what Frege called the application of logical *functions* to *arguments*. (These were ideal types: 'one' representation may have both analogue and Fregean features.)

Sloman's distinction was clearer than most accounts of analogue representations—a term used in many different ways in any case (Haugeland 1981*b*). That was largely due to his experience of AI. The AI researcher who provides a representation of something must also provide a clearly specified way for the system to use ('understand') it. In the 1970s, however, the need to specify not only what a particular representation is, but also *how it can be used*, still wasn't appreciated by most psychologists (or philosophers, either).

A lot of ink was spilled on the relative merits of iconic versus descriptive representations (for a review, see Boden 1988: 27–44; for Paivio's most recent version, see Paivio 2005). The debate continued well into the 1980s, with critiques of Pylyshyn's methodology and further Shepard–Kosslyn experiments to the forefront. John R. Anderson even put the cat among the pigeons by arguing that it's impossible to discriminate empirically between various forms of representation. The busy programme of experimentation on imagery was a waste of time. Or rather, it might—it had—come up with intriguing new data. But it would never settle the question *iconic or descriptive?*

Anderson agreed with Pylyshyn that propositional representations *could* underlie what we call imagery, and that the concept of iconic representation was unacceptably vague (Anderson and Bower 1973, sect. 14.5; Anderson 1978, 1979). But no experimental evidence could decide between them, because a propositional code could (in principle) represent anything. What was needed, then, was a judicious use of Occam's razor to identify horses for courses:

> At best . . . this is what the imagery-versus-propositional controversy will reduce to—that is, a question of which gives the more parsimonious account of which phenomena. . . . [There] is *no way to prove one is correct and the other wrong*. Moreover, even if parsimony could yield a decision between propositional and imagery theories, there would still remain the possibility that there are *other, fundamentally different, representations* which are as parsimonious as the preferred member of the propositional–imagery pair. (1976: 13; italics added).

(A few years later he relented, up to a point. It was the notations used to express representations which were indistinguishable, he now said, not the computational processes defined in relation to them—1983: 46.)

Anderson's suggestion that there may be other forms of representation would eventually be widely accepted (Chapters 10.iii.a, 12.v–vi and x, 13.iii.a, and 14.viii). However, that took time. In 1980 the first edition of his textbook declared:

> Interestingly, researchers seem to be strongly divided into two camps: those claiming that there is only a propositional-code representation and those claiming that there is only a dual-code [iconic plus verbal] representation. Out of what is probably a false sense of parsimony, no one seems willing to admit that there might be three codes—abstract, verbal, and visual [though] a proposal that comes close to this eclectic point of view [is Kosslyn's] computer simulation for visual imagery. (Anderson 1980: 126)

Evidently, his message hadn't yet been heard.

As his final comment implies, however, it wasn't only experiments on human subjects which had flourished. Computer simulation had forged ahead too. Kosslyn's model (1980, 1981) was soon announced to the world in a trade book (1983). He'd accepted Pylyshyn's point that descriptive knowledge can influence imagery. But he still posited a base of quasi-spatial "depictive" representations in the FFA, transformed in 'spatial' ways (e.g. 'rotation'). These were stored in a *matrix* which "functions like a space"—with a limited extent, a specified shape, and a "grain of resolution" (highest at the centre). Like Anderson himself (and Grossberg too: 14.vi.c–d), Kosslyn continually adapted his model to include new experimental findings.

He optimistically described the most recent version (Kosslyn 1996) as "the resolution of the imagery debate". One doesn't have to share that optimism to allow that psychologists' ideas about the possible types of visual representation have come a long way since the 1970s.

For the record, Shepard (1984) eventually used his epochal 1971 data as the seed of a theory of perception *in general*. That theory became so influential that *BBS* recently devoted a special issue to it (Shepard 1994/2001). (His own historical perspective on this progression is given in Shepard 2004.)

The mental-rotation experiments had suggested that the mind/brain models external reality to a remarkably precise degree. Shepard's explanation of this appeared as the very first paper in the new journal *Cognitive Psychology* (Shepard and Chipman 1970). He argued that the phenomenon was due to "second-order isomorphism" between similarities among shapes and similarities among their internal representations:

> [The] isomorphism should be sought—not [as Craik, and most of his followers, had assumed: 4.vi.a] in the first-order relation between (a) an individual object, and (b) its corresponding internal representation—but in the second-order relation between (a) the relations among alternative external objects, and (b) the relations among their corresponding internal representations. Thus, although the internal representation for a square need not itself be square, it should (whatever it is) at least have a closer functional relation to the internal representation for a rectangle than to that, say, for a green flash or the taste of a persimmon. (Shepard and Chipman 1970: 2)

In other words, this was "a call for the representation *of* similarity instead of representation *by* similarity" (Edelman 1998: 450). (This idea was later developed into a computational theory of vision in general by Shimon Edelman. As he put it, the task of the visual system is to "*represent similarity between shapes, not the geometry of each shape in itself*" (1998: 451)—which is why people are much better at recognizing similarities between shapes than at perceiving shape as such.)

As for how the brain is able to recognize similarities, Shepard later claimed that a number of cognitive universals act as "the evolutionary imprint of the physical world". Just as our biological circadian rhythms have internalized the diurnal movement of the earth so, he said, have our perceptual mechanisms internalized other aspects of the world and our bodily relations to it: the geometry of observed objects-in-motion, for example.

These mechanisms were comparable to Kantian intuitions and schemas—but placed in us by evolution, not God. That is, Shepard's 1984 theory was an instance of evolutionary psychology—a nativist enterprise that had hardly begun when he did his initial experiments on imagery (see Section vi below, and Chapter 8.ii.d–e).

b. Vision from the bottom up

A huge change in the way that psychologists thought about vision was brewing in the late 1970s. Some bubbles were formed as a result of Marr's first papers on vision (1975*a*,*b*,*c*), and of his general statement in MIT's book on home-grown AI (D. C. Marr 1979). But it boiled over in 1982, with the publication of Marr's book on the subject. The book was widely read even by people with no special interest in vision, because of its challenging claims about the nature of psychological explanation *in general* (Section iii.b, above). But visual psychologists, of course, had even more reason to be interested.

Marr had cut his computational teeth in the late 1960s, by pioneering formal theories of the brain (Chapter 14.v.b–e). By the mid-1970s, however, that work couldn't be carried any further. Thanks largely to Minsky, whom he met in May 1972, he now turned his attention to vision—and to computer modelling, as opposed to abstract formal modelling. (The story of his eventful encounter with Minsky is told in Chapter 14.v.f.)

Despite his intellectual debt to Minsky, he didn't approach computer vision in the way that was dominant in Minsky's AI Lab at the time. Instead of favouring GOFAI work, he drew inspiration from MIT's Berthold Horn, who was doing detailed research on optics and psychophysiology. Marr's computer models were connectionist (but not PDP), consisting of many simple processing units operating in parallel and communicating locally (see Chapter 12). As such, they were very different from GOFAI models. Indeed, much of Marr's energy through the 1970s, until his premature death in 1980, would be devoted to criticizing GOFAI scene analysis (10.iv.b), and the New Look assumptions that informed it.

In particular, he avoided talk of top-down processing guided by learnt concepts, or high-level expectations. He frequently pointed out that we can see (locate, distinguish, manipulate) things we *don't* expect to see, because we've never seen them before.

When Captain Cook's sailors first met a kangaroo, they were instantly able to pick it out from the bush, see the soft texture of its fur, and locate it precisely—whether to stroke its flanks or to shoot it in the head. And the ship's artist was able to draw it. On his return to England, many people thought his picture a flight of fancy, so different was this animal from any they'd seen before. But the artist had 'used his eyes'. Of course, he'd used his brain too (remember Gregory?)—but the visual more than the associative cortex.

Marr's view was that for this sort of thing to be possible, we must be able to rely on *general* image-processing mechanisms. He wasn't denying that top-down perceptual influence occurs: how could he, given the intriguing playing-cards data described in Chapter 6.ii.a? But his prime interest was in the automatic, bottom-up, processing that enables us to see even unfamiliar things.

It's because this processing is automatic that, as Gregory had shown, many visual illusions are culture-neutral and persist even when we *know* they're illusory. The processes involved aren't modifiable by conceptual knowledge (in Pylyshyn's terms, they're not cognitively penetrable) because they're built-in, honed by millions of years of evolution. In other words, Marr was positing what Chomsky and Fodor would later call visual *modules* (see Section vi.d, below).

His research at MIT was "computational" in three senses:

* First, it was grounded in his three-level theory of psychological explanation (iii.b, above).
* Second, it offered specific algorithms intended as descriptions of visual processing—for detecting (for instance) edges, texture, or depth.

* And third, it involved extensive computer modelling, on MIT's state-of-the-art machines. (At first, even these weren't powerful enough for what he wanted to do: see 14.v.f.)

Crucially, his theory—and his computer models—assumed that visual information is passed bottom-up rather than top-down. (Remember the kangaroo.)

It was already known that light-intensity gradients are computed (or, as some people still preferred to put it, detected) in the retina. In other words, the 2D image on the retina, caused by the light entering the eye, is coded by a series of nerve impulses representing the 2D-changes between light and dark. That's done by cells in the retina itself. But Marr argued that the mapping between raw light intensities, or even intensity gradients, and a 3D description of the object concerned is far too complex to be computed in one step. There must be a *series* of visual representations, of increasing abstractness, between the retinal image and the final perception (see subsections b–c).

It follows, he said, that a theory of vision, in identifying these, must specify distinct representational primitives at each stage, showing how they might be constructed from the primitives of the stage before. And because information that will be required for later computations mustn't be lost at earlier representational stages, the theorist should prove (*sic*) that specific sorts of information can be implicitly preserved at a given level, even if they're not explicitly coded by the primitives of that level.

By the time Marr turned to the study of vision, the neuroscientists had spent twelve years discovering the rich variety of feature detectors in visual cortex, and even longer on studying the retina (Chapter 14.iii.a and iv). In addition, the psychophysicists (such as Horn) had related what was known about the retina to theoretical optics. Unlike most connectionist modellers (of whom he had a low opinion: 14.v.f), Marr took these facts seriously, keeping as close to them as he could. And in doing so, as in his earlier work on the brain as a whole, he *made sense* of the neuroscience. Christopher Longuet-Higgins later described neurophysiology pre-Marr as "a theoretical vacuum", despite the profusion of intriguing data (Chapter 14.v.f).

Besides making sense of the neurophysiology, Marr set a new standard of rigour for the psychology of vision. He offered an integrated set of precisely specified, and computer-testable, hypotheses about visual representations and processes. Even if they turned out to be wrong (as his first theory of stereopsis, for instance, soon did: 14.v.f), the onus was on others to replace them with theories of equal clarity. (Stephen Grossberg was attempting something similar at much the same time, but for lack of rhetorical crispness his work wasn't widely read: see 14.vi.a.)

Sutherland, ever contemptuous of vagueness in psychology (Chapter 5.ii.a) and a committed computationalist to boot, was full of admiration. He'd been won over even before Marr started publishing on vision in the late 1970s, due to Marr's papers on the brain—and their personal friendship. He described Marr's volume *Vision* (1982) as "perhaps the most important book on the subject to appear since Helmholtz's *Physiological Optics*" (1982: 692).

Unfortunately, Marr himself didn't live to see this encomium. With typical English understatement, his Preface declared that "in December 1977, certain events occurred that forced me to write this book a few years earlier than I had planned". He was referring to a diagnosis of leukaemia. By the time the book went to press, overseen by his MIT colleagues, Marr was dead.

c. Maths and multimodels

Marr had set his sights high. He'd aimed to revolutionize the psychology of vision, not by replacing one highly respected theory by another but by providing *the first theory worthy of any respect.* (Helmholtz, to be sure, was an intellectual hero: but the mid-nineteenth century was far too early for *theories.)* This was why one reviewer said:

> I urge psychologists to read [his book], *if only to ponder in a spirit of self-criticism whether Marr is justified in his largely dismissive remarks about the contributions of psychology so far* to the study of visual perception. His view is widely shared by many scientists from different disciplines, so it deserves close attention. (Morgan 1984: 165; italics added)

In his critique of previous accounts, Marr was uncompromising. We need not plausibility, not persuasion, but mathematical proof:

> extreme care is required in the formulation of theories because nature seems to have been *very careful and exact* in evolving our visual systems . . . Having to formulate the computational theory of a process introduces *a great and useful discipline* into the subject. No longer are we allowed to invoke a mechanism [based on correlations, or Fourier transforms, for instance] that seems to have some features in common with the problem [e.g. stereopsis] and to assert that the mechanism works *like* the process. Instead, we have to *analyze exactly* what will work and be prepared to *prove* it. (1982: 75; all italics added except "*like*")

Just what sort of beast was this 'first ever' visual theory? In general, Marr argued from first principles—the optics of the image-forming process (where the 'image' is the pattern of excitation on the retina)—in suggesting how the visual system interprets light intensities. As he put it:

> From an information-processing point of view, our primary purpose now is to define a representation of the image of reflectance changes on a surface that is suitable for detecting changes in the image's geometrical organization that are due to changes in the reflectance of the surface itself or to changes in the surface's orientation or distance from the viewer. (1982: 44)

He identified six very general (physical) constraints on physical surfaces, including the fact that they're organized on different levels of detail—so requiring "a number of different [representational] processes, each operating at a different scale" (p. 46). The basic elements in the retinal image are intensity changes, but these are structured (organized) in a way that "yields important clues about the structure of the visible surface" (p. 51). Therefore, the image structure "needs to be captured by the early representations of [it]". If it weren't, then the information about organization would be lost, and couldn't be reconstructed later.

Marr posited several levels of visual representation, each building on the one before. First, there's the Primal Sketch. Next, the $2\frac{1}{2}$D Sketch. And finally, the 3D model (or 'object model'). And he did what he'd said should be done (see above): at each stage, he defined the representational primitives precisely, and related them to the primitives of the stage below.

Distinct algorithms were defined for computing particular aspects of visual perception—depth, orientation, motion, colour, and so on. These were automatic, and procedurally separate from each other. In other words, they were "informationally encapsulated": Section vi.d below. (Later, it was discovered that the different aspects are dealt with by distinct anatomical regions in visual cortex: Zeki 1993.)

These algorithms, in turn, could consist of several sub-levels of processing. The Primal Sketch, for example, is built up in several hierarchical stages.

First, retinal cells identify the changes in light intensity ("zero-crossings") in the image. At base, that's all that the visual system has to go on. The "raw" Primal Sketch codes these in terms of (2D) shape, size, orientation, and discontinuities. That is, it distinguishes blobs and bars of various widths and lengths, and locates bar-ends and curves.

Next, *groups* of bars or blobs are distinguished, on grounds of common shape, size, or orientation. Finally, the full Primal Sketch represents the overall disposition of the groups: so, for example, a boundary is constructed between two sets of groups sharing a common orientation, as in Figure 7.4. (Notice that Figure 7.4 already includes much of the information relevant to the Gestalt principles of "good form".) In short, the Primal Sketch explicitly codes information about (2D) *textures*.

To go from 2D textures to 3D surfaces requires further computation. So the system now constructs the $2\frac{1}{2}$D Sketch. This represents the 3D surfaces in the scene, and their orientation and depth—all described *relative to the viewer*. In effect, the $2\frac{1}{2}$D Sketch tells you (for instance) that a certain surface is 3 feet away from you, now, and oriented at the same angle as you are, now.

But what you usually need to know is what objects (not surfaces) are out there, and where they are in 3D space, irrespective of your viewpoint. A fully 3D coordinate frame, which explicitly relates visible surfaces to (partly invisible) objects, is constructed at the third level: the 3D model.

This codes volumetric as well as surface properties, using the information about edge and surface contours contained in the $2\frac{1}{2}$D Sketch. Now, it's possible for the visual system to represent the fact that two different retinal images depict one and the same physical object, whose shape, size, volume, and location are all *independent* of viewpoint. (This progression from visual subjectivity to objectivity would later be used to explain how concepts develop from non-conceptual content: Chapter 12.x.f.)

Marr posited a multitude of internal 3D models, to explain the objective interpretation of unfamiliar 2D silhouettes. But (unlike the models highlighted by the New Look) these are automatically constructed bottom-up, not learnt and applied top-down. In describing them, he drew on work done by GOFAI's Thomas Binford and the mathematician Harry Blum. Binford had spoken of "generalized cylinders" in discussing the computer recognition of shape *in general* (1971; Agin and Binford 1973; Nevatia and Binford 1977). Blum (1973) had focused on biological shapes, and had defined a bottom-up method for identifying the *axis* of an unfamiliar body or body part.

Marr argued that certain optical constraints on the image-forming process can be exploited *only* by a representation based on "generalized cones". This is the class of volumes generated by moving a closed curve along an axis, where the curve may change its size—and, in a fully generalized cone, its shape. (Think of the in-and-out mouldings

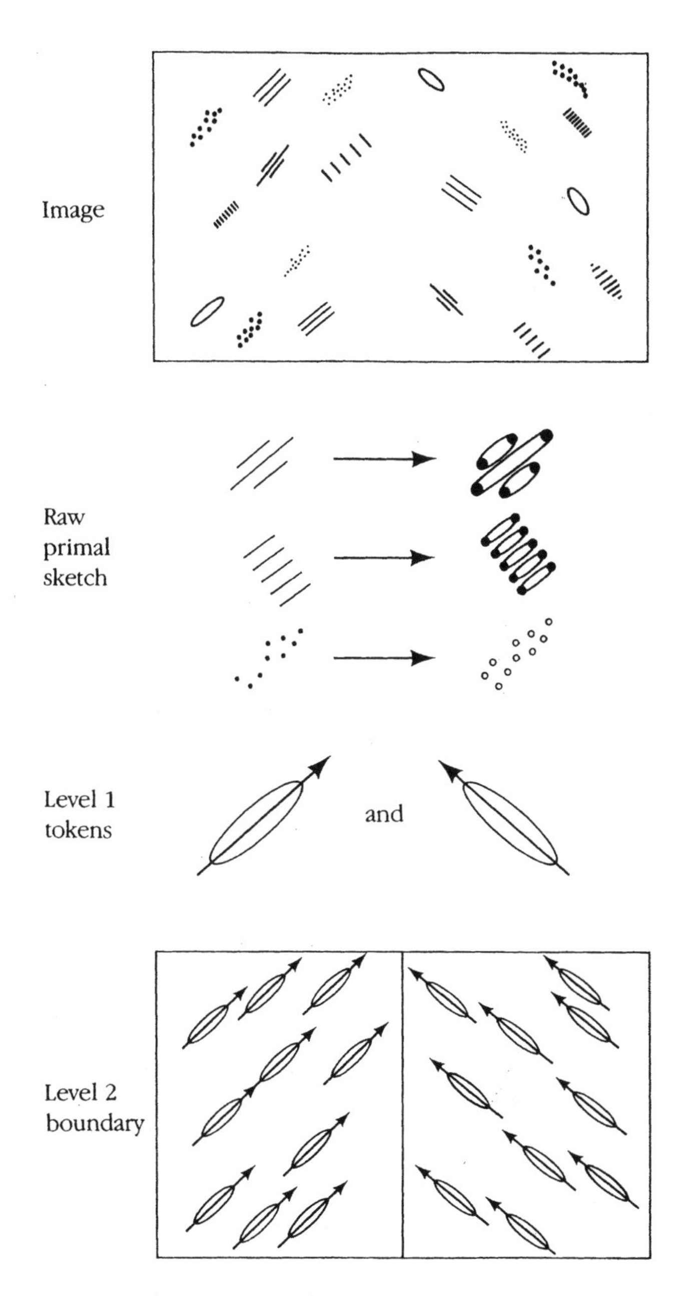

FIG. 7.4. A diagrammatic representation of the descriptions of an image at different scales which together constitute the Primal Sketch. At the lowest level, the raw Primal Sketch faithfully follows the intensity changes and also represents terminations, denoted here by filled circles. At the next level, oriented tokens are formed for the groups in the image. At the next level, the difference in orientations of the groups in the two halves of the image causes a boundary to be constructed between them. The complexity of the Primal Sketch depends upon the degree to which the image is organized at the different scales. Reprinted with permission from D. C. Marr (1982: 53)

on the vase in Figure 5.1.) And he suggested, though without offering a proof, that animal bodies can be perceived from their silhouettes by identifying the hierarchical organization of their axes (limb, thigh, foot, toes, arm, head, trunk . . .). Since each body part can be coded as a generalized cone, Binford and Blum could be united.

Blum had remarked that a mammalian body can be described in terms of one main axis (usually, horizontal), or as a major axis with four minor axes representing the limbs. A crawling human differs from a horse, and a rearing horse from a walking man, in the spatial relations between the constituent axes—such as the angle between head and body, and/or the length of the neck. Of course, one must learn to *recognize* these species differences. (Kangaroos seemed strange in eighteenth-century England not only because of the babies peeping out from their pouches, but also because of their unusual pattern of body axes.) But Marr insisted that animal bodies can be *seen* by the untutored eye, thanks to bottom-up axis-finding mechanisms evolved by the visual system.

Marr's theory of how vision progresses from 2D subjectivity to 3D objectivity was seen as hugely exciting by many psychologists. But his views on the multitudinous 3D models made less of a splash. They were too 'mathematical' to be biologically plausible.

In principle, perhaps, every 3D shape consists of one or more generalized cones. In practice, many can't be economically described in this way. Think of a piece of crumpled paper, or a velvet scarf draped carelessly over a pile of books. A finite visual system couldn't function in real time if it had to construct 'cone descriptors' for every visible object. (Shape description in the general case is still an unsolved problem, even in technological AI where one can 'cheat' as much as one likes; for two ingenious, and very different, efforts to deal with it see: Pentland 1986; Kass *et al.* 1987.)

Very likely, animals don't bother with overall shape descriptions but use special-purpose cues, including IRMs (Chapter 5.ii.c), to recognize certain objects quickly. (As Gigerenzer might put it, they employ "fast and frugal heuristics" for visual recognition.) In that sense, theories of 'situated' cognition are more realistic than Marr's mathematical purism (see Section iv.a above, and 15.viii.a).

d. The fashion for Mexican hats

Marr's mathematization of vision was evident, too, in his fondness for Mexican hats—not on the head, but inside it.

"Mexican hat" is the name given to a particular mathematical curve, which looks like a sombrero: see Figure 7.5. In Marr's theory, Mexican hats (or DOG functions: see below) were used to code the intensity array when constructing the "raw" Primal Sketch. What's interesting here is that Marr, working with Ellen Hildreth, used what would previously have been thought a very *non*-psychological way of coming up with them (Marr and Hildreth 1980; Marr 1982: 53–73). That is, he relied on wholly a priori argument to give Mexican hats pride of place.

He started by assuming—perhaps wrongly: see below—that evolution, over millions of years, must have come up with the mathematically *optimal* way of coding changes in light intensity. And he argued (at length) that the best such operator must be computationally efficient, mathematically simple, insensitive to differences in edge orientation, and sensitive to differences in size. A number of mathematical operators, some of which had been used by other psychologists of vision, were defined and

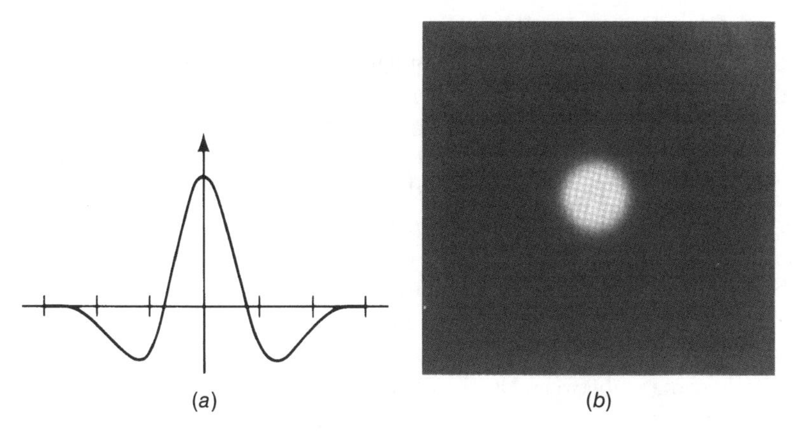

Fig. 7.5. (*a*) shows the Mexican hat function (defined as the second-order differential of a Gaussian) in one dimension. (*b*) shows it in two dimensions; the light intensity indicates the value of the function at each point. Adapted with permission from D. C. Marr and Hildreth (1980: 191); first reprinted in D. C. Marr (1982: 55)

compared. And the winner, on these purely a priori grounds, was what's known as the Laplacian operator.

(I'll omit the equations, even though for Marr they were of the essence. Mathematical readers won't need them, and others might not understand them. Indeed, they might not even get to see them: when I used Marr's crucial equation elsewhere, it appeared with an egregious misprint—and no erratum slip—*despite* my having forewarned the copy-editor/compositor against making that very mistake, and *despite* my having pointed it out at two stages of proofs, and on publication—Boden 1988: 64; for the correct version, see D. C. Marr 1982: 54.)

However, a pure Laplacian wouldn't do, because (lacking any size variable in its definition) it can't produce size sensitivity. Marr and Hildreth needed to define a class of filters that would *blur* the image at different spatial resolutions, by getting rid of all lower-scale detail. They chose what's known as the Gaussian distribution, which does have a size variable in it.

This had two advantages. First, although in principle it extends to infinity, in effect it's localized to a specific (circular) part of the image. So different Gaussians could be functioning in different places at one and the same time, which could account for our ability to see fine detail in only *part* of the image. Second, a Gaussian function is smooth at the edges: its value doesn't fall suddenly at the boundaries of its frequency (scale) or localized field. That meant that it was unlikely to introduce spurious intensity changes that weren't present in the original image.

The final step was to define a brand-new mathematical operator, by *combining* the Laplacian and Gaussian functions. It's this operator whose curve is the Mexican hat—and which was misprinted despite my repeated warnings. It's localized to a specific part of the image. And by giving a specific value to the size variable (inherited from the Gaussian), it would pick up light-intensity changes *at a given scale.* Since a large group of Mexican hats could be simultaneously applied to different points in the image, operating at different scales (i.e. tuned to different spatial frequencies), they could together detect a very rich variety of light-intensity changes.

All this, you'll notice, without a single reference to biology. In other words, this was the "computational" level of explanation. But Marr had long been interested in real brains, and was no less interested in real vision. His next step, however, *wasn't* to go straight to the neurophysiologists and ask them what they'd found. Instead, he used a priori argument again, asking how Mexican hats *could* be embodied in a physical mechanism (whether evolved or engineered).

At this point, the DOGs entered the picture. Marr and Hildreth proved mathematically that the Mexican hat operator is near-equivalent to a DOG function (Difference of Gaussians). This compares two Gaussians—one positive and one negative—at different scales. The best match between DOG and hat, they showed, would occur when the two Gaussians in the DOG function have space constants in the ratio of 1:1.6. In that case, the two curves are almost identical (see Figure 7.6).

It followed, they said, that Mexican hats could be near-perfectly approximated by physical mechanisms—whether natural or artificial—capable of computing DOG functions at differing scales, where the ratio between the scales is 1:1.6. Indeed, Marr suggested that specific cells of the retina and the lateral geniculate body (the second 'way station' in the visual system) were actually DOG detectors (1982: 64).

The DOGs hadn't been conjured out of nowhere. They'd long been barking in neurophysiology, as well as mathematics. As far back as the 1950s, the ON-centre and OFF-centre retinal ganglion cells, discovered twenty years before that, were found to be sensitive to changes in light intensity (see Chapter 14.iii.c). By Marr's time, various theories, and even computer models, of vision had already incorporated them. But

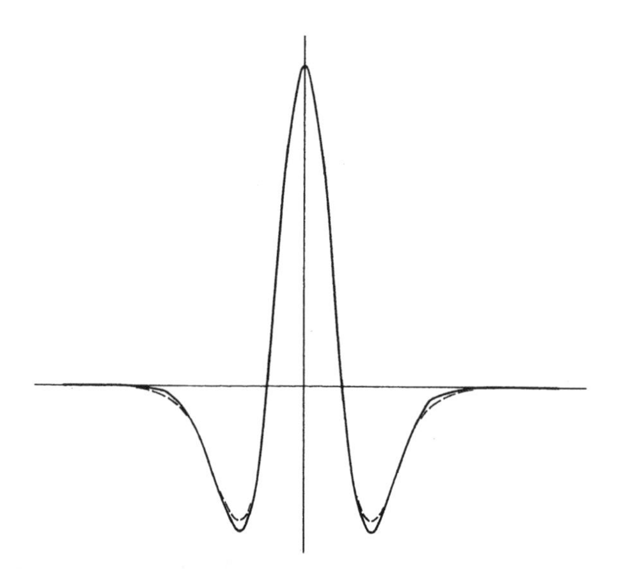

FIG. 7.6. The best possible match between the Mexican hat operator and the DOG function, shown by the continuous and dotted lines respectively. Adapted with permission from D. C. Marr and Hildreth (1980: 217); previously reprinted in D. C. Marr (1982: 63)

those psychologists had been working *from the physiology to the mathematics.* Marr, by contrast, was (notionally) working in the opposite direction. (Notionally, because—as remarked in Section iii.b—the physiology can, and in this case probably did, provide a hint that *such-and-such* mathematical operator might be the one worth thinking about.)

As with the Marr–Binford cones, Marr's purist mathematical approach was questioned here too. As Sloman (1983) pointed out, the assumption that evolution *must* have found the computationally optimal solution for a visual task was a dangerous one. Genetic algorithms in AI sometimes find the optimal solution—but not always (15.vi). In biological evolution too, satisficing may often win out over optimizing—as we saw in Section iv. Especially where speedy reaction is needed, quick-and-dirty methods (reflexes, heuristics . . .) may be more adaptive than purist ones. This applies to vision no less than problem solving.

In other words, Marr was closer to GOFAI than he was willing to admit. Quite apart from borrowing specific GOFAI ideas, such as Binford's work on generalized cylinders and Alan Mackworth's (1973) on "gradient space", he shared the early-GOFAI emphasis on *general* methods. And for Marr the mathematician, general meant not only universal but optimal. It's not clear that biological organisms always have that luxury.

There were countless other objections to his views. These concerned every level of his theory, even including the very first stage of zero-crossings. One critic, for instance, argued that the retina can't be using Mexican hats, because the spatial ratio of the centre and surround functions of the relevant cells is larger than 1:1.6 (Robson 1983). Others gave an even deeper mathematical analysis of the task of stereopsis than Marr had done (Longuet-Higgins and Prazdny 1980). And John Mayhew (1983), a researcher at Sheffield University, went so far as to say that Marr's theory of stereopsis was "almost completely wrong".

Yet within the very same breath, Mayhew acknowledged "an immense intellectual debt" to him. Even then, it was already clear that Marr had initiated what Imre Lakatos (1970) had called "a scientific research programme": a series of studies extending over many years, generated by a theory that's amended as discovery proceeds but whose central insights remain. Various aspects of his original theory have dropped by the wayside, as we've seen. But Marr's influence—his search for precisely defined (and neurophysiologically located) algorithms, computing many-levelled visual representations—is still strong today.

e. Direct opposition

The people who niggled about Mexican hats and DOG ratios, or who charged that Marr was "almost completely wrong", were nevertheless on his side. Indeed, they often presented themselves as his intellectual disciples (e.g. Mayhew and Frisby 1984). They thought he was doing psychology in the right way, even if he didn't find all the right answers. Some others didn't: namely, the Gibsonians.

The Gibsonians had moved into Psychology's House in the 1940s, long before Marr crossed the threshold (Chapter 5.ii.c). They'd been hugely influential. Gregory, for instance, had been "saturated" by their views as a student (see Chapter 6.ii.e). And although their carpets were similar to Marr's (like him, they were grounded in the detailed physics), their self-assembled theoretical furniture was very different. For one thing, they had an alternative view of what vision is *for*. For another, they didn't speak in terms of computation, or representation. When other people did so, they bridled.

As regards what vision is for, we've seen that Marr defined the task of vision as converting 2D information to 3D information. (He didn't say much about the fourth dimension, except on how 'optical flow' could aid 2D-to-3D mapping.) But isn't that an overly abstract, disembodied, view? After all, the *point* of vision, its biological function, isn't to emulate a geometer's knowledge of 3D space, or a sculptor's either. It's to navigate 3D space, to manipulate 3D objects, and in general to recognize opportunities for (3D-located) actions of various kinds: approaching, fleeing, fighting, eating, courting, grooming, mating . . . and so on. Also, and especially in social species such as *Homo sapiens* and chimpanzees, it includes the recognition of the emotional state and action-readiness of others.

These opportunities, termed "affordances", were highlighted in Gibson's later work, his "ecological" theory (1977, 1979). This excited many biologically minded psychologists because of its explicit linking of perception and (appropriate) *action*. Its general focus was on how the organism's sensory systems are adapted to the environment it lives in, and how they serve—or even prompt—the behaviour needed to survive in that environment.

Gibson wasn't a behaviourist—but he wasn't a mentalist, either. He had no time for talk of "images" on the retina, and even rejected apparently more innocent vocabulary:

> The function of the brain is not even to *organize* the sensory input or to *process* the data, in modern terminology. The perceptual systems [of all the senses, not just vision], including the nerve centers at various levels up to the brain, are ways of seeking and extracting information about the environment from the flowing array of ambient energy. (1966: 5)

He saw affordances as *properties of the environment*, even though they were defined by reference to the organism's needs. For example, a clear passage really is a pathway, a cherry really is edible, and a solid floor really is safe to walk on. In other words, Gibson wasn't focusing on the subjective meaning of object properties, but on their objective ability to be biologically relevant (useful/dangerous).

This interpretation of affordances was all of a piece with his philosophical realism (see below.) Nevertheless, it presented perception as inherently *meaningful*. Some philosophers, notably Edward Reed (1954–97), who soon became one of Gibson's leading disciples, pricked up their ears accordingly (E. Reed 1996*a*,*b*; Reed and Jones 1982). How it's possible for organisms/humans to support meaning is a fundamental philosophical problem (see Chapter 16).

When Marr turned from the brain to vision in 1972, the ecological approach was already making waves. Gibsonians were preparing for a three-week conference, to be held at the University of Minnesota's Center for Research in Human Learning in 1973 (R. Shaw and Bransford 1977). Gibson's key book appeared a few years later, in 1979 (the year he died).

By that time, however, Marr was terminally ill. In so far as he learnt from Gibson, it was the earlier work which he'd considered. For ecological psychology hadn't come out of the blue. It was grounded in Gibson's pioneering mid-century research on perception—first vision (1950), but then generalized to the other senses (1966).

In those early writings, done at Cornell, Gibson had sought to turn psychologists' attention *away from* the organism. Instead of focusing on behaviour, or even neurophysiology (for instance, the retina), he'd looked to pervasive features *of the environment.* Specifically, he'd considered the ambient light that impinges on the organism as it moves through the world.

In short, Gibson—like the maverick biologist D'Arcy Thompson before him—took the implications of *material embodiment* seriously. Much as D'Arcy Thompson had seen basic physical forces as moulding bodily forms and behavioural repertoires (Chapter 15.iii), so Gibson saw the physics of the environment, such as the details of light and sound, as moulding perception.

Even then, Gibson was interested in how vision can be *useful.* Like Broadbent, he'd worked extensively with pilots in the Second World War. He'd been particularly struck by the fact that pilots seemed to control their planes, in landing and take-off, by constant reference to the ground and the horizon. But 'ground' and 'horizon' weren't terms recognized by orthodox psychophysics. In short, visual psychologists had been missing the point:

> As I came to realise, *nothing of any practical value* was known by psychologists about the perception of motion, or of locomotion in space, or of space itself. The classical cues for depth referred to paintings or parlour stereoscopes, whereas the practical problems of military aviation had to do with takeoff and landing. (1982: 15)

In discussing the nature of the ambient light, Gibson pointed out that it contains a rich array of information about the world surrounding the organism, and the various objects within it. And that information is *structured,* being constrained by environmental "invariants" in how light is reflected from physical things.

For example, textured surfaces "expand" and "contract" in a lawful manner as we move towards and away from them (or even as we move our heads), or as *they* move, in relation to us. Similarly, the apparent size of an object changes constantly as it moves, if the movement has a constant speed; if the speed varies, the non-constancy of the change in apparent size provides information about the change in velocity. Again, the ratio of an object's height to the distance between its base and the horizon is constant—and this, said Gibson, explains the puzzling phenomenon of size-constancy. It's puzzling because, as Descartes had remarked, the size of the relevant part of the retinal image varies as the object moves near or far—yet we don't *see* the object as changing in size.

Descartes, and Helmholtz too, had explained this in terms of what we'd now call top-down (conceptual) inferences. Gibson, by contrast, was saying that information about size constancy is *already there,* in the light itself.

This approach was revolutionary: previous theories of psychophysics had concentrated on *local* features. They'd considered independent 'stimuli' or 'signals', such as the difference in light intensity between two adjacent points. Gibson was 'anti-Newtonian' in rejecting those atomistic categories.

He wasn't ready for nativism, however. At mid-century, that was still a step too far. Granted, he held that we don't need to *learn* to deal with visual information: visual systems have evolved to pick it up directly—and to act accordingly. In that sense, he was bordering on nativism. But what we've inherited is the ability to respond to information that's actually out there, not the ability to use unconscious inference to "go beyond" the information given. Where the New Look psychologists, and soon Chomsky too, constantly stressed the poverty of the stimulus, Gibson insisted on its richness.

One famous demonstration, designed by his wife Eleanor (a developmental psychologist), after a visit to the Grand Canyon, involved the "visual cliff" (E. J. Gibson and Walk 1960). This was a sudden change in depth, which babies (and kittens too) can perceive even before they can walk (see Figure 7.7). The evidence—remember: the baby can't speak yet, either—is that he/she refuses to crawl across the drop. (Of course, it's covered by a strong glass floor!) Yet no such chasm has ever been seen by the child before. Moreover, the baby doesn't just *see* the drop: he/she *does* something (stops crawling) as a result. This, said the Gibsons, is an example of an evolved connection between sensitivity to what's out there—i.e. rich visual information—and environmentally appropriate action.

When the visual cliff was first reported, it seemed that learning was *in no way* involved, and that the baby *simply would not* cross the drop. Later, it became clear that things are more complicated—and the visual cliff spawned a still-continuing experimental 'industry' accordingly.

For instance, 9-month-old babies, who can't yet crawl, may show fear (increased heart rate) if lowered over the 'deep' end of the cliff, but in 2-month-olds the heart rate decreases, suggesting raised interest (Campos *et al.* 1970, 1978). Recognition of the drop, then, appears earlier than fear of it. There's evidence that the degree of wariness depends on the baby's past experience of locomotion and of heights (Bertenthal *et al.* 1984; Campos *et al.* 1992), but counter-evidence that it depends on maturation (Richards and Rader 1981). Moreover, the mother may be able to coax the baby across (her voice being even more persuasive than her smile), but if she puts on a frightened expression then the baby won't move—presumably, because it picks up her fear (Sorce

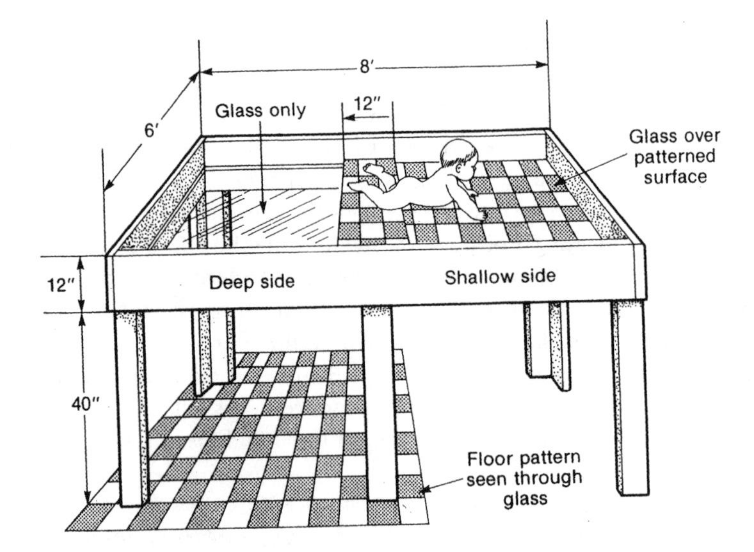

Fig. 7.7. Gibson's visual cliff. If the baby turns round and crawls towards the 'cliff', it stops at the edge of it—although the invisible glass floor continues. Reprinted with permission from Dworetzky (1981: 187)

et al. 1985; Campos *et al.* 2003; Vaish and Striano 2004). And mother-enticed babies look at the mother continually while crossing the cliff: only gradually do they become able to cross 'autonomously'.

In short: vision, locomotion, emotion, and social behaviour develop in subtle interaction with each other (for a recent review, see Campos *et al.* 2003). In Jean Piaget's terminology, they develop "epigenetically", thanks to an interplay between evolution (nature) and experience (nurture).

All that was for the future, however. Most psychologists still took it for granted that 'nature versus nurture' is a sensible opposition—see Section vi.g, below. As for Gibson, he was less concerned with how vision develops than with how it works. (Notice the then typical assumption that the latter can be understood without considering the former.) And he stressed automatic, environment-driven, mechanisms.

Marr did that too, of course. So what was the disagreement between them?

f. Let battle commence!

Marr had some kind words to say about Gibson—moderated by an important criticism:

> [Gibson asked] How does one obtain constant perceptions in everyday life on the basis of continually changing sensations? This is exactly the right question, showing that Gibson correctly regarded the problem of perception as that of recovering from sensory information "valid" properties of the external world. (Marr 1982: 24)

> In perception, perhaps the nearest anyone came to the level of computational theory was Gibson (1966). However, although some aspects of his thinking were on the right lines, he did not understand properly what information processing was, which led him to seriously underestimate the complexity of the information-processing problems involved in vision and the consequent subtlety that is necessary in approaching them. (p. 29)

Reading those words cold, one might infer that Gibson was a computational psychologist in the usual sense of the term, but that his theory of algorithms and internal representations left something to be desired. One couldn't be more wrong. The word "computational", here, was intended in the (atypical) sense defined in Section iii.b. Broadly, Marr meant only that Gibson had taken the constraints of physics seriously in trying to say what vision does. As for Gibson's theory of algorithms and representations—there wasn't any.

Gibson was adamant, right from the start, that perception is *direct.* Indeed, he devoted many pages to philosophical argument (well, he was a maverick!) in favour of direct realism. He had no time for Descartes's representational theory of perception, nor for Kant's hidden world of noumena either (Chapter 2.vi.a). Indeed, his commitment to realism led him to criticize Gregory for concentrating on illusions—by definition, *unrealistic.* (Hence his dismissive remark about "parlour stereoscopes".) According to Gibson, the information in the ambient light is there to be discovered, and is effortlessly (though actively) "picked up" by the visual system. It's neither altered (processed) nor interpreted by the eyes and brain. As he often put it (e.g. 1979: 54), we don't see *light,* we see *objects.*

For Marr, by contrast, we don't see objects until after 'seeing' (processing) light—and many other things besides, namely the representations making up the Primal and $2\frac{1}{2}$D

Sketches. For Gregory too, seen objects are the result of unconscious inference, or interpretation. So when the New Look psychologists and (later) Marr started talking about internal representations, and the computational processes involved in constructing and interpreting them, Gibsonians rejected the whole mentalistic story. *There are no representations*, they said—and no computations, either.

The result was a running battle between Gibsonians and computational psychologists in general. Sometimes, Gibson won new converts. Neisser, for instance, switched camps in the early 1970s. He abandoned his "analysis by synthesis" approach to perception, replacing it by a more ecological, realist, theory—developed in a book he dedicated to Gibson (see Chapter 6.v.b). But that was pre-Marr. After Marr, the stakes were upped. For his theory, which appealed to some similar (physics-based) insights, had enthused people at least as much as Gibson's had done. Moreover, it had helped them to realize how very difficult it is (that is: how much information processing actually has to be done) to see an edge, never mind an object.

One computationalist attack on Gibson was launched by Fodor and Pylyshyn (1981). They were less interested in the specifically visual issues than in his concept of affordances—which they castigated as "a dead end" (1981: 192). The place to which his pathway was supposedly leading was an understanding of meaningful behaviour in terms of environmental properties (see above). But they despaired of his philosophical compass.

In particular, they complained that he offered no principled way of deciding *what counts* as an affordance, nor of *how* affordances can be grasped. Mental states and processes (beliefs, goals, inferences . . .), they said, are necessary to understand how affordances are exploited—and even how they're recognized in the first place. Some animals may have a few automatic links between perception and action (e.g. IRMs). But intentional behaviour can be explained only in terms of computations over representations (Chapter 16.iv.c). Admittedly, the philosophy of intentionality was a mare's nest. But ecological psychology couldn't help.

Almost immediately, four of Gibson's closest colleagues rose to the bait (Turvey *et al.* 1981). Even if high-level human cognition involves mental states and inferences, they said, perception doesn't. It can be explained by "natural laws", which focus on observable organism–environment regularities rather than hypothetical mechanisms inside animals' heads. And indeed, many animals had already yielded to the Gibsonian treatment. Ecological psychologists had studied a range of species, showing how they found different affordances (in effect, different environments) in the real world. But it was the objective properties of that world (for example, optics), together with the bodily potential of the animals concerned (their "effectivities"), which accounted for their behaviour. Meaning, in the sense of adaptiveness, can be given a naturalistic explanation. Human-level semantic meaning might be another matter—but that's not what Gibson had been talking about.

Another explosive skirmish was staged in the interdisciplinary pages of *BBS*. Soon after founding the journal (while Marr was still fighting for life), Harnad circulated a target article by Marr's MIT colleague Shimon Ullman. The title announced the challenge: 'Against Direct Perception'.

Ullman (1980) expressed the mystification common to all computational psychologists: that the Gibsonians seemed to believe in some unexplained, quasi-magical process of "information pickup" (Gibson 1966, ch. 13). Gibson had used the metaphor of a radio, which doesn't interpret the radio waves but picks them up directly and "resonates" with them. A radio isn't magical, to be sure. But it's far less complex than even 'low-level' vision. If one looks hard at the neurophysiology, one finds ample evidence that specific parts of the eye/brain do (compute) specific things. Moreover, the engineering task of building a visual system had turned out to be different in kind from—and vastly more difficult than—building a radio. In short, said Ullman, whether or not one accepts the details of Marr's theory, one must allow that vision is far from "direct".

Among the many peer commentaries—*Marr* v. *Gibson* was a fight that everyone wanted to join—was a judicious piece by Geoffrey Hinton (1980). Hinton agreed with Ullman that there's no such thing as a-procedural information pick-up. Low-level vision is indeed automatic—but that means that it invariably happens, and that it can't be influenced by conscious effort or other top-down factors. It doesn't mean that it 'just happens', as if by magic. A lot of information processing has to go on for us to see objects, or even edges and surfaces. Anyone who doubts this, he said, should try getting a computer to distinguish those things.

Gibson's mistake, Hinton continued, was to think that "computation" must mean either conscious deliberation (as the Wittgensteinians were saying) or formal–symbolic processing of the type favoured in GOFAI. Now, in 1980, we had a better sense of various types of bottom-up visual processing—of which Marr's was one example and Hinton's another. (He might have added, but didn't, that "computation" is a word whose meaning isn't fixed but gets richer as AI progresses: 16.ix.) In short, one could be a computationalist without falling into the traps rightly feared by Gibsonians.

Hinton's vision research had been strongly influenced by Sloman, when they'd collaborated at the University of Sussex (12.v.h). And Sloman (1989) soon sought to bring Gibson's *later* ideas inside the computational camp. His own work on vision, like his later discussions of personal psychology, had always been done in an 'architectural' spirit (see 10.iv.b). In other words, he was concerned with the role that sight, our most well-developed sense, plays in the mind as a whole. In outlining the potential for "a Gibsonian computational theory of vision", he argued that affordances are hugely important—and entirely consistent with a computational approach. In particular, he pointed out that there might be learning mechanisms which could link specific visual cues with all manner of associations, including widely diverse opportunities for action.

Gibson and Marr died within a year of each other, but their spirits still live. Researchers in what's sometimes called nouvelle AI (that is, situated robotics and enactive perception) favour Gibson, as do those neuroscientists and philosophers who stress "embodiment" as a crucial ground of our psychological abilities. Michael Turvey and Peter Kugler, at the University of Connecticut, have been especially influential in discussing the nature—and emergence—of intentional movement and perceptual "measurement" (Kugler and Turvey 1987, 1988; Kugler *et al.* 1990; Turvey and Shaw 1995). Like their late mentor Gibson, they stress the physical properties of interacting dynamical systems—so wouldn't want to be called "computational" psychologists.

In short, the battle between ecological realism and representationalism continues today. That's so not only in the psychology of vision but also in A-Life (Chapter 15.vii–viii), neuroscience (14.viii.a–c), and philosophy (14.viii.d and 16).

But perhaps, as Sloman (among others) had said, one shouldn't be engaging in a *battle*. A recent target article in *BBS* (again!) argues that both these flowers can blossom in the same flowerbed. Joel Norman (2002) tries to "reconcile" the Marrian and Gibsonian perspectives, largely by relating them to neuroscientific discoveries that happened after both men's deaths. (I'm using "Marrian" here as code for "constructivist": Marr himself isn't cited by Norman.)

The peer commentaries—including one from Neisser, who switched from constructivist to ecological psychology in the early 1970s but is now more catholic in his tastes (Neisser 1994)—show that there's been significant advance. Yes, there are still people who regard one approach as more significant, more interesting, than the other. (Neisser is one: see Chapter 6.v.b and Neisser 1997.) But there's a common realization that vision can't be fully captured by either.

7.vi. Nativism and its Vicissitudes

The fourth behaviourist assumption identified in Chapter 5.i.a was that one must look *outside* the organism to understand why it does what it does. In particular, nativist psychologies are wrong-headed: there is no innate knowledge, and there are no inborn psychological powers (apart from a few very simple reflexes). That's largely why the ethologists (5. ii.c) had been relegated to the back corridors of Psychology's House.

The earliest cognitive science made no difference. Neither informational psychology nor the New Look gave any hint of nativism—and nor did 1950s Chomsky. *Syntactic Structures* put many cognitivist cats among the behaviourist pigeons: remember Jenkins's rueful "Everybody in the meeting jumped on me." But nativism wasn't one of them. Chomsky was still keeping quiet about that. Even when he excoriated Skinner in 1959, he didn't use nativism as part of his armoury (Chapter 9.vi.e and vii.b).

In the mid-1960s, however, Chomsky came out of the closet—and with a bang, not a whimper. In a flurry of books within only four years (1964, 1965, 1966, 1968), he made a scandalous psychological claim, which put nativism at the very heart of his linguistics. An adequate theory of language, he now declared, "incorporates an account of linguistic universals, and *it attributes tacit knowledge of these universals to the child*" (1965: 27; italics added).

Many psychologists—and philosophers and linguists, too—baulked at this. Some attacked his a priori arguments for nativism (innate ideas), while others criticized the empirical evidence, or lack of it, for linguistic universals (Chapter 9.vii.c–d). The fact that many major thinkers in centuries past would have been less shocked (9.ii–iv) didn't save his theory, in these critics' eyes.

However, early work in developmental linguistics, and unsuccessful attempts to teach language to apes, persuaded some others that he was right. This increased people's interest in the inherited capacities of different species, whether linguistic or not. Over the last twenty-five years of the century, then, the ethologists moved into more capacious quarters within Psychology's House. Moreover, much of the extra space was allotted to their new close cousins, the evolutionary psychologists (see 8.iv–v).

The Piagetians, by that time, needed expanded house room too—and this made for complications. For Jean Piaget, "Nativism: *yes* or *no*?" had never been an acceptable question (5.ii.c). Eventually, that Piagetian view prevailed. By the turn of the millennium, nativism *as commonly understood* was almost as discredited as it had been in behaviourist days.

Nevertheless, our biological inheritance was being acknowledged more strongly than ever before. The fourth tenet had been rejected—but with an interesting twist.

These changes weren't due to cognitive scientists alone, for some of the crucial evidence described below (especially in subsections b–c) was garnered by people with very different concerns. An ethologist who chose to live with a band of gorillas, or a psychologist who tried to teach language to chimpanzees, might have no interest whatever in computational ideas. But others, who did, used their findings as evidence in theorizing about the types of information processing going on in infant and adult human minds.

In short, nativism could have been resuscitated without any conceptual impetus from AI, or from Chomsky either. As a matter of historical fact, however, it owed a great deal to both.

a. The words of Adam and Eve

Those psychologists who were willing to take Chomsky's nativism seriously looked first to the development of language in children. Here, the social psychologist Roger Brown (1925–97) had a head start. For as a denizen of Bruner's Center, he'd known of Chomsky's heresy even before it was published.

In the early to mid-1950s he'd studied the development of language (R. Brown 1958), in connection with the Whorfian hypothesis (Chapter 9.iv.c). But that work had focused on words/concepts (not grammar), and had been done in the usual way: many different children, of various ages, were studied simultaneously, and their behaviour was compared. A few years later, he pioneered a very different methodology: developmental psycholinguistics as it's understood today.

In 1962 Brown began a long-running and scrupulously detailed study of the development of language—initially in just two children, "Adam" and "Eve", but "Sarah" soon joined them. This work was to be hugely influential. Linguistic development became a thriving speciality within psychology, the techniques becoming more discriminating as the years passed. Indeed, by the time that Sarah was included in the pioneering study, the investigators had already learnt that they should record certain subtle syntactic/morphemic distinctions which (with Adam and Eve) they'd formerly ignored.

Brown's team, which included the young Ursula Bellugi (who later did influential studies of American Sign Language and Williams Syndrome), discovered many intriguing facts about how the children's utterances became longer and more grammatically complex. Sometimes, they got "a shock" (R. Brown 1973: 284), when their naturalistic observations *didn't* match expectations based on late 1950s experiments exploring the grammatical capacities of 4- to 7-year-old children. (They managed to 'explain away' almost all of these surprises: pp. 284–93.)

For instance, they noted when two-word utterances first appear (at between 20 and 28 months) in the "Stage I" child, and how some words come to be used as hooks on which to hang many others (*milk all-gone, Daddy all-gone . . .*). And they studied the development of particular grammatical constructions, such as plurals and the past tense. With respect to the past tense, Brown's team found that the irregular forms (ate, went, came, did) preceded the regular ones (*walked, lifted, laughed*)—pp. 311–12. (For some other verb changes, such as third-person inflections, the regular preceded the irregular.) Some irregular pasts (*came, fell, broke, went, sat*) occurred very early indeed, and irregular pasts were always more frequent than regulars in the infants' speech, as they are in the speech of adults (p. 260).

Brown's general conclusion was that there are many structural regularities in the development of language, and that these don't depend primarily on the input (the mother's/carers' speech). The order in which specific grammatical structures (plurals, definite descriptions, possessives . . .) arise doesn't match the mother's speech: even her 'baby-talk' includes all these features from the start. It must, therefore, depend on sources within the child's own mind. In other words, there was seemingly some evidence for what Chomsky called a "Language Acquisition Device" providing innate knowledge of, or preparedness for, grammatical rules.

The Chomskyans soon claimed additional evidence for their case. Psycholinguists found that when the child comes to possess a large number of regular forms, the (correct) irregulars disappear—or anyway, occur less reliably—for a while. So an infant who'd previously said *went* appropriately would now say *goed*. Combined with the fact that young children will also say *mouses* and *sheeps*, this seemed to be strong evidence for the nativist position. (The past tense, in particular, was put under a developmental microscope. The details, excruciatingly boring in themselves, would suddenly become hugely interesting in the early 1980s: Chapter 12.vi.e.) Chomskyans saw such over-regularizations as proof that their guru was right. For, they said, these quasi-plurals and quasi-pasts weren't *learnt*: the child never heard its carers saying "goed" or "sheeps". They could have been generated only by applying GOFAI-type rules, such as *add '-ed' to the verb stem*. Indeed, any regularities that didn't reflect the speech of the carers, including the two-word to three-word progression, seemed to imply some inborn, language-specific, 'program'.

The Piagetians weren't convinced. They argued that Chomsky's nativism was over-simple, that language develops epigenetically from both active experience and biological maturation (Piatelli-Palmarini 1980). But there was much mutual misunderstanding, not least because Piaget's own writings were notoriously vague. Conrad Waddington, the biologist from whom Piaget had borrowed the concept of epigenesis, was unknown to psychologists (and regarded as a maverick by other biologists: see 15.iii.b). Most non-Piagetians of the 1960s–1970s saw epigenesis as even more mysterious than Chomsky's nativism (see subsection g, below).

Many 1970s psychologists, then, believed that Chomsky had been vindicated—or, at least, that there was strong evidence in his favour. Part of that evidence was the success of various GOFAI programs in parsing natural language (9.xi.b). Clearly, formal rules *could* be used to interpret grammatical structure. And phenomena like the infant's over-regular *goed* seemed to show that this is what goes on in human heads.

Twenty years after Chomsky's nativism shocked the psychological world, another scandal would arise—thanks to connectionist AI. David Rumelhart and James McClelland (1986) argued that *mouses* and *goed* might result from learnt statistics, not innate rules (see Chapter 12.vi.e). In the early 1990s, others showed that a connectionist system might be able to "start small" (compare: two-word utterances before three-word ones), enabling it eventually to learn a complex structure *without* being primed on what to look for (12.viii.c–d). And, also in the 1990s, developmental neuroscientists would give some firm empirical content to the notion of epigenesis (subsection i below, and 14.ix.c).

But all that was in the future. Meanwhile, some psychologists decided to focus on the second half of Chomsky's nativism. This was his claim that *Homo sapiens* is the *only* species to possess language.

b. Some surprises from ethology

Psychologists had two ways of confronting Chomsky's humanism. They could seek evidence that non-human animals possess language naturally, or they could try to teach it to them in the laboratory.

Informal studies of both kinds had been done for centuries, and people had also investigated feral children (9.ii.c). What was different now (from the mid-1960s on) was that Chomsky's formal theory was being used—by some people, anyway—as a criterion of language. In a word: no grammar, no language. So the central question was no longer "Do animals communicate?", nor even "Do they develop/learn vocabulary?" Rather, it was "Can they develop/learn syntax?" I say "develop/learn" because Chomsky had insisted not only that non-human animals *don't* develop language naturally, but also that they *can't* acquire it artificially (9.vii.c–d). As he put it:

> [All] normal humans acquire language, whereas acquisition of *even its barest rudiments* is quite beyond the capacities of an otherwise intelligent ape. (Chomsky 1968: 59; italics added)

Whether animals in the wild develop language naturally was a matter for the ethologists. Previously unfashionable, in the last third of the century they became highly visible. As late as 1971, they were still being sidelined. Pribram remarked:

> [At] the International Congress of Psychology in Tokyo [in 1971], comparative psychologists held a symposium on the state of their field. *The tone was rather gloomy.* They had not heard of the chimpanzee's challenge to the uniqueness of man. Washoe, Sarah, and Lana—and a whole new generation of scientists and apes—are attempting to spell the end of Cartesian dualism [i.e. the split between animals and man]. (Rumbaugh 1977, p. xvi; italics added)

Soon afterwards, the gloom was dispelled. The ethologists attracted attention from professional psychologists and philosophers—and from the media-influenced public (including 'animal rights' movements) around the world.

Some 1960s ethologists were more interested in *animals* than *theory*. For example, in 1960 Jane Goodall—then 26 years old (having paid her boat passage to Africa in 1957 with saved waitressing tips) and working as secretary to the anthropologist Louis Leakey—went to live with a group of chimpanzees by Lake Tanganyika. She revolutionized people's view of these creatures, for she discovered that their social skills were hugely more diverse and complex than had been thought. And, by implication, so were the cognitive skills required to generate them.

Her Ph.D. was published in 1968, by which time she'd already returned to Africa: a fuller account appeared some twenty years later (Goodall 1986). When her work, and comparable 'living research' with gorillas by Dian Fossey (1983), entered the public arena the opposition to vivisection on primates soared accordingly. Some countries tightened their rules as a result (see 2.ii.f).

But Goodall and Fossey were special. Most ethologists studied their chosen species either by relatively brief trips to the wild, or by observing them in quasi-natural captivity. In the latter case, they typically did experiments to discover the bounds of their behavioural capacities, much as Konrad Lorenz had done years before with his greylag goslings. And with the rise of cognitive psychology, some started asking questions about animals' *cognitive* capacities too. Specific computational processes and models were rarely mentioned (but see Boden 1983). Nevertheless, animals were seen as information-processing organisms, whose psychology—in some cases, anyway—was broadly comparable to ours.

By the mid-1970s, Donald Griffin (1915–2003) had amassed enough evidence to publish a book called *Animal Awareness* (1976)—a phrase which, even then, still had the capacity to shock. Strictly, and despite the reference to 'Mental Experience' in the subtitle, he was talking about cognition, not consciousness. (He remedied that later: Griffin 1992.) In other words, he was asking whether intentional concepts in general can be ascribed to some animals, and if so what the limits of their intentional capacities are.

The self-styled "cognitive ethologists" (Griffin 1978) went far beyond IRMs. Seeking to get "inside the mind of another species" (the subtitle of Cheney and Seyfarth 1990), they looked for evidence of language—and representation, intention, planning, imitation . . . all distinctly non-behaviourist concepts—in various species.

A cautionary 'aside': Here, let's ignore Thomas Nagel's (1974) famous argument that we simply can't get inside the mind of a non-human animal, if this means *knowing what it's like to be* that animal. And let's ignore the question of whether any creature capable of intentionality ("aboutness") must also have consciousness: after all, it's possible to discuss the former while ignoring the latter (see, for instance, Searle 1980). Both intentionality and consciousness are philosophical minefields. Ascribing either to animals is doubly problematic, but that's what the cognitive ethologists were prepared to do. Their work often contained some philosophical argument, and as the years passed the philosophers discussed cognitive ethology in their turn. (Some early interdisciplinary discussions are in the *BBS* peer commentary on Griffin 1978 and Premack and Woodruff 1978; later examples include Heyes and Dickinson 1990, 1995; Dennett 1991, chs. 7, 14; 1996; and Allen and Bekoff 1997.)—Of course, we shan't be able to ignore this philosophical can of worms for ever. We'll open it later (in Chapters 12.x.f; 14.viii.d, ix.b, and x–xi; and 16 *passim*).

The cognitive ethologists found many surprises, such as the alarm calls of vervet monkeys. These were observed in 1977–8 by the husband-and-wife team Robert Seyfarth and Dorothy Cheney, then at UCLA (now, at the University of Pennsylvania). They insisted that these warning signals have *semantic*, not just emotional, content—referring (*sic*) specifically to snakes, or to leopards, or to eagles (Seyfarth *et al.* 1980). They said this not only because of patterns of co-occurrence (snake calls happen only in the presence of snakes), but also on the basis of the monkeys' behaviour. When the calls were recorded and played back in the absence of any predators, the vervets did the 'appropriate' thing. For instance, they took to the trees to escape from a snake on the ground—who, of course, wasn't there at all.

Moreover, Cheney and Seyfarth (1990) discovered that vervet babies have to learn the alarm calls, gradually restricting the scope of reference. At first, they'd produce the eagle signal when any bird was present; eventually, it would occur only for eagles (alias predators). Towards the end of the century, similarly reference-restricted warning cries were found in other species. Even more surprisingly, the ethologists sometimes observed 'modifier calls' which, when produced in combination with some other call, apparently altered its meaning. In addition, previously unsuspected forms of communication were discovered in dolphins and whales. (Many of these findings fascinated the general public: whale "songs" even entered the pop charts.)

Predictably, these phenomena were often described—even by ethologists—as "the rudiments of language". However, the ethologists hadn't found *grammar*.

Most animal communications are discrete signals, like the cries of the vervets. To be sure, birdsong often consists of long strings of sounds. But there's no evidence of compositional structure or 'creative' generativity. Mature nightingales can sing many different songs, some of impressive length and complexity. But this isn't *generative* complexity. The nightingales are more like a pianola with a dozen different (sometimes cut-and-pasted) rolls in it than a person, or even a Stage 2 child, coming up with a new sentence/word-string every few minutes. Moreover, although greater complexity in an individual bird's song correlates with greater success in attracting mates, there's no evidence of semantic content, nor any hint that the *order* of the cries in the 'combination signals' makes any difference to their functionality.

Thirty years of this type of ethological research have shown that the cognitive capacities of many animal species are far greater than psychologists had previously thought. That's true even if one believes that *representation* isn't the right way to explain them (see 14.vii and 15.vii–viii). Indeed, even crickets and frogs possess information-processing mechanisms of surprising power, inflexible though they may be (14.iv and vii, and 15.vii). But the natural linguistic (i.e. syntactic) endowment of non-human animals is, apparently, nil.

c. From Noam to Nim

Even before Goodall first ventured into Africa, an entire experimental industry had arisen alongside human psycholinguistics: the increasingly ingenious effort to teach language to non-human animals in captivity. (Mostly primates, but eventually also parrots: Pepperberg 2000.) For, given Chomsky's claim that even the "barest rudiments" of language are beyond the capacities of apes, the challenge was clear: was there anyone out there who could coax chimpanzees, our closest cousins, to speak?

At first, it seemed—on biological grounds—that the answer must be "No!" For Eric Lenneberg (1921–75), a biologist admirer of Chomsky, had shown in intriguing detail not only that human bodies are nicely prefigured for the spoken word, but that the chimpanzee's oral cavity and musculature make speech (i.e. pronouncing phonemes) impossible. The very *anatomy* of language, it seemed, is species-specific (Lenneberg 1964, 1967).

However, Elizabeth McCall had recently argued in her MA dissertation that American Sign Language (ASL), which uses finger movements, may be a 'real' language—that is, one with a grammar (McCall 1965). Spontaneous signing by the deaf had been the subject of speculation for centuries: Descartes was just one of many who'd discussed it (Chapters 2.iii.c and 9.ii–iv). But again, Chomsky's work had sharpened the relevant questions. McCall's suggestion was later taken up by Bellugi (with Edward Klima), who found that ASL is indeed a syntactically structured language (Klima and Bellugi 1979; Bellugi and Studdert-Kennedy 1980).

So maybe all the past attempts to coax language from apes had failed simply because they can't *speak*? Even Viki, a chimp whose lips and tongue had been constantly 'shaped' by the fingers of her human foster parents in the 1940s, had learnt to produce only four word-ish—very '-ish'—sounds (K. J. Hayes and Hayes 1951). However, chimps have fingers nearly as nimble as ours. Perhaps one should try to teach them sign language, instead?

The first psychologists to attempt this were the husband-and-wife team of R. Allen and Beatrice Gardner, at the University of Nevada (1969, 1971, 1974). Like Winthrop and Louise Kellogg a quarter-century earlier (Kellogg 1933), they took an infant chimp into their house to be raised much as a human child. Unlike the Kelloggs, however, they used not speech but ASL (the version in which each handmade sign stands for its meaning 'directly', rather than spelling out the English word). They used all the teaching methods they could think of, including showing their own ASL signs to the animal; gradually shaping its behaviour by rewards (operant conditioning); moulding its hands into the required position; and simply 'chatting' to it while playing, feeding, and so on.

Their protégée Washoe (named for Washoe county, Nevada) eventually learnt to recognize and produce about 240 ASL signs. (More accurately, her trainers *reported* that she did so: see below.) Most of these stood for *classes* of object, being correctly applied (for example) to a new bird, never seen before. On at least 300 occasions, Washoe produced novel combinations that 'made sense'. (The same caveat applies here: see below.) Some of the sign combinations that she was specifically taught were five signs long.

The Gardners were persuaded that Washoe's performance matched that of a 2-year-old child, and they set out to persuade the world. By the early 1970s their chimp-child was famous, featured in newspapers and TV programmes in many different countries. (*Hello!* magazine hadn't yet been founded: perhaps its camera bulbs would have been flashing too.)

But had Washoe, in learning to sign, learnt a sign *language*? The Gardners said "Yes!" They pointed out that Washoe sometimes came up with new combinations of signs, thus—so they claimed—paralleling the "creativity" of language which Chomsky had emphasized so strongly (9.iv.f and vii.b). One famous example was her signing *water bird* on first seeing a swan. The awkward question, however, was whether she was signing *water-bird*, or whether she was merely signing *water*, quickly followed by *bird*. After all, both the water and the bird were clearly visible. And what if she'd happened to sign *bird* before *water*? For native speakers of English, "bird-water" isn't nearly so persuasive a combination (though it might be interpreted by enthusiasts as the sentence *There's a bird on the water*).

It turned out on closer examination that Washoe's 'two-word utterances' didn't respect word order, whereas Stage 1 children do. Although the two-word child can't yet say either "Dog bites cat" or "Cat bites dog", she can distinguish these situations by saying *dog bite* or *bite dog*, respectively (R. Brown 1970). It doesn't follow that Washoe couldn't 'tell the difference' between the two situations. But she couldn't *tell* that difference. In English, word order is an important grammatical cue—one of the "barest rudiments" of the language. There was no evidence that Washoe could deal with it.

Some years later, in 1982, the media reported worldwide that Washoe, by then with Roger Fouts at the University of Oklahoma, had taught her adopted son Loulis to sign. Or, less tendentiously, that—being raised in her company—he'd learnt to sign spontaneously (Fouts *et al.* 1982). However, Loulis didn't respect word order either.

Washoe's contemporary Sarah, though less suitable as chat-show material, also became famous. Unlike Washoe, she didn't 'make remarks' spontaneously, nor use a 'human' method of communication. Indeed, it wasn't clear that she was *communicating* at all. Rather, she responded to the tasks/questions set to her by her trainer David Premack (1925–), aided by his wife, Ann, by moving plastic tokens, or 'symbols', on a magnetized board. (She'd been taught by means of operant conditioning.)

Sarah's 'linguistic' accomplishments, it appeared, were far greater than Washoe's. Indeed, they were staggering. According to Premack (1971), they included *yes–no* interrogatives, *wh*-interrogatives, negatives, word compounds, plurals, quantifiers . . . and more, many more.

For instance, she was given tokens for *same* and *different*, and trained to pick the right one when presented with two objects from the same or different classes. Next, a token for the question mark, *?*, was placed between two objects, and Sarah was trained to replace it with one—the right one—of the *same/different* tokens. Premack interpreted this as a simple *wh*-question, namely, "*What* is the relation between the two objects?" Many other tasks, more complicated than this one, were also mastered by Sarah. Seemingly, she'd learnt a great deal more than the barest rudiments of language.

But had she, really? Psychologists sympathetic to Chomsky suggested, in effect, that Sarah's superficially impressive behaviour was Newtonian at heart (e.g. R. Brown 1973: 32–51). With respect to the more complicated tasks, her 'successes' might have been due to chains of conditioned reflexes, as seen in pigeons playing ping-pong (Chapter 5.iii.a, above). In this context, it was significant that her success rate, about 80 percent, *didn't* vary with what people see as the 'difficulty' of the task. Her performance, one might say, was too good to be true.

Moreover, Sarah's behaviour was elicited in very narrowly defined situations. Brown (1973: 48) compared her grasp of "Tokenese" with his own of Japanese. He'd received some lessons ingeniously "programmed . . . in an almost Skinnerian way", and had learnt to reply appropriately to sentences like those in the training set. He'd thought he was doing well. Faced with 'real' Japanese, however, he was virtually helpless. He could neither interpret nor generate sentences he hadn't come across before (and the problem wasn't merely his lack of vocabulary).

Finally, many of Sarah's 'successes' could have been due to unconscious signalling on the part of the human experimenters. Robert Rosenthal had recently shown that rats and horses could learn to respond to such signals (R. Rosenthal and Fode 1963; Rosenthal 1964; Pfungst 1911/1965, introd.). In some cases, the signals were so slight (e.g. a postural change caused by increased muscular tension) that they were invisible *even to other humans highly motivated to observe them.* If rats and horses could do this, then presumably people could do it too. It followed, for instance, that any experiment on telepathy/ESP where *someone perceptible by the subject* already knows the answer is worthless. And the implications didn't stop there. As Rosenthal wryly put it:

> That many experimenters over the years may have fulfilled their experimental prophecies by unintentionally communicating information to their subjects may be a disquieting proposition. (Rosenthal, in Pfungst 1911/1965, p. xxii)

For chimp-language research, Rosenthal's findings were potential dynamite. Over the years, the trainers' experimental methods were altered so as to minimize experimenter bias. But Herbert Terrace (1979*a*,*b*, 1981), for instance, had the honesty to admit that films of his ape subject—a cousin of Washoe's, mischievously named Nim Chimpsky as the experiment got under way—showed that the animal was often unknowingly cued by the psychologists working with him. Unlike Sarah (but like Washoe), Nim sometimes signed spontaneously. Nevertheless, Terrace concluded that his stalwart efforts to train Nim to 'talk' had been *unsuccessful* (Terrace 1979*a*,*b*, 1981; Terrace *et al.* 1979). The sarcastic baptism had backfired: Chimpsky had vindicated Chomsky.

Besides the type of experimenter bias described by Rosenthal, there's the type of bias highlighted by the New Look psychologists. One sees what one expects to see—and what one *wants* to see. This explains the highly sceptical remarks of the only person on the Washoe team who, being profoundly deaf, was raised with ASL as their native language:

> Every time the chimp made a sign, we were supposed to write it down in the log... They were always complaining because my log didn't show enough signs. All the hearing people turned in logs with long lists of signs. They always saw more signs than I did... [They] were logging every movement the chimp made as a sign... Sometimes [they'd] say, "Oh, amazing, look at that, it's exactly like the ASL sign for *give*!" It wasn't. (quoted in Pinker 1994: 338)

So where I said (above) that Washoe learnt 240 signs and produced 300 sensible combinations, it would have been more accurate to say that she was *perceived* as doing so by her trainers. Just how overgenerous their interpretations were, it's difficult to know. But that they *were* overgenerous is sure.

Other efforts to refute Chomsky's view on apes and language included two conducted at Georgia State University. These were Duane Rumbaugh's (1977) work with Lana, and its successor: Sue Savage-Rumbaugh's still-continuing training of Kanzi (1991; Savage-Rumbaugh and Lewin 1994; Savage-Rumbaugh *et al.* 1998).

By the late 1990s, the Georgia team were confidently claiming that their long-time pupil had acquired linguistic and cognitive skills equal to those of a $2\frac{1}{2}$-year-old child. Their work bristled with intriguing experiments, in which their animals—unlike Sarah—used 'language' to help further their own purposes.

For example, one chimp learnt to use abstract 'symbols' to ask another chimp for a specific tool. (The second couldn't see what the first was doing, so couldn't work out for itself which tool was needed at the time.) And in January 2003 the *New Scientist* reported that Kanzi had made up four 'words' for himself. Apparently, he was emitting sounds meaning *banana*, *grapes*, *juice*, and *yes*—and a team member insisted, "We haven't taught him this. He's doing it on his own." The next steps, the researcher continued, would be to find out whether Kanzi was trying to imitate human speech, and whether other chimps would learn to treat Kanzi's sounds as *communications*.

Like Sarah, however, Savage-Rumbaugh's chimps were accused by sceptics of being well-conditioned rather than well-spoken. Kanzi's newly minted 'words' could have been due to the Rosenthal effect. Even his seeming (limited) comprehension of syntax, or word order, could have been closer to Rosenthal than to Chomsky. And the tricky question remained: *why*, if chimps can emulate a 2-year-old, can't they emulate a 3-year-old or 4-year-old . . . ?

The progress—or lack of it—made in a decade of dedicated work can be gauged by comparing two highly detailed reviews (Ristau and Robbins 1982; Wallman 1992). Strictly, the jury's still out. Who knows what may be achieved tomorrow? But don't hold your breath. As yet, there's no clear evidence that any non-human animal can learn a syntactically structured, indefinitely creative, language. Descartes, it seems, was right.

But maybe Chomsky wasn't. For besides agreeing with Descartes that no ape could ever learn language, he'd also posited a dedicated "Language Acquisition Device" providing the human baby with an innate knowledge of universal grammar (9.ii.c and vii.c–d). His disciple Steven Pinker argued for that position in a best-selling book on *The Language Instinct* (1994). However, not everyone was convinced.

Non-Chomskyans suggested that whatever it is which enables babies to acquire language may be some combination of mechanisms evolved for more general purposes—including visual perception and problem solving. The psycholinguist Elizabeth Bates (1947–2003), for instance, famously said: "Nature is a miser. She clothes her children in hand-me-downs, builds new machines in makeshift fashion from sundry old parts" (E. Bates *et al.* 1979: 1). If these "sundry old parts" *make language possible*, that's not to say that they're dedicated to it alone (cf. 12.x.g). More recently, the neuropsychologist Arbib (2003, 2005) has named seven general capacities for which neuroscientific evidence exists, and which may underlie the evolution of language. (For an overview of current ideas on its development, see Tomasello and Bates 2001.)

d. Modish modules

Nativism is nativism: why stop at language? Chomsky himself had speculated that the mental "organ" for language is only one of several such organs (1980*a*). Now, in *The Modularity of Mind* (1983), Fodor picked up Chomsky's mental-organs idea, and Marr's work too, and ran with them very far and fast indeed.

He didn't merely say that inherited mental organs ("modules") exist. He argued that *only* behaviour which is grounded in them can be grist for psychology's mill: "the limits of modularity are also likely to be the limits of what we are going to be able to understand" (1983: 126). To be sure, he added the words ". . . given anything like the theoretical apparatus currently available". But he didn't expect to hear of a sufficiently bright idea some time next week. Instead, the 'higher mental processes' were being ruled out of bounds (see Section iii.d, above).

Provocative as ever, he increased the shock by subtitling his new book 'An Essay in Faculty Psychology'. The nineteenth-century faculty psychologists had been ousted from Psychology's House long ago (hence no mention of them in Chapter 5.ii). Was he inviting them back in? Not quite: he wasn't heralding 'organs' for Will and Imagination. But he was resuscitating their notion that we're born with various mental powers, which are—and remain—distinct.

By a module (alias a faculty), Fodor meant a biologically evolved input system triggered by domain-specific ("proprietary") information, whose computations are automatic and "encapsulated". That is, they can't be influenced by each other nor, top-down, by concepts or beliefs (Chapter 16.iv.d). He was using the distinction that he and Pylyshyn had made earlier, between processes defined by the FFA and cognitively penetrable phenomena (Section v.a, above). But the FFA was now seen as comprising a significant number of domain-specific components.

If Fodor's book carried a negative message (*Hands off the higher mental processes!*), it also carried a positive one: *Inborn faculties exist, and can be found—and understood.* Finding them, he said, involves "carving nature at its joints" (p. 128). He discussed only language and vision, because computational theories—in Marr's sense—had been outlined only for those. His own account of language included David Lewis's (1969) convention of truth telling as one of the constraints:

> We say of *x* that it is *F* only if *x* is *F*. Because that convention holds, it is possible to infer from what one hears said to the way that the world is [much as we can infer from visual stimulations what the 3D layout of the scene is]. (Fodor 1983: 45)

(Lewis's term *convention* wasn't well chosen: "truth telling" isn't like wearing a hat, or even driving on the left/right of the road, for we can't decide not to follow it.) This, by the way, enabled Fodor to escape from the trap of methodological solipsism (Fodor 1980*a* and 16.iv.d). The logic of Fodor's argument left space for other, as-yet-undiscovered, domain-specific faculties. He speculated in passing about possible motor modules. But some people went much further: namely, the evolutionary psychologists (cf. Chapter 8.ii.d–e).

Evolutionary psychology was ethology made good. It was both more cognitive and—by 1980—more 'respectable' than the work mentioned in Chapter 5.ii.c. The new data about the mental powers of chimps and gorillas had helped, as had the growing appreciation of Simon's ant. Many evolutionary psychologists specifically highlighted computational ideas (Tooby and Cosmides 1992: 64–9, 94–108, 112–14). Moreover, Marr and Fodor were often cited in support of evolutionary accounts of sociocultural behaviour (see Chapter 8.ii.e). And Pinker (1997) wrote a popular trade book in which Chomsky and evolutionary psychology were celebrated together.

A wide variety of modules were posited by evolutionary psychologists, each supported by its own adaptationist story. For instance, Leda Cosmides (1957–) and John Tooby (1952–) attributed the recognition and punishment of "cheaters" to an inborn faculty evolved to detect and sanction people "violating a social contract" (Cosmides and Tooby 1992: 180; see also Cosmides 1989).

Such a faculty, they argued, would be very useful to a social species, for without it various types of antisocial parasitism would emerge. Whereas formal logic is a general-purpose inference mechanism, cheat detection—they said—involves a dedicated, and largely automatic, system. They claimed that this explains the startling results of Wason and Johnson-Laird's experiments (Section iv.d). Indeed, these were seen as strong evidence in their favour, since they showed that people are far better at detecting violations of if–then rules if (besides being familiar) they involve *cheating on a social contract* (pp. 180–206).

Given that *Homo sapiens* is a social species, cheat detection wasn't the only candidate for 'interpersonal' modules: a number of others were posited too (A. W. Byrne and Whiten 1988). The most important, a faculty for attributing intentional states to other people, is discussed at length below (subsection f). Three more suggested modules—a body-ratio assessor, a symmetry spotter, and a viable-habitat recognizer—are mentioned in Chapter 8.iv.a; and a fifth, a face recognizer (for which there is strong neuroscientific evidence), is described in Chapter 14.ix.c.

But the candidates for modularity numbered many more than these. Indeed, the mind was repeatedly compared to a Swiss army knife, a tool consisting of many little tools—each of which does a different job, and each of which functions separately.

One of the most extreme expressions of this viewpoint came from the neuropsychologist Charles Gallistel. He even argued, at a time when Hebb's name was echoing loudly in psychology's halls, that there's *no* content-neutral learning mechanism (see 14.ix.g). As for modules, he asked:

> Did the human mind evolve to resemble a single general-purpose computer with few or no intrinsic content-dependent programs? Or does its evolved architecture more closely resemble an intricate network of functionally dedicated computers, each activated by different classes of content or problem, with some more general-purpose computers embedded in the architecture as well? ... [In other words, does] the human mind come equipped with any procedures, representational formats, or content-primitives that evolved especially to deal with *faces, mothers, language, sex, food, infants, tools, siblings, friendship, and the rest of human metaculture and the world*? (Gallistel 1990: 94; see also Gallistel 1995)

His answer—no prizes offered!—was a resounding *Yes!*

Just how separately the different modules function was a bone of contention. For instance, some people stressed our ability to mix concepts from different domains. The developmental psychologist Karmiloff-Smith showed that such mixing is impossible for infants, but becomes possible later (subsection h, below). And the archaeologist Steven Mithen (1996*a*,*b*) argued that the evolution of modern man from earlier hominids involved a move from a 'pure' Swiss-army-knife mind to one of "cognitive fluidity", where concepts evolved for one domain can be applied in another. Members of *Homo sapiens* can think of people also as animals, or even as objects. These types of racism were impossible for our phylogenetic cousins *Homo neanderthalis*, said Mithen, because for them people could only be *people*.

As the example from archaeology may suggest, modules became fashionable way beyond the pages of psychology books. By the turn of the century, even theologians of a distinctly traditional cast of mind were using the term en passant, without comment or explanation. So the Chancellor of York Minster, while bemoaning the unremitting rise of secularity, remarked:

> [The] decline of organized religion ... does not necessarily mean that people are losing an inclination to seek personal identity through *modules of understanding* which they would probably call 'spiritual'. (E. Norman 2002, p. viii; italics added)

This was in a little book written for a general audience, and utterly lacking in jargon—theological or otherwise. Clearly, the author assumed that his readers would know what he meant. (Whether there really are modules for spiritual understanding is discussed in Chapter 8.vi.b–f.)

That's why, when Pinker—in the very same year—published yet another trade book putting the case for modules, some of his own colleagues accused him of flogging a dead horse. In *The Blank Slate* (2002), he ridiculed fourth-tenet externalism and championed an inborn "human nature". A number of psychologists (in reviews, and in Internet discussion groups) complained that he was wildly exaggerating the extent to which people still believed in John Locke's "blank slate" (6.i.c).

Pinker's critics were exaggerating, too. On the one hand, postmodernism in the humanities—and anthropology (8.ii.a–c)—still had a strong influence. Besides their anti-realist take on science (1.iii.b), the postmodernists still insisted that human minds are culturally constructed *rather than* (not *as well as*) biologically grounded. On the other hand, there was some resistance to nativism within cognitive science itself. The connectionist McClelland was now generalizing his notorious mid-1980s attack on innate grammar (12.vi.e) to cover the concepts—animals, plants . . . and so on—believed by evolutionary psychologists to be innate semantic modules (McLelland and Rogers 2003; T. T. Rogers and McClelland 2004). Even "Theory of Mind" (see below) was said by some to arise from general connectionist roots.

The point of historical interest, however, is that many psychologists felt that Pinker was attacking a straw man. Nativism, it seemed, had graduated from professional heresy to near-orthodoxy in a mere thirty years. (It was still anathema for many of the general public, however: so Pinker's book didn't entirely miss its mark.)

e. But how many, exactly?

Speculations about Neanderthal man were bound to raise eyebrows, not to say blood pressure. But they weren't the only ideas to be scorned by critics as Just So stories.

Rudyard Kipling's (1902) tales of how the elephant got its trunk, and why we indulge the cat who walks by himself, were hardly less fanciful—so these critics thought—than many claims made in evolutionary psychology. The modules posited were so numerous, and so socioculturally 'sexy' (going way beyond boring Mexican hats), that they threatened to undermine Darwinian psychology's hard-won respectability.

One could be bothered by this proliferation of modules even as a modularist oneself. Fodor, for instance, refused to accept all the examples posited by the evolutionary psychologists. He countered Pinker's Swiss-army-knife account of *How the Mind Works* (Pinker 1997) with his own *The Mind Doesn't Work That Way* (Fodor 2000*b*), a longer version of his stinging squib in the *London Review of Books* (1998*c*).

Fodor's problem wasn't the nativism: "I'm a committed—not to say fanatical—nativist myself" (2000*b*: 2). (He even took this opportunity of reiterating his belief in an innate Language of Thought: 16.iv.c.) Nor was it the commitment to computationalism, which he still thought "far the best theory of cognition that we've got". Rather, it was Pinker's "ebullient optimism" about what a computational psychology could actually achieve:

> I would have thought that the last forty or fifty years have demonstrated pretty clearly that there are aspects of higher mental processes into which the current armamentarium of computational models, theories, and experimental techniques offers vanishingly little insight. And I would have thought that all of this is common knowledge in the trade. How, in light of it, could anybody manage to be so relentlessly cheerful? (2000*b*: 2–3)

His own sympathy was with A. A. Milne's Eeyore: "'It's snowing still' said Eeyore. '...And freezing...However,' he said, brightening up a little, 'we haven't had an earthquake lately.'" That, Fodor remarked, captured his mood exactly.

In part, Fodor was repeating his pessimism about explaining the higher mental processes (Section iii.d, above). He did this by attacking what the anthropologist Sperber (1994) had called the "massive modularity thesis", or MM thesis (Fodor 1987*a*; 2000*b*, ch. 4). The MM thesis (in Fodor's words) is the view that "the mind is mostly made of modules". In Sperber's account, modules came in many formats and sizes, including micro-modules the size of a single concept. In short, there are lots of them.

Fodor was happy to allow that there may be more modules than he'd originally expected. But how are their outputs integrated? Where, as Descartes would have put it (2.iii.c), is the *common* sense? How is it that, faced with perceptually intractable illusions such as the Müller-Lyer diagram, we aren't actually fooled by them—because we *know* they're illusions? What domain-specific module could possibly explain that? None, said Fodor, explicitly aligning himself with "the New Rationalists" (see Chapter 9.ii.a). In short, *many* modules needn't imply *massive* modularity. (Sperber eventually admitted that his MM thesis had been "extremist". Nevertheless, he insisted that it was true enough to justify further "speculation" about modules: Sperber 1994.)

As for just how many modules there may be, Fodor was relatively miserly. Marr's work on vision offered the clearest examples, but recent research on human infants had plausibly suggested others—not only in folk physics, but in folk psychology too (see below). However, Sperber and others had gone overboard: "the modularity thesis gone mad" (Fodor 1987*a*: 27).

The adaptationist arguments used by the evolutionary psychologists, said Fodor, were largely bogus. For one thing, tiny changes in brain structure may have huge (i.e. relatively sudden) effects on the behavioural phenotype. Indeed that seems to be so, for chimps' brains are much more like ours than their behaviour is. It follows, given our current ignorance about the brain, that we don't *have to* posit a gradual evolutionary shaping of each aspect of human behaviour to explain 'how we got here from there'.

For another, adaptationists often commit "fallacies of rationalization" (Freud's word). For example, Cosmides and Tooby had argued backwards from their recognition that it would be useful for a social species to possess a cheat-detection faculty. But, said Fodor, from the fact that *if* someone's motive were to increase his chances of survival/reproduction *then* the best thing to do would be X, it doesn't follow that he *actually* did X because he (or his selfish genes) consciously or—*pace* Freud—unconsciously 'wanted' that evolutionarily happy result. The person's conscious avowal that he did X *because he wanted something else* is an alternative, and often more plausible, explanation. In other words, cognitive science can be nativist without being Darwinist—if that means massively, reductively, adaptationist.

Fodor wasn't the only philosopher contributing to the modularity debate. Many others queried *just what was meant* by "module", as well as asking whether this or that version of the modularity hypothesis was coherent and/or plausible (useful collections are: J. L. Garfield 1987; Segal 1996; Carruthers and Chamberlain 2000).

For example, Samuels (1968–) noted a threefold ambiguity (Samuels 1998, 2000; Samuels *et al.* 1999). Modules, he said, were assumed to be:

(1) innate stores of information about a specific domain, ranging from organized sets of items to single concepts, or
(2) information-processing mechanisms, applied only in a given domain, or
(3) both.

Samuels defined a compromise, the "Library Model of Cognition", according to which modules are:

(4) stores of domain-specific information processed by general-purpose mechanisms. (An example of the latter might be the general preference for moderate novelty, discussed in Chapter 8.iv.c.)

The Library Model, he argued, is just as evolutionarily plausible as MM.

Tony Atkinson and Michael Wheeler (2004) offered a further alternative. Besides showing that there's rarely a clear distinction between "domain-specific" and "domain-general" processes (because domains can be defined at different hierarchical levels), they pointed out that there may be

(5) innate general information-processing mechanisms *as well as* specific ones.

Each of these alternatives (Samuels and Atkinson–Wheeler) is a form of MMM thesis, the three Ms here positing *moderately* massive modularity (Carruthers 2003).

For our purposes, what's interesting is the *computational* flavour of almost all of these philosophers' discussions. Anyone could grumble about the spurious plausibility of Just So stories, of course—and many did. (It was a favourite gambit of the postmodernist opponents of evolutionary psychology, for instance.) But the arguments about informational integration, and about general versus domain-specific information-processing mechanisms, could have been suggested only in an intellectual climate which took for granted that the causes of behaviour can/should be conceptualized in functionalist (computational) terms. The philosophy of mind had changed unrecognizably since the 1950s (see Chapter 16.i).

(By the end of the century, the notion of "module" had changed almost unrecognizably too. Many psychologists and neuroscientists now argued that modules are "inborn" only in a highly qualified sense. Instead of being preformed, they *develop* epigenetically: see subsections g–i, below.)

f. Theory of Mind

Developmental psychologists' research on babies' understanding of the physical world—movement, for instance, or solidity, or size-constancy—made some astounding discoveries in the 1980s. For Piaget's claim that real-world knowledge arises only from active manipulation of it was being cast in doubt. Manipulation, once it has appeared, may well enrich the infant's object concept. But it was now becoming clear that Piaget had hugely underestimated babies' abilities.

For example:

* newborns have some sense of size-constancy (Slater *et al.* 1990),

* and various aspects of the physical world become salient one after another (Spelke 1991).
* Babies only $2\frac{1}{2}$ months old show surprise when one (hidden) solid object apparently passes through, or jumps over, another instead of moving along a connected, unobstructed pathway.
* Four-month-olds assume that a hidden object will maintain a constant size and shape.
* And by six months they also expect unsupported objects to move downwards, and moving objects to continue to move in the absence of obstacles.

In short, even very young babies have grasped a significant portion of what the AI scientist Hayes was calling "naive physics" (13.i.ii). Comparable discoveries were being made about many other concepts, too (Carey 1985; Carey and Gelman 1991). But some of the most high-profile nativist research concerned "Theory of Mind" (ToM).

The idea of ToM originated in 1970s studies of chimpanzee behaviour, and was buttressed by 1970s discoveries in neuropsychology. However, two of the seminal publications appeared in 1983 and 1985, when Fodor's book on modularity was making intellectual waves, and it became strongly associated with that concept. Indeed, Cosmides and Tooby—two leading proponents of the Swiss army knife—later wrote the introduction for a widely read book on ToM (Baron-Cohen 1995).

The label "Theory of Mind" was coined in the late 1970s by the primatologists Premack and Guy Woodruff (P&W). After training the long-suffering Sarah to use quasi-linguistic tokens, Premack had turned to ask what intentional understanding she had. For example, could she attribute goals and plans to other agents? P&W's answer was *Yes!*

In one experiment, for instance, they familiarized Sarah with heaters—and how they were lit. Then, they showed her a video of a shivering experimenter next to an unlit heater. They gave her two pictures to choose from: one showing a lit match, the other some unrelated scene. Almost always, she chose the match. Now they raised the stakes. Sarah was given three pictures: a lit match, an unlit match, and a burnt-out match. Again, she almost always chose the lit match. A similar experiment, with comparable results, involved a person trapped in a cage, and a key (ratcheted up to a key, a twisted key, and a broken key).

These results showed, said P&W, that Sarah was applying a theory of mind. She realized that the human wanted to be warm, that a sub-goal was to light the heater, and that a lighted match was a means to that sub-goal. Moreover, P&W interpreted their label literally: "A system of inferences of this kind is properly viewed as a theory" (1978: 515).

Published in *BBS*, P&W's paper received very wide attention. One item of the extensive peer commentary was especially fruitful. The philosopher Dennett (1978*d*) pointed out that their results, like Premack's previous work on Sarah's language, could be explained as due to conditioning. (P&W had actually admitted this, but said an intentional explanation was simpler.) But he added some constructive comments.

Based on his account of the intentional stance (16.iv.a–b), Dennett listed five necessary conditions for attributing ToM—"beliefs about beliefs"—to an agent. We can't be sure that chimps meet these conditions, he said, because they lack language.

But children do have language. Moreover, children evidently appreciate the subjectivity of belief, for they can attribute *false* beliefs to others, and make reliable predictions accordingly. For instance, their enjoyment of Punch and Judy involves their knowledge that Punch believes—mistakenly—that Judy is trapped in the box he's about to throw off the cliff.

By the summer of 1979, Dennett's analysis was being applied in work on children, not chimps (J. Perner, personal communication). It inspired an ingenious experimental design involving pretend play with puppets, developed by Heinz Wimmer and Josef Perner (1948–)—then at the universities of Salzburg and Sussex respectively. They already knew that infants have "meta-representation" (beliefs about beliefs) and even understand that different individuals—for instance, people sitting on opposite sides of a screen—may have different beliefs. Now, they discovered that this meta-representation is severely limited: the full subjectivity of belief isn't yet understood (Wimmer and Perner 1983).

In a nutshell, they found that 3-year-olds can't grasp the fact that people may have beliefs that are false. A different version of their original test was soon defined by Simon Baron-Cohen (1958–), at University College London (UCL). He showed, for instance, that a 3-year-old will say that Sally, who left the room just before the marble she'd put into a basket was moved into a box by Ann, will look for it *in the box* on her return (Baron-Cohen *et al.* 1985). That's why infants don't lie: they *can't* lie, for they can't comprehend that a statement they know to be true might be thought by someone else to be false. (It turned out later that pygmy 3-year-olds can't conceive of false belief either, and become able to do so at around 5 years just like their Western counterparts: Avis and Harris 1991—and see Chapter 8.vi.d.)

As for when subjectivity (and deception) are at last understood, Wimmer and Perner pointed to 5/6 years of age. The Sally–Ann ToM test put it at 4 years old. But the central 'nativist' point was the same: some developmental (not purely experiential) factor enables children to realize, eventually, that people have *minds*.

ToM became a major experimental industry in the late 1980s. (For a few of the products, see Astington *et al.* 1988; Wellman 1990; Perner 1991; Whiten 1991; Leslie 2000. And for an account of how ToM-endowed children develop *trust* appropriately, see P. L. Harris and Koenig forthcoming.) Here, two points are especially relevant. On the one hand, there was a dispute about *how ToM functions*. On the other, there was the suggestion that ToM is a *module*.

Regarding the first, some psychologists—such as Perner himself (1991)—followed P&W in taking the "T" in ToM literally. They believed that the 4-year-old develops a *theory* that people have subjective beliefs and goals. Thenceforth, the child/adult interprets and predicts behaviour by inferring hypotheses from that theory. In effect, the New Look's notion of perception-as-science (6.ii) was being generalized to the perception of mind. Others disagreed. ToM, they said, is the ability to *simulate* the other person's behaviour. One puts oneself in the place of the other, empathizes with them, adopts their point of view. Then, one 'runs through', or imagines, the thoughts/behaviour that would result if starting from *that* place.

This approach was originated by the Cambridge philosopher Jane Heal (1986) and the Missouri philosopher Robert Gordon (1986). Heal spoke of "replication", Gordon of "simulation"—and it was Gordon's term which was generally adopted. Later, after some years of heated dispute, Heal clarified what's meant by simulation and showed that it could be understood in two different ways (Heal 1994, 1996, 1998).

She pointed out that the general assumption that 'theory versus simulation' is an *empirical* dispute was misleading. Theory theory (in a "strong" sense) is a priori false, and simulation theory a priori true. Or rather, it's true if one takes simulation theory to mean "co-cognition": the ability, at the personal level, to think about the subject matter of the thoughts of others. The reason is that thinking about Anita's thoughts about opera, or coffee, *necessarily* involves thinking about (exploiting our own knowledge of) opera, or coffee. (According to theory theory, all we have to consider is folk psychology, or the nature of *thoughts*, never mind opera and coffee—Heal 1998: 485–8.) The empirical question is what "sub-personal cognitive machinery" is involved in implementing such co-cognition. One possible answer (there are others) is that:

> when we think about others' thoughts we sometimes "unhook" some of our cognitive mechanisms so that they can run "off-line" and then feed them with "pretend" versions of the sorts of thought we attribute to the other. (Heal 1998: 484)

The "theory theory" versus "simulation theory" debate filled many pages through the 1990s. Because it concerned the nature of folk psychology, it intrigued philosophers as well as psychologists. Lively interdisciplinary discussions resulted, in *BBS* and elsewhere (e.g. P. L. Harris 1991; Goldman 1993; M. Davies and Stone 1995*a*,*b*; Carruthers and Smith 1996; Peacocke 1994: 99–154). The epistemologist Alvin Goldman, at Rutgers University, even developed a "hybrid" account, in which both simulation and 'theorizing' played a part (Goldman forthcoming).

Recently, a third possibility has been proposed by Alan Leslie (1951–). An ex-student of Bruner's at Oxford, Leslie was originally employed at UCL but is now at Rutger's Center for Cognitive Science. As required by Dennett's five conditions for ascribing second-order beliefs, he considers both true/false belief and positive/negative desires. And he claims to have provided "the first information-processing model of belief–desire reasoning" (2000: 1235).

Leslie sees ToM as a dedicated frame, or schema, associated with a processing mechanism that fills the frame slots in a particular way (see Chapter 10.iii.a). The child "is endowed [*sic*] with a representational system that captures cognitive properties underlying behavior" (p. 1235). This ToM representation marks four things: an agent; an agent's attitude (e.g. pretends-true, believes-true, believes-false); an aspect of the world that "anchors" the attitude (e.g. a telephone, a banana); and the content of the attitude (e.g. "it's a telephone"). The processing mechanism operates whenever attention is directed onto an agent's behaviour. It *selects* a content for the agent's belief, which by default is assumed to cohere with reality. (This makes evolutionary sense, for most beliefs have to be true if we are to survive.)

In false-belief tasks, the default assignment has to be inhibited. That's done, Leslie says, by *general* information-processing mechanisms, which take at least four years to develop. But if the non-factual content is somehow made more salient (for instance, by asking "Where will Sally look for the marble *first*?"), then 3-year-olds can pass the ToM test. Indeed, even a 2-year-old can understand his/her mother pretending, with exaggeration, that a banana is a telephone (Leslie 1987).

As for the second point remarked above, we've seen that ToM research didn't need the concept of modularity to get started. But the two became closely associated in the mid-1980s. This wasn't due only to the 'modishness' of modules. To the contrary, there was strong evidence that ToM is an inherited faculty. Modules (by definition) function—and so can malfunction—independently. If ToM is a module, there should be cases where it fails to develop, even though other psychological powers are normal. And this, it now appeared, was so.

The evidence came from a trio at the MRC's Cognitive Development Unit at UCL: Baron-Cohen, Leslie, and Uta Frith (1941–). They'd been working on ToM, pretend play, and autism—and in the mid-1980s they brought these three ideas together. They showed that ToM can fail to develop even though other aspects of intelligence are normal. And the result, they said, is autism—a combination of poor social skills, lack of eye contact, delayed language, and no pretend play—or its milder form, Asperger's syndrome. (The early papers are Baron-Cohen *et al.* 1985; Leslie and Frith 1987; see also U. Frith 1989, 1991; Baron-Cohen 1995, 2000*a*.)

"These *three* ideas" (ToM, pretend play, autism): modularity wasn't one of them. The UCL work was partly 'prepared' by recent work in clinical neuropsychology—which strongly suggested nativism. Two London-based colleagues, Elizabeth Warrington and Timothy Shallice, had discovered surprisingly specific behavioural deficits, often associated with damage to specific brain regions:

* In cases of language deterioration, for instance, low-level concepts were typically lost before high-level ones (Warrington 1975; cf. Quillian 1969);
* long-term and short-term memory could be damaged independently, and were seemingly stored in different parts of the brain (Shallice and Warrington 1970);
* and motor action could be impaired by different flaws in planning, which mapped onto different brain lesions (Shallice 1982).

Frith (personal communication) now remembers these discoveries as having encouraged her group to pursue ToM, whereas the Chomsky/Fodor writings on modularity had little or no influence on them.

The UCL view was revolutionary—and highly controversial. Autism had previously been blamed on emotional deficits, or on inadequate mothering. The revolution might have happened a few years earlier, for it was already known that autistic children wouldn't engage in pretend play. Indeed, that was part of Leo Kanner's (1943) original definition of the syndrome. What Leslie had offered was a retrospective explanation, based on the fact that pretend play involves temporarily accepting/attributing false beliefs.

However, his ideas were so "radical" at that time that he had "incredible difficulties" publishing them (U. Frith, personal communication). Moreover, a lack of pretend play was the least of sufferers' problems: who cared? As it turned out, Leslie's paper on play didn't appear in print until two years *after* the trio had published their suggestion that autistics don't develop ToM (Leslie 1987).

Their first experiments, done by Baron-Cohen as a Ph.D. student, showed that the 25 per cent of autistic children who have normal causal/mathematical intelligence cope very badly with false-belief tests. Twelve-year-olds fare far worse than *much* younger controls, and also than 10-year-old intellectually retarded Down Syndrome children (success rates were 20 per cent, 86 per cent, and 85 per cent respectively). Even Frith, already an expert on autism, was taken aback by these findings: "I was amazed by Simon's first results—I had to go along and see for myself" (personal communication).

Over the following years, further evidence for ToM modularity emerged. (And evidence *against* it, too: see subsection h.) Some came from brain scanning done by Frith's husband's team at Hammersmith Hospital (C. D. Frith and Frith 2000; Baron-Cohen *et al.* 1994). They reported brain areas that 'light up' when normal people are using intentional concepts, but don't seem to be activated in Asperger subjects.

Other evidence was behavioural. Baron-Cohen—now co-directing Cambridge's Autism Research Centre (the MRC's UCL Unit was disbanded in 1998, when its Director John Morton retired)—recently showed that autistics' weakness in the domain of folk psychology can coexist with very high intelligence in the physical domain (2000*b*). Indeed, he has suggested that Sir Isaac Newton's personal failings, remarked at the opening of Chapter 5, may have been due to Asperger's (Baron-Cohen and James 2003).

In addition, there was new evidence of sex linkage. It was already known that autism is more than twice as common in boys than in girls. Now, it was found that the testosterone level before birth (in the amniotic sac) varies inversely with the amount of eye contact—and the vocabulary size—during the first two years of life. (This applies to normal children: data aren't yet available for autistics.) Baron-Cohen (2002, 2003) now contrasts "empathizing" people/brains with "systemizing" people/brains, and even describes autism as 'The Extreme Male Brain'.

Despite the autism evidence, whether ToM could strictly be counted as a *module* wasn't clear. The more it was assimilated to theoretical inference in general, the more problematic this was. Leslie allowed general processing mechanisms to play a role, but he did define a four-slot *dedicated* ToM representation. Even so, this version of ToM was very different from the fully automatic processing and highly specific input information of modules for Mexican hats.

The murkiness was increased by the unclarity of just what "module" means (see subsection e, above). Moreover, a close colleague of the UCL trio, the developmentalist Karmiloff-Smith, had formulated a theory that not only undermined the MM thesis but also complicated the concept of nativism itself (see subsection i). And she'd specifically argued that the autistic's ToM deficit might be due to *general computational* rather than *specific representational* deficits (1992: 168).

While the developmental psychologists were energetically studying ToM, its originators—the primatologists—were doing so too. Many of them attributed ToM to non-human primates. They did so on the basis of studies (and 'natural' observations) of imitation, self-recognition, social relationships, deception, role taking, and perspective taking.

Sometimes, the conclusion was that these animals had some ToM, but not much. Specifically, they apparently don't understand beliefs, but do understand some psychological processes, such as seeing (Tomasello 1996; Tomasello *et al.* 2003*a*,*b*). They—and some non-primate species, too—can also recognize "basic" emotions such as fear and anger (see i.d, above) from their conspecifics' facial expressions (Sripada and Goldman 2005). So to ask whether chimps have a theory of mind, *Yes or No?*, is "exceedingly misleading": there's no such thing as "a monolithic 'theory of mind' that species either do or do not have" (Tomasello *et al.* 2003*b*: 239, 240). To put the point in another way,

ToM has evolved gradually, rather than springing into existence fully formed (Povinelli and Preuss 1995).

But such caution was rare—especially in the early days, and especially in the newspapers. Predictably, media accounts of ToM in the jungle proliferated in the 1990s. The public were entranced—and the animal-rights movement further enraged (see 2.v.a). Indeed, their guru—the philosopher Peter Singer—founded the Great Ape Project, whose aim was to persuade governments and individuals that some non-human primates have rights comparable to ours and should be treated accordingly (Cavalieri and Singer 1993). (Today, the World Transhumanist Association recommends ascribing moral rights also to man-machine cyborgs, and to some computer agents, too: see <http://www.transhumanism.org>.) It seemed, then, that Goodall and Fossey had been vindicated—with knobs on.

However, all was not as it seemed. Cecilia Heyes (1960–), a philosophically sophisticated psychologist at UCL, argued in *BBS* that in every reported case the results could have occurred (1) by chance, or (2) by non-mentalist association, or (3) by inference based on non-mental—e.g. causal—categories (Heyes 1990, 1998; cf. Povinelli and Vonk 2003). Modulo the latter concession to cognitivism, this was Sarah-and-the-tokens all over again.

Heyes noted that primatologists (including P&W, as we've seen) usually supported their ToM explanations by appeal to theoretical parsimony—including Dennett's pragmatic defence of the intentional stance (Heyes 1998: 109–10). Occam's razor was taking precedence over Lloyd Morgan's canon (see 2.x.b). But Heyes favoured the canon. Moreover, she said, these experimenters had misinterpreted the notion of parsimony. Their ToM explanations might appear "simpler" than long-winded associationist accounts, but they weren't. For one thing, the so-called convergent evidence rarely rested on independent (i.e. non-ToM) assumptions.

She wasn't saying that ToM in animals is impossible. (She even proposed an experiment, based on conditional discrimination training and transfer tests, to find out whether chimpanzees really do have the concept *see*: 112 ff.) And she admitted that many would think her a "killjoy", making "carpingly narrow objections" to "elegantly bold ideas". Her own instinct, indeed, was *not* to shout for the nit-picking methodologists. Nevertheless:

> [It] is precisely because Premack and Woodruff's question is important and intriguing that it warrants a reliable answer; and without some sober reflection, acknowledging the limitations of current research, we may never know whether nonhuman primates have a theory of mind. (p. 114)

g. The third way

Given the many meanings of modularity sketched above, it was abundantly evident by the end of the century that nativism wasn't a clear concept. In truth, it never had been.

For example, Chomsky's mid-1960s attempt to resuscitate rationalist doctrines of innate ideas had been criticized at the time by leading empiricist philosophers (9.ii.c and vii.c). They'd shown that nativism could be interpreted in many different ways, some of which weren't supported by Chomsky's claims.

Even more to the point, an entire school of psychology, which had been moving beneath the mantle since the 1920s (5.ii.c), was committed to a compromise position. Piaget had repeatedly described himself as taking a "middle way" between nativism and environmentalism, rationalism and empiricism . . . and other tempting philosophical dichotomies (Boden 1979: 5 ff.). He saw the question *Is behaviour x innate or isn't it?* as ill-posed. Instead of pure nativism or pure environmentalism, he favoured *epigenesis*: a self-organizing dialectic between biological maturation and experience.

But "self-organization" and "dialectic" weren't buzzwords in the ears of the fashionable. So although by the 1960s Piaget's ideas had been hugely influential in the theory and practice of education, they were much less so in experimental psychology as such. Indeed, he'd been ignored, not to say scorned, by most tough-minded psychologists—including early cognitive scientists such as Sutherland (5.ii.a and c).

Piaget was well into his sixties by that time—not a good age for adopting a brand new technology/methodology. To be sure, he'd previously approved of cybernetics and now approved of the GOFAI approach to psychology too (Boden 1979, ch. 7). So he was quick to say: "I wish to urge that we make an attempt to use it" (Piaget 1960). On reaching his seventies, he remarked that the techniques of computer simulation promised to be "the most decisive for the study of structures" (Piaget 1967: 318). And in 1969, in a Radio Canada TV interview with Thérèse Gouin-Decarie, he "wished strongly" that the AI approach would be explored (Jean Gascon, personal communication). But it was much too late for him to do so himself.

Younger bloods, however, leapt in. Owing to the widespread indifference to his work remarked above, they were rather few. Nevertheless, in the early 1970s several computationalists tried to implement aspects of his theory by way of production systems. These attempts were scattered across the USA (Klahr and Wallace 1970, 1976), Canada (Baylor *et al.* 1973; Baylor and Gascon 1974; Baylor and Lemoyne 1975), and UK (Richard M. Young 1974, 1976).

For instance, Richard Young (1944–), at the MRC Unit in Cambridge, studied the micro- and macro-developmental changes involved in children's grasp of seriation—specifically, arranging blocks to make a staircase (see 5.ii.c). True to Newell and Simon's inspiration, Young relied on an abstract task analysis, and also gathered behavioural protocols in the laboratory. Both were reflected in his computer model. Besides the 'correct' and 'incorrect' picking-up and placing of blocks (which Piaget had described), it included details such as the child's stretching out a hand to pick up a block but withdrawing it before the block was actually touched; touching a block without picking it up; and double-tapping one block with a finger moving along the staircase.

Young unified his various models—which represented both different Piagetian "stages" and different children—by way of a theoretical matrix defining a space of seriation skills. The three dimensions of the space were *selection* of the next block, *evaluation* of the structure built so far, and *placement and correction* of individual blocks. Any type of seriation behaviour, including characteristic errors as well as successes, could be defined by locating it within this space. Adding and/or removing certain rules (with appropriate prioritizing) would take the system from one child to another, from one experimental session to another, or from one stage to another. For example, the transformation from the concrete to the formal operational stage was effected by adding *IF you want to add a block to the staircase THEN pick up the biggest block.*

That was all very well—but it wasn't clear that Young was simulating what Piaget had been talking about. Whatever "self-organization" might mean, it didn't include an extra rule's being added by the external hand of the programmer. Moreover, stage transformations were supposed by Piaget to be holistic, so that visual and tactile seriation (for instance) develop simultaneously. But because Young's systems were so nicely matched to his specific experimental protocols, they weren't readily generalizable to other domains (although he did generalize from blocks to discs).

In any event, Young—like his fellow modellers in the USA and Canada—wasn't aiming to advance Piaget's theory in any fundamental way. The seriation matrix was new, to be sure, but it added system not substance. Trying to clarify the theory was challenge enough.

Throughout the 1980s, the challenge was taken up by others—including some at Carnegie Mellon, the birthplace of production systems (Klahr *et al.* 1987; Siegler 1983, 1989; Newell 1990). Although most of these models focused on behaviour at a given developmental stage, a few tried to model development as such (e.g. Wallace *et al.* 1987).

Just for the record: In the 1990s, production systems were still going strong (e.g. Klahr 1992). But they'd been joined by a constructivist approach developed by Minsky's student Gary Drescher (1989, 1991). Drescher modelled Piaget's notion of 'schema' in a GOFAI program which, by simulating bodily action in the physical world, gradually built up concepts of persisting, perceiver-independent objects. Since it didn't start out with any specific world knowledge, Drescher claimed to have solved the chicken-and-egg problem of empiricism: how to learn the "relevant" environmental regularities, given innumerable candidates, without already knowing which ones they are. (In one sense, he needn't have bothered: as we'll see, Piaget's view that newborns aren't predisposed towards specific environmental patterns was mistaken.) Later still, GOFAI was joined by connectionism, when Thomas Shultz modelled seriation in a network using "cascade correlation" (12.ix.c).

While the production-system Piagetians were still at work in the 1970s–1980s, others *were* trying to move the 'third way' forward. Eventually, they succeeded—although some of Piaget's key claims were modified/abandoned in the process. Throughout the last quarter-century, the concept of epigenesis was enriched by new theoretical insights and empirical findings.

In the 1990s epigenesis became prominent in A-Life, where Waddington was 'officially' promoted from maverick to guru (see 15.iii.b and Boden 1994*a*, introd.). It had revived in biology too (Piaget's first love), being applied—for example—in understanding bone anatomy (Lovejoy *et al.* 1999).

But the main impetus had come from psychology and neuroscience—with connectionist modelling playing a role as well. Indeed, an influential six-authored book called *Rethinking Innateness* (Elman *et al.* 1996) spelled out epigenesis from each of these disciplinary bases. (And from a base of dynamical theory too: Esther Thelen's account of the A-not-B error in terms of bodily skills, not the object concept, is outlined in Chapter 14.ix.b.)

No one reading that book could rest easy with the simplistic interpretations of nativism that had been held for so long, by defenders and opponents alike. There were some critical reactions (e.g. Fodor 1997; Samuels 2002). Nevertheless, the huge complexity of development, and its importance for understanding *adult* minds, could no longer be gainsaid.

This interdisciplinary revival was carried out by many different people. I'll discuss three in particular (one here, two in later chapters), but the caveat spelled out in Chapter 1.iii.e–f applies here, as elsewhere. None of them was a lone spirit, theorizing in isolation from their developmentalist colleagues.

One of the earliest prime movers in the reawakening of epigenesis was the neuroscientist Jean-Pierre Changeux, whose approach will be outlined in Chapter 14.ix.c–d. Others included Jeffrey Elman, one of the six co-authors of *Rethinking Innateness*. We'll see in Chapter 12.viii.b–c that he contributed two important ideas to connectionism, namely recurrent networks and "starting small"—where the latter was an insight into epigenesis *in general*. In this chapter however (and also in 12.viii.d), I'll focus on yet another of the six co-authors, the psychologist Karmiloff-Smith (1938–).

Karmiloff-Smith (1990*b*) aimed to cement a "marriage" between Chomsky and Piaget, and between Fodor's anti-constructivist nativism and Piaget's anti-nativist constructivism (Karmiloff-Smith 1992: 10). (Fodor was unwilling, and tried to make the cement crumble: 1997, 1998*d*.) She saluted Fodor for having had "a significant impact on developmental theorizing" (1992: 1). But she challenged his two major claims: that 'higher' thought can't be explained, and that adult modules are already preformed in the infant. (The first challenge is discussed in subsection h, below, the second in subsection i.)

Like her fellow enthusiasts for the third way, she didn't see developmental psychology as a specialism, safely ignored by 'straight' psychologists. On the contrary, she said, cognitive science *as such* needs a developmental perspective. So she was always irritated by well-meaning people who gushed "You must love babies!" (This would happen even more often after her popular TV series and best-selling books, some written with her daughter Kyra: Karmiloff-Smith 1994; Karmiloff and Karmiloff-Smith 1998, 2001.) Besides the fact that they wouldn't have said that to her male colleagues, or anyway not in the same tone of voice, this betrayed a fundamental misunderstanding. She studied development not because she loved babies but because she wanted to understand human minds.

Karmiloff-Smith became a student of Piaget's in Geneva in 1967, and was a member of his research group in the 1970s. From 1982 to 1988 she worked alongside the ToM trio at UCL, and then founded the Neurocognitive Development Unit at the Institute of Child Health (attached to UCL and to the Great Ormond Street Hospital for Children). In the 1960s, she'd been a professional simultaneous translator for the United Nations, including two years working in Beirut's Palestinian refugee camps. She's one of those rare people who came to science late but made a significant contribution anyway.

How significant? Well, we'll see in the rest of this section that she provided a rich trove of empirical data showing that Piaget's theory of "stages" was mistaken, and his focus on "accommodation to error" greatly overdone. She defined a computational theory of the development of the higher mental processes. She offered an epigenetic account that turned the MM thesis upside-down. And she helped to institute a novel rapprochement between psychology (developmental, cognitive, and clinical) and neuroscience (ditto, ditto, and ditto).

h. What makes higher thinking possible?

Karmiloff-Smith was no orthodox Piagetian. For one thing, she didn't posit holistic stage changes. Rather, she spoke of a series of "phases" within each domain (e.g. language or drawing) and micro-domain (e.g. using pronouns, or the definite/indefinite article). And her data showed that the different domains don't march in step. For another thing, she allowed, even "insisted", that

> there are *some* innately specified, domain-specific predispositions that guide epigenesis. Young infants have more of a head start on development than Piaget granted them. (1992: 172)

This had become clear in the recent explosion of research on very young babies, sometimes only a few hours—or minutes—old. In the mid-1970s, studies of early turn taking, eye-gaze following, and pointing had shown that these activities "scaffold" the growing intersubjectivity between baby and mother, including language (Scaife and Bruner 1975; Bruner 1978; C. B. Trevarthen 1979; see 6.ii.c). By the early 1990s there was a mountain of evidence that babies can do much more than Piaget had thought.

For instance, newborns pay more attention to human speech than to other sounds; within a few days, they prefer their mother's language; and soon afterwards they can recognize some of its syntactic properties. Further examples included seeing things as continuously existing objects (5.ii.c); looking at faces, especially the eyes (14.ix.c); and anticipating 'realistic' physical movements. (See Carey 1985; Sophian and Adams 1987; Mehler *et al.* 1988; Gallistel 1990; Spelke 1991; Butterworth 1991; Carey and Gelman 1991; Leslie 1991.)

However, since babies can't use language, or pencils, or pianos . . . as adults can, the nativist has a problem:

> The more complex the picture we ultimately build of the innate capacities of the infant mind, the more important it becomes for us to explain the flexibility of subsequent cognitive development. (Karmiloff-Smith 1992: 9)

The most influential nativist of the day, namely Fodor, had announced this to be impossible: the higher mental processes were out of bounds (Section iii.d, above).

Karmiloff-Smith disagreed. In the mid-1970s, when she was still in Geneva, she'd begun a long-continuing programme of experiments—on language, weight balancing, drawing, making maps, and writing—designed to test her ideas about "representational redescription", or RR (e.g. Karmiloff-Smith and Inhelder 1975; Karmiloff-Smith 1979*a*,*b*,*c*, 1984, 1986). She believed that RR is found only in *Homo sapiens*, and identifies what's specifically human about human cognition.

According to RR theory, adult intelligence is achieved by successive representational changes, whereby information that was previously *implicit in* the system becomes *explicit to* the system. (The earlier representation is supplemented, not replaced: it's still available for use.) These changes aren't driven by error—Piaget's "accommodation" to the external world (5.ii.c)—but by autonomous internal development. That is, they happen only *after* mastery has been achieved at the lower level. The child constructs increasingly explicit "theories" of its own skills, which it uses to vary those skills in increasingly flexible ways. Eventually, skills learnt in one domain or micro-domain can be used in others too.

A similar progression happens when adults learn to read late in life, or to play/improvise on the piano (Karmiloff-Smith 1986: 97–8; 1992: 16–17; cf. Sudnow 1978/2001; Hermelin and O'Connor 1989). David Sudnow's subtle account of the subjective experience of learning to improvise on the piano is especially interesting here. He made no claim to identify the psychological mechanisms involved. But he did indicate the development of previously undescribed motor, perceptual, and musical schemata—which psychology (e.g. RR theory) should be able to explain.

RR isn't the work of an instant: constructing a new theory takes time. Indeed, the child who previously performed successfully starts making mistakes, which disappear when the new representational phase has been consolidated (so graphs plotting task success are U-shaped). But although much of the observable behaviour is the same as before—the beam is balanced, or the appropriate word pronounced—the cognitive processes that generate it are very different.

Karmiloff-Smith posited four levels of (meta-)representation, culminating in verbalizable consciousness. Each new level describes the skill more explicitly, adding opportunities for self-monitoring and (eventually) voluntary self-control—that is, freedom (Section i.g, above). The order of skilled movements, and—later—their organization into hierarchical (sub)routines, is marked. Once an aspect has been marked, it can be varied. Steps (subroutines) can now be omitted, iterated, inserted at different points, or even smoothly integrated into a routine developed to deal with a very different category. In other words, the child (or skill-learning adult) becomes successively more imaginative, or creative (Boden 1990*a*: 63–73).

The many detailed experiments which put empirical flesh onto these RR-theoretical bones included a series on drawing (Karmiloff-Smith 1986; 1992: 155–61). Children between 4 and 11—all of whom could fluently draw an *ordinary* man (or animal, or house)—were asked to draw "a funny man" (or animal, or house), or "a man that doesn't exist". Karmiloff-Smith found, as she'd predicted, that their ability to do this develops phase by phase.

The youngest children couldn't do it at all. They'd learnt how to draw a man (animal, house) successfully, and unvaryingly (e.g. head or roof first), but that was it. The first changes affected the shape and/or size of elements (e.g. a bubble-shaped arm, and/or a tiny head) or of the whole thing (e.g. a house shaped like an ice-cream cone). Then, elements could be deleted, although at first only the last part normally drawn. (Compare: novice pianists can halt a piece at mid-point before they're able to start it from there.) The 5-year-olds could draw a cross-category figure (e.g. a house with wings) if asked, but only by adding the cross-category elements *after* the main picture had been drawn—and very few of them did this 'of their own free will'.

Only from about 8 or 9 years of age were extra elements smoothly, and spontaneously, inserted into the drawing while it was being done (e.g. a second head, with no 'unnecessary' lines between it and the single neck). And only then was an element's position/orientation changed (e.g. the head put on the side of the shoulder), or elements from different categories inserted (e.g. man and animal, to give a 'centaur').

Comparable results were observed in the other domains, in each of which the children showed a growing ability to comment on what they were doing. With respect to speech, for instance, Karmiloff-Smith (1986; 1992, ch. 2) found that when children first use the words *the* and *a* (or their French equivalents) correctly, they have no insight into their speaking skill. If asked to pick up "a watch", they will (correctly) pick any of the several watches, whereas if asked for "the red watch" or "my watch", they pick a particular one; and they use the right word in describing what they picked up. But they can't reflect on their speech, to say what the relevant difference is. They can't do this even when they start correcting themselves for wrong usage ("the watch" when there are two watches, for example). At first, they can rectify the mistake only by providing the right form of words, not by giving the general principle. Later, however, they can explain that if there'd been two watches on the table it would have been wrong to say "I picked up the watch" (or "Pick up the watch"), because the listener wouldn't know which one was meant. At last, they've achieved a conscious grasp of the structure of their own linguistic skill.

Or rather, they've consciously grasped *some* aspects of it. Even highly articulate adults don't always know just what they're doing when they speak. For instance, they can't explain why they use a pronoun rather than a full noun phrase in particular communicative contexts (Karmiloff-Smith *et al.* 1993). We can all use *she* to track the role of the main character in a story—but without realizing that we're doing so, never mind how we do it. (That's why pronouns are so difficult to handle in NLP *text* generation: see Chapter 9.c and f.)

Karmiloff-Smith suggested that we lack metalinguistic awareness here because decisions about pronominalization are taken rapidly on-line, depending on what's just been said to the listener. In other words, redescription doesn't happen because there's nothing in LTM to be redescribed. (In her later work she found that "notational" skills are sometimes varied earlier than language or music making; for instance, children can sometimes delete parts of a "funny man" earlier than she'd expected. Her explanation was that the physical mark provides the child with cues for self-monitoring which otherwise would be made available only by RR.)

As her remark about text pronominalization implies, she thought about thinking in computational terms. This had been true almost from the start. She'd learnt LISP programming by the late 1970s, and used GOFAI as a source of concepts, metaphor, and inspiration. Her account of what the meta-representations enable the child to do recalled GOFAI planning, including Gerald Sussman's self-correcting HACKER (Chapter 10.iii.c). And she compared the lowest level of skill to a compiled program and post-RR skills to interpreted programs, employing "procedural" and "declarative" knowledge respectively (1990*a*; but see 1992: 161–2).

Later, she turned to connectionist AI (1992, ch. 8). She was drawn to PDP because it showed that global, stagelike, transformations can occur as a result of continuous local changes. For example, the development of the past tense (sketched in subsection a), which Chomsky and Fodor had ascribed to general computational *rules*, appeared to happen as a result of cumulative changes in system statistics (see 12.vi.e). What's more, PDP networks initially capable of learning many different things became more specialized, less plastic, with training. This was analogous, she felt, to the process of modularization (see below).

Karmiloff-Smith's connectionist work is described in Chapter 12 (Sections viii.d and e, and ix.a)—as is Elman's and Kim Plunkett's, both also co-editors of *Rethinking Innateness*. As we'll see, GOFAI is (still) better at modelling what meta-representations enable the system to do once they've arisen, connectionism—perhaps—at modelling how they arise in the first place. Having seen that RR was "precisely what is missing thus far [i.e. in about 1990] from connectionist simulations" (1992: 179), Karmiloff-Smith sketched some ways in which RR might come about (A. J. Clark and Karmiloff-Smith 1993). However, her suggestions were highly programmatic. Others used cascade correlation to model cognitive development (12.ix.c). But their success, too, was limited—hence that cautionary "perhaps". From the computational point of view, the development of RR is still opaque.

What of RR theory's challenge to Fodor? It outlined how certain structural types of high-level thought become possible—which is largely what a psychological science is about (see Section iii.d). But it couldn't explain, still less predict, particular cases in detail. (No Mexican hats.) It couldn't say *just why* an 11-year-old decided to draw a house with wings, as opposed to something else equally bizarre.

Compare what Fodor had said about analogy. He'd remarked that it's often involved in scientific discovery—in the Rutherford–Bohr solar-system model of the atom, for instance. But he'd despaired of its being understood:

> [Analogical] reasoning would seem to be . . . a process which depends precisely upon the transfer of information among cognitive domains previously assumed to be mutually irrelevant. By definition, encapsulated systems [i.e. modules] do not reason analogically . . . It is striking that, while everybody thinks that analogical reasoning is an important ingredient in all sorts of cognitive achievements that we prize, *nobody knows anything* about how it works; not even in the dim, in-a-glass-darkly sort of way in which there are some ideas about how [scientific] confirmation works. I don't think that this is an accident either. In fact, I should like to propose a generalization; one which I fondly hope will some day come to be known as "Fodor's First Law of the Nonexistence of Cognitive Science". It goes like this: the more global . . . a cognitive process is, the less anybody understands it. (Fodor 1983: 107; italics added)

His charge that "nobody knows anything" about analogy had been rebutted by RR theory. For scientific discovery is houses-with-wings with knobs on. RR had even indicated some very general types of self-monitoring that can twiddle the knobs to good effect. How the Rutherford–Bohr atom *could possibly arise* was thus somewhat less obscure than before.

But how it *actually* arose wasn't. For the reasons outlined in Section iii.d, we'll probably never know—not even with the relevant diaries, notebooks, and recollections available from the scientists concerned. Moreover, AI models of analogy still fell short of the human reality in many ways (see 13.iv.c and 12.x.a). Despite the necessary cavils about "nobody knows anything" then, Fodor's First Law remained standing.

i. The modularization of modules

In "insisting on" the post-1970 evidence of babies' abilities, Karmiloff-Smith had accepted a form of nativism—but not Fodor's. Indeed, she turned the MM thesis on its head. Instead of "prespecified modules" she posited "a process of modularization" (1992: 4). To the extent that the adult mind/brain is modular, this happens—she said—not because it was born that way but as a result of development.

For many years, this was an interesting speculation rather than a proven fact. To be sure, it made sense from the epigenetic point of view. Indeed, Piagetian neuroscientists were already discovering aspects of brain development that made it broadly plausible

(14.ix.c–d). But specific proof, for specific cases, required advances in cognitive neuroscience which didn't happen until the late 1980s.

Karmiloff-Smith was waiting for them (1992: 5). Meanwhile, and alongside her RR studies of normal development, she'd begun to study abnormal development too. In Geneva she hadn't had links with a hospital. But at UCL's Cognitive Development Unit (funded by the *Medical* Research Council) she did. Moreover, her UCL colleague Frith was already working on autism when she arrived, and Frith's student Baron-Cohen was about to discover some surprising comparisons between autism and Down Syndrome. Those comparisons bore on the issue of modularity, as we've seen (subsection f, above). Her interest aroused, Karmiloff-Smith too started working with congenitally brain-damaged people.

She became especially interested in Williams Syndrome (1990; 1992: 168–71; Karmiloff-Smith *et al.* 2003). Besides some physical abnormalities, such as 'elfin' faces and SVAS (a narrowing of the aorta where it leaves the heart), Williams patients have an unusual cognitive profile. They (seem to) develop language normally, though up to a year late, and face recognition too. However, they're gravely retarded in general intelligence, including number, problem solving, spatial skills, and planning (their IQs range between 40 and 85: Tassabehji 2003).

For instance, Karmiloff-Smith reported a conversation with an 18-year-old Williams patient who said she was very interested in vampires. This young woman, unable to tie her shoelaces or perform simple spatial-alignment tasks, spoke about vampires thus:

EXPERIMENTER: What do vampires do?
WILLIAMS PATIENT: They break into women's bedrooms in the middle of the night and sink their teeth into their necks.
EXPERIMENTER: Why do they do that?
WILLIAMS PATIENT: (Clearly never having asked herself the question) Maybe they are inordinately fond of necks.

(Karmiloff-Smith 2001)

The grammar and vocabulary were impressive. But the conceptual understanding was not.

So far, so Fodorian. Williams Syndrome seemed to provide strong evidence for modularity. And MM theorists said so: Pinker, for instance, spoke of language being "unimpaired" in several congenital conditions while other faculties were damaged (1994: 51).

However, by the early 1990s it was clear—from neuroscientific as well as behavioural evidence—that face recognition, for example, isn't a preformed mechanism (14.ix.c). To the contrary, it gradually becomes more discriminating and domain-specific. In general, it now seemed that innate predispositions do no more than lead the baby/child to attend to certain broadly specified types of input: three-blob 'face' stimuli, for instance, or human speech sounds, or the mother's eyes. But the child eventually processes more richly specified inputs, in more encapsulated and domain-specific ways. In short, modules aren't there from the start. They're *produced* by epigenesis.

If that's so, one would expect that still-developing 'modules' are less separate than Fodor had believed. Damage to one would probably be reflected in some way by damage to another. From the epigenetic point of view, the notion that one or two developing modules could remain intact while others are grossly deficient was highly implausible.

Moreover, there was growing evidence that if a module can't develop normally—because of lack of input, for example—the brain's plasticity compensates to some extent (14.ix.d). So in ferrets whose auditory nerve is cut, the auditory cortex may develop systematically structured *visual* feature detectors (Sur *et al.* 1988, 1990). Similarly, brain scanning had shown that the auditory cortex in congenitally deaf children comes to be used for computing *visuo-spatial* input in linguistically relevant ways (Karmiloff-Smith 1992: 172—and see Chapter 14.ix.d). In short, much the same function (though not exactly the same, as we'll see) comes to be served by different parts of the brain.

Accordingly, a number of psycholinguists got involved in interdisciplinary studies of congenital brain damage. The ASL specialist Bellugi, for instance, had done so with respect to Williams Syndrome in the 1980s (Bellugi *et al.* 1988, 1991). Karmiloff-Smith's work, too, became increasingly interdisciplinary.

At the Neurocognitive Development Unit, she was collaborating with a wide variety of clinicians and neuroscientists, including her partner Mark Johnson (1960–): see 14.ix.d. Soon, she was working with clinical geneticists too, including a team at Manchester's St Mary's Hospital. In addition, she helped to build new connectionist models (and reviewed older ones, including the past-tense learner: 12.vi.e) to show that the 'same' behavioural deficiency in child and adult need not necessarily be due to the same underlying causal mechanisms (M. S. C. Thomas and Karmiloff-Smith 2002).

(As yet, those "causal mechanisms" have been modelled largely by ignoring issues of timing and synchrony in the nervous system. But Karmiloff-Smith's team expect this to change:

> There are likely to be temporal parameters that have a material effect on the ability of a network to represent relational information; variations in such parameters would be new candidates for mechanisms to drive cognitive development....
>
> Only computational implementation will clarify what notions such as processing capacity, speed of processing, and inhibition involve, and how they may be related both to each other and to components of implemented systems. For example, in considering the notion of speed of processing, computational specification forces certain issues to be addressed. Of what is there a higher rate per second? By what precise mechanism does this rate alter the quality of computations? Are changes in speed the outcome of changes in other parameters, in the way that increasing the discriminability of processing units might allow recurrent circuits to settle more quickly into stable states? Or is it a primitive, on a par with the velocity of neural conductance? (M. S. C. Thomas and Karmiloff-Smith 2003: 148)

The omission of timing in most connectionist modelling has often drawn criticism from neuroscientists, and others. Recently, however, people have started taking these issues on board: see Chapter 14.ii.d and ix.g.)

The combination of these disciplinary approaches strongly suggested that the Residual Normality (RN) assumption is mistaken. According to RN, atypical development can produce selective deficits while the rest of the system develops normally. That RN is untrue is, of course, just what the third way would lead one to expect.

Karmiloff-Smith's mature position was presented to a general audience—and to the media, who featured it widely—in her BPS Centenary Lecture in November 2001. She showed that all the disciplines just mentioned must be integrated if we're to understand intelligence, whether normal or abnormal. We must consider the "complex pathways from gene-to-brain-to-cognitive-processes-to-behaviour" (2001: 206).

That may sound like a platitude. But what she meant by integration wasn't simply inter-level consilience. Rather, it was consilience-with-epigenesis.

She pointed out that recent neuroscience had described many cases of brain plasticity, in which deficits in one region affect others (see Chapter 14.ix). At the psychological level, this meant that even apparently 'pure' syndromes aren't as pure as all that. So for example, when I said (above) that Williams children develop language "normally", that wasn't quite accurate. On closer examination, their language—vocabulary, semantics, grammar, pragmatics, and reading—had turned out to be abnormal after all (Paterson *et al.* 1999; Karmiloff-Smith 2001; Karmiloff-Smith *et al.* 2003). For instance, English-speaking Williams infants have trouble learning words with the first syllable accented (e.g. teddy) but not the last (e.g. guitar); and they acquire a fairly extensive vocabulary *before* they're able to use it for name-based categorization (Nazzi 2003).

Much the same applied to their apparently 'normal' face recognition. Yes, they could recognize faces successfully. But *the processes by which* they recognize faces are unusual. Instead of considering the face (or the car, or the monkey . . .) as a whole, with the parts interrelated within it, they 'count out' the individual facial features. Moreover, brain scans indicated that they mostly use the left cerebral hemisphere, whereas 'global' face recognition uses the right.

Further experiments showed that Williams patients have impairments in low-level activities such as eye-movement planning, and in patterns of neural firing in the brain. For instance, the brain-firing of Williams adults resembles that of normal 3-month-olds. They may also have abnormal *neurones*, since laboratory mice with 'Williams genes' deleted—see below—sometimes develop unusual dendritic spines (Tassabehji 2003). Such neurological details may underlie the behavioural impairments which led to Williams Syndrome's being noticed in the first place. Yet other studies showed that the differences in proficiency (spatial, language, social . . .) between a Williams adult and a Down adult *are not* the same as those between a Williams infant/child and a Down infant/child.

In a sense, these new findings were more of the same. The latter discovery, for example, was a special case of the epigeneticist's general claim that the adult isn't simply a larger version of the child. (So dissociations in adult life, after strokes perhaps, are *not* reliable evidence for 'inborn' modularity: see 14.x.b, and M. S. C. Thomas and Karmiloff-Smith 2002.) And as for the separation of mental powers, Karmiloff-Smith had already pointed out that ToM isn't so untouched in Down Syndrome as the modularists had believed: although Down children can pass the Sally–Ann test, they can't cope with the more difficult ToM tests which normal 7- to 9-year-olds can manage (1992: 170). By the turn of the century however, the degree of psychological and neuroscientific detail had been hugely increased.

What's more, genetics had now entered the picture. Where modularists—and journalists—were quick to speak of "the" gene for behaviour X, Karmiloff-Smith was one of the many who criticized this. (Another, who gave a host of examples from non-human animals, was the biologist Matt Ridley: 2003.) Her Centenary Lecture reported new illustrations of the fact that single genes don't have single effects. So the gene trumpeted by the Associated Press—admittedly, with some scepticism and much mockery—as "the grammar gene" is no such thing (Pinker 1994: 297–8; Gopnik and Crago 1991). Not only does it affect language in a *number* of ways (whereas some had linked it to a specific syntactic phenomenon), but it also affects other intellectual abilities.

As for Williams Syndrome, this had recently been shown to involve the deletion of sixteen genes on one copy of chromosome 7. (Two years later, the count had risen to twenty-five: Tassabehji 2003.) However, Karmiloff-Smith's Manchester collaborators discovered that people can lack a subset of those genes without showing an 'equivalent' subset of symptoms. Even with 70 per cent missing, a person could show (similar physical symptoms, such as SVAS, but) *no* impairment of intelligence. Yet with only two genes deleted (both included in the 70 per cent just mentioned), Williams Syndrome appears.

Why did Karmiloff-Smith choose that strange lecture title, 'Elementary, my Dear Watson . . . '? Well, although system complexity makes life difficult for the psychologist, it can also ease it a little. Any clinical syndrome has major effects (i.e. those we happen to be most interested in), but it will have side effects too. If some of these are observable more easily and/or earlier, diagnosis may benefit. This 'scattering' is useful also for theoretical psychology. If we know that a variety of effects tend to be found together, we can ask *What underlying computational process* could give rise (in cooperation with others, of course) to *just these* effects? So tiny clues, sometimes found in unexpected places, can help cognitive scientists to understand the full range of human intelligence—but they'll need the detection skills of a Sherlock Holmes.

By the new millennium, then, the MM thesis had been given a very different twist. Yes, "there are *some* innately specified, domain-specific predispositions that guide epigenesis" (1992: 172). But also, "the brain progressively sculpts itself, *slowly becoming more specialized* over developmental time" (2001: 207; italics added). In short, the third way was being driven further through the interdisciplinary forest.

7.vii. Satellite Images

The six psychological stories told above described very different topics, each covering many sub-topics. Most of the researchers involved chose to focus on only one of them—often, indeed, on sub-sub-(. . .)-topics that I haven't even hinted at. That was unavoidable: they were down on the ground, doing the detailed work.

Something that's unavoidable may nevertheless be regrettable. So Cronbach, for instance, had complained about the "fragmentation" of 1950s psychology. Years later, Newell would make similar complaints:

> Allen Newell, typically a cheery and optimistic man, often expressed frustration over the state of progress in cognitive science. He would point to such things as the "schools" of thought, the changes in fashion, the dominance of controversies, and the cyclical nature of theories. One of the problems he saw was that the field had become too focussed on specific issues and had lost sight of the big picture needed to understand the human mind. (John R. Anderson and Lebiere 2003: 587)

His own view of what "the big picture [of the mind]" would have to be like is sketched in subsection c, below. But one can also ask about the big picture *of computational psychology itself.*

If, like a latter-day Cronbach, we were to look down from a great height on this large area, what would we see? And how would the picture have altered over the last half-century?

a. A telescopic vision

The change that would be most evident, given a camera orbiting far above, is *a growing realization of the subtle complexity of the human mind.*

Intuitively, this complexity was already recognized. Cronbach's "philosophers or artists" had a good grasp of it (5.ii.a). So did his "scientific psychologists", when gossiping with/about their friends over coffee. But in their explicit theorizing, they didn't. Most ignored mental processes and semantic content entirely and/or tried to reduce them to highly abstract general principles, such as 'Newtonian' S–R laws (5.i and iii).

The early cognitive psychologists avoided S–R laws, but their theories were often hardly less simplistic. For they were frequently posed as dichotomies: serial versus parallel processing, analogue versus digital representation, single-trial versus continuous learning . . . and the like. This might have delighted Claude Lévi-Strauss (8.vi.c), but it didn't delight Newell. He argued that 'You Can't Play Twenty Questions with Nature and Win' (1973*c*), because mental processes can't be corralled within such crude theoretical categories. Gradually, more and more psychologists came to agree with him on that general point, even if they didn't accept his specific theories.

The computationalists of the 1960s, having accepted the mentalist baton from their non-Newtonian predecessors (Chapters 5 and 6), tried to delve deeper into the specifics of human minds. They even tackled personality and emotion (Section i.a–b, above). But their efforts, especially in the more personal areas, fell woefully short of the human reality. They didn't yet have powerful computational tools, whether machines or programming languages (10.v). More importantly, they didn't have appropriately detailed computational *ideas.*

But that changed, as the orbiting camera would show. Over the next forty years, this late twentieth-century version of *verum factum* (1.i.b) produced a huge crop of fruit. New computational concepts enabled psychologists to conceive theories about *precisely what may be going on* when we think. And not just when we "think": the satellite evidence would support the claim made at the outset of Chapter 1.ii, that cognitive science isn't concerned only with cognition.

The successes, failures, and limitations of programs actually run on a computer enabled psychologists to hone their ideas in a way that no verbal speculations could (iii.c, above). Similarly, they enabled them to express theories far more detailed than had previously been possible.

* Compare, for instance, what Abelson said about "attitude change" with what Dienes and Perner said about why it's more difficult for a hypnotist to persuade someone to hear voices than to make their arm rigid (Section i.a and h).
* Compare work in scene analysis (6.ii.e and 10.iv.b) with post-1980 theories of low-level vision (Section v.b–f above, and 14.vi).

* Compare Colby's part-semantic, part-numerical, account of different types of anxiety with Sloman's architectural analysis and nursemaid program (Sections i.a and f).
* Compare LT and GPS with Newell and Simon's production systems, or with Johnson-Laird's mental models (Section iv).
* Compare Markovian models of language with grammatically structured ones, and 'perfect' speech with Clippinger's anxiety-ridden pronouncements (Section ii and 9.vi–xi).
* Compare Skinner's simplistic accounts of language learning with the mini-programs for language acquisition offered by Miller and Johnson-Laird (Section iv.d–e), or with models of past-tense learning (12.vi.e).
* Compare early theories, and computer models, of rationalization and the attribution of agency (Section i.c) with Sperber and Wilson's theory of communicative relevance (iii.d), or with recent neuroscientific work on schizophrenic delusions about bodily 'takeover' (i.i).
* Or compare the all-too-familiar fluff about creativity (tact forbids . . .) with cognitive–computational theories of it, and with computer models of musical composition or of the generation and comprehension of familiar types of joke (Chapter 13.iv.c).
* Finally, compare Morton's confession in 1981 (p. 232) that "Experimental psychology has a disastrous history with respect to its relevance" with recent computational work on emotion (i.d–f), reasoning (iv), and various clinical syndromes (vi.i and 12.ix.b).

The point is not that the theories named in the second half of each of these comparisons are *correct*. Maybe they are, maybe they aren't. More likely, they are *and* they aren't. In other words, each of them suggests some promising directions—and indicates that theories *of at least that level of complexity* will be needed to explain the mental phenomena concerned. That realization in itself is a genuine advance.

Some readers may be shifting uneasily in their seats at this point. "What about the ant?", they may be saying (Section iv.a). "The satellite pictures show us Simon's ant, crawling on the ground—but it *reduces* the mental complexity needed to explain behaviour." Well, yes. But much of the "missing" complexity is transferred to the environment—especially to the cognitive technologies made available by human cultures. And ant-theory as a whole is more discriminating than the previously popular Twenty Questions dichotomies.

The camera-on-high would record that no one took much notice of the ant, at first. Or if they did, they pictured it crawling mostly *inside* the brain. In Newell and Simon's production systems, for instance, most of the *conditions* were supposed to be mental, not environmental, events (iv.b). By the 1990s, however, psychologists—especially those sympathetic to A-Life (Chapter 15)—were more ready to give embodiment its place, and the environment its due. One example was Gigerenzer's work on "simple heuristics that make us smart" (Section iv.g, above), but the renaissance of Gibsonian psychology counts here too (v.e–f). So does low-level Marr, despite his battles with the Gibsonians.

Crawling alongside the ant, and visible even from the sky by the end of the millennium, was the concept of *epigenesis* (Section vi.g). This was imported into psychology by Piaget

(5.ii.c). But his writings weren't taken seriously by the early computationalists (5.ii.a), partly because they seemed too philosophical to be respectable (Boden 1979, chs. 1 and 5). It was difficult to think about epigenesis in detail: one needed to consider both broad developmental trends and micro-developmental changes contingent on the environment (vi.h). And one needed to consider neurology, too. It wasn't until very late in the century that the epigenesis of (a few) abilities could be described in both psychological and neurological terms (vi.i above, and 14.ix.c–d).

Also visible from a great height would be the continuing and (despite challenges from the ant-lovers: 13.iii.b) deepening concern with *representation*. Craik had kicked this ball into play in the 1940s (4.vi). So had McCulloch and Pitts—not once, but twice. Their two representational balls were differently shaped, much as soccer and rugger balls are: one logical, the other probabilistic (4.iii and 12.i.c–d). Since then, psychologists' studies of representation (aka schemas) have burgeoned. Abelson's earliest notion of "scripts", for instance, concerned the mental representation of familiar interpersonal concepts (i.c). And representation *as such* was discussed with respect to vision (v.a–f), rationality (iii.d and iv.d–e), development (vi.f–h), consciousness and disturbances thereof (i.h), and psychological explanation in general (iii).

From the satellite, empirical research on representation that's described in other chapters would be visible too. This was focused, for example, on conceptual prototypes (8.i.b); on whole-animal integration (14.vii and 15.vii); on connectionist and distributed representations (12 *passim*); and on their implementation in the brain (14 *passim*, but especially 14.viii). In addition, people disputed the psychological relevance of GOFAI representations in general (10.iii.a and iv.a, and 13.iii.b).

Last but not least, the sky-high camera would record pertinent work being done by philosophers. Some were trying to buttress various versions of the representational theory of mind (12.x and 16.ii–iv), others to question it (16.v), and others to declare it fundamentally absurd (16.vi–viii). But one didn't have to be a professional philosopher to raise, and explore, such issues. Indeed, the camera's earliest takes would record that Craik himself had discussed representation in philosophical terms, noting for instance that *similarity* between X and Y isn't enough to make X a representation of Y (4.vi.b).

As for where the representations (and the responses) come from, even the grainiest photographs would show an increasing interest in evolution over the last sixty years. Craik had spoken of inbuilt/evolved models for making adaptive sense of perception (4.vi.b–c). The "evolutionary psychologists" of the last quarter-century had a great deal more to say about such matters (Sections i.i, iv.g, v.b–f, and vi above, plus nearly all of Chapter 8). Psychologists now have a much better sense of the similarities between *Homo sapiens* and other species, and of the huge—mostly language-based—differences too (Sections i.g, vi.a–c, and vi.h, and Chapters 8 and 9 *passim*).

Much of the relevant evidence was garnered as a result of asking specifically computational questions. But data gathered from non-computational sources, such as primatology (smudgy images of Washoe, Sarah, Lana, and Kanzi here), were woven into a computational fabric.

Closely associated with the evolutionary movement was the notion of "modules". This gained added respectability from Chomsky's linguistics and Fodor's philosophy of mind, both of which came out of the same Massachusetts stable (vi.d above, 9.vii.c–d, and 16.iv.c–d). However, it was countered to some extent by the growing emphasis on epigenesis (vi.i).

The satellite images would indicate a continuing background of interdisciplinarity, despite the foreground fractionation of computational psychology into multiple specialisms. In some areas, indeed, the interdisciplinarity deepened with the passage of years. One illustration of that is the developmental work described in Section vi.i. Another, as yet less detailed, is the rapprochement between cognitive and neuroscientific accounts of delusional states (i.h–i, above).

In the most recent years, the celestial camera would also pick up myriad activities of brain scanning. Occasionally, these would illuminate some specific psychological claim. However, and despite the enthusiastic pronouncements of their discoverers, most brain-scan data don't have much psychological relevance—except in the very broadest sense. So it's not clear that they should be counted as "discoveries" at all (see 1.iii.f). The theoretically alert satellite-image interpreter would see them less as *message* than as *noise* (14.iii.a and x.b). (In future decades, they might indeed form part of the message—but only if the relevant psychological theory has been developed.)

b. Forking footpaths

Different footpaths afford different flowers. That's as true in psychology as it is in field botany. The general methodologies visible from on high have changed over the years. Information theory and GOFAI were the first. For it was the early 1950s mathematical and New Look psychologists, soon joined by Newell and Simon, who got computational psychology off the ground and provided its first, intoxicating, successes (6.i–iv).

The satellite images transmitted in that mid-century period would also show connectionism. The early models of Hebbian learning (which illustrated very general features of neuropsychology) aroused late 1950s psychologists' interest (5.iv.b–f and 12.i–ii). In particular, they were intrigued by Selfridge's Pandemonium, and even more by Rosenblatt's perceptrons.

But that flurry of interest didn't last. For one thing, connectionism would be near-crushed under a huge obstacle thrown across its path in the late 1960s (snarling close-ups of Minsky and Papert, here: 12.iii). For another, connectionism couldn't yet be used to address specific psychological topics—such as the past tense, for example (12.vi.e). So it dropped out of the picture, remaining invisible for some twenty years. (Set at a higher resolution, the satellite camera would have detected important work, on associative memory, going on behind the scenes: 12.iv–v.)

Over the years, the camera would show GOFAI persisting as a major influence. It was used—and is still being used—to address many long-standing psychological topics, from personality to problem solving. Every section of this chapter showed 1960s work with the GOFAI imprimatur. These researchers theorized in GOFAI terms, and sometimes drew inspiration from GOFAI as such—for example, from work on knowledge representation (10.iii.a) and from computer-vision research on "scene analysis" (6.ii.e and 10.iv.b).

From the mid-1980s, the satellite images of GOFAI would jostle alongside images of connectionism (see Chapter 13). So viewers would see part-GOFAI theories of psychopathology (e.g. clinical disturbances of speech and everyday action), and of hypnosis and absent-mindedness too (Section i.g, and 12.ix.b). In addition, they'd see part-GOFAI work on the society of mind and/or mental architecture in general (Section i.e–g).

The camera would confirm that psychologists concerned with *mind as machine* have followed two different pathways (1.ii.a). (Or perhaps two different games: GOFAI soccer and connectionist rugger.) At first, the two weren't clearly distinguished—and anyway, most people who were interested in the one were also interested in the other (Chapter 4 *passim*). When a schism eventually occurred (4.ix), most psychologists took the GOFAI route. But the 'cybernetic/connectionist' road remained intact, and would later be heavily trodden (12 *passim*). One track within it would reflect interest in "dynamical" systems, now being studied in psychology, neuroscience, and A-Life (14.ix.b and 15.viii.c, ix, and xi).

Finally, what of the famous Zeitgeist? Would a sky-high viewer of computational psychology's history be as aware of this as Edwin Boring was in 1957? Probably not, since his canvas was stretched over many more than fifty years (although in Chapter 2 we skated lightly over many centuries). Within just one half-century, there's less time for the rise of new Zeitgeists.

However, our period did see one important example. This new viewpoint was chronicled and celebrated by Theodore Roszak (1969) as "the counter culture" (i.iii.c; see also Toulmin 1999, ch. 5). Postmodernism in the humanities was one expression of it, and the popular New Age movement was another. In general, Enlightenment optimism about the all-encompassing relevance of science, and about the likely success of technologies based on it, was roundly rebutted. For many people now, science was the enemy, and technology deeply suspect.

The most obvious effect that this cultural development had on professional psychology was in social psychology and certain parts of clinical psychology. (Psychology-in-anthropology was affected even more disastrously: see 8.ii.b–c.) 'Scientific' approaches were firmly rejected—and explicitly insulted. Presumably, Cronbach was appalled. Certainly, Broadbent was. So much so that he wrote a book, and prompted a professional conference, specifically to defend "empirical" psychology, including its informational and computational versions, from this counter-cultural attack (6.i.d).

Many members of the general public, doubtless including actual or potential psychology students, turned away from computational ideas as a result of their sympathy with the counter-culture. While some were driven primarily by philosophical worries, others were motivated, at least in part, by the militaristic context of computational psychology and AI in general (Chapter 11.i; see also Edwards 1996, ch. 6; Haraway 1986/1991). But with the advent of computational theories of self-organization and distributed cognition, some former opponents were part-mollified (e.g. Kember 2003). Hard-headed businessmen were affected, too: in theories of business management in the 1990s, there was a drift from the formal to the concrete, and from hierarchical to distributed control (Toulmin 1999, ch. 5).

Even *within* computational psychology, there were repercussions. From the satellite in outer space, one would be able to see that 'ant-friendly' theories were in general viewed more favourably by people who sympathized with Roszak. Decentralization, distributed control, bottom-up processing, emergence, epigenesis, embodiment, self-organization, 'simple' heuristics . . . all these concepts were attractive to people with counter-culturalist leanings (1.iii.d). And all crept into psychology in the last quarter-century, as well as informing A-Life and much of neuroscience.

They weren't usually put there by card-carrying New-Agers (though sometimes they were: Varela *et al.* 1991; Agre 1988, 1997). Indeed, almost all of them had been hovering in the background since the very earliest days. And some, such as Newell and Simon's decentralized blackboard architecture, were provided by very un-New-Age-ish people (iv.b). But with the scent of the counter-culture wafting in the air, their reception was eased. (And for professionals seeking to seduce the media, the way was clear: Chapters 12.vi.a, vii.a, and x.a, and 15.x.a.)

In sum, the satellite film would show widening, and deepening, activity over the last fifty years. Both major footpaths would be prominent, and various minor tracks would be seen to branch off them. Countless detailed examples of what John Ziman (2000*a*) called "reliable knowledge" have resulted from this research. If the sub-sub-sub- . . . topics could have been mentioned here, that would have been unmistakable. As it is, the six psychological stories told above, along with the relevant portions of other chapters, will have to suffice.

c. The Newell Test

In leafing through the satellite pictures to get an overall view of what's happened in the field, one could do worse than bear in mind Newell's criteria for psychologizing about the mind as a whole. He claimed that a theory of the human mind must consider, and a plausible computer model of it should implement, twelve functional properties—and his own SOAR system attempted to do just that (Section iv.b, above).

John ('ACT') Anderson and Christian Lebiere have recently suggested using these twelve properties as the "Newell Test" for cognition (Anderson and Lebiere 2003). This isn't intended as a post-millennial version of the (computer versus human) Turing Test, but as a way of comparing two computer models with each other. In a historical context, it offers a way of tracking research interests, and of estimating progress, over the last half-century.

The Newell Test assumes that we're considering an AI model with psychological, not merely technological, pretensions. And it advises us to ask whether it can do the following things:

1. Behave as an (almost) arbitrary function of the environment
   —Is it computationally universal, as opposed to (wholly) dedicated?
2. Operate in real time
   —Given its timing assumptions, can it respond as fast as humans?
3. Exhibit rational (that is, effective) adaptive behaviour
   —Does the system yield functional adaptation in the real world?
4. Use vast amounts of knowledge about the environment
   —How does the size of the knowledge base affect performance?
5. Behave robustly in the face of error, the unexpected, and the unknown
   —Can it produce cognitive agents that successfully inhabit dynamic environments?

6. Integrate diverse knowledge
   —Is it capable of common examples of sensory and intellectual combination?
7. Use natural language
   —Is it ready to take a test of language proficiency?
8. Exhibit self-awareness and a sense of self
   —Can it produce functional accounts of phenomena that reflect consciousness?
9. Learn from its environment
   —Can it produce the variety of human learning?
10. Acquire capabilities through development
    —Can it account for developmental phenomena?
11. Arise through evolution
    —Does the theory relate to evolutionary and comparative considerations?
12. Be realizable within the brain
    —Do the components of the theory exhaustively map onto brain processes?

(Adapted from John R. Anderson and Lebiere 2003: 588)

Applying this test to classical connectionism (Chapter 12) and to ACT-R (Section iv.c above), Anderson and Lebiere judge that—by and large—ACT-R is better, but that connectionism wins out on several points.

They point out, however, that to satisfy what psychologists in general are usually aiming for, Newell's twelve *functional* criteria would need to be supplemented by at least two more:

13. Match human mental processes.
    —Given that a computer model 'works', and fulfils the twelve demands listed above, does it also match the actual details of human psychology?
14. Be useful for practical applications.
    —Can the theory be used to improve everyday practice in psychotherapy, education, entertainment, and other areas of applied psychology?

These fourteen criteria have all been touched on in this "psychological" chapter. But—as a satellite image of *cognitive science as a whole* would show—they involve matters touched on in other chapters too. Here's a summary concordance:

1. Freedom from the environment:
   Sections i.h, iv.b–c, and vi.h.
   Chapters 8.v–vi, 9.iv.f, 9.vi.c–d, and 12.ix.b.
2. Real-time operation:
   Sections ii.c and iv.g.
   Chapter 14.vi.c–d and ix.g.
3. Rational/effective adaptive behaviour:
   Sections iv and vi.i.
   Chapters 4.viii, 5.ii.c. 5.iii, 6.iii, 8.iii–vi, 10.iii.c, 13.iii.b–c, 14.vi–vii, and 15.vii–ix and xi.
4. Vast environmental knowledge:
   Section iii.d.
   Chapters 9.iv.a–b, 10.iii.e, 13.i, and 14.vi.c–d.

5. Robustness in face of error/uncertainty:
   Section iv.g–h.
   Chapters 12 *passim*, 13.iii.b, and 14.vi.
6. Integration of diverse knowledge:
   Sections i.e–f, iii.d, and iv.c.
   Chapter 14.vi.c–d and vii.
7. Language:
   Sections ii, iv.d–e, vi.a–c, and vi.i.
   Chapters 4.iii.c, 6.i.e, 8.i.a–b, 9 *passim*, and 12.vi.e.
8. Self and self-awareness:
   Sections i and vi.f.
   Chapters 5.ii.a, 12.iii.d, 13.vi.e, and 14.x–xi.
9. Learning:
   Sections iv.c and vi.g.
   Chapters 5.iv.b–f, 6.ii.b–c, 12 *passim*, and 15.v.
10. Development:
    Section vi.
    Chapters 2.vi.b, 5.iii.c, 9.vii.c–d, 12.viii.d, 14.ix.c, and 15.iv.
11. Arise by evolution:
    Sections iv.g, v.b–f and vi.
    Chapters 8.ii.d–e, 8.iii–vi, 9.iv.e, 14.vii, and 15.vi–vii.
12. Realizable in the brain:
    Sections iv.b, v.b–f, and vi.i.
    Chapters 4.viii.c–d, 5.iv, 12.i.c–d, 14 *passim*, and 15.vii and xi.a.
13. Match human psychology:
    Sections i–vi *passim*.
    Chapters 8, 9, 12, 14, and 16 *passim*.
14. Be practically useful:
    Sections i.a and d–f, iv.a, and vi.i.
    Chapters 9.x–xi, 10.iv.c and v.g, 12.vi–vii, 13.v–vi and vii.b, and 14.x.b.

What would our judgement be, were we to look at today's computational psychology with these questions in mind? We'd have to admit that we're still a very long way from a plausible understanding of the mind's computational architecture, never mind computer models of it (Section i.e–g, above). And as remarked elsewhere (especially Chapters 8, 9, 13, and 16), we're still a long way from human-level modelling of language and the recognition of relevance—even supposing this to be in principle possible. In other words, psychological modelling is still bedevilled by the open texture of language, and by the frame problem in general.

Nevertheless, if we bear the Newell Test (or even common sense) in mind, many lines of progress can be seen. For instance, a half-century of work on nativism has hugely increased our understanding both of the facts and of the concept itself (Section vi—and see also item 10 in the concordance above). We're much more sophisticated about what "nativism" means, so less ready to assume that a particular ability must be *either* innate *or* learnt (dichotomies, again). We realize that learning can take place before birth, as well as after it. We've got a better understanding of the distinction between learning and development, and of how specific developmental changes are needed to support higher-order thought. And we've got a better sense of how (animal and human) behavioural studies can be integrated with neuroscience on the one hand and computer modelling on the other.

These advances happened in the context of cognitive science in general, and computational psychology in particular. Without the new forms of non-Newtonianism that arose in the 1950s (Chapters 5 and 6), they wouldn't have occurred.

d. Low focus

Even the low-resolution images described above picked out a fair amount of detail. What if the satellite camera were set at a lower resolution still?

At the low-focus extreme, the overall story would relate a gradually growing recognition by psychologists, especially cognitive psychologists, that a new form of explanation was available. (Or, if you prefer, two new forms—reflecting the two pathways to mind-as-machine.) And not merely available but, in many people's view, superior to anything that was available before.

The Oxford cognitive psychologist D. Alan Allport, for instance, didn't mince his words. The advent of AI models, he said some twenty-five years ago, was the "single most important development in the history of psychology" (1980: 31). As if that weren't praise enough, he added that "artificial intelligence will ultimately come to play the role vis-à-vis the psychological and social sciences that mathematics, from the seventeenth century on, has done for the physical sciences". In effect, Longuet-Higgins was saying the same thing, if less dramatically:

> It is perhaps time that the title "artificial intelligence" were replaced by something more modest and less provisional... *Might one suggest, with due deference to the psychological community, that "theoretical psychology" is really the right heading under which to classify artificial intelligence studies of perception and cognition.* (1981: 200; italics added)

Those comments are notable not only for their rhetorical power but also for the high professional standing, and scrupulousness, of both Allport and Longuet-Higgins. These men weren't playing to the gallery. Nor were they so closely identified with the field, as Newell was, that *of course* they'd champion it (Newell had said long ago that AI might as well be called "theoretical psychology": 1973*b*: 25–6). But nor were they alone. By 1981, at which time computational psychology had existed for a quarter-century, there were plenty more such remarks scattered in the literature.

That's not to say, however, that AI models were universally accepted by psychologists. At much the same time as those two encomia were being declared, another British cognitive psychologist said to me that the field was "too specialized" to be the theme of the first sub-meeting of the History and Philosophy of Psychology group within the BPS.

If interpreted as an intellectual judgement, that was a mistake. The computational approach (symbolic and connectionist combined) is held by its practitioners to be potentially applicable to all mental processes, so—if they're right—it's not "specialized" but universal. In other words, Neisser's fear that computational psychology would become "a narrow and uninteresting specialist field" was misguided (see Chapter 6.v.b).

However, if interpreted as a statistical observation about people's concerns at the time, her comment was correct. (The general professional reluctance to take computational theorizing on board was partly due to the lack of hands-on experience discussed in Chapter 6.v.a–b.)

It wouldn't be correct today. The interest aroused by the work described in this chapter has been—and is—increasingly strong.

Consider, for example, the BPS conference held in London in 2004. An entire day—one plenary session, plus half a dozen sessions running in parallel with others, plus two lengthy 'slots' for symposiasts' discussion—was devoted to the psychology of creativity *considered from the computational point of view*. If even creativity can be looked at in this way (cf. 13.iv), it's hardly surprising that many of the other 2004 conference sessions reflected this approach too. Indeed, they didn't even bother to say so: whereas session titles like 'Creativity and Computers' still have power to shock, computational ideas about perception, memory, reasoning, and language are ten tens a penny. Professionally speaking, computational psychology has arrived.

That's not to say that *computer models* are scattered like autumn leaves on the ground. Even psychologists who are sympathetic in principle may not be able to use that methodology (as opposed to the computational concepts driving it) to address their chosen research topics. In current studies of psychopathology, for instance, some people are using explicitly computational analyses and/or models. (Remember the research on autism and Williams Syndrome in Section vi.f and vi.i; and consider the analyses of pathological action errors mentioned in Chapters 12.ix.b and 14.x.b.) But many cognitive theories of psychopathology are computational only in a very broad sense, with no attention to detailed mental processing—still less, to computer modelling (e.g. D. M. Clark 1996).

Even so, the computational movement—or if you prefer, the cognitive revolution—has influenced the general form of these clinicians' theorizing. To that extent, even MGP's most flamboyant promises have been partially met (6.iv.c; and cf. 17.iv).

e. The bustling circus

In the late 1950s, Alan Baddeley feared for the future of psychology:

> [With respect to the controversy over Edward Tolman and Clark Hull: see 5.iii] I was very struck by the capacity of psychological theory to change direction rapidly, apparently as a result of fashion rather than evidence. This led to the question, commonly raised at the time, as to whether psychological theory could ever be cumulative. (Baddeley 2001: 345)

His own response was "to favour theorisation based on relatively simple generalisations tied to robust phenomena", and to opt for Stephen Toulmin's (1953) philosophy of science, which allowed for a pragmatic clutch of separate theories, rather than the then popular axiomatic approach (Chapter 9.v.a).

Professional psychology today is closer to Toulmin's vision than to Hull's. It's not "cumulative" in the sense that physics is. (Although even physics, notoriously, can't yet integrate relativity with quantum theory.) But the various topic areas are closer in spirit to each other, and to the other cognitive sciences, than they were when Baddeley was a young man.

Nevertheless, there are persisting areas of what Cronbach called "desert". Many psychologists, still, ignore the computational viewpoint. Some of them are more concerned with *What?* than with *How?* They don't deny that there are how-questions to be asked. They may even allow that computational ideas would be needed to provide the answers. But they aren't interested in asking them. These people include some primatologists (but see Section vi.e–f above), some evolutionary psychologists (but see 8.ii.d–e, iv, and vi), and most social psychologists and psychometricians—in other words, what Cronbach termed the descriptive/comparative psychologists.

Others spurn *any* 'scientific' approach, computationalism included. They believe that meaning, intentionality, personality, purpose, and consciousness simply *cannot* be explained naturalistically (Chapter 16.vi–viii).

These desert dwellers include postmodernists such as Kenneth Gergen (1994, 2001), who thirty years ago was among the first to attack empiricist social psychology (6.i.d). They also include the personality theorists and therapists whose predecessors were dismissed by Cronbach as "philosophers or artists". Some of them are specifically offended by what they see as the dehumanizing, technologistic, influence of computer models—even more alien than rats (1.iii.c–d, 6.i.d, and 8.ii.b–c). In general, they agree with humanists such as Giambattista Vico (1.i.b) and Wilhelm von Humboldt (9.iv) that knowledge of human minds is essentially different from, and even superior to (better grounded than), knowledge gained by science.

This particular area of intellectual "desert" is less scorpion-ridden in professional psychology than in anthropology, but it exists nonetheless (6.i.d and 8.ii.b–c). Hardly anyone attempts to cross it, from either side. That's not surprising: it's a special case of a fundamental split in the philosophy of mind and epistemology (14.xi and 16.vi.b).

A few brave souls have recently ventured forth to reconcile the seemingly irreconcilable (15.viii.b, 16.vii.b–d, ix.e–f, and x.c; and see M. W. Wheeler 2005). The most ambitious is the AI-based cognitive scientist Brian Cantwell Smith. However, his account of computation and mind is idiosyncratic to a startling degree (16.ix.e–f). For sure, it's not going to convince the masses. In any event, it's more concerned with metaphysics than with psychology. The specifically psychological desert-crossers include Francesco Varela, James Thompson, and Eleanor Rosch, whose recent writings on mind and cognition stress the importance of *embodiment* (Varela *et al.* 1991; see 15.viii.b). But despite the occasional date tree (13.iii.e), their pathway hasn't led to many significant oases of experimental/computational research.

In short, Cronbach's remarks on the fragmentation of psychology are still apt. The 2004 BPS conference, like its US equivalent half a century before, was "a circus, but a circus far grander and more bustling than any Barnum ever envisioned" (Cronbach 1957: 671). Given the differences of topic interest and of philosophy just remarked, the circus will continue to bustle for many years yet. Computational psychology, however, will be a star performer in the ring.

8

THE MYSTERY OF THE MISSING DISCIPLINE

Anthropology studies the cultures constructed by human minds, and by which they're largely constructed in turn.

That's shorthand, of course. Anthropologists have defined "culture" in many different ways (Kuper 1999). Moreover, some "cutting-edge cultural anthropologists" have "virtually eliminated" the term (Strauss and Quinn 1997: 1). Besides problems of definition, postmodern anthropologists today see "culture" as an unacceptably essentialist notion (they prefer terms like "discourse", "interest", and "strategy"). But I take it that we know, roughly, what we mean by culture—and that, as the cognitive anthropologist Bradd Shore (1996: 9) has said, we need to refine the notion rather than discard it.

I take it, too, that culture is paradigmatically human. It's not purely linguistic, since it covers material artefacts and rituals, and even bodily posture and gestures too. Indeed, the archaeologist Colin Renfrew (2003) argues that the external symbol storage of monumental architecture (e.g. Stonehenge) was the next important step in cultural evolution after language—an honour sometimes granted to cave-painting (Mithen 1996*b*) or writing (M. Donald 1991). Nevertheless, culture is very close to, and often expressed in, language—which is essentially human (see 7.vi.c).

So anthropologists *don't* normally include "fledgling birds mimicking their species-typical song from parents", or "rat pups eating only the food eaten by their mothers" as examples of culture, even though some psychologists do (Tomasello 1999: 4). Nor do they see "young chimpanzees learning the tool-use practices of the adults around them" as culture—although promoters of the Great Ape Project, which stresses the *similarities* between the higher primates, typically do (Cavalieri and Singer 1993). If the young birds, rats, or chimps didn't engage in that early learning, most other remarks about their minds—or, if you prefer, their lived worlds—would remain much the same. But that's not true of people. In Jakob von Uexküll's terminology (see 5.ii.c), the *Umwelten* of *Homo sapiens* are overwhelmingly cultural, or semiotic: human beings don't just live *with* culture, but in it, through it, and by means of it. Or in John McDowell's terminology (16.viii.b), culture—and in particular, language—is our "second nature": it's part of our ethology, without which we wouldn't be recognizably *human* at all.

One might expect, then, that anthropology would form part of cognitive science. For language has been a central concern of the field ever since the late 1950s (Chapter 6.i–ii), and the importance of "cognitive technologies" in general—which include not only language but also cave painting, architecture, and writing—was made much of by Jerome Bruner from the early 1960s (6.ii.c). Yet it's usually missing from lists of the disciplines involved. For instance, it wasn't included among "the disciplines contributing to cognitive science" in a leading recent textbook on *Understanding Intelligence* (Pfeifer and Scheier 1999: 39). Nor did it appear in the titles of the six chapters on individual disciplines that opened the huge *MIT Encyclopedia of the Cognitive Sciences*, or *MITECS* (R. A. Wilson and Keil 1999). If it's mentioned at all, it's apt to be downplayed as "a trace substance" (Simon 1994*b*: 1).

When cognitive science began, things were very different. Many people then thought it obvious that anthropology should play a significant role. By the 1980s, however, it had become near-invisible.

Anthropological questions were continuing to feature nevertheless. Indeed, an expert on one African culture declared in the 1990s that "The study of cognitive processes, however incipient in cognitive psychology, *allows us to reformulate many classical problems of anthropology*" (Boyer 1994, p. xi; italics added). Strictly speaking, then, the discipline isn't "missing" so much as unacknowledged.

This chapter explains why that is. It starts by outlining the origins and development of cognitive anthropology. Section ii shows why the field dropped out of sight, even though highly relevant research was still going on. Sections iii and iv discuss teamwork in navigation and cross-cultural aspects of aesthetics, respectively. The history of ideas about cultural evolution is sketched in Section v. Finally, Section vi considers religion: its universality, and its underlying psychological mechanisms. It also asks why some 'chains' of communication are stable enough to avoid turning into a game of Chinese Whispers: were that not so, culture would be impossible.

8.i. Anthropology and Cognitive Science

Work labelled "cognitive anthropology" started in the early 1960s, and the closely related "ethnoscience" even before that. It was seen as a key aspect of the incipient cognitive science and an inescapable part of its future.

Admittedly, there were sceptics right from the start: we've already noted Ulric Neisser's charge that a computational approach must be "indifferent to culture" (Chapter 6.v.b). Moreover, anthropology wasn't one of the disciplines listed on the dust jacket of the first book devoted to "cognitive science" (Bobrow and Collins 1975). Nevertheless, the editor of the series—entitled Language, Thought, *and Culture*—was himself an anthropologist (Eugene Hammel). And Marc de Mey, writing in the early 1970s, included it in *The Cognitive Paradigm* (published ten years later), pointing out that rituals and institutions are the expression of representational systems (belief systems) that define a world and a way of life (1982, p. xv).

Moreover, in 1978 anthropology was officially identified as one of the six major dimensions of the field. Its place in the cognitive pantheon seemed assured.

a. The beginnings of cognitive anthropology

The scent of cultural questions was wafting in the air from the very earliest days of cognitive science. The cybernetics movement at mid-century had attracted the anthropologist Gregory Bateson, an ex-student of the schema theorist Frederic Bartlett (5.ii.b). He described cultures as self-balancing systems capable of containing 'contradictory' multilevel relationships (e.g. Bateson 1936). And the maverick Gordon Pask did some work on criminal subcultures that one might describe as nascent urban anthropology (4.v.e).

What's more, anthropologists in the 1950s were starting to move away from purely descriptive ethnography towards more theoretically based approaches—some cognitive, some not (D'Andrade 2000). So the ethnoscientists of the late 1950s tried to unite anthropology with formal linguistics.

They sought semantic equivalents of the finite set of phonemes which—as had recently been discovered—constitute all human languages (see Chapter 9.v.d). So, for instance, cultural anthropologists such as Ward Goodenough (1956, 1965), Floyd Lounsbury (1956), Benjamin Colby (1966), and David Schneider (1965) asked whether a componential analysis could capture kinship terms in every natural language—including the terminologies used by their fellow Americans, from Yankees to Pawnees.

In the early 1960s, psychology entered the equation. This wasn't unexpected. Bruner's Harvard mission station was founded in 1960 with the intention that anthropology should feature strongly in it. He'd been interested in the subject ever since his days at Duke University, where he'd heard William McDougall's tales of the famous 1898 Anthropological Expedition to the Torres Strait (Bruner 1983: 133–6). And on his arrival at Harvard, he'd encountered several leading anthropologists—notably Clyde Kluckhohn (1905–60), who'd done pioneering work on the Navaho.

Kluckhohn also helped found the Department of Social Relations, Bruner's interdisciplinary 'home' (Parsons and Vogt 1962). The Department would train many now-famous anthropologists. These ranged from Clifford Geertz (1926–), who eventually became strongly *anti-cognitive* (Geertz 1973*a*), to his close contemporary A. Kimball Romney (1925–), one of the first to encourage cognitive anthropology (Romney and D'Andrade 1964). Indeed, Romney trained and/or influenced "nearly a third of the people who have done significant work" in the field (D'Andrade 1995*a*: 245).

So Bruner's Center for Cognitive Studies included anthropologists as valued members from the start. During my stay there (1962–4) Franz Boas, Edward Sapir, and Benjamin Whorf were frequently mentioned, and Humboldtian questions about culture, language, and thought were rife (Chapter 9.iv).

Another name in the air was A. Irving Hallowell (1892–1974), an expert on Native American tribes who'd long considered not only how cultures appear from the outside but how they're experienced by—and how they affect—the people living in them. That is, he'd generalized 'Humboldt–Whorfism' from language to personality. A few years earlier, in a book of essays celebrating his sixtieth birthday, he'd called for a new field of *ethnopsychology*, which would study "concepts of self, of human nature, of motivation, [and] of personality" in different cultures (Hallowell 1955: 79). The Department of Social Relations contained a number of people—not least, Gordon Allport, Henry

Murray, and David McClelland—deeply interested in personality theory, and some of them had regular contacts with the Center.

So the many 'cultural' questions heard in the corridors hadn't *originated* from the mind-as-machine hypothesis. But they'd been *sharpened* by the formalist ethnoscientists. And they were soon to be *broadened* with the help of the computational ideas of the Center's co-founder George Miller.

Miller's brainchild, the MGP manifesto (see 6.iv.c), inspired one of the first examples of cognitive anthropology, and one of the most widely cited (A. F. C. Wallace 1965). Indeed, Miller *et al.*'s *Plans and the Structure of Behavior* was the only work mentioned in this paper's References, except for three by the author himself.

Instead of MGP's "hammering a nail", Anthony Wallace (1923–) used "driving to work" as his main example. He described this activity as guided by plans, or rules, of different types—some drawing on very general knowledge, some on the cultural context, and some on personal experience. The Image, one might say, was being attended to at last (see Chapter 6.iv.c).

The driver's "cognitive map", according to Wallace's account, is many-levelled. It represents (for instance) routes and landmarks. It codes stop signs, traffic lights, and places—such as schools and shopping malls—where children may dart out into the road. It covers visibility, weather, traffic level, and road conditions. And it includes the tools of the trade (clutch, indicators, gear shift . . .), the bodily actions required to operate them, and the driver's comfort when doing so.

"The simplest model of how this total process operates", said Wallace, "is to consider the driver as a cybernetic machine" (1965: 287). He was noting the need for monitoring (of self, vehicle, and environment) and feedback—both of which had been included within the TOTE unit. His nine rules for "Standard Operating Procedure", seven outside-of-car dimensions (allowing 216 combinations), and five car controls (allowing forty-eight combinations of actions for "unitary response") in effect defined a TOTE hierarchy, with "choice points" at specified junctures.

Wallace suggested that analysis in terms of Action Plan, Action Rules, Control Operations, Monitored Information, and Organization could explain other tool-using activities, even including hunting and warfare (p. 291). Whether it could also capture "behavior in social organizations", he said, "remains to be seen" (p. 292).

Wallace wasn't the only one to embark on a cognitive anthropology. A multi-authored collection of papers under that label appeared in 1967, and was soon reissued (Tyler 1967), and others appeared in due course (e.g. Dougherty 1985). "Mathematical" cognitive anthropology was heralded too (Ballonoff 1974), and information/feedback theory used to distinguish the idea systems and subsystems—religious, economic, kinship, marriage . . .—accepted in Punjab (Leaf 1972). (The use of computers in anthropology had been the topic of a conference as early as 1962; however, this discussion included some decidedly *non-cognitivist* approaches, such as HOMUNCULUS: see Hymes 1965 and Chapter 7.i.c.)

The early 1970s saw a few anthropologists sketching "culture grammars" (B. N. Colby 1975). These were intended to capture underlying structural possibilities, much as computational psychologists were trying to do for stories (Rumelhart 1975). And in 1977 the Society for Psychological Anthropology (SPA) was founded, under the umbrella of the American Anthropological Association (AAA).

By this time, however, Wallace's concerns for material culture (traffic lights and gear shifts) and social institutions (shopping malls and traffic laws) had faded into the background. They'd be resuscitated later, as we'll see in Section iii. Meanwhile, the focus had turned to language, and especially to classification.

Wilhelm von Humboldt's view (9.iv) that culture is primarily represented in, and learnt from, *language* had seemingly triumphed. So much so, indeed, that when one Wallace sympathizer in the early 1980s started writing a book on his field, he became "disillusioned" and lost interest in the project (E. L. Hutchins 1995, p. xii; see Section iii, below). However, that was because he favoured a different focus of research, not because there was no cognitive anthropology being done.

In fact, a lot of work had been done—and cognitive psychology was a close partner in the enterprise.

b. Peoples and prototypes

In the late 1960s and 1970s, cross-cultural studies of the structure and processes of thought blossomed (e.g. Warren 1977). Most of the interest was in whether some ways of conceptualizing the world are psychologically 'natural', and therefore universal, or whether categorization is basically arbitrary and culture-dependent.

The novelist Jorge Luis Borges had presented an "ancient Chinese" taxonomy of the animal kingdom:

> [Animals] are divided into (a) those that belong to the Emperor, (b) embalmed ones, (c) those that are trained, (d) suckling pigs, (e) mermaids, (f) fabulous ones, (g) stray dogs, (h) those that are included in this classification, (i) those that tremble as if they were mad, (j) innumerable ones, (k) those drawn with a very fine camel's hair brush, (l) others, (m) those that have just broken a flower vase, (n) those that resemble flies from a distance. (Borges 1966: 108)

This was obviously a joke: no such taxonomy could exist. But why? What, exactly, was 'unnatural' about it?

An answer was offered by the Berkeley psychologist Eleanor Rosch (1938–) (Rosch 1978: 27), and by two of her Berkeley colleagues: the anthropologists O. Brent Berlin and Paul Kay. (Kay later transferred from the Department of Anthropology to Linguistics; both he and Rosch had been trained in the interdisciplinary halls of Social Relations at Harvard.) Their work was hugely influential, and brought psychology and anthropology closer together.

Berlin (1936–) and Kay (1934–) had been working on colour names throughout the 1960s. They hoped to use colour to resolve the debate about linguistic relativity, aka the Whorfian hypothesis.

Comparing vocabularies in nearly a hundred different languages, they'd identified startling differences in colour classification—and in the experience of colour (O. B. Berlin and Kay 1969). For example, native English speakers see blue and green as 'obviously' similar—but some cultures don't. So far, it might have seemed, so Whorfian. But they'd also found interesting similarities, presumably due to inborn biological mechanisms—a very non-Whorfian idea.

For instance, when asked to identify the just-noticeable differences between patches of colour, people show no cultural differences. Even more strikingly, some colours seem to be 'basic' to all languages—and some are more basic than others. As Berlin and Kay put it, their "major findings" were:

> (1) the referents for the basic color terms of all languages appear to be drawn from a set of eleven universal perceptual categories, and (2) these categories become encoded in the history of a given language in a partially fixed order. (1969: 4–5)

Accordingly, the subtitle of their 1969 book on colour terms referred to *Their Universality and Evolution*. Both key hypotheses were later refined, thanks to the—continuing—explosion of interdisciplinary research inspired by their work. But universality and evolution remain crucial to anthropologists' understanding of colour (P. Kay and Maffi 1999).

Rosch (then publishing as Eleanor Heider) was one of the first flames in that explosion. At the start of the 1970s she continued Berlin and Kay's line of thought by doing cross-cultural experiments (Rosch/Heider 1971, 1972; Rosch/Heider and Olivier 1972). Working with the Dani tribe of New Guinea (now West Papua)—who have only two colour terms, 'translated' as *cool/dark* and *warm/light*—she found that colour grouping and memory for colours are indeed influenced by language. Nevertheless, her results implied—as Berlin and Kay had suggested—a universal base underlying the cultural variety. Certain ("focal") colours are perceptually salient: they attract the infant's attention more readily, their names are more easily learnt and remembered, and they're more likely to have special names in the language.

She soon extended her research to categorization *in general* (Rosch 1973, 1975, 1978; Rosch and Mervis 1975). Her theory of "prototypes" identified various ways in which Borges's imaginary taxonomy was psychologically absurd. And it provided a more realistic view of concepts than BGA had done (Chapter 6.ii.b).

For example, it confirmed the common-sense intuition that a robin is a 'better' example of a bird than is an owl or an ostrich. She even measured the difference: her experiments showed that a robin has a typicality rating of 1.02, an owl 2.96, and an ostrich 4.12 (1978: 36). On the BGA property-list view, this made no sense: a bird is a bird, is a bird... Robins and ostriches are *equally* birds, even if robins are much the more familiar.

Prototypes were relevant to AI too, for they explained various knotty aspects of knowledge representation. Consider the oft-cited AI example "Birds fly, but emus don't". Rosch's description of prototypes could defuse the apparent contradiction, and justify the default assumption that—in the absence of information to the contrary—if something is a bird, then it flies. Similarly, the theory helped to explain why the frame problem can arise in respect of virtually any natural-language word, or common-sense reasoning (see 10.iii.e and 7.iii.d).

But as well as describing prototypes, Rosch offered an explanation of why they are as they are. The categories used within any culture, she said, have been developed according to two universal psychological principles. One is "cognitive economy", whereby the strain on memory and information processing is minimized. (Bounded rationality, again: see 6.iii and 7.iv.)

For example, hierarchical concepts prevent discriminations being coded unnecessarily. Why think/speak of *cats and dogs and birds and fish and flies and*... when one can think/speak simply of *animals*? If the properties one's interested in can be predicted by *animals*, then the high-level term is an economical way of storing that information. (A warning: Rosch often said she was talking about the "structure" of information, not its "processes". She wasn't a *computational* psychologist, and later co-authored a radical critique of mainstream cognitive science: Varela *et al.* 1991.)

The other general principle is "perceived world structure". Whereas BGA's concepts were defined by arbitrary sets of geometrical features, everyday concepts are based on our perception of real structures in the real world. If we see feathers we expect wings, because feathers and wings really are highly correlated. So the concept of *bird* is 'natural' (and culturally widespread), whereas a concept defined as *suckling pigs and stray dogs* isn't. Granted, culture can sometimes provide the correlations. If it were the custom, perhaps justified by a particular myth, to provide a portion of suckling pig to every stray dog encountered on the street, a concept covering the two animals might well develop.

"Prototypes" were Rosch's explanation for a host of psychological phenomena that had been noticed previously (by Ludwig Wittgenstein for instance: 9.x.d) but which she detailed more fully. In summary, concepts don't have clear-cut boundaries—again, compare BGA. A concept is actually a 'cloud' of slightly different concepts (each one defined within a large set of discriminatory features) centred on a paradigm case, or prototype. More accurately, there may not be a single paradigm case: *several* items might have typicality ratings very close to 1. Poker and tennis, for example, are both more typical examples of *game* than patience is, so are more likely to act as paradigms (cf. Wittgenstein 1953: 31 ff.). A prototype is a statistically central measure, which may or may not correspond to an actual representation.

Prototypes aid cognitive economy, because prioritizing *one* (relatively central) case within a continuously varying set helps us to think quickly and decisively. For example, Rosch showed that people can judge whether something is or isn't a bird more quickly if it's a robin than if it's an owl or an ostrich.

To put things the other way around, investigating the judgement speeds of people in a given culture provides one with a key to the semantic structure of their conceptual networks—and their practical priorities. The hierarchical level at which the prototype is found is the one which provides the most useful information—namely, the most inclusive level that captures the structure of attributes perceived in the world. ("Perceived" is important: city-dwelling children were found to have a less differentiated concept of *tree* than rural children do.)

In short, Rosch claimed that the basic discriminatory features for all concepts are shared. It's their grouping into classes which is culture-dependent. So by the late 1970s she felt that Whorf's linguistic relativity (9.iv.c) must be taken with a very large pinch of salt:

> When many of us first came in contact with the Whorfian hypothesis, it seemed not only true but profoundly true . . . (Rosch 1977: 95)

> At present [however], the Whorfian hypothesis not only does not appear to be empirically true in any major respect, but it no longer even seems profoundly and ineffably true. (Rosch 1977: 119)

Many anthropologists were intrigued. Her theory raised a host of questions about which features are basic, what prototypes are favoured by different cultures, and how their concepts are structured.

Berlin, for instance, was by this time gathering data on "folk biology": how various cultures classify plants and animals (O. B. Berlin 1972, 1974; Berlin *et al.* 1973). He saw Rosch's work as highly relevant in asking to what extent biological categories are culture-dependent (O. B. Berlin 1978: 9, 24). That opinion is still widely shared by cognitive anthropologists today.

c. Hopes and a hexagon

In the late 1970s, when Berlin and Rosch were prime names to conjure with, the early hopes for anthropology as a key player within cognitive science were still strong.

That's clear from Howard Gardner's (1943–) widely read book *The Mind's New Science: A History of the Cognitive Revolution* (1985), which contained an entire chapter on anthropology. First, Gardner outlined the contributions of Edward Tylor (1832–1917), Boas, Sapir, and Lucien Lévy-Bruhl (1857–1939). Then, he discussed more recent work—especially the structuralism of Claude Lévi-Strauss (1908–) and the componential analysis of ethnoscience (pp. 236–57). He also touched briefly (pp. 242 ff.) on the views of Geertz and Daniel Sperber (1942–)—of whom, more below.

This highlighting of anthropology wasn't an idiosyncratic decision on Gardner's part. His book was sponsored by the Sloan Foundation, being one of the many projects they funded under the banner of "cognitive science" (6.iv.f). As such, it would have been unthinkable not to give space to anthropology. For the Report submitted to the Foundation, when they were considering (from 1976) whether they should give hugely generous support to the field, had picked out anthropology as one of the six key disciplines (State of the Art Committee 1978).

This Report was written by experts, not bureaucrats. Those responsible included Michael Arbib, George Miller, Donald Norman, Zenon Pylyshyn (all mentioned in Chapters 6 and/or 7), the psycholinguists Joan Bresnan and Ronald Kaplan (see Chapter 9.ix.b)—and the San Diego anthropologist Roy D'Andrade (1931–), yet another graduate of Harvard's Department of Social Relations.

The common research objective of cognitive science, they declared, was *to discover the representational and computational capacities of the mind and their structural and functional representation in the brain* (p. 76). Their 'Sloan hexagon' (Figure 8.1) represented anthropology as no less crucial to this quest than Psychology, Philosophy, Linguistics, Neuroscience, and Computer Science.

Moreover, "Anthropological linguistics" and "Cognitive anthropology" were identified as essential subdomains. And both featured in the key example. This example, which comprised some 50 per cent of the Report, described the recent "transdisciplinary" findings on "the names we give to colors"—including, of course, those due to Berlin (pp. 76 ff.).

To be sure, not everyone accepted the Sloan Report's definition of cognitive science. In fact, it aroused such virulent disagreement that it wasn't officially published until some years later (6.iv.f). This isn't surprising: as remarked at the outset of Chapter 1, there's *still* no agreed definition of the field. In the late 1970s as now, different basic assumptions and methodologies were favoured by different individuals.

Moreover, this was hardly a disinterested debate. With $15 million at stake, people were seeking to cultivate their own garden patch. The committee members were all

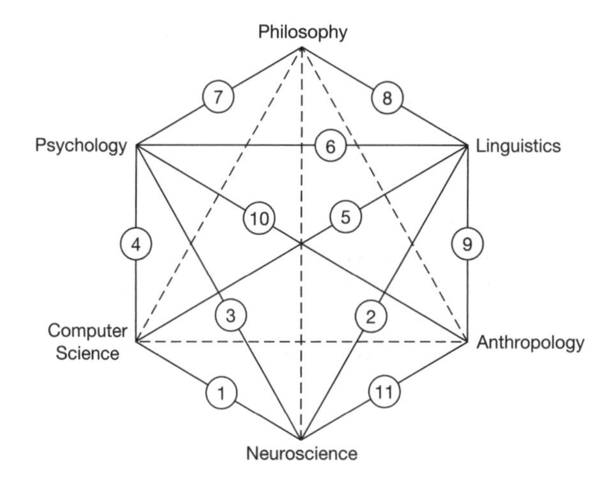

Fig. 8.1. The Sloan Hexagon (designed in 1978). Original caption: Subdomains of Cognitive Science: 1. Cybernetics. 2. Neurolinguistics. 3. Neuropsychology. 4. Simulation of cognitive processes. 5. Computational linguistics. 6. Psycholinguistics. 7. Philosophy of psychology. 8. Philosophy of language. 9. Anthropological linguistics. 10. Cognitive anthropology. 11. Evolution of brain. Redrawn with permission from Machlup and Mansfield (1983: 76)

highly respected cognitive scientists, and they'd taken formal advice from over twenty others—not to mention their many informal contacts and Sloan-relevant meetings. But they didn't include any specialist in straight computer science, or even AI. Disgruntled computer scientists felt that their discipline had been underplayed, even though AI was included in the hexagon.

When the money started flowing, disagreements—and jealousies—multiplied. There was a widespread (though not universal) welcome for Sloan's opening strategy of distributing a large number of small grants to many different places. But when the few large-scale grants were made, tempers rose. (The key AI departments, of course, had become used to receiving ARPA/DARPA largesse, originally provided by Joseph Licklider: Chapter 10.ii.a.)

But if the Sloan Report was controversial, it wasn't perverse. Anthropology was seen in this important document as a key part of the overall endeavour—and very few, if any, of the intended audience were unhappy with that. They might want more money for their own pet discipline, but they wouldn't have dreamt of arguing that anthropology didn't deserve any.

d. More taxonomies (and more Darkness than light)

Encouraged by the Sloan accolade, work in cognitive anthropology continued throughout the last quarter-century. Research on classification was the first to advance.

Kinship—a traditional focus of anthropologists' concerns—was joined as an object of interest by many other classificatory systems. Studies of ethnobotany, ethnozoology, ethnopsychology, folk medicine, and folk theology (for example) burgeoned. Each of these involved fieldwork in various cultures, to gather the necessary data.

Today, Rosch's theory still provides a theoretical basis for analysis and cross-cultural comparison (cf. Ellen 2003: 52). However, some anthropologists now believe that her results weren't due simply to statistical patterns in the environment: more robins than ostriches, and *always* feathers. For they see classifications and folk taxonomies as resulting also from inborn predispositions to think in terms of "essences" for natural kinds. In this, they're picking up on the current fashion for positing "modules" in evolutionary psychology (see Chapter 7.vi.d–e, and iv.a below).

(That fashion is regarded as mistaken by some psychologists. For instance, the influential connectionist James McClelland argues—backed up by computer modelling—that apparently "innate" domain theories could arise from familiar types of connectionist learning: McClelland and Rogers 2003; T. T. Rogers and McClelland 2004. His argument is closely comparable to his connectionist assault on Chomskyan innate grammar, which caused a sensation in the mid-1980s: see Chapter 12.vi.e.)

The anthropologist Scott Atran (1952–), for instance, says:

> Humans, let us suppose, are *endowed* with highly articulated cognitive faculties for "fast-mapping" *the world they evolved in, and for which their minds were selected.* The "automatic" taxonomic ordering of phenomenal species, like the spontaneous relational ordering of colors, would then be a likely product of one such faculty. (Atran 1990: 65; italics added)

> Humans appear to be inherently disposed to classify living things according to presumptions about their underlying physical natures. Cross-cultural evidence indicates that people everywhere spontaneously organize living kinds into rigidly ranked taxonomic types despite wide morphological variation among those exemplars presumed to have the nature of their type. (pp. 70–1)

The remark about "rigidly ranked taxonomic types" is over-optimistic. For it's now clear that Berlin underplayed the variation among folk biologies. He tried to assimilate all taxonomies of living things to a hierarchical, quasi-Linnaean, model. But it's turned out that not all classificatory systems are like that (Ellen 2003, esp. 53–5). Atran's other claim, however, may be better grounded. Certainly, his reference to "the world [we] evolved in" is an example of a form of thinking that's recently become common within cognitive anthropology (see Sections ii.d–e and iv–vi, below).

The explosion of research in ethnobiology would probably have happened anyway, given Berlin's provocative start. But it benefited from funding led by medical/pharmaceutical interests, and sometimes by the pharmaceutical companies themselves. This high-tech industry was now anxious to consider the previously scorned indigenous knowledge of local flora and fauna—not least, because the relevant species were rapidly disappearing from the rainforests.

Whether such research involved exploitation, deceit, political knavery, and worse became a topic of controversy worldwide at the millennium. For at that time Patrick Tierney published his outraged—and outrageous—book *Darkness in El Dorado* (2000).

He charged two anthropologists who'd worked with the Yanomami in the Amazon forests—James Neel (1915–2000) and Napoleon Chagnon (1938–)—with all the sins mentioned above. He even added deliberate genocide, claiming that Neel and Chagnon had knowingly caused "hundreds, perhaps thousands" of deaths through improper use of a measles vaccine. Most scandalous of all, this was said to be not a tragic mistake but a deliberate experiment to test Neel's "fascistic eugenics".

Tierney wasn't chasing minnows. Chagnon was described by the University of California's (Santa Barbara) Center for Evolutionary Psychology as "perhaps the world's most famous living social anthropologist" (Tooby 2000*a*). Neel had recently died, as had Tim Asch, the ethnographic film-maker attached to the team. But he'd been a physician and "a founder of modern medical genetics". (And, for the record, a fierce opponent of eugenics for sixty years: Tooby 2000*b*.)

This caused "the most sensational scandal to emerge from academia in decades" (Tooby 2000*b*). Tierney's accusations were vociferously spread by many anthropologists, including some officials of the AAA—which had known about them from the early 1980s. But they were just as hotly disputed by others. John Tooby, in his role as President of the Human Behavior and Evolution Society, checked them thoroughly and found them to be utterly "fraudulent" (Tooby 2000*b*). Admittedly, he was *parti pris*, since Chagnon was a leading member of the Society. In the end, however, they were officially declared groundless in a report published by the University of Michigan, Neel's former employer (Cantor 2000).

Moreover, the AAA admitted that they'd been "unable and unwilling . . . to address them in a fair and open manner" (Albert *et al.* 2001). (They didn't actually rescind their criticisms, however: in February 2005, yet another attempt was made by a few AAA members to make them do so.) Meanwhile, they'd split the anthropological 'community' right down the middle. Or rather, since the two sides were by that time highly unequal in terms of numbers, it accentuated the pre-existing gulf between the 'scientific' (marginal) and the 'interpretative' (mainstream) anthropologists: see Section ii.a–c below.

The specialist ethnographers who did the wide-ranging post-Berlin research on cultural categories weren't the only anthropologists to develop an interest in patterns of concepts. For it became increasingly evident that the various classifications found within a given culture overlap significantly.

Computational explanations were given for this. It had been suggested in the early 1980s that, for reasons of informational economy, similar concepts are likely to be used for classifying both human social relations *and* non-human domains (B. N. Colby *et al.* 1980; Ohnuki-Tierney 1981). Today, there's a rich store of ethnographic data, concerning both lexical codes and 'wordless' cultural practice, to support that view. As Roy Ellen (1947–), a cognitive anthropologist at the University of Kent, has recently said:

> [All] human populations apprehend the social in terms of the natural world and the natural in terms of metaphors drawn from the social world. The two are intrinsically complementary, although in certain neurological pathologies they may conflate in unusual ways [see Chapter 7.vi.d–e] . . . The classificatory language we use for plants and animals is derived from the way we talk about genealogical relations, and we understand the functional dynamics of both organisms and ecological systems in terms of our experience of participating in social systems . . . (Ellen 2003: 50)

The topic of this book, *mind as machine*, is of course no exception. Indeed, Ellen immediately went on to point out that "technology provides many productive analogies", one of which is to see "the brain [or the mind] as a computer".

e. And modelling, too

Seeing the mind/brain as a computer became even more attractive for cognitive anthropologists in the mid-1980s, when PDP (parallel distributed processing) connectionism entered the picture (see Chapter 12.vi). For this methodology offered a way of talking about concepts without having to talk about language (grammar), and without having to adopt static, cut-and-dried definitions. Rosch's "fuzzy" concepts and "central" prototypes could be modelled at last—and with some (limited) neurological plausibility.

Maurice Bloch (1939–), at the London School of Economics, was one of the first anthropologists to welcome connectionism. He argued that it was better suited than GOFAI to model culture-specific categorizations, which typically happen very quickly:

> Since much of culture consists of the performance of these familiar procedures and understandings connectionism may explain what a great deal of culture in the mind–brain is like. It also explains why this type of culture cannot be either linguistic or "language-like". Making the culture efficient requires the construction of connected domain-relevant networks, which by their very nature cannot be stored or accessed through sentential logical forms such as govern natural language. (M. E. F. Bloch 1991: 192)

Besides rebutting the symbolist school in anthropology, who'd assimilated cultural knowledge to linguistic codes, Bloch was also defending the traditional method of participant observation. In this approach, "knowledge" (of foreign categories, for instance) is picked up by the visiting anthropologist more or less effortlessly, and typically without the informants' being able to justify their own judgements explicitly. Similarly, he said, connectionist learning doesn't depend on verbal (programmed) instruction, but on the system's tacit recognition of statistical patterns in the input examples.

But it wasn't all about connectionism, for Bloch also approved other ideas from cognitive science. In particular, he picked up on "mental models" and schema theory: the spirit of Bartlett—accompanied by those of Kenneth Craik, Robert Abelson, Marvin Minsky, Michael Arbib, and Philip Johnson-Laird—was hovering in the background. Although he criticized psychologists for downplaying the sociocultural and non-linguistic aspects of "meaning", Bloch unambiguously recommended cognitive science to his colleagues (M. E. F. Bloch 1991, 1998).

That shouldn't have been too shocking. After all, Bartlett had been dealing with 'anthropological', cross-cultural, material (see 5.ii.b). And Abelson's IDEOLOGY MACHINE had been an attempt *to distinguish and also to unify* distinct (political) mini-cultures within the USA in formal–schematic terms (7.i.c).

Schema theory and connectionism were soon used to model fieldwork done in a community closer to 'home' than usual. Claudia Strauss and Naomi Quinn, in an SPA meeting held in 1989 and in their later book (1997), provided original ethnographic data on contemporary American understandings of *marriage* and *success*. These data were considered in a new way, well fitted to describe shifting-yet-stable cultural meanings. Specifically, they were analysed in terms of schema theory, connectionism, and distributed cognition ("shared task solutions").

Quinn (1987, 1991) had earlier found that all the many metaphors used by her hundreds of informants fell into one of eight classes: sharedness, lastingness, mutual benefit, compatibility, difficulty, effort, success/failure, and risk. The relations between those classes (e.g. the potential contradictions between lastingness and mutual benefit) guide Americans' reasoning about marriage. It's guided also—and complicated—by three different ways of thinking about individuals:

* as a human being, with rights, obligations, and capacities;
* as a bearer of a social role, such as spouse, parent, or teacher;
* and as someone with personal characteristics, such as generosity or idleness.

Further complexities are added by the varying emotional commitments to the eight features on the part of different individuals (and of the same individual at different times). In short, there are multiple, and sometimes mutually conflicting, constraints to be borne in mind.

It's no accident, then, that when PDP connectionist models became widely known, in the late 1980s, Strauss and Quinn chose them as a way of exploring their hypotheses. For PDP is specifically concerned with multiple constraint satisfaction (Chapter 12.v–vi).

This approach could be used to model the thoughts—and emotional weightings—of individual informants (Quinn had recorded many verbatim quotations). It could also be used to explore possible constellations of American 'marriage thinking' in general. In short, connectionism offered a way of testing the authors' hypotheses that *these* schemas, inferential patterns, and beliefs were in play. (The informants themselves weren't usually able to identify them by introspection.)

Another study of shared task solutions, not in marriage but in seamanship, also used connectionist modelling (see Section iii.a, below). Here, PDP was valued for its ability to model not only fuzzy judgements, but also different patterns (and persuasiveness) of communications between one person and another.

By the 1990s, cognitive science—sometimes, with psychodynamic theory—was providing much of the theoretical inspiration for psychological approaches to culture (D'Andrade and Strauss 1992). For instance, Strauss (1992) described how differentially salient marriage-relevant schemas such as *success* and *breadwinner* can direct behaviour in American blue-collar workers. She showed that what directs one worker isn't exactly the same as what directs another, because even people in the 'same' subculture think about their life-world somewhat differently. Accordingly, she argued that anthropologists must recognize semantic organization at the personal (as well as the cultural) level.

At much the same time, at Emory University, Georgia, cognitive science was being called on to provide yet another definition of "culture". Shore (1945–), a long-time ethnographer of Samoa (Shore 1982), redefined cultures as distinct sets of mental models (Shore 1996). In explaining what he meant by this, he drew inspiration from Bartlett's work on schemas (5.ii.b), Erving Goffman's (1959, 1967) descriptions of the tacit conventions of everyday life, Roger Schank and Abelson's concept of scripts (7.i.c, 9.xi.d), and Johnson-Laird's mental models (7.iv.d–e).

It's mental models, said Shore, which distinguish the construction of meaning from mere information processing (1996: 157). They cover every aspect of cultural life, whether linguistic or non-linguistic. So he specified examples at several different scales of experience, running from bodily postures and table manners all the way to myth, marriage, and religion. They weren't seen as 'purely' cognitive. To the contrary, they were said to underlie, and help construct, the specific motivations and emotions typical of the culture concerned. (In that sense, they resembled William McDougall's notion of *sentiment*: see 5.ii.a.)

Samoan *alofa*, for example, is an attachment emotion whose socio-psychological functions differ significantly from those of Westerners' *personal love* (Chapter 7.i.f). For instance, it's *alofa*—according to what the Samoans themselves say—which motivates them to adopt relatives' children, and even to give up their own babies willingly for adoption by kinfolk. (My own Polynesian friends, in the Cook Islands, often speak of someone's "birth mother" and "feeding mother".) This widespread Polynesian custom may have evolved, as functionalist anthropologists would put it, to distribute scarce resources efficiently. But the point here is that the concept, or mental model, of *alofa* plays an explanatory and motivational role that's intriguingly different from the Western concept of personal love.

Similarly, Samoan grief at the death—specifically, the murder—of a loved one is different from ours, though recognizable as grief (Shore 1982, ch. 1). Samoan 'apology' or 'reparation' for a murderous act on the part of a family member is not what one would see in the West. And Samoan anger at a loved one or authority figure is different from ours. As in other parts of Polynesia, it's not socially acceptable for such anger to be expressed directly. Moreover, Eleanor Gerber (1985) has shown that it's not conceptualized or experienced in the same way that 'Western' anger is. (Gerber's work is a fascinating example of the 'ethnopsychology' called for in the 1950s by Hallowell: see above.)

In Shore's terminology, the mental models of attachment and anger that are prevalent in Samoa differ from ours. Individuals in a given culture share a particular set of mental models. The cultural community may be a Samoan village (the focus of his earlier work), a Hopi or Navaho gathering, or a US high school or baseball team. The last four examples were among the communities studied by his Center for Myth and Ritual in American Life (MARIAL), set up at the millennium and funded—like so much else in cognitive science—by the Alfred P. Sloan Foundation. Like Strauss and Quinn, then, he was now looking close to home for much of his ethnographic data.

As well as studying specific cultures or subcultures, Shore *compared* them, in terms of the extent to which their members share the same mental models. He even claimed that *a common biological context* underlies emotions (such as love and *alofa*) conceptualized—and practised—by different cultures in different ways. In short, he was highly unorthodox.

He'd already criticized the dominant (Geertzian) anthropological approach for being too "particularistic": its methods of *local* interpretation, he said, can't show the extent to which meanings are, or aren't, cross-culturally shared (Shore 1988). If there is any universal base to human cultures (a suggestion firmly avoided by most of his fellow professionals, as we'll see), these methods wouldn't find it. At the millennium, in an essay on globalization (Shore 2000), he not only made many cross-cultural comparisons but committed the cardinal sins of rejecting cultural relativism, and of criticizing certain aspects of some non-Western cultures. He even said, in opposition to the sort of 'identity' politics that perfused the simultaneously published *Darkness in El Dorado*, that some of these aspects aren't worthy of preservation, and would be better lost.

Just how unorthodox this was will become clear in Section ii.b–c. Here, it's worth noting that Shore's plea for a more scientific anthropology—note his 1988 title: "Interpretation Under Fire"—didn't spring from a lack of knowledge or sympathy. His BA had been in English at Berkeley, and his early ethnographic work (based on a two-year stay in Samoa undertaken even before he turned to academic anthropology) was done in an interpretative spirit (Shore 1982). Moreover, he was deeply influenced by Bruner, who moved towards the interpretative side of anthropology in the last quarter-century (6.iv.d).

But Shore's interpretations, and his teaching, were based on huge amounts of data:

> [The students were expecting] a lot of anecdotes about my personal experiences, and then a vague account about empathetic insights, what Weber called *Verstehen*, that comes from just living with people. Interpretation as a kind of intuitive speculation. Well there is that of course, but what I brought in were piles of genealogies, demographic data, detailed data on political titles and their holders in a Samoan village, hundreds of questionnaires and the data analyses that resulted from them and of course thousands of pages of interview transcripts. I just went on and on until the students looked at me completely baffled and said, "This is symbolic anthropology?" And I said "Well", I said, "You know, symbolic anthropology does have data, and because of the problems of interpretation and the ease with which we can read into things—I mean, there's no question that human beings with a great imagination can read any interpretation they want into virtually any text—we hold ourselves all the more accountable to data. (Shore 2004: 72)

In addition, he insisted that we sometimes need what Bruner called "the recognition of the right answer versus the wrong answer". To be sure, Samoan *alofa* and Western *personal love* form part of different discourses, which can't be neatly mapped onto each other. But anthropology needs *theories* as well as "conversations":

> Rather than trying to sort of resolve disagreements, the tendency [among anthropologists] is to reframe the issues as part of different discourses, and wall them off from one another. [... This] strategy leads to a kind of rejection of theorizing itself... It's one thing to show a kid that her wrong answer is the right answer to another question, and quite another to suggest that all issues of disagreement are really simply to be handled by a kind of post-modern bunker mentality in which each question is simply dismissed as part of a different discourse. At an earlier time we would have said that we were simply playing different [Wittgensteinian] language games. More recently we use Richard Rorty's image of simply different conversations. They all share the notorious post-modern resistance to meta-narratives [theories], or in a way to going meta [generalizing] at all. (p. 154; cf. 1.iii.b)

Shore, then, had a respect for "scientific" methods (and even for "scientific" explanations) as well as hermeneutic ones. In fact, he was a rare case of someone sympathetic to both sides of the intellectual divide within the profession (see ii.a–c, below.)

(Another rare case is the philosopher of science Nicholas Maxwell, whose account of science as "wisdom" provides a strong counterpoint to value-blind scientism: 1984, 2004. Significantly, he considers science as a whole—his own scientific expertise is in quantum physics. If the self-styled scientific anthropologists had been familiar with his work, they might have been able to deflect some of the odium of their interpretative colleagues without risking a descent into postmodernism: see Section ii.c, below.)

Science-based discussions were going on *outside* professional anthropology, to explain how cultures are possible in the first place. For instance, the primatologist Michael

Tomasello (1999) cited Theory of Mind and RR theory (representational redescription) as essential prerequisites of "culture" properly so-called (Chapter 7.vi.e and h–i). These mechanisms, he argued, enable humans to acquire, develop, and transmit culture—which non-human animals can't do. Many other evolutionary psychologists (*sic*) were trying to explain cultural phenomena (see Sections ii.d–e and iv–vi, below). Despite many disputes about detail, they had the support of their professional peers. Shore didn't: he was swimming against the tide.

However, he wasn't the only one. By the 1990s, there'd been enough work in cognitive anthropology to justify a history of it—written by a one-time member of the Sloan Committee (D'Andrade 1995*a*). (There'd also been a survey of computer-based research, but this dealt more with ethnographic data processing and mathematical modelling than cognitive simulation: Fischer 1994.)

To cap it all, the relevance of computational theorizing—and of computer models—was confirmed in a mid-1990s encyclopedia article. Colby, a pioneer of culture grammars in the early 1970s (B. N. Colby 1975), advised intending cognitive anthropologists to study various AI skills. These included knowledge representation, NLP, connectionism, and symbolic logic (B. N. Colby 1996: 214–15).

Today, encyclopedia articles on this topic may mention a recent addition to the AI-modelling stable: agent-based distributed cognition (see Chapter 13.iii.d–e, and cf. Section iii.b below). One such, written by David Kronenfeld (1941–), the first anthropologist to try swimming in these waters (Kronenfeld and Kaus 1993), concludes by saying that "agent-based computational models seem to offer one very promising method" for understanding culture, and especially the relation between individual and shared schemas (Kronenfeld 2004*a*). Another, by the same author (2004*b*), also ends by recommending computer models—and mentions the online publication *Journal of Artificial Societies and Social Simulation.*

Known as *JASSS*, this was begun in 1998. Another online journal in this area, founded just a year later, is *MACT*: *Mathematical Anthropology and Culture Theory*. *MACT* has organized sessions on 'Cultural Systems' and 'Cognitive Clarity' at Vienna's European Meeting on Cybernetics and Systems Research since 2002, and many *MACT*-initiated papers appear in the EMCSR journal *Cybernetics and Systems*.

Apart from the online journals (for no one could have predicted the Web), that was just what the early cognitive scientists would have expected. For links with both AI and cybernetics had been on the cards since the beginning, as we've seen.

Maybe that counts as a happy ending. But the intellectual journey that led there was far from happy. Meeting in a smoky Viennese café during their first EMCSR conference, the *MACT* editors helped launch both the Salon des Refusés and SASci (Paul Ballonoff, personal communication)—of which, more below. Neither of these would have been needed if things had gone to plan. In other words, if the expectations implicit in the Sloan hexagon had been fulfilled there'd have been no Salon, and no SASci.

8.ii. Why Invisibility?

One might imagine, given the history sketched above, that anthropology and cognitive science hopped into bed together in the late 1950s and remained happy bedfellows ever after. They didn't. Hopes were high at the beginning, but as time went by the two were mentioned in the same breath ever more rarely.

Why? How did anthropology come to be described by a leader of cognitive science as a mere "trace substance"? What happened, to make it vanish?

There are two answers. On the one hand, the anthropologists themselves—well, most of them—decided to give the cold shoulder to psychology in general, and cognitivism in particular. On the other, much of the relevant research was renamed—and the new label didn't include the word "anthropology".

These changes were due to two very different intellectual movements of the late twentieth century, each of which spread way beyond anthropology. The first started in the 1960s and blossomed in the early 1970s. The second came soon afterwards. By the end of the century, cognitive anthropology was travelling in the shadows, largely incognito.

a. Psychology sidelined

Anthropology's current invisibility can be superficially explained by the fact that cognitive anthropologists remained rare beasts. The history of the field mentioned above reported that "most of the work has been carried out by a shifting core which has never been larger than about 30 persons" (D'Andrade 1995*a*, p. xiv). If more British names had been included, the number would have risen—but not much. Evidently, departments of anthropology weren't good breeding grounds for this species of animal.

Few university researchers means few students to pick up the baton. So in the absence of a critical mass, cognitive anthropology couldn't flourish. But why was there no critical mass? The excitement in other areas of cognitive science was enormous, so why not in anthropology too?

The reason is indicated by two key chapter titles in Strauss and Quinn's end-of-century book. This contained a defensive early chapter on 'Anthropological Resistance' and an optimistic final chapter called 'Beyond Old Oppositions'. Clearly, the authors felt that they were walking through a minefield.—And it was their anthropology colleagues who'd laid the mines.

That is, the profession's accepted career path had taken a very different route from the one endorsed by D'Andrade and his fellow authors of the Sloan Report. In the last quarter of the century, most anthropologists had little sympathy with their cognitive colleagues. They didn't want anthropology to be part of Donna Haraway's "cyborg science" (Chapter 1.i.b). Indeed, they didn't want it to be seen as intellectually dependent on (individualistic) *psychology* of any sort. That being so, studies of folk taxonomies (for example) went out of favour. As for cognitive science, anthropologists in general had no interest in being part of an interdisciplinary endeavour in which psychology was a crucial dimension.

The flight from psychology began in the early 1970s, and was led by Geertz (since 1970, at Princeton's Institute for Advanced Study). He scorned what he called "the cognitive fallacy", the idea that "culture consists of mental phenomena" (Geertz 1973*b*: 12). Goodenough's definition of culture as knowledge ("whatever it is one has to know or believe in order to operate in a manner acceptable to its members") was specifically rejected (Geertz 1973*b*: 11).

As Geertz put it to his biographer Fred Inglis (a scholar not of anthropology, but of English literature):

> I want to do more, attempt to use some of the techniques that literary critics use, that historians use, that philosophers use, to explicate cultural matters. (Inglis 2000)

The "philosophers" he had in mind weren't the analytically minded philosophers of cognitive science, but the neo-Kantians—including the postmodernists (Chapters 1.iii.b–d, 2.vi, and 16.vi–viii). Since Geertz fell within the long-standing hermeneutic tradition of social science (Sherratt 2005), his preferred method was "interpretation":

> The concept of culture I espouse . . . is essentially a semiotic one. Believing, with Max Weber, that man is an animal suspended in webs of significance he himself has spun, I take culture to be those webs, and the analysis of it to be therefore not an experimental science in search of law but an interpretive one in search of meaning. (Geertz 1973*b*: 5)

To go "in search of meaning" was to try to give a "thick" description of someone's behaviour (a term he borrowed from Gilbert Ryle: 1949). A thick description reports (for example) not a movement of a hand or an eyelid, but a greeting, or an expression of complicity.

Moreover, a wink or wave can mean very different things in different cultures: knowledge of *local* meanings is what's required. So Geertz insisted that reliable interpretation could result only from cultural immersion. The anthropologist, he said, must be a *participant* in the culture, not an *observer* of it. Most cognitivists were (and remain) appalled: "we deplore the rejection of explanation in favor of interpretation in anthropology at present" (Strauss and Quinn 1997: 9, 259).

Instead of calling for the fruitful collaboration between psychology and anthropology which the early cognitive scientists had hoped for, Geertz declared cognitive psychology to be *persona non grata*. Or rather, psychology *considered as the study of the processes going on inside the individual's head* was dismissed. Geertz's position had nothing to do with hostility to computers. (Because of their position in the science wars, outlined below, interpretative anthropologists would naturally be suspicious of computational psychology; but in principle that's a separate matter.) It was rooted, instead, in an unorthodox view of the mind.

Specifically, Geertz persuaded his professional colleagues to accept an externalist philosophy of mind (a position which became newly popular in the philosophy of mind itself a quarter-century later: 16.vii.d). He insisted that "mind" is naturally located *outside* the head, in the cultural institutions, conventions, and artefacts with which so-called individual minds are imbued. (As a one-time member of Bruner's Center at Harvard, he was in effect generalizing Bruner's views on "cognitive technologies": Chapter 6.ii.c.)

As he put it: "Thinking consists not of 'happenings in the head' (though, happenings there and elsewhere are necessary for it to occur) but of a traffic in . . . significant symbols" (Geertz 1966: 45). So not only should anthropology not be psychological, but psychology should be anthropological:

[The evolution of humans was very extended, since] the phylogenetic history of man took place in the same grand geological era—the so-called Ice Age—as the initial phases of his cultural history. Men have birthdays, but man does not.

What this means is that culture, rather than being added on, so to speak, to a finished or virtually finished animal, *was ingredient, and centrally ingredient*, in the production of that animal itself. . . .

Most bluntly, it suggests that *there is no such thing as a human nature independent of culture* . . .

[. . . Chartres cathedral can't be understood merely as stone and glass, but only as a creation of a specific religious culture.] It is no different with men. They too, every last one of them, are *cultural artifacts*. (Geertz 1966, sect. iii; italics added)

In other words, cultural variations are constitutive of human minds, not superficial additions to them.

Many cognitive scientists would have rejected Geertz's "most blunt" expression (that there's "no such thing" as a human nature independent of culture). But one didn't have to be a postmodernist to agree with him to a large extent. The cognitive anthropologist Shore refers to "Geertz's profound but elementary insight", and puts it like this:

The study of human nature minus culture does not produce a more basic understanding of the human but an understanding of a *protohuman*, a creature that is all bioessence but *lacking recognizable qualities of human existence*. (1996: 33; italics added)

He deliberately speaks of "culture in mind", to suggest "both an ethnographic theory of mind and a cognitive theory of culture" (1996: 13). In other words, Shore sees Geertz as having gone too far, posing a false dichotomy between "private" mind/psychology and "public" culture/anthropology (1996: 51). Rather, anthropology is "the study of human nature *through* the study of human difference" (1996: 379).

One example of the way in which Shore applies his notion of "culture in mind" is his recent definition of "techno-totemism" (1996, ch. 6). He argues—like Haraway, but in a less antagonistic spirit—that we have ways of thinking, and even concepts of humanity and of self, that are deeply informed by specific aspects of our culture's technology. Cognitive science itself, of course, is one expression of techno-totemism (cf. also Gigerenzer 1991*b*). Indeed, Shore sees "modularity" as a foundational schema of modernity (1996: 117–18), without which techno-totemism couldn't arise. And by modularity he means functional separation in general, as stressed by Herbert Simon's *The Sciences of the Artificial* (of which evolutionary "modules" are a special case: Chapter 7.vi.d–e).

By the mid-1990s even Bruner himself no longer saw anthropology and computational psychology as part of the same enterprise—and he favoured the former. Although in 1983 he'd "regretted" not having got more involved with computational theorizing (see 6.ii.d), and although he still allows that computational scientists have a role to play in "the study of man" (in Shore 1996, p. xiii), he also complains that AI has "trivialized" psychology (Bruner 1997). The remedy, he says, is to make psychology more like anthropology—and both of them less "impersonal" (2002: 111–12). Specifically, he asks for a "narrative turn", as exemplified by Geertz's interpretative anthropology—and by the Annales school of historians too.

Geertz was recently described by Bruner as "surely . . . the most cited and most often attacked anthropologist of our era" (2002: 118). But he sees him as being on the winning side: "one increasingly widespread view is that culture is as much a prod to the development of human cognition as human cognition is to the development of culture" (p. 52). He'd said that in the 1960s, of course (Chapter 6.ii.c). But his work on cognitive technologies was done in an experimental context. Now, he prefers interpretation to experimentation.

So, too, do most anthropologists. For Geertz's arguments revolutionized the field. That's partly why anthropological linguistics, so prominent in Kluckhohn's vision for the Department of Social Relations, has gone out of fashion—especially in the USA. Chomsky's influence, including his excessive emphasis on English (Chapter 9.vii.d), led to detailed studies of "exotic" languages becoming much rarer in departments of linguistics. And because of Geertz's 'literary' approach in anthropology, they're rarer there too. The focus is less on nouns than on narrative. (In Europe, this effect is less marked. Nijmegen's Max Planck Institute for Psycholinguistics, for instance, set up a "Cognitive Anthropology Research Group" to study neo-Whorfian questions about language and cognition: e.g. D. Hill 1993; Pederson 1993.)

Within anthropology (and other disciplines too), interpretation and experimentation—or anyway, objective observation—were destined to clash. As Jerry Fodor has recently said, in discussing evolutionary psychology:

> Cultural relativism is widely held to be politically correct. So, *sooner or later*, political correctness and [nativist] cognitive science are going to collide. *Many tears will be shed, and many hands will be wrung in public.* (Fodor 1998*c*; italics added)

In anthropology, as we'll see next, it was sooner rather than later. The collision happened years ago, and the tears have already flooded the floors.

b. Skirmishes in the science wars

Gardner, some twenty years ago, had seen the writing on the wall. While the Sloan monies were being generously distributed and thankfully spent, he remarked:

> [Much] of anthropology has become disaffected with methods drawn from cognitive science, and there is *a widespread (and possibly growing)* belief that the issues most central to anthropology are better handled from *a historical or a cultural or even a literary perspective.* (1985: 44; italics added)

He was right. An ill-tempered schism was developing within the profession, between those who were suspicious of scientific approaches, including computationalism, and those who weren't.

Anthropologists had long been used to having colleagues with different interests: the bones or the beads, one might say. But this wasn't a matter of choosing between physical and cultural anthropology. It was a fundamental philosophical disagreement about how to study culture. Specifically, it was a special case of the "science wars" mentioned in Chapter 1.iii.b–c.

Initially, the conflict was played out privately, at anthropological conferences and within university departments of anthropology. Some groups split into two as a result. Even when they didn't, the psychologically minded anthropologists sometimes fled to more welcoming homes (at Harvard, for example, largely to Education).

The battle was especially fierce in America, where many felt that an intellectual commitment to the discipline should feed a practical commitment to 'identity' politics too. (Hence the passion stirred up by *Darkness in El Dorado.*) D'Andrade sees the political radicalization of US campuses in the Vietnam war as a crucial factor in the transformation of anthropology (2000: 221 ff.). The sociologist Adam Kuper (1999) agrees, regarding the US debate as fuelled by hangovers from the civil rights movement and Vietnam. Europeans, he said, with their bitter memories of the two world wars, were less open to a politics celebrating cultural difference, less ready to regard universalist approaches as banal.

(Most non-cognitive anthropologists avoid universalism. But proponents of "reflexive" anthropology do claim that there's a universality in human emotions. For instance, one Westerner mourning his wife said he could empathize with the Ilongot headhunter in the Philippines who told him that "rage, born of grief, impels him to kill his fellow human beings": Rosaldo 1989; cf. Behar 1996. Indeed, he added that *his task as an anthropologist* was to learn to empathize with the 'alien'. Many of his colleagues argued that this approach is self-indulgent and self-deceptive, and obscures important cultural differences. The discussion of grief in Chapter 7.i.f suggests that they're right. Although there may well be some universal emotional core, grief will be experienced—not just expressed—in significantly different ways across cultures with different concepts of love, marriage, loyalty, obligation, and so forth. Indeed, the grief felt by Strauss's American informants will likely differ, to the extent that their concepts of marriage differ: see Section i.e, above.)

The fight didn't stay private for long. When the Stanford department bifurcated in 1998, and also during the rancorous run-up to the fracture, reports from the war zone reached the general public. They appeared in widely read media such as *Science* and *The Nation*, for instance.

This was a sign of the times. The growing disaffection with science, chronicled (and encouraged) by Theodore Roszak's cult book *The Making of a Counter-Culture* (1969), was evident not just in the public arena but in the academy as well. A similar professional split was threatening psychology too, although somewhat less public (and, as it turned out, less damaging: Chapter 6.i.d). Moreover, new currents in the theory of *literature and history* prompted many highly publicized battles on both sides of the Atlantic.

One such battle was triggered by a spoof article written by the physicist Alan Sokal, subtitled 'Towards a Transformative Hermeneutics of Quantum Gravity'. This was innocently published by a postmodernist journal, and reprinted with a hilarious commentary by the author (Sokal and Bricmont 1998, esp. 199–258).

The *casus belli*, here, was the shameless—and ignorant—misuse of scientific terminology in some postmodernist literary/philosophical writings. (If you doubt this, just look at the quotations given in the book.) Specifically, the problem was that these writers were treating physics as what Sperber calls a "symbolic" communication, when in fact it isn't (see Section vi.c, below).

The *New York Review of Books* ran commentaries by the physicist Steven Weinberg and the literary scholar George Levine, among others. And the *London Review of Books* published a desperate attempt by John Sturrock (1998), an author on structuralism and translator of Proust (and the *LRB*'s consulting editor), to defend the indefensible. Sokal, who'd admittedly made some injudicious remarks in his post-publication commentary, was mocked as "le pauvre Sokal". This, said Sturrock, was what Jacques Derrida had called him, with "a seen-it-all-before sigh". Many other protagonists entered the ring.

It was a fierce fight, and a dirty one. The pens were filled not with ink, but venom. (It didn't help that the disagreements fuelling the science wars were political as much as philosophical: see I.iii.b–c.)

Other clashes were rooted in the hostilities between Leavisites and literary theorists at the University of Cambridge. This conflict hit the international presses twice: in the 1980s when the structuralist Colin McCabe wasn't promoted to a permanent post in English, and in the 1990s when an honorary doctorate was offered to Derrida. A letter to *The Times* (9 May 1992) signed by many Cambridge philosophers protested that Derrida's work consisted of "little more than semi-intelligible attacks upon the values of reason, truth, and scholarship".

If that was the version for public consumption in the staid columns of *The Times*, in private the invective was vicious—and it was returned in kind. Anyone even touching on the dispute risked being insulted by one side or the other. (I received an abusive letter from a Cambridge-based fellow Fellow of the British Academy, a scholar of English based in Cambridge, merely for inviting him to take part in an Academy discussion of post-structuralism.)

In that atmosphere, whatever value there was in the positions of either side was ignored, or not even recognized, by the other. If this was true in the context of literary studies, it was also true in the context of science. The fact (noted in Chapter 1.iii.b) that the social constructivists had made some important points about the history and sociology of science was forgotten by those outraged by their relativism. And the more someone subscribed to "the Legend", the more hostile they would be. John Ziman's even-handedness was much needed, but missing.

For instance, Derrida's point that because a text is written in a public language one can (contra John Austin: 1962*a*) interpret it without reference to the intentions of the author, was expressed more provocatively (and much less clearly) than it might have been (Derrida 1977*a*,*b*), and countered more forcefully too: "*not just a series of muddles and gimmicks [but] a large mistake*" (Searle 1977, 1983*a*: 77; italics added).

As for cultural examples, the sociologist Pierre Bourdieu (1977, 1990) likened human life to a "game" (with influential "umpires") whose systematic rules—covering both bodily practices and associated "beliefs"—we don't choose, but are born into. (Hence his scare quotes around *beliefs*.) Also, he highlighted the near-inescapable constraints implicit in social conventions (such as gift giving), and in culturally approved artefacts (such as cutlery). For instance, if someone receives a gift there are—in any given situation, in any given culture—well-understood conventions about what type and value of gift they should give in return, and when that should be done. These constraints aren't strictly *rules*, he said, but psychological mechanisms which have much the same effect:

> The active presence of past experiences . . . deposited in each organism in the form of schemes of perception, thought, and action, tend to guarantee the "correctness" of practices and their constancy over time, *more* reliably than all formal rules and explicit norms. (Bourdieu 1990: 54; italics added)

Such work was potentially relevant to anthropology, but was ignored by the 'scientific' camp. They weren't merely being perverse: the postmodernists' jargon, and their often *deliberate* unclarity, made their insights difficult to find. But the high emotions aroused by the very public hostilities precluded most opponents from searching for them in the first place.

In Cambridge, after a hugely divisive and ill-tempered debate, Derrida's degree was awarded. The neo-Kantians had won that battle. Within anthropology, however, they'd won the war.

c. Top dogs and underdogs

In a mini-history tellingly titled 'The Sad Story of Anthropology 1950–1999', D'Andrade reported the rout of traditional anthropology (including cognitivism) by the postmodernists:

> By the mid-1980s, critical anthropology had become mainstream. [Its goal] was to critique hidden and open oppressions of Western bourgeois culture: its racism, sexism, nationalism, homophobia, *and scientism*. The Enlightenment . . . came to be seen as a well of poison . . . The World Bank and the International Monetary Fund were enemies, *science was an enemy*, and *rationality was a destructive force*. (D'Andrade 2000: 223; italics added)

In the USA, the cognitivists "struggled" to get their work included in degree programmes in anthropological theory and history—and that's still true today (Strauss and Quinn 1997: 255). Across the pond, it's taught in several UK and Scandinavian universities (the latter, thanks to the influence of Fredrik Barth at Oslo's Ethnographic Museum: T. Hall, personal communication). But the degree in Anthropology with Artificial Intelligence at Sussex exists no more (see Chapter 6.iv.e). Initiated in 1975, it was withdrawn in the early 1980s because the interested students were increasingly being dissuaded by their Anthropology tutors from taking it. And there's still only one M.Sc. in the UK dedicated to cognitive anthropology, founded by Bloch at LSE. So although Bloch's 1991 paper (see Section i.d) was accepted by—or anyway, published in—a leading professional journal, he's still outside the mainstream.

The victory was officially recognized in the early 1980s, when the anti-science faction gained the ascendancy in the all-powerful AAA. The 'scientists' bitterly accused them of mounting a takeover, and peppered their communications with searing criticisms of the party line. But to no avail.

At the AAA's business meeting in Atlanta in 1994, a lengthy motion attacking the postmodernist hegemony was presented by Berlin and Romney, among others. It began:

> For the last ten or fifteen years, anthropology has been engaged in a struggle for its disciplinary life. Established notions of scientific evidence and scholarship have been jettisoned as unwarranted baggage in anthropology departments across the country. Post-modernist ethnography, post-processual archaeology, and cultural studies rhetoric have become the dominant canon in much anthropological writing [including, now, our flagship journal the *American Anthropologist*].

It ended by calling for the journal's editorial policy to be re-examined, so as to include "the diversity of approaches that have always been the hallmark of our discipline". The motion failed, by 112 votes to 86.

The following year saw a debate on 'Objectivity and Militancy' between D'Andrade (1995*b*) and Nancy Scheper-Hughes (1995), a prominent feminist anthropologist at Berkeley. D'Andrade presented "a defense of objectivity and science", saying that ethical/political commitments (like those which would underlie the *Darkness* controversy a few years later) aren't a matter for anthropologists as professionals, but as private citizens. The widespread anthropological emphasis on "oppression, demystification and denunciation" was fundamentally—and often explicitly (e.g. Rosaldo 1989; Scheper-Hughes 1992)—opposed to an objective approach. It was less an example of anthropology than a transformation of it:

> Originally, I thought these attacks [on traditional anthropology and ethnography] came from people who had the same agenda I did, just different assumptions about how to accomplish that agenda. I now realize that an entirely different agenda is being proposed—*that anthropology be transformed from a discipline based upon an objective model of the world* to a discipline based upon a moral model of the world. (1955*b*: 399; italics added)

This, he argued, was not only inappropriate but self-defeating, since any respectable moral position—and "any moral authority that anthropologists may hold"—must rest on an objective understanding of the world. Moreover, the current moral model was ethnocentric and even colonial in its vision of "evil" as a lack of equality and freedom: "In my opinion these are not bad values, but they are very American. These are not the predominant values of [many modern-day societies]."

D'Andrade's defence of objective anthropology appeared, alongside an unrepentant riposte from Scheper-Hughes, in a leading professional journal. So did Stephen Reyna's (1994) and Charles Lindholm's (1997) similarly aimed salvos. Ernest Gellner (1988) addressed a more general audience, but backed up his journal article with a characteristically punchy little book waving the banner for "reason" (1992). None of these defences took Ziman's measured stand on postmodernism versus the Legend: the odds were far too high for that.

The journal editors were more accommodating to the profession's underdogs than were the officials of the AAA. In 1997–8 an attempt was made to initiate a "Computational Anthropology" Section within the AAA. This was thwarted, and the web pages were deleted by the 'parent' Anthropology Department at UCLA. (This was largely because another section imagined that their territory was being threatened: "Their mission was to study people who used computers. Ours was to use computers to study people"—Nick Gessler, personal communication.)

Taking things outside the family onto the public stage, neo-Kantian anthropology was criticized at length in a stand-alone book (Kuznar 1997). It was also attacked in papers included in widely read collections on the science wars (Gross and Levitt 1994/1998; Gross *et al.* 1996). Predictably, these hostile writings, like D'Andrade's contribution to the debate in the peer review journal *Current Anthropology*, cut no ice with the opposition.

The new millennium saw the 'scientists' fighting back once more. Much as the Impressionists in Paris had mounted the Salon des Refusés in 1863, because they couldn't get their canvases hung in the main exhibition, so in November 2002 the disaffected anthropologists held a Salon des Refusés (they borrowed the name) in New Orleans. This new Salon, too, had a European provenance, having been mooted in Vienna by the *MACT* editors and colleagues six months earlier. It was held because the AAA had rejected several of their suggestions for symposia, including two on distributed cognition (Section iii below).

This wasn't a matter of spurning substandard papers by individuals, but of ruling entire topics and/or methodologies out of court. The reason was the holism and anti-psychologism of neo-Kantian approaches. Most of the profession had no time for the notion of distributed knowledge, which—thanks to PDP connectionism—had entered cognitive anthropology in the 1980s. As D'Andrade had pointed out a few years before the Salon:

> questions concerning . . . distribution [of knowledge] did not even make sense within much of the mainstream framework of cultural anthropology. One has to have a notion of separate units before the study of their distribution has any meaning. (D'Andrade 1995*a*: 247)

The Salon was advertised as being "concurrent with the AAA conference". A press release announced the foundation of a Society for Anthropological Sciences (*sic*), or SASci, and complained of "the anti-science bias of much contemporary anthropology". The founders hoped that their Society would be accepted as a special-interest group within the AAA, but were prepared to make it a breakaway group if necessary.

Meanwhile, one of their number—Nick Gessler of UCLA—had initiated the so-called Lake Arrowhead conferences: annual meetings on 'Human Complex Systems', the first being held in 2000. The Call for Papers for the 2003 conference invited submissions on (for example) agent modelling; artificial societies and artificial cultures; simulating social intelligence; and emergent social structure. It could have been mistaken for the CFP for an A-Life get-together, or even one on distributed processing in 'straight' AI. (It probably will be—if so, it may be another example of renaming leading to invisibility: see below.) In fact, the primary motivation was to advance anthropology.

Lake Arrowhead was an oasis of sanity, in the opinion of SASci members. The draft manifesto for SASci was part-written by Gessler, and distributed on the cognitive anthropologists' email list. It called for anthropologists to seek general principles, whether psychological or not: detailed fieldwork on individual societies was necessary, of course—but as initial data for scientific explanation, not as a final 'story'. And besides stressing SASci's commitment to scientific method, it declared: "We expect that effective computer simulation of cultural models will eventually lead to more rigorous mathematical representations."—*Anathema!*

If cognitive anthropologists are today near-invisible within the AAA, it's not too surprising that they're barely visible to their fellow cognitive scientists. Indeed, their discipline as a whole will very likely be scorned: one well-known experimental psychologist recently complained of "a clearly postmodernist atomization and often trivialization of 'the human science' " (G. Mandler 1996: 26).

As for people outside cognitive science, they can't be expected to place anthropology in it. After all, most of the anthropologists they know, or know about, wouldn't want to be there.

d. What's in a name?

Despite the opposition from the AAA, cognitive anthropology grew through the 1980s–1990s. A few examples, such as the research of Strauss and Quinn, were mentioned in Section i.d; another will be discussed at length in Section iii. But many more developed in a way that left anthropology largely unseen. The reason, in two words, was *evolutionary psychology*.

The last two decades of the century saw an explosion of research in this new field. In general, it focused on the information-processing mechanisms underlying animal and human behaviour (see Chapter 7.vi.d–f). Some of these helped to explain our cognitive abilities, some our social life, and some were relevant to culture as such (Buss forthcoming). One and the same mechanism could even affect all three, as we'll see. Indeed, culture's being "given its due" was one of the five 'Good Thing' bullet points listed by the linguist Steven Pinker (1954–) in *The Language Instinct* (Pinker 1994: 411).

The "explosion" wasn't just in research, but occurred in the publicly accessible media too. Over the next decade, Pinker made continuing efforts to bring this approach to a wide audience. His books on *How the Mind Works* (1997) and *The Blank Slate* (2002) were highly readable paeans to evolutionary psychology. He wasn't the only one. Henry Plotkin, Professor of Psychobiology at UCL, published a culture-centred introduction to the field that reached far beyond dusty library shelves (1997). Lionel Tiger—Charles Darwin Professor of Anthropology at Rutgers—was even more visible. From 1998 on, he wrote on a wide range of cultural topics in media such as *The New Yorker*, *The Wall Street Journal*, and New York's *Daily News* (Tiger 2003). And a non-technical book by two leaders in the field—one based in an anthropology department—is being readied for the press as I write (Boyd and Richerson forthcoming). Its elegantly written chapters should put evolutionary anthropology (*sic*) firmly on the public's map.

One shouldn't imagine that the newborn evolutionary psychology was primarily concerned with 'cultural' evolution. It wasn't. A few people were interested from the start in the evolution of culture as such: development in science, technology, or religion, for instance (see Sections v–vi). But most weren't. At first, there was more interest in how cultures are made possible than in how they change.

Even so, more evolutionary psychologists focused on mechanisms relevant to *sociality* than to *culture*. The universal interest in gossip, for instance, or the disapproval of cheating, were said to be mechanisms evolved to aid the efficiency of social groups (Barkow 1992: 627–31; Cosmides 1989; Cosmides and Tooby 1992). The more culturally relevant examples included mechanisms said to underlie aesthetic preferences of various kinds (Section iv), and others said to give rise to religion (Section vi).

In any event, the name of the game (literally) was *evolutionary psychology*. This new label didn't mention anthropology. And what is, in effect, the *MITECS* anthropology chapter is called 'Culture, Cognition, and Evolution'. "Culture" had survived, but "Anthropology" hadn't—*even though* the chapter was written by two anthropologists (Sperber and Hirschfeld 1999).

In short, the older discipline was largely hidden within the younger one. The new terminology had caused it to vanish.

e. Barkow's baby

Things could well have happened otherwise. The anthropologist Jerome Barkow (1944–), an expert on West African societies based at Canada's Dalhousie University, had called for a 'Darwinian Psychological Anthropology' in a leading anthropological journal as far back as the early 1970s (Barkow 1973).

That had been a daring thing to do: Darwinism wasn't generally favoured in the social sciences. That was mostly for philosophical reasons (Winch 1958). Also, many associated it with the horrors of Nazi Germany (Campbell 1965: 25). Indeed, an equally daring *psychologist*, Donald Campbell, had recently complained about the "overwhelming" rejection of evolutionary theories of culture (Campbell 1965: 19). Anthropologists' suspicion was strong, as detailed in a paper published on the centennial of the *Origin of Species* (White 1959*a*). The few brave souls willing to mention Darwin and anthropology in the same breath weren't easily heard (e.g. White 1959*b*; Sahlins and Service 1960).

However, Barkow's approach seemed less outrageous to his colleagues then than it would some years later. For the virulent anti-science movement in anthropology (described above) hadn't yet got off the ground. And Geertz was only just setting out on his anti-psychology crusade. So Barkow's approach persuaded some of his readers. Indeed, the 'evolutionists' were able to mount no fewer than two symposia at the AAA meeting of 1976 (Chagnon and Irons 1979). No need for a Salon des Refusés, yet.

But Barkow's suggested label was a mouthful. If he'd abbreviated the fourteen syllables to a punchy "DPA", it might have caught on. In that case, his home discipline would be less invisible now.

(To say that the label might have caught on, among those who were interested, isn't to say that the activity might have caught on, among anthropologists in general. Thanks to the science wars, there was no chance of that. As much as twenty years later, a 'scientific' anthropologist remarked, with admirable understatement, that "in anthropological theory, the very mention of evolutionary constraints is bound to trigger some hostility, not always of a strictly rational nature"—Boyer 1994: 15.)

Besides foreseeing a Darwinian anthropology in the early 1970s, Barkow did a great deal to initiate it in the early 1990s—albeit under another name. He was the senior editor of what became the seminal publication in the field of evolutionary psychology: a volume of specially commissioned papers on *The Adapted Mind* (Barkow *et al.* 1992).

Like MGP's manifesto for computational psychology (Chapter 6.iv.c), this manifesto for evolutionary/computational psychology was partly prepared at the Stanford think-tank. Subtitled 'Evolutionary Psychology *and the Generation of Culture*', it introduced many non-specialists to evolutionary psychology—and to some of its cultural implications. Indeed, Barkow wasn't the only anthropologist in the driving seat. One of his two co-editors, Tooby, was an anthropologist also, as we've seen. Only Leda Cosmides (Tooby's wife) was officially a psychologist. However, the content of the book showed how misleading these disciplinary distinctions had become.

The three co-editors made a point of allying evolutionary psychology with cognitive science:

* Their 'statement of intent' repeatedly assimilated the two (Tooby and Cosmides 1992: 64–9, 94–108, 112–14).
* In their opening pages, they praised cognitive scientists for adding rigour to psychology, and for asking *how* things happen in human minds (Cosmides *et al.* 1992: 7–11).
* They adopted Jerry Fodor's talk of "modules" (7.vi.d).

* They claimed to be following Marr's advice about abstract "computational" explanation (7.iii.b and v.b–d), since (they said) evolutionary psychology bases its hypotheses about cognitive mechanisms on the *environmental constraints* faced by our hunter–gatherer ancestors and our cultural forebears.
* And, tacitly calling on Sperber's work on relevance (7.iii.d, and v.d below), they stated that:

> evidence about the structure of memory and attention [in general] can help cultural anthropologists understand why some myths and ideas spread quickly and easily while others do not. (Barkow *et al.* 1992: 12)

In other words, they deliberately blurred the boundaries between psychology and anthropology—which Barkow himself had done in his DPA paper. This was even more of a professional heresy in 1992 than when Barkow first suggested it. Most late-century anthropologists agreed with Geertz that

> The main source of theoretical muddlement in contemporary anthropology is [the view that] culture [is located] in the minds and hearts of men.
>
> Variously called ethnoscience, componential analysis, and cognitive anthropology . . . this school of thought holds that culture is composed of psychological structures by means of which individuals or groups of individuals guide their behavior. (Geertz 1973*b*: 11)

In some quarters, however, the heresy spread. SASci members complained that although Geertz's followers were happy to speak of the self, meaning, and identity, they weren't prepared to talk about "the psychological processes and structures that help explain these" (Strauss and Quinn 1997: 9). And in his provocative Malinowski Memorial Lecture given in London in 1984, Sperber (based at Paris's CREA, or Centre de recherche en épistémologie appliquée) declared:

> There exists . . . no threshold, no boundary with cultural representations on one side, and individual ones on the other. Representations are more or less widely and lastingly distributed, and hence more or less cultural. (Sperber 1985: 74)

He even offered *psychological* reasons why certain representations rather than others are "widely and lastingly distributed" (see Section v.d).

Barkow's colleagues accepted the traditional concept of "culture", which many of their contemporaries had spurned. But they decomposed it into three types, defined—provocatively—in terms of different sorts of *information processing* (Tooby and Cosmides 1992: 121). As they saw it:

> Culture is not causeless and disembodied. It is generated in rich and intricate ways by information-processing mechanisms situated in human minds . . . To understand the relationship between biology and culture one must first understand the architecture of our evolved psychology. (Cosmides *et al.* 1992: 3)
>
> [Our] central premise . . . is that there is a universal human nature, but that this universality exists primarily at the level of evolved psychological mechanisms, not of expressed cultural behaviors. On this view, cultural variability is not a challenge to claims of universality, but rather data that can give one insight into the structure of the psychological mechanisms that helped generate it. (p. 5)

It followed from this that "the socially constructed wall that separates psychology and anthropology (as well as other fields) will disappear" (Tooby and Cosmides 1992: 121).

In arguing that anthropology and psychology were—or anyway, should be—closely related, Barkow wasn't saying anything he hadn't said twenty years before. He'd written his DPA paper partly because he'd been trained as an interdisciplinary animal (in Chicago's "Committee on Human Development" programme), so was wary of Chinese walls.

Moreover, he'd always used ideas from cognitive science. In the early 1970s he'd posited internal representations mediating social status (Barkow 1975, 1976). Later, he did so in respect of gossip too (1992: 627–31). And he'd explained culture and consciousness in terms reminiscent of MGP, speaking of goals and sub-goals, plans and sub-plans, and codes and sub-codes—at base inborn, but hugely varied and complicated in acculturated human adults (1989).

As a corollary, Barkow had always insisted on the "unity" of science. He didn't mean this in the logical positivists' sense: reducibility to physics (see Chapter 9.v.a). Rather, he meant that explanations at various levels should be *compatible* even though they're not *reducible* (Barkow 1983, 1989). His theory of consciousness, for instance, accommodated anthropological, psychological, neurological, and evolutionary perspectives.

Within cognitive science, the need for such inter-level consilience is a platitude. But it's not so regarded throughout anthropology, as we've seen. With respect to the mainstream of the profession Barkow was, and remains, a maverick.

The same is true of cognitive anthropologists in general. In the eyes of the AAA, they were studying forbidden topics in forbidden ways. The rest of this chapter describes some examples of these professional heresies.

8.iii. Minds and Group Minds

One of the forbidden topics featured at the New Orleans Salon des Refusés was distributed cognition, a concept prominent in cognitive science since the 1980s (Chapters 13.iii.d–e and 12.v–vi). Here, knowledge is represented across many different places, where this involves not duplication (copying) but distribution: i.e. sharing, in the sense of division of labour. The "many places" may lie within an individual brain (or network), or within a group of intercommunicating organisms (or networks, or robots).

Most anthropologists are interested in distributed cognition, though they may call it role assignment, kinship relations, shared/individual cultural schemas, division of labour, or task sharing. But the scientifically minded can now call on computational concepts to help clarify the issues involved. Indeed, we saw above that the online journal *JASSS* has been founded specifically to describe "social" computational theories and simulations.

People working on this topic sometimes speak informally of "group minds", although that term is usually avoided in print. More often, they speak of "the cognitive properties of groups". They see "planning" not as an in-the-head phenomenon, but as a dynamical engagement with the environment, including other people. The anthropologist Lucy Suchman (1951–), for example, described the "interactive" planning she'd observed in the operations room at San Jose airport—an account that would influence research in AI (Suchman 1987; see 13.iii.b).

Similarly, the philosopher John Searle distinguished "collective intentions" from "individual intentions", arguing that the former can't be analysed in terms of the latter. A pass in a game of football, for instance, isn't made up of the kicks/headers of the individual players. He was well aware of the strangeness of such a claim:

> How, one wants to ask, could there be any group behavior that wasn't just the behavior of the members of the group? After all, there isn't anyone left to behave once all the members of the group have been accounted for. And how could there be any group mental phenomenon except what is in the brains of the members of the group? How could there be a "we intend" that wasn't entirely constituted by a series of "I intend"s? (Searle 1990*c*: 402)

His own answer was in terms of a definition of (and notation for) "collective intentionality", which assumed "a preintentional Background of mental capacities that are not themselves representational" (p. 401).

Other cognitive scientists appeal to a background of a different kind, namely the ready-made technologies provided by one's culture (including language as just one example). Bruner had done this as early as the 1960s (6.ii.c). But some of today's writers make this approach the ground of a highly unusual concept of the "individual". Specifically, they *deny* that the enculturated individual's mind, or self, stops at the skull or the skin.

One such denier is the philosopher Andy Clark (Chapter 16.vii.d). Another, whose research on navigation is often cited by Clark, is Edwin Hutchins (1948–). Hutchins is an anthropologist at San Diego. Besides being a colleague of D'Andrade, he's close to San Diego's pioneering connectionist group (Chapters 6.iv.e and 12.vi–vii).

Hutchins has also worked with Norman on interactive interfaces, an interest partly fuelled by his concern for how naval crewmen use the instrumentation aboard ship (E. L. Hutchins *et al.* 1986). Indeed, his account of distributed cognition on ships has recently been generalized into an analytical tool for thinking about the design of workplaces in general, of which interfaces—and human collaboration via computer interfaces—are important examples (Hollan *et al.* 2000; Y. A. Rogers forthcoming). In addition, he has studied the interactions within teams of programmers cooperating on complex software maintenance (Flor and Hutchins 1991). So besides learning from non-anthropological cognitive science, he has contributed to it too (see 13.iii.d–e).

a. Models of seamanship

Hutchins started doing cognitive anthropology thirty years ago, but his chief concerns have developed in an interesting way.

In the mid-1970s, he spent two years studying reasoning strategies in Papua New Guinea. His book on *Culture and Inference* (1980) used schema theory to describe conversations he heard on PNG's Trobriand Islands—primarily, disputes about land tenure. Despite the 'exotic' subject matter and cultural presuppositions, his conclusion was that there's no essential difference between the reasoning of Trobriand Islanders and Westerners. By implication, then, reasoning is a human universal. There was a caveat, however: since psychologists still don't fully understand 'home-grown' reasoning (see

Chapter 7.iv), anthropologists aren't standing on firm ground when they compare it to reasoning in other cultures.

His later work moved beyond language and reasoning, to include material artefacts and non-linguistic behaviour. (Wallace revivified, one might say.) Indeed, it was he who was disillusioned by the linguistic—not to mention the 'literary'—turn when trying to write a survey of cognitive anthropology (see Section i.a, above).

By far the most of Hutchins's post-PNG fieldwork has been done in the community of commercial airline pilots (personal communication). Like Craik and Donald Broadbent, but at the cultural not individual level, he has studied the operation and design of airline cockpits and the nature of pilot training (he holds a commercial pilot's certificate). In cockpits, he says, one can literally "step inside the cognitive system". Unlike the processes inside human heads (about which Hutchins is agnostic), much of the internal organization and operation of this cognitive system is *directly* observable (1995: 128–9).

What's of special interest here is Hutchins's broadly similar research on navigation at sea (1995). Most of the fieldwork for his book on this topic was done on US Navy ships, but some involved canoe navigation in Micronesia; besides, he's an accomplished amateur sailor. His work is relevant not only because he uses computational ideas, but also because he challenges the assumptions underlying much current cognitive science (and virtually all of the computational psychology discussed in Chapter 7). His extension of "the mind" *beyond* the person's skull/skin is one indication of this.

Like the evolutionary psychologists mentioned above, Hutchins pays homage to Marr's views on explanation (7.iii.b). But because of the nature of the task, he's able to take Marr more literally. He specifies four abstract computational constraints that must be satisfied by any successful (adaptive) method of navigation. He describes a variety of algorithms (navigational procedures), and a variety of implementations (in brains and/or artefacts), in 'Marrian' terms. And he provides computer models—connectionist networks-of-networks—confirming some of his claims.

The four abstract constraints (too complex to summarize here, but see Hutchins 1995: 52–8) are:

(1) The Line of Position constraint.
(2) The Circle of Position constraint.
(3) The Position–Displacement constraint.
(4) The Distance–Rate–Time constraint.

Different combinations of these four constraints determine the answers to the navigator's key questions: "Where am I?", "How far have I come?", "Where is my intended port-of-call?", and "How do I get there from here?"

As for *how one can compute* those answers while crossing vast expanses of ocean, there are two fundamentally different ways of generating navigational algorithms that satisfy the constraints. One is to think of the vessel and stars as moving, while the islands remain stationary. The other is to think of the vessel as fixed, while the islands and stars move towards it and then pass it by. (It's not just thinking, but *seeing* too: Hutchins and Hinton 1984.) The first representation is used by Western sailors, the second by Pacific islanders.

From the *computational* point of view (in Marr's sense), and considering only navigation (not how these pictures fit with other theories about the world), there's nothing to choose between them. But the *algorithms* derived from the one or the other will differ. Indeed, some 'obvious' questions posed from within one point of view simply won't make any sense to someone familiar only with the other. (Compare the fact remarked in Chapters 1.iii.a and 2.vii.f, that some scientific questions posed in the past make little or no sense today.)

(Also, although Hutchins doesn't discuss this, navigational practices which might seem to be 'obviously' implied by others may nevertheless be absent. For instance, *north* in Icelandic is judged by careful reference to the stars when at sea, but not on land. There, the judgements are radial; so a land traveller headed for the northern quadrant is said to be travelling north even if they're 'really' going north-east: Haugen 1957.)

Moreover, any instruments devised to *implement* navigational information processing will embody the option favoured by the culture using them. Once implemented in instruments, the conceptual scheme appears further objectified. Alternative schemes therefore look even more like groundless myths and mind games, mysteriously capable of compensating for gross ignorance.

Specific aspects of alien practices may be misunderstood accordingly. David Lewis's fascinating *We the Navigators* (1972), for instance, is trustworthy when he describes *what the Pacific islanders do* during their voyages. But it's not (or not always) reliable when he tries to explain *why they do it* (E. L. Hutchins 1995: 78–83).

The actions of the US Navy sailors could in principle be implemented (in their heads) as a global top-down plan, *à la* MGP—or Wallace. Alternatively, they could be implemented as 'antlike' production systems (Chapter 7.iv.b), wherein "each crew member only needs to know what to do when certain conditions are produced in the environment" (Hutchins: 199; cf. Fararo and Skvoretz 1984; Boden 1984*a*,*b*).

Hutchins shows that, in essence, the second of these is how it's actually done. The organization of specific duties ensures that the appropriate sequence of actions can be performed efficiently and flexibly, without burdening even the lowliest rating with a cognitive map of the entire navigational process. ("Ensures" is over-optimistic: accidents do happen, and some are tellingly analysed by Hutchins—e.g. pp. 241–2.)

In human societies, many tasks are shared, in the sense that they're distributed across several different individuals. This doesn't include a tug-of-war, where every man does the same thing and all have access to the same information about progress/success. Rather, people holding distinct roles do different things, and have different types of information. Often, specific communicative procedures have to be instituted (*sic*) in order to keep their diverse activities "in step" and to monitor progress in the overall task. (Compare Wallace's traffic lights and stop signs, for instance.) The communication may pass (heterarchically: see 10.iv.a) both upwards and downwards in the status hierarchy, and sideways too. And there are many ways of communicating besides speaking: pointing, for example, or raising a flag.

Hutchins illustrates these points in detail, describing elaborate shipboard procedures for finding and communicating relevant information. In a real sense, the sailors interact not only with each other but also with the ship and its instrumentation—including the naval charts. For Hutchins, knowledge amassed over many centuries by mariners, scientists, shipbuilders, and instrument makers is distributed *also* in these naval artefacts.

Even the placement of the chart table is significant, and coheres with the organization of activity aboard the ship.

b. Networks of navigation

The need for communication, given differential access to information, is where Hutchins's computer models come in.

Solving navigational problems is often difficult. The evidence may be incomplete (e.g. weather conditions may preclude use of some instruments), and inconsistent (e.g. instrument readings and/or crew members' beliefs don't match). The strong point of PDP-connectionist AI, as compared with GOFAI, is that it 'naturally' provides multiple constraint satisfaction (Chapter 12.v). Connectionist systems can reach *the most coherent* interpretation, even if some evidence is missing or contradictory. Just which they settle on depends on the evidence available/communicated at the time.

Hutchins uses *networks made up of networks* (12.ix.a) to explore "the relationships among properties of individuals and properties of groups" (p. 247). Put more provocatively, he's studying the relations between minds and group minds.

His networks model both individual sailors (e.g. the captain, helmsman, or radar operator) and the teamwork involved when the crew members act as a community. The sailor networks show what psychologists call confirmation bias: once a hypothesis has been formed, people pay more attention to evidence which confirms it than to contrary data (cf. 7.i.c). And the crew networks can model situations in which sailors disagree (about what they see, or what they think should be done), fail to make their opinion known to some other relevant person, or fail to convince that person of its truth.

A system composed of two or more networks has at least seven parameters that aren't found in a single network (p. 248). Three concern the distribution of structure and activation-state across the members of a community of networks; four concern communication among them. So Hutchins's computer models represent a sailor's schemata for interpreting what he sees/hears; his access to environmental evidence; his predispositions and current beliefs; who talks to whom, about what, and when; and how persuasive they are.

For instance, *current beliefs* are represented by the initial pattern of activation across the units of a sailor net. The *persuasiveness* of sailor *x*'s communications to sailor *y* is coded by the strength of the connections between the two nets concerned. The *topic* is coded by the pattern of interconnectivity among the units of the communicating sailor nets. And *who talks to whom* is captured by the pattern of interconnections among the nets.

Hutchins studied how systematic variations in these seven parameters affected the performance of the multi-network as a whole. For instance, a real-life situation in which two ships collided was modelled partly by degrading the 'visual' information available to the 'captain' (pp. 241–2, 249). And he showed that it's not always true that the best way to improve the performance of a group is to improve the communication among its members. In some conceivable circumstances, improved communication between individual minds reduces the effectiveness of the group mind. The confirmation bias typical of individuals is echoed—indeed, strengthened—at the group level, making the group interpretation even more hasty and intransigent (pp. 253 ff.).

The "attitude change" effected by one person communicating their belief to another was studied in the very early days of computational psychology (7.i.c). It was recognized, then, that the second person's level of confidence in the belief concerned *is not* always raised, and can even be lowered—and the computer simulations were designed accordingly. Hutchins's models are hugely less crude than those 1960s examples. But they assume—what he admits is "the most problematic simplification in the system" (p. 250)—that communication does always have this confidence-raising effect.

How much the second sailor network's belief activation is raised, however, depends (among other things) on the persuasiveness of the first. Persuasiveness was varied in different runs (pp. 251–9), and could have been modelled in even more detail (p. 251).

A main determinant of persuasiveness is the individual's social status. In hierarchical organizations (such as the crew of a navy ship), the leader's opinion counts very heavily, and can overrule the opinions of inferiors even if they have more direct access to the evidence. In consensus societies, such as a Quaker meeting, that can't happen: everyone—eventually—has to agree.

Hutchins's models illustrate two drawbacks of such (consensus) organizations. First, impasse can result if conflicting opinions have settled before communication begins. Second, if attempts are made to unblock the system by changing the pattern of access to evidence, this may prevent *any* reasonably coherent opinion from being reached. Finally, in systems where the solution is reached by voting, the result may differ from what it would have been had further communication been allowed.

"The cognitive properties of groups", Hutchins concludes, "are produced by interaction between structures internal to individuals and structures external to individuals" (p. 262). The social organization is crucial:

> Doing without a social organization of distributed cognition is not an option. The social organization that is actually used may be appropriate to the task or not. It may produce desirable properties or pathologies. It may be well defined and stable, or it may shift moment by moment; but there will be one whenever cognitive labor is distributed, and whatever one there is will play a role in determining *the cognitive properties of the system* that performs the task. (p. 262; italics added)

Four things follow from Hutchins's account:

* Anthropology and psychology are intimately connected.

* Psychology isn't enough. Even *social* psychology, if this means how individuals are affected by their social context, isn't enough. The matters highlighted by Abelson, for example, are important (Chapter 7.i.c)—but they leave a crucial dimension of analysis untouched.

* Classical AI, and all computational psychology inspired by it, is incapable *in principle* of capturing the cognitive properties of groups—aka cultural phenomena. (Compare the critique of orthodox cognitive science outlined in 13.iii.e.)

* And distributed AI, by contrast, can counter the excessive individualism of which many critics of GOFAI complain. (Hutchins favours PDP, as we've seen; however, distributed AI isn't necessarily connectionist: see Chapter 13.iii.d–e.)

In short, computational anthropologists can consider not only minds but also group minds—and they can do this without mystery mongering.

8.iv. Mechanisms of Aesthetics

Anthropologists know better than anyone that different cultures dress, dance, and decorate in different ways. Music, for example, has diverse styles and social significance for distinct societies (Nettl 2000; Mache 2000). Even within a given society, distinct subgroups may value very different genres. Indeed, a new way of doing art will typically involve explicit defence of a new aesthetic—as happened with computer-based "interactive" art, for instance (see 13.vi.c and Boden forthcoming). In short, arts and crafts vary greatly across the world. (The arts more than the crafts, because of the different psychological mechanisms involved: crafts depend on universal Gibsonian affordances, whereas fine art depends on culture-specific themes and stylistic references—7.v.e–f and Boden 2000*a*.)

Neo-Kantians, following Humboldt (9.iv.b), are likely to stress the *uniqueness* of each culture's aesthetic. However, some—such as Boas (1927)—have argued that aesthetic values in all cultures are rooted in an appreciation of technical skill and patience. And a fortiori those, like Barkow, who hope for a *Darwinian* psychological anthropology will seek some universal basis here. Beauty is in the eye of the beholder—but perhaps, like the three Graeae sisters of Greek mythology, all human beings share a single eye?

In the last twenty years of the century, there was a surge of interest in this question. It had been revived partly because of evolutionary theory in general, and partly because of cognitive science's stress on specific information-processing mechanisms.

a. From Savanna to Sotheby's

The sociobiologist Edward Wilson (1929–) was one of the first proponents of the single eye (E. O. Wilson 1984). His influential "biophilia" hypothesis was applied to all members of *Homo sapiens*, not just three. Indeed, it was applied even more widely: all the way from the art experts at Sotheby's to the busy members of his favourite species, ants.

The biophilia hypothesis claimed that animal species, including humans, have evolved to find certain aspects of their habitat attractive. They tend to pay attention to them, and to prefer them—in the behaviourists' language, to be reinforced by them. The relevant aspects of the environment, Wilson said, range from specific chemicals in water or food to 'friendly' conspecifics, and include many visible features.

Wilson himself saw this not merely as a disinterested scientific hypothesis but also as the empirical basis for a "conservation ethic" (E. O. Wilson 1993). To the extent that anthropologists favoured conservation—of mountains, flora, and fauna, and of human *experience/knowledge* of these things too—they could find support in Wilson's writings.

More to the point for present purposes, the biophilia hypothesis implied that species-universal aesthetic preferences underlie our culturally diverse practices. The question naturally arose as to just what those preferences might be. In the decade following Wilson's book (and Fodor's *Modularity of Mind*, published just a year earlier), evolutionary psychologists posited various "modules" supposed to underlie aesthetic judgements.

It was found, for instance, that—despite varying cultural preferences for 'fattipuffs' and 'thinnifers'—there's a universal attraction to certain ratios of body measurements in males and females (Singh 1993; G. F. Miller 2000: 246 ff.). In a nutshell, people are aesthetically valued for their probable efficiency as reproductive partners (Hersey 1996). From an evolutionary point of view, that made perfect sense. But other aesthetic values weren't so obvious. Let's consider just two examples: landscape and symmetry.

The appreciation of landscape—including people's responses to city squares and to offices or hospital wards with/without windows—was studied by environmental psychologists (e.g. Ulrich 1983, 1984, 1993; Kaplan and Kaplan 1982; S. Kaplan 1992). Evolutionary assumptions often lay in the background.

The zoologist/ecologist Gordon Orians (1932–), in particular, saw these matters in an evolutionary context (1980, 1986; Orians and Heerwagen 1992; Heerwagen and Orians 1986, 1993). His core hypothesis was that people would prefer certain types of scenery because of information-processing mechanisms evolved for selecting advantageous habitats.

Consider animals forced to flee because of predators, or lack of food. When should they stop, and stay put? Only when it's safe to do so. If they make camp far from water, or with no hiding places and/or no escape route, they're less likely to survive. So a disposition to stop running when one comes to a place having these valuable features is likely to be selected. It will become 'natural' to prefer such places: to tarry when one reaches them, to explore them (to gather information about them), and perhaps to settle comfortably in them. The first stage is crucial, for the other two can't happen without it.

So far, so speculative. But Orians reported that (in various species) the decision to tarry is taken automatically and very quickly, whereas exploration takes longer and is more 'cognitive' in nature. Moreover, clues predicting advantageous future states should be more useful than indicators of current, and perhaps transient, states—and some species behave accordingly. For instance, many birds use general patterns of tree density and branching as "primary settling clues", rather than trying to assess food supplies directly (Orians and Heerwagen 1992: 555–6).

Findings like this had promoted Orians's idea from a Just So story to a hypothesis for which there was some specific evidence. Now, he looked for comparable evidence in humans too.

Just how the relevant features are recognized wasn't at issue (but remember the discussion of affordances in Chapter 7.v.e–f). However, he called his approach "computational" in the Marrian sense that its purpose was "to guide research into the psychological mechanisms that promote adaptive functioning in different environmental contexts" (p. 561). Specifically, he predicted that the clues concerned would typify the resources of the African savannas where humans originated:

> If we assume that [we've evolved] psychological mechanisms that aid adaptive response to the environment, then savanna-like habitats should generate positive responses in people ... This is because the savanna is an environment that provides what we need: nutritious food that is relatively easy to obtain; trees that offer protection from the sun and can be climbed to avoid predators; long, unimpeded views; and frequent changes in elevation that allow us to orient in space. Water is the one resource that is relatively scarce and unpredictably distributed on the African savannas [so it's likely to function as a 'direct' settlement clue]. (p. 558)

Borrowing a method already in use (Balling and Falk 1982), Orians showed photographs of landscapes to people from widely differing geographical habitats and cultures. Subjects were asked which images they found most attractive, or which places they'd most like to visit or live in. These questions aren't equivalent, of course: modern tourists visit many places they wouldn't want to live in, much as fashionable late eighteenth-century gentlemen hung 'sublime' Romantic landscapes on their walls depicting places they'd never willingly inhabit. Nevertheless, Orians's results, whatever the question, showed a clear preference for all the clues listed in the quotation above (and for the presence of animals, too).

Aesthetic preferences for landscape art, from garden design to painting, were studied in these terms. For example, Orians saw what Stephen and Rachel Kaplan (1982) had called "mystery" as inviting the second stage: exploration. And indeed, roads winding behind hills, bridges affording (*sic*) safe access to distant terrain, and even partially blocked views were all found attractive by his human subjects.

Moreover, nineteenth-century garden design apparently confirmed one of his more startling discoveries. Orians found that people around the world are attracted by images of trees shaped like those on the African savanna (Orians and Heerwagen 1992: 559). Even more amazingly, they prefer those on *high-quality* savanna (his photographs showed only one species of tree, whose shape varies markedly with the availability of water). The favoured trees had a spreading canopy, moderately dense foliage, and a trunk that branches close to the ground. Humphrey Repton (1752–1818), the renowned English landscape designer (and author of three books on garden theory), was quoted as recommending trees of that very type (p. 561).

Orians's results were later replicated by others—at least in part (Ruso *et al.* 2003: 280–1). These follow-up researchers included two environmental psychologists at UC Davis: Robert Sommer (Sommer and Summit 1995; Sommer 1997) and Richard Coss. Coss, in particular, had long studied the role of perceptual and emotional mechanisms in art (e.g. 1965, 1977, 1981).

On turning to Orians's savanna hypothesis, he found that 3- to 5-year-old children (across cultures) find wide-crown trees less "pretty" than adults do. However, they see them as the best trees "to climb to hide", "to feel safe from a lion", and "to find shade" (Coss and Moore 1994; Coss and Goldthwaite 1995; Coss 2003: 82–5). These results, from children far too young to be experienced climbers, suggested "a precocious understanding of tree affordances useful historically and currently to avoid predators and to prevent dehydration" (Coss 2003: 84).

As for preventing dehydration, Coss also argued that the widespread (universal?) aesthetic preference for shiny things, including lustrous fabrics, is based in visual mechanisms evolved to identify water (Coss and Moore 1990; Coss 2003: 86–90). A sensitivity to reflectance could have helped early *Homo sapiens* to find expanses of still water at a distance and/or partly hidden by trees or shrubs, or water dispersed as dewdrops on—lick-affording—plants. Today, 7-month-old babies are twice as likely to lick or suck glossy things as dull ones.

The water example is highly plausible. Think of silver lurex, or of Lord Leighton's sensuous portraits of satin-clad women; or consider the chromium and glistening paint on new cars. And meditate on the fact that bone artefacts more than 70,000 years old show signs of 'useless' polishing, as do 14,000-year-old ivory figurines too (Coss 2003:

88–9). (And recall that the German *schön*, beautiful, has the same root as *scheinen*, to shine.)

The tree example, however, may seem mind-boggling. But the discovery of IRMs had already shown that visible shapes can be coded in the genome (e.g. the moving hawklike cross that prompts the gosling to crouch: Chapter 5.ii.c). And it was discovered at much the same time as Orians published his first 'habitat paper' that birds who naturally prey on snakes, but have never seen one, show strong aversion to wooden dowels painted with wide yellow and narrow red rings—like the skin of a venomous coral snake (S. M. Smith 1977). So it's not impossible that a particular tree shape is somehow favoured by our genes.

Humboldtians will bridle, pointing out that few people regard the savanna as their ideal environment. Woody Allen presumably prefers New York; and many favour the flamelike cypresses of Provence over the spreading canopies of Africa. But Orians knew that. He allowed that experience of other habitats can result in emotional attachments that make *them* seem ideal. His point, rather, was that such preferences have to be learnt. By contrast, everyone responds positively to savanna clues, whether they've experienced them or not.

b. The seductiveness of symmetry

Another feature valued by cultures all around the world is symmetry. Boas (1927) listed it as one of the universal aesthetic values in "primitive" art. But unlike savanna trees and shininess, it has no obvious adaptive advantage. Indeed, it's highly abstract: delightful to mathematicians, perhaps, but why should the rest of us care? Can it *really* be as important as Boas implied?

Its power is seen, for example, in the *maneaba*—a sacred meeting house, one per island—of the Kiribati, the people of the Gilbert Islands (Maude 1980: esp. 15–16, 20, 24). The floor plan of every *maneaba* is almost perfectly rectangular (exactly fitting one of nine length/breadth combinations), and the two main posts supporting the ridge-pole are almost perfectly vertical. "How clever of them", you may condescendingly think, "to get so close to Euclid—they might have got even closer if they'd had tape measures." You'd be way off the mark. The eastern side is deliberately made slightly longer than the western side. Moreover, at one point during the highly ritualized construction, which measures not by metre rules but by limb-lengths and pandanus fronds, the posts *are* vertical. But the Kiribati then push them towards the west by a hand's width, so as to overhang the territory of the evil spirits. This cultural practice couldn't have arisen if symmetry weren't a very special value indeed.

It was appreciated even in prehistoric times. When *Homo ergaster*, some 1,400,000 years ago, first made stone hand-axes to recognizable standards—as opposed to merely knocking flakes off rocks, to use as cutting edges—symmetry was a key feature (Kohn and Mithen 1999; Mithen 2003). And when *Homo sapiens*, perhaps 50,000 years ago, started to vary hand-axe designs, symmetry remained a key characteristic in the tools and art of all prehistoric societies.—But why?

In answering that question, evolutionists noted that symmetry is *present* in nearly all known species, and is *perceptible*—and even *preferred*—by many of them. For instance, psychologists have discovered that people can discriminate nicely between symmetric and asymmetric faces and body shapes, and that they find the former more attractive (for reviews, see Etcoff 1999 and Thornhill 1998). (Much of this work was done in the 1990s, by David Perrett's group in Edinburgh: see Chapter 14.iv.d.) Indeed, symmetry is attractive to monkeys too—and even to carrion crows. As long ago as the late 1950s, these animals had been seen to prefer items carrying regular patterns over irregularly patterned ones (Sutterlin 2003: 134), and the evidence has mounted since then (e.g. James R. Anderson *et al.* 2005).

Three evolutionary explanations were offered. The first suggests that animals will tend to be symmetric. The others suggest that they're likely to be able to recognize symmetry in other animals.

First, bodily symmetry is efficient, requiring less genetic coding than asymmetry. (A centipede with a hundred differently shaped legs would need a much-enlarged genome to develop them.)

Second, visible symmetry/asymmetry is a useful clue to fitness (P. J. Watson and Thornhill 1994). All manner of things can go wrong in the development of egg into adult. Symmetries, then, can act as indicators of 'good genes'—which is to say, a good choice as a mate.

And third, symmetrical patterns (especially those with several axes of symmetry) are relatively easy to spot, since they look much the same despite variations of orientation and/or viewpoint (Enquist and Arak 1994). Their usefulness for identifying conspecifics would favour sexual selection—although their usefulness to predators, in identifying a potential lunch, would be a selective disadvantage. (The third hypothesis was seemingly supported by a computer model that evolved increasingly complex symmetries. It turned out, however, that this result was an artefact of the simulation design: Bullock and Cliff 1997.)

The third explanation was offered as an alternative to the second. In principle, however, all three could be jointly true. That is, symmetry might be favoured by natural selection for coding efficiency and/or for ease of recognition, and by sexual selection also (see below). But whatever the evolutionary explanation/s may be, the fact remains that human beings have a remarkably good eye for symmetry—and also tend to prefer it.

An information-processing mechanism capable of recognizing symmetries *in general* is a different kind of beast from one which recognizes only hawklike crosses. But we know that visual matching of the images in the two eyes is 'built in' for stereopsis (14.v.f). Some broadly comparable matching mechanism could have evolved to cope with symmetries.

c. Universality in variety

The claim that there's a universal mechanism for appreciating savanna, or symmetry, raises a problem for any 'massively modular' aesthetics (see Chapter 7.vi.e). For artists sometimes deliberately break symmetries, or paint landscapes that are far from inviting.

Contrast the classicism of Andrea Palladio with the asymmetrical architecture of Daniel Libeskind, for instance, or the symmetry of a diamond tiara with the challenging jewellery of the American goldsmith William Harper. (Even *classicism* has its variants: Palladio was a more faithful respecter of symmetry than his fellow classicist Nicholas

Hawksmoor.) And where landscape is in question, contrast the rural scenery of John Constable with the unnatural panoramas of Salvador Dali.

One doesn't have to go to Western art to find examples. No cultural pattern is safe from variation, not even in communities which value change much less than industrial societies do.

What's more, such variation is often admired as "interesting", or "creative"—where what this means, put very crudely, is that it surprises us. But why value surprises? Wouldn't life be more comfortable without them? Why evolve a capacity (and a preference) for surprise, if current behaviour suffices for survival? In short, how can evolutionary psychology explain this pervasive aspect of human cultures?

An answer was recently suggested by Geoffrey Miller (1965–), in a 500-page tour de force drawing on a treasure chest of evidence from psychology and anthropology, and half-a-dozen other disciplines besides (G. F. Miller 2000). (He was well aware that evidence is needed to turn a Just So story into a plausible scientific hypothesis.) He'd been working on this topic for some ten years, at Stanford, Sussex, LSE, and UCL—with a brief sojourn at the Max Planck Institute in Munich. Now, he's at the University of New Mexico.

Miller is a cognitive scientist who has worked on A-Life computer models, and on perceptual representation in humans (Cliff and Miller 1995; G. F. Miller and Freyd 1993; see Chapters 15.vii.c and 14.vii.c, respectively). He's wary of the old-fashioned computer metaphor being taken too literally:

> [To] psychologists who pride themselves on their seriousness... the mind is obviously a computer that evolved to process information. Well, that seems obvious now, but in 1970 the mind as a computer was just another metaphor... The mind-as-computer helped to focus attention on questions of how the mind accomplishes various perceptual and cognitive tasks. The field of cognitive science grew up around such questions.
>
> However, the mind-as-computer metaphor drew attention away from questions of evolution ... creativity, social interaction, sexuality, family life, culture, status, money, power... As long as you ignore most of human life, the computer metaphor is terrific. Computers are human artifacts designed to fulfill human needs, such as increasing the value of Microsoft stock. They are not autonomous entities that evolved to survive and reproduce. This makes the computer metaphor very poor at helping psychologists to identify mental adaptations that evolved through natural and sexual selection. (G. F. Miller 2000: 153)

But although he scorned GOFAI, Miller was committed to mind-as-machine, understood in the catholic sense defined in Chapter 1.ii.a. And he saw evolutionary theory as the key to cultural variation.

Miller's book pointed out that Charles Darwin posited two principles of evolution: natural selection (for survival) in *On the Origin of Species* (1859) and sexual selection (for mate choice) in *The Descent of Man* (1871). Darwin used the second to explain courtship behaviour in animals, and anatomical features such as the hugely expensive peacock's tail (Cronin 1991). Miller, over 100 years later, applied it also to the runaway increase in brain size of *Homo sapiens*.

This didn't happen, he argued, because it enabled us to solve survival problems more efficiently: killing sabre-toothed tigers, for instance. Rather, brain size increased because it allowed us to represent—to generate and/or recognize—increasingly complex motor/perceptual patterns. In a word, it offered more opportunities for surprise. (Surprise, by definition, involves deviation from some recognized pattern.)

A preference for a moderate degree of novelty has evolved in various species. Even very young babies (and monkeys) can be rewarded simply by giving them new images to look at. This novelty preference isn't an input-triggered *module*, like Orians's settlement mechanism, but an aspect of information processing in many different domains. In general, then, one can expect that moderately surprising behaviour on the part of a potential mate will be attractive.

If 'surprisingness' is combined with fitness indicators such as intelligence, determination, and muscular control, so much the better. Someone who can carve a symmetrical hand-axe, paint a geometrical design onto a pot or a T-shirt, or dance and drum to a complex beat, must have all of these. (Hence the fact noted long ago by Boas, that all cultures find skilled craftsmanship beautiful.) So someone capable of making a deliberate *departure* from the familiar style (e.g. by breaking the symmetry) isn't only advertising their intelligence and self-control, but is also surprising—and thereby attracting—the perceiver.

The departure mustn't be too great. If there's no intelligible connection with the previous style, the surprise won't be welcomed—as the Parisian Salon des Refusés showed (Boden 1990*a*/2004, ch. 4). But moderate pattern breaking is favoured. (What counts as moderate varies: 'traditional' cultures tolerate a lesser degree of change than 'modern' cultures do, and even these include more and less change-tolerant subcultures.)

Miller didn't claim that sexual selection is the *only* principle underlying cultural variation and individual creativity. It certainly can't explain why a particular society favours a particular style. The geometrical nature of Islamic art, for instance, has nothing directly to do with biology but is explained by theological objections to representing human figures. In other words, Miller was no more a cultural reductionist than Barkow was.

He even allowed that it's not easy to assess just how strong the evidence for his position is. For example, if mate selection were the *only* factor, then males should write more books, compose more music, win more Nobel Prizes, and lead more political and religious campaigns than females do; and their interest in such cultural patterns should rise markedly at puberty. And indeed, that's so: "Demographic data show not only a large sex difference in display rates for [creative] behaviors, but male display rates for most activities peaking between the ages of 20 and 30" (82–3). But there are many *cultural* factors operating too. If women (still) produce fewer aesthetically valued behaviours than men, that's partly because their cultures (still) offer them fewer opportunities, fewer role models, less self-confidence, and sparser encouragement when they do break cultural moulds.

In sum: this was a form of the MMM thesis (7.vi.e). Cultures have developed largely in the service of mate selection, thanks to domain-general psychological mechanisms for generating, detecting, and valuing changes in behaviour. But domain-specific mechanisms, such as a preference for certain landscapes or body ratios, have evolved too. That's why *universal* savanna preferences coexist with a huge *variety* of cultural patterns. The description of this extraordinary variety is what anthropology is about.

These explanations of aesthetic preferences have drawn on examples of *biological* evolution. But people sometimes speak of *cultural* evolution, too. What have cognitive scientists said about that?

8.v. Cultural Evolution

Darwin himself, in *The Descent of Man*, suggested that cultures evolve. And clearly they do, in some sense. After all, "evolution" is loosely used today to mean any putatively progressive series of changes. It can even be used to describe putatively *regressive* changes, such as the current secularization of Anglicanism bemoaned by the Chancellor of York (J. Norman 2002; see Chapter 7.vi.d). The word was used with various meanings in Darwin's time too—for instance by Herbert Spencer (1820–1903), who wrote at length about social evolution (Robert M. Young 1967). That's why Darwin avoided the term, speaking of "descent with modification" instead (2.vii.e).

Unlike some of his predecessors, he wasn't merely describing evolution. He was explaining it—and in a new way. The question for Darwinian anthropologists, then, is whether *descent with modification* can explain the process of cultural change.

The first person to explore this idea seriously was William James, who opened a lecture to the Harvard Natural History Society by saying:

> A remarkable parallel, which I think has never been noticed, obtains between the facts of social evolution on the one hand and of zoological evolution as expounded by Mr. Darwin on the other. (James 1880: 441)

He went on to praise Darwin for separating "the causes of production" and "the causes of maintenance" of variations. Social evolutionists, he said, can ignore the first—much as Darwin ignored them in the physiological domain. Their question, rather, is how sociocultural *selection* (maintenance) happens.

In modern terms, James's distinction is between *how cultural ideas arise* and *how they spread*. Cognitive science has had something to say about both.

Today, there are five 'schools' of cultural evolution, each offering slightly different answers to James's questions (Laland and Brown 2002).

* The first three have been mentioned already: sociobiology (e.g. Wilson on biophilia); then human behavioural ecology (including the work of Orians and the Kaplans); and evolutionary psychology (e.g. Barkow—but see also subsection b, below).
* The last two to emerge were gene-culture coevolution and memetics (subsections b and c, respectively).

It has taken some fifty years for those five schools to develop. The beginnings lay in the mid-twentieth century. By the 1960s, several people—from several disciplines—were using Darwinian ideas in trying to answer James's two questions. Initially, however, the second question was more prominent than the first.

a. Evolution in the third world

One mid-century Darwinian was the cultural anthropologist George Murdock (1956), then at Yale (where he'd recently taught Goodenough). He'd already had an impact on anthropology in the 1940s, having provided a systematic empirical "coding" of social institutions which he'd applied to some 300 cultures. This classification encompassed a society's customs (such as rituals, punishments, and rewards), beliefs, and collective ideas. Now, he suggested using these cross-cultural comparisons to study cultural change, and the extent to which it resembled evolution.

However, Murdock was a methodologist rather than a 'grand theorist'. He was more interested in showing how specific cultural claims could be stated and tested than in making general claims himself. A more widely influential mid-century voice was the philosopher Karl Popper (1902–94).

Popper wasn't interested in anthropology as such, caring little about what different peoples happen to believe about this or that. But he was interested in how a society can develop a viable political system (1944/1957, 1945), and in how science can progress in a rational manner (1935, 1963). Ever since the 1930s–1940s he'd explained both these things in terms of carefully constrained trial and error, or "conjectures and refutations"—but without mentioning Darwin.

The initial ideas or hypotheses, he said, could come from anywhere. Politicians can draw on common sense, social science, historical example, and the like. Scientific ideas can originate not only in the laboratory but also in dreams, fables, religion, or even metaphysics. (This was an unorthodox view in the 1930s, for the logical positivists of the Vienna Circle had declared metaphysics to be meaningless: Ayer 1936.) In short, the context of discovery (what James had called "the causes of production") was very different from the context of justification ("the causes of maintenance"). In the first, anything goes. But the second is ruled by Popper's criteria of falsifiability and resistance to falsification.

In the late 1960s, he put his position in more explicitly Darwinian terms (1965, 1968). The unconstrained context of discovery was now a source of 'variation'. Falsifiability and lack-of-falsification were scientific 'fitness' and 'survival'. And epistemology in general was to be thought of in an evolutionary way. So, for instance:

> The growth of knowledge . . . is not a repetitive or a cumulative process but one of error-elimination. It is Darwinian selection . . . Classical epistemology . . . can only be described as pre-Darwinian. (1968, sect. 8)

(A digression, which relates to Section iv.b above: At that stage, Popper thought Darwin's theory of natural selection—the survival of the fittest—to be "almost tautological": a piece of unfalsifiable metaphysics, albeit prompting questions of great scientific interest—1965, sect. xviii. Later, he recanted, saying that the theory was not only testable but in some cases falsified—1978: 344 ff. For, as Darwin himself had pointed out, the peacock's tail isn't "useful". It's the result of sexual selection, not utility-driven natural selection—cf. Cronin 1991.)

Popper argued that human evolution in general is largely driven by our use of "exosomatic organs", comparable to wolves' lairs or beavers' dams. These included not only material artefacts—weapons, paper, computers (used to do sums: "to support argumentation")—but *intangible* objects too, namely publicly accessible ideas. He even posited a "third world", alongside the physical and mental worlds, to hold those ideas. Scientific discovery, he said, happens not (subjectively) within the minds of individuals but (objectively) within the scientific community. A new idea can't become part of science unless it's communicated, made available for criticism. (This was a new interpretation of the Cartesian "cooperation" necessary for science: see 2.ii.b–c.)

Popper's evolutionary ideas had an effect. Other philosophers, such as Stephen Toulmin (1961, 1972), applied them to a wide range of historical examples within the 'culture' of science. The biologist Richard Dawkins (1941–) would later acknowledge Popper's position on the third world as an anticipation of "memes" (1976: 204). And as we'll now see, they entered psychology too—reinforcing a Darwinian account that had been embarked on independently.

b. A new mantra: BVSR

Science—and cognition in general—was seen in evolutionary terms also by the psychologist Campbell (1916–96). A student of Edward Tolman and Egon Brunswik in the 1940s (5.iii.b), Campbell spent most of his professional life at Northwestern University. He started out as a social psychologist, with a special interest in cross-cultural psychology. He soon became a leading expert in experimental/statistical methodology, which helped to earn him the presidency of the American Psychological Association in 1974.

But Campbell was much more than a methodologist. His professional interests ran all the way from cybernetics, through perception in New Guinea, to practical pedagogy. His William James Lectures in 1977 were on 'Descriptive Epistemology: Psychological, Sociological, Evolutionary'. His *New York Times* obituary was headed 'Master of Many Disciplines'. And he died as Emeritus Professor (at Lehigh University, Pennsylvania) of Sociology, Anthropology, Psychology, and Education.

The root of this unusual interdisciplinarity *wasn't* a commitment to mind-as-machine: Campbell was a cognitive psychologist, not a cognitive scientist. However, in his Presidential Address to the APA, he dated his "fascination" with evolutionary theory from his reading of W. Ross Ashby's *Design for a Brain* in 1952. In that book, he said, "the formal analogy between natural selection and trial and error learning is made clear" (Campbell 1975: 1105). That formal analogy underlay his interdisciplinarity, for he became convinced that, where Darwinism is concerned, one size fits all. In other words, he used evolutionary theory as what Daniel Dennett (1995*a*) would later call a "universal acid", capable of solving/dissolving every problem within the social and biological sciences.

By the early 1960s, he'd already sketched 'Darwinian' accounts of perception and creative thinking (Campbell 1956, 1960). Like the New Look psychologists (Chapter 6.ii), he'd compared perception to scientific reasoning. But unlike them, he described both in evolutionary language. That vision and science can provide us with world knowledge, he felt, is just as amazing as the intricate anatomy of the eye itself. (As he put it later, "Does the power of visual perception to reveal the physical world seem so great as nearly to defy explanation?", and "Do you marvel at the achievements of modern science, at the fit between scientific theories and the aspects of the world they purport to describe?"—1974*a*: 139.)

He'd originally been inspired by Brunswik. But when Popper's philosophy of science was translated in 1959, it found a ready reader in Campbell. (He cited it, for instance, in his paper of 1965: 27). Soon, he started sprinkling his Darwinian acid over other aspects of culture too (Campbell 1965).

Like all acids worthy of the name, Campbell's had a formula: BVSR. This stood for "Blind Variation and Selective Retention". That was his way of expressing what Darwin had called "descent with modification". But instead of applying it to physical morphology and behaviour, he—like Popper, at around the same time—applied it to the anatomy and influence of cultural ideas.

There was nothing metaphorical about this, for Campbell had defined BVSR in abstract terms: blind variation and selective retention of items within category *x*. Category *x* might be biological (species, feathers, kidneys...), or cultural (artefacts, dance, kinship systems, science, religion...), or even computational (cellular automata, robots...). It's because evolution can be understood abstractly that A-Life workers can sensibly say that their systems evolve (15.vi).

John von Neumann had realized this in the mid-1950s (15.v). But when Campbell defined BVSR twenty years afterwards, von Neumann's later work was still largely unknown even in the AI community. For psychologists and anthropologists, BVSR was a new idea. (Many would have read Popper, of course; but Campbell made more of an effort to match specific concepts of Darwinian biology.)

A respectable theory of BVSR, he said, requires one to specify "mechanisms for introducing variation", a "consistent selection process", and "mechanisms for preserving and/or propagating the selected variants" (1960; cf. 1965: 27). More specifically, it requires one to ask to what extent biological concepts can find analogues in cultural contexts. And in a move to placate the "cultural relativists" in anthropology, who "emphasize the excellent adaptedness and internal coherence of simple cultures as of complex ones" and who reject notions of 'progressive' evolution accordingly, he said:

> [One] may be a cultural relativist within the framework of a correct [i.e. non-'progressive'] social evolutionary theory. Indeed, the relativist's frequent emphasis upon the adaptiveness or functional validity of customs that seem bizarre to outsiders *implies* a selective retention and elimination process. (1965: 39; italics added)

Despite his disclaimer regarding evolution-as-progress, Campbell's approach was most avidly taken up by historians of science and, especially, technology (e.g. Basalla 1988). For in the latter case, detailed information about the structure and development of artefacts was relatively accessible. Moreover, it seemed—to some extent (see below)—to fit. So Ziman has said:

> In my own limited reading [anyone who knew Ziman will take that phrase with a pinch of salt: M.A.B.] I have come across suggested technological analogues of what an evolutionary theorist would term *diversification, speciation, convergence, stasis, evolutionary drift, satisficing fitness, developmental lock, vestiges, niche competition, punctuated equilibrium, emergence, extinctions, coevolutionary stable strategies, arms races, ecological interdependence, increasing complexity, self-organization, unpredictability, path dependence, irreversibility*, and *progress*. (Ziman 2000*b*: 4–5)

Like Popper before him, Campbell saw epistemology in general through Darwin-tinted spectacles. He spoke of "a continual breakout from boundaries", followed by testing and further development/breakouts towards increasingly "reliable" knowledge. In short, he used BVSR as the core of an "evolutionary epistemology", or EE (Campbell 1974*a*,*b*).

This was a descriptive enterprise, not a normative one. It was therefore "undertaking a different task from that of traditional analytic epistemology" (1974*a*: 140). In Fregean terms, it wasn't philosophy, but psychology (see 2.ix.b). However, some leading philosophers, notably Toulmin (1961, 1972) and Willard Quine (1969), were engaged in a similar activity—what Quine called "naturalized epistemology".

Campbell allowed that *within professional philosophy* this was "a minor heresy" (1974*a*: 140). Within psychology, it began as a heresy too. But it grew apace. For when "evolutionary psychology" surged in the early 1980s, Campbell was there waiting. His EE approach now got favourable attention, for example, from the biologist Plotkin (1982), from philosophers of biology such as David Hull (1982, 1988*a*), and from the multi-talented duo of Robert Boyd and Peter Richerson (Boyd and Richerson 1985).

Boyd (1948–) was a member of UCLA's Department of Anthropology, though his initial training had been in physics and ecology (including energy management). Richerson (1943–) was an environmental scientist and zoologist, at UC Davis. As ecologists, they favoured theories of biological evolution focused at the level of *genetic populations*—which led on naturally to concerns about *cultures*. They developed a "dual inheritance" approach, showing how genes can influence cultural evolution and how culture can affect biological evolution. (They were later criticized by some anthropologists for neglecting cultural evolution *as such*, i.e. change wholly within the realm of cultural constructs—much of which is neutral with respect to biological fitness—Durham 1991: 437 ff.)

The Stanford geneticists Luigi Cavalli-Sforza and Marcus Feldman had been modelling the co-evolution of biology and culture even earlier (1973, 1981). Initially, they'd focused on matching DNA to human populations, and later moved on to the evolution of languages. The work on language has been roundly criticized by the Sussex linguist Larry Trask (Trask 1996: 376–404). I mention it here not because it was valuable but because it was an influential example of the evolutionary Zeitgeist. Boyd and Richerson were informed by that Zeitgeist too. However, they were looking at examples of specific cultural practices described by anthropologists.

They'd published their first joint paper comparing cultural and biological evolution in the mid-1970s (Richerson and Boyd 1976), and many more had appeared since then. Their 1985 book, which pulled their previous work together, was even more interdisciplinary than most exercises in cognitive anthropology. Perhaps for that very reason, it wasn't generally thought of under the label "anthropology" (see Section ii.d). But it threw new light on the nature of culture.

The key questions concerned what sort of mind/brain is capable of imbibing (and contributing to) culture, and how/why it evolved. The main influences on cultural evolution, they said, include "biased transmission": the fact that the process of cultural transmission itself can favour certain variants over others. This can take place in three ways.

"Direct" bias happens when people adopt an idea, or practice, because they judge it to be valuable. (Jogging is 'good for you'.) "Indirect" bias occurs when a variant is selected because it's associated with a valued variant. (Role models admired for their success are often copied in ways that have nothing to do with the feature which made them valued in the first place.) Third, "frequency dependent" bias is in play when the spread of a cultural variant among a population affects its probability of adoption by later generations. (A tourist resort can become increasingly fashionable merely because it's already fashionable; but it may later lose custom because people choose to swim against the tide.) These factors, and especially the second, explain why so many cultural traits have no clear advantage—and why some can even be maladaptive.

Boyd and Richerson identified various types of cultural subsystem, correlated with particular psychological mechanisms for 'deciding on' and effecting their transmission. The correlations weren't merely superficial, still less accidental. On the contrary, they were explained as adaptations to distinct kinds of environment—not rocks or trees or expanses of water, but different statistical patterns of variation.

The two authors drew on rich ethnographic data in expressing, and exploring, their theories. But they had another way of doing this too, namely, mathematical modelling. They defined a number of models detailing how ideas ("information in brains") might be conserved and altered, and how they might be affected by different environmental conditions—including the variability of other animals' behaviour.

For example, they showed that whether imitation ("social learning") is more adaptive than working out how to do something for oneself depends on how "cost-effective" that learning would be. Different statistical properties in the environment favour different types of social learning. One is blind copying of some—any—conspecific. Others are guided by heuristics, such as "copy dominants" or "go with the majority" (which require the capacity to recognize dominants and majorities). High variability in the environment is correlated with the ability to make complex cognitive analyses of the problem situation. And human behaviour, of course, is hugely variable.

In other words, some results of their formal models underscored the more intuitive arguments of Geoffrey Miller (Section iv.b, above). Human culture is not only made possible by human intelligence, but drives it to evolve further. (In their most extensive discussion of intelligence, they attributed increase in brain size to the evolution of cognitive strategies adapted to environmental deterioration: Richerson and Boyd 2000.)

Boyd and Richerson weren't merely expressing vague analogies between biology and culture. They were doing a systematic study of well-defined cultural possibilities, grounded in what was known about computational psychology—and anthropology, ecology, and biology. Much as Hutchins would later show how different patterns of communication emerge from different network models, so they showed how distinct cultural phenomena would result from distinct computational assumptions.

Today, some twenty years later, EE is a thriving industry (e.g. Radnitzky and Bartley 1987; Campbell and Cziko 1990; Heyes and Hull 2001). It became an "industry" partly because there was much disagreement about just how relevant BVSR really is to cognition and culture. For example, was Ziman right to claim that those twenty-one neo-Darwinian concepts could all be applied to cultural (technological) change? Answering such questions turned out to be a very lengthy, and still ongoing, exercise.

Here, let's just note two main dimensions of discussion:

* How blind is Blind?, and
* What are the units of cultural variation?

Darwin hadn't insisted on absolutely blind variation, for his occasional willingness to consider "panspermia" allowed for the inheritance of acquired characteristics. (Panspermia was the idea that the sex cells are formed by the agglomeration of tiny 'copies' of cells sent from all parts of the body.) However, he believed that descent with modification could be blind, and often said that it is. That is, variations don't happen because of their likely biological usefulness. In particular, he rejected Jean-Baptiste Lamarck's view that *conscious desire and effort* can result in (acquired) heritable variations.

His views prevailed. Thanks to the work of August Weismann in the 1890s (Chapter 2.vii.e), plus modern genetics, twentieth-century neo-Darwinism taught that "Blind" really did mean *blind*. Hence the title of Dawkins's book (the one in which he introduced his A-Life "biomorphs"), *The Blind Watchmaker* (1986). Admittedly, recent research suggests that there may be tiny glimmers of 'vision' in the origin of some biological variations (Jablonka 2000, esp. 33 ff.). In general, however, blindness rules.

This posed a problem for cultural BVSR, because cultural/conceptual variations are rarely generated blindly. To the contrary, they're highly Lamarckian. People—from mechanical engineers to Popper's "experimental" politicians—typically have a specific goal and/or fitness criterion in mind when they try to come up with new ideas. The anthropologist William Durham, for instance, intimated as much in the 1970s and spelt it out at length some years later (Durham 1976, 1979, 1991).

BVSR devotees, however, argued that this didn't make Campbell's formula irrelevant. The way he'd put it was to say that "In going beyond what is already known, one cannot but go blindly" (1974*b*). That was true *in principle*, given a strict interpretation of "beyond what is already known". But how true was it *in practice*?

Detailed studies by historians of technology provided an answer, for they showed that even the most carefully designed artefacts have some unpredictable features. Their fitness for their intended purpose therefore has to be tested—and developed—in the real world (Ziman 2000*b*,*c*; Wheeler *et al.* 2002). That's even more true of politics, as Popper had taken pains to point out. And it's true also of creative ideas in art, which typically develop by continual interaction between the artist and his/her still-uncompleted artefact or text (A. Harrison 1978). In short, although people evolving new ideas aren't completely blind, their vision is only limited.

The second dimension of disagreement about EE concerned the units of variation: what are they? Here, Campbell's writings were largely eclipsed in the late 1970s by Dawkins's. This wasn't because Dawkins was saying anything essentially different, but because his lucid, even racy, prose demanded less from the reader than Campbell's did. In particular, he provided a seductive new label: *memes*.

c. The meme of memes

Dawkins's phenomenally successful *The Selfish Gene* (1976) kicked off the popular surge of interest in Darwinism in the last quarter-century. It sold to all manner of people, and was translated into many languages. One person enthused by it was the Microsoft billionaire Charles Simonyi, who began recommending its ideas to his co-workers in the early 1980s (A. Lynch, personal communication). In the 1990s, Simonyi founded the Chair of the Public Understanding of Science at Oxford, specially designed for Dawkins.

Most of the book concerned the evolution of physical phenotypes: flowers, feathers, eyes, and so on. The final thirteen-page chapter, however, dealt with 'Memes'. And the penultimate sentence declared: "We are built as gene machines *and cultured as meme machines* . . ." (1976: 215; italics added).

Like genes, said Dawkins, memes are "replicators", whose success requires three properties: longevity, fecundity, and copying fidelity (p. 208). They are the units of variation in evolving cultures: a meme is "a unit of cultural transmission, or a unit of *imitation*" (p. 206). Examples included "tunes, ideas, catch phrases, clothes fashions [e.g. stiletto heels], ways of making pots or of building arches". In other words, any concept, belief, or cultural practice counted as a meme. Interlinked groups of memes were "meme-complexes", which included whole cultures, specific belief systems, and socially established styles of doing things.

The meme-complexes attributed to "Socrates, Leonardo, Copernicus, and Marconi" have survived hundreds of years longer than the four men themselves (p. 214). So far, so banal. But now Dawkins made a startling claim:

> [When] we look at the evolution of cultural traits and at their survival value, we must be clear *whose* survival we are talking about. Biologists . . . are accustomed to looking for advantages at the gene level (or the individual, the group, or the species level according to taste). What we have not previously considered is that a cultural trait may have evolved in the way that it has, simply because it is *advantageous to itself.* (p. 214)

"Advantage", for memes, meant survival in the minds of mankind. And the more the merrier: a meme's fitness was its capability of being imitated, so as to enter human minds as quickly, widely, and stably as possible.

It followed, he said, that we don't *need* to explain prominent cultural traits—music and dancing, for example—in terms of biological survival value. To be sure, many of them (even including his *bête noire*, religion) may actually have such value. They may aid group cohesion, for instance. Or they may increase fitness by way of Darwin's second evolutionary principle, sexual selection (Section iv.b, above). But the basic evolutionary concern is the meme's "selfish" ability to survive and spread in the minds/brains of enculturated individuals.

Dawkins described memes as "viruses of the mind". Mostly, he compared them to biological viruses:

> When you plant a fertile meme in my mind, you *literally* parasitize my brain, turning it into a vehicle for the meme's propagation *in just the same way* that a virus may parasitize the genetic mechanism of a host cell. (1976: 207; italics added)

But sometimes he spoke of "informational" computer viruses, too:

> For data on a floppy disc, a computer is a humming paradise just as cell nuclei hum with eagerness to duplicate DNA . . . Computers are so good at copying bytes . . . that they are sitting ducks to self-replicating programs: wide open to subversion by software parasites. (Dawkins 1993)

The analogy had originally been drawn by the psychologist Nicholas Humphrey. On reading a draft of his friend's book, Humphrey had remarked that memes are like viruses, since they propagate themselves by "parasitizing" human brains (Dawkins 1976: 206–7).

That applied to *all* memes, so Dawkins often described memes-in-general as viruses. Sometimes, however, he seemed to be thinking of viruses as disease bearers—in which case beliefs "backed by good reasons" such as science weren't included, but religious beliefs were (see below). (A later writer described memes more neutrally, as "contagious", pointing out that laughter, happiness, and joy are contagious too: Lynch 1996, 2002.)

If Dawkins's view was "startling" for most of his readers, it was less so for some anthropologists. For (as he pointed out in passing), one of their number—F. Ted Cloak (1931–) at the University of North Carolina—had already said much the same. Dawkins cited a mid-1970s paper by Cloak (1975*a*), in which he'd used the software/hardware distinction in speaking of the "programming" of the nervous system. But there'd been much earlier ones, too (Cloak 1966, 1968, 1973). In subsection d, we'll look more closely at just what Cloak had said, and ask why it still hadn't caught on—as Dawkins's term immediately did.

Six years later, in the more academically oriented *The Extended Phenotype*, Dawkins gave a clearer definition. And he cited Cloak in doing so: "a meme should be regarded as an item of information residing in a brain (Cloak's 'i-culture')" (1982: 109). But the damage had been done. His loose definition of memes on the word's introduction had already led, and would lead again, to various conflicting accounts, some more catholic than others. To add to the confusion, some of those would occur in books almost as popular as Dawkins's—so the notion spread ever more widely.

Dennett, for example, broadcast it far and wide in his best-selling *Consciousness Explained* (Dennett 1990; 1991: 199–226). He used "meme" to include not only neural representations of ideas but also the ideas themselves; the virtual machines comprising mind, self, and consciousness; cultural artefacts; and socially accepted ways of behaving. And, by the way, he mentioned neither Campbell nor Cloak—who'd recently published a fuller version of his "instructional" theory (Cloak 1986). Dawkins got all the credit.

Dennett was one of many. The historian Peter Munz, an ex-pupil of Popper who'd already published books on myth and religion as well as science, offered a version of EE expressed in the language of memes (1993). Memes spawned "memetics", and a new (online) journal was founded in 1997 by Manchester University's "Centre for Policy Modelling", the *Journal of Memetics: Evolutionary Models of Information Transmission*. Two years later, Susan Blackmore (1999) used Dawkins's phrase "meme machine" as the title of a best-selling popular book announcing "a new science": memetics.

For many, that was a step too far. If "meme" had become a buzzword for some cognitive scientists (and for many members of the public), it was regarded with suspicion by others (Aunger 2000). There were two problems: vagueness, and lack of match to psycho-cultural reality.

Those who'd followed Campbell in championing evolutionary accounts of culture didn't necessarily favour talk of memes. The cognitive anthropologist Bloch (2000), for instance, accused the memeticists not only of reinventing anthropological wheels but also of repeating some of the mistakes previously made by his professional colleagues. (The 'invisibility' of anthropology discussed in Section ii presumably hadn't helped.) And the philosopher Hull defended cultural evolution—that is, a focus on heritable variation and competitive selection of cultural features—but had few kind words for memes (D. L. Hull 1982, 1988*b*).

The same was true of Boyd and Richerson. They did sometimes use the word "meme", for convenience. More often, they spoke (technically) of "cultural variants" or (informally) of "ideas", "beliefs", "values", and "skills". They defined culture as "(mostly) information in brains" (forthcoming, ch. 3). They granted that some cultural information is stored in artefacts: pots, for instance. Perhaps, they said, some youngsters learn to decorate pots by looking at old pots rather than talking to old potters. In general, however, cultural knowledge is linguistic.

But memes, they argued, weren't helpful in thinking about cultural change. Dawkins had defined memes as discrete entities, faithfully transmitted from one brain to another. On the contrary, they said, the actual units of cultural variation and inheritance are neither. They denied that Darwinian evolution *necessarily* involves "copying fidelity", arguing (and demonstrating by their models) that blending inheritance is compatible with BVSR evolution. For example, learning (and historical change) in speech pronunciation could conceivably be caused by alternative sets of computations in the infant's mind, one of which computes average (i.e. blended) phonological values in adults' speech. Provided that there are also sources of heritable variation (slight differences in the anatomy of the larynx, for instance, or mishearings on the part of the child), evolution can occur. In sum, cultural variants are *not* close analogues to genes, but "different entities entirely, perhaps a rather diverse class of entities, about which we know distressingly little".

Boyd and Richerson weren't alone. The general view was that the analogy between genes and memes isn't nearly as close as Dawkins had suggested. Sceptics pointed out that memes—unlike genes—can't be crisply identified. They appear to include concepts, schemata, scripts, theories, cultures, and cultural practices, none of which are easily defined. (Consider Rosch's work on concepts, for example: Section i.b, above.)

They aren't always copied faithfully, but are often contaminated by new cultural associations, or by metaphor. Indeed, such contamination can be a matter of 'Lamarckian' choice. They don't have separate lineages, for while one can imitate one's parents' or neighbours' ideas one can also (thanks to books and other media) imitate ideas far away in time and space. They can be discarded and replaced by others, whether voluntarily or not: Paul on the road to Damascus is one example, acceptance of a new scientific theory another. And whereas genes are clearly distinct from the phenotypic characters they 'cause', memes are seen sometimes as the brain mechanisms that underlie a concept or cultural practice and sometimes as those phenomena themselves.

None of this controversy denied that anthropologists had been on the right lines when they'd used certain types of pottery, or specific elements of myth, to indicate past cultural movements and influences. Such evidence was just what an evolutionary approach would favour. Similarly, anthropologists' accounts of one tribal group's gaining dominance over another could now be understood as describing "group selection in action" (Sober and Wilson 1998: 191). But, quite apart from the non-blindness of Blind, BVSR should be applied to cultural matters only with care—and it didn't gain from pseudo-scientific talk of memes.

d. Cloak uncloaked

Dawkins's theory of memes had been anticipated, and in some detail—but neither he nor his non-anthropological readers realized it. Indeed, not all of the anthropological readers did. For the relevant work wasn't well-known. Cloak, one might say, was cloaked—so much so, as to be near-invisible.

His first two papers on cultural evolution had appeared in a highly obscure source, not held today by either the British Library or the Library of Congress. (I was given copies by his follower Aaron Lynch.) And an important lecture given at an anthropological gathering in 1973 had—by accident—remained officially unpublished, although it was available in the conference preprints and from Cloak himself. Indeed, he was eventually denied tenure—and at least one intending graduate student left the profession in disgust as a result (Aaron Lynch, personal communication).

This setback may have been part-grounded in the scantiness of Cloak's publications list. But perhaps his colleagues also felt that his "radical" suggestions for restructuring the whole of anthropology were too ambitious, not to mention too uncomfortable. In addition, his grandiose claim that his theory could account for "all known biological, social, and cultural phenomena" was likely to invite scepticism, not to say mockery.

What was radical about Cloak's approach wasn't any commitment to computational theorizing. His first relevant paper appeared only a few months after Wallace's, but was very different. It didn't rely on ideas about computational processes, whether from MGP or anyone else. However, it did focus on the transmission of *information* ("instructions"), both cultural and genetic; and by the early 1970s he was talking about the "software" and "programming" of the nervous system. In that sense, he was aligned to the early cognitive scientists, Wallace included—and working at a higher level of abstraction than most anthropologists.

In an address written for the AAA's annual meeting in 1966, Cloak had argued that previous comparisons between biological and cultural evolution had picked the wrong biological units (i.e. species) on which to base the analogy. The important likeness, he said, was to gene flow (the diffusion of instructions) between geographically separate but reproductively linked populations. And he continued:

> I must beg leave to enter a provisional concept here, *analogous to the gene*, defining it as "that which is transmitted when cultural diffusion takes place". I call it the *unit of cultural instruction*, or *UCI for short, putting an asterisk before it as a reminder that it is provisional. (Cloak 1966: 8; first italics added, second in original)

As in the biological case, he said, an idea (*UCI) survives if it has some adaptive advantage—"technological, social, or moral"—for its bearer. (He would retract this later: see below). But unlike the biological case, cultural diffusion requires that the new ideas be "selected" and "fixed" not by chance but according to specific structural features of the receiving culture.

Cloak ended his address by referring gnomically to unspecified "empirical evidence" showing that the sequence of cultural transmission is "predictable". He wanted his anthropological audience to focus on theory, not ethnographic detail:

> [I intend to continue my empirical studies of cultural change, but] I think that more attention should be directed to *the theoretical problem of the nature of the unit of cultural instruction* (the *UCI) and the principles by which *UCI's and cultural structures interact in the ongoing process of cultural microevolution. (1966: 10; italics added)

Two years later, his new theoretical unit had become less "provisional". Cloak (1968) used the work of Niko Tinbergen and Konrad Lorenz (Chapter 5.ii.c) as the inspiration for a "cultural ethology", which distinguished "behavioral propensities" from "modes of acquisition". Cultural learning, he said, includes imitation and receipt of verbal instructions—which can affect the manner in which even inborn propensities are exercised. So women flirt differently in different cultures (1968: 39). The "inborn" components (the smile, the enlarging of the eyes, and the wrinkling of the skin at the bridge of the nose) occur in different orders, and some components—such as the flourishing of the nail varnish, or the fluttering of the fan—are wholly cultural.

In general, he said, studying the distribution, replication (*sic*), and if possible the natural order of adoption and loss, of behavioural variations within and between cultures will "broaden our understanding not only of the history of particular cultures but of *the mechanisms which control culture change*" (p. 41; italics added). He'd already done this, in relation to behavioural replication in immigrants to Trinidad (this was the empirical evidence he'd hinted at in his AAA address). The cultural patterns he'd studied ran from 'meaningless' phrases in ancient songs (compare: *Ring-a-ring of roses, a pocket full of posies* originally referred to the plague) to stock answers to genuine questions, answers that often were not only incomplete but two-thirds false: *Q. What do you grow in your garden? A. Corn–peas–cassava* (1968: 43).

The *corn–peas–cassava* stock phrase soon acquired a thirteen-syllable label (with no defensive asterisk). Three years before Dawkins's book appeared, Cloak (1973) addressed an international gathering of anthropologists, saying:

> Elementary self-replicating instructions [ESRIs, for short] are the constructors of life and, given life, of culture (in the sense that coral polyps are the constructors of a reef).

Crucially, those ESRIs could be interpreted materially (as genes, or neural structures) or informationally (as ideas). So he'd prefaced his talk by declaring:

> A reconstruction of Darwinism, the proposed theory . . . *brings genetic and cultural evolution into a common conceptual framework, including a common system of notation*, and it reconciles hitherto opposed viewpoints in cultural anthropology . . . For example, the evolution of functional *social* "structures", as well as of functional *material* structures, is rendered explicable.
>
> Materialistic, naturalistic, mechanistic, deterministic, the theory purports to account for all known biological, social, and cultural phenomena and to contain no terms or concepts not reducible, in principle, to physico-chemical terms. . . .
>
> *Key terms developed in the communication include*: 'determinant', 'behavioral event', 'event of natural selection' ('ENS'), 'replication', 'self-replication', 'exploitation', 'domestication', 'system (of instructions)', 'function (in a system)', 'environmental sub-region', 'frontier (between sub-regions)', 'evolutionary event', 'organism'. (Cloak 1973: 1; italics added)

In short, Cloak had outlined "a radical reconstruction of general anthropology" in Darwinian terms. And he'd done so by trying to define cultural equivalents of over a dozen key biological concepts.

In his Congress talk, and in his 1975*a* paper too, Cloak had distinguished between "i-culture" and "m-culture". The i-culture consisted of the cultural "instructions" carried in the heads of the members, while the m-culture was the material artefacts and physical practices produced by the i-culture. Dawkins mentioned that distinction. But he didn't point out that Cloak had also said this:

> [The] survival value of *a cultural instruction* is the same as its function; it is *its value for the survival/replication of itself or its replica(s), irrespective of its value for the survival/replication of the organism which carries it* or of the organism's conspecifics. (Cloak 1975*a*: 168; italics added)

That is, Cloak had come up with the very same idea which Dawkins expressed by calling memes "selfish". By this, Dawkins explained, he meant: "Once the genes have provided their survival machines [i.e. animals/humans] with brains which are capable of rapid imitation, the memes will automatically take over" (Dawkins 1976: 214).

"Meme" itself was a widely imitated meme, as we've seen. Cloak's alternatives, "unit of cultural instruction", "elementary self-replicating instruction", and "elementary cultural replicator", weren't. Like Barkow's "Darwinian psychological anthropology", they were non-memorable mouthfuls. By contrast, Dawkins's pithy (and gene-rhyming) "meme" was a jewel of rhetorical skill—and, as such, more likely to enter the history books (see Chapter 1.iii.h).

What's more, Cloak's 1973 discussion (though not his two 1975 papers) was highly formal—and very dry, with nary a concrete example. Even the leading anthropologists sitting in the Congress hall, or perusing it at their leisure in the preprints, were likely to find it tough going. For 99 per cent of the readers of *The Selfish Gene*, it would have been altogether too taxing. So one can't reasonably wish that the general public had been subjected to the anthropologist Cloak, rather than the biologist Dawkins. Cloak's work would have left them cold.

One can reasonably regret, however, that few academic readers had looked at it. For as Cloak's list of "key terms" (quoted above) suggests, his account was much more precise than Dawkins's notion of memes. Indeed, in the year when *The Selfish Gene* appeared, the luminary Campbell (1976: 381) was already describing Cloak as "one of the most meticulous and creative thinkers about social evolution".

Campbell had been one of the reviewers for Cloak's 1975*a* paper, received by the journal editor as early as February 1973 (A. Lynch, personal communication). He was so enthusiastic that he wrote to ask Cloak for copies of his earlier publications. And the remark just quoted was made in an APA Presidential Address which drew many published comments, including one from the sociobiologist Wilson.

What's more, Cloak himself was in the habit of sending his unpublished work out quite widely. (The final section of his 1975*a* paper said his ideas on culture should be spread "into such areas as ethology, anthropology, human ecology, philosophy and logic of science, computer science, statistics, experimental psychology, and neurophysiology".) And his 1975*b* paper had already been presented to the 1974 meetings of the Animal Behavior Society *as well as* the Central States Anthropological Society.

In sum, it's possible that Cloak's ideas (though not necessarily his name) had spread quite widely among sociobiologists and evolutionary psychologists by the time that Dawkins wrote *The Selfish Gene*. If so, that would help explain the rapid take-up of the rhetorically attractive meme of *meme*.

8.vi. The Believable and the Bizarre

Ever since Tylor's *Primitive Culture* (1871) and James Frazer's even more widely read *The Golden Bough* (1890), anthropologists have reported exotic beliefs about all manner of things. These range from cookery and kinship, through flora and fauna, sickness and death, all the way to shamans and religion.

The cautious have done so with some trepidation. As Humboldt pointed out long ago, identifying another culture's concepts and beliefs is problematic in principle (Chapter 9.iv.b). It's not even straightforward in *one's own* culture—which is why some philosophers argue that folk psychology can't be the basis of a science (see Chapter 16.iv.b).

For instance, do the Micronesian navigators really believe that the islands move? They *see* them that way, to be sure (E. L. Hutchins and Hinton 1984). But just what does that prove? And should one assume that all the members of a given culture share the same beliefs? Do all Polynesians, from Tuvalu to Samoa, Aoteoroa to Rapa Nui, hold the same beliefs about their ancestral homeland Havaiki? (No: for one thing, it's less celebrated in western Polynesia—Bellwood 1987: 68.) Do all modern Americans think of *marriage* in the same way? (No, despite an eightfold similarity: see Section i.d, above.) Simply *defining* "culture" in terms of (universally?) shared beliefs merely begs the question.

Gallons of ink have been spilt on these issues, by philosophers and methodologists alike (e.g. Winch 1958 and Clifford and Marcus 1986, respectively). Let's ignore them here, however, and assume that foreign ideas can be pretty well understood. Let's focus instead on three undeniable facts.

First, some alien beliefs are not just unfamiliar but—to us (i.e. the likely readers of this book)—bizarre in the extreme. And, no doubt, vice versa.

Second, others can strike us as deeply *familiar*. That's so even though they may be accompanied, in their home culture, by practices or pronouncements which we find very odd indeed.

And third, 'strange' ideas are accepted even within one's own society—and some spread like wildfire. Think of backwards-facing baseball caps. Or if that's too trivial, and too hideous, to contemplate, think of religious cults: the Moonies, perhaps, or the interpreters of messages from the planet Clarion (Chapter 7.i.c).

A number of questions arise, some of them very old. For instance:

* Why do some new ideas spread faster than others? Certainly, cults often employ ingenious brainwashing techniques to make conversions. But are they more likely to succeed with some ideas rather than others—and if so, why?
* Why do 'meaning-less' rituals (behaviour and dress) survive for centuries, even in some parliamentary democracies?
* Why do religious people, everywhere, make claims which they know to be highly counter-intuitive? Pick your own bugbear, and ask: How could anyone believe *that*?
* Finally, are there any limits? Or could *any* belief, no matter how strange, thrive in some culture, somewhere?

a. An epidemiology of belief

Dawkins had taken it for granted that some ideas (memes) will spread easily and others won't. However, he hadn't said why that is. Or rather, he'd explained it in terms of common-sense psychology. For instance, he'd attributed the spread of Christianity to its threats of hellfire, its promises of salvation and immortality, and its self-protective meme of *faith*—which disarms rational criticism (1976: 212). What he hadn't done was to ask *what cognitive mechanisms* underlie the varying survival of ideas.

The anthropologist Sperber tried to fill the gap. In his Malinowski Lecture at LSE in 1985, he said that "Representations are more or less widely and lastingly distributed, and hence more or less cultural" (see Section ii.d above), and went on to ask why *this* representation becomes "widely and lastingly distributed" while *that* one doesn't. To find the answer, he said, we need an "epidemiology" of ideas, or beliefs, which would explain why some ideas are more "contagious" than others.

In his audience, he might have seen the LSE sociologist Eileen Barker. For she'd long been engaged in a study of late twentieth-century religious cults, including the Moonies for instance (1989; Barker and Warburg 1998). Her work gave fascinating details about their conversion practices, and useful advice on how to resist them. But Sperber wasn't concerned with such brainwashing techniques.

Nor was he concerned with the question of which ideas people, whether cult leaders or anyone else, would actively *choose* to communicate. (He was criticized, accordingly: the metaphor of epidemiology downplayed the Lamarckian aspect of cultural evolution, because it ignored people's valuations of one idea as opposed to another: Durham 1991: 197–201.) Sperber's question, rather, was whether there are some ideas which would be especially easy, or difficult, to disseminate—and if so, why?

In calling for an epidemiology of ideas, he was calling for something he'd already done much to develop. His book on symbolism had appeared ten years earlier, and the one on communication was already in press when he gave this lecture (Sperber 1975; Sperber and Wilson 1986).

That second book was an interdisciplinary mix of linguistics and psychology, with anthropology hovering in the background. Sperber saw cognitive psychology as crucial here. It could illuminate the psychological processes involved in metaphor—a topic that had recently received much attention, some of it relevant to anthropology (Ortony 1979; Lakoff and Johnson 1980). Even more to the point, it could help discover which concepts are remembered well in all cultures (J. M. Mandler *et al.* 1980). His new theory of "relevance" sought to explain *what makes a concept memorable*, whether it's widely dispersed cross-culturally or not.

Sperber saw his theory of cultural evolution as an alternative to Dawkins's account, not an elaboration of it. He insisted that ideas aren't "replicators", since they don't satisfy Dawkins's criterion of copying fidelity. As outlined in Chapter 7.iii.d, he saw communication as an activity that *changes* the hearer's mental representations.

Dawkins seemed (by default) to agree with John Locke (1690), who'd said that communication involves ideas in one mind being passed on to another. But Sperber held that new ideas are actively assimilated, not passively received. Understanding is a matter of making inferences, guided by "cost–benefit" computations concerning one's own cognitive economy. In other words, the information processing involved virtually guarantees that transmission produces variants, not copies:

> What human communication achieves in general is merely some degree of resemblance between the communicator's and the audience's thoughts. Strict replication, if it exists at all, should be viewed just as a limiting case of maximal resemblance. (Sperber 1990: 30)

Moreover, the variants aren't random: the nature of the information processing puts bounds on what they will be.

Dawkins had allowed for meme variation too, of course (1976: 209). After all: no variation, no evolution. But he'd apparently assumed that faithful copying is the norm, as it is in genetics. If the double helix works properly, the new gene is identical to the old one. With respect to the copying *mechanism*, mutations are faults. Analogously, the successful communication of ideas was assumed to involve perfect copying.

As a student of culture, however, Sperber was well aware that the change in the communicated ideas can't be too great:

> [Cultural beliefs differ from personal ones in that] those representations that are repeatedly communicated *and* minimally transformed in the process will end up belonging to the culture. (Sperber 1990: 30; italics added)

What James called the "maintenance" of culture depends on the fact that cultural transmission, despite the inevitable variations, isn't a game of Chinese Whispers. This is the "cultural ratchet" which—together with Theory of Mind (7.vi.f)—enables human creativity, unlike innovations by animals, to endure and even develop through the generations (Tomasello *et al.* 1993).

But how does it work?

In the book that was in the press as he spoke, Sperber and Deirdre Wilson had concentrated on the transmission of banal sentences such as those previously discussed by H. Paul Grice with respect to "speech acts" and "conversational conventions". So *Can you pass the salt?, He's a snake!*, and even—an example close to my heart—*It took us a long time to write this book* (pp. 122–3) were more to the point than exotic claims about snakes as totems or 'soul twins'. Cross-cultural examples were mentioned only in passing. In the next ten years, however, he would apply the theory to many such instances, based in a wide range of ethnographic data (Sperber 1990, 1994, 1997*a*).

One of the distinctions made in Sperber's epidemiological theory was between "intuitive" and "reflective" beliefs (1990, 1997*b*). Broadly, an intuitive belief is one which someone holds on the basis of everyday experience, underlain by universally shared cognitive modules. Reflective beliefs, by contrast, are "believed in virtue of second-order beliefs about them". Examples of intuitive beliefs include *That house offers shelter, The stone is about to hit the water*, and *She wants me to speak to her.* Examples of reflective beliefs include *Water consists of hydrogen and oxygen, Baseball caps are cool*, and *God is everywhere.* Each of these is held on other people's say-so. (So are most scientific, historical, political, economic, legal . . . beliefs—Shapin 1994, ch. 1.) "X says it" is crucial in all three cases. But who counts as an acceptable "X" differs greatly.

With respect to the belief about water, "X" is held to be a *reputable* scientist, or a *trustworthy* journal or institution. That's why the word of a gentleman was so crucial to the foundation of modern science, and why peer reviewing and institutional affiliation are so important today (see Chapter 2.ii.b–c).

This second-order aspect of scientific culture is unavoidable. Even if one decided to check the constitution of water for oneself, one would have to rely on what scientists say about the theoretical reasoning (and instruments) involved. The reflective belief couldn't be turned into an intuitive belief without doing the same for all the scientific propositions involved in its justification. In practice, this isn't possible. So as Popper (and René Descartes) said, science is—and has to be—a *cooperative* enterprise.

The better grounded a reflective belief is, the more likely it is to be transmitted and maintained. Science has become a global epidemic as a result. In Sperber's words:

Well-understood reflective beliefs . . . include an explicit account of rational grounds to hold them. Their *mutual consistency* and their *consistency with intuitive beliefs* [including experimental observation] can be ascertained, and plays an important, though quite complex, role in their acceptance or rejection. (Sperber 1990: 37; italics added)

In the case of science, *inconsistencies* with intuitive beliefs (experimental observations) are an embarrassment. They're usually disregarded as the result of 'faulty procedure', especially if they disappear on repeating the experiment. However, intractable inconsistencies do sometimes occur. These may be quietly swept under the carpet, or admitted as intriguing "anomalies" awaiting some future theory to cope with them (Kuhn 1962). These examples show that mutual consistency between reflective beliefs can *compensate* for their inconsistency with intuitive beliefs.

(One might expect this compensatory effect to be especially important for religion, where the inconsistency with daily experience is very great. And indeed, theological arguments—including Charles Babbage's analysis of miracles: Chapter 3.i.b—are largely devoted to maximizing internal consistency, and justifying the inconsistencies in the process. So theology may assist the spread of religious ideas among highly educated people. But there's a puzzle here. Many highly literate believers ignore theology, seeing it as an optional extra—or, worse, a trivial pastime of logic-chopping monks and scholars. They don't spend much effort in trying to make their ideas systematic. Yet the religion spreads, nevertheless—and its adherents even *glory* in the inconsistencies. Later, we'll consider one cognitive anthropologist's explanation of why that is.)

What of *Baseball caps are cool*? The second-order beliefs, again, are of the form "X says it"—or perhaps "X does it". But now, "X" can be the boy-next-door, the members of one's gang, a film star, or a football player. This reflective belief has spread rapidly, far and wide. But it's limited to people who respect, and want to emulate, the role models concerned (whose numbers have grown as the epidemic spread) and/or who value *being in fashion* or *appearing young* as a general practice. (Or as an occasional gimmick: the British politician William Hague, when leader of the Conservative Party, was widely mocked when he wore a baseball cap to meet the press photographers.) However, role models can come to be seen as less attractive: maybe they misbehave, or simply become boring. And fashions in clothing change, not least for commercial reasons over which the average person has no control. So this epidemic may fade away.

A more lasting, because officially sanctioned, clothing epidemic is the khaki *sulu* worn by Fijian policemen. *Sulus* are the traditional form of dress for Fijian men. Often patterned and/or brightly coloured, they're more similar to Western women's straight skirts than to the looser (unisex) Polynesian sarong. When the British colonialists imported the role of *policeman* into Fijian culture, they designed the uniform to be 'appropriately' low-key and neat (no patterns, no bright colours, no swirling folds of fabric). But instead of replicating the British bobby's trousers (as the French in Tahiti replicated the uniform of Parisian *gendarmes*), they included aspects of the local dress. This variant was far from "blind". It was a canny choice of mixed memes, intended to foster a higher degree of acceptance of the new cultural role than would otherwise have been the case.

As for *God is everywhere* (Sperber 1990: 90), this is a glaring example of inconsistency between reflective and intuitive beliefs. (How can anyone be everywhere? If He's everywhere, why can't we see Him?) And there are plenty more where that came from. What has cognitive anthropology had to say about them?

b. Religion as a cultural universal

In the late nineteenth century, Tylor (1871) coined the term *animism*, meaning "belief in spiritual beings", declaring this to be "the minimal definition" of religion. Soon afterwards, Frazer (1890) defined religion as "the propitiation or conciliation of powers superior to man which are believed to direct and control the course of nature and of human life". Since then, many alternative definitions have been offered.

Some don't focus on the content of religious beliefs, as Tylor and Frazer did, but on the attitudes of awe and mystery that attend them—as in Émile Durkheim's (1912/1915) contrast between the "sacred" and the "profane". Increasingly, suggested definitions have recognized the *social* aspects of religion, including not only rituals but also political/priestly power relations. And Geertz defined religion in terms of "symbolism":

> [A religion is] (1) A system of symbols which acts to (2) establish powerful, pervasive and long-lasting moods and motivation in men by (3) formulating conceptions of a general order of existence and (4) clothing these conceptions with such an aura of factuality that (5) the moods and motivations seem uniquely realistic. (1963/1966: 31)

One could nit-pick over definitions ad nauseam. For present purposes let's take it, following Pascal Boyer (1994), that:

* A religion is a culturally accepted set of representations (ideas, beliefs, attitudes) and ritual practices which is counter-intuitive *even to its proponents*.
* In other words, these ideas and practices posit unnatural, often supernatural, entities and causal powers.
* The relevant concepts—which differ greatly in detail, from culture to culture—are largely inscrutable, in that their meaning and implications can't be pinned down.
* The entities/powers may be associated with sacred artefacts (icons, idols, vestments, the bread and the wine . . .).
* They're believed to influence the daily lives—and sometimes the afterlives—of individuals in the culture concerned.
* And some of those individuals (shamans, priests, gurus) are thought to have special knowledge of and/or access to the supernatural beings—how to avoid them, communicate with them, or placate them.

Boyer's definition isn't wholly new. Gellner, for example, arguing that psychoanalysis is in effect a religion, said:

> A compelling, charismatic belief system . . . must engender a *tension* in the neophyte or potential convert. It must tease and worry him, and not leave him alone. It must be able to tease and worry him with both its promise *and* its threat, and be able to invoke his inner anxiety as evidence

of its own authenticity. Thou wouldst not seek Me if thou hadst not already found Me in thy heart! . . . *Demonstrable or obvious truths do not distinguish the believer from the infidel, and they do not excite the faithful. Only difficult belief can do that.* (Gellner 1985: 40; final italics added)

(The Catholic Church agrees: hence Gregory XI's discomfiture at Ramón Lull's reasoning machine—see Chapter 2.i.b.) However, the "difficulty" here could be mere implausibility, or counter-intuitiveness. What Boyer adds to most definitions of religion is the aspect of inscrutability, wherein the meaning and implications of religious concepts and claims are not just *hard to accept* but *hard to pin down.*

Thus defined, there's no doubt that religion is universal to cultures (though not to individual human beings). But why? In other words, why do such inscrutable beliefs occur over and over again?

(For current purposes, our focus is on semantic *content*: the concepts and beliefs, or what Boyer calls the religious "representations". Other important aspects of religion, including the rituals and the priestly power relations, will be ignored. Boyer himself did discuss them, but he too focused primarily on the cognitive representations involved.)

Many believers and theologians will answer that religion reflects reality. They may admit that some cultures grasp this truth more fully than others, much as some people see or hear more acutely than others. But *truth* it is. Even if that's so, however, a psychological question arises. Just as we can ask what mechanisms enable human beings to see, or hear, physical reality, so we can ask what mechanisms—provided, perhaps, by the grace of God—enable us to glimpse religious reality.

If religion is *not* a reflection of reality, then the psychological question is sharpened. If religion is illusory, why is it there? Indeed, why is it everywhere? And why is this *universal* phenomenon also hugely *diverse*?

As in medical epidemiology, there are two sub-questions here: *Where do religious ideas come from?* and *How do they spread?* Dawkins had asked both, but offered an answer only to one. He explained the spread of religious memes in terms of people's susceptibility to threats (of hellfire or excommunication), promises (of salvation or immortality), and priestly instructions (to have faith, to go to synagogue or Church, and to send the children to *shul* or Sunday School). But he didn't know where the idea had originated:

Consider the idea of God. We do not know how it arose in the meme pool. *Probably* it originated many times by independent "mutation". In any case, it is very old indeed. (Dawkins 1976: 207; italics added)

That "probably" was a con trick. He had no specific reason, *over and above* its frequency around the globe, to think that religion is likely to arise. If it has arisen many times, independently (which experimental science, for example, has not), then that fact itself requires an explanation.

The irreligious may ask these questions in a spirit of amazement that anyone could ever accept such absurdities. Freud, for instance, had no doubt that religion is an "illusion", essentially comparable to a childhood neurosis (1927). The fact that (in his view) it was a psychosocial necessity didn't prevent its being illusory. And Dawkins became notorious for his militant atheism. His discussions of religious memes—usually drawn from Christianity, but now increasingly from Judaism and Islam also—typically included vitriolic denouncements of their idiocy and evil effects (e.g. Dawkins 1993). When he said that religion is a "virus", he meant that term in the nastiest sense possible.

Bertrand Russell would have cheered him on. For besides his careful philosophical arguments, he too had a nice line in anti-religious polemic (Russell 1957, chs. 1–2, 13–14). The meme enthusiast Dennett was less ready with the polemic, but defined religion as a "maladaptive" meme, because it "discourages the exercise of the sort of critical judgment that might decide that the idea of faith [i.e. belief despite counter-evidence] was, all things considered, a dangerous idea" (1995: 349). Russell would have cheered him on, too. For he'd argued, similarly, that the contentment and resignation encouraged by Christianity prevents people from seeking scientific solutions—the only solutions possible—for the various evils that plague them.

However, one doesn't have to be a Freud, a Russell, or a Dawkins to wonder why people accept religion. Religious followers themselves admit, indeed insist, that their ideas are very strange—hence the awe and mystery surrounding the sacred. And nobody subscribes to every religious idea found across the world, not even in an attenuated, or through-a-glass-darkly, fashion. It follows that even the devout believer needs a naturalistic explanation of why *those people out there* accept *that.*

At least eight explanations have been suggested:

(1) *Perhaps it's because we use our intellect?*

This reply underlies claims that religion is, at least in part, prescientific explanation. Darwin, for example, said:

> [The belief in unseen or spiritual agencies] seems to be universal with the less civilised races. Nor is it difficult to comprehend how it arose. As soon as the important faculties of the imagination, wonder, and curiosity, together with some power of reasoning, had become partially developed, man would naturally crave to understand what was passing around him, and would have vaguely speculated on his own existence. (1871: 143)

Twenty years later, Frazer (1890) announced that "it is the imperious need of tracing the causes of events which has driven man to discover or invent a deity". Indeed, St Thomas Aquinas himself had claimed that his 'Five Ways'—including the Argument from Design and the Cosmological Argument—could prove the existence of a creator God by reason alone (but see subsection d, below). The same aim underlay the "natural theology" of the *Bridgwater Treatises*, or the Gifford Lectures (see Chapters 2.vii.b, 3.i.b, and 7.i.g).

(2) *Could there be some human ability to receive supernatural revelation—perhaps universal, perhaps confined to a priestly elite?*

Whether one answers "Yes" or "No", one must allow that there are many varieties of so-called religious experience (James 1902; G. W. Allport 1951; Bourguignon 1976: Ferrari 2002). These include visions and involuntary actions induced in states of 'possession' (Chapter 7.i.i), and the observation of unique or wonderful events supposed by the religious person to be caused and/or preordained by supernatural agency (see 3.i.b). Many cases of religious experience, including hallucinated voices and delusions of alien control of one's own body, can now be explained neuroscientifically and/or investigated by brain scanning (some references are given in 14.x.c; see also Atran 2002: 174–96).

But those who answer "Yes" will contrast "genuine" miracles, visions, or states of possession with illusory and/or self-deceiving ones. (Consider the role of the Devil's Advocate, who puts the counter-arguments when the Vatican assesses candidates for beatification. Or eavesdrop on the Fang people of Cameroon, when they argue about whether a supposed shaman really does have a hotline to the "ghosts" in the forest—Boyer 1994, ch. 6.)

(3) *Or perhaps religion satisfies an inner need for comfort and guidance?*

An especially complicated—and universalist—psychological story was told about that by Sigmund Freud (1913, 1927, 1930, 1932). It's difficult, today, to swallow it whole. Besides relying on the scientifically shaky Oedipus complex (Chapter 5.ii.a), he developed his own mythology of "the primal horde", to explain the historical origins of religion. But his insight that religion can offer various kinds of psychological security was sound. His account fitted monotheism best, but he did discuss totemism as well. Indeed, his work inspired Bronisław Malinowski's (1884–1942) pioneering studies of the Trobriand Islanders—who, said Malinowski, *didn't* develop an Oedipus complex, because in Melanesian culture the father's role wasn't as authoritarian as in Freud's Vienna (Malinowski 1915–18, 1922).

(4) *Possibly, our mental/spiritual health depend on it in some other way, linked perhaps to self-integration?*

This view was held by Carl Jung (1933), and by many Third Force psychologists (G. W. Allport 1951; Maslow 1962; see Chapter 5.ii.a). And since Freud saw religion as a protection against neurosis, he held a version of it too.

(5) *Perhaps religion aids social cohesion?*

A view favoured by many of those evolutionary psychologists who see religion as an *adaptation*, and by some sociologists and anthropologists. Thus Durkheim:

> [Religion is] a system of ideas with which the individuals represent to themselves the society of which they are members... The god of the clan, the totemic principle, [is] nothing else but the clan itself, personified and represented to the imagination under the visible form of the animal or vegetable which serves as totem. (1912/1915: 225–6)

(6) *Maybe there's some special faculty, or instinct, which leads us to religious belief?*

Darwin said not: he insisted that there's no "aboriginal endowment" for belief in "an Omnipotent God" or even "one or more gods" (1871: 143). (Some religious popularizers today, though not the evolutionary psychologists themselves, disagree: they even speak of "the God module".)

(7) *Or perhaps there are some more general mechanisms in human minds, evolved for other purposes but prone to generate religion?*

In other words, religion isn't an adaptation but a (possibly useful) side effect of non-religious adaptations.

Item (1), above, is a special case of this view. Recent cognitive science, however, has added further detail—as we'll soon see.

(8) *Finally, perhaps religion is learnt from the surrounding culture, just as table manners are?*

Despite its chicken-and-egg air, this is the answer that most anthropologists would give (see subsection f, below).

Each of these answers is consistent with at least one other. But three are mutually exclusive—namely (6), (7), and (8). The first two of these are nativist: they ascribe religious belief to inborn psychological mechanisms, either religion-specific or more general. The last is environmentalist—and universalist, too. That is, it respects the third and fourth 'Newtonian' tenets identified in Chapter 5.i. By implication, anything can be learnt—and whatever is learnt, is learnt in the same way.

It's these three answers which have most relevance for cognitive science. Some late twentieth-century cognitive anthropologists criticized their more traditional colleagues for their commitment to (8). They themselves were flying the flag for (7): the generation of the *extraordinary* out of the *ordinary*.

c. Symbolism

The sense in which religions are extraordinary was discussed by Sperber, in the mid-1970s. On his view, *the way in which we interpret* religious concepts is different from the way in which we interpret remarks about tables and chairs. He was the first anthropologist to explain religion in terms of the mind's underlying cognitive *processes*, and to take cognitive psychology seriously in so doing.

Abstract cognitive *structures*, to be sure, had been posited at mid-century by the then fashionable anthropologist Lévi-Strauss. He'd employed a binary semiotic "code", or symbol–meaning mapping, to analyse (for instance) different kinship systems—such as the "avunculate" pattern, in which the maternal uncle holds the family authority over a young boy. Similarly, myths and religions were supposed to be structured around basic semiotic oppositions and themes. Dichotomies such as *men/women*, *raw/cooked*, *village/forest*, *good/evil*, *elder/younger*, and others, were seen as fundamental to all cultures (Lévi-Strauss 1958/1963, 1964/1969).

Lévi-Strauss had been revolutionary in the early 1950s. Instead of focusing on cultural practices and/or artefacts, he claimed to be describing the human mind's classificatory rules, which structured those practices. That's why he was given prominence, thirty years later, in Gardner's Sloan-sponsored history (Section i.c, above). But he'd relied on abstract armchair analysis of ethnographic data, not on discoveries in cognitive psychology.

Sperber (1975), by contrast, argued that anthropologists should consider the processes by which people actually interpret symbols. In a word, this wasn't decoding, but inference. (That insight was later developed, in his co-authored book on *Relevance*: see 7.iii.d.)

As we've seen, Sperber called *God is everywhere* a "reflective" belief, accepted on the say-so of others: especially priests. But it's notoriously tricky. It's not clear just what it means, nor how we can know whether it's true *independently* of the words of the priests. (Nor is it clear how to recognize the priests in the first place, as we'll see.) Even Aquinas admitted that no particular observational belief follows from it, or conflicts with it. The same applies to the ritual sacrament of the Eucharist: the theologians tied themselves into knots over the bread and the wine.

Thomas Hobbes had little patience with them accordingly. He put the Ancient Greeks' mythical "Satyres, Fawnes, Nymphs, and the like" in the same class as his neighbours' "Fayries, Ghosts, and Goblins; and the power of Witches". The Church, he claimed, had encouraged or tolerated these beliefs "on purpose . . . to keep in credit the use of Exorcisme, of Crosses, of holy Water, and other such inventions of Ghostly men" (1651: 7). As for the dogmas pronounced by the theologians,

> [Such statements are among] those we call *Absurd*, *Insignificant*, and *Non-sense*. And therefore if a man should talk to me of a *round Quadrangle*; or *accidents of Bread in Cheese*; or *Immaterial substances*; or of *A free Subject*; *A free-Will*; or any *Free*, but free from being hindered by opposition, I should not say he were in an Errour; but that his words were without meaning; that is to say, Absurd. (1651: 20)

Three centuries later, the logical positivists agreed with him: religious statements are meaningless. As Antony Flew (1950/1955) famously put it, they suffer "death by a thousand qualifications".

If that's so, however, they take a long time a-dying. Most people might grant Hobbes the round quadrangle, but not the other examples. Religious concepts, they feel, do have some meaning. Atheists who reject *God is everywhere* normally say that it's false ("in an Errour"), not meaningless. They're happy to agree with Flew, that the meaning is highly elusive. But meaning *of some sort* there seems to be. How is that possible?

Sperber suggested an answer, by defining meaning in psychological—not logical—terms (Chapter 7.iii.d). In his sense, countless facts are potentially "relevant" to *God is everywhere*. For in the mind of a Christian believer, almost anything can be wheeled in to support, or illustrate, God's ubiquity, goodness . . . and so on (A. J. T. D. Wisdom 1944). Much the same is true, anthropologists tell us, in other cultures. The precise import of informants' remarks about their clan's totem, or the spirits in the forest, is as hard to pin down as anything Aquinas puzzled over.

For Sperber (1975), this was true of "symbolism" in general—religious concepts being a special case. The elusiveness—"inscrutability"—of symbolic meaning, he said, results from an asymmetry in the way it's interpreted. Whereas non-symbolic communication (including science) draws each claim from as many inferences as possible, symbolic language draws as many inferences as possible from each particular claim. It follows that a potentially unmanageable cloud of schema-based inferences—some symbolic, some not—attends every symbolic communication.

Think of how one reads certain types of poetry, for instance, or how one interprets certain types of visual art: ambiguity and unexpected associative richness are considered virtues, here. On Sperber's analysis, the slipperiness of religious statements, and the awe and mystery they arouse, results from this skewed interpretative process.

On Sperber's account, we don't have to be *told* to interpret the poem, picture, or prayer in this sort of way (though cultural training can lead us towards certain interpretive paths rather than others). If and when we find that we can't interpret a communication in the "rational" mode, we *automatically* switch to the "symbolic" mode. So by "rational", Sperber didn't mean consciously deliberated or argued, still less scientific. After all, science isn't/wasn't prominent in some cultures. But people everywhere make statements based on the evidence of their senses, from "The baked beans are on the top shelf" to "The totem pole is broken". Granted, culture-specific concepts inform such everyday perceptions (Chapters 6.ii and 16.iv.e). But Sperber's position was that non-symbolic interpretation kicks in first, as the (evolutionarily grounded) default procedure.

Just why we can't interpret religious statements rationally was considered at greater length by Sperber's follower Boyer. His answer relied on the contrast between the ordinary and the extraordinary.

d. The extraordinary out of the ordinary

Consider Darwin's dog. This creature, "a full-grown and very sensible animal", growled and barked fiercely when an unattended parasol left open on the lawn moved slightly, because of the breeze. That was unusual, said Darwin, since if anyone had been standing near the parasol the dog would have "wholly disregarded" it. In explanation, he attributed a primitive animism to his pet. But this animism was of type (7), not type (6):

> He must, I think, have reasoned to himself, in a rapid and unconscious manner, that movement without any apparent cause indicated the presence of some strange living agent, and no stranger had a right to be present on his territory. (Darwin 1871: 145)

In other words, the dog's ordinary reactions—the ascription of causes, the recognition of animate beings, and the assertion of the territorial imperative—were the root causes of his extraordinary ferocity.

Over a century later, the anthropologist Boyer (1959–) would offer a similar account of human religions. This didn't make him popular with his professional colleagues (Boyer 1994: 11). For most of them, as we've seen, psychology was irrelevant and human "universals" *tabu*.

Cognitive anthropologists, however, were less hostile. One of them, Scott Atran at the University of Michigan, even tried to take Boyer's theory further. He formulated "the Mickey Mouse problem", asking what distinguishes beliefs about the Disney rodent from beliefs about the gods (Atran 1998: 602). In a nutshell, his answer was: passionate personal commitment. So his evolutionary explanation drew on inherited mechanisms underlying motivation and social relations *as well as* the more strictly cognitive mechanisms stressed by Boyer (Atran 2002). Here, however, our focus is on Boyer—for whom Atran was a complement, not a rival.

By the mid-1990s, nativism had reappeared in psychology after its temporary eclipse, and had grown real teeth besides (Chapters 7.vi and 14.ix.c–d). So where Darwin could only hand-wave towards inborn mechanisms, Boyer (then at King's College, Cambridge) could call on developmental psychology, psycholinguistics, and neuroscience to locate them more closely. In addition, he could lean on Sperber's work on the cognitive mechanisms of symbolism and communication.

He'd already discussed culturally authorized "persuasion" by people regarded as truth-sayers (Boyer 1990). Now, he turned especially to religion, using many detailed examples from the African Fang people, with whom he'd lived for several years in the early 1980s (Boyer 1987, 1992, 1993, 1994).

His definition of religion (given in subsection b) was a psychological one. That's unusual: most anthropologists speak, rather, of religious *cultures*, *belief systems*, or *world-views*. Boyer regarded such terms as vague, mystical, and misleading, since they reify high-level abstractions instead of focusing on the underlying reality (1994, e.g. 51, 296). So although he often mentioned "culture", he did so as a convenient shorthand, not as a technical term. (For the record, he's now located in a Department of *Psychology*.)

He was similarly wary of theology. Religious representations, even religious experiences (naturalistically understood), are more real than religious "beliefs". Indeed, most people don't worry about the things that concern theologians. Boyer agreed with Meyer Fortes that even highly elaborate religious beliefs and practices "can be carried on perfectly well without a doctrine or lore of the nature and mode of existence of the 'beings' to whom they are ostensibly directed" (p. 94).

In short, it's a mistake to assume that people care greatly about the consistency of their religious beliefs. But that fact in itself is puzzling. Why do people seek consistency in respect of everyday trivia, but blithely ignore it in the seemingly important area of religion?

Boyer tried to give an answer. With respect to the *universal* aspects of religion, he suggested that the inferences that people link to religious statements aren't wholly unconstrained. (So when I said, above, that almost anything can be wheeled in to support religious claims, that "almost" was important.) The constraints he was talking about weren't further examples of symbolism, not even carefully argued theo-logical symbolism. Rather, they were cognitive schemas found in every human mind. These ordinary mechanisms are what give rise to the extraordinary claims of religion:

> Human evolution has resulted in the development of particular intuitive principles, geared to particular domains or aspects of the natural and social environment. Whatever cultural input can trigger some of these intuitive principles is likely to be more easily acquired and communicated, thereby giving rise to the recurrent features of cultural phenomena. Cultural recurrence (including the recurrence of religious representations) is therefore the outcome of a runaway process in the iterative application of epigenetic constraints. (Boyer 1994: 294)

The "intuitive principles" concerned cover a variety of ontological and relational schemas, in a number of different domains. They constitute our naive physics, naive biology, and naive psychology—several aspects of which had been discovered in pygmy children as well as in Westerners (e.g. Avis and Harris 1991). They've evolved for recognizing and making inferences about inanimate objects; causal connections; plants and animals; natural kinds such as water, or gold (conceptualized as having some hidden "essence"); human agency; and social status.

In making this list, Boyer was drawing not only on ethnoscientists' work on folk taxonomies (Section i.a–b, above), but also on recent developmental psychology (7.vi.i). Susan Carey (1978, 1985), for example, had noted that children overextend the words they're learning (so that "daddy" is applied to all men, and "doggie" to all four-legged animals), and suggested that animism results from this sort of overextension. Indeed, an interdisciplinary conference held at Ann Arbor in 1990 had focused on how universal cognitive mechanisms may give rise to various "cultural" phenomena (Hirschfeld and Gelman 1994).

Theory of Mind (7.vi.f), for instance, underlies anthropomorphic views about gods and spirits, from totems and witches to God our heavenly father. In using ToM (intentional) concepts to describe the supernatural, we help ourselves to the many default inferences associated with them. A father can wish, plan, love, protect, admonish, punish, and reward. And a witch, or a ghost, or a god can (by default) do all these things too.

Ghosts are especially interesting here, as is the awe/dread we often feel in the presence of a dead body—especially if we'd loved the dead person during their life. Regarding dead bodies in general, Theory of Mind naturally (*sic*) induces us to treat them as intentional systems—which, unfortunately, they no longer are. Boyer sees this mismatch as the source of the uncanniness experienced in the presence of the dead. And this in turn gives them their religious significance, and underlies the *interpersonal* dimensions of funeral rituals.

As for the religious significance of dead loved ones, that can be explained in comparable terms. Personal love is rooted in Theory of Mind, though it has weighty cultural aspects too. It's a highly complex set of cognitive, motivational, and emotional dispositions, all related to the purposes and attitudes of the loved one (Chapter 7.i.f). As a crucial computational structure within the lover's mental architecture, it can't simply be switched off at the loved one's death. Hence the inevitability of grief, and of mourning (again, see 7.i.f). (The same considerations imply that a committed believer in a personal God who loses their faith will suffer the tortures of hell in this world, if not in the next.)

And hence, too, the likelihood that some small comfort may be drawn from positing an *invisible* ghost, or soul, with whom some of the previously established personal relations can still be preserved. In other words, this isn't just superficial 'wishful thinking', but an attempt to maintain a central aspect of one's own mental structure.

Similarly, if priests or shamans or religious healers are a special *kind* of person, then they possess some hidden essence which gives them their priestly powers. (For the Fang people, this is an invisible magical organ, called an *evur*—Boyer 1994: 34–5, 45, 86–7.) However, since they look the same as everyone else, doubts can arise about their status—doubts that are *not* necessarily settled by their ancestry or gender, nor by any 'ordination' ritual they may have undergone (Boyer 1994, ch. 6, pp. 237–40, 252–3). It follows that doubts can arise also regarding the reflective beliefs about the supernatural which people are invited to accept on their say-so.

Religions, then, are natural phenomena: their key concepts are 'attractors' in our biologically rooted mental space. (In his later writing, Boyer related them to certain aspects of neuroscience: Boyer 2003.) But they're "un-natural" too (they're very *strange* attractors, a punster might say), for they *deny* some of the assumptions normally activated by these universal mechanisms. So an animal can be of only one species, but a totem may be a bear–snake. Nothing can pass through a solid wall—but Marley's ghost could, and so can the "ghosts" feared by the Fang people (Boyer 1994: 113–14). Every person is limited in power—but God isn't. Every physical event has a cause—but a creator–god doesn't. A block of wood or brass has no intentional states, nor any causal powers beyond gravity—but a graven idol does. And Everyman will die—but God won't; even animistic gods are ageless, or have an "age" that stays constant.

Communications using supernatural concepts are (up to a point) intelligible, contra Hobbes and Flew, because they automatically activate a host of familiar inferences in our minds. That's even true of 'weird' beliefs from other cultures—which is why Bartlett's subjects in Cambridge could make some sense of the alien story 'The War of the Ghosts' (Chapter 5.ii.b).

What makes such communications near-unintelligible, or slippery, is that one or more default assumptions is modified, or dropped. If witches fly on banana leaves (as claimed in the Fang religion), they'll have no difficulty in getting to wherever they want. But does it follow, or doesn't it, that they'd be 'grounded' if a forest fire destroyed the banana trees? If God is everywhere, and omniscient, then—unlike Ann's friend Sally (Chapter 7.vi.f)—He knows everyone's thoughts. But does it follow, or doesn't it, that (given that He's omnipotent too) He's part-responsible for the evil that humans do? And if He's wholly benevolent too, then what about other miseries? Darwin was one of many bothered by this:

I am bewildered... There seems to me too much misery in the world. I cannot persuade myself that a beneficent and omnipotent God would have designedly created the *Ichneumonidae* [a type of wasp] with the express intention of their feeding within the living bodies of caterpillars, or that a cat should play with mice. (letter to Asa Gray, 22 May 1860; in Darwin 1887)

And if God is omniscient, just how all-encompassing is that little prefix "omni"? Darwin, again:

[Do] you believe that when a swallow snaps up a gnat that God designed [or even foresaw] that that particular swallow should snap up that particular gnat at that particular instant? (letter to Asa Gray, July 1860; in Darwin 1887)

Such questions are inevitable, for they arise *naturally* from the conceptual/inferential schemas concerned. But they're also largely unanswerable, because those schemas have been fundamentally altered. (This is a psychological version of Immanuel Kant's "antinomies": metaphysical questions that we find ourselves compelled to ask, but which can never find an answer—Kant 1781.)

This, by the way, explains the phenomenon of heresy. If a powerful religious hierarchy has laid down a dogma, certain unorthodox answers may be officially denounced as "heresies". They're the ones which are likely to occur again and again, due to their close links with the intuitive schemas. Unorthodox answers that tempt only very few people aren't worth labelling as heresy.

It also helps explain the power of the priests, and the survival of religions once they've arisen. For the person who's socially sanctioned to interpret important life-relevant concepts which to the flock are unintelligible possesses a significant degree of social power. To that extent, natural religion (attained by method (1) defined above) *threatens* the status of the religious hierarchy. It's not surprising, then, that when Aquinas defined his Five Ways to the creator God, he pointed out that they said nothing about the personal God. For that, he said, we need not reason but revelation (guided, of course, by the Catholic Church). Claims to revelation, in turn, can be buttressed by religious experiences: item (2) on the list. Someone who's especially susceptible to—and/or especially skilled at inducing—florid states of possession, or strange visions, will very likely be revered as a religious authority (Chapter 7.i.i).

At this point, you may be thinking "What about quantum physics?" (thanks to Ron Chrisley for reminding me of this). Quantum physics, too, plays fast and loose with basic everyday concepts: cause, time, location, identity—not to mention Schrödinger's cat. (In that sense, quantum physics is a better candidate for Atran's Mickey Mouse problem than Mickey Mouse himself, or Karl Marx either—cf. Atran 2002: 13–15.)

The physicists themselves find this stuff weird (even "crazy": Chapter 17.i), and many others—myself included—simply can't begin to get their heads around it. If one's unable to cope with the mathematics, one can have recourse only to natural language—which is transgressed over and over again. So is quantum physics a religion?

Cynical remarks about power play and one-upmanship aside, no it isn't. Quantum physics does attract attention by (and even glory in) its inscrutability to common sense. (Sometimes, even non-physicists arrogantly trumpet the anti-commonsensical nature of science in general: L. Wolpert 1992.) But it grounds its counter-intuitive claims in as much empirical evidence as it can, justifying them by scientific experiments and by successful technologies. That is, it doesn't go in for Sperber's "symbolic" communication. Indeed, a key reason for the physicist Sokal's impatience with post-modernist interpretations of his field was that the postmodernists were doing just that (Section ii.b, above).

Moreover (recalling the other criteria in Boyer's definition), quantum physics doesn't buttress cultural rituals outside the laboratory, nor disseminate and validate sacred artefacts (with the possible exception, as I write this in mid-2004, of the *iPod*!). They don't claim the existence of strange forces that can influence specific events in the personal lives of those individuals who accept quantum physics. Similarly, they don't encourage prayer or propitiation as a way of avoiding unfortunate life events. Nor do they feel drawn to die, or kill, for physics. They're as envious as the rest of us of the status of the shamans, the power of the priests. So they may (they often do) 'play God' at cocktail parties, and even at scientific conferences involving non-physicists. But that's as far as it goes.

e. Anything goes?

Cognitive science has told us something about what sorts of bizarre ideas will arise—in James's terms, the production of religion. But are there any ideas *so* bizarre that they won't ever form the basis of a religious cult? To answer that question, we need to consider how religion spreads and how it's maintained.

(Boyer focuses on the cognitive aspects of *how* it's maintained, not the social/motivational reasons *why*. But he can allow, for example, that those cognitive aspects which buttress priestly power will be stressed by the priesthood in their own interests, and highlighted—largely thanks to them—in religious rituals. Similarly, items (3), (4), and (5) on the list of putative explanations above aren't inconsistent with Boyer's account, unless they're offered as the *only* explanation of the origin of religion. As for item (1), we've already seen that this is a special case of item (7)—which is Boyer's explanation.)

Religions spread and survive, according to Boyer, largely thanks to the flouting of the common-sense categories concerned. It's the counter-intuitive claims—about witches flying on banana leaves, or God being everywhere—which make religion so fascinating. (And, as just remarked, quantum physics too.) Like a 'face' with three eyes, they *naturally* attract even more attention than the unmodified schemas do. (They're further examples of behavioural novelty, which we've evolved to find attractive: see iv.b, above.)

That's why religious adherents themselves acknowledge, even revel in, the strangeness of their beliefs. A Fang adult takes it for granted that there are witches, never questioning this for a minute—but he does realize that witches are pretty strange. And the Church father Tertullian, converted by the courage of the Christians thrown to the lions in second-century Carthage, is famous for declaring *Credo quia absurdum est*: "I believe it because it's absurd." If his remark weren't so seductive, it wouldn't have survived for almost 2,000 years. (In fact, it hasn't survived unaltered. There's been a micro-evolutionary change, for his actual words were slightly different. Of the death of the son of God, he said *Credibile est, quia ineptum est*; and of the resurrection, *Certum est, quia impossibile*—Moffatt 1916: 170. However, saying "I believe it" is more personal—that's to say: more inference-arousing, more relevant—and so more memorable than saying "It's believable", or even "It's certain".)

It's no surprise, then, that religion is found everywhere. If it also happens to satisfy our needs for explanation, mental health, and/or social cohesion—items (1) and (3–5), above—its maintenance will be that much more likely. (Some psychological mechanisms making 'full-blown' religious belief possible are sketched in Boden 2002.) The recurrence of religion's universal core *isn't* due to cultural transmission.

But the recurrence of its superficial diversities *is*. Fang witches ride on banana leaves, Christian angels on clouds or wings. The survival of these representations from generation to generation is due to cultural transmission—or in a word, learning. For only by being a member of that culture, or an attentive visitor, can one acquire these concepts. To *that* extent, item (8) on our list, above, is vindicated.

It doesn't follow, however, that cultural learning is wholly unconstrained—a tacit assumption of item (8). Perhaps there are limits, even to religion? Are there some things which can't be believed, or can't even be thought?

Philosophers may approach this question in an 'absolutist' spirit. For instance, consider Colin McGinn's discussion of brain and consciousness (Chapter 14.x.d). He argued that we'll never find mind–body explanations. The problem (he said) isn't that the mind–body relation is *essentially* mysterious, or that minds are divorced from the material world: he assumed that consciousness arises from the brain. The problem is that our cognitive capacities aren't up to understanding it. Analogously, there's nothing mysterious about arithmetic, or the January sales—but dogs and goldfish will forever be oblivious to them. McGinn's claim, then, is that there are some truths that can never be known, or even entertained, by human minds. Some things, eminently worth thinking, simply can't be thought.

Possibly, that's so. For anyone who accepts a realist philosophy of science, this worry is inescapable. (For a non-realist, the problem can't arise.) It would be hubris to assume that we're potentially capable of understanding *every* aspect of the universe. But that's not what's at issue here.

For cognitive anthropologists, the question is different. It's this: are there some things that are thinkable in principle, but *not* capable of being widely accepted and/or reliably transmitted? Or can human cultures, differing greatly as we know they do, maintain just any idea?

Someone might cite multiply nested sentences here (Chapter 7.ii.b). For it's virtually impossible to understand them, especially without paper and pencil. We can easily get our heads around the linear version of *This is the house that Jack built*. But we couldn't cope with its hierarchically embedded equivalent. So a linguistic representation that's possible (grammatical) in principle is unintelligible in practice. However, its semantic content can be captured by a different representation, which can be understood even in the nursery. Indeed, Anglophone culture has already included Jack's house in its mythology—and the malt, the rat, the cat . . . right up to the farmer sowing his corn—by describing these things in the linear way.

A more telling example is Borges's joke taxonomy (Section i.b). In a weak sense, to be sure, he could understand it. He could describe it at his leisure, in written prose. But he couldn't have used it. He couldn't even have remembered it, except parrot-fashion—or perhaps with the help of a specially concocted mnemonic.

Some of the reasons were provided by Rosch, as we've seen. She showed that the structure of concepts, and the processes of conceptualization, simply don't fit the bizarre example imagined by Borges.

Even more to the point is Sperber's work on relevance. He and Wilson (1986) showed how inescapable constraints on information processing both bound and enable our rationality. Certain interpretations of what another person says won't arise, even though in principle they could. That's why, in Chapter 7.iii.d, I described relevance theory as addressing "the frame problem" for high-level thought. What goes for understanding goes for memory too. Something that's difficult to grasp in the first place will be difficult to retain—and to transmit. In other words, the mind's information processing influences not only James's "causes of production" of variation in ideas, but the "causes of maintenance" too.

Notoriously, beliefs can be shared in one culture which in others are literally incredible. Members of that culture can rely on this fact, in making sense (*sic*) of their fellows' communications. For instance:

> [Every] Freemason has access to a number of secret assumptions which include the assumption that all Freemasons have access to these same secret assumptions. In other words, all Freemasons share a cognitive environment which contains the assumption that all Freemasons share this environment. (Sperber and Wilson 1986: 41)

So what's irrelevant in one culture may be highly relevant in another. If a web of conceptual relations has already been learnt, a new communication may be effortlessly interpreted in a way that would never even occur to someone from a different cultural background. That's why one person can hold a belief which another can hardly understand, and certainly not credit: "How on earth can they believe *that*?"

But what of the 'Tertullian factor'? We've seen that the un-natural aspects of religious ideas help make them interesting, and therefore memorable. So perhaps the more bizarre the better?

Boyer thought not:

> [Certain] combinations of intuitive and counterintuitive claims constitute a cognitive optimum, in which a concept is both learnable and nonnatural (Boyer 1994: 121).

> Religious ontological assumptions generally comprise a culturally transmitted [i.e. explicit] part, which violates intuitive expectations, and a tacit, schematic part, which confirms them. Assumptions of this type *are likely to be more recurrent if they reach a cognitive optimum*, in which (1) the violation clearly marks off the putative entity or agency from ordinary objects and beings

and (2) the confirmation imposes maximal constraints on the range of inferences that can be drawn from cultural cues. (p. 287; italics added)

In short, the cognitive equilibrium between the familiar and unfamiliar must be maintained. If the balance is thrown too strongly towards the latter, the potential belief won't be accepted—or anyway, won't be widely communicated.

For instance, remember the flying saucers promised in the message from the Guardians on the planet Clarion (Chapter 7.i.c). They'd been expected in the first place only by the cult members. When they failed to arrive, the charismatic leader's second "message"—that the whole town had been saved from the flood at the last minute by the faith of the cult—was credited only by the few members still cooped up with her in her suburban house, offering each other mutual support. And when those few faithful finally tried to spread their "good news" over the media, nobody listened—except to laugh.

At first glance, rescue-by-flying-saucer seems no more bizarre than bodily ascensions or virgin birth, both of which are accepted by millions of people in "the same" culture. To be sure, the latter beliefs have two millennia of tradition behind them. But even that's not accidental. Boyer's analysis helps us understand why the idea of bodily ascension spread widely, and is still maintained by millions, whereas the vision of flying saucers didn't, and isn't. And even flying saucers and other-worldly Guardians are intelligible, having strong roots in our intuitive physics and theory of mind. Had Borges aspired to lead an 'animal' cult, he'd have found no followers: he'd have been on his own, tangled up in his artificial mnemonics. As Boyer put it, "the variability of cultural ideas is not unbounded" (1994: 5).

Most anthropologists, however, seem to assume that it is. The Borges example should give us pause. And Boyer suggested a simpler one: that *everyone* can be a medium, but only *every other day* (1994: 6). There are two things wrong with this. (In fact, they're two sides of the same coin.) On the one hand, a medium is supposed to be a special kind of person, whose essential properties (including access to the supernatural) can't be switched on and off. On the other hand, the natural (*sic*) assumption that "same causes have same effects" can't be—*can't be*, not just *isn't*—given up without strong justification. Conceivably, a medium might be able to commune with the dead only on the anniversary of the deceased's birthday, marriage, or death; or a Christian saint might be able to work miracles only at Christmas or Easter, or only on Sundays. But *every other day* . . . ? That's not merely bizarre. It's almost meaningless, since this imaginary limitation is utterly irrelevant to other 'spiritualist' representations. Even if *per impossibile* a tiny cult were to accept this every-other-day belief, it would never sweep the world.

Once an inferential constraint has been loosened, the way is left open for inferences that would otherwise be blocked. If angels are immaterial agents, they have no weight: so why shouldn't they fly around on golden wings? And once they're doing that, why shouldn't they stop to take a rest on a passing cloud? Clouds or rest, in turn, will trigger other mental schemas—some of which may be highly culture-specific. In short, this is epigenesis: the universal core can develop in many differently detailed ways (Chapters 7.vi.g and 14.ix.c). Hence the enormous cultural diversity that anthropology describes: given the creative imagination of *Homo sapiens*, weird stories about supernatural beings can blossom to our hearts' content.

However, our hearts—or rather, our minds—can't be contented by *just anything*. If too many default assumptions are flouted, the "relevant" inferences become so computationally unconstrained that they're not merely slippery but unmanageable—and unmemorable, too. In other words, they aren't really relevant at all. And since communication depends on the (mostly tacit) recognition of relevance, the problematic representation can't be transmitted either.

f. The impurity of induction

Anthropologists in general are highly suspicious of evolutionary approaches, as we've seen (Section ii.b). Chagnon, one of the two men slandered by *Darkness in El Dorado*, co-edited a large evolutionary volume soon after the sea change in anthropology had got under way (Chagnon and Irons 1979). Clearly, given the AAA's initial support for *Darkness* twenty years later, this did nothing to enhance his reputation in the eyes of the mainstream. And Boyer remarked defensively at several points in his 1994 book that universalist accounts still weren't considered *comme il faut* by most of his peers, because of "the occupational disease of relativism" (p. 111)—their chief weapon in the science wars (Chapter 1.iii.b).

Besides disliking evolutionary accounts (as scientistic), most anthropologists think they don't need them. For they generally assume that *cultural* transmission can work for all representations. This assumption depends on their faith in pure induction, item (8) on the list above. That is: any pattern, any similarity, in the input can be learnt if it's encountered often enough.

Some early cognitive scientists made the same assumption. (Not Warren McCulloch, however: see 4.iii.b and 14.ii.b.) Now, they don't. They believe, with Boyer, that

> The more people pick up available information from the environment, the more they are working on (and constrained by) implicit hypotheses about what is to be picked up. (Boyer 1994, p. x)
>
> The variety of situations that a subject experiences supports indefinitely many possible inferences, of which only a small subset are ever entertained by the subject. Unless we have a good description of the mechanisms that constrain the range of hypotheses, the empiricist account is insufficient. (p. 25)
>
> [There] is no such thing as a simple inductive device, and in fact there cannot be such a device. (p. 95)

One might say they've become more Kantian, for Kant argued that sensory experience must be informed by structuring "intuitions", or "categories" (Chapter 9.ii.c). But that doesn't mean they've all been reading Kant. (More likely, though also not necessarily true, they've been reading Noam Chomsky: 9.vii.c.) Their initial optimism was damped by difficulties, not to say failures, in *implementing* pure induction:

> This is, again, a very familiar point in cognitive modelling; unless one wants to accept indefinitely long searches, one must assume that knowledge structures [mental schemas] provide some information to constrain inferential processes. (Boyer 1994: 59)

In brief, inductive learning requires some initial guidance and/or restriction. Without that, the computational system would be overwhelmed by a host of similarities in the input, most of which are of no significance whatever. That's been shown independently by work in GOFAI, in connectionism, in linguistics, in developmental psychology, in adult cognitive psychology, in the philosophy of science, and in neuroscience.

The system may start out with relatively specific guidance—such as a module for face recognition, a "Language Acquisition Device" (not necessarily Chomskyan), or a scientific hypothesis (Chapters 7.vi.d–e, 9.vii.c–d, and 6.b–d, respectively). And/or initial processing restrictions may prevent it from being swamped by the input, being gradually relaxed thereafter (see 12.viii.c–d). Or it may benefit from a helpful input history (12.viii.e). There may even be different neural learning mechanisms for different domains: the domain-neutral associative bond may be what Randy Gallistel has called "the phlogiston of psychology" (14.ix.g).

But whatever the details turn out to be, the empiricist model of learning adopted by most anthropologists is vacuous:

> [Many] anthropological accounts of recurrence and transmission [have] a strong magical flavor . . . [Saying] that "cultural models are somehow transmitted through socialization" is not very different from saying that "performing the ritual somehow makes the rains fall". Or, to be less polemical, it is very much like saying that "turning the ignition key somehow makes the engine start", which certainly constitutes a reliable principle, but hardly a theory of thermal engines. (Boyer 1994: 265)

The theory of *mental/cultural* engines that we need is "a *rich* psychology", not an artificially simplified one (p. 288).

The banana leaves, and the angels' wings, are representations that can be easily transmitted (learnt) because they're inferentially linked to intuitive schemas about animate agents travelling through space. We don't bother to mention that if a witch/angel wants to speak to someone, they need to move to the place where that person is. Probably, we don't even explicitly think it: wings and banana leaves are much more interesting. In some languages, we might not even be able to make it explicit. But that doesn't mean we wouldn't be tacitly influenced by it: if a culture has no word translatable as *cause*, *belief*, or *intention*, it doesn't follow—contrary to some anthropologists' claims—that they don't possess those concepts tacitly (Lee 1949; Boyer 1991).

It's true that if we're challenged—or if the witch/angel is replaced by God, who we're told is *everywhere*—we're prepared to drop the commitment to location change. That's just one of Flew's death-dealing qualifications. They number "a thousand" because, where religion is concerned, no single inference is safe from dropping. Without any inferential guidance at all, however, we'd be lost. We couldn't acquire the concepts in the first place. (We couldn't even acquire religious *rituals* if they weren't rooted in intuitive intentional schemas, like those which underlay Robert Abelson's account of beliefs and actions: Chapter 7.i.c; cf. Boyer 1994: 202–5.)

Admittedly, highly abstract religions do exist: certain forms of Buddhism, for instance. However, they appeal only to highly literate specialists. The populace understands them in more concrete ways. Benedict Spinoza's (1632–77) austerely non-anthropomorphic vision of God has intoxicated many individuals, myself included. The Spinozahuis in The Hague where he died teems with his admirers, eager to sign the visitors' book as an act of homage. But his closely argued *Ethics* (1677), which rejects every familiar predicate in monotheism (even including "God is *one*"), will never become the basis of a culture-wide religion. It can be the topic of philosophy, or of psychology—but never of anthropology.

In sum, even religion has its limits. It's not the case that—in some culture, somewhere—anything goes. Cole Porter was wrong. But if Cole Porter was wrong, his contemporary Bartlett was right. In his book on memory and social psychology, Bartlett pointed out that anthropologists ignored the mechanisms that make culture possible:

> [Very little has been said about] social conduct in any genuinely observed sense, except by anthropologists *whose interests are apt to stop short at detailed description.* (Bartlett 1932: 239; italics added)

That's no longer true. Or anyway, it's no longer true of the entire profession. Thanks to the cognitive anthropologists, we now have computational accounts of a wide range of cultural phenomena. These are mostly schematic outlines, not nitty-gritty computer models. Nevertheless, the sixth point on the Sloan hexagon has been respected, polished, and sharpened.

9

TRANSFORMING LINGUISTICS

Theoretical linguistics has been enormously important in the history of cognitive science *as a whole*. But it might not have been. After all, why should psychologists, or philosophers, or anthropologists, or neurophysiologists, or computer scientists be interested in linguistics? If they happen to be concentrating on language, then perhaps it's relevant. If not, why should they bother with it?

As late as 1950, most of them didn't. And, in fact, most of them don't today. But in the late 1950s to 1970s all cognitive scientists had to pay some attention to it. This chapter explains why that was. (Why the interest then dwindled to near-zero is explained in Chapter 7.iii.a, and in Section ix.g below.)

The cyberneticists gave no thought to linguistics—and very little to language. The nearest they got was Warren McCulloch's 1920s search for a "logic" of verbs (4.iii.c), a pastime foreign to his cybernetic fellows. Even Gregory Bateson and Gordon Pask, who both championed language-based psychotherapy, weren't interested in the structure of language. When Alan Turing and others in the UK wrote the first computer programs involving language, they ignored linguistics (see x.a, below). And the (few) psychologists who focused on language paid scant attention to linguistic theory, and—crucially—didn't see their work as relevant for the study of mind *in general*. Nor was linguistics needed to spark off the cognitive revolution: that originated in psychology and AI, not linguistics (4.iii.e, 5.iii–iv, and 6.i–iii).

It turned out, however, that theoretical linguistics—specifically, the study of *syntax*—stoked the fire sparked off by other disciplines. (The contemporary work on phonetics was less relevant in this regard, and I shan't discuss it.) By the early 1960s no one involved in cognitive science, or in the philosophy of mind, could ignore it. The reason, in just two words: Noam Chomsky (1928–).

Linguistics turned computational with Chomsky. He tried to formulate a mathematical definition of language *as such*: a "computational" account, in the abstract sense later stressed by David Marr (7.iii.b). His work, including a pioneering paper on computer languages, encouraged others to persevere with, or to embark on, the computer modelling of language. But he himself didn't join them. He was highly sceptical about what's normally called computational linguistics, or natural language processing (NLP). He dipped his toes into the water only once (G. A. Miller and Chomsky 1963: 464–82)—by which time many others were already swimming in the shallows (see Section ix).

Chomsky's early publications—up to the mid-1960s—had a huge, and lasting, effect on cognitive science overall. This needn't have happened, even if his *linguistics* was, as one disciple later said, "*the* existence proof for the possibility of a cognitive science" (Fodor 1981*b*: 258). But Chomsky generalized his philosophical and methodological claims. He even defined linguistics as "part of psychology", focused on "one specific cognitive domain and one faculty of mind" (Chomsky 1980*b*: 1). Granted, he cared little about the psycholinguistic processing details—especially when they turned out to be less consilient with his theory than had at first appeared (Chapter 7.ii.a). But he saw his linguistics, in effect, as a theory of *mind*.

Specifically, he revived the then hugely unfashionable doctrine of *nativism*. This is the view that the mind/brain of the newborn baby is already equipped with knowledge of, or dispositions towards, language—and various other things, too. Modern proponents of nativism, such as his ex-pupil Steven Pinker (1954–), cite many biological data that were unknown to Chomsky (Pinker 1994, 2002). But he remains their key inspiration. Indeed, a very recent book by a leading anti-nativist has acknowledged that his "arguments [for linguistic nativism] are still widely read, *and carry as much weight today as any newly-produced publications*" (Sampson 2005: 7; italics added). In brief, he's still a major force.

Chomsky's influence on cognitive science was beneficial in many ways. In particular, he deepened the nascent questioning of behaviourism, and encouraged "mentalist" theories couched in terms of internal rules and/or representations (Chapter 6.i.e). He revivified research on nativism (7.vi), and indirectly encouraged the growth of evolutionary psychology (8.ii.d–e and iv–v). He offered a vision of theoretical rigour which inspired linguists and non-linguists alike. And, despite his own scepticism, his work encouraged others to attempt the computer modelling of mind. In short, for the field as a whole, he was a Good Thing.

Nevertheless, this chapter starts, unusually, with a health warning: *beware of the passions that swirl under any discussion of Chomsky* (Section i).

Next, Sections ii–v locate Chomsky in relation to the three centuries of language scholarship—including linguistically inspired philosophies of mind—that preceded him. The ancient writings are discussed in some detail here because he himself made a point of comparing his own work to them in a provocative way. I begin by distinguishing predecessors from precursors (in Section ii), and then consider the various writers named by Chomsky as his "rationalist" precursors (see Sections iii–iv). Section v sketches the state of theoretical linguistics during Chomsky's youth.

In Section vi, I describe how he first sought to change it. Section vii deals with his nativism, and the associated attack on behaviourism, while Section viii says a little about the subsequent changes in Chomskian theory. Various fundamental critiques of his account of language are outlined in Section ix, which also shows the need for the health warning.

Finally, the last two sections describe the early days, and the eventual maturation, of work in the computer modelling of language. These matters could have been discussed in Chapter 10, for NLP was central to GOFAI research. I decided to place them here instead, largely because the problems experienced in NLP recalled theoretical questions highlighted by the scholars featured in Sections iv–v, below.

A word on demarcation: this chapter deals with linguistics rather than psycholinguistics. The focus is on the nature of language *as such*. Psychological questions about how we understand, communicate, and remember linguistic meanings do crop up here, of course. But they're addressed more directly in many other parts of the book (e.g. Chapters 7.i.c, ii, iii.d, and iv.d–e; 8.i.a–b and vi; 10.iii.a and e; 12.x; and 16.i.c and iv.c–d). Indeed, Chomsky's influence led many linguists to ignore such questions: see Section ix.g, below.

9.i. Chomsky as Guru

Many people, including youngsters, take Chomsky as their political guru. Indeed, it's mostly because of his courageous and uncompromising political writings—which far outnumber his publications on linguistics—that he's the most-quoted living author (Barsky 1997: 3). I'll say a little about his politics in Section vii.a, below. But it's his role as a *scientific* guru that's relevant here.

I said, above, that Chomsky was a Good Thing for the development of cognitive science. So he was. But to acknowledge this isn't to accept the tenfold Chomsky myth.

a. The tenfold Chomsky myth

Those who uncritically take Chomsky as their scientific guru share ten beliefs:

* They think (1) that besides giving a *vision* of mathematical rigour, he always achieved it in his own work.
* They believe (2) that his linguistic theory was, or is now, indisputably correct in all essentials.
* They hold (3) that the (nativist) psychological implications he drew from it were convincingly argued at the time, and
* (4) that they are now empirically beyond doubt.
* On matters of history, they suggest (5) that his work of the 1950s was wholly original.
* And they suppose (6) that his writings of the 1960s were—as he himself claimed—the culmination of an august tradition of rationalism dating back 300 years.
* Some also assume (7) that without Chomsky's grammar there would have been no computer modelling of language.
* Many believe (8) that Chomsky was responsible for the demise of behaviourism.
* Most take for granted (9) that he reawakened and strengthened the discipline of linguistics.
* And many assume (10) that linguistics is as prominent in cognitive science today as it was, thanks to Chomsky, in the late 1950s to 1970s.

Each of these beliefs, strictly interpreted, is false. But all, suitably qualified, approach the truth. (The least defensible are nos. 9 and 10: in some ways, as we'll see, he *set back* the field of linguistics, and *undermined* its relevance for cognitive science in general.)

Moreover, many people in Chomsky's heyday accepted all of them (except no. 10, of course)—and a significant number still do, with no. 10 now added. That's why

Chomsky is not only a pivotal thinker in theoretical linguistics, but also a hugely important figure in the wider story of cognitive science.

However, "important" means *influential, provocative, fruitful* . . . perhaps even *largely beneficial*. It doesn't necessarily mean *right*.

b. A non-pacific ocean

The tenfold myth attracts passions of surprising depth. Surprising, that is, to anyone who believes the Legend: that science is a disinterested enterprise (see 1.iii.b).

While doing the research for this chapter, I was astonished to discover the lack of mutual respect between the two camps in linguistics, Chomskyan and non-Chomskyan. I'd known there were unpleasant tensions of course, but I hadn't realized their degree. They range from unscholarly ignorance of the very existence of important competing theories (as I discovered in a recent conversation with a young MIT linguist) to vitriolic hostility and public abuse.

As one illustration, a leading American linguist who (unusually) has an open mind on the issue of nativism has recently bemoaned "what is, I am afraid, an often disturbingly unserious, indeed irresponsible approach to the innateness position on the part of its major advocate" (Postal 2005, p. viii). Paul Postal (1936–) then quotes some "revealing", "outrageous", and "grotesquely untrue" remarks by Chomsky, the burden of which is that all the arguments against innateness are such confused nonsense that nobody ever bothers to answer them. And his verdict is that "These statements [by Chomsky] are not scientific comments but rather the analog of awful, partisan political discourse" (2005, p. ix).

As another illustration, a non-Chomskyan complained to me about Chomsky's theorizing about syntax:

> Anything that any human being actually says is at once defined by Chomsky out of the discipline, leaving him and his followers free to pursue their strange little version of navel-gazing, free from any contamination from the real world. *And I can't forgive Chomsky for perverting my discipline in this sick manner.* (Larry Trask, personal communication)

Larry Trask told me, a few months before his tragically early death in 2004, that he was happy for his remark to be quoted, since he'd said much the same in print. (And I'm confident that he'd have been happy with the italics, which are mine.)

Third, the computational linguist (and AI–vision researcher) Shimon Edelman (1957–) has recently referred to "the Bolshevik manner of [Chomsky's] takeover of linguistics", and the "Trotskyist ('permanent revolution') flavor of the subsequent development of [his] doctrine" (S. Edelman 2003). To be fair, he allows that there were cultural reasons for the "takeover" in early 1960s America (Koerner 1989). These were the widespread counter-cultural rejection of the views of the older generation (see 1.iii.c), and the post-Sputnik increase of research funding—especially at MIT (see 11.i). Thanks to that money, many new departments of linguistics were set up in the USA at that time, and it was "inevitable" that they'd be staffed by young lecturers trained at MIT (James McCawley, quoted in Koerner 1994). So Edelman isn't suggesting that Chomsky was an all-powerful tyrant, solely responsible for what happened. By the same token, the Chomskyan "revolution" wasn't caused by purely *intellectual* factors (S. Edelman 2003: 675).

And last, an anonymous linguist who reviewed this chapter accused me of pursuing "a personal vendetta" against Chomsky instead of "doing intellectual history". He, or possibly she, couldn't have been more wrong. I'm not a professional linguist, so have no axe to grind either way. (And like Postal, I have an open mind about linguistic nativism: see Chapter 7.vi.) Moreover, it's precisely because I'd done the intellectual history carefully that I'd realized just how questionable, and in certain respects just how damaging, some of Chomsky's claims were.

It's a telling sign that my account—which isn't written in a polemical spirit (though it does quote polemics, from both sides), and which gives chapter and verse for every criticism—should have been misinterpreted in that way. So be warned: these are turbulent waters. When it comes to Chomsky, few linguists, if any, are emotionally neutral.

(The same reviewer wanted me to change the chapter because "it will upset people". Well, I've no doubt that it will upset some people—though it will probably hearten others. But I haven't changed it: it's an honest account of what I found in the literature, and of what I was told in conversations with several leading linguists.)

Far from having mellowed with time, the passions have risen over the years. In Sections viii and ix, we'll see why this is so. But before then, we'll consider Chomsky's early work (up to the mid-1960s), its immediate reception, and—first of all—its intellectual background.

9.ii. Predecessors and Precursors

Let's start with some banalities:

A predecessor is a person who discussed the same topic as some later scholar.

A precursor is a predecessor who said similar things about it.

A precursor needn't have influenced the scholar concerned: intellectual anticipation isn't the same as intellectual ancestry.

Identifying precursors requires one to decide just what one means by "similar".

Similarity isn't merely a question of degree, but of nature. Two œuvres may have more or less of a certain feature, and/or they may have partly different features.

Finally, someone who interprets "similarity" too broadly not only obscures important differences, but casts all the scholars concerned in intellectual roles for which they aren't best suited. So reflected glory may be more distorting than illuminating.

a. Why Chomsky's 'history' matters

The points just listed, which relate to the caveats given in Chapter 1.iii.f, are pertinent in any discussion of the history of ideas. But they're especially relevant here. They concern not only Chomsky's intellectual pedigree, but also his rhetorical efforts to share in the glory associated with great names of the past.

Those efforts weren't mere self-seeking. Rather, they were a form of philosophical defensiveness. Chomsky had deliberately suppressed his (nativist) ideas on language-and-mind when writing his first book (1957), judging them to be "too audacious" for publication (see Section vi.e). When he did finally announce them, he defended them partly by locating them in the context of respected thinkers of the past (1965, 1966). In so doing, however, he misrepresented those thinkers in various ways.

"So what?" you may ask. "Why bother to rap Chomsky over his historical knuckles? He wasn't a historian, after all." Well, no. But historical accuracy is important here for two reasons.

He himself, in defending nativism, gave his Cartesian–rationalist labelling a high profile (and many Chomsky fans uncritically took him at his word). Today, his long-time champion Jerry Fodor (1998*c*) speaks of "the New Rationalists", of whom he counts himself one (7.vi.d). We need to know just how literally this tag should be taken.

Even more to the point, we must try to grasp the shifting subtleties of the long-standing debate about "innate ideas" if we're to understand nativism *in general.* In this context, we need to know what the philosophers he referred to actually said—which often differed, more or less, from what he implied that they said.

On the one hand, Chomsky himself repeatedly named as precursors people—dating back to the early seventeenth century—whose views differed significantly from his own, and implied that they were more similar to each other than is the case. (In so far as they are indeed his precursors, they're anticipatory rather than ancestral: he studied most of them only after developing his own ideas.)

On the other hand, the fact that linguists today situate themselves—in agreement or opposition—with respect to Chomsky might suggest that his work was a complete break with what went before. In other words, it suggests that his teachers were mere predecessors, not precursors.

In discussing these matters, we must distinguish his views on language-and-mind from his theory of language (primarily, of syntax) as such. The "one hand" (above) concerns the former. The "other hand" concerns the latter.

With regard to his grammatical theory, one might say that Chomsky had no precursors. More accurately, the sense in which others before him had "said similar things" was an attenuated one. He himself claimed that the transformational model of language was "not at all new", having been anticipated by the Port-Royal logicians of the seventeenth century (Chomsky 1964: 15–16). But that was true only in the most general sense. Even those of his immediate predecessors who developed formal theories of syntax did so in much less detailed ways. Chomsky's grammatical work, although clearly dependent on contemporary advances in formal linguistics and information theory, was largely novel (see Sections v–vi).

As for his views on the relation between language and mind, Chomsky was the first to say that he had many precursors (Chomsky 1964: 8–25). But most, he says, were long dead and largely forgotten:

> These early modern contributions were scarcely known, even to scholarship, until they were rediscovered during the second cognitive revolution, after somewhat similar ideas had been independently developed. (Chomsky 1996*a*: 7)

That remark apparently applies to Wilhelm von Humboldt (1767–1835), whose work one historian sees as having suffered "a pattern of recurrent, or successive, amnesia" in American linguistics (Hymes and Faught 1981: 98). And it applies also to the Port-Royal *Grammar* (see Section iii.c), which another historian of linguistics has described as "forgotten, [and] often misunderstood" by the end of the nineteenth century (Koerner and Asher 1995: 174). But it doesn't apply so clearly to the Port-Royal *Logic*, which was being translated into English for the fourth time just as Chomsky's first book appeared. And I know from my own experience that Herbert of Cherbury (on innate ideas) was routinely recommended to Cambridge philosophy undergraduates in the 1950s. Perhaps professional linguists weren't reading these people, but others certainly were.

These examples suggest that some of Chomsky's remarks about his precursors may need to be taken with a pinch of salt. Perhaps a sackful: a third specialist historian has gone so far as to say that Chomsky's version of the history of linguistics is "fundamentally false from beginning to end" (Aarsleff 1970: 583).

Even if that judgement is too harsh, there are systematic problems with regard to Chomsky's account. I noted, above, that talk of precursors depends on what one counts as similar. Chomsky interpreted the similarities so broadly that his descriptions of his forerunners were often misleading. In particular, his frequent uses of "rationalist" and "Cartesian" were inaccurate. And it's not at all clear that what his forebears had meant by the "inner form" of language, or by "creativity", was what he meant by these words.

The central doctrine of "Cartesian linguistics", said Chomsky, is that "the general features of grammatical structure are common to all languages and reflect certain fundamental properties of the human mind" (Chomsky 1966: 59). In other words, "Cartesian" linguistics holds that there is some inborn grammatical core of all natural languages.

Chomsky did attach a caveat to his inclusive term ("Cartesian"), pointing out that the rationalist/nativist writers he had in mind fell into more than one philosophical school. Nevertheless, he implied that they comprised a coherent tradition of humanist rationalism, one that had been forgotten but had now been revived by his own research (Chomsky 1966, 1968).

If we're to situate that research in its historical context, we need to understand how his nominated precursors differed—from him and from each other—as well as how they were alike. In this section, we'll consider their general philosophical background; specific examples are discussed in Sections iii and iv.

b. The rationalist background

In broad outline, the writers Chomsky named as his "rationalist" precursors agreed on two things. They believed that language is the origin of human thought, and that it's innate in every child.

Some—rationalists in the strict sense—interpreted "thought", here, to mean access to necessary truth. Some, but not all, remarked that the second (nativist) belief implies the existence not only of some specifically human propensity for language, but also of some common core of language as such. Some believed in, or hoped for, a past (or future) universal language, intelligible to everyone because it was (or would be) wholly or largely natural, not purely conventional. Some saw language as the essential core of humanity. And some stressed the role of language in creativity and culture.

As we'll see in Section vii.d, Chomsky wasn't a rationalist in the strong sense. He didn't posit innate access to necessary linguistic principles, still less to necessary truth. Nor was he interested in projects aimed at designing a universal language, or at locating the origins of language in prehistory. But he did believe that language is what makes us human, that language and thought are intimately connected, that language is essentially creative, that all languages share a common grammatical core, and that linguistic ability is innate in (only) the human species.

In Chomsky's own judgement, his most important precursor was René Descartes. As we saw in Chapter 2.iii.c, Descartes taught that language is uniquely human. We don't share it with any animal, nor could it be provided to a machine:

> If you teach a magpie to say good-day to its mistress, when it sees her approach, this can only be by making the utterance of the word the expression of one of its passions. For instance, it will be the expression of the hope of eating, if it has always been given a tidbit when it says the word. Similarly, all the things which dogs, horses and monkeys are taught to perform are only the expression of their fear, their hope or their joy [all of which are entirely mechanical: see Chapter 2.ii.d] ... But the use of words, so defined, is peculiar to human beings. (Descartes 1646: 207)

No machine, according to Descartes, could "arrange words variously in response to the meaning of what is said in its presence, as even the dullest men can do". The human ability to arrange words variously was to be heavily stressed by Chomsky, who both sought to explain it and used it as a stick to beat behaviourism. However, what Chomsky meant by this ability wasn't what Descartes meant by it (see Section iv.f).

As for where language ability comes from, Descartes believed that it's provided to all human minds by God. His reason was that, because of its variability, it can't be explained in material terms. He pointed out that deaf-mutes—fellow human beings—spontaneously use gestures and make up signs that other people can learn to understand (1637, ch. 5).

It's not at all clear, however, that Descartes was a "Cartesian" according to Chomsky's definition (above). He didn't specifically say that language has a core structure, of which we have innate knowledge. Indeed, he described languages as unpredictably diverse, to be learnt from the environment much as other social customs are. And he attributed the variability of language to the innate power of reason (the ability to come up with new and appropriate thoughts), not to any property of language as such. He did believe, however, that some human abilities rest on innate knowledge.

c. The puzzle of innate ideas

In general, for Descartes, our knowledge depends heavily on innate ideas. These exist in the mind prior to any experience.

The idea of God must be innate, he said, since the concept of infinity could be neither derived nor—as Thomas Hobbes had argued (see 2.iii.b)—extrapolated from experience. The true proposition "God exists" is also innate (said Descartes), because it is self-contradictory to suppose its falsity. Similarly, he saw the necessary principles of logic and mathematics as present in the mind a priori.

For Descartes, no true *propositions* about the empirical world are innate: only experience can teach us that roses are red and violets blue. Here, he was less nativist than Chomsky, for whom knowledge of certain empirical facts about language—facts which might have been different—is inborn (see Section vii.c–d). But Descartes sometimes argued that the *sensory concepts* themselves must be innate, since no genuine causal relation between brain and mind is possible:

> [Besides the ideas of movements and figures] so much the more must the ideas of pain, colour, sound and the like be innate, that our mind may on occasion [*sic*] of certain corporeal movements, envisage these ideas, for they have no likenesses to the corporeal movements. (trans. Haldane and Ross 1911: i. 42)

Descartes wasn't the first to posit innate ideas, whether concepts or propositions (McRae 1972; Chomsky 1966: 59–72). They'd been spoken of for centuries. Plato's doctrine of *anamnesis*, introduced in *The Meno* to explain the slave boy's intuitive grasp of geometry, is a well-known example. Most rationalist writers followed Plato in attributing necessity to innate ideas. Descartes's occasional suggestion that even sensations are innate was unusual.

But familiarity didn't bring clarity. There was much controversy about just what this doctrine amounted to, in Descartes's writings as in other people's. This was so, irrespective of whether it concerned only concepts or also propositions.

Did it mean that we have *actual* ideas—of number, extension, God, or even colour—prior to any experience? Or did it mean that our *potential* to do arithmetic and geometry, to think about God, or to perceive red and blue, is a natural capacity, or disposition, possessed from birth by all human minds—as opposed to a learnt skill? And if the latter, *just what* is a "potential", "capacity", or "disposition" to have certain ideas? Are these expressions anything more than philosophically misleading ways of saying that we can, in fact, do so?

Many of Descartes's correspondents, including the arch-empiricist Hobbes, raised these questions. Comparing innate mental dispositions to a family's inborn propensity for a certain disease, Descartes replied:

> I never wrote or concluded that the mind required innate ideas which were in some sort different from its faculty of thinking; but when I observed the existence in me of certain thoughts which proceeded, not from extraneous objects nor from the determination of my will, but solely from the faculty of thinking which is within me, then . . . I termed them INNATE. (trans. Haldane and Ross 1911: i. 442)

But his reply didn't end the debate. On the contrary, it raged on long after he died.

John Locke (1690) interpreted the doctrine literally, and rejected it. Or rather: "he formulated every clear version he could think of and showed each to be either trivially true or obviously false" (Goodman 1969: 138). He likened the newborn mind to a blank wax tablet, or tabula rasa (which, trivially, has certain potentialities rather than others). Locke's view was very widely accepted at the time—not least, for its 'enlightened' political implications. Indeed, the entry on "Idea" in Ephraim Chambers's *Cyclopaedia* of 1728 stated that "our great Mr. Locke seems to have put this Matter out of dispute".

But Chambers was over-optimistic. Gottfried Leibniz (1690), specifically targeting Locke's discussion, interpreted the doctrine dispositionally, and defended it. He compared the mind to a veined block of marble, which both enables and constrains the figures that can be sculpted from it.

Empiricists accepted Locke's interpretation. They saw knowledge in general as built from passively received sensations, and language as the learnt ability to use certain words to stand for certain ideas.

In justification, they pointed out that languages are very different, and that children learn to speak as the people surrounding them do. They remarked that feral children, much discussed in the seventeenth and eighteenth centuries (Itard 1801), don't use language spontaneously, but will learn it if rescued while still young (for a modern discussion, see Curtiss 1977). They argued that dispositional accounts of innate ideas were merely pretentious ways of stating a commonplace: that people can learn language and mathematics. And they complained that positing innate ideas prevents one from asking more searching questions about the origins of knowledge. These empiricist arguments were still prominent in the twentieth century, and were marshalled by some of Chomsky's critics (see Section vii.c–d).

Those of a rationalist cast of mind favoured Leibniz's interpretation. But it was unclear just what cognitive dispositions might be involved and/or just how they contribute to knowledge. Rationalists claimed, for instance, that necessary truths are innate, but they couldn't explain why geometry and arithmetic seem to provide genuinely new knowledge.

That changed after the late eighteenth century, thanks to Immanuel Kant's *Critique of Pure Reason* (1781). The problem of innate ideas—at least as regards our knowledge of the physical world—seemed, to many, to have been solved by Kant's distinction (sketched in Chapter 2.vi.a) between empirical intuitions and categories on the one hand and sensory experience on the other. The former are (he said) innate, but are merely structuring principles. Only the latter can give actual content to our knowledge. In Kant's words: "Thoughts without content are empty, intuitions without concepts are blind" (Kant 1781: 61).

It was more difficult, however, to gloss language as innate in this sense. For there seem to be no universal structuring principles. Whereas we all acknowledge causality and Euclidean space, our languages appear chaotic. Not only do we use different words for the same thing (*children, enfants, tamariki . . .*), a fact linguists call "the arbitrariness of the sign", but we put the words together in very different ways. In short, languages differ not only in vocabulary, but also in grammar.

This seemingly incontrovertible truth, coupled with the growth of empiricism outlined in Chapter 2, led—in scientific circles—to the demise of the rationalist view of language. By the end of the 1940s, when Chomsky was a student, the predominant school of scientific linguistics favoured environmentalist accounts of language learning (see Section v.b).

That situation was partly due to the dominance of behaviourism in American psychology. But even anti-behaviourists fought shy of nativist theories of language. Thus the Gestalt psychologists, who developed neo-Kantian explanations of the perception of causation and identity, didn't do so for language. And Jean Piaget, who recognized universal features of language such as temporal order and hierarchy, saw them as based in the universals of logic. These in turn, he said, are constructed by the infant's sensori-motor activity in the physical world (Piaget 1952*b*; Piatelli-Palmarini 1980). In short, he saw language as neither unique nor innate.

In humanist circles, by contrast, the nativist viewpoint on language and mind survived. In the three centuries separating Descartes's birth from Chomsky's, it was elaborated in various ways. Literally Cartesian versions of innate ideas gave way to neo-Kantian and Romantic ones. And, in place of Descartes's mix of pure mentalism and mechanistic biology, there arose a holistic view of mind as grounded in self-organization, similar to the autonomous development of the embryo (see 2.vi). The diversity of languages wasn't denied: quite the contrary (see Section iv.b). But the basic humanist commitment to species-specific linguistic nativism remained.

9.iii. Not-Really-Cartesian Linguists

When is a door not a door? And when is a Cartesian not a Cartesian? The first question is more swiftly answered than the second.

The people named by Chomsky as Cartesians (and undeniably influenced by Descartes) differed between themselves about matters of interest to us—including some of the questions listed in Chapter 1.i.a. Many of those questions are still (largely) unanswered. So, quite apart from evaluating Chomsky's accuracy as a historian, there's good reason to consider just what they did say and just what they didn't say.

a. Descartes's disciple

Géraud de Cordemoy (*c.*1620–84), sometime Director of the Académie française, is the only person named as a precursor by Chomsky (e.g. 1996*a*: 3) who was an orthodox Cartesian. Born four years after Descartes, he was one of his leading disciples. Indeed, he was a member of the party who travelled to Stockholm in 1666 to bring Descartes's exhumed remains back to France—minus the writing forefinger, which the French ambassador was allowed to cut off, and the skull, which had been stolen. (After being sold several times, the skull ended up in the Musée de l'homme—Lindeboom 1979: 13–14.)

Descartes had died of pneumonia sixteen years earlier, having risen daily at four o'clock in the Swedish winter to teach Queen Christina (aka Greta Garbo) philosophy in her freezing castle. Cordemoy outlived his master by over thirty years. For much of that time, he was spreading, and elaborating on, Descartes's views.

In promulgating the Cartesian philosophy, Cordemoy devoted an entire essay to language, the *Discours physique sur la parole* (Cordemoy 1668: 196–256). As the title implies, he had much to say about the physical aspects of language (the lungs, speech organs, ears, and nerves), pointing out that the human bodily machine is naturally suited for producing and hearing it. He even discussed the bodily aspects of writing. Nevertheless, he argued that language is possible only for beings with souls (minds), which is to say, human beings. It can't be explained in purely physical terms, nor mimicked by artificial machines—and it isn't acquired by the learning methods used by animals.

Cordemoy marvelled at the child's acquisition of language, and remarked that it doesn't seem to depend on reinforcement:

It's true that one usually tries to excite some emotion (such as joy) in infants, by exclaiming when one shows them something while speaking its name, so that they are more attentive and, being more affected by this method of teaching, retain the words better.

But, no matter what trouble one takes to teach them certain things, one often finds that they know the names of thousands of other things, which one has not attempted to show them. And the most surprising thing about this is that, by the time they are two or three years old, and by the power of their attention alone, they are able to pick out the name of something from all of the constructions that one uses to talk about it.

Next, and with the same concentration and discernment, they learn the words that signify the properties of the things whose names they know.

Eventually, extending their knowledge further, they notice certain actions or movements of these things; and simultaneously observing people talking about them, they distinguish—by their ability to pay attention to the movements, and to hear people repeat certain words in combination with the names of the things or of their properties—those words which signify actions. (Cordemoy 1668: 213–14; my trans.)

Adverbs, he said, are acquired much later, because modifications of actions are less important to the infant than the actions themselves. And grammarians teaching adults a foreign language concentrate first on nouns, then adjectives, and then verbs—"in imitation of the precepts given by nature to children" (p. 215).

Words are partly physical, in so far as they are sounds produced by our vocal apparatus. In that, they resemble the cries of animals. But animal 'signs' are communications without meaning. Their communicative effect results from the harmony between beast and beast, and between animal and environment, established by God as a system of interlocking reflexes. Words, by contrast, have meaning as well as sound, for (as Descartes had taught) they are the voluntary expression of conscious thoughts guided by reason. This is evident from the unpredictability and aptness of the words produced by human beings.

Moreover, said Cordemoy, souls are universally the same. Differences in linguistic ability, and in intelligence in general, are due to imperfections in the brain, not the mind. Freed of the body, all human souls would be equal, because equally rational. (To deal with the philosophical mind–body problem lurking here, Cordemoy developed a neuropsychological version of occasionalism: Chapter 2.iii.b.)

Whether Cordemoy was a "Cartesian" as well as a Cartesian is questionable. (We've already seen that Descartes himself wasn't.) The quotations given above might be read as implying that some specifically grammatical capacity leads the infant to concentrate first on nouns, then on adjectives, and finally on verbs. And perhaps Cordemoy did believe that. However, his remarks about the modifications of actions being less important to the child than the actions themselves suggest that it is the child's general interests, not its specifically grammatical preparedness, that guide the order of language acquisition.

In sum, Cordemoy may not have believed that grammar as such is foreshadowed in the infant's mind. But he certainly believed that humans have a natural capacity for language, which has no parallel in either animals or automata.

b. Arnauld and the abbey

The belief in some universal basis of language was promoted also by Antoine Arnauld. He was a highly valued correspondent of Descartes (2.ii.g)—and of Leibniz, too (Arnauld 1662, p. xxxv).

Although he was influenced by Descartes's philosophy, Arnauld wasn't a Cartesian but a (highly controversial) Jansenist. He was known primarily for holding unorthodox views on grace and the Eucharist. More to the point, here, he rejected Descartes's claim that we have innate knowledge of God. He didn't think that God's existence can be proved by the light of reason, but taught that Christian theism is—literally—our best bet (Arnauld 1662: 357). That is, he accepted the famous "wager" argument of his friend Blaise Pascal—who spent his last years at Arnauld's home base, the Abbey of Port-Royal.

Port-Royal was a Cistercian abbey for women, allowed by papal decree to accept male seculars in retreat. Arnauld was associated with it from early childhood: one sister was appointed Abbess in 1602, and four more—as well as his widowed mother—were nuns. In maturity, he was a key member of the influential 'Port-Royal logicians', and the main author of their most famous publications: the *Grammaire générale et raisonnée* and its successor treatise on reasoning, *La Logique; ou, L'Art de penser* (Lancelot and Arnauld 1660; Arnauld 1662). These texts openly adopted some of Descartes's ideas. But, as many of their pietistic examples showed, they weren't promoting pure Cartesianism. In particular:

> Port-Royal, thanks to the excellent philosophical tools employed by Arnauld, in general grammar developed a branch of Cartesianism which Descartes himself hadn't emphasized: namely, the study and analysis of language in general, assumed to be founded in reason alone. This Cartesian branch, planted and nurtured at Port-Royal, to some extent advanced the ideas familiar in the seventeenth century, and anticipated the labours of the eighteenth century, in which it would be taken up directly by [among others] du Marsais [and] Condillac... (Sainte-Beuve (1867), quoted in Cornelius 1965: 121; my trans.).

The scholastic logic of the medieval period had leant strongly on grammar when it was first developed. Indeed, medieval treatises contain many statements broadly consonant with Arnauld's (and Chomsky's) views. Roger Bacon (1214–94), for example, taught that "Grammar is substantially the same in all languages, even though it may vary accidentally" (cited in Lyons 1991: 172).

But most medieval logicians—as opposed to grammarians—weren't much interested in the grammar of real sentences. Rather, they used natural language to construct a semi-formal language with its own grammar, in order to demonstrate certain formal inference patterns.

By the opening of the seventeenth century, then, the links between logic and language were primarily with rhetoric (I. Thomas 1967). The Port-Royalists changed that.

c. The Port-Royal *Grammar*

In their fresh approach to logic, Arnauld and his colleagues revived its relation to grammar—which characterizes human speech, they said, but not the mimicking of word sounds by parrots (Lancelot and Arnauld 1660: 21). They aimed to state a "rational grammar". By this they meant one that is universally shared, and based on necessary principles of reason rather than contingent (and diverse) facts of linguistic usage.

The Port-Royal group distinguished a sentence's underlying propositional structure from its final expression. In this, they were following the medieval and Renaissance logicians. They claimed that words have a "natural order", in the sense that they provide the natural expression of our thoughts. Or rather, words considered as parts of speech (Subject, Verb, Object, Copula) have a natural order, although words as constituents of phrases or idioms may not. The natural logic underlying the various parts of speech may differ from the common-sense view: all verbs, for instance, were analysed as the Copula and an attribute (so that *Peter lives* became *Peter is alive*). Language and mind (rationality) were thus seen as intimately connected.

Moreover, just as reason is universal, so are some grammatical features. These weren't formal syntactic rules, as they were for Chomsky. Rather, they were features which—because of their natural relation to thought—aid the function of language: communication. (Significantly, the two subtitles of the *Grammar* advertised 'The Fundamental Principles of the Art of Speaking' *before* 'The Reasons of the General Agreement, and the Particular Differences of Languages'.)

For instance (the group argued), all languages that distinguish singular and plural must *naturally* ensure that nouns and adjectives, and verbs and nouns/pronouns, agree in their number (Lancelot and Arnauld 1660: 148–9). For to do otherwise would be to confuse the hearer, who hopes to discover the speaker's thought.

The Port-Royalists were well aware that languages don't always behave 'naturally'. English, for instance, distinguishes singular and plural in nouns, but not in adjectives: both *man* and *men* can be *learned*. Greek is puzzling, too, since plural neuter nouns take singular verbs. Arnauld dealt with such counter-examples by saying that a "figure of discourse" or idiom (which may lend a superficial elegance to the language) will have "some word [implicitly] understood", and that the hearer copes "by considering the thoughts more than the words themselves" (Lancelot and Arnauld 1660: 149). The clear implication was that some natural languages might be more natural (closer to rational thought) than others. No prizes need be offered for guessing which one was favoured by the denizens of Port-Royal:

> [There] is scarce any language, which uses these [superficially unnatural] figures less than the French: because it particularly delights in perspicuity, and in expressing things as much as possible, in the most natural and least intricate order; tho' at the same time it yields to none in elegance and beauty. (p. 154)

Arnauld distinguished between agreement, which is universal (except in a figure of discourse), and government, which is "almost entirely arbitrary" (Lancelot and Arnauld 1660: 148). By government, he meant the fact that using one word can cause some alteration in the form of another. An example discussed at some length is the phenomenon of case: genitive, dative, accusative, and so on. The case, which is a logical relation in thought, may or may not be apparent in the form of the noun (compare the Latin *terra/terram* with the English *ground/ground*). There may or may not be a special particle for a certain case (such as the French *de*). And a verb may or may not always take the same case (consider *servir quelqu'un* and *servir à quelque chose*). Moreover, since "cases and prepositions were invented for the same use", many prepositions govern the noun they are used with—thus *ad terram*, but *ab terra* (chs. 6 and 11, and pp. 149–52).

The various languages differ greatly in whether, and if so how, they mark this or that case by changes in the form of words and/or by the insertion of extra words. Occasionally, different governments alter the sense of the governed word. So in Latin (Arnauld pointed out), *cavere alicui* and *cavere aliquem* mean *to watch over a person's safety* and *to beware of him*, respectively. Such meaning shifts are grounded not in logic, but in arbitrary linguistic convention (p. 152). (As these examples show, the Port-Royalists did consider several European languages. But their main focus was on French, and some of their universalist claims were inconsistent even with English or Greek.)

With respect to those grammatical constructions where they believed that logic is involved, the Port-Royalists gave many detailed examples of the natural order they had in mind. In discussing relative clauses, for instance, Arnauld related specific grammatical forms to their "subordinate propositions" in thought (Arnauld 1662: 118–21). He distinguished "explicative" and "restrictive" uses of words like *which* and *who*, and argued that a particular grammatical conversion (ellipsis) can produce the underlying thought in one case but not the other.

For example, the subordinate proposition expressed by the words "men, who were created to know and love God" can be attained by substituting the antecedent for the pronoun itself, thus: "men were created to know and to love God". The reason is that the relative clause here is (for Arnauld) explicative: according to his theology, the idea of "man" includes the idea of a being created to know and love God. By contrast, the restrictive clause in the sentence "men who are pious are charitable" doesn't express the subordinate proposition (arrived at by a similar pronominal substitution) that "man as man is pious". Rather, it states that men may be pious ("the idea of piety is not incompatible with the idea of man"), and that the idea "charitable" can be affirmed of the complex idea "pious man".

Chomsky described the Port-Royalists' work as an anticipation of his own contrast between "deep" and "surface" structure. Further, he endorsed some of their specific hypotheses about how these are correlated (Chomsky 1966: 33–51, 97).

He claimed, for instance, that they employed a type of phrase-structure grammar (see Section vi.c). They held that the deep structure, or thought, consists of one or more elementary (subject–predicate) propositions, whereas the surface structure may be syntactically very complex: recursively nested, for example. Moreover, expressing the thought as an actual sentence may require more than just expressing each elementary proposition in words. It may also involve rearranging, replacing, or deleting parts of the first-stage (unspoken) sentences—which is to say, transforming them (Chomsky 1966: 40).

Broad similarities certainly exist. But these concern aspects of Chomsky's thought that weren't especially original or controversial. Moreover, some of Chomsky's comparisons are questionable in detail.

For instance, he says that Arnauld's treatment of relative clauses anticipated various "modern" logical–semantic insights, including the distinction between meaning and reference "in pretty much their contemporary sense" (Chomsky 1996*a*: 7).

It's true that the explicative–restrictive contrast rests on a distinction between the meaning of the term concerned (in this case, "man") and those objects that are picked out by it (actual men—who may or may not be pious). However, some distinction between meaning and reference had already been made by medieval logicians. Moreover, the Port-Royal contrast between the "comprehension" and "extension" of a general term wasn't equivalent to any distinction made in modern logic. It muddled singular and general terms, and also gave a definition of meaning—according to which it's part of the meaning of "triangle" that the internal angles equal two right angles—which is now regarded as over-inclusive (Kneale and Kneale 1962: 318 ff.).

Indeed, modern logicians see Arnauld's definition of meaning as an example of psychologism (2.ix.b). They therefore criticize the Port-Royal group as "the source of a bad fashion of confusing logic with epistemology" (Kneale and Kneale 1962: 316). That's why, by the 1950s, the Port-Royalists were only a minority taste in philosophical logic.

Nonetheless, the Port-Royal *Logic* was a creative advance at the time, and dominated logic for the next 200 years. And their *Grammar* was regularly reprinted until 1846. Consequently, their view that grammar reflects thought was widespread throughout this period, and still respected in the late nineteenth century.

d. Deaf-mutes and Diderot

One of the thinkers who accepted this Port-Royalist view was Denis Diderot (1713–84). He was editor-in-chief of the *Encyclopédie*, the 'bible' of the French Enlightenment—a project that had started out as a scheme to translate Chambers's two-volume *Cyclopaedia* (see Chapter 2.ii.c), but which ended up as twenty-eight newly written volumes (Porter 2001: 8). And he, too, was someone named by Chomsky as a precursor.

Descartes, as we've seen, had referred briefly to deaf-mutes, in making a point about the universality of reason and communication. But Diderot wrote a 17,000 word essay about them, exploring the relations between gesture, language, and thought (Diderot 1751). Part of his aim was to discuss the origins of language, and to ask whether there is some "natural" universal language, intelligible to all human beings.

Both questions were prominent topics of debate in the eighteenth century. They featured in several novelists' tales of imaginary voyages—some of which described "universal" artificial languages and symbol systems (Cornelius 1965).

Some people argued that the spontaneous signing of deaf-mutes represented the universal gestural language from which conventional languages had originally developed, and which might now be deliberately extended to form a worldwide communication system (Knowlson 1975; Large 1985). This 'handmade' system would be naturally intelligible to all, because it would be logically transparent (much as the Russell–Whitehead logical notation was later intended to be: 4.iii.c). Others had suggested new sign languages and/or alphabets, designed for universal use (Large 1985: 19–63).

The most interesting of these projects was that of John Wilkins (1614–72), Bishop of Chester and a founder member—and first Secretary—of the Royal Society. More interested in phonetics than in grammar, he proposed a universal alphabet, systematically designed to encompass "all such kinds of simple sound, which can be framed by the mouths of men" (J. Wilkins 1668: 357–62).

His reference to "the mouths of men" was no mere rhetorical flourish. For his notation was comprised of iconic symbols depicting various positions of the speech organs: root/tip of tongue, one/two lips, top/root of teeth, back/front of palate, parts of mouth, and nose. It also marked whether or not the sound involved breathing. In addition, Wilkins designed a more easily usable alphabet. This had some highly stylized iconic features, and some coherence in using similar symbols to represent similar sounds.

The supposed intelligibility of this language wasn't grounded only in the universal ability to make speech sounds and mouth movements. For the systematic coherence spread into semantics, too. Wilkins gave every letter a significance, based on the forty classes into which, according to him, *everything* could be fitted. Each class was subdivided into differences, and these into species. To each class, he gave a two-letter syllable, and a consonant and a vowel were assigned to each difference and species, respectively. So flame was called *deba*: the *de* meant 'element', the *b* meant 'fire' (the first of the four elements), and the *a* meant a small part of the element: namely, a 'flame'. In short, semantic hierarchies were explicitly reflected in phonetics, and in spelling too.

Another aim of Diderot's essay, regarded by some as "one of the outstanding examples of literary criticism in the eighteenth century" (A. M. Wilson 1972: 123), was to distinguish various rhetorical genres. So he compared prose, poetry, drama, scientific writing, and so on.

Here, Descartes was lurking in the background. His clear prose, which reflected his views on the best "method" of thinking, had caused a revolution in scientific and philosophical writing. This is evident, for example, in the striking stylistic differences between the three versions of Joseph Glanville's *Vanity of Dogmatizing* (Glanville 1661–76). These arose because, between the first and second drafts, Glanville adopted the Royal Society's recommendation of Descartes's prose over the more florid writing of the early Renaissance. (One aspect of this was the rejection of interpretative accounts of Nature, considered as God's text, in favour of empirical ones: see Chapter 2.iii.b.)

Diderot went further, arguing that authors should choose not only the most appropriate genre for the type of thought they wished to express, but also the most fitting language. He had in mind not just choice of vocabulary (*le mot juste*), but selection among natural languages themselves. And here, Chomsky has remarked (1965: 7), he illustrated the lasting influence of Port-Royal. For he declared, as Arnauld had done before him, that French surpasses all other languages in the degree to which it matches the order of our thoughts. Everyone, in so far as they are rational, must think their thoughts in the same order. Because French matches this rational order best, he argued, it is the natural language for philosophy and science. Greek, Latin, Italian, and English are more appropriate for less reasoned enterprises, such as the theatre—and in general for persuasion, emotionalism, and deceit.

Only his French readers, no doubt, found that particular claim compelling. But Diderot's view that grammar has some principled basis in thought was widely shared.

9.iv. Humboldt's Humanism

Ten years after the final publication of the *Encyclopédie*, the Enlightenment criticisms of traditional institutions and beliefs were followed by Kant's radical critique of human knowledge itself. Accordingly, some late eighteenth-century philosophers, described as "Cartesians" by Chomsky, conceptualized language in neo-Kantian terms.

That is, they posited—however vaguely—some linguistic equivalent of the empirical intuitions and categories (2.vi.a). This was supposed to be a formative principle shared by all humankind, which moulds all conceivable languages much as the Kantian intuitions inform our ideas about physical reality.

Typically, this 'grammar' wasn't thought of as a static set of given principles—like the Port-Royal logic, or even Kant's space and time. Rather, it was a developing, self-organizing, inner form. As such, it was broadly comparable to the organic forms then being posited in *Naturphilosophie* and Johann von Goethe's biological morphology (see 2.vi.e).

One of these neo-Kantian writers was Johann Herder (1744–1803), who in a prizewinning essay on the origins of language argued that language wasn't given to us in its full form by God, but isn't wholly invented either (Herder 1772). It's part of our innermost nature, and very different from the narrowly focused and unvarying instincts of animals. Rather, it's the ground of mankind's freedom and self-development. Indeed, Herder compared the genesis of language to the mature embryo pressing to be born. He also described a nation's mother tongue as an aspect of the national "soul", or "spirit".

All these notions were to be taken up by the Romantics in general, and by Humboldt in particular. Humboldt, in turn, would have a strong influence on nineteenth-century Romanticism. (In the early 1800s, he spent four years in Paris at a time when William Wordsworth, for instance, was living there too.)

a. Language as humanity

Herder's essay was read, about ten years later, by the young Humboldt—in his teens at the time. Eventually, Humboldt became a renowned humanist scholar (and sometime Prussian ambassador to England). But he wasn't just a fine scholar: he added some enormously influential ideas of his own.

Humboldt's philosophy of language, like Kant's epistemology, has been described as a Copernican revolution (Hansen-Love 1972). For it presents the human mind as the creative origin of knowledge and culture, instead of being dependent on them for its ideas. Accordingly, his position (following the *verum factum* tradition of Giambattista Vico: see 1.i.b) promised a deep knowledge of language and culture—much deeper than scientific knowledge of material things could ever be.

Humboldt saw human beings as *set apart* from nature (including animals) by the mental faculty of language, through which the human mind or spirit is both developed and expressed. Like Herder, he thought of language in terms of organic development, describing it as a "living seed" and "an internally connected organism" (Humboldt 1836: 61, 21). He was thinking not only of the development of language in infancy but, even more, of the historical development of language from its earliest roots:

> *Language*, regarded in its real nature, is an enduring thing, and at every moment a *transitory* one... In itself it is no product (*Ergon*), but an activity (*Energeia*). Its true definition can therefore only be a genetic one. For it is the ever-repeated *mental labour* of making the *articulated* sound capable of expressing *thought*... To describe languages as a *work of the spirit* is a perfectly correct

and adequate terminology, if only because the existence of spirit as such can be thought of only in and as activity. (Humboldt 1836: 49).

Humanity as such was continually celebrated by Humboldt. He often did so in spiritual terms, as when he said that "thought at its most human is a yearning from darkness into light, from confinement into the infinite" (1836: 55). This causes a recurring difficulty when one tries to relate his work to modern linguistics and cognitive science (and when one translates his term *Geist* as "mind" rather than "spirit").

For instance, he pointed out, as Cordemoy had done before him, that human anatomy seems to be especially apt for the production of speech (a fact to be explored in fascinating detail over 100 years later: Lenneberg 1967). But he *didn't* explain these anatomical phenomena as due to the benevolent design of God. Instead, he saw them as aspects of the activity of the creative human spirit. He even linked bipedalism to—or explained it in terms of?—our language and humanity:

And suited, finally, to vocalization is the upright posture of man, denied to animals; man is thereby summoned, as it were, to his feet. For speech does not aim at hollow extinction in the ground, but demands to pour freely from the lips towards the person addressed, to be accompanied by facial expression and demeanour and by gestures of the hand, and thereby to surround itself at once with everything that proclaims man human. (1836: 56)

In short, although Humboldt's observations were often highly astute, many of his scientific questions were fundamentally different from those posed in the empiricist tradition. His work therefore sets one of the "traps" mentioned in Chapter 1.iii.a. To interpret Humboldt's questions and/or answers in terms of the explanatory categories of modern science—which Chomsky frequently did—is often to distort them.

It would be misleading, for instance, to praise Humboldt's naturalistic (i.e. non-theological) explanation of bipedalism as being about as close as a pre-Darwinian scientist could have got. For Humboldt was not only pre-Darwin: he was also pro-Goethe. In other words, his conception of "science" was very different from the empiricist's (see 2.vi).

Like Goethe (a close friend, and a neighbour), Humboldt favoured organic holism over analytic science:

The comprehension of *words* is a thing entirely different from the understanding of *unarticulated sounds*... [The word] is perceived as articulated, [which perception] presents the word directly through its form as part of an infinite whole, a language. (1836: 57–8)

It is thus self-evident that in the concept of linguistic form no detail may ever be accepted as an *isolated fact*, but only insofar as a method of language-making can be discovered therein. (p. 52)

Connected discourse ... is all the more proof that language proper lies in the act of its real production. It alone must in general be thought of as the true and primary, in all investigations which are to penetrate into the living essentiality of language. The break-up into words and rules is only a dead makeshift of scientific analysis. (p. 49)

To be sure, language involves "a system of rules" (p. 62). But that system is an organic whole, which "grows, in the course of millennia, into an independent force", restricting how we can express (and think) our thoughts. Its generative principle is humanity itself ("the unity of human nature"):

Language belongs to me, because I bring it forth as I do; and since the ground of this lies at once in the speaking and having-spoken of every generation of men, so far as speech-communication may have prevailed unbroken among them, it is language itself which restrains me when I speak. But that in it which limits and determines me has arrived there from a human nature intimately allied to my own, and its alien element is therefore alien only for my transitory individual nature, not for my original and true one. (p. 63)

The similarity between this passage and Chomsky's nativism is evident. But the differences are arguably even greater (see subsections e–g, below).

b. Languages and cultures

Drawing partly on Diderot's *Encyclopédie*, and in particular on the three volumes of its summary (the *Encyclopédie méthodique*) that were devoted to grammar and literature, Humboldt studied all the European languages, ancient and modern. But he didn't stop there.

He also described several native American languages, relying on novel data brought back by his younger brother Alexander—a noted scientist much admired by Charles Babbage (Hyman 1982: 72–4), who spent five years exploring the Americas. In addition, he was familiar with many oriental examples, including Chinese, Sanskrit, and Hebrew. And, thanks to the interest in non-Indo-European languages that was aroused by European colonialism in the eighteenth century, he also had knowledge of Melanesian and Polynesian—and the various dialects of Kawi, the ancient literary and sacred language of Java.

Kawi was especially interesting to him largely because it seemed to be a mixed language: although its vocabulary was Sanskrit, its grammar was Malayan. Moreover, its geographical position suggested (to him) that it might be linked to virtually all the known languages of the world. Given Humboldt's passion for intellectual synthesis, and his view that "language" is the fundamental unifying force within all humankind, Kawi seemed to offer the hope of integrating "old world" and "new world" languages—and cultures (Humboldt 1836, p. xii).

His three volumes on Kawi, prepared in the closing years of his life, included a 350-page introduction that gives the clearest statement of his views on *language in general* (Humboldt 1836). As well as considering syntax, morphology, and (especially) phonetics, Humboldt studied the various literatures and orally transmitted poetry and prose. When he thought of language, then, he had in mind both linguistics and literature.

These wide-ranging studies weren't mere scholarly stamp collecting: Humboldt was no Dr Casaubon. They were driven by Humboldt's novel ambition to distinguish language groups in terms of their grammatical (and phonetic) structure, rather than their historical origins. They were intended to show, too, that all languages are essentially alike:

My aim is [not to gather the external details of individual languages, but] a study that treats the faculty of speech in its inward aspect, as a human faculty, and which uses its effects, languages, only as sources of knowledge and examples in developing the argument. I wish to show that what makes any particular language what it is, is its grammatical structure and to explain how the grammatical structure in all its diversities still can only follow certain methods that will be listed one by one . . . (Humboldt 1836, p. xiv)

His studies also underpinned Humboldt's distinctive position on the relation between language and culture, and on the individuality of cultures.

If thoughts can be expressed only in language, difference in language implies difference in culture. Humboldt argued that each natural language is unique. *Perfect* translation is impossible, since apparently equivalent words will have different associations in the minds of speakers of the two languages (and even in individual speakers of "one" language). Nevertheless, *good* translation—on which he had many interesting things to say (Novak 1972)—can expand the language and mentality of the reader.

Although he held that everything can in principle be expressed in every language, he pointed out that in practice languages (cultures) differ in what they choose to express. Translations of literary masterpieces can thus awaken latent potential in the receiving nations. (Scientific thought, he said, varies less across languages.)

Each culture, for Humboldt, has its own *Weltansicht*, its unique manner of regarding the world. Human individuals are deeply informed by their culture, as well as by their unique life experience. And their cultural identity must be thought of holistically. For a language, like an organism, is a self-organizing whole, not a collection of independently definable parts.

c. Humboldt lives!

Humboldt's stress on the diversity, and the holism, of cultures informed some early twentieth-century work on topics relevant to cognitive science. So, too, did his views on the near-identity of language and thought.

The holistic view of language was to be taken up, for example, by the linguists Ferdinand de Saussure (1857–1913) and Roman Jakobson (1896–1982). Although his birth-date locates Saussure firmly in the nineteenth century, his general intellectual presence dates from well into the twentieth. For his influence blossomed only after his death, with the posthumous publication of his lecture notes on "general" linguistics (Saussure 1916).

After a lifetime of detailed research in historical philology, Saussure had finally considered language as such. (It was he who coined the phrase "the arbitrariness of the sign"—see ii.c, above.) He described language as a holistic system, in which the meaning of one word, or sign, can be defined only in relation to others. To include a word, such as *cat*, in a language is not to provide an inert pointer referring to some pre-existing class, but to identify some class by distinguishing it from others picked out by different words (*kitten*, *dog* . . .).

Jakobson was a neo-Humboldtian whom Chomsky has acknowledged as a major influence in the origination (not just the *post hoc* justification) of his own thinking. Based at Harvard while Chomsky was a Junior Fellow there, Jakobson concentrated on phonology, not meaning, and combined holism with nativism (compare Humboldt's "inner form", discussed below). He claimed—as it turned out, wrongly (Sampson 1974)—that the seemingly chaotic speech sounds heard around the world can be defined in terms of twelve oppositional "distinctive features" (Jakobson 1942/3; Jakobson *et al.*

1952). These, he said, comprise a hierarchical system of phonological capacities shared by all human beings. Each language selects only some of the possibilities available.

As for linguistic diversity, Humboldt's ideas were echoed by the anthropological linguist Franz Boas (1858–1942), who set in hand the systematic description of the fast-disappearing native American tongues (Boas 1911). (It was he who'd translated the story used in Frederic Bartlett's experiments on memory: see Chapter 5.ii.b.)

Whereas previous accounts of these languages had focused largely on vocabulary, Boas focused on grammar. He argued that each natural language has a distinct grammatical structure. Moreover, he said, the structural differences can't always be expressed by using shared grammatical categories, as when one says that adjectives are usually put after the noun in French but before it in English. Rather, they may need grammatical categories specifically designed to describe the particular syntax concerned.

Edward Sapir (1884–1939) and Benjamin Whorf (1897–1941) went further, claiming that virtually none of the concepts of one language can be understood (thought) by people brought up to speak another. Whorf even held that each language conceals a specific metaphysics, or ontology, unintelligible to people with different mother tongues (Carroll 1956). On this view, there is no 'true' ontology: languages differ arbitrarily from each other, and views of reality differ likewise (cf. 1.iii.b).

This claim—that a person's thought can't escape their language—was hotly disputed through the 1960s and beyond (Hoijer 1954; R. Brown 1958; Hymes 1964: 149–53; Rosch 1977; Lucy 1992). Empiricist philosophers complained that the Sapir–Whorf 'hypothesis' is slippery and vague, and if clarified becomes either obviously true or obviously false (M. Black 1959). But many social scientists, doubting the obviousness, garnered evidence to test it.

Some of these people, influenced by Chomsky's nativism (see Section vii.c–d), claimed that apparently arbitrary differences in colour vocabulary are in fact underlain by universally shared colour categories (O. B. Berlin and Kay 1969; Rosch/Heider 1972; Rosch 1973; see Chapter 8.iia–b). This early work had glaring methodological faults (Sampson 1980: 95–102). But Chomskyans still reject the Sapir–Whorf hypothesis, saying "there is no scientific evidence" for it and attributing it to "a collective suspension of disbelief" (Pinker 1994: 58).

Others followed Sapir and Whorf, suggesting that cognitive differences may be associated with, or even grounded in, syntactic variations as well as vocabulary. For instance, the Navaho language uses a different form of the verbs for handling things (such as *throw*), depending on the shape of the object handled. This grammatical distinction isn't found in Indo-European languages, and suggests that speakers of Navaho may have a relatively keen awareness of shape (Carroll and Casagrande 1966: 26–31; Lucy 1992: 198–208). Only "suggests": independent tests need to be done to confirm any specific instance of the Whorfian hypothesis.

d. A fivefold list

There's a puzzle here, however.

To concentrate on the syntactic differences between languages seems decidedly un-Chomskyan, given that Chomsky posited a universal grammar. Indeed, Humboldt's aim of describing the grammatical structure of as many languages as possible was defined as the major goal of linguistics by the very people against whom Chomsky reacted in the 1950s (see Section v.b, below). And Chomskyans, as remarked above, reject neo-Humboldtian views on the boundless diversity of languages and concepts.

Yet Humboldt was repeatedly named by Chomsky as an important precursor, who had essentially similar aims. In his first act of homage to the "Cartesian" linguists, Chomsky devoted more space to Humboldt than to anyone else (Chomsky 1964: 17–25; 1966: 19–28).—Why is this?

The answer is fivefold:

* First, Humboldt insisted on the universality of language in the human species.
* Second, he saw it as an innate mental faculty, or force, distinct from other psychological abilities and from the instincts of animals.
* Third, he saw language as inexplicable in terms of its ultimate origins (although he explained many features of specific languages by reference to earlier ones).
* Fourth, he contrasted the finite number of objects in the physical world, including the fixed number of speech sounds, with the infinite creativity of language, which can always come up with some new thought. Language, therefore, implies freedom.
* And fifth, he saw language as having an organic "inner form" that unites humanity, irrespective of culture, in a fundamental way.

Stated briefly, as I've just done, the similarities between Humboldt and Chomsky seem to be clear—indeed, overwhelming. In truth, however, they're neither.

We needn't waste time cavilling over the first two items. These (nativist) beliefs are evident in Humboldt's reply to the familiar empiricist argument that children will learn quite different languages, depending on who brings them up:

> Man is everywhere one with man, and development of the ability to use language can therefore go on with the aid of every given individual [i.e. caretaker]. It occurs no less, on that account, from within one's own self; only because it always needs an outer stimulus as well, must it prove analogous to what it actually experiences, and can do so in virtue of the congruence of all human tongues. (Humboldt 1836: 59)

He also argued (like Chomsky) that the infant's linguistic capacity must be shared with everyone else—otherwise, how could they learn so quickly (pp. 58–9)?

So far, so similar. But the last three items need fuller discussion.

e. Origins

The third item is included in the list because Chomsky claimed that there could be no evolutionary explanation of language—or at least, that we could never find one. Language, he argued, is a holistic system with an all-pervasive structure, not a collection of independently acquired habits or tricks. Humboldt had said much the same thing. But Humboldt, born almost 100 years before *On the Origin of Species*, couldn't consider the possibility of a Darwinist evolutionary account. Chomsky could, and did.

Chomsky was no less pessimistic than Humboldt about our ever understanding the origin of language as such:

> There seems to be no substance to the view that human language is simply a more complex instance of something to be found elsewhere in the animal world. This poses a problem for the

biologist, since, if true, it is an example of true "emergence"—the appearance of a qualitatively different phenomenon at a specific stage of complexity of organization. (Chomsky 1986: 62)

In fact, the processes by which the human mind achieved its present stage of complexity and its particular form of innate organization are a total mystery .... It is perfectly safe to attribute this development to "natural selection", so long as we realize that there is no substance to this assertion, that it amounts to nothing more than a belief that there is some naturalistic explanation for these phenomena. (p. 83)

Admittedly, "evolution" was listed as one of only five keywords for Chomsky's official précis of one of his books (1980*b*: 1). However, this was due not to the author but to the editor, on the grounds that talk of universal grammar predictably raises questions about origins (A. N. Chomsky, S. Harnad, personal communications).

Chomsky's pessimistic remarks of the 1980s, cited above, implied that "emergence" is inexplicable. They implied, also, that the naturalistic explanation of language would refer to some single genetic mutation, or possibly some small set of mutations, that occurred early in our species' history, whose effect on the "emergent" phenotype is opaque. If that were so, then we couldn't hope to find it/them, nor to understand its/their significance if we did.

Similar anti-evolutionary scepticism has often been associated with the complex structure of the eye. Darwin himself admitted that to explain the eye in terms of natural selection "seems, I freely confess, absurd in the highest degree" (quoted in Cronly-Dillon 1991: 15). Nevertheless, he rebutted this scepticism in general terms, and evolutionary biology can now do so in detail. It turns out that the eye is neither a miracle nor even a singularity: eyes have evolved independently a number of times. (Or perhaps not: evidence that all types of eye have evolved from the same root is given in Gehring and Ikeo 1999.) Possibly, then, we shall one day understand the evolution of language much as we now understand the evolution of the eye.

Some Chomskyans, indeed, have speculated at length about the evolution of "the language instinct" (Pinker 1994). As for Chomsky himself, he now allows that we're beginning to understand how a qualitatively different phenomenon can arise spontaneously from a simpler base (see Chapters 14.x.a–b and 15). By the new millennium, he had even suggested that language might be based in spontaneous physical self-organization:

There is little doubt that the human language faculty is a core element of specific human nature ....

Recent work suggests more far-reaching possibilities. The minimal conditions on usability of language are that languages provide the means to express the thoughts we have with the available sensorimotor apparatus. *One far-reaching possibility* is that, in non-trivial respects, the language faculty approaches an optimal solution to these minimum design specifications (with "optimality" characterised in natural computational terms). If true, that would suggest interesting directions for the study of neural realization, and perhaps for the further investigation of the critical role of physical law and mathematical properties of complex systems in constraining the "channel" within which natural selection proceeds.

What the work on optimal design of language seems to suggest is that language *might* be more like the appearance of familiar mathematical structures in nature such as shells of viruses or snowflakes than like prey/predator becoming faster to escape/catch one another [see Chapter 15, below]. When the brain reached a certain state, some small change *might* have led to a reorganisation of structure that included a (reasonably well-designed) language faculty. *Maybe.* (Chomsky 1999: 31; italics added)

However, if natural selection must somehow underlie language, as it underlies every other biological phenomenon, that's not to say that Chomsky expects a detailed evolutionary explanation. On being charged (on the evidence of the passage just quoted) with changing his views on the role of natural selection, he replied:

[I presuppose] that natural selection is operative in this case (as in others). [And I have raised some tentative] questions about the role of "the 'channel' within which natural selection proceeds." That natural selection proceeds within such a "channel" is too obvious to have been questioned by anyone. Only the most extreme dogmatist would produce a priori declarations about its role in any particular case, whether it is virus shells, slime molds, bones of the middle ear, infrared vision, or whatever.

As before, I take no particular stand on the matter, for the simple and sufficient reason that virtually nothing significant is known—even about vastly simpler questions, such as the evolutionary origins of the waggle dance of honeybees. (Chomsky, email to "evolutionary psychology" list, 19 Oct. 1999; italics added)

Chomsky's agnosticism, here, is well judged. It's not clear that one can go much further than speculation on this topic. Languages don't fossilize. Nor do soft structures such as vocal cords (although oral bones and skulls do). Moreover, animal communication systems seem to be very unlike ours, having only finite productive power—as both Humboldt and Chomsky were keen to point out.

On the other hand, mammalian movements are highly variable—and recent work suggests that gestural language is grounded in neuroscientific mechanisms (such as mirror neurones; Arbib 2005) originally evolved for controlling other bodily actions (14.vii.c). Even the logic of *subject–predicate* has been attributed to specific neural circuits in the brain, embodying mechanisms necessary for many tasks performed by non-human primates (Hurford 2003).

In sum, if scepticism about the very possibility of linguistic evolution is out of place, pessimism about our ever being able to detail it may not be (Maynard Smith and Szathmáry 1995, ch. 17; 1999, ch. 13).

f. Creativity of language

The fourth item is included on the list (in subsection d, above) because Chomsky sees Humboldt as anticipating his own notion of the creativity of language. He first argued for this in 1962, at the International Congress of Linguists in Cambridge, Massachusetts; his paper was revised and expanded before publication (Chomsky 1964: 17–22).

Certainly, Humboldt constantly described language—alias the human spirit—as creative, indeed as self-creative. And he spoke of language making infinite use of finite means:

But just as the matter of thinking, and the infinity of its combinations, can never be exhausted, so it is equally impossible to do this with the mass of what calls for designation and connection in language. In addition to its already formed elements, language also consists, before all else, of methods for carrying forward the work of the mind, to which it prescribes the path and the form.

> The elements, once firmly fashioned, constitute, indeed, a relatively dead mass, but one which bears within itself the living seed of a never-ending determinability. (Humboldt 1836: 61)

Whether Chomsky is right to see Humboldt, and the other "Cartesians", as a precursor in this respect is another matter.

Chomsky's definition of the creativity of language referred to syntax, not thought (see Section vi.b). Admittedly, when (in the early 1960s) he included semantics within his grammar, novel meaning would automatically accompany novel form. But even then, he still prioritized syntax (see Section viii.c).

Descartes, by contrast, had focused on the power of (language-informed) reason or imagination to generate new thoughts, expressible in language. And Humboldt, too, had spoken of the power of language to define new thoughts. Chomsky stressed the purely formal fact that his grammar can generate infinitely many new sentences and deep structures, saying that "this 'creative' aspect of language is its essential characteristic" (Chomsky 1964: 8), and that "recursive rules ... provide the basis for the creative aspect of language use" (Chomsky 1967: 7). His later complaint that "A number of professional linguists have repeatedly confused what I refer to here as 'the creative aspect of language use' with the recursive property of generative grammars, a very different matter" was thus misdirected (Chomsky 1972*a*, p. viii). In short, his notion of linguistic creativity was not that of his august predecessors.

Moreover, Chomsky's concept of creativity fails to capture all the cases ordinarily covered by the term, including those which Humboldt seems to have in mind here:

> At every single point and period, therefore, language, like nature itself, appears to man—in contrast to all else that he has already known and thought of—as an inexhaustible storehouse, in which the mind can always discover something new to it, and feeling perceive what it has not yet felt in this way. (Humboldt 1836: 61)

What Chomsky calls creativity is the exploration of an unchanging conceptual space, or thinking style (namely, generative grammar). Some human creativity is like this: run-of-the-mill jazz improvisation, for instance, or mundane examples of what Thomas Kuhn (1962) called normal science. But the most interesting cases are not.

These cases involve either unfamiliar combinations of familiar ideas, or the transformation of an existing conceptual space by altering one or more of its defining dimensions (Boden 1990*a*/2004, 1994*b*; cf. 13.iv). Such transformations make it possible to generate structures that were previously impossible (Humboldt's "things not felt this way before"?). Successive transformations over days, years, or even centuries (Humboldt, again?) can evolve the thinking style in complex and unpredictable ways: the development of post-Renaissance tonal music, for example (C. Rosen 1976; Boden 1990*a*: 59–62).

Both these types of creativity go beyond what Chomsky means by the term. And both, arguably, were what the Romantics were interested in. The poet Samuel Taylor Coleridge (1772–1834), who popularized neo-Kantianism in England, highlighted the mind's propensity for combinational creativity (Livingston Lowes 1930; Boden 1990*a*, ch. 6; see also the preamble to Chapter 12, below). And transformational creativity is perhaps comparable to the more structured, developmental, creativity posited by Humboldt.

I say "arguably" and "perhaps", here, because this point depends on how we understand the fifth item on our list. Just what did Humboldt mean by the "inner form" of language?

g. The inner form

In his Kawi-introduction, Humboldt described this inner form as a creative human force that is unfurled and developed in life, a vital linguistic instinct whose expression can be shaped, favoured, and hindered by external circumstances (Humboldt 1836, esp. 48–53). The environment doesn't impose itself on a passive tabula rasa, but elicits an active response from a mind inherently suited to make linguistically relevant distinctions—between speech sounds and other sounds, for example. And, he said, even (indeed, especially) the languages of "so-called savages" show a typically human diversity of expression.

The broad similarity to Chomsky's approach is, again, evident. And Chomsky declared that Humboldt's notion of form as generative process was "his most original and fruitful contribution to linguistic theory" (Chomsky 1964: 17). But the similarity between the two linguists becomes more fuzzy on closer inspection.

Humboldt's definitions of the inner form of language were both various and vague:

> The constant and uniform element in this mental labour of elevating articulated sound to an expression of thought, when viewed in its fullest possible comprehension and systematically presented, constitutes the *form* of language. (1836: 50)

So far, so (apparently) Chomskyan. However, Humboldt immediately added a gloss that doesn't sound Chomskyan at all:

> In this definition, form appears as an *abstraction* fashioned by science. But it would be quite wrong to see it also in itself as a mere non-existent thought-entity of this kind. In actuality, rather, it is the quite individual *urge* whereby a nation gives validity to thought and feeling in language. (p. 50)

Two pages later, having remarked that "the characteristic form of languages depends on every *single* one of their smallest *elements*", and that "language, in whatever shape we may receive it, is always the spiritual exhalation of a nationally individual life", he passed from nation states to the ultimate nature of language:

> From the foregoing remarks it is already self-evident that by the form of language we are by no means alluding merely to the so-called *grammatical form* . . . The concept of form in languages extends far beyond the rules of *word order* and even beyond those of *word formation*, insofar as we mean by these the application of certain general logical categories, of active and passive, substance, attribute, etc. to the roots and basic words. It is quite peculiarly applicable to the formation of the *basic words* themselves, and must in fact be applied to them as much as possible, if the nature of the language is to be truly recognizable. (p. 51)

In that quotation, Humboldt seemed to have historical philology in mind. The crucial questions would then concern (for instance) how the Indo-European languages developed, and how they differ from the Malayan group, rather than (for example) the relations between active and passive, or the hidden depths of *John is easy/eager to please*. But Chomsky interprets inner form as embracing "the rules of syntax and word formation as well as the sound system and the rules that determine the system of concepts that constitute the lexicon" (Chomsky 1966: 26–7). And Chomsky's interpretation finds support in the following remark of Humboldt:

> [We must engage in] a laborious examining of fundamentals, which often extends to minutiae; but there are also details, plainly quite paltry in themselves, on which the total effect of languages is dependent, and nothing is so inconsistent with their study as to seek out in them only what is great, inspired, and pre-eminent. Exact investigation of every grammatical subtlety, every division of words into their elements, is necessary throughout, if we are not to be exposed to errors in all our judgments about them. It is thus self-evident that in the concept of linguistic form no detail may ever be accepted as an *isolated fact*, but only insofar as a method of language-making can be discovered therein. (Humboldt 1836: 52)

Because Humboldt defined "inner form" in so many different (and imprecise) ways, it's not clear whether he really is Chomsky's precursor in this fifth sense. Indeed, one might say that he never *defined* it at all: "what it means is never revealed either by way of explanation or example, let alone definition which is a device he seems to have spurned" (Aarsleff, in Humboldt 1836, p. xvi). It's hardly surprising, then, that "for a hundred years all discussion has failed to converge on any accepted meaning" (p. xvi). ("*All* discussion" includes scholars who can read Humboldt in the original German, as I myself cannot; so I don't think the translation can be to blame, here.)

When Humboldt was more specific about just what these formal relations might be, he was often less plausible. Thus he remarked that certain stories have developed independently in different cultures, explaining this not as the result of shared human experiences but as the unfolding of innate ideas (Novak 1972: 128). He even claimed that similar sounds are used by numerous languages to express similar ideas—where he wasn't thinking of onomatopoeia, as in *buzz*, *miaow*, or perhaps *crack* (Jespersen 1922: 57).

In short, Humboldt's "inner form" of language was specified only a little more clearly than Leibniz's veins of marble. Even his friends criticized him for his irremediable vagueness (Humboldt 1836, p. xv). And one of his twentieth-century admirers, the Danish linguist Otto Jespersen (1860–1943), put it in a nutshell: "Humboldt, as it were, lifts us to a higher plane, where the air may be purer, but where it is also thinner and not seldom cloudier as well" (Jespersen 1922: 57). If such a remark could be made even by an admirer, less sympathetic critics would clearly need some persuading.

A twentieth-century linguist wishing to revive the unfashionable humanist/nativist ("Cartesian") approach would face several challenges:

* They'd have to clarify the relation between linguistic structure and meaning.
* They'd have to explain just how linguistic variability is possible, and how it differs from the changes (learning) found in other types of behaviour.
* They'd have to specify what the universal innate ideas of language might be, and reconcile them with the diversity of actual tongues.
* And they'd have to show that nativist theories of language aren't mere optional alternatives to empiricist accounts, but inescapably more appropriate.

In Sections vi–viii, we'll see how Chomsky tried to do all those things. And in Section ix we'll see that by the 1980s, despite his enormous influence over the preceding quarter-century, some of his fundamental arguments were being rejected.

First, however, we must understand why it was that "Cartesian" linguistics became unfashionable in the first place.

9.v. The *Status Quo Ante*

Chomsky's immediate predecessors had no interest in the tasks just listed. The professional linguists he encountered as a student dismissed questions about inner forms of grammar, although some did engage with Jakobson's inner forms of phonology. They assumed that languages are arbitrarily diverse systems, wholly learnt from the environment.

If Humboldt's influence was still strong (whether acknowledged or not), it was in his wish to describe the many distinct grammars exemplified by natural languages, and to avoid historical bias in doing so. These two aims suffused the tradition in which Chomsky was trained: American "descriptive", or "structuralist", linguistics.

Following Boas's lead (see Section iv.c), the structuralists produced detailed descriptions, or taxonomies, of the grammatical categories of many little-known languages. They saw linguistics as an autonomous enterprise based firmly in the present, an example of what Saussure (1916) had called "synchronic" linguistics. That is, they aimed to describe speech, morphology, and grammars by a scientific method not tainted by history—nor, as we'll see, by introspection.

Like any intellectual movement, this one included a variety of positions. With respect to structuralism as a whole, it's been said that Chomsky's linguistics "continued some fundamental traits of its predecessor, recovered others, and unwittingly rediscovered still others" (Hymes and Faught 1981: 1). What's most relevant for our purposes is that the dominant school in the late 1930s–1940s—Bloomfieldian linguistics—adopted a broadly behaviourist methodology, and that it included theorists who tried to analyse linguistic structure in formal terms. Chomsky's work would undermine the former and emphasize the latter.

a. Two anti-rationalist 'isms'

The Bloomfieldians' rejection of introspection, and their striving for formalism, were inherited from two closely related movements in early twentieth-century thought. These were behaviourism in psychology and positivism in the philosophy of science.

Introspection was ousted from (American) psychology long before being banished from linguistics. Boas's follower Leonard Bloomfield (1887–1949) started out as an admirer of Wilhelm Wundt, whose mentalistic folk psychology informed his first book (Bloomfield 1914). As a professor of German, he could read Wundt in the original: the first five (of ten) volumes of the *Volkpsychologie*, had been published by the time Bloomfield wrote his text. (A brief summary appeared in English two years later: Wundt 1916.) His book also paid homage to Humboldt, as the founder of general linguistics (1914, ch. 10).

With hindsight, it's intriguing that Bloomfield recounted Wundt's theory of "inner forms". A meaning, or inner form, may be expressed in various grammatical sequences, as in *Mary ate the apple* and *The apple was eaten by Mary*—plus, of course, equivalent sentences in other languages. Wundt had argued that linguistics should focus on *sentences*, where a 'sentence' can't be defined in terms of words or phrases, but is recognized intuitively by native speakers. He'd also held that processes of transformation (*Umwandlung*) produce grammatical sequences from inner forms.

But Bloomfield didn't follow up these ideas. In the very year in which he praised Wundt's mentalistic work in a leading psychological journal (Bloomfield 1913), another such journal published John Watson's seminal paper on 'Psychology as the Behaviorist Views It' (J. B. Watson 1913). That paper swept the board in psychology: Watson was elected President of the American Psychological Association only two years later. Eventually, Watson would defeat Wundt in linguistics too.

Bloomfield soon abandoned Wundt, and his pre-Watsonian talk of inner forms and hidden transformations. Indeed, thanks to a "shocking" book very different from his first one, he became the leader of "scientific" structuralist linguistics, which downplayed meanings and had no place for inner forms (see below). But this didn't happen overnight. Two decades would pass before the behaviourist approach conquered in linguistics.

Watson's iconoclasm had been fuelled by frustration with the experimental impasse in psychology, *not* by independent philosophical argument. But the later behaviourists justified their approach by appeal to logical positivism, which swept the philosophy of science in the 1920s and 1930s.

Several leading positivists emigrated from Europe to America shortly before the Second World War. They included Hans Reichenbach (1891–1953) and Rudolf Carnap—one of whose protégés was Walter Pitts (see 4.iii–iv).

Besides stressing observational evidence and operational definition, these philosophers tried to axiomatize physics (an aim which intrigued the young Herbert Simon: Chapter 6.iii.a). In this, they were still following Descartes's formalist dream, which had inspired modernism—in science, art, and politics—since its origin in the seventeenth century (Toulmin 1999: 152–60). Believing that other sciences could then be defined in terms of physics, they hoped to "unify" science within a single deductive system.

Some behaviourists tried to do much the same thing, expressing their theories in terms of psychological definitions and quasi-axioms. The most detailed examples were developed by Clark Hull in the late 1930s and early 1940s (C. L. Hull 1943) (see 5.iii.b). But perhaps the most ambitious appeared much earlier, when logical positivism was relatively new: Albert Weiss's 'One Set of Postulates for a Behavioristic Psychology' (1925).

Weiss listed ten "postulates", or axioms, of psychology, outlining its potential unification with physics. These ranged from the motions of electrons and protons, through behaviour (including language, as "the characteristic factor in human behavior"), to civilization. And, strange as it may seem, civilization was defined by Weiss partly by reference to "electron–proton movements" (Postulate 9). The underlying thread supposedly linking electrons to civilization was Weiss's assumption that "human behavior and social achievement are ... forms of motion", as opposed to "psychical or mental phenomena".

As for language, this was defined as a system of "sensorimotor and contractile effects" linking stimulus and response with environmental changes.

If that were so, then sentences about propositional attitudes, using (intentional) verbs such as *know*, *believe*, *want*, and *hope*, could be analysed in purely truth-conditional (extensional) terms. But it's not clear that this is possible. The truth of *Mary believed that the bracelet had been lost*, for instance, is independent of the truth of the embedded sentence about the bracelet.

Philosophers were aware of this difficulty. So much so, that many saw intentionality as an insuperable logical–metaphysical boundary between psychology and natural science (Chisholm 1967; cf. Boden 1970). (What's more, many still do: see Chapter 16.) But the behaviourists ignored it. They either avoided psychological terms entirely or assumed—with Weiss—that they could be interpreted as forms of motion (stimulus and response).

Not all behaviourists went so far as trying to axiomatize psychology, but they did accept the positivists' general viewpoint. Indeed, they shamelessly took it for granted. Psychology textbooks before the 1970s typically opened with a chapter first exulting that psychology had at last 'escaped' from philosophy, and then uncritically offering a positivist account of science.

b. The shock of structuralism

Bloomfield was influenced by both behaviourism and positivism. He was converted to the former by Weiss (Bloomfield 1931), and invited by the philosopher Otto Neurath to contribute one of the first essays in a positivist encyclopedia of "unified science" (Bloomfield 1939). (Not that one can put too much weight on that: Thomas Kuhn's (1962) devastating attack on positivism—and Popperianism—would later appear as an annexe to the same encyclopedic review.)

At one stage, Bloomfield followed Weiss's example by formulating a set of fifty postulates for scientific linguistics and, for comparison, twenty-seven postulates for historical linguistics (Bloomfield 1926). More to the point, he referred repeatedly to "stimulus" and "reaction" in his most influential book, *Language* (Bloomfield 1933, e.g. 29–30).

This was the volume which tipped the linguistic profession—at least, in America—into the behaviourist mould. To be sure, the neo-Humboldtians Boas, Sapir, and Whorf weren't converted. They continued to study language (and meaning) in much the same way as before. But the mainstream shifted.

Bloomfield's new account was seen on its publication as "a shocking book: so far in advance of current theory and practice that many readers, even among the well-disposed, were outraged by what they thought a needless flouting of tradition" (B. Bloch 1949: 92). Nevertheless, it soon became "almost orthodoxy" (p. 92).

Part of this new orthodoxy was the *rejection* of speculation on the relation between language and mind:

> [Many linguists accompany their] statements about language with a paraphrase in terms of mental processes which the speakers are supposed to have undergone. The only evidence for these mental processes is the linguistic process; they add nothing to the discussion, but only obscure it. (Bloomfield 1933: 17)

Another part was the rejection of *meaning* as a theme for linguistics. Meaning, Bloomfield argued, can't be studied scientifically. Meanings are in practice defined (even by mentalists) "in terms of the speaker's situation and, whenever this seems to add anything, of the hearer's response" (p. 144). Rigorously defining the speaker's situation is impossible, however, for we don't know what associations the speaker has learnt (a point with which Humboldt would have agreed).

Accordingly, the Bloomfieldians focused on phonology and syntax, not semantics. They identified their theoretical categories without reference to meaning, by attending to physical measurements and statistical distribution patterns. They all tried to define their terms clearly. But many wished to replace verbal definitions by formal–symbolic ones. If that could be done, they thought, they might even identify scientific (meaning-free) methods by which phonological, morphological, and grammatical categories could be reliably—perhaps automatically—discovered.

In this journey towards formalism, they were running in parallel with empiricist philosophy of language—though with less delicate tread. During most of Bloomfield's life, the dominant movements were logical atomism and logical positivism, both of which drew heavily on Bertrand Russell's logic (Urmson 1956).

Some philosophers focused on meaning. For instance, they tried to define class concepts by lists of necessary and sufficient conditions; and they discussed the semantics of definite descriptions (noun phrases starting with *the*), and of the English words *and*, *if–then*, and *some*. (There's more to the English *and* than meets the logician's eye: according to logic, *London is the capital of England and fish have gills* is unproblematically true, but these two conjuncts would never be linked by *and* in normal conversation.)

It was this general approach which led the young McCulloch to try to formulate a logic of verbs: distinguishing past, present, and future while ignoring grammatical questions, such as where to place the past participle (4.iii.c). But it didn't affect mainstream linguistics, which had outlawed studies of *meaning*. The philosophy of language that was potentially relevant for structuralism was concerned rather with *grammar*.

Some positivists discussed logical/linguistic syntax at length (Carnap 1934; Reichenbach 1947, ch. 7). Indeed, Carnap explicitly argued that the best way to understand natural grammar is to compare it with the syntax of an artificially constructed language. But this positivist work attracted only the bravest structuralist souls. It was highly abstract (Carnap's formal notation was notoriously obscure, as remarked in Chapter 4.iii.f), and said very little about the specifics of natural languages. Not until the late 1940s was there a structuralist grammar that could compare in sophistication with logical studies of syntax.

Bloomfield himself, in his "shocking" book, offered a very simple formalism to explain the appearance of new word forms. His "proportional formulae" represented the selection of new forms by analogy with previously experienced ones: see Figure 9.1. (Bloomfield 1933: 406).

Bloomfield used these formulae to explain both correct uses of regular forms, such as the plural *cows*, and overgeneralizations of irregulars, such as the past-tense *dreamed* instead of *dreamt* (cf. Chapters 7.vi.a and 12.vi.e). He applied proportional formulae also to syntactic innovations in the history of a given language: for instance, the change in sixteenth-century English when people started to introduce subordinate clauses by the word *like* (as in *to do like Judith did*) (Bloomfield 1933: 407).

sow : sows = cow : x

scream : screams : screaming : screamer : screamed
= dream : dreams : dreaming : : x

FIG. 9.1. Proportional formulae for creating new word forms. Reprinted with permission from Bloomfield (1933: 406)

However, to call these expressions "formulae" was to claim too much. One can't reasonably complain that they had to be interpreted intuitively, for in the early 1930s the notion of automatic computation had yet to be defined (see 4.i.c), still less implemented. Even so, they were highly informal. Apart from the proportionality sign borrowed from mathematics, there was no symbolization here, nor even an explicit separation of the words into constituent morphemes (such as *scream* and *-ed*). This wasn't a new notation, nor a grammatical calculus, merely a mnemonic for summarizing certain morphemic patterns.

Bloomfield's gloss on Figure 9.1 showed that it wasn't a description of conscious (verbally reportable) processes, either:

> Psychologists sometimes object to this formula, on the ground that the speaker is not capable of the reasoning which the proportional pattern implies. If this objection held good, linguists would be debarred from making almost any grammatical statement ... Our proportional formula of analogy and analogic change, like all other statements in linguistics, describes the action of the speaker and does not imply that the speaker himself could give a similar description. (Bloomfield 1933: 406)

Unconscious (physiological) mechanisms were doubtless responsible for the speaker's behaviour: Bloomfield didn't believe in magic. But the linguist's job was to describe the abstract structure of the linguistic behaviour, not to speculate on the processes by which it was achieved.

c. The formalist Dane

A more ambitious linguistic formalism was published a few years later by Jespersen. Situated as he was in Europe, Jespersen wasn't a card-carrying structuralist. Indeed, Chomsky (1996*a*) names him as a "Cartesian" precursor. But his early attempts at formalism attracted praise from Bloomfield (1933: 86), and were cited by Reichenbach in his positivist analysis of "conversational" language (Reichenbach 1947, ch. 7).

Initially, Jespersen used his notation to describe only phonemes, which he analysed in considerable detail. His "analphabetic" phonetic symbolism, a twentieth-century version of Wilkins's project of the 1660s, represented many distinct states of the vocal organs. In his *Analytic Syntax* (1937), Jespersen extended his notation to cover grammatical structure too. His work was not only ambitious, but original:

> So far as I know, this is the first complete attempt at a systematic symbolization of the chief elements of sentence-structure, though we meet here and there with partial symbolizations ... (Jespersen 1937: 97)

He compared his notation with the formulae of chemistry and logic (p. 13). And (without digressing to explain the individual symbols) we can see that it was an analytic tool far in advance of Bloomfield's proportionalities. The phrases *burning hot soup* and *wide open windows* were both rendered as "2/321", and *curious little living creatures* was "21(21(21))"—(see p. 19). The sentence *I saw the soldiers, some of them very young indeed* was represented as "S V O [1^qp1 323]"—(see p. 47), and *I've found the key that you spoke of* as "S V O (12(3^c/$1^{c*}S_2$ V p*))"—(see p. 76). And the two superficially similar sentences *The story is too long to be read at one sitting* and *The story is too long to read at one sitting* were represented respectively as "S V P (32p1(I^b3))" and "S* V P(32p1(IO^{o}*3))"—(see p. 64). As for more complicated examples, the sentence *Brother Juniper, forgetting everything except the brother's wishes, hastened to the kitchen, where he seized a knife, and thence directed his steps straightway to the wood where he knew the pigs to be feeding* became this: "S(1 1 2(Y O p1($1^2$1)) Vp1 2(3^c S V O & 3 V O($S^2$1) 3p1 2(3^c S V O(S_2 I)"—(see p. 93). As Jespersen laid it out, it was as follows:

Brother Juniper, forgetting everything except the brother's wishes,
S(1 1 2(Y O p1(1^2 1))
hastened to the kitchen, where he seized a knife, and thence directed
V p 1 2(3^c S V O & 3 V
his steps straightway to the wood where he knew the pigs to be
O(S^2 1) 3 p 1 2(3^c S V O(S_2 I)
feeding. [NB Opening/closing brackets are unequal in number.]

Jespersen stated his general policy as "to follow the sentence or word combination that is to be analyzed word for word". (Even without the detailed explanations, one can see this strategy at work in the examples given above.) However, he allowed many exceptions to this rule. Specifically, "such combinations as *the man*, *a man*, *has taken*, *will take*, *is taking*, etc. (generally also *to take*), even *can take*, are reckoned as one unit" (Jespersen 1937: 15).

Some of Jespersen's remarks about language recall Humboldt, whom (as remarked in Section iv.g) he admired greatly—and, with hindsight, Chomsky too. For example:

> What is certain is that no race of mankind is without a language which in everything essential is identical in character with our own . . . (Jespersen 1922: 413)

> The complexity of human language and thought is clearly brought before one when one tries to get behind the more or less accidental linguistic forms in order to penetrate to their notional kernel. Much that we are apt to take for granted in everyday speech and consider as simple or unavoidable discloses itself on being translated into symbols as a rather involved logical process, a fact that is shown, for instance, by the number of parentheses necessary in some of the examples. (Jespersen 1937: 15)

The reference to the "notional kernel" betrays Jespersen's belief, previously argued in his *Philosophy of Grammar* (1924), that behind the syntactic categories which describe the partly "accidental" form of an actual language there are deeper categories that are independent of existing languages. He saw the linguist's task as to investigate the relations between notional and syntactic categories, and to discover in what way all natural languages are "essentially identical".

Despite these Chomskyan overtones, Jespersen's work was significantly different from Chomsky's. Most importantly, his grammatical formalism wasn't generative. It described the syntactic structure of a sentence as given, not how it could in principle be built. (The term "generative" is here intended in its timeless mathematical sense, indicating the class of structures defined by some set of derivational rules.) As a corollary, Jespersen didn't explicitly ask what makes word strings grammatical or ungrammatical, since his theory was applied only to sentences actually present in the corpus.

(A "corpus" is a representative sample of sentences, or phrases, gleaned from some natural source: books, conversations, interviews . . . In Jesperson's time it might have held only a few score items. Today, it may hold hundreds of thousands. The CHILDES corpus, for instance, is a database containing over 150 megabytes of speech-exchanges between parents and children of various ages, some using a second language and some the mother tongue; sub-corpora can be extracted which hold only utterances of parents/carers, or only utterances of 4-year-olds . . . and so on.)

Jespersen failed to consider sentence generation partly because this typically requires syntactic procedures to be performed recursively. For instance, one noun phrase can be nested inside another, and that in another . . . and so on. But the mathematics of recursion wasn't yet understood.

Even fifteen years later, when functioning machines using recursion were available, the scientific potential of recursion wasn't obvious. The philosopher Carl Hempel (1905–97) observed in 1952 that "recursive definitions which play an important role in logic and mathematics . . . are not used in empirical science" (Hempel 1952: 11 n. 11). As Chomsky put it in the 1960s:

> The fundamental reason for the inadequacy of traditional grammars is a . . . technical one. Although it was well understood [by traditional grammarians] that linguistic processes are in some sense "creative", the technical devices for expressing a system of recursive processes were simply not available until much more recently. In fact, a real understanding of how a language can (in Humboldt's words) "make infinite use of finite means" has developed only within the last thirty years, in the course of studies on the foundations of mathematics. (Chomsky 1965: 8)

d. Tutor to Chomsky

The first linguist to take recursion seriously was Zellig Harris (1909–92) at the University of Pennsylvania—where the ENIAC was built in the mid-1940s (3.v.b). Harris taught Chomsky as an undergraduate and as a graduate student, and they kept in close contact after the younger man moved to Massachusetts.

Chomsky later acknowledged Harris as his main inspiration, and a frequent discussant of his own work (Chomsky 1975: 4). Harris was no more interested than Bloomfield in the humanist/nativist ('language-and-mind') questions listed at the close of Section iv.g. But if anyone was a precursor of Chomskyan grammatical analysis, it was he.

In 1951 Harris published a book widely received as the most important contribution to descriptive linguistics since Bloomfield's *Language*. His *Methods in Structural Linguistics* was shown to Chomsky in proof, on joining Harris's seminar in 1947. It contained a detailed discussion of how to use distribution patterns to identify phonemes, morphemes, morpheme sequences (phrases), and other linguistic classes. In addition, it offered a formalism for representing such classes. This formalism—which Chomsky adopted and modified—was used by Harris to state a finite set of grammatical rules that could describe an infinite set of sentences.

When electronic computers appeared on the horizon, Harris hoped also to define a formal method which, if expressed as a computer program, would automatically identify the grammatical classes and patterns within a set of sentences drawn from an unknown language with an alien grammar (Z. S. Harris 1951, 1968). In this, a computer-age version of a long-standing structuralist goal, he didn't succeed.

That should have been no surprise, for the task is in principle impossible. It assumes that scientific categories and hypotheses can be derived from the data by theory-free induction—a view that was already being questioned by Popperian philosophers (K. R. Popper 1935; Goodman 1951). One can't even identify phonemes and morphemes without some guiding grammatical theory (Chomsky 1962: 125). However, Harris's system was "the most ambitious and the most rigorous attempt that has yet been made to establish what Chomsky was later to describe as a set of 'discovery procedures' for grammatical description" (Lyons 1991: 34).

It turned out, many years later, that the discovery procedures needed may be less specifically linguistic than some Popperians might expect. Recent work in the machine learning of language shows that relatively "light" theoretical assumptions can go a long way, if combined with powerful statistical techniques (see Section xi.a, below). These assumptions aren't language-specific in the sense that interested Chomsky, for they aren't grammatical.

Nevertheless, the general Popperian point still stands. In the 'pure' statistical analysis of the linguistic data, words are given for free by the spaces between the input letter strings. Moreover, phonetic categories—in the form of the alphabet—are taken for granted. From the point of view of psycholinguistics, the latter is entirely defensible. Although nativism with respect to syntax is highly questionable (see Section vii.c–d), nativism with respect to phonetics isn't: there's a finite set of human speech sounds, and newborn babies pay more attention to them than they do to sounds in general. Refutation of the Popperian claim would require successful induction over purely acoustic features—and even then, the acoustically structured anatomy of the cochlea could be seen as providing an implicit theory. (Although whether, and if so in what sense, acoustic feature detectors are really "innate" is debatable: cf. Chapter 14.vi.b and x.a–b.)

Harris introduced Chomsky not only to the project of rigorous formalism, but also to the notion of transformations (Z. S. Harris 1952, 1957, 1968). This idea wasn't mentioned in *Methods*, but was already prominent in Harris's mind. Indeed, by 1947 (when the book was completed) he was discussing it in his seminars, and in conversations with other linguists (Bar-Hillel 1970: 292; Z. S. Harris 1952: 18–25). The young Chomsky was a research assistant for some of the necessary analyses (Z. S. Harris 1952, n. 1). By the time of his own first publications, Harris's work on transformations was well known.

"Discourse analysis" was, in effect, a scientific study of the phenomenon of literary genre that had so fascinated Diderot. It showed how to give a structuralist description of entire connected texts, including spoken conversation. Its first public presentation (in 1949) was to an audience of anthropologists, for Harris's motivation in defining transformations was to apply linguistics to culture and personality. One might almost say that Humboldt's assimilation of culture and linguistics was being revived—but from a very different philosophical base, and with a very different methodology.

Instead of asking a priori what type of language should be used for a certain type of task, as Diderot had done, Harris asked what sentence structures actually occur within discourses devoted to that task. He answered this question in terms of inter-sentence distribution patterns, defined in terms of grammatical "transformations" specifying relations between entire sentences or morpheme sequences. Two syntactically complex sentences might be transforms of one, simpler, kernel sentence. And he found that "often it is possible to show consistent differences of structure between the discourses of different persons or in different styles or about different subject matters" (Z. S. Harris 1952: 1).

Rhetoricians and literary critics, Diderot included, had known for centuries that some such differences exist. And they had identified various genres and authorial signatures accordingly. But discourse analysis could describe these structural differences more precisely, as well as discovering many that hadn't been noticed before.

Harris's Presidential Address to the Linguistic Society of America in 1955 summarized his approach. Two sentence structures, related (for instance) as *active/passive*, or even as *question/answer*, were classed as transforms of each other on the basis of their observed distribution patterns. Specifically, they must share the same set of individual co-occurrences of morphemes. For example, *The kids broke the window* and *The window was broken by the kids* share the same verb and two nouns, although in reverse order. (The sentence *The young people destroyed the pane of glass* wouldn't count as a transform: Harris approached meaning only indirectly, through word distributions.) Problems arise with the morpheme pair *-ed/will*, since the first co-occurs with *yesterday* whereas the second does not (*The cliff crumbled yesterday*, *The cliff will crumble tomorrow*). Accordingly, just what is to count as the relevant "sentence environment" is not a straightforward matter. Harris (1957) gave a detailed discussion of how this problem can be addressed. (I can't resist remarking that the cliff actually crumbled on 11 January 1999, a few weeks after this passage was written, when Beachy Head—a famous landmark, and suicide point, near my home—sent tens of thousands of tons of chalk crashing into the English Channel.)

e. Not quite there yet . . .

If Harris's work was a crucial inspiration for Chomsky, it was nonetheless different from his in two—maybe three—important ways.

First, Harris defined his transformations by reference to distributional (statistical), and purely surface, features of sentences, as we've seen. Chomsky, by contrast, defined transformational grammar in terms of deep and surface structures. Or rather, he did so in his publications in the 1950s and 1960s, which are the crucial ones for our purposes. (More recently, he has dropped the deep/surface distinction: see Section viii.b.)

Another often remarked difference is that Chomsky's theory was generative whereas Harris's wasn't. Chomsky certainly saw his grammar as being original in this way, and Harris apparently agreed: "Noam Chomsky has combined transformational analysis [which features also in my own system] with a generative theory of sentence structure [which does not]" (Harris 1968: 4).

This is puzzling, however. It's true that Harris's aim was to describe sentences, not to show how they could be generated. But, as noted above, his formal theory potentially covered an infinite set of sentences. In that sense, it was a generative system—even though Harris didn't describe his rules as "derivational", nor use them to discuss sentence generation (cf. Hymes and Faught 1981: 166).

The third difference is uncontestable, and the most significant of all. As Harris put it: "Chomsky [unlike myself] has produced a mathematical specification of context-free languagelike systems within a spectrum of languagelike systems" (Z. S. Harris 1968: 4). In other words, Harris didn't locate his formalism within *the class of all grammars*.

That idea was developed by Chomsky. In so doing, he brought theoretical linguistics into the ambit of information theory, computer science, AI, and computational psychology.

9.vi. Major Transformations

The major transformation in twentieth-century linguistics dates back to 1957, with the publication of Chomsky's *Syntactic Structures*. It wasn't a high-profile event. On the contrary, the volume appeared in an obscure specialist series, from a small publishing house in Holland. Chomsky himself said later that they accepted it only because it was recommended by Jakobson, and that it wouldn't have been noticed but for a long, and near-simultaneous, review in a leading linguistics journal by his student Robert Lees (1957).

Lees didn't merely recommend the book, he championed it. Chomsky's "first established convert", he "went around to linguistics conventions and got the ball rolling" (Weimer 1986: 300). As Howard Gardner put it, he was "playing Huxley to Chomsky's Darwin" (H. Gardner 1985: 189). This was to become an all-too-common role for Chomsky's students, as we'll see (Section viii.a).

Nevertheless, Lees and Jakobson weren't the only ones to be impressed. This little book, of only 118 pages (and by an author only 29 years old), revolutionized the subject.

a. Chomsky's first words

If linguists had had their eyes open, the revolution might have happened a few years earlier. For Chomsky's first book wasn't his first publication. A handful of technical papers had already appeared, most in journals not read by linguists, such as the *Journal of Symbolic Logic*.

One of these was Chomsky's paper on 'Three Models for the Description of Language' (1956). This was a hierarchical classification of grammars, defined in purely abstract (non-linguistic) terms. With hindsight, it's clear that this paper, not *Syntactic Structures*, was Chomsky's most original and most well-founded contribution—even if many of Chomsky's fans have never read it (cf. also Chomsky and Miller 1958; Chomsky 1959*a*).

'Three Models' did what Harris admitted he'd never thought of doing: it located the grammar of natural language within the class of all possible grammars. At the top of Chomsky's formal hierarchy were the "recursively enumerable" languages: namely, all languages having some finite definition. A subset of those were the "context-sensitive" grammars, which in turn contained the "context-free" grammars (see below). At the lowest level were the "regular" grammars, definable in very simple terms. (Twenty years later, other linguists would insert "indexed grammars" between the context-sensitive and context-free levels: see Section ix.d.)

Computer scientists were interested immediately. Similar research was already being done by some of them, but Chomsky's was a highly valued contribution.

His distinction between context-free and regular grammars (Chomsky 1956) helped them to codify, respectively, the lexical structure of artificial languages and the high-level structure of programs. Moreover, he proved in 1962 that different types of computational device—again, abstractly defined—would be needed to deal with the various classes of grammar. A universal Turing machine (see Chapter 4.i.c) can compute *any* definable language. Push-down stacks (see Chapter 10.v.b) can compute context-free languages (an insight that helped establish the theoretical basis of modern parsers, and of compiler design). And, since regular grammars are lower down in the hierarchy, they can compute those too. But push-down stacks aren't necessary for dealing with regular grammars. Such grammars can be computed by finite-state machines (described below).

The linguists, by contrast, didn't notice these early papers. For they weren't published in 'their' journals. (Chomsky's initial connection with MIT was the Research Laboratory of Electronics, from which the Linguistics Department later emerged: see 10.ii.a.) In other words, the scientific cooperation dreamt of by Descartes (2.ii.b–c) was—temporarily—prevented by disciplinary specialism.

Nor did they read the mimeograph/microfilm versions of the typescript of Chomsky's voluminous background research, which had been available to the cognoscenti since 1955. Even the cognoscenti had found it hard going, and very few (perhaps none?) devoured it from first page to last. This wasn't mere laziness on their part: the typescript was neither easily legible nor readily intelligible. Many long passages consisted of sequences of definitions expressed in algebraic symbolism. Indeed, in 1956 MIT Press had refused to publish it in its original form. (It was eventually published twenty years later: Section ix.d.)

In short, most professional linguists, if they'd encountered this work at all, weren't interested. They saw Chomsky's line of enquiry as "unpromising and exotic" (Chomsky 1975: 2). They weren't unaccustomed to formalism, as we saw in Section v.c–e. But whereas Harris (and Jespersen) had employed formalism in the service of meticulous empirical description, Chomsky's early research focused on formalism *as such*.

Most psychologists weren't interested either. Some mathematical psychologists realized the crucial implication of the 'Three Models' theme: namely, that human minds must have a level of computational power capable of dealing with whatever class of grammar is appropriate to natural languages. But, working within the information-theoretic paradigm, they assumed that regular grammars would suffice.

It was *Syntactic Structures*, described by Chomsky as a "sketchy and informal outline" of this background material, which set the linguistic world afire. Its impact for linguistics was immediately likened—in the review that Chomsky mentioned (above)—to the transition from alchemy to chemistry (Lees 1957: 375–6). And as we'll see, it had a revolutionary effect outside linguistics too.

b. The need for a generative grammar

The research summarized in *Syntactic Structures* was inspired by Harris's discourse analysis. It also reflected Chomsky's interdisciplinary education: at Harris's suggestion, he'd studied mathematics and philosophy as well as linguistics. Formalism had flourished in the philosophy of science for some years, as we've seen. But Chomsky went beyond mere formalism. His approach was fundamentally mathematical in a sense in which Harris's wasn't.

Chomsky asked two new questions about natural-language syntax. On the one hand, he sought to state a generative grammar—as opposed to a merely descriptive one—for a given language (such as English). On the other hand, he aimed to compare the computational power of various types of grammar.

Readers of *Syntactic Structures* weren't burdened with the bristling technicalities of his 1956 paper, but those ideas underlay the argument. Although he didn't locate his new grammar at a particular point within the grammar hierarchy, Chomsky's focus on differences in computational power largely explains the book's wide influence: across linguistics, computer science, psychology, and the philosophy of language.

The brief text of *Syntactic Structures* took a fresh, not to say shocking, approach. It denied two fundamental structuralist (Bloomfieldian) assumptions: that the goal of linguistics is to define a discovery procedure for grammars, and that language can be explained in probabilistic, behaviourist, terms. Moreover, it started out not from a rich description of actual usage but from dry definitions mentioning not a word of English, Sanskrit, or Kawi.

In the opening pages, Chomsky defined language, and the goal of linguistics, in abstract terms:

> I will consider a *language* to be a set (finite or infinite) of sentences, each finite in length and constructed out of a finite set of elements. All natural languages are languages in this sense... Similarly, the set of "sentences" of some formalized system of mathematics can be considered a language. The fundamental aim in the linguistic analysis of a language L is to separate the *grammatical* sequences which are sentences of L from the *ungrammatical* sequences which are not sentences of L and to study the structure of the grammatical sequences. The grammar of L will thus be a device that generates all of the grammatical sequences of L and none of the ungrammatical ones. (Chomsky 1957: 13)

A generative grammar is a set of (timeless) derivational rules. It doesn't tell us just how to build or parse a specific utterance. In other words, it isn't a computer program. However, it is a computational notion. The "device" mentioned here—and the "machine" mentioned elsewhere (e.g. p. 52)—was a purely mathematical one, as is a Turing machine (see 4.i.c–d).

But if Chomsky wasn't talking about computers he was, indirectly, talking about people:

> Any grammar of a language will *project* the finite and somewhat accidental corpus of observed utterances to a set (presumably infinite) of grammatical utterances. In this respect, a grammar mirrors the behavior of a speaker who, on the basis of a finite and accidental experience with language, can produce or understand an indefinite number of new sentences. Indeed, any explication of the notion "grammatical in L"... [offers] an explanation for this fundamental aspect of linguistic behavior. (p. 15)

Even so, his grammar wasn't a set of instructions for psychological processing. Rather, it represented the underlying grammatical competence of the speaker (see Chapter 7.iii.a).

In offering his new vision of what a theoretical linguistics should be like, Chomsky asked four interrelated questions:

* How can we discover a grammar for a given set of sentences?
* How can we know that a grammar generates all and only these sentences?
* How can we evaluate alternative grammars?
* And how can we express the differences between grammars, of either artificial or natural languages?

His answer to the first question wasn't what structuralists expected. On his view, no scientific ("practical and mechanical") discovery procedure can be found (1957, ch. 6). Even Harris, whose work was more careful than most, hadn't succeeded in defining one. The best we can hope for (and what Harris had actually provided), said Chomsky, is an evaluation procedure: a method for deciding, given a finite set of sentences, which of two candidate grammars is preferable. Even this is difficult to define formally, for grammars—like all scientific theories—are evaluated largely in terms of their simplicity, a notoriously elusive notion.

A candidate grammar for a natural language must generate all and only the grammatical sentences. But how do we know which these are? No search of an existing corpus can identify them. Statistics are inappropriate, since most sentences in any reasonable corpus will be one-offs: uttered on only one occasion. And if a sentence isn't there at all, so what? Any native speaker can, and frequently does, come up with new ones.

This is a matter not merely of observation, but of syntax. (At this stage, it was syntax, not semantics—still less, reason or imagination—which was given the credit for the creativity of language: see Section iv.f, above.) One can produce new sentences, for instance, by conjoining two or more old ones, or by recursively nesting one inside another. (Whether the result will always be intelligible was addressed in Chapter 7.ii.b.)

It follows, said Chomsky, that judgements of grammaticality must be left to the intuitions of native speakers. For they can reliably say whether a given word sequence is grammatical even if they can't explain their judgement.

Data-respecting critics soon complained—and still do—that by *native speakers* he meant *a single native speaker whose intuitions one trusts*—in Chomsky's case, Chomsky. Such a source hardly seems reliable: Chomsky himself changed his mind twice about his famous example "Colourless green ideas sleep furiously". And how were these armchair intuitions to be monitored? The structuralists, too, had used native informants to assess grammaticality, but had insisted that the conditions in which evidence is elicited from them be carefully controlled, to prevent experimental bias of various kinds (Z. S. Harris 1957, sect. 4). Disagreement arose, in psychology as well as linguistics, over the role of extensive sampling as against the consideration of individual cases: see Chapter 7.iii.d. Chomsky himself, however, remained content to rely on his natural, untutored, intuitions.

Assuming that we've agreed on a set of grammatical English sentences, we must ask "what sort of device can produce this set". This requires us, said Chomsky, to compare the power of different conceivable grammars. He addressed this task by starting from Claude Shannon's information theory (4.v.d).

c. Beyond information theory

Shannon had defined a number of artificial languages in the 1940s, suggesting that:

> These artificial languages are useful in constructing simple problems and examples to illustrate various possibilities. We can also approximate to a natural language by means of a series of simple artificial languages. (Shannon and Weaver 1949: 13)

Chomsky agreed that artificial languages are useful comparison points for natural language. But, with his grammar hierarchy lurking in the background, he saw the prospects for Shannon's serial "approximations" as hopeless.

The simplest example Chomsky considered, lying at the lowest ("regular") hierarchical level, was a *finite-state grammar* (*language*, *machine*), or FSG. This is a grammar—of which there are infinitely many—that describes a device with a finite number of states, including an initial state and a final state, and a finite set of rules for passing from one state (considered in isolation) to the next (cf. Shannon and Weaver 1949: 15–16). In a linguistic context, "passing from one state to the next" would involve either emitting or accepting a word, depending on whether the FSG was being used to produce word strings or to parse them, respectively.

Taking up Shannon's discussion, Chomsky pointed out that some FSGs could produce only a finite number of "sentences"—in some cases, only two (see the first state-diagram in Figure 9.2) (Chomsky 1957: 19). Clearly, such examples couldn't model the apparently infinite potential of natural languages. But a more powerful type of FSG could be defined by adding one or more closed loops, so that instead of passing directly from state n to state $n + 1$, the device may—optionally—return to state n again, and again . . . before finally leaving it. As a linguistic analogy, the sentence *The man comes* can be extended to *The old man comes*, *The nasty old man comes*, etc. by inserting one or more adjectives at the relevant point. This type of FSG is recursive, so can produce infinitely many sentences (compare using five adjectives, or fifty, or . . .).

Because of the potentially infinite number of sentences of English (French, Kawi . . .), Chomsky had the extended type of FSG primarily in mind (cf. Chomsky and Miller 1958). Of the two FSGs shown in Figure 9.2, then, he was less interested in the first than in the second, which contains a closed loop. Indeed, the simpler type is now sometimes forgotten, or anyway ignored (as an uninteresting special case). So FSGs *as such* are sometimes said to be capable of producing infinitely many word strings, or sentences.

Shannon had pointed out that a specific probability could be assigned to each possible state transition allowed by an FSG, and therefore that different artificial languages could be defined by assigning distinct sets of probabilities. Such probabilistic FSGs are called "Markovian". In natural languages, too, the successor probabilities for both letters and words vary. In English, though not in Polish, the state *z* is highly improbable as a successor to *c*. Similarly, the state *man* is fairly probable, given the state *old*, whereas the state *because* is highly improbable as a successor to *the*. If realistic probability measures were added for each state transition, Shannon said, something resembling natural language would result.

To illustrate this point, Shannon suggested a number of simple rules to produce various Markovian letter strings and word strings. These strings approximated English more and more closely as the number of previous letters or words considered (zero,

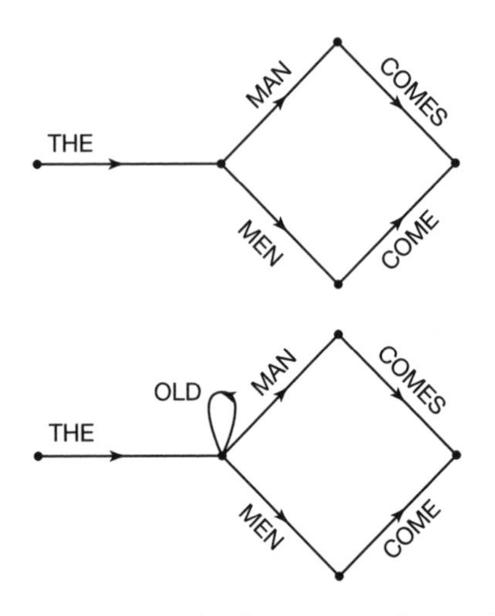

Fig. 9.2. State-diagrams for two FSGs. The first can produce only two sentences, whereas the second–because of the loop–can produce infinitely many. Redrawn with permission from Chomsky (1957: 19)

one, or two) increased. For instance, choosing words independently of each other (but guided by their individual frequencies) resulted in this:

REPRESENTING AND SPEEDILY IS AN GOOD APT OR COME CAN DIFFERENT NATURAL HERE HE THE A IN CAME THE TO IF TO EXPERT GRAY COME TO FURNISHES THE LINE MESSAGE HAD BE THESE. (Shannon and Weaver 1949: 14)

Not all of the two-word sequences included here would actually occur in English, and only one of the four-word sequences (AND SPEEDILY IS AN). By contrast, allowing the choice of each word to depend upon its immediate predecessor produced this:

THE HEAD AND IN FRONTAL ATTACK ON AN ENGLISH WRITER THAT THE CHARACTER OF THIS POINT IS THEREFORE ANOTHER METHOD FOR THE LETTERS THAT THE TIME OF WHO EVER TOLD THE PROBLEM FOR AN UNEXPECTED. (p. 14)

This passage contains many sequences of four or more words that could occur in real sentences. (Anyone but an information theorist would say "... that could make sense".) Indeed, the twelve-word sequence from FRONTAL to POINT could occur in a guidebook on country walks, in a sentence such as "It is because of the savage frontal attack on an English writer that the character of this point on the pathway has changed."

The method Shannon used to compose the latter word string was a simple one. Taking a book from his bookshelves, he opened it at random, and wrote down a word selected randomly from the open page. Next, he opened the book at another randomly chosen page, and read on until he encountered the previously recorded word. At that point, he wrote down the immediately following word. Then he repeated the process again and again, always using the most recently recorded word as his guide.

Clearly, he didn't pick *Alice's Adventures in Wonderland.* Indeed, he may have deliberately avoided picking a 'literary' text as his sample source. The reason is that this method would grind to a halt if any newly chosen word occurred only once in the entire book.

(i) *Sentence* → *NP* + *VP*
(ii) *NP* → *T* + *N*
(iii) *VP* → *Verb* + *NP*
(iv) *T* → *the*
(v) *N* → *man*, *ball*, etc.
(vi) *Verb* → *hit*, *took*, etc.

Fig. 9.3. Rewrite rules for a simple phrase structure grammar. Reprinted with permission from Chomsky (1957: 26)

Even this simple method was time-consuming. It would be interesting to go further, said Shannon, by considering more than one previous word. But the labour involved in hand-searching the sample text would become enormous even at the two-word stage. (George Miller, instead of searching texts, asked people to volunteer 'the next word' after having been given one, two, or . . . six words: the higher-order, more redundant, strings were remembered more easily: G. A. Miller and Selfridge 1950.)

If computers had been available at mid-century, to collect the necessary statistics, many structuralists would have been interested. Think of the textual studies that might have been done on the basis of Harris's discourse analysis, for instance. Today, such studies are possible. But in the 1950s, computing technology was still in its infancy. In practice, then, Shannon's approximations weren't followed up by those linguists whose theoretical sympathies were similar to his.

Chomsky, in any case, wasn't one of their number. He saw the statistical approach as a dead end in theory, not just in practice. He believed he could prove, as a matter of principle, that FSGs, although they can generate infinitely many new sentences, cannot produce all the examples of everyday English. Specifically, they cannot generate sentences involving nested expressions, where the selection (in Shannon's terms, the probability) of word $n+1$ depends not on its immediate predecessor, word n (nor on the m immediate predecessors), but on some word indefinitely many places earlier. Consider this sentence, for instance: *The cat with black fur and long whiskers was sitting on the mat.* The ninth word here depends not on the eighth, but on the second: replace *cat* by *cats*, and you must change *was* to *were*. It's structure which is doing the work here, not statistics.

Such nested dependencies, Chomsky argued, can in principle occur on infinitely many levels. (We saw in Chapter 7.ii.b, however, that *in practice* they cannot.) They require a "phrase-structure grammar", in which sentences are derived by a series of ordered "rewrite rules". Rewrite rules *as such* were invented not by Chomsky, but by the logician Emil Post (1943: see 10.v.e). But Chomsky was the first to apply them to natural language. He used them to produce phrases, including terminal elements (words), on various hierarchical levels (see Figure 9.3).

Unlike Jespersen twenty years earlier, Chomsky aimed to account for every single word in the sentence. So he analysed nominal phrases like *the man* and *a man* into two parts—and when he came to discuss transformations (see below), he analysed auxiliary phrases such as *has taken*, *will take*, and *is taking*, and showed the underlying relations between them.

The complete grammatical structure of the sentence, though not the order in which the rewrite rules were applied, can thus be represented by a tree diagram in which every

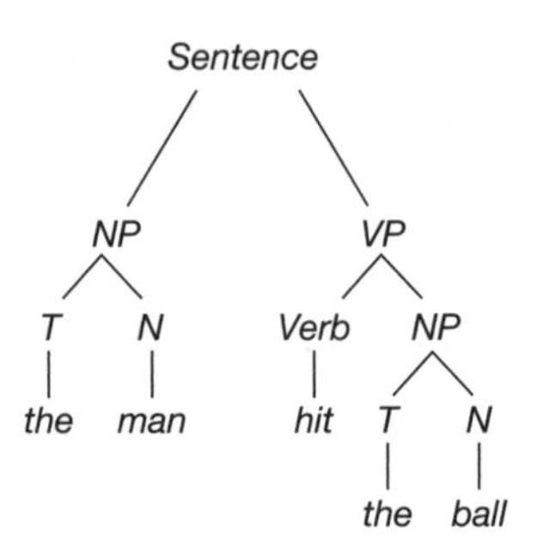

FIG. 9.4. Phrase-structure diagram showing the derivation of "The man hit the ball". Redrawn with permission from Chomsky (1957: 27)

word is separately considered (see Figure 9.4). This diagram shows how the grammatical categories are linked: from top to bottom, not—as in Markovian grammars—from left to right.

d. Transformational grammars

Phrase-structure grammars weren't new. They implicitly underlay immediate-constituent analysis, in which a sentence is recursively broken into contiguous constituents. This method had been used by many people, from the Port-Royal logicians to Harris, not only to parse unproblematic sentences but also to explain ambiguities. Linguists had long known, for instance, that the phrase structure of *The old men and women* could be either *The (old men) and women* or *The old (men and women)*, and that the meanings differed accordingly.

Similarly, the intuitive notion that different grammars might have different computational power wasn't new. Shannon, after all, had already contrasted FSGs having finite or infinite string sets.

The novelty in Chomsky's argument—as considered so far—was to define phrase-structure grammars as a general class, to prove that they're more powerful than Markovian (probabilistic) approaches, and to show that *natural* languages require grammars whose computational power is at least as great as that of a phrase-structure grammar. Significantly, he did the latter by relying, not, as Shannon had done, on some intuitive notion of computational power, but on his new classification of grammars. Each level in the hierarchy could generate all the strings generated by the level below, and more. FSGs lay at the lowest level, and phrase-structure grammars lay above them.

Chomsky was speaking here about *context-free* phrase-structure grammars. In his previous mathematical analysis, he'd distinguished between context-free (CF) and context-sensitive (CS) phrase-structure grammars.

In a CF grammar, each rewrite rule can be applied whenever the one-and-only symbol on the left-hand side is present. In a CS grammar, more than one symbol occurs on the left-hand side of the rewrite rules; a rule may rewrite one of them, and carry the others over unchanged. In the hierarchy of computational power, CS grammars had been located much higher than CF grammars. It was already clear that CS grammars included many "languages" of interest only to mathematicians, not to linguists. For linguists, the more interesting, more restrictive, and more plausible hypothesis was that natural languages have CF grammars. Indeed, the term "phrase-structure grammar" is often used to mean CF grammar.

At this point in the argument, Chomsky went further: he claimed that grammars of even greater computational power than phrase-structure grammars are needed for natural language. (My use of the word "claimed" here, instead of "proved", will be justified in Section ix.d–e.) In other words, natural languages involve a grammar located somewhere between phrase-structure grammars and universal Turing machines.

Chomsky admitted: "I do not know whether or not English is itself literally outside the range of [context-free phrase-structure] analysis." But he claimed that it could be so described (if at all) "only clumsily", by a complex, ad hoc, and unrevealing theory (p. 34). The reason he gave was that certain simple, and intuitively obvious, ways of describing sentences can't be expressed in terms of phrase structures, as then understood.

Everyday examples that require something more than a phrase-structure grammar, said Chomsky, include sentences involving auxiliary verbs (*The man has read the book*, *The man has been reading the book*); passive sentences (*The book is read by the man*); interrogatives (*Does the man read the book?*, *Will the man have been reading the book?*); *and* wh-questions (*Why/where/when/how does the man read the book? Who reads the book?*). It's intuitively obvious that there are close grammatical relations between the sentence *The man reads the book* and all these other sentences. But, Chomsky claimed, these relations can be simply expressed only by rules of the general form: *If the derivation of a grammatical sentence S-1 is such-and-such, then another grammatical sentence, S-2, can be produced by altering S-1 thus-and-so.* (*Sentence*, here, is a technical term, that covers any legal structure within a grammatical derivation. So *The man VP* is a sentence in this sense, for it is a structure derivable on the way to the terminal structure, *The man reads the book*.)

Intuitively, all these sentences seem to be related to the last one: *The man reads the book*. What's therefore required, said Chomsky, is a set of rules that can start with this sentence and derive all the others. Because rules like this define ways of transforming one sentence, or tree structure, into another (S-1 into S-2), grammars that contain them are called transformational grammars, or TGs.

A TG consists of a "base component"—a phrase-structure grammar that generates the initial trees—plus one or more transformational rules. This general definition covers many examples, including Harris's work and perhaps even Port-Royal grammar (see Sections iii.c and v.d–e). But the term TG is widely used to refer only to Chomsky's grammar, or variations of it—a fact that has aroused angry criticism, as we shall see (in Section ix.d).

TGs in general can't be located at any specific point in the grammar hierarchy, because the base component needn't be context-free (though it usually is), and because it's not clear how much—if any—computational power is added to it by the various transformations. The latter point makes it difficult to place even a particular TG in the hierarchy. The assumption implicit in *Syntactic Structures* was that a TG must lie somewhere between Turing machines and CS grammars—but just where was unclear.

In Chomsky's 1957 terminology, the simplest of the English sentences given above, and the point of reference for all of the others, is a "kernel" sentence. In other words, *The man reads the book* is derivable from its underlying grammatical form (which is derived by phrase-structure rewrite rules) by using only obligatory transformations (p. 45). These are the transformations that must be applied if the terminal sentence is to be grammatical at all. Non-obligatory transformations are applied either to the forms underlying kernel sentences, or to previous transforms thereof. In deciding which transformations are obligatory in English, and which optional, Chomsky relied on grammatical intuitions such as the one just remarked.

Chomsky's general claim, here, was that sentences of natural language must be represented on more than one theoretical level. In broad terms at least, many linguists were quick to agree. One later recalled Chomsky's argument as having had a "dramatic input" at the time:

> By God, native speakers DID associate active and passive counterparts... Yet the two were different and even unrelated in a phrase-structure analysis. His ideas seemed salient, dramatically salient, from the outset. (Hymes 1972)

As a result, "within a few years, quite a number of linguists were working on transformational grammar" (Chomsky 1975: 4).

Within a few years, also, the philosopher Hilary Putnam noticed that transformational grammars as defined above have the same power as a Turing machine (Putnam 1961: 101 ff.). If we allow just anything to count as S-1 and S-2, then this class of grammars has potentially infinite power. Any conceivable language, whether natural or artificial, recursive or non-recursive, could be derived by some TG (if it's computable at all). The set of prime numbers, for example, can be TG-generated.

This point was awkward for linguists claiming that TGs underlie natural languages. A scientific theory should not only explain what *does* happen, but also restrict what *could* happen (see 7.iii.d).

A grammar, for instance, should generate all *and only* the acceptable sentences of the language concerned. If any conceivable language, even including the (highly un-natural) set of prime numbers, can be 'explained' by some TG, then the class of such grammars can't exclude any possibilities. To be sure, a given TG might explain English and exclude French. (To someone seeking a TG to fit *all* natural languages, as Chomsky later did, this would be a weakness rather than a strength: see Section vii.d.) But Chomsky's supposedly surprising claim in 1957 was that each natural language requires *some TG or other*, and that *some TG or other* will suffice to explain it. Putnam's work suggested that this amounted merely to saying that natural languages are Turing-computable. That had been assumed for years by logicist philosophers (see Chapter 4.iii.c), so was hardly exciting.

The Turing-equivalence of transformational grammar was to be discussed in detail in the 1970s (Peters and Ritchie 1973). Partly because of this, by 1980 some linguists had rejected transformations altogether. They offered more restrictive scientific theories of natural language, relying on grammars low down in Chomsky's hierarchy.—But that was for the future (see Section ix.d–f, below).

Meanwhile, Putnam's theoretical point was shunted into the background for over twenty years, as linguists tried to discover *which transformations in particular*, and in which order, are needed to represent a given natural language—usually, English. The specific transformations that Chomsky hypothesized for English included, for instance, rules for converting a kernel sentence into a question either by moving the auxiliary verb to the beginning (so *The man will read* becomes *Will the man read?*), or by inserting the appropriate form of the auxiliary *do* at the beginning, and altering the form of the main verb (so *The man reads* becomes *Does the man read?*).

The details don't concern us (pp. 61–84). Nor need we discuss the rather different account of transformations that Chomsky was to give in his third book, in which kernel sentences, with their compulsory/optional transformations, were nowhere to be seen (Chomsky 1965). (Transformations as such became rarer in Chomsky's later work, as many functions previously effected by transformations were transferred to other rules of the grammar: see Section viii.b.)

More important, for our purposes, is the general form of the argument in *Syntactic Structures*. For this had implications far beyond linguistics. It alerted philosophers of language, psycholinguists, and the early computational psychologists. (One might say that it didn't alert psycholinguists so much as *create* them; the countless studies of "verbal behaviour" were focused on learning, not language—word strings, not sentences: see Chapter 6.i.e.) But it *didn't*, yet, alert psychologists in general.

e. So what?

Chomsky's analysis had radical—and to many, highly unwelcome—implications. For it followed, as he took pains to point out, that utterances (i.e. "sentences" in the everyday, non-technical sense) must be theoretically represented on several levels.

On the one hand, there is the surface structure: namely, the grammatical form of the sequence of words (morphemes, phonemes) that constitutes the sentence. The surface structure is interpreted phonetically, to produce a spoken utterance. On the other hand, there is the deep structure: the derivational history (including rewrite rules, kernel forms, and transformations), which will usually have many levels. The deep structure is interpreted semantically, to give the sentence's meaning.

(Although the technical terms "deep/surface" were introduced in Chomsky's second book, a related distinction was clearly made in the first. Similarly, the terms "competence/performance" didn't feature there, but the relevant distinction did.)

From the point of view of Bloomfieldian linguists (and behaviourist psychologists, too), this added insult to injury. Having already been told that statistical theories can't describe language, they were now being told that the observable aspects of language are underlain by a complex unobservable structure that must be represented in any adequate theory. Mentalism had returned, with a vengeance.—Mentalism, but not introspectionism: Chomsky agreed with Bloomfield that the speaker has no direct access to inner mechanisms.

One might argue that grammar is a mathematical exercise having nothing to do with mentalism. Indeed, theoretical computer scientists, who were much influenced by Chomsky's analysis of the computational power of different types of "language", gave no thought to psychology. And Chomsky himself said nothing, here, about the broader mentalistic implications that were to be so controversial in the 1960s (see Section vii.c–d).

In other words, *Syntactic Structures* contained no explicit comment on behaviourism, although Chomsky's anti-Markovian arguments were clearly relevant to it. Moreover, there was no mention of universal grammar, and therefore no hint that linguistics has nativist psychological implications, nor any relevance for the philosophy of mind. The theory of generative grammar wasn't yet being explicitly presented as "a particular sub-branch of cognitive psychology", still less as a study of "the human language faculty as such" (Chomsky 1975: 9).

The reason these "Cartesian" hypotheses weren't mentioned wasn't that Chomsky hadn't yet thought of them, but that he was deliberately suppressing them. They would have clashed with structuralist assumptions about the diversity of language and the emptiness of talk about language-and-mind—and the book was radical enough, even without them. Indeed, its ideas were so unconventional that, as he later recalled: "The one article I had submitted on this material to a linguistic journal had been rejected, virtually by return mail" (Chomsky 1975: 3). Understandably, then, he felt it to be "too audacious to raise the psychological issues which were in the background of my mind" (p. 35).

However, psychology wasn't ignored entirely. Linguistics, he said, is relevant to psycholinguistics—specifically, to the study of meaning. He was concerned not with vague claims like the Sapir–Whorf hypothesis, but with clearly specified theories about the psychological abilities (though not the processing details) involved in understanding:

> [My] formal study of the structure of language . . . may be expected to provide insight into the actual use of language, i.e. into the process of understanding sentences. (p. 103)

> [The] notion of "understanding a sentence" must be partially analyzed in grammatical terms. To understand a sentence it is necessary (though not of course sufficient) to reconstruct its representation on each level, including the transformational level where the kernel sentences underlying a given sentence can be thought of, in a sense, as the "elementary content elements" out of which this sentence is constructed. (p. 107–8)

The last remark occurred in the context of a discussion of the prospects for semantics. Here, the book was rather more acceptable to Chomsky's Bloomfieldian predecessors. For meaning was sidelined. Any theory of syntax, Chomsky argued, must be independent of semantics. Indeed, semantics is secondary to syntax, for understanding the meaning of a sentence rests on grammatical analysis.

Even so, structuralists were being challenged. Although Chomsky agreed with them that a science of meaning can't be based on vague appeals to introspection, or on unknown past associations, he did say that a (syntax-based) science of meaning is in principle possible. Indeed, semantics would play a fundamental role in his next book, in which he described a grammar as a set of rules relating the meaning of a sentence to its phonetic structure (Chomsky 1965).

These passages alerted psychologists, some of whom were aghast and others inspired. The repercussions in psycholinguistics, and the relevance of Chomsky's distinction between competence and performance, were discussed in Chapter 7.ii–iii.

In Chapters 5 and 6, we considered the growing revolt against behaviourism—to which Chomsky was an enthusiastic latecomer. Now, let's look more closely at what he had to say about the then prevailing psychological orthodoxy.

9.vii. A Battle with Behaviourism

Chomsky initiated a battle with behaviourism, but not the war itself. Others were fighting at the same time, and some had reached the battlefield earlier (see 5.iv and

6.i–iii). But it was Chomsky who flourished the sword most cuttingly and wounded his opponents most bloodily. And it was he who returned to the attack over and over again, in both his scientific and his political writing. Consequently, he's been awarded most of the victory medals, even though others merited them too.

(Not being one for half-measures, he also rejected Piaget's "epigenetic" compromise between nativism and behaviourism: see Chapters 7.vi.g and 14.x.c. Contributions to the Chomsky–Piaget debate are collected in Piatelli-Palmarini 1980.)

Chomsky's training in descriptive linguistics had been informed by behaviourist assumptions, as we've seen (in Section v.a). But in the early 1950s, he ignored this psychological movement. He looked at it in detail only after he'd finished his basic theoretical research on language, in 1956–7. Indeed, he said later that the critique of finite-state languages with which he'd opened *Syntactic Structures* was "an afterthought", not included in his larger manuscript (Chomsky 1975: 40).

From the late 1950s on, however, he pursued the behaviourists relentlessly. By the mid-1960s, he had rejected—or rather, expanded—the definition of linguistics given in *Syntactic Structures*. Now, he identified the goal of linguistics in explicitly nativist terms—with radical implications for psychology, and for the philosophy of mind.

a. Political agenda

In his pursuit of behaviourism, Chomsky was driven at least as much by political passion as by abstract argument. For he saw it in political terms, and didn't like what he saw.

He linked behaviourism with the social-scientific 'experts' whose advice to the US government, and especially to the military, he—and countless others—regarded as political anathema. AI scientists were suspect too, despite their non-behaviourist approach, for their work had made 'smart' bombs and aerial reconnaissance possible (Chapter 11.i).

Chomsky expressed this complaint on many public platforms at the height of the Vietnam war. Indeed, he spoke in the very first public demonstration against the war, held on the Boston Common in October 1965. He spoke, but he wasn't heard:

> the mobs were so hostile that none of us could say an audible word. The only reason we weren't killed, I suppose, was that there were hundreds of police—who didn't like what we were saying one bit, but didn't want to see anyone murdered on the Common. (quoted in Swain 1999: 29)

Besides braving mobs on the Boston Common, he was put in prison on more than one occasion (Swain 1999: 24). He was thus a powerful voice in the counter-culture (Chapter 1.iii.c).

He still is. For, unlike his MIT colleague Joseph Weizenbaum (11.ii.d), he continued—and intensified—his complaints long after the 1970s. He repeated them, with meticulous documentary evidence, all around the world in connection with various other events—most recently, those of 11 September 2001 in New York, and their grisly aftermath in Afghanistan and Iraq. He still berates "respected intellectuals" for justifying the actions of "violent and murderous states"—including the USA (Chomsky 2001: 73; 2005). And he still sees political engagement as a responsibility of the intellectual (1996*c*).

His speeches continue to draw a multitude of listeners: his Royal Institute of Philosophy Lecture in 2004, for instance, saw about 3,000 youngsters queuing along the London streets (Chomsky 2005). And his writings have attracted adulatory praise from political sympathizers (Barsky 1997). They doubtless drew some people to his scientific writings who wouldn't otherwise have been interested in them.

Given the combination of his political and linguistic work, he eventually became the most-quoted living writer, and the eighth most quoted in history (Barsky 1997: 3). (The other seven were Marx, Lenin, Shakespeare, Aristotle, Plato, Freud—and, in fourth place, the 'author' of the Bible.)

In making these politico-scientific links, Chomsky repeatedly criticized what he saw as the philosophical assumptions underlying the science. Specifically, he claimed that empiricist and rationalist views of humanity inherently tend to promote political oppression and freedom, respectively (e.g. Chomsky 1972*b*: 45–54 and *passim*; cf. N. V. Smith 1999).

The justice of this general claim isn't obvious; some have argued that the reverse is true (Sampson 1979*a*). For instance, Locke's political philosophy—which underlay the English, French, and American revolutions—hardly suggests that empiricism is incompatible with liberty. Indeed, Locke saw doctrines of innate ideas as politically dangerous, since they encourage "blind credulity" whereby the populace becomes "easily governed" by a "dictator of principles and teacher of unquestionable truths" (J. Locke 1690: i. iv. 25). One of his followers, Lord Chesterfield, even wrote to his son that "A drayman is probably born wyth as good organs as Milton, Locke, or Newton" (Strachey 1924–32, pp. ii–136). True to his time, however, when Locke framed the constitution of Carolina in 1669 he gave slave-owners full jurisdiction over their slaves (Porter 2001, pp. xxii–xxiii). Lockean liberty had its limits.

However that may be, this dual significance helps explain the virulence with which Chomsky attacked behaviourism in general, and Burrhus Skinner in particular. Skinner—who wrote the first book on "verbal behaviour"—not only theorized language in a manner Chomsky saw as scientifically vacuous, but also rejected traditional assumptions about the way to the good life (Skinner 1948) and humanist conceptions of freedom and dignity (Skinner 1971).

These views didn't endear him to the counter-cultural Chomsky (1.iii.c). He attacked them in typically uncompromising terms:

> Consider a well-run concentration camp with inmates spying on one another and the gas ovens [visibly] smoking in the distance, and perhaps an occasional verbal hint as a reminder of the meaning of this reinforcer. It would appear to be an almost perfect world... In the delightful culture we have just designed, there should be no aversive consequences... Unwanted behavior will be eliminated from the start by the threat of the crematoria and the all-seeing spies... Nevertheless, it would be improper to conclude that Skinner is advocating concentration camps and totalitarian rule (though he also offers no objection). Such a conclusion overlooks a fundamental property of Skinner's science, namely, its vacuity. (Chomsky 1973: 129–30)

The familiar vocabulary of freedom and dignity, said Chomsky, far surpasses Skinner's terminology in discussions of politics. Indeed, it's all we have: there's no scientific theory of politics, and perhaps there never can be. In refusing to use such humanistic expressions, therefore, Skinner was being not scientifically scrupulous but politically irresponsible.—Perhaps merely politically neutral? No, Chomsky insisted, for that comes to the same thing: to "offer no objection" is, in effect, to support the status quo.

Later, Chomsky widened this attack to include all quietist academics. He contrasted Albert Einstein's "very comfortable life" of research, interrupted by "a few moments for an occasional oracular statement", with Russell's passionate espousal of nuclear disarmament and vociferous critiques of the Vietnam war, which landed him in prison on more than one occasion (Barsky 1997: 32–3). (Russell was one of the first public critics of the H-bomb, and in 1958 became founding president of the Campaign for Nuclear Disarmament; his pacifist speeches and writings around the time of the First World War had also got him jailed—see Monk 1996: 525–40; 2000: 367–90, 405–13.) In general, Chomsky argued that intellectuals should actively, if perforce riskily, confront "privilege and authority"—both on public platforms and in their personal lives. Certainly, he himself has never stopped doing so.

For the rest of our discussion here, let's ignore the politics. If this brands us as irresponsible quietists, so be it. For Chomsky also gave non-political arguments against behaviourism.

b. That review!

Politics aside, Chomsky made three main claims in his ongoing critique of behaviourism:

- First, that behaviourist theories are in principle inadequate to represent grammatical structure, which exists on several theoretical levels.
- Second, that Skinner—the most influential behaviourist in this area—used covertly mentalistic terms to explain language, while having no theoretical right to do so.
- And third, that human babies, and *only* human babies, can acquire language only because they have innate knowledge of its fundamental structure.

One could coherently admit the first two claims while rejecting the (more controversial) third. Nevertheless, by the end of the 1960s, many cognitive scientists had accepted all three.

The first claim was the theoretical core of Chomsky's dispute with behaviourism, and rests on his argument in *Syntactic Structures* (presented there without explicit reference to behaviourism). The structure and "creativity" of language can't be described by Markov processes, but must be represented as a hierarchy of theoretical levels (see Section vi). The central idea here wasn't new: Karl Lashley had expressed it in the 1940s, in relation to speech and other motor skills (Chapter 5.iv.a). But Chomsky put it in mathematical terms.

He took this criticism to apply to behaviourism in general. As regards those behaviourist theories which (unlike Skinner's) posited internal Ss and Rs, he was justified: mathematically, it doesn't matter whether the Ss and Rs are observable or not. (Compare the objection that production systems are merely an internalist form of behaviourism: Chapter 7.iv.ii.)

Whether he was justified as regards all possible forms of behaviourism was less clear. Some empiricist philosophers argued that behaviourism, or (as they put it) scientific psychology, shouldn't be defined so restrictively as to include only Markovian theories based on conditioned response (e.g. Quine 1969: 96–7). (It turned out, later, that simple statistical methods can enable a connectionist network to learn the past tense, even generating the detailed errors made by children: see Chapter 12.vi.e and x.d.)

The second prong of Chomsky's attack was unsheathed in his notorious review of Skinner's book on the psychology of language (Skinner 1957; Chomsky 1959*b*). This book was published in the same year as *Syntactic Structures*, but had been circulating in draft for some time. It was based on Skinner's William James Lectures, given at Harvard in 1947, and in his own mind represented his crowning achievement. (He'd been working on it for twenty-three years—Skinner 1967: 402.)

Chomsky allowed that in his animal research Skinner had been scrupulous in defining his terms, and had made genuine scientific discoveries. On turning to language, however, he had abandoned both experiment and scruple. His central theoretical concepts of *stimulus*, *response*, and *reinforcement* were being applied so loosely as to be vacuous. And his new, specifically linguistic, concepts (such as *tact* and *mand*), though defined in putatively observational terms, were actually being understood mentalistically (*mand* as an indiscriminate amalgam of *command*, *request*, *question*, *prayer*, *advice*, and *warning*, for instance).

This was hardly surprising, Chomsky said. For if one struggled to interpret Skinner's vocabulary in a strictly behaviourist fashion, his theory instantly lost all plausibility. Moreover, Skinner had made no serious attempt to test his theory of language experimentally, relying instead on anecdote and speculation.

Chomsky did acknowledge that some operant-conditioning experiments on language had come up with surprising and non-trivial results. For instance, speakers can be—unknowingly—conditioned to produce more or fewer plural nouns, to use the personal pronoun more or less often, or to talk about one topic rather than another (Krasner 1958). The first of these is especially interesting, since it's defined in terms not of a single response, like uttering the word *I* or *we*, nor even of a phonemic pattern such as words ending in *-s*, but of a syntactic class. But Chomsky was sceptical: "Just what insight this gives into normal verbal behavior is not obvious" (Chomsky 1959*b*, n. 7).

This review, one-third as long as Chomsky's little book, attracted enormous attention. Its many readers were not only gripped by the argument, which was detailed and careful, but also amused and/or outraged by the rhetorical style. For this essay was very different from the dry and symbol-ridden *Syntactic Structures*.

Chomsky's pen had been dipped in vitriol, as well as in ink. Words such as *pointless*, *confused*, *gross*, *absurd*, *delusion*, *dogmatic*, *arbitrary*, *useless*, and *empty* leapt out from the pages. Skinner's fundamental concept (reinforcement) was scornfully said to have "totally lost whatever objective meaning it may ever have had", and Skinner himself to be "play-acting at science". There was delectable ridicule, too. According to Skinner, said Chomsky, the best way to show one's appreciation of a prized work of art owned by a friend is to shriek "Beautiful!" repeatedly in a loud and high-pitched voice, without ever pausing for breath.

This was welcome light relief from the daily academic grind. Small wonder that the piece was so widely read. Only Skinner avoided it—or so he claimed: "I have never actually read more than half a dozen pages of Chomsky's famous review of *Verbal Behavior*. (A quotation from it which I have used I got from I. A. Richards)" (Skinner 1967: 408).

Rhetoric apart, however, Chomsky's second point was unassailable. Skinner was guilty as charged: he was misusing his own terminology and betraying his own scientific method, by tacitly relying on everyday mentalistic intuitions. In that skirmish, Chomsky won hands down.

c. Nativist notions

Chomsky's third salvo caused even more sound and fury—and is still echoing today (see 7.vi, 12.vi.e, and 14.ix.c–d). More intellectually daring than the first two points, and arguably less well grounded, it was a combination of a negative claim and a positive one.

Negatively: that the child's language acquisition can't be fully explained by experience, even if this is conceptualized in non-Markovian terms. Positively: that babies must therefore have innate knowledge of the structure of language—which is to say, of the rules of some universal grammar.

The positive corollary, that there's some inborn "language acquisition device" specially apt for learning grammatical patterns, was even more obviously heretical than its negative ground. It had been "too audacious" to be included in *Syntactic Structures* (see above), and was merely hinted at in the notorious review, where Chomsky was more concerned to criticize Skinner's theory than to present his own. But it was made fully explicit—and defensively related to his august "Cartesian" precursors—in Chomsky's publications of the 1960s (Chomsky 1964, 1965, 1966, 1968). Indeed, he now built it into the very definition of linguistics as such:

> A theory of linguistic structure that aims for explanatory adequacy incorporates an account of linguistic universals, and it attributes tacit knowledge of these universals to the child. (Chomsky 1965: 27)

> Given a variety of descriptively adequate grammars for natural languages [which can be achieved if one sees the goal of linguistics as I defined it in *Syntactic Structures*], we are interested in determining to what extent they are unique and to what extent there are deep underlying similarities among them that are attributable to the form of language as such. *Real progress in linguistics* consists in the discovery that certain features of given languages can be reduced to universal properties of language, and explained in terms of these deeper aspects of linguistic form. Thus *the major endeavor of the linguist* must be to enrich the theory of linguistic form by formulating more specific constraints and conditions on the notion "generative grammar". (Chomsky 1965:35; italics added)

The negative nativist claim rested partly on Chomsky's anti-Markovian arguments, and partly on his assertion that the language the infant hears is inadequate to allow grammar to be induced so quickly. The child, he said, may hear a certain syntactic construction only very rarely. Moreover, everyday speech abounds with unfinished sentences, restarts, and grammatical errors. (People may get lost inside subordinate clauses, for instance.) So the child very often hears utterances that are grammatically flawed. Given such patchy and contaminated data, rapid learning of a perfect grammar simply couldn't happen.

This negative assertion was immediately criticized by those of an empiricist cast of mind, as being an argument from ignorance. The fact that no one has yet come up with an explanation of how children learn language, they said, doesn't mean that it's impossible to do so. Induction (as remarked above) needn't be conceptualized in the simple terms of traditional empiricism (Quine 1969: 96–7; Harman 1967, 1969). Moreover, the notion of "experience" was said to be richer than Chomsky allowed. It was no longer thought of as passive or atomistic, and with further psychological research might become richer still (M. Black 1970: 458).

This neo-empiricist shaft hit its mark, but left Chomsky standing. Facing only promissory notes from his opponents, his argument from ignorance remained in play. (It was soon followed by a formal proof that 'nested' languages can't be learned on the basis of *positive* evidence only: Gold 1967.)

As for Chomsky's comment that the infant's heard corpus is largely degenerate, this seemed obviously true to many people. But—like Cordemoy's remarks (see Section iii.a) about nouns preceding adjectives, preceding verbs—it hadn't actually been tested.

When it was, psycholinguists reported that mothers typically use a restricted dialect ('motherese') when speaking to babies and young infants, employing sentences both syntactically simpler and more grammatically correct than those of adult speech (R. Brown 1973). On the other hand, they also found that the child's syntax develops in predictable stages, two-word utterances appearing first, and the complexity increasing with age (7.vi.a). This was generally interpreted (perhaps wrongly: see Chapter 12.viii.c–d) as evidence of an innately driven developmental process, supporting the spirit, if not the letter, of Chomsky's nativism.

The positive nativist claim caused huge controversy. Much of the initial scepticism concerned what Chomsky—or indeed anyone—could mean by "innate knowledge" or "innate ideas". All of the questions, and most of the answers, that had been hurled at Descartes, Locke, and Leibniz resurfaced.

Some philosophers objected that innate knowledge is impossible, because "knowledge" implies justification (Edgley 1970; Harman 1969). Chomsky replied that unconscious knowledge (competence), unlike what Gilbert Ryle (1946) had called "knowledge that", requires neither verbalization nor justification (Chomsky 1969; 1980*a*, ch. 3).

Even when the debate was put in terms of innate ideas instead of innate knowledge, or dispositions instead of ideas, the problems sketched in Section ii.c (above) arose—and were debated with passion. High-visibility symposia pitting Chomsky against leading epistemologists and philosophers of language occurred from the mid-1960s on, and innate ideas became a hot topic of scientific and philosophical debate (e.g. *Synthese* 1967; Hook 1969, pt. ii; Stich 1975).

The discussions were often ill-tempered, and many of Chomsky's adversaries rivalled his rhetorical armoury in their attack. One leading philosopher, Gilbert Harman, said that Chomsky "misrepresents [my comments] at almost every point", to which Chomsky testily retorted: "I see no reason to try to trace the various confusions Harman develops, none of which have any relation to the views I actually hold . . . " (Harman 1969: 143; Chomsky 1969: 154). His irritation was doubtless fuelled by the fact that, some years earlier, Harman had denied the need for transformations (Harman 1963). Another influential philosopher dubbed Chomsky's theory 'The Emperor's New Ideas', and remarked:

> The theory of innate ideas is by no means crude. It is of exquisite subtlety, like the gossamer golden cloth made for that ancient emperor. But the emperor needs to be told that his wise men, like his tailors, deceive him; that just as the body covered with the miraculous cloth has nothing on it, the mind packed with innate ideas has nothing in it. (Goodman 1969: 141–2)

Willard Quine, in characteristically gentlemanly fashion, mildly declared that the behaviourist is "knowingly and cheerfully up to his neck in innate mechanisms of learning-readiness" (Quine 1969: 95–6). Reinforcement, for example, depends on "prior inequalities in the subject's qualitative spacing . . . of stimulations". And "unquestionably much additional innate structure is needed, too, to account for language learning".

But there was a sting in the gentleman's tail. Quine ended his paper thus:

> Externalized empiricism [which makes no reference to "ideas"] or behaviorism sees nothing uncongenial in the appeal to innate dispositions to overt behavior, innate readiness for language-learning. What would be interesting and valuable to find out, rather, is just what these endowments are in fact like in detail. (Quine 1969: 98)

Chomsky could answer that question only by saying: universal grammar, whatever that turns out to be.

d. Universal grammar?

What universal grammar will turn out to be—if it exists at all—is still unclear. Indeed, the very notion of *innateness* is now interpreted in a much more sophisticated way (Chapters 7.vi.g and 14.ix.c). Added to that is the fact that Chomsky himself changed his mind more than once about just what the "universal" base of language is.

Chomsky's initial postulation of universal grammar had depended only on argument, the main premiss being the supposed inadequacy of learning theory to explain language acquisition. But, as Quine had said, it needed empirical evidence also.

From 1965 onwards, Chomsky suggested a number of abstract grammatical principles that appeared to apply to several languages and might apply to all. Some concerned aspects of the base component (for example, that all languages share the same syntactic categories: noun, verb, NP, etc.). Others concerned the transformational rules.

For instance, in *Aspects of the Theory of Syntax*, the first major publication in which he made his nativism fully explicit, Chomsky said:

> Although a language might form interrogatives, for example, by interchanging the order of certain categories (as in English), it could not form interrogatives by reflection [reversing the order of words in the sentence], or interchange of odd and even words, or insertion of a [syntactic] marker in the middle of the sentence. (Chomsky 1965: 56)

Later, he posited the *A-over-A* principle. This forbids any language-specific (English, French, or . . .) transformation from picking an embedded noun phrase out of the noun phrase in which it is embedded—or more generally, from extracting any phrase of category *A* out of some larger phrase of the same category. Another example was a "principle of cyclic application" which Chomsky supposed to govern the way in which noun phrases are deleted and replaced by pronouns.

However, each of Chomsky's suggestions was challenged as applying only (if at all) to English, or to a small set of languages. Thorough testing would require detailed descriptions of a multitude of tongues.

This, of course, is what the out-of-fashion structuralists had aimed to provide. But Chomskyans didn't. One critic has described them as being as obsessed with English as the Port-Royalists were with French (Itkonen 1996: 487; cf. Section iii.c). Indeed, Chomsky himself would defiantly declare that "I have not hesitated to propose a general principle of linguistic structure on the basis of a single language"—an admission dubbed "preposterous" by this critic (Itkonen 1996: 487).

Some of Chomsky's "universals" were criticized even with respect to English. The *A-over-A* principle, for instance, was soon decomposed into several independent "island constraints" (J. R. Ross 1968), which covered the data better (and some of which hadn't been covered by the original principle). In the years that followed, Chomsky tried to integrate the island constraints into a single rule. He came up with the "Subjacency Principle", but even this didn't work for all of the island constraints. Again, Chomskyans tried to find some even more abstract principle underlying the variety of surface forms.

But non-Chomskyan linguists weren't persuaded. Worse (for Chomsky), they would eventually accuse him of constantly redefining his position over the years so as to make it, in effect, unfalsifiable. Sometimes, this would be said in waspish exasperation, caused by frustration at the still-continuing dominance of Chomskyans in the linguistic profession (see Section viii.a):

> I [have become] convinced that there is nothing, absolutely *nothing*, that could make Chomskyans admit that there is or has ever been anything amiss with their theory. In the game of linguistics, truth is a secondary consideration: not to lose face has top priority. (Itkonen 1996: 497; cf. pp. 490–4)

On this view, the passage from A-over-A to island constraints wasn't a step forward in a healthy Popperian progression of conjectures and refutations, but one of many intellectual retreats in a degenerating research programme (K. R. Popper 1935; Lakatos 1970). Today, half a century after Chomsky came on the scene, linguists are divided between these two judgements (see Sections viii & ix).

In short, Chomsky's universal grammar of the 1960s was, at best, a promissory note. Twenty years later, and largely due to Chomsky's influence, nativism would be expressed more acceptably and backed by considerable *non-linguistic* evidence (see Chapter 16.iv.c–d and Fodor 1981*b*). Twenty years later still, 'straightforward' nativism would be replaced by epigenesis, and modular theories of the mind reinterpreted, or rejected, accordingly (Chapters 7.vi and 15.x.a–c). Meanwhile, Chomsky couldn't meet Quine's challenge.

It's important to recognize, however, that (whatever one might say about his later theories) Chomsky's early suggestions on this matter had the Popperian merit of being mistaken. They weren't vague claims unamenable to testing—which isn't to say that the Chomskyans were eager to test them.

Still less were they a priori arguments from first principles, with no obvious relation to reality. On the contrary, they were empirical, falsifiable, *scientific* claims. Chomsky was saying "Languages are like *this*—but they could, conceivably, have been like *that*." As he put it:

> There is no *a priori* consideration that lends plausibility to a theory of [universal] syntactic operations [such as mine] that precludes the formation of interrogatives by left–right inversion of the corresponding declarative, or innumerable other perfectly conceivable, and quite efficient operations. It is just this fact that constitutes the scientific interest of such a theory. (Chomsky 1970: 468)

In other words, Chomsky wasn't a rationalist in the strict sense. The Cartesian philosophers of the seventeenth century took language to be close to reason, being informed by principles of necessary truth. And even Kant held that no alternative mental structure (system of intuitions) is conceivable by us (see Section ii.c). But Chomsky shared neither of these beliefs. Whether a language-universal grammar exists, and if so what it is like, are empirical questions.

On those points, Chomsky's behaviourist opponents were willing to agree. But even if he had identified a grammar suited, as a matter of fact, to all natural languages, they wouldn't have conceded defeat. For universality doesn't guarantee innateness. Max Black put the point thus:

> In science in general, reference to dispositions and powers is usually a promissory note, to be cashed by some identifiable internal structure... But there is no serious question at present of finding such internal configurations in human organisms: we hardly know what we should be looking for, or where to look. So far as stimulation of research goes, 'nativism' looks to me like a dead end... (M. Black 1970: 458)

Black's charge that no interesting research can be prompted by nativism fell wide of the mark. Cognitive scientists excited (whether for or against) by Chomsky's nativism did much interesting work as a result, the many studies of the development of grammar in children being just one example (see Chapter 7.vi). But Black was right to insist that, strictly, the nativist requires some independent evidence of internal mechanisms. Not until the 1990s would the debate on innate ideas, and inborn "dispositions", be illuminated by reference to specific neurological processes (see Chapter 14.vi.b and ix.c).

Not even Chomsky, of course, suggested that babies are born knowing French—or perhaps English, if destined for a life of persuasion, emotionalism, and deceit? (see iii.d, above). But their knowledge of universal grammar, he said, acts as a framework which guides them to attend to certain features, certain distinctions, in the language spoken around them. In syntax as in phonetics, the specific value of those distinctive features varies across natural languages: hence their apparent diversity.

On this view, the child is, in effect, like a scientist. Instead of collecting data by pure induction (which is impossible), the scientist formulates theories and hypotheses which suggest what to look for, and where to look for it. This, essentially, is what the baby has to do in learning its mother tongue.

The view of the speaker/hearer as a hypothesis tester had always been implied by Chomsky's work, quite irrespective of nativism. It was implicit in *Syntactic Structures*, in his claim that the speaker/hearer must ascribe unobservable, many-levelled, grammatical structure to utterances in order to understand them, and was made explicit in the first review (Lees 1957: 406–7). As he himself admitted, his theory that language users generate grammatical hypotheses owed much to Jerome Bruner's "New Look" in perception (6.ii), which swept cognitive psychology in the late 1940s and 1950s (Bruner 1980: 81).

What Chomsky's nativism added (in 1965) to the widely current idea of hypothesis testing was the claim that babies may not have to create their linguistic hypotheses out of thin air, nor even out of infantile dreams and visions. Rather, he said, they can produce them on the basis of a powerful theoretical framework—a "Language Acquisition Device"—already present in their minds.

9.viii. Aftermath

Chomsky's intellectual footsteps were greatly amplified in the two decades after his appearance, and are still clearly audible today.

His early papers had a huge, and lasting, influence on pure computer science. Virtually every introduction to compiler design cites Chomsky's hierarchical classification of languages (Edmund Robinson, personal communication). And a leading textbook on theoretical computer science remarks that "[Chomsky's] notion of a context-free grammar and the corresponding push-down automaton has aided immensely the specification of programming languages" (Hopcroft and Ullman 1979: 9; cf. 217–32). That's why I said (in Section vi.a) that the 'Three Models' paper was his most original and most well-founded contribution.

For cognitive science in general, his crucial writings were the formalist *Syntactic Structures* and the nativist *Aspects*—plus, for light relief, his coruscating review of Skinner. These three publications had a huge impact on psychology and philosophy, and a significant influence on AI (see 6.i.e, 7.vi.a–d, 10.i.g, and 16.iii–iv). In effect, cognitive science in the 1960s had a love affair with Chomsky: one historian of linguistics has remarked on "the sweeping and unrequited optimism of the honeymoon years of cognitive psychology and transformational grammar" (R. A. Harris 1993: 257).

With respect to linguistics as such, the sequel was clear: all linguists came to situate themselves with respect to Chomsky (Lyons 1991: 9). This was true even of those whose specialism was in some other area, such as sociolinguistics, historical linguistics, descriptive linguistics, and anthropological linguistics—all of which, and especially the last two, declined in status post-Chomsky (Sampson 1980: 78 ff., 146–7). All linguists had to decide whether to adopt some version of his account in their own work. And all had to allow that he'd set new standards of clarity for theoretical linguistics.

Whether his many followers lived up to those standards is questionable. Trouble was brewing even in the 1960s. A champion of formalization in science—echoing Leibniz's "calculemus" as the only way of achieving objectivity (see Chapter 2.ix.a)—praised Chomsky for seeding "one of the healthiest and most important controversies in psychology in this century", but complained that his leading disciples couldn't *prove* their claim that behaviourism is in principle inadequate (Suppes 1968). Twenty years later, things were even worse. One critic pointed out that, of the "thousands" of papers published on Chomsky's grammar between 1970 and 1986, *only one* (in a relatively obscure journal) attempted a mathematical statement (Gazdar 1987: 125). Indeed, Chomsky himself would eventually be fiercely criticized for lack of rigour (see Section ix.e).

In addition, he was accused of being mistaken in various ways even about *syntax*, never mind arcane philosophical topics such as innate ideas. This fact should be no great surprise. It's obviously possible to be influential without being right. What is surprising is the extraordinary emotional tone of many of those criticisms. Disinterested discussion of Chomsky's views was—and still is—hard to come by, as we'll now see.

a. Polarized passions

For linguists who specialized in syntax, semantics, or phonetics, Chomsky soon became a constant presence. Too much so, perhaps: the sociology of science hindered the science itself, as Chomskyan cliques came to dominate various conferences and journals.

Bruner, who as a non-linguist has no professional need to 'take sides' (and who'd influenced Chomsky deeply in the late 1950s: see vii.d, above), is one of many who've complained about this. He recently remarked:

[Chomsky is a] systems builder, and ruthless systems at that. And it's so funny, because the fact of the matter is he claims to be a deep believer in democratic values. But the ways in which he goes about it, and the ways in which they, the Chomskians, have gone about it by a kind of taking over of the apparatus of linguistic scholarship in America is nothing less than just a hostile takeover of the whole damned system. Everybody else is out. (Shore 2004: 51)

Naturally, this phenomenon didn't go unchallenged within the profession. Eventually, linguists' complaints about Chomsky's hegemony in linguistics became even more heated than philosophers' debates about his theory of innate ideas (see Section vii.c). One observer recently remarked that "the level of enmity is truly stunning". As he put it, "there are people who genuinely wish that Chomsky would die, or retire, or move exclusively into political or philosophical domains, and just leave poor little linguists alone" (R. A. Harris 1993: 256).

This "blood-boiling animosity" arose from professional frustration. An English linguist surveying the international scene in 1980 saw it like this:

The discipline of linguistics seems to be peopled by intellectual Brahmanists, who evaluate ideas in terms of ancestry rather than intrinsic worth; and, nowadays, the proper caste to belong to is American. The most half-baked idea from MIT is taken seriously, even if it has been anticipated by far more solid work done in the "wrong" places; the latter is not rejected, just ignored . . . To the young English scholar of today, the dignified print and decent bindings of the *Transactions of the Philological Society* smack of genteel, leather-elbow-patched poverty and nostalgia for vanished glories on the North-West Frontier, while blurred stencils hot from the presses of the Indiana University Linguistics Club are invested with all the authority of the Apollo Programme and the billion-dollar economy. Against such powerful magic, mere common sense . . . and meticulous scholarship (in which the London School [e.g. Firth, Halliday, Hudson] compares favourably, to say the least, with the movement that has eclipsed it) are considerations that seem to count for disappointingly little. (Sampson 1980: 235)

Comparable complaints have been made more recently, if less colourfully. A Finnish linguist in the mid-1990s remarked that Chomskyans often identify "linguistics" with "generative linguistics", and that "there are nowadays relatively few publications which subject generative linguistics to explicit critical scrutiny". The lack of publication was attributed to sociological factors:

A lengthy discussion [on the Internet] concerning "mainstream linguistics" . . . was started by the observation that people who had to criticize the generative paradigm preferred to stay anonymous. This was interpreted as reflecting the opinion that open criticism of "mainstream linguistics" might jeopardize a person's career prospects and possibilities for publication. Although such a view was heatedly denied by representatives of the generative paradigm, at least my own experience confirms it. In several informal meetings I have found out that linguists agree with some or even all aspects of the criticism [that I am about to present]; they just do not want to say it publicly. (Itkonen 1996: 471)

In a nutshell, the "representatives of other schools seem anxious to maintain what might, depending on one's point of view, be called either 'peaceful coexistence' or a 'balance of terror' " (p. 471).

Similarly warlike language was occasionally heard on Chomsky's side of the Atlantic, too. One leading American linguist, a colleague of Chomsky's at MIT in the late 1950s and 1960s (see Section x.b), had lost patience with him by the 1980s. Victor Yngve (1920–) , arguing that linguistics should focus on people (communicators) rather than language, referred to "puffery, empty polemics, intellectual bullying, and the great-man syndrome", and bemoaned "unscientific" a priori reasoning "in support of philosophical preconceptions and maintained by early ecstatic reviews, political acceptance, largely unquestioned personal authority, and the abuse and contempt . . . poured on other positions" (Yngve 1986: 7, 43, 108). Chomsky's name wasn't mentioned. Clearly, however, the cap fitted and he was being asked to wear it.

In short, Chomsky was the high priest of a new orthodoxy, almost a new paradigm (Kuhn 1962). Most young linguists aspired to be his devoted acolytes. Heretics (if noticed at all) weren't treated lightly—a situation due largely to Chomsky's personality, which thrives on embattlement.

For example, the MIT linguist who defined island constraints (see Section vii.d) was systematically reviled and excluded by his so-called colleagues (R. A. Harris 1993: 245–6). And George Lakoff, a pioneer of generative semantics (see below), was venomously accused by Chomsky himself of (among other things) having "discussed views that do not exist on issues that have not been raised, confused beyond recognition the issues that have been raised and severely distorted the contents of virtually every source he cites" (quoted in R. A. Harris 1993: 157). Skinner, then, wasn't the last to be lacerated by Chomsky's pen.

Nevertheless, heretics there were. Not everyone agreed with him. Indeed, even Chomsky disagreed with Chomsky. In other words, he had second thoughts—and third, and fourth . . . and more. Before considering his linguistic critics (in Section ix), let's look briefly at his own revisions of his ground-breaking theory of 1957.

b. Revisions, revisions . . .

Revisions weren't slow to come, for the grammar of *Aspects* differed significantly from Chomsky's earlier approach. The main differences were its terminology of deep and surface structure, its new treatment of transformations, and its inclusion of semantics. (Besides the new treatment of *grammar*, there was also a new emphasis on nativism, as we've seen.)

The 1965 account—known as "standard" theory—lasted, with relatively minor revisions, for fifteen years, while Chomsky devoted much of his time to politics. These revisions reduced the number of transformations in his theory: in 1970, for example, he suggested that lexical rules should replace morphological transformations (Bresnan 1978: 5). But the revised standard theory wasn't his last gasp. In the early 1980s, he provided a new approach, the theory of "government-binding" (GB) (Chomsky 1980*a*,*b*, 1982).

GB theory was even more unlike *Aspects* than that had been unlike *Syntactic Structures*. Chomsky had already suggested, in the late 1970s, that what had previously been thought of as a number of different transformations (involving wh-questions and various other matters) could be seen as special cases of a more general one. By the early 1980s, only a single, highly general, transformation remained. (Known as "move alpha", its pivotal notion was subjacency.) Much of the explanatory weight was now carried by innate "principles" and "parameters" (Chomsky and Lasnik 1977).

Principles are unvarying linguistic universals. Parameters are like variables: each has a limited number of possible values, which are partly interdependent. The diversity of actual languages is explained by differences between their sets of parameter values. So infants, in acquiring language, must discover which values characterize their mother tongue, and set their internal parameters accordingly (Lyons 1991: 184–8). As Chomsky put it:

> If the system of universal grammar is sufficiently rich, then limited evidence will suffice for the development of rich and complex systems in the mind, and a small change in parameters may lead to what appears to be a radical change in the resulting system. (Chomsky 1980*a*: 66)

This theory is reminiscent of claims made by biologists who see the range of possible adult forms as highly constrained, yet different, with no "great chain of being" filling the gaps. Conrad Waddington, and his student Brian Goodwin, have argued that small changes in epigenesis can tip the embryo into one developmental pathway rather than another: see Chapters 14.ix.c and 15.ix.a–c.

Eventually, in the early 1990s, GB theory gave place to "minimalism" (Chomsky 1993, 1995). "Deep structure" was abandoned. Still more shockingly, the previously pivotal notion of grammaticality was dismissed as being "without characterization or known empirical justification" (Chomsky 1993: 44). And several new principles—including Greed, Procrastinate, and Last Resort—were introduced (Chomsky 1995: 200–12).

But minimalism was too much for some of Chomsky's critics. One reviewer remarked that it was so very different from its GB predecessor that Chomsky was here "contriving to reclaim the role of the lone revolutionary", without acknowledging that his "new" position owed much to the generative semanticists of twenty years before (see below)—Pullum 1996: 137. As for the core theoretical ideas of minimalism, these were "sketched so hesitantly as to be hard to describe", and afforded "a level of explication [that] is risible" (Pullum 1996: 140–1). He mocked the new principles of "Greed" and the like, and complained about "anthropomorphic" accounts of "phrases moving in a bid to get their needs satisfied, and abstract nodes yearning to discharge their feature burdens" (Pullum 1996: 142). Minimalism, this critic declared, involved "a complete collapse in standards of scientific talk about natural language syntax".

These heretical judgements cut no ice with Chomsky's followers. The mantle of intellectual guru had settled firmly on his shoulders in the 1960s, and sits there comfortably still. Late in 1998, for example, his theory of minimalism was reported as a quarter-page news item—not a book review, nor even a feature article—in a British national newspaper. This largely explains his reviewer's ill-tempered use of vocabulary such as "risible".

In truth, the rot had set in long before minimalism. Chomsky's own standards of rigour had always been much lower than was generally believed. This fact, which became clear (to those with eyes to see) in the mid-1970s, is important for cognitive science in general, and is discussed in Section ix.e.

The many theoretical changes in Chomsky's later writings, whether rigorous or not, don't concern us. The huge impact of Chomsky's work on cognitive science resulted from his first three books, and was already integral to the field by the 1980s.

However, two of Chomsky's post-1957 theory changes do have a wider relevance, so merit mention here. The first was his increasing stress on "modularity". His style of linguistic nativism always implied that the language "faculty"—or, as he also says, "organ"—is a self-contained part of the mind, functioning in independence of others. Over the years, he made this more explicit (now, he analyses the language faculty itself into distinct sub-modules: Chomsky 1986). Moreover, he speculatively generalized it to other mental "faculties" (Chomsky 1980*a*).

The idea of modularity didn't originate with Chomsky. Conceptualized as "hierarchy", it had already been stressed by Simon, in his widely read discussion of the design (and evolution) of complex computational systems (Simon 1962/9). Chomsky's championing of it influenced (and drew support from) research in the psychology of vision and developmental psychology (Chapter 7.v and vi); ran in parallel with the incorporation of expertise (as opposed to general reasoning methods) in AI (see Chapter 10.iv.b–c); and inspired an influential philosophy of mind developed by Fodor (16.iv.c–d).

Nevertheless, the general relevance of modularity is controversial (see Chapter 7.vi.d–i). It has been challenged, for instance, by wide-ranging research in developmental psychology (Karmiloff-Smith 1992). And Simon's arguments about the evolutionary necessity of modularity have recently been questioned by the philosopher Tadeusz Zawidzki (1998). He argues that Simon's desideratum that evolution pass through "stable intermediate stages" is met by Stuart Kauffman's models of genetic regulation (see 15.ix.b), which are *not* decomposable into separate, hierarchical, parts.

The second broadly influential change was that Chomsky included semantics in *Aspects*, having previously ignored it.

c. Semantics enters the equation

By introducing semantics into the discussion, Chomsky was trying (for example) to account clearly for the deviance of sentences like "Colourless green ideas sleep furiously", and to decide whether they should be regarded as truly grammatical.

This word string *seems* to be grammatical: two adjectives and a noun, making a nice NP, and a verb and adverb providing the necessary VP. From the syntactic point of view, then, it appears to be fine. (Some people even claimed to be able to give it a meaningful interpretation.) Nevertheless, there does seem to be something wrong with it—but what, exactly?

To deal with that question, among others, Chomsky now (in *Aspects*) said that any generative grammar must have both a syntactic and a semantic component. And he regarded the latter as "purely interpretive". That is, each deep-structure (syntactic) expression is sent to a semantic rule system, which gives it a formal description in terms of (universal) semantic primitives. These include the distinction between mass and count nouns, and between human, animate, inanimate, or abstract things.

Because the semantic component was supposed to be only interpretative, "it follows that all [this] information . . . must be presented in the syntactic component of the grammar" (Chomsky 1965: 75). But there is no a priori answer, he said, to the question whether—and if so, to what extent—it should influence the generation of syntactic structures. He did include various semantics-based "selectional rules" restricting the structures that his syntax could generate. But he pointed out that these could be transferred to the semantic component; conversely, all of the semantic interpretative rules might be applied before some of the syntactic generative rules, "so that the distinction between the two components is, in effect, obliterated" (Chomsky 1965: 158–9). In short, Chomsky regarded questions about the relation of syntax to semantics as a matter of judgement, not principle.

His close MIT colleagues Jerrold Katz and Fodor (1963) agreed that syntax and semantics are to some extent intertwined. Nonetheless, they insisted that we need a specifically semantic theory to describe and explain the role of meaning in language. Moreover, they said, such a theory would contribute to (and draw on) the psychology of language. Our ability to understand the meanings of novel sentences must depend on the meanings of the constituent words (morphemes).

This implies, they argued, that we need a semantic "dictionary" of morphemes, and also an understanding of the principles by which meanings can be sensibly combined. A dictionary and grammar alone cannot explain understanding. The prime reason for this is that many words are multiply ambiguous (*bank*, *bill*, *seal*, *bachelor*), and the dictionary must include all the possible senses. Syntax cannot always do the disambiguating for us: *He banked the bill* isn't ambiguous, but *They were approaching the bank* is. In addition, then, we need "projection rules" for using the dictionary and grammar, which take account of the meaning relations between morphemes in generating semantic interpretations of sentences.

The burgeoning interest in semantics took many forms. One of these was "generative semantics". This label, originated by Lakoff in 1963 (before the publication of *Aspects*), was attached to a number of theories in the late 1960s to mid-1970s (Dean Fodor 1977; R. A. Harris 1993, chs. 5 and 6; McCawley 1995). The single label obscured various differences, both theoretical and political.

Some generative semanticists criticized Chomsky's reliance on mathematical models, as opposed to detailed linguistic data. Some even saw Chomskyans as "scientific Calvinists" out of touch with the anti-elitist counter-culture of the 1960s (McCawley 1995: 344–5). Politics aside, the charge of mathematical affluence combined with data poverty would be levelled by many other critics too. But those generative semanticists who did focus on mathematical models argued that the basic generative system was semantically driven and interpreted by syntactic rules, rather than the other way around (see above).

It was as though they were saying "Meaning first, syntax afterwards"—which, at least on first hearing, is intuitively plausible. However, after "some years of involved and at times acrimonious argument", it was eventually agreed that there was less difference between the two positions than had been supposed (Lyons 1991: 93). Both "generative" and "interpretative" semantics concerned (timeless) linguistic competence, not performance. Giving the generativity to the syntax or to the semantics made little difference in the types of sentence that could be (non-procedurally) accounted for.

In general, a semantic theory should give us a principled account of synonymy, and explain the (non-syntactic) ambiguity of *The bill is large*, and its resolution in *The bill is large but need not be paid*. Katz and Fodor assumed that concepts are analysable into distinct semantic components, and posited semantic primitives universal to all natural languages—and all human minds. They explained synonymy in terms of shared entries in a semantic dictionary, and paraphrase in terms of shared semantic structure between sentences that might be lexically and syntactically very different (*The oculist examined me/I was inspected by the eye-doctor*). Different senses of one word (such as *bachelor*) falling in the same syntactic category (*noun*) were identified by differing "semantic markers" (primitives such as *Human, Animal, male*) and "distinguishers" (for instance, *never married, young knight, first academic degree, young fur seal*).

The componential analysis of meaning is a tenet of classical empiricism—found in Locke, for instance. But there are difficulties about how to identify semantic primitives (for example, how to differentiate Katz–Fodor markers from distinguishers—Dean Fodor 1977: 144–55). Worse, it's not obvious that the componential assumption is true. Even in apparently simple cases, such as the concept of *cousin*, it's not clear what the analysis should be. If one uses the 'obvious' constituents *mother, father, son, daughter* (and perhaps *brother* and *sister*), one can't represent someone's being a cousin without committing oneself on the sex of the various people involved (unless disjunctions are included). This problem disappears if one analyses in terms of *parent, child, sibling, male, female*; but then one can accept only blood cousins, not cousins by marriage (which would require *spouse*). Moreover, suppose the word *cousin* actually was used in this restrictive way: so what? Someone who creatively used it (in an appropriate context) to include cousins by marriage might raise some eyebrows, but would nevertheless be understood. This example illustrates what the philosopher Friedrich Waismann (1955) had called the "open texture" of language.

Scepticism about componential analysis and the identification of semantic primitives arose in linguistics, philosophy, and AI (e.g. Fodor 1970; Kempson 1977; Wittgenstein 1953: 31 ff.; Wilks 1978; Sampson 1979*b*). It was one reason why generative semantics went out of fashion. Fodor, for instance, abandoned his "primitivist" account, and later suggested—following the psychologist Eleanor Rosch (1978—see Chapter 8.i.b)—that certain psychologically salient concepts may be the semantic focus of language (Fodor 1983).

Besides these criticisms, however, a very different account of semantics—an account based in mathematical logic—began to gain ground by the mid-1970s (Dowty *et al.* 1981). Componential analysis, in the most general sense, was retained. But the components were now conceptualized as abstract ontological entities and mathematical relations, not as familiar notions such as *Human, Animate, parent, child,* or *male*. The assumption that the distinction between syntax and semantics is a matter of judgement and emphasis was roundly rejected. Moreover, the psychologism of Chomsky and his followers was rejected too.

In short, the Chomskyan paradigm was being challenged. This approach to semantics is outlined in the next section, along with some other important anti-Chomskyan work. Some of this would cast doubt not only on his *grammar*, but on his position on cognitive science in general.

9.ix. Challenging the Master

Some of the new theories challenging Chomsky would undermine only his linguistics, strictly so called. People not especially enamoured of NPs and VPs might not be much interested. Indeed, they might be downright bored.

But other challenges would also reject his close juxtaposition of language and mind, with its supposed implications for both philosophy and psychology. And one would drive a coach and horses through his claim that phrase-structure grammars are inadequate—so hugely improving the power of computer models of language.

In this section, I'll highlight these points of general interest, rather than concentrating on strictly linguistic nit-picking. Some nit-picking, however, there will have to be.

a. Linguistic wars

The grammarians who disagreed with Chomsky did so for a variety of reasons. Accordingly, the "linguistic wars" of the 1970s and early 1980s involved many different skirmishes (R. A. Harris 1993). One linguist in the mid-1980s identified forty-two competing grammars by name, ominously adding "among others" at the end of the list (Yngve 1986: 108).

Broadly speaking, the non-Chomskyan combatants fell into two armies. On the one hand, there were people who, perhaps even before Chomsky's first publications, took a fundamentally different approach to syntax. They might be spurred to greater clarity by his highly formal example. But they didn't accept his theory—or even his goal.

The London School, for example, didn't aim at a generative grammar from which all possible sentences could be derived (Sampson 1980: 212–35). Founded in the 1920s by J. R. Firth (1890–1960), these linguists focused on the functional properties of language as communication.

Where grammar was concerned, they asked how syntactically different types of sentence relate to different communicative contexts. They were much less interested in (Chomskyan) questions concerning word-by-word derivations of individual sentences, or decisions about whether a particular word string is grammatical. Like the Port-Royalists, they saw grammar as a communicative system that happens to be formalizable. Chomsky, by contrast, saw grammar as a formal system that happens to be used for communication. (His prioritizing of formal syntax was criticized also by philosophers working on speech acts, whose research dealt with issues—of meaning and speaker's intention—he left untouched; Searle 1972.)

Michael Halliday (1925–) was one of the London group. A specialist in Chinese, he had an early interest in machine translation, and was associated with Richard Richens and the Cambridge Language Research Unit, or CLRU (see Preface, ii). But his main contribution to linguistics was "systemic" grammar—later used in one of the first impressive automatic parsers (see Section xi.b).

Halliday classified sentences into a limited set of syntactic types. The speaker has to choose one of these, in deciding how to communicate what they want to say. Each type was defined in terms of the syntactic "features" included in it, and each feature was chosen from a feature set of mutually exclusive alternatives. The features were largely interdependent (hence the name, systemic grammar): choice of a feature not only ruled out its immediate alternatives, but also constrained the allowable feature choices from other feature sets.

Whereas the speaker must make the choices, the hearer uses them to aid interpretation. (An example of how syntax can help understanding is discussed in Section xi.c.) And both speaker and hearer are guided by context: if one syntactic type is more appropriate in context than another, then it will (or should) be chosen. Some linguists applied Halliday's approach to practical problems of language teaching, rhetoric, and literary criticism—hardly grist to the Chomskyan mill.

On the other hand, there were those who followed Chomsky onto his own ground and challenged him there. Unlike the London School, whose general approach pre-dated Chomsky, these linguists couldn't have done their work without him. They agreed that the goal of linguistics is to find some generative grammar—but not necessarily his.

Many merely quibbled about this transformation or that one. Someone could do that, yet still be playing on Chomsky's team. Others risked the charge of heresy by offering a very different transformational grammar.

But some went much further. They excommunicated themselves, by rejecting transformations altogether. And some of these, in turn, criticized Chomskyans' mathematical claims—and even Chomsky's formal skills—in uncompromising, often contemptuous, terms.

b. Who needs transformations?

Chomsky had argued that context-free phrase-structure (CFPS) grammars can't generate natural language, so must be supplemented by transformations. This strategy was theoretically inelegant, especially when—as in the early versions of Chomsky's linguistics—a long list of transformations was involved. But, for a quarter of a century, it was widely assumed to be unavoidable.

Some murmurings, however, were afoot. Even before *Aspects*, Harman had outlined a generative grammar without transformations (see Section vii.c). And the suspicious fact—first pointed out in 1961 (see Section vi.d)—that the class of transformational grammars is unconstrained, being equivalent to a Turing machine, was clarified a decade later by the Stanford logician Stanley Peters (1941–)—Peters and Ritchie (1973). Chomsky himself started to rely less on transformations (see Section viii.b). Eventually, then, some theorists stopped asking which transformations are needed, and asked instead whether they're needed at all.

One person who posed this question was Joan Bresnan (1945–), initially a student of Chomsky's at MIT. Bresnan's linguistics was clearly an example of cognitive science, and was highly influential at Stanford's (early 1980s) interdisciplinary Center for the Study of Language and Information (CSLI). Together with the computational linguist Ronald Kaplan (1946–) she developed "lexical-functional grammar", or LFG—so named because it stressed lexical rules for word morphology, and functional rules governing relationships such as *agent* and *passive* (Bresnan 1978; Bresnan and Kaplan 1982).

These two types of rule replaced some of the transformations defined in *Aspects*. Bresnan still felt some structural transformations to be necessary, and sometimes described her grammar as "transformational" accordingly (Bresnan 1978: 36–40). But she conceptualized them procedurally as online operations on a single string, not as changes in one (deep) string to turn it into another (surface) one. That is, transformations in the literal sense had disappeared.

This was no accident, for the main motivation underlying LFG was to avoid altering already built structural descriptions during the derivation or parsing of a sentence. Bresnan and Kaplan saw this strategy as computationally clumsy and psychologically implausible. Instead of (sequentially ordered) transformations, LFG provided several levels of description that apply simultaneously. In generating a sentence structure, each partial description could only add to, not cause changes in, the partial descriptions already accepted on other levels.

Bresnan took Chomsky's psychologism even further than he had done. In accordance with his structuralist training, Chomsky had developed his grammar first and only then suggested that it had psychological implications. Moreover, when psycholinguistic evidence went against his theory, he retreated into grammar as such: "[my] generative grammar does not, in itself, prescribe the character or functioning of a perceptual model or a model of speech-production" (Chomsky 1965: 9). But LFG was conceptualized from the beginning as the ground of detailed psycholinguistic hypotheses. As Bresnan put it:

> [A realistic grammar requires that we] define for it explicit realization mappings to psychological models of language use. These realizations should map distinct grammatical rules and units into distinct processing operations and informational units in such a way that different rule types of the grammar are associated with different processing functions. (Bresnan 1978: 3)

A realistic grammar, she also said, should be tested by computational methods, as well as by psychological experiments. A prime reason for her collaboration with Kaplan, who also saw grammars as "mental representations of language", was that he'd written computer programs—using ATNs (see Section xi.b)—intended to simulate what goes on in people's minds when they understand a sentence (R. M. Kaplan 1972, 1975). Similar models, and a related programme of psycholinguistic studies, were developed by two of Kaplan's co-researchers—who showed, for instance, how relative clauses could be parsed without backtracking (Wanner and Maratsos 1978; Boden 1988: 100 ff.).

Some years later, the computational linguist Martin Kay (a one-time member of CLRU: see Preface, ii) located Bresnan's work—and that of others discussed below—within the overall space of possible human parsing strategies (M. Kay 1980). He argued that the psycholinguist needn't be strictly constrained by considerations of computational elegance, still less by a devotion to step-by-step (sequential) algorithms. There are many reasons for thinking that people process language in parallel, in the sense that a number of different parsing strategies are at work at the same time, and are somehow integrated in generating the completed parse tree. And there are reasons for thinking that both top-down and bottom-up methods may be involved.

Each point in Kay's abstract space defined an *algorithm schema*, identifying a function to be computed but not saying just how this should be done. Actual algorithms could be generated in a principled way by combining a schema with an *agenda* listing various psychological strategies. These could direct both depth-first and breadth-first search (see Chapter 10.i.d). In addition, Kay discussed in general terms how parsers (using *charts*) could be constructed to avoid unnecessary duplication of effort, such as repeating prior computations during backtracking. When computer parsers proliferated in the 1980s and 1990s, this seminal attempt to see the wood as well as the trees was to be highly influential.

c. Montagovian meanings

Whereas Bresnan had insisted that grammar and psychology should be very closely related, some others who challenged Chomsky's account of language made a point of avoiding that claim. Richard Montague (1930–71) at UCLA was an important example (Dowty *et al.* 1981).

Montague was a logician and philosopher of language, working in the strongly anti-psychologist tradition of Frege (see Chapter 2.ix.b), and inspired by Carnap rather than Chomsky. His work wasn't an instance of cognitive science. He had no interest in the mind or mental processes, nor in computational linguistics. And his theoretical concepts were drawn not from computer science, but from mathematical logic. However, his approach appealed to some linguists whose views were highly relevant to cognitive science, as we'll see. In addition, it was taken up by a highly influential psychologist working on language and reasoning (Chapter 7.iv.e).

His influential paper on "universal grammar" had nothing to say about the relation between language and mind, still less innate ideas (Montague 1970). By invoking a universal grammar he meant, rather, that all languages, both natural and artificial, should be described in the same way. In this belief, he was echoing Carnap (see Section v.a). What was novel about his approach, and what commended it to some linguists, was the unusually close theoretical relation between syntax and semantics.

His system was called "intensional logic", because it dealt with meanings as well as logical formulae. It contradicted Chomskyan assumptions about the autonomy of syntax. It also rejected all semantic theories that merely offered translation into a more basic language—whether this was some form of logic, or a language of semantic primitives. For Montague, to translate one language into another isn't to address the central problem. The crucial questions of semantics concern the relation between linguistic expressions (of whatever kind) and entities in the world.

Logicians had long defined semantics in terms of truth conditions: what would have to be the case in the world for the expression to be true. And they had long assumed compositionality, so that the truth conditions of a complex expression depend recursively on those of its component parts. So far, so good—but just how were truth conditions to be conceptualized?

The logical positivists had answered this question in terms of (pure) observation protocols, but by mid-century it had become clear that this approach had intractable problems. In the 1960s (though not published until after his death in 1970), Montague developed a more abstract answer, and applied it to various types of sentence in English (Montague 1970, 1973).

Criticizing Chomskyan grammar as mathematically imprecise and unsystematic, he went back to Russell's set theory. With this as his ground, he defined "the set of all possible worlds" in terms of an abstract ontology of individuals, properties, *n*-ary relations etc. The semantics of a linguistic expression was then defined as its denotation (its model) in some possible world. That model was found by applying recursive mapping rules, such that each syntactic form (NP, VP, Det, etc.) had a corresponding semantic form. A proper name, for instance, always maps onto some individual.

This 'rule-for-rule' approach coupled syntax and semantics much more closely, precisely, and systematically than Chomsky and his followers had done. The theory of semantic markers, for example, had posited "primitives" such as *Animate* and *Human* (see Section viii.c). But to capitalize a familiar term of English isn't to display its meaning. Moreover, it wasn't clear just how to interpret the bracketed lists of semantic primitives offered by such theories: their syntax didn't establish their meaning. They were understood intuitively, much as the higher-level words (such as *bachelor*) were. In Montague's hands, by contrast, some specific—albeit highly abstract—meaning was definitively implied by the syntax.

Montague's theory was mathematically elegant, but couldn't solve most of the problems about meaning that had long plagued linguistics. He held, for example, that if a rule-for-rule mapping of a given English sentence succeeds in all possible worlds, it is necessarily true—and semantically equivalent to every other such sentence. Yet the component words of these distinct sentences may arouse very different associations in the hearer's mind. If, with Humboldt (or even Bloomfield), one regards these as part of the "meaning" of language, then Montague's theory is inadequate.

Again, semantic relationships such as synonymy, implication, sub/superordinacy, and consistency/inconsistency were rigorously defined by Montague. Synonymy, for instance, requires that both expressions be true in the same set of possible worlds. However, this abstract definition may not help us to decide whether two actual words are synonymous. (Montague agreed with Humboldt that accurate translation from one natural language to another is impossible—not because of different mental associations, but because it's unlikely that any two sentences in the different languages will share the same mapping in model theory.) Nor did his approach help in defining *cousin*: all the problems noted in Section viii.c, including the open texture of natural language, remained.

The new semantics offered clear definitions for everyday words such as *the*, *every*, *some*, *all*, *any*, *never*, *ever*, all of which are notoriously difficult to define in dictionary style. They were interpreted by Montague as set-theoretic expressions defined over possible worlds, whose logical–mathematical relations could be precisely stated.

However, most of the long-standing problems about assimilating natural language to logic had been brushed aside, rather than solved (see Chapter 4.iii.c). As for sentences about propositional attitudes, these were as intractable as ever (Dowty *et al.* 1981: 170–5). No semantic mapping for *Mary believed that the bracelet had been lost* could be completed, because the bracelet's fate still had to be left open (see Section v.a).

Model-theoretic semantics was a further step along the road of formalism pioneered by Frege, Russell, and the early Wittgenstein. This road had been followed also by Carnap and Reichenbach. But whereas in the 1930s–1940s few linguists had paid much attention to Carnap's work on language, by the 1970s linguistics itself was largely formalized, and had been strongly linked (by Chomsky) with the analysis of artificial languages. In other words, the ground had been prepared for Montague.

Even so, persuading linguists to read Montague took some time and effort. His work was so fiercely abstract, and so tersely written, as to border on unintelligibility. Most linguists tackled it by way of some introductory account, preferably one relating set theory (more familiar to philosophers of language than to linguists) to transformational grammar. Barbara Partee (1975) was especially influential in introducing Montague's ideas to other linguists. But even four years later, professional journals were still prepared to publish explanatory notes on Montague rather than original research, defending their decision by reference to his forbidding writing style (Halvorsen and Ladusaw 1979: 185).

After the mid-1970s, when (thanks to Partee) linguists in general became aware of him, Montague's work soon came to dominate theoretical semantics, and various theories of syntax too. (Ironically, Chomsky himself now puts much less stress on formalization than he used to, and even queries the relevance of "most of the results of mathematical linguistics"—Lyons 1991: 199.)

Psycholinguists, however, virtually ignored Montague. (Although some would take up his theories in the early 1980s: see 7.iv.e.) Being ignored by the psycholinguists wouldn't have bothered him at all. As remarked above, his theory was a purely logical–ontological one, making no claims about psychology.

d. Transformations trounced

Among the linguists inspired by Montague in the mid-1970s was Gerald Gazdar (1950–), of the University of Sussex. Together with Ewan Klein, Geoffrey Pullum, and Ivan Sag, Gazdar developed a new syntactic theory: generalized phrase-structure grammar, or GPSG (Gazdar 1979*a*, 1981*a*, 1982; Gazdar *et al.* 1985). GPSG adopted Montague's model-theoretic approach, and applied it to a much wider range of English syntax than he had done.

Like Montague, Gazdar treated language as a purely formal system, not as a psychological phenomenon. Nevertheless, he was inspired also by the psychologically oriented Bresnan. What drew him to Bresnan's work wasn't her use of grammar to relate language and mind, but her rejection of syntactic transformations.

Another formative influence was Peters—who in 1973 had expanded Putnam's insight about the Turing equivalence of transformations in general (see above). Gazdar took this mathematical discussion still further. Besides co-developing GPSG, he directly challenged some of Chomsky's formal claims about the power of various types of grammar. In so doing, he implicitly cast doubt on arguments that had influenced cognitive science in general (see Section ix.e–f).

In formulating GPSG, the prime aim of Gazdar and his associates was to develop a single-level (non-transformational) tree-based grammar that would describe English elegantly. That task could be, and often is, approached without any concern for the abstract mathematical properties of language. But two of the authors (Gazdar and Pullum) had already claimed that the widely accepted arguments against CFPS grammars were either formally invalid or based on false empirical premisses. Indeed, their work had revived interest in this long-neglected question, and led to some improved arguments for the opposing view (G. Gazdar, personal communication).

Partly because of this interest (which not all syntacticians shared), and partly as a matter of methodological discipline, the GPSG group took pains to prove that their new account of syntax was mathematically equivalent to a CFPS grammar.

The GPSG authors were careful not to claim that GPSG can describe every natural language, although they pointed out that it seemed to fit about twenty diverse tongues (Gazdar *et al.* 1985: 15). But they did assume that it could describe the whole of English.

Soon afterwards, it became clear that this was almost true, but not quite (see below). Nevertheless, it was evident by the early 1980s that GPSG could successfully describe a host of English sentences which Chomsky had said require appeal to transformations. Moreover, no extra rules had to be added to link each syntactic form with its semantic interpretation. These relations were transparent, by virtue of a model-theoretic account of the mapping between syntax and semantics.

GPSG captured the grammatical intuition noted in Section vi.d, that *The man reads the book*, *The book is read by the man*, and so on, are closely related. Nevertheless, it considered only the surface string, so didn't need transformations (which were introduced to derive surface-level from deep-level structures). It specified only a set of phrase-structure rules. GPSG was more flexible than the phrase-structure grammars considered by Chomsky, because in addition to the familiar syntactic categories (S, NP, VP, Det, and so on) it contained three other formal devices: slash-categories, metarules, and linking rules.

A slash-category marks a constituent with a syntactic gap: so S/NP, for instance, represents a sentence with a noun phrase missing. As this example shows, the nature of the gap was defined in terms of the basic (and familiar) phrase-structure rules. Given slash-categories, the problem arose of defining how the grammar should deal with them. Consider S/NP: this must denote either *an NP-missing-an-NP followed by a VP*, or *an NP followed by a VP-missing-an-NP*. These are the only possible interpretations, given that the basic grammar prescribes that S $\rightarrow$ NP + VP. Other slash-categories may allow more than two possibilities.

One might have expected, therefore, that GPSG would include an inelegant profusion of extra rules, several for each slash-category. Not so: instead, Gazdar's grammar contained a small number of metarules. Each of these was a general rule schema, from which (together with the basic rules) the new rules required for a given slash-category could be derived. Finally, the (more numerous) schematic linking rules dealt with the housekeeping: they defined how to introduce and delete slash-categories, and enabled the gap to be carried down to lower levels of the parsing tree as necessary.

This new grammar not only dealt with all the sentences generated by transformational grammar, but also explained some things simply which Chomsky could explain only in tortuous ways. For instance, consider these two sentences: (1) I think *Fido destroyed the kennel*, and (2) The kennel, I think *Fido destroyed*. Most linguists (including Chomsky) had assumed that the two italicized expressions fall into the same syntactic category—namely, a (subordinated) sentence. But because the second expression isn't a complete sentence (it cries out for a noun phrase telling us what Fido destroyed), they had to give a complicated account of (2) in order to show this. GPSG, by explicitly labelling the gap, placed the italicized strings in different syntactic categories. But it enabled one to give equally simple descriptions of how to generate the two whole sentences (Gazdar 1981*a*).

GPSG had a significant influence on linguistics, being taken up by a number of syntacticians in the early 1980s. Clearly, it could achieve a great deal—but how much? It seemed to encompass English, and at least twenty other languages too. But there were growing doubts. It was already known that some European languages contain phrase-structure trees lying outside GPSG, and in 1984 two further languages—one from West Africa, another a Swiss dialect of German—were found to contain recalcitrant sentences (Gazdar *et al.* 1985:16; Gazdar 1998). Indeed, English itself, as Klein had discovered, contains some sentences—involving nested comparatives—that couldn't be described by GPSG.

Accordingly, Sag went on to develop an improved, but no longer context-free, version: "head-driven" phrase-structure grammar, or HPSG (Pollard and Sag 1994). And Gazdar showed that both GPSG and HPSG were special cases of "indexed" grammar. This is a general type located just above context-free grammar in the abstract Chomskyan hierarchy, but still appreciably below context-sensitive grammar (Hopcroft and Ullman 1979: 389–92).

Even within linguistics, however, GPSG didn't attract attention as widely as it deserved. This was largely due to the sociological factors cited in Section viii.a—including the *not invented here* syndrome that still pervades MIT. Stanford, to be sure, was open to Gazdar's theory—but CSLI didn't rule the world.

Today, much of the best work on non-Chomskyan syntax (including LFG, GPSG, and HPSG) is done in departments of computational linguistics, not of linguistics as such. This is only partly because the theorists concerned are interested in computer modelling. More to the point, they are "locked out" of the many departments of linguistics still dominated by the Chomskyan paradigm. One indication of this dominance is the fact that the term "generative grammar" is sometimes used as if it referred only to Chomsky's work—the current version of which is arguably not generative at all (Gazdar *et al.* 1985: 6).

e. Why GPSG matters

In cognitive science *as a whole*, GPSG was even less well known than in linguistics. In a sense, it was its own worst enemy. For it was yet more fiercely abstract than Chomsky's early publications had been. Non-specialists would need plenty of black coffee.

Moreover, the GPSG linguists, unlike Chomsky, had no desire to discuss broad issues of "language and mind". And, unlike Bresnan and Kaplan, they had no interest in psycholinguistics. They drew no psychological morals from their grammar, and no psychological discovery could either support or falsify their (purely formal) work. Some psycholinguists, including Janet Fodor and Rosemary Stevenson, did experimental research inspired by GPSG (G. Gazdar, personal communication). But because of the GPSG group's explicitly anti-mentalist stance (and their limited visibility in professional linguistics), most psychologists were unaware of their work.

Nonetheless, it was—and is—potentially important for cognitive science in five different ways. (Four are discussed in this subsection, and the fifth in the next.)

First, GPSG questioned Chomsky's fundamental critique of CFPS grammars. He had admitted that he couldn't strictly prove that English lies beyond any conceivable phrase-structure grammar. Now, his claim (his hunch) that it could be so analysed, if at all, "only clumsily" had been challenged. GPSG, which covers most if not quite all English sentences, is a formally elegant theory, not a clumsy one. It didn't deserve Chomsky's contemptuous accusation that it was "simply a needlessly complex variant" of a transformational theory, supplemented by "a class of superfluous devices" (Gazdar 1981*a*: 280–1).

Moreover, if natural languages can indeed be described by some context-free phrase-structure grammar (a CFPSG, for short), then a huge mathematical literature becomes potentially relevant. This literature was developed (partly as a result of Chomsky's papers of the early 1950s) in theoretical computer science, to deal with the parsability and learnability of programming languages.

Compiler programs, for instance, have to parse context-free programming languages in order to translate them into machine code. This is "perhaps the most researched and best understood area of formal language theory" (Gazdar 1981*a*: 275). In other words, and as Montague had already shown, the mathematical analysis of language—and, significantly, the comparison of natural and artificial languages—can be taken much further than Chomsky took it.

Third (a corollary of the previous point), the standards of rigour set by Harris and raised by Chomsky had been raised further still. Indeed, Chomsky's own standards had already been criticized by people outside the GPSG group. The Stanford-based logician Montague didn't even deign to criticize them, but contemptuously ignored "the attempts emanating from the Massachusetts Institute of Technology" (Montague 1973: 247). He was presumably offended by (among other things) Chomsky's basing his core claim about the necessity of transformations not on a proof, but on a hunch.

Others agreed, and accused Chomskyans in general of not having done their mathematical homework. As remarked in Section vi.a, the technical evidence behind *Syntactic Structures* wasn't published until 1975. Meanwhile, the famous mimeograph ("three thick volumes of faint purple typewriting with handwritten annotations") had probably been read only by a handful of those who had bothered to get hold of it (Sampson 1979*c*: 356).

One reviewer of the published version confessed that he himself had for years taken on trust Chomsky's hints that the theoretical lacunae in *Syntactic Structures* had already been filled in by the work presented in those purple pages. He confessed also that when he did finally get hold of a copy (while waiting to receive the published book for review), he "rapidly laid it aside" because of its near-illegibility. And he acidly commented:

> The publication of this book [in 1975] symbolizes a remarkable situation in the recent history of linguistics, a situation which can hardly have many parallels in other disciplines... [For the best part of two decades], during the heyday of Transformational Grammar, its hundreds of thousands of advocates in the universities of the world not only didn't but *couldn't* really know what they were talking about. (Sampson 1979*c*: 355–6)

Part of the explanation for this unhealthy state of affairs was that most linguists couldn't have understood Chomsky's manuscript even if they had read it. Realizing this, they'd been content to leave its arcane mathematics to the "high priests" at MIT, assuming that it answered all their theoretical questions. But they were mistaken. Those with mathematical eyes to see would find, when they did actually read it, that Chomsky's argument was often "maladroit" and "perverse", and sometimes "just plain wrong" (Sampson 1979*c*: 366).

(I must confess that I can't justify this claim at first hand, for I *don't* have mathematical eyes to see. Several specific examples have been pointed out to me, but I have to take them on trust. However, I'm a colleague of two of Chomsky's major 'mathematical' critics, Gazdar and Sampson, and I've pressed both of them hard on whether they really mean what they say in the passages I've quoted. They do.)

Many of Chomsky's mistakes, according to Sampson (1979*c*), were "elementary", and some were crucial. For example, he'd repeatedly claimed that the syntactic phenomenon of "Affix Hopping" can be stated as a single transformation (though he never showed just how to do so). That claim had persuaded many linguists into his camp. Yet it now appeared, Sampson complained, that one of his own formal definitions rendered it impossible. In other words, Affix Hopping had arguably been "a standing refutation of [his] theory throughout its history". Yet Chomsky had never discussed this problem (Sampson 1979*c*: 365).

In light of such points, the professional adulation of Chomsky was—or anyway, should have been—distinctly embarrassing. To be sure, most linguists weren't mathematical animals. But their misplaced admiration for Chomsky's formal work was due to more than mere ignorance. It was encouraged by sociological factors: his by-then-established grip on departments of linguistics, and his towering status as a worldwide guru.

Similarly negative judgements of Chomsky's work and influence have been expressed for twenty years by Gazdar and his colleagues. Indeed, it was Pullum who recently dismissed Chomsky's theory of minimalism as "risible" (see Section viii.b). Admittedly, Gazdar had earlier declared—and still does (personal communication)—that *if* any work in syntactics is to be called "foundational", as Frege's work was foundational for mathematical logic, then it is Chomsky's—especially his hierarchy of grammars (Gazdar 1979*b*: 197). But that's not to say that he saw all of Chomsky's "proofs" (in Chomsky 1975) as equalling Frege's in respect of precision and accuracy. To the contrary, he didn't.

Commenting on Chomskyan linguists in general, Gazdar's group pointed out that "mathematically respectable" proofs of general claims about grammars are very difficult to obtain. For instance, new versions of transformational grammar had been proposed (by Chomsky and others) which were held to be formally equivalent to, though more elegant than, some previous version; and linguists often claimed that such-and-such a structure cannot be generated by this or that grammar. But proving it was a different matter. Since most Chomskyans didn't even try to do so, the GPSG theorists were less than impressed:

> Such claims [about the formal equivalence and computational power of different grammars] are made in many contemporary linguistic works without there being any known way of demonstrating their truth. In some works we even find purported "theorems" being stated without any proof being suggested, or theorems that are given "proofs" that involve no definition of the underlying class of grammars and are thus empty. (Gazdar *et al.* 1985: 14)

Rigour was necessary, they said, not just for clarity and a good mathematical conscience, but for providing explanation. Thus any proposed "universal" should be a necessary consequence of some formal definition of the class of natural grammars. Otherwise, it is (at best) a description of some feature found, as a matter of fact, to be universal—without any understanding of why this is so (pp. 2 ff.). This point was similar to Marr's claim that we can explain visual processing only if we can relate it to (ideally, derive it from) the abstract constraints on vision as such: Chapter 7.iii.b.

A fourth point of interest for cognitive science as a whole is that GPSG implied that Chomsky's nativist 'argument from ignorance' against behaviourism must be reconsidered. Even if the simple learning theories popular at mid-century are inadequate to explain language acquisition, because they cannot parse phrase structure, some more sophisticated—but still purely surface-level—version might succeed. In other words, GPSG might be learnable (without the need for any dedicated acquisition device), even if transformational grammar isn't.

To be sure, the framework for this more powerful type of 'induction' might be innate: even Quine had allowed that "unquestionably much additional innate structure is needed . . . to account for language learning" (see Section vii.c). Even so, and although linguistic nativism hadn't strictly been disproved by Gazdar's work, the empirical evidence demanded years before by Black had become even more necessary.

One reason why GPSG didn't spread widely throughout cognitive science was that evidence for some sort of language-specific nativism was by then more abundant. Besides the developmental studies mentioned in Section vii.c, it included analysis of the many detailed features of human anatomy and physiology seemingly evolved specifically (*sic*) for language (Lenneberg 1967), and the apparent impossibility of teaching even simple grammatical structure to non-human primates (Chapter 7.vi.c).

In addition, neuroscientific evidence for *visual* nativism had—seemingly—been accumulating since the 'Frog's Eye' discoveries in the late 1950s (Chapter 14.iv & vi). And the new popularity of modular theories of mind (see above) discouraged reliance on general learning mechanisms to explain language acquisition.

That's not to say that nativism had been conclusively proved, for it hadn't. Despite the best efforts of Chomskyans to marshal their evidence (Pinker 1994), the facts didn't—and still don't—show that language, even linguistic universals such as coordination and hierarchical structure, can't be learned and/or evolved (Sampson 1997, esp. chs. 3 and 4). Furthermore, even if the nativists were right, Black's point still stood: the nature of the innate dispositions supposedly concerned, and of their neurophysiological implementation, were—and still are—unclear.

Indeed, the notion of nativism is itself unclear. Recent research in both psychology and neuroscience has shown how over-simple it is to label behavioural or anatomical features as 'innate' (see Chapters 7.vi.g and 12.viii.c).

The fifth important aspect of GPSG is perhaps the most significant of all. It has deep implications for both psychology and software technology, and especially for NLP.

f. Computational tractability

Because of its equivalence to a CFPS grammar, GPSG is less computationally demanding than transformational grammar. And that means, in a nutshell, that more can be achieved by relatively simple systems—whether minds or machines—than Chomsky believed.

With respect to psychology, GPSG suggested (for instance) an explanation of why people can parse natural language so rapidly (Gazdar 1981*a*: 276). And it lessened the difficulty of seeing language as the product of evolution, since context-free parsing algorithms are less evolutionarily improbable than transformational ones. More generally, if natural languages are indeed CFPS structures, then psycholinguists don't need—so shouldn't propose—psychological theories with the power of a context-sensitive grammar.

This is a special case of Lloyd Morgan's canon, framed in the late nineteenth century to promote intellectual hygiene in comparative psychology (see Chapter 2.x.b). Conwy Lloyd Morgan had forbidden explanation in terms of "a higher psychical faculty" if the behaviour in question could also be explained in terms of "one which stands lower in the psychological scale" (Lloyd Morgan 1894: 53).

For sixty years, this hugely influential advice could be followed only by relying on some intuitive sense of what counts as a "higher" psychical faculty. But Chomsky's hierarchical classification of grammars had, in effect, defined "higher" in terms of distinct levels of computational power, requiring processing mechanisms of specific types—with or without push-down, for example. (This enabled 'extra' levels to be inserted in the hierarchy: indexed grammar, for instance, lying above CFPSGs but still significantly below context-sensitive phrase-structure grammars, or CSPSGs.)

In short, the linguistic hypothesis that natural languages can be described by CFPSGs implied a psychological hypothesis about the minimum computational power required by the human mind/brain. Humboldt's "creative spirit of humanity" would have to rise as far as CFPSGs to generate language, but—if the GPSG theory was correct—need rise no further.

As for the software implications of GPSG, these were quickly recognized and highly influential. By the early 1980s, this grammar had already been adopted in various NLP projects in Europe, Japan, and the USA.

One of these was the "Alvey Tools Grammar", developed for the UK's Alvey Project (11.v.c) and still in use in various places today. And Sag's HPSG, which shares many features with GPSG although it isn't context-free (it permits unbounded recursion in certain circumstances), is now probably the most widely employed grammar formalism in NLP, especially in Europe (G. Gazdar, personal communication). (Very recently, Gazdar (1998) has defined a class of grammars, which includes GPSG, that is mathematically equivalent to a type—known as "linear tree adjoining grammars"—now gaining popularity in NLP circles.)

These NLP projects of the 1980s fell clearly into "technological" rather than "psychological" AI (see Chapter 1.ii.b). GPSG's psychological plausibility was irrelevant. So was the 'pure' linguist's question whether it could describe *every* syntactic construction in *every* human language. From the point of view of NLP research, that question is comparable to theological disputes about angels dancing on pinheads. What was important about GPSG, besides its success in describing almost the entire range of English (and some other languages), was its computational tractability.

Even devotees of GPSG, however, don't deny Chomsky's seminal importance. His work, besides affecting computer science as such, profoundly influenced NLP. Or rather, his classification of different types of grammar did so. His ever-changing syntactic theories have had less influence here. Partly because of his own lack of interest in, even hostility to, computer modelling, very few Chomskyan programs have been written (but see Chapter 7.ii.b). And, for reasons remarked above, many computational linguists are explicitly anti-Chomskyan. However, to be anti-Chomskyan isn't to be uninfluenced by Chomsky: LFG, GPSG, and HPSG were all developed in reaction to his pioneering efforts.

Moreover, Chomsky's formal–generative approach to linguistics in the late 1950s (and his linking of different grammars to different automata) buoyed the spirits of people already doing computer modelling of language, and encouraged others to start trying. (Section x will recount the early development of this field.)

g. Linguistics eclipsed

The last item within the tenfold Chomsky myth (i.a, above) was that linguistics is as prominent in cognitive science today as it was in the 1950s–1960s. That's not true. Ironically, its eclipse was largely due to Chomsky himself.

Part of the problem was his continual recasting of his own theory of syntax, described above. But that alone wouldn't have sufficed: if the price of doing good cognitive science had been keeping up with Chomsky, then that price would have had to be paid. More important than his inconvenient changes of mind was his mid-1960s distinction between *competence* and *performance* (Chapter 7.iii.a). This put a theoretical firewall between the psychology of language and linguistics as such. The firewall prevented commerce in both directions. Linguists, in effect, had been instructed to take no heed of psychology. But by the same token, psychologists were tempted to regard abstract linguistics as largely irrelevant to their concerns.

Researchers in non-linguistic disciplines didn't stop working on language, of course. But they stopped looking to Chomskyan theory for their key inspiration. As for the people who weren't working on language, they no longer gave linguistics a second thought. The general lessons (about formalization and generativity) that Chomsky had taught had been well learnt, and were no longer thought of, at least by the youngsters entering the field, as having any special connection with linguistics. By the 1990s, the discipline was "arguably far on the periphery of the action in cognitive science" (Jackendoff 2003: 651).

Recently, things have started to change. For instance, Pinker has put linguistics in the context of cognitive psychology, and in particular of evolutionary psychology (Pinker 1994, 1997, 2002). Moreover, the late 1990s saw a new field being named: cognitive linguistics. Its adherents are *linguists*, but linguists who take special note of the cognitive processes involved in using language (see 7.ii, preamble, and 12.x.g).

Similarly, the Brandeis (now Tufts) linguist Ray Jackendoff (1945–) has recently tried to integrate linguistics with psychology, neuroscience, and evolutionary theory (Jackendoff 2002, 2003). An ex-pupil of Chomsky, Jackendoff is still wedded to the idea of a universal grammar. But he claims to have shown (what Chomsky questioned) that it's possible for this to have evolved *incrementally*. He also argues that semantics doesn't need to be "paralysed" by a failure to solve the philosophical problem of intentionality (2002: 280). And he couches his argument in computational terms, describing a parallelist architecture (blackboard-based: see 10.v.e, and xi.g below) in which the rules of grammar are directly involved in language processing.

Jackendoff's book is described as a "masterpiece" by one *BBS* commentator, namely the philosopher Daniel Dennett (2003*b*: 673). Others disagree. For instance, another philosopher (F. Adams 2003) complains of its cavalier attitude to semantics and intentionality (although Dennett defends Jackendoff on this point—2003*b*: 673–4). And a neuroscientist objects to its assumption that the brain has evolved specialized components of language, as opposed to a small number of general capacities (he names seven) that make it possible to discover such "components" over the course of many millennia and/or to learn them within a few years (Arbib 2003: 668). Right or wrong, however, Jackendoff's theory is a far cry from Chomsky's chastely firewalled linguistics.

Or perhaps not? Employing the revolutionary analogy cited in Section i.b, Edelman has described Jackendoff's book as

> one of several recent manifestations in linguistics of the Prague Spring of 1968, when calls for putting a human face on Soviet-style "socialism" began to be heard (cf. the longing for "linguistics with a human face" expressed by Werth 1999: 18). (S. Edelman 2003: 675)

Edelman's prognosis for Chomskyan theory isn't hopeful:

> In a totalitarian political system, this may only work if the prime mover behind the change is at the very top of the power pyramid: Czechoslovakia's Dubcek in 1968 merely brought the Russian tanks to the streets of Prague, whereas Russia's Gorbachev in 1987 succeeded in dismantling the tyranny that had sent in the tanks. In generative linguistics, it may be too late for any further attempts to change the system from within, seeing that previous rounds of management-initiated reforms did little more than lead the field in circles. . . . *If so, transformational generative grammar, whose foundations Jackendoff ventures to repair, may have to follow the fate of the Communist bloc to clear the way for real progress in understanding language and the brain.* (S. Edelman 2003: 676; italics added)

To be sure, Edelman isn't a disinterested critic. For he argues, contra the Chomskyans, that the inductive learning of hierarchical structure is in principle possible. He and his colleagues have devised a pioneering "ADIOS" algorithm (Automatic DIstillation Of Structure), which can induce both context-dependent and context-free grammars (Solan *et al.* forthcoming). It uses a combination of statistics and structured generalization to extract hierarchical regularities from input sequences without knowing what patterns to expect, and without being given clues as labels on sequence parts (e.g. part-of-speech tags on the input words). Its only presupposition is that the sequence contains "partially overlapping strings at multiple levels of organization" (p. 3).

ADIOS has been applied to large sets of 'grammatical' sentences (e.g. the Bible and the *Wall Street Journal*), and also to the CHILDES corpus of spontaneous—and largely 'ungrammatical'—speech. (Those scare quotes are needed because Edelman rejects Chomsky's view that only strings fitting some set of abstract syntactic rules can be termed grammatical: language, for Edelman, is what people actually speak.) The system works even better with CHILDES than with the *WSJ* corpus, which Edelman (personal communication) explains in terms of Jeffrey Elman's stress on the importance of "starting small" (see Chapter 12.viii.c): whereas CHILDES contains many short word strings, the *WSJ* contains many longer ones.

In addition, ADIOS has coped well with six natural languages, including English, French, and Chinese. It has also been tested on artificial context-free languages, some having thousands of rules. And it has even been applied to DNA and protein sequences from several species. In each case, it has learnt to recognize the hierarchical patterns involved, and to reject anomalous sequences.

As its success with CHILDES illustrates, it has done this *despite* having ungrammatical strings in the input (which Chomsky saw as a huge barrier to inductive learning). Moreover, simulating the "creativity" of language, it has generated new (and legal) structures too—the first learning algorithm to do so. (The patterns it classifies in the learning phase are transformed into rewrite rules, which are then used for the generation phase.)

It doesn't follow that this is what infants do. Indeed, the authors point out that ADIOS has no access to conceptual knowledge, nor any "grounding" of the verbal input in current action and/or environment. (They also remark, however, that babies appear to use both statistics and rule-learning in acquiring language: Saffran *et al.* 1996; Seidenberg 1997; G. F. Marcus *et al.* 1999.) But ADIOS's success does dispatch arguments for the *necessity* of linguistic nativism.

In sum, linguistics may regain its place in cognitive science—if not as a *root* of theorizing in the other disciplines, at least as an equal, and collaborative, partner. If that happens, however, linguistics will be a very different enterprise from that engaged in by Chomsky.

9.x. The Genesis of Natural Language Processing

At mid-century, Harris wasn't the only person hoping to apply computer technology to linguistic problems. On both sides of the Atlantic, there were people "enchanted by the idea that [machine-based experiments to test linguistic hypotheses] could now be performed in the flesh rather than only in the spirit" (Bar-Hillel 1970: 293). And some were already doing so.

The late 1940s and early 1950s saw the first computer models for NLP. These attempted machine translation (MT), automatic classification, information retrieval (IR), text generation, and (for a while) the induction of grammars.

As it happened, these topics weren't the focus of the early research in any of the three pioneering AI labs at MIT, Stanford, and Carnegie Mellon (see 10.ii.a). Consequently, when people talk about the origins of AI they often ignore work in MT which pre-dated their foundation. By the same token, however, when MT suddenly fell into disfavour in the mid-1960s (a story told in Section x.e–f, below) AI as a whole—although it suffered—wasn't so badly affected as one might have expected.

a. Ploughman crooked ground plough plough

Hopes for automatic translation long pre-dated the computer age. It had been envisaged in some of the "universal language" projects of the seventeenth century (see Section iii.d, above). And the first (purely mechanical) MT patents had been taken out, in Russia and France, in 1933.

Whereas the French patent merely described a simple paper-tape dictionary that stored equivalent words in several languages, Petr Smirnov-Troyanskii's (1894–1950) machine could transform a grammatical analysis of a sentence in one language into its equivalent in another. Moreover, he claimed that it should also be possible to automate the (two-way) conversion between these analyses and actual sentences. But he wasn't able to demonstrate this, for lack of funds.

As late as 1944, the Institute of Automation and Telematics refused his bid for support; and his 1948 design for an electromechanical machine similar to the Harvard Mark I (see 3.v.b) was similarly ignored. (He died two years later, so had no chance to develop it.) This lack of appreciation was perhaps inevitable, for his ideas about MT were far ahead of his time:

> In retrospect, there seems to be no doubt that Troyanskii would have been the father of machine translation if the electronic digital calculator had been available and the necessary computer facilities had been ready. History, however, has reserved for Troyanskii the fate of being an unrecognised precursor; his proposal was neglected in Russia and his ideas had no direct influence on later developments; it's only in hindsight that his vision has been recognised. (W. J. Hutchins 1986: 24)

With the development of computers in the UK and USA, their use for various language-related purposes soon followed. In 1946, for instance, Andrew Booth (1918–) of Birkbeck College, London, suggested using computers for translation—and had designed and built a computer by 1947 (Lavington 1975: 5). In the late 1940s he started work with Richard Richens on what would become a lifelong project (Richens and Booth 1955; Richens 1958; Booth 1967, p. vi), and in the early 1950s he co-edited the first volume on MT (W. N. Locke and Booth 1955).

Richens, who was later attached to Margaret Masterman's Cambridge Language Research Unit (CLRU), has been credited with "the first serious contribution" to MT (Bar-Hillel 1959: 303; 1970: 303). His work (initially carried out by hand simulation using punched cards) was interesting not least because it considered grammar as well as vocabulary: the input verbs were split into stem and affix, so that *start*, *started*, and *starting* would have only one vocabulary entry, while *-ed* and *-ing* could be attached to indefinitely many different verbs. Booth and Richens's first experiments on MT were mentioned at a conference (on 'Science Abstracting') in Paris in June 1949, and featured briefly in the *Scientific American* six months later. But they became more widely known thanks to Warren Weaver's memo on MT, written after Booth's visit to the USA in 1947.

Booth's discussions in 1947 with the information theorist Warren Weaver, then Director of the Natural Sciences Division of the Rockefeller Foundation, indirectly influenced the initiation of MT research in America. MT programmes were set up at several US universities in the early 1950s, as a result of an influential memorandum written by Weaver two years after Booth's visit (Weaver 1949; W. J. Hutchins 1986: 24–30).

Financially, the burgeoning of NLP was largely driven by Cold War funding. This was especially true of machine translation. For military purposes, the prime languages were English and Russian, but in the USA they were soon joined by Vietnamese (Newquist 1994: 115).

The first public demonstration of an MT system, in January of 1954, translated forty-nine Russian sentences into English (using a 250-word vocabulary and six rules of grammar). Despite the six rules of grammar, this was basically a word-for-word translator. As such, it was hugely fallible. Nevertheless, it was used right up to the mid-1960s by the USA's Atomic Energy Commission (which was monitoring nuclear sites in Italy, as well as America), and by other governmental agencies too. In other words, it was *still* state-of-the-art: nothing better (for practical use) than word-for-word translation had been achieved.

That program, developed with CIA funding at Georgetown University, was one of many MT projects initiated in the USA in the 1950s. Even in the UK, much of the money for MT and other types of NLP came from various parts of the American defence programme. For instance, CLRU (founded in 1954) was partly supported by the US Office of Naval Research.

Predictably, NLP was developed in the Soviet Union too. Following the visit of a Russian computer scientist to the USA in 1954, for whom the Georgetown demonstration was proudly repeated, several research programmes were set up in the USSR (Hutchins 1986: 37, 133–45). By 1959, there were many hundreds of Soviet researchers in MT alone.

Intellectually, the field was driven by linguistics, by various logical and philosophical theories of language, and by the recent discussions of communication in cybernetics and information theory (see Chapter 4.v). Much early NLP research made a serious attempt to think about language systematically and/or to work towards practically useful ends.

But to say it was serious isn't to say it was successful. The well-known story—or rather, the group of mutually inconsistent stories—about "The spirit is willing but the flesh is weak" being rendered (after translation and retranslation) as "The whisky's fine, but the meat is rotten" is almost certainly apocryphal (W. J. Hutchins 1986: 16–17). However, there were certainly many such early-MT howlers. In the mid-1950s, for instance, Virgil's precisely inflected Latin sentence *agricola incurvo terram dimovit aratro* was once translated at CLRU as *ploughman crooked ground plough plough* (Masterman *et al.* 1957/1986). As remarked in the Preface, ii, this happened because CLRU used a thesaurus, not a dictionary.

With no computable theory of syntax—as opposed to a few simple programmed grammatical rules—available at the time, that's hardly amazing. (A fairly modest dose of syntax would have sufficed to deliver *A/The ploughman ploughs the ground with the crooked plough.*) And with no adequate theory of semantics either, it's not surprising that the CLRU's computational work on meaning didn't prevent this infelicity. Not until the early 1970s would there be powerful computational models of syntax, or of syntax allied with semantics (see xi.b, below).

b. Shannon's shadow

Shannon's theory of information was an important early influence on NLP. He himself used it in the late 1940s to produce "approximations" to English, as we saw in Section vi.c.

With hindsight, it's easy to see that these were word strings rather than sentences, because neither grammatical structure nor meaning were represented in information theory. At the time, this wasn't so obvious. Karl Lashley's (1951*a*) arguments against Markovian theories (see Chapter 5.iv.a) were unknown to most linguists and logicians, and Chomsky's more formal critique hadn't yet been published. Accordingly, many people were enthused by Shannon's ideas about the automatic generation of language.

Besides attempts to generate English text probabilistically, a number of people in the 1950s tried to induce grammars by statistical programs. In effect, they were aiming at the automatic discovery procedure sought by structuralism (see Section v). They focused on artificial rather than natural languages, not least because natural syntax was both complex and unclear.

Chomsky, for instance, cooperated with the psychologist George Miller (then at Stanford University, but soon at Harvard) in describing the formal properties of finite state grammars (Chomsky and Miller 1958). They also did experiments to find out what strategies people use, consciously or otherwise, to discover them (G. A. Miller 1967). Letter strings were labelled by the experimenters as grammatical or ungrammatical, and the subjects tried to learn to distinguish the two classes. If they could actually state the regularities involved, so much the better.

Their Cambridge colleague Bruner had investigated the information strategies used in concept formation (see Chapter 6.ii.b). Moreover, when Miller and Chomsky first described their experimental work (at the University of Michigan in 1957), they alerted the information scientist Ray Solomonoff, a participant at the Dartmouth meeting in the previous year, to define ways of inducing (artificial) grammars automatically (Solomonoff 1958, 1964). (These methods underlie the 1990s work on statistical grammar induction mentioned in Section v.d.)

That there should be linguists, psycholinguists, and information theorists at one and the same venue in 1957 was no accident. Communication between these groups had been growing throughout the 1950s. The Dartmouth meeting had caused an extra upsurge of interest, especially in AI models of learning and problem solving (see Chapter 6.iv.b).

But even earlier, in 1951, the information-theoretic approach to language was already predominant at Harvard and MIT—where Chomsky and his wife, Carol, respectively, were then working:

> Few linguists in Greater Boston at that time dared not use freely "message" and "code", "information" and "bits" in their shop talk, and nobody was "in" if he did not master, or at least professed to master, a good amount of probability theory and statistics. Everybody who was somebody in the field would sooner or later show up at MIT... All of us were enormously impressed by Shannon's well known [probability-based] experiments in what he called "approximations to ordinary English" and were convinced that speech, in English or any other language, was a Markov process. From this to the conviction that the English language, i.e. the set of all English sentences, can be generated by a Markov source, was only a small step, and I am not sure that we noticed at the time that this was a step at all. (Bar-Hillel 1970: 294–5)

The person reminiscing here, and quoted above as initially "enchanted" by the possibilities of the new technology, was the logician Yehoshua Bar-Hillel (1915–75). He'd encountered the logical positivists' approach to language in 1935, while studying at the Hebrew University in Jerusalem. He experienced their work as "nothing short of a revelation", later describing Carnap's volume on the logical analysis of language as "the most influential book I read in my life" (Bar-Hillel 1964: 1). And Bloomfield's linguistics "kept the fire of my interest burning".

So Bar-Hillel shared the structuralists' dream of an inductive discovery procedure for grammars. His hope was intensified when Harris, on a visit to Palestine in 1947, remarked that he was considering using the new electronic computers for this purpose. And, at close of the 1940s, he became "the first prophet in Israel" of cybernetics, including its potential relevance to the mind–body problem (Bar-Hillel 1964: 5).

With these views firmly in place, in 1950 Bar-Hillel accepted a fellowship at the University of Chicago, where he got to know Carnap personally. In the following year he visited MIT, well primed to be drawn into the enthusiastic information-theoretic community described above. Indeed, he was very soon appointed the first director of MIT's MT research. His remit was to survey the nascent field, to assess its prospects, and to plan MIT's research accordingly. His initial survey—of which, more below—was submitted in 1951, and in 1952 he organized the first MT conference (Hutchins 1986: 34–6).

Bar-Hillel didn't stay at MIT very long. On his return to Jerusalem in 1953, he was succeeded by Yngve, also from the University of Chicago. Yngve, now in his eighties, has been described by one NLP expert as "still the most original of all computational linguists" (Y. Wilks, personal communication).

In the 1950s, Yngve used both cybernetic and symbolic approaches. Besides doing early work on MT (Yngve 1955), he studied statistical methods for generating English (Yngve 1961) and developed the COMIT programming language, the first to be designed specifically for NLP (Yngve 1958). In the 1960s, he wrote a *Scientific American* article describing MT to the general public, and a technical survey defending MT against a damaging attack from an official committee (see below) (Yngve 1962*c*, 1967).

Yngve also made an early, and influential, attempt to relate syntactic theory to specific hypotheses about how people process language (Yngve 1960). That was an indication of the difference between his interests and those of the more austere grammarians, whether Chomsky or Gazdar. Over the years, his interests moved more towards "people" and away from "language" (Yngve 1986). And as remarked in Section viii.a, he attacked Chomsky on sociological as well as intellectual grounds.

During Bar-Hillel's brief stay in Massachusetts in the early 1950s, he met Chomsky. The two men interacted regularly, and Chomsky expressed his gratitude to Bar-Hillel several times in his early manuscript on the logical properties of language (e.g. Chomsky 1975: 31).

By 1955, Chomsky had accepted an appointment at MIT's Research Laboratory of Electronics (jointly with Modern Languages). In 1957, as remarked above, he and Miller were doing psychological experiments that immediately prompted information-theoretic work aimed in part at the development of NLP. And his hierarchy of grammars, as we've seen, helped to suggest what type of computational device would be needed to compute natural language.

Nevertheless, Chomsky never shared Bar-Hillel's enchantment with NLP. On the contrary, he was deeply sceptical from the outset:

> My personal reaction to this particular complex of beliefs, interests, and expectations [that information-theoretic computer modelling, combined with empiricist psychology, would make NLP possible, and further the positivist unification of science] was almost wholly negative. The behaviorist framework seemed to me a dead end, if not an intellectual scandal . . . I had no personal interest in the experimental studies and technological advances. [I felt the latter to be] in some respects harmful, [leading people to focus on] problems suggested by the available technology, though of little interest and importance in themselves. As for machine translation and related enterprises, this seemed to me pointless, as well as probably quite hopeless. As a graduate student interested in linguistics, logic, and philosophy I could not fail to be aware of the ferment and excitement [in the early 1950s]. But I felt myself no part of it and gave these matters little serious thought. (Chomsky 1975: 40)

When he did become interested in them, soon after completing his *magnum opus*, he wrote his fierce review of Skinner (see Section vii.b). His only foray into NLP work was a program schema that used transformational grammar (rewrite rules), not statistics, in its processing (G. A. Miller and Chomsky 1963: 464–82).

Nor was he a fan of AI in general. Quite apart from any intellectual reasons for scepticism, there appears to have been some unpleasant rivalry between him and Marvin Minsky—both young, brilliant, famous, and ambitious. In the mid-1960s, when the *Alchemy* scandal was at its height (see 11.ii.b), he went over to the MIT AI Lab, to sit in on one of the heated debates between Weizenbaum and Hubert Dreyfus on one side and Minsky and Seymour Papert on the other. David Waltz, an AI student at the time, now recalls:

> I remember once Noam Chomsky came over [to these debates] also. There was personal animosity between him and Minsky, and people were very rude to Chomsky when he came to the AI lab. They booed when he talked, and he was very miffed. (interview in Crevier 1993: 123–4)

As for Bar-Hillel, and partly because of his discussions with Chomsky, his love affair with—though not his interest in—NLP was to be short-lived, as we shall see.

c. Love letters and haikus

Despite the "ferment and excitement" aroused at mid-century by information theory and statistical cybernetics, some NLP pioneers chose instead to use the newly available digital computers as symbol manipulators.

Even very early on, these machines could be programmed with rules (written in binary notation) for selecting words from distinct grammatical classes. But with the introduction of list-processing languages, such as IPL and COMIT, in the mid-1950s (see 10.v.b), hierarchical syntactic structures could be represented more easily.

Probably the earliest computerized text generator was Turing's playful love-letter program. This was written in the late 1940s for MADM, the world's first electronic stored-program digital computer (Chapter 3.v.b). Using random numbers to choose the words, MADM produced this (Lavington 1975: 20):

Darling Sweetheart,

You are my avid fellow feeling. My affection curiously clings to your passionate wish. My liking yearns to your heart. You are my wistful sympathy: my tender liking.

Yours beautifully

M.U.C. [Manchester University Computer]

The program hasn't survived. But presumably the last two lines, apart perhaps from the adverb *beautifully*, were provided in 'canned' form, while some overall template instructed the machine to select an adverb here, a noun there, and perhaps to pick two endearments (*Darling, Dearest, Sweetheart, Love* . . .) for the first line. Template fillings of this type were soon to be used at CLRU to generate texts fitting the strict constraints of a Japanese haiku (Masterman and McKinnon-Wood 1968; Masterman 1971). Two examples are:

All green in the leaves
I smell dark pools in the trees
Crash the moon has fled

All white in the buds
I flash snow peaks in the spring
Bang the sun has fogged

As the CLRU team realized, the apparent meaningfulness of these mechanically generated haikus was due primarily to the human reader, whose "effort after meaning" (Bartlett 1932: see 5.ii.b) projected significance onto the initially puzzling lines. But if the CLRU realized this, many others didn't. The program was exhibited at the first international exhibition of computer art, and impressed the visiting public more than it should have done (J. Reichardt 1968).

d. Wittgenstein and CLRU

Some of the intellectual sources of early NLP were surprising, being very far in spirit from structuralist linguistics and information theory, and from symbolic computing too. This applies, for instance, to the pioneering research done at CLRU under the direction of Masterman (1910–86).

Their computational work was informed by the anti-positivist philosophy of the later Ludwig Wittgenstein. In scientific circles, this approach to language was nothing if not unorthodox. Wittgenstein himself had likened it to Goethe's views on morphology—see Chapter 2.vi.e–f (Monk 1990: 303–4, 501).

Although these new ideas weren't published until after his death, Wittgenstein had lectured on them—and his lecture notes had circulated—in Cambridge for many years. Moreover, he'd dictated some of them to Masterman herself (see Preface, ii). The lectures provided a radical critique of his own earlier (logicist) position (Wittgenstein 1953), and a very different view of language. Whereas he had once declared that "Everything that can be said at all can be said clearly" (Wittgenstein 1922: 27, 79), he now saw an indefinable penumbra of meaning surrounding every statement.

One of his main points was that words don't have cut-and-dried (essentialist) definitions, in terms of necessary and sufficient conditions. Rather, the items falling under a given word (*game*, for example) are linked by a network of "family resemblances" (Wittgenstein 1953: 31 ff.). Analogously, relatives may share the family nose, the family eyebrows, the family fingernails... without any one person having all these characteristics. Admittedly, some items are better examples of the concept concerned, so can function as "paradigm cases". (These have Rosch-typicality ratings close to 1: see Chapter 8.i.b.) But there's no clear cut-off point.

Masterman was convinced by the doctrine of family resemblances, and disturbed by the all too evident weakness of word-for-word dictionary translations (first attempted by Richens and Booth 1955 and Yngve 1955). Her critique of the early word-for-word efforts (Masterman 1967) was backed up by her own thesaurus-based approach to NLP (Masterman 1957, 1962). One of the early AI workers influenced by her research was M. Ross Quillian, who visited King's College, Cambridge, in 1961 to give a talk about his early ideas on semantic networks (Chapter 10.iii.a).

The thesaurus was envisaged as a semantic interlingua, through which the translator could pass from one natural language to another. (For an early application of CLRU's interlingua approach, see the discussion of "catataxis" in Parker-Rhodes 1956.) Many examples of semantic ambiguity, which would defeat a word-for-word MT procedure, could be resolved by this means. The semantically ambiguous word would fall under at least two different heads in the thesaurus, and the one actually selected by the program would be the one under which some other words in the sentence fell also. So *I put my money in the bank* and *I know a bank whereon the wild thyme blows* would be recognized as dealing with financial and rural matters, respectively. (Syntactic analysis would be needed to deal with *I drew some money from the bank to buy some wild thyme.*)

Initially, Masterman used a thesaurus derived by human intuition, namely, a cut-down version of Peter Roget's pioneering work published 100 years earlier. But her group also developed procedures for constructing thesauri automatically. Formal clustering methods were used to identify word classes, in terms of the co-occurrence of various words. Those word classes determined lexical substitutability for the purposes of machine translation. In addition, they aided in the automatic classification, indexing, and retrieval of documents (Sparck Jones 1988). (The pioneering librarian Gabriel Naudé would have been intrigued: see Preface.)

CLRU's thesaurus approach offered flexibility of various kinds. It could be applied to texts of any length, not just to isolated sentences. It was a source of semantic primitives that could be used to define meaningful messages (or "gists"), based on *actor–action–object* patterns, that were helpful for single-language interpretation and précis as well as for translation. And, not least, it enabled Masterman's group to take some account of the unpredictable extensibility of language. This had been stressed by Wittgenstein and Waismann—and, as Masterman often remarked, by the philosopher Ernst Cassirer (1874–1945), who followed Humboldt in highlighting the unceasing development of language (Cassirer 1944).

The extension of language was modelled, for instance, by Yorick Wilks (1939–). Like Masterman, he was especially interested in the interpretation of metaphysical texts—dismissed as "meaningless" by the logical positivists fashionable at mid-century. He wrote a program in the 1960s to assign semantically coherent representations to textual snippets drawn from the writings of a wide range of philosophers.

Wilks's program could do this successfully for some test paragraphs from Descartes, Leibniz, and David Hume, *without* any need to extend word meanings. But other test paragraphs did require that. Accordingly, his program would locate the semantically problematic word in a chunk of text (not a trivial matter), and then use thesaurus-driven methods to identify its meaning with some other term or phrase within the passage (Wilks 1972: 166–72). If the problematic word was an extension of sense relative to some interpretation already stored in the (thesaurus-based) dictionary, Wilks's EXPAND algorithm could develop the rules for 'literal' interpretation so as to make sense of the anomalous word in context.

This procedure wasn't foolproof. What if there were no other suitable word in the text, and/or some major sense of the problematic word had been omitted from the dictionary? And what is to count as a "major" sense, anyway? (Bar-Hillel, clearly, was hovering in the background.) Nevertheless, it gave acceptable readings of some notoriously tricky texts, including passages drawn from Wittgenstein's *Tractatus Logico-Philosophicus* and Spinoza's *Ethics*. Short of torturing it with extracts from *Finnegans Wake*, one can hardly imagine a more taxing test. In short, this early NLP model tried to escape the logicist straitjacket by acknowledging the creative extensibility of language. (Some of Wilks's later work did so too: his "preference semantics" was used in automatic parsing, interpretation, and inference—Wilks 1975.)

The methods developed by CLRU weren't intended as theories of how human beings translate, interpret, or classify. As remarked in Chapter 1.ii.b, this NLP group was doing technological, not psychological, AI. (Quillian, by contrast, was at least as interested in simulating human psychology as in producing useful technology: 10.iii.a.) But the practical demands were taken seriously from the start. One of the strengths of CLRU was its emphasis on careful testing and experiment to compare IR (information-retrieval) systems (Sparck Jones 1988: 14–17, 25–6). Mere plausibility wasn't good enough—and was sometimes found to be misleading. In short, Drew McDermott's well-deserved criticisms of much early AI didn't apply to Masterman's group (see 11.iii).

Whether Wittgenstein himself would have approved of any computerized approach to translation or classification, even one (such as Masterman's) part-inspired by his own ideas, is another matter. He didn't explicitly discuss any of the budding AI work. Nor do we know of any informal meeting between him and Turing, still less any discussion between them about the possibility of AI. (For an entertaining account of an imaginary meeting, see Casti 1998.) But when Turing attended some of Wittgenstein's lectures on the philosophy of mathematics, the two men repeatedly disagreed on fundamental points (Monk 1990: 417–22). The whole tenor of Wittgenstein's mature philosophy suggests that he wouldn't have welcomed computational work on language.

In part, this follows from his view of language as a network of social conventions rooted in the human "form of life", a notion that can hardly be applied to a computer. But in part, it follows from his stress on the flexibility and open texture of language (and concepts), which seems to lie beyond any scientific explanation.

This flexibility may be approximated by some computational methods, such as CLRU's thesaurus. But even if anti-logicist NLP is in practice less inadequate than logicist approaches, the philosophical problems remain. For Wittgenstein, presumably (S. Shanker 1998)—and for his follower Dreyfus, certainly (H. L. Dreyfus 1972; Dreyfus and Dreyfus 1988)—neither type of NLP can escape the fundamental limitations of AI in general. The Wittgensteinian position agreed with what Descartes had argued long ago (2.iii.c): machine mimicry of language use can only be highly imperfect.

e. Is perfect translation possible?

However, in this area of AI as in others, the question whether some task could in principle be performed (or mimicked) *perfectly* by a machine must be distinguished from the question whether it could in practice be performed (or mimicked) *well enough to be useful.*

The first question is especially difficult where translation is concerned, because it's not even clear what—if anything—would count as perfection in the human case.

Humboldt had claimed that no human translator can achieve perfect equivalence (see Section iv.b, above). In scientific texts, he allowed, translators could hope to do their work with no important loss of meaning. But in literary contexts, they couldn't (Novak 1972). In discussing the (many) possible criteria for a good literary translation, he suggested that the translator should choose words and sentence structures that somehow indicate the original language. The more one does this, however, the less acceptable (in the derivative language) the new sentence will become. Mark Twain's spoof of the German habit of delaying the verb illustrates the point:

> I am indeed the truest friend of the German language—not not [*sic*] only now, but from long since—yes, before twenty years already. And never have I the desire had the noble language to hurt; to the contrary, only wished she to improve—I would her only reform. It is the dream of my life been . . . I would only some changes effect. I would only the language method—the luxurious, elaborate construction compress, the eternal parenthesis suppress, do away with, annihilate; the introduction of more than thirteen subjects in one sentence forbid; the verb so far to the front pull that one it without a telescope discover can. With one word, my gentlemen, I would your beloved language simplify, so that, my gentlemen, when you her for prayer need, One her yonder-up understands . . .
>
> After all these reforms established be will, will the German language the noblest and the prettiest on the world be. (Twain 1897)

Moreover, it's not even clear that this particular criterion of fidelity to the source language should be considered at all. Humboldt himself was well aware that there are conflicting views on what counts *in principle* as a high-quality translation. In short, "perfect" translation is a chimaera. And even high-quality translation is often very difficult.

If this is true for expert human translators, the difficulties facing machines (or rather, their programmers) are so much the greater. Both meaning and grammar stand in their way.

Meaning is notoriously elusive: if it can be pinned down in principle, which even empiricists often deny (Quine 1960), it doesn't follow that it can be captured in practice. Semantic ambiguity and vagueness, not to mention metaphor, are obvious problems. Moreover, a translation machine (eschewing word-for-word translation and statistical methods) needs a universal grammar or an artificial interlingua to mediate between languages, and/or full descriptions of the relevant pair of natural-language grammars. But defining a grammar to cover a single language, never mind a universal grammar, is no mean feat, as we saw in Section ix. And to define a grammar isn't to program it: successful parsers weren't available until the early 1970s (and true computational effectiveness had to await GPSG, more than a decade later).

Most of these theoretical points were made in the 1950s by an increasingly disillusioned Bar-Hillel (1951, 1953, 1960). Despite having been enchanted by the linguistic promise of the new technology in the 1940s, he later expressed deep pessimism—not only about machine translation, but about computational linguistics in general (Bar-Hillel 1970: 298 ff.). He even apologized for his own early influence on the field:

> I myself was led into confusion by the strong surrounding currents [of cybernetics] and was responsible to a degree for increasing this confusion and leading others into blind alleys from which some never returned. (Bar-Hillel 1970: 289)

In recanting his earlier beliefs, Bar-Hillel didn't merely lament the unexpectedly slow progress. He claimed to prove that certain things, their feasibility largely taken for granted by the young NLP community, simply cannot be done by automatic means.

For example, Harris's suggestion that texts made up of syntactically complex sentences can be represented as transformations of several much simpler kernel sentences had led many to hope that NLP could proceed by normalizing texts to some simpler form. Bar-Hillel proved, by contrast, that "Not everything that can be said at all can be said, in a given language, by using syntactically simple sentences exclusively" (Bar-Hillel 1963: 31).

As for semantic ambiguity, this often calls for world knowledge as well as knowledge of language as such. The simple sentence *The box is in the pen*, for instance, can be understood only if one knows that playpens are large enough to contain boxes whereas writing pens are not (Bar-Hillel 1960, app. iii). An MT program concerned specifically with such topics could of course be given this information beforehand. But, Bar-Hillel insisted, such priming can't be done in the general case.

In short, what he called FAHQT (fully automatic high-quality translation) is impossible.

f. Is adequate translation achievable?

Bar-Hillel was complaining about the "science-fictional claims" made by some proponents of NLP, namely those who were promising FAHQT. He wasn't arguing that computer models of language are wholly useless.

In his first report to MIT, he'd said that whereas high-accuracy, fully automatic, MT isn't feasible, "there appear to be less ambitious aims the achievement of which is still theoretically and practically valuable" (Bar-Hillel 1951: 154). Almost a decade later, when he surveyed the progress of MT in the USA and UK (and USSR), Bar-Hillel again stressed the impossibility of human-quality MT (Bar-Hillel 1960). Instead of aiming for this unattainable goal, he said, AI should aim for imperfect-but-intelligible automatic translations, possibly supplemented by human post-editing.

Alongside this potentially hopeful advice, however, he argued that even this less ambitious project should be undertaken in a different way. In his second report, after MT had been under way for some years, he rejected approaches based on statistics and information theory, which he'd previously thought so promising. Nor did he have much time for the then-fashionable methods of GOFAI (Good Old-Fashioned AI: see Chapter 10). He criticized virtually all the major MT groups by name, and attacked the most common methodologies.

The immediate effect of Bar-Hillel's second report was almost to destroy faith in MT on the part of the general public, including scientists not working in the field. A best-selling popular book on *Computers and Common Sense* by Mortimer Taube (1910–65), for instance, took up Bar-Hillel's negative remarks while ignoring his positive ones (Taube 1961).

Taube's book announced in highly emotional terms that MT was impossible and MT research—indeed, AI in general—a waste of taxpayers' money. The nation's monetary resources, Taube said acerbically, would be better devoted to "research looking toward the Second Coming" than to research aiming to build machines with "intelligence, knowledge, and knowledgeability". His dismissive conclusion was that "One may wonder why reputable scientific journals publish material of this sort and why it should have an audience beyond the readers of the Sunday supplements."

Taube, an internationally famous librarian, knew some of his computer onions: he'd pioneered powerful new forms of indexing and information retrieval. Even so, his attack cut little ice with the AI community. Being called "ignorant" and "jejune", and being accused (repeatedly) of "scientific aberration", led to so much offence that they hardly bothered to marshal a defence. Moreover, the book was—as one early cognitive scientist put it—"neither reliable nor responsible", being largely comprised of "allegations presented as facts, of misunderstanding, of debaters' tricks identical to those he decries in others, and of statements about the work of others which are simply untrue" (Reitman 1962: 718).

However, Taube's attack encouraged an attitude of scepticism, not to say ridicule, in the general reader. And the "general readers", in this case, included not only counter-culturalists such as Theodore Roszak (1969: 295) but also people who might at some point have to make decisions about funding policies for MT.

As if that weren't damaging enough, Bar-Hillel's critique also dampened the optimism of MT researchers who were aiming to approach human-quality results. Those who were willing to tolerate many linguistic infelicities, provided that these didn't compromise understanding, were less set back. Even so, and even though significant (military-led) funding continued for over a decade, many MT workers lost heart. At Harvard, for example, almost all research in this field had ceased by 1964.

In that year, the scepticism deepened further. A survey had just been conducted by the US government's Automatic Language Processing Advisory Committee (ALPAC), chaired by the Bell Telephones executive John Pierce (Hutchins 1986: 164–7). Their report was thirty-four pages of dynamite—plus ninety pages of smouldering appendices. It concluded that previous MT research had been a failure, that fully automatic MT was impossible, and—even worse—that there was "no immediate or predictable prospect of useful machine translation".

This negative judgement wasn't a purely scientific one. It was partly based on ALPAC's view that not enough customers would want to use MT to make it commercially viable. Machine aids for human translators, however, might well be feasible.

The effect on informed (or rather, "semi-informed") public opinion was predictable. Yngve described it like this:

> The [general] public is now where the pioneers were many years ago, vaguely aware of the difficulties but allowing that the thing is possible, while the workers, with more than a little knowledge now, see quite clearly that there is very much yet to be done. And the semi-informed public gives indications of swinging back to [their] original position, that the task is virtually impossible, therefore no proper area for research. In this they may be echoing a few despairing noises coming from some of the workers who had expected a quick victory. (Yngve 1967: 453)

Despite criticisms that the ALPAC report was "narrow, biased, and shortsighted", and imbued with a "hostile and vindictive attitude", it had a devastating effect on MT research in the USA. The activity was virtually halted for over a decade. And the devastation was near-instantaneous: "Funding for [MT] projects dried up within weeks of the ALPAC report, and several research labs shut down soon thereafter" (Newquist 1994: 116).

It had a negative influence in the UK and Soviet Union, too. Work on translation at CLRU, for instance, ceased in the late 1960s—even though the Unit itself continued, and did valuable work on various aspects of information retrieval (Sparck Jones 1988; Wilks forthcoming).

If the negative aura of ALPAC could cross oceans, it could cross some intellectual boundaries too. It threw a shadow over AI in general, and NLP in particular. Although the three main AI departments weren't much affected (they weren't doing NLP anyway), several US universities that had been planning to start departments of AI decided not to do so (Newquist 1994: 117).

Nevertheless, ALPAC didn't end NLP as a whole. On the contrary, it favoured increased support of *basic* research in computational linguistics. If the embarrassingly prolific MT cousins of *ploughman crooked ground plough plough* were ever to be avoided, this would require a better understanding—and effective computational implementation—of theoretical linguistics.

9.xi. NLP Comes of Age

NLP would come of age (in the 1970s) with the acceptance of more realistic goals for MT, with the appearance of effective computer parsers, and with the recognition that, besides processing single sentences, one must also model the rhetorical coherence of entire texts. The last two lines of development—which required advances in general AI—are still under way.

Moreover, it's now clear that improved statistical methods, let loose on huge samples of real-life linguistic data, can go much further than Shannon's early critics imagined. Even as late as 1987, that was less clear. At a statistics/linguistics conference in Oxford, sponsored by the Engineering and Physics Research Council (EPSRC), the atmosphere was civilized but there was no real communication (G. Gazdar, personal communication). At a similar meeting held at the Royal Society in 1999, there was (I witnessed that for myself). What's more, while the EPSRC may have expected all the benefit to go from the statisticians to the linguists, in fact it went in both directions.

For instance, Markovian three-word chunks can now be used to make word predictions almost as well as people can, although they can't generate grammatical sentences as reliably (Pereira 2000). These probabilistic models typically induce "hidden" rules from huge databases, and some rely on symbolic rules or tree structures derived from theoretical linguistics (Gazdar *et al.* 2000). Many are trained on collections made up of several hundreds of millions of words. Even so, not all sentences (or newly coined words) can be included. Chomsky's complaint that Markovian theories can't allow for the creative use of language is countered by "smoothing algorithms" that avoid assigning zero probabilities to unencountered events.

Some advances, whether symbolic or statistical, are of primarily technological interest. They will affect our daily lives, but not necessarily our understanding of the mind. Others are of interest only to highly specialized linguists. Below, I outline a few post-1960 landmarks, selected for their relevance to cognitive science *in general*.

Strictly, a certificate of maturity for NLP would also require advances in speech processing—both understanding and synthesis. The former involves the analysis of continuous speech into individual words, and the classification of acoustically different sounds as one and the same phoneme. It had to await the development of vocal-acoustic technology, and also required an integration of several theoretical levels: acoustic, phonetic, morphemic, syntactic, and semantic.

Since I've virtually ignored phonetics throughout this chapter, I'll ignore speech systems too. An early machine for learning to recognize phonemes was mentioned in Chapter 6.ii.c, but this was merely an exploratory toy. The pioneering HEARSAY program of the early 1970s will be mentioned in Chapter 10.v.e, because of its innovative "blackboard" architecture, which integrated processes based in syntax and semantics with processes grounded in phonetics. And the 1980s connectionist system NETtalk is discussed in Chapter 12.vi.f, because of its methodological interest. But speech processing as such isn't crucial for our purposes. Apart from a few remarks at the close of the chapter (subsection g), it won't feature here.

a. MT resurrected

As for machine translation, the ALPAC authors were shown in the long term to have been over-pessimistic. But their suggestion that basic research in computational linguistics would be helpful was justified. Bar-Hillel, by contrast, turned out to be correct on both main counts: although human-quality MT isn't achievable in the general case, practically useful MT is. Masterman's key problem of word-sense disambiguation, for example, can be approached today by methods fundamentally similar to hers but much more powerful (Ide and Veronis 1998).

There are now many MT systems in daily use (W. J. Hutchins 1994, 1995). Most, as Bar-Hillel had predicted, deal only with a highly restricted area, but some can cope with a fairly wide range of topics. They are used, for example, to scan foreign-language news media or scientific publications, or to express political debates and legislation.

Their translations are typically good enough to be understood by informed human readers, and/or to show whether it would be worthwhile to ask a human translator to improve them. Sometimes, only minimal post-editing is required. But sometimes (for instance, when Japanese is involved), significant pre-editing and post-editing are needed.

The SYSTRAN program is one well-known example (P. J. Wheeler 1987). A descendant of an early Russian–English system funded by the US Air Force, it was written in the late 1960s by a team led by Peter Toma, a participant in the famous Georgetown project. The European Community adopted it for English and French in 1976, since when it has been broadened to include all the official languages of the European Union. It is used also by NATO and the International Atomic Energy Authority, and by commercial companies such as Xerox and General Motors.

Some of these applications appear at first glance to evade Bar-Hillel's criticism. For many documents in English can be translated by SYSTRAN into other European languages with near-perfection. The trick is to use only a small subset of English for the original document. A number of other commercially available programs can also provide near-perfect translation of technical manuals and the like, by using highly restricted vocabulary and syntax.

What might well surprise Bar-Hillel is that some very recent MT systems rely heavily on statistical pattern matching (W. J. Hutchins 1995: 440–1; 1994). Rule-based approaches have given way to corpus-based ones, and the emphasis has shifted from syntax to the lexicon (Hutchins 1994, sect. 9). Instead of careful syntactic analysis, these systems rely on probabilistic methods based in some huge corpus of paired source/translation texts. Shannon, one might say, has returned with a vengeance.

IBM's CANDIDE system, for instance, draws on the voluminous (and multi-topic) French–English 'Hansard' of the Canadian Parliament. In the mid-1990s this program contained no grammatical rules whatsoever, relying purely on statistical matching of words and phrases. Its performance amazed most observers, given that the analytical approach had been largely taken for granted in NLP for thirty years. Even ALPAC had seen detailed computational linguistics as essential for automatic translation. Almost 50 per cent of phrases were acceptably translated, being matched either with their database originals or with equivalent expressions.

However, one must not forget the other 50 per cent. To deal with those, future versions of CANDIDE will include some morphological and syntactic information, as well as more sophisticated statistical methods. But grammar and morphology will be used as sparingly as possible. In other words, the team expect the statistics to trump the syntax. It remains to be seen whether they are right, and to what extent MT can afford to ignore theoretical linguistics.

Certainly, grammatical correctness is MT's Achilles' heel. The general assumption, *pace* CANDIDE, still is that this demands painstaking attention to syntax (gender, number, tense, and so on). Measures of intelligibility and correctness have shown a dramatic improvement in the former, but much less progress on the latter.

For instance, in the two-year period 1976 to 1978, the intelligibility of translations generated by SYSTRAN rose from 45 to 78 per cent for the raw text, and from 95 to 98 per cent for the post-edited text (W. J. Hutchins 1986: 261). Human translation, it's worth noting, gave not 100 per cent but only 98–9 per cent.

SYSTRAN's correctness, where every tiny error counts, scored much lower. In 1976 the measure for grammatical agreement was only 61–80 per cent; and in 1978, only 64 per cent of words were left untouched by the human post-editors. Even so, human post-editing of a page of SYSTRAN output took only twenty minutes in the mid-1980s, whereas normal (fully human) translation would have taken an hour. However, over half of the amendments involved substituting a different word, not simply altering a word ending. In both cases, the main source of errors seemed to be the dictionary, which in 1978 contained only 45,000 entries. Accordingly, the computerized dictionaries have been gradually enlarged.

The more ambitious EUROTRA system was multilingual as opposed to severally bilingual, dealing with forty-two different language pairs (King and Perschke 1987). It held 140,000 words in its prototype stage, in 1988: 20,000 words for each of seven languages, with the expectation of at least two more to come. But no operational version was ever built. EUROTRA was influential, nevertheless. It was recently recognized in an official report as providing much of the basis for later work in MT across Europe (*Euromap Report* 1998: 59).

More ambitious still would be an MT system that could switch between languages in very different language groups, such as English, Japanese, and Hindi. In Japanese, for example, the words aren't segmented as they are in English (no equivalent of our *post-ed*, for the past tense of *post*), and the phrase orderings in the parse trees are reversed (Powell 2002: 112). Although there are MT programs combining Japanese and English, and also for switching between several of the many languages of India, they aren't as efficient as the best European-language systems.

(I've mentioned corpus-based work only in MT, but similarly statistical approaches are increasingly being used in other areas of computational linguistics too: see Sampson and McCarthy 2004.)

b. Automatic parsing

If computerized dictionaries are relatively easy to improve, automatic parsing is a different matter.

NLP in the 1950s and early 1960s, as we've seen, was hampered by the lack of a powerful theory of syntax. Computational work on syntax started to show results in the early 1970s. This development was largely due to the post-Chomsky interest in formal grammars. (This is *not* the same thing as an interest in Chomsky's formal grammar: as remarked in Section ix.f, very few people used Chomsky's theory as the basis of their NLP programs.)

The first impressive computer parser was written by William Woods at Bolt, Beranek & Newman (BBN), using "augmented transition networks", or ATNs (W. A. Woods 1970, 1973). The computational advantage of ATNs, which soon became widely used in NLP, was that they considered the input sentence in one pass, working left to right on one word or syntactic constituent at a time. Nevertheless, they could deal—recursively—with nested sentences or noun phrases (such as *the judgement of the court*), and with nouns embellished by several adjectives.

In effect, they worked top-down, continually making syntax-driven predictions about grammatical form and testing the input word to see whether it fitted the current prediction (see Boden 1988: 91–102). On encountering the word *the*, for example, the ATN would predict either an adjective or a noun, in that order; but if an adjective were found, it would immediately predict another one before looking for a noun. This simple procedure would enable it to recognize, for instance, *The girl*, *The clever girl*, and *The clever French girl*.

Woods used his parser in his LUNAR program, which answered questions about lunar geology, drawing on NASA's data about the mineral samples collected from the moon on the Apollo 10 mission. Its answers were in 'written' English (though Woods soon worked also on the HEARSAY speech system). This early version couldn't handle relative clauses in a theoretically elegant manner. Nevertheless, LUNAR was a huge advance on the question-answerers of the 1960s (surveyed in Simmons 1965, 1970).

These included the BASEBALL program, co-authored by Carol Chomsky (B. F. Green *et al.* 1961). This could search and reason about its database so as to answer questions such as "How many games did the Yankees play in July?" (see Chapter 10.iii.a). To deal with the English-language input, BASEBALL and its contemporaries had relied on a menu of ELIZA-like pattern matches (10.iii.a), not on parsing. Simple rules enabled ELIZA to switch pronouns so that (for example) the input *I* ***** *you* would elicit *WHY DO YOU* ***** *ME*. Woods's program, by contrast, modelled syntactic structure in some detail.

ATNs had originally been defined by a research group in Scotland (Thorne *et al.* 1968). Their NLP program was successfully tested on Lewis Carroll's 'Jabberwocky', among other things. It was based on transformational grammar, and built syntactic representations of the input sentence on both deep and surface levels. Woods adopted the general approach of ATNs, but eschewed Chomsky's grammar.

Nor was he the only one to do so. Because of the computational load involved in mapping between two entire syntactic structures, and thanks to the development of non-Chomskyan grammars (see Section ix), very few NLP projects have implemented transformations. The most interesting exception is Mitch Marcus's PARSIFAL, a program inspired not only by Chomsky's syntax but also by his views on language-and-mind (see Chapter 7.ii.b and M. P. Marcus 1980).

The Scottish team, like Marcus, had been doing psychological AI. They saw Chomsky's theory, and their left-to-right ATNs, as modelling some of the mental processes involved in language. Woods wasn't much interested in whether his parser was psychologically plausible. But some people used his approach to formulate specific psychological hypotheses about how people use language—and memory, too (see Chapter 7.ii.d and iv.d–e). The first to do so was Kaplan (1972), who later went on to co-author a psychologically motivated grammar, LFG, that has been borrowed extensively for technological purposes (see Section ix.b). Another was David Rumelhart, who used ATN models of grammar to study the psychology of reading errors (Stevens and Rumelhart 1975).

This swinging pendulum of intellectual influence illustrates a general point made in Chapter 1.ii.b. The boundary between psychological and technological AI, and therefore between what is cognitive science and what is not, is often fuzzy. ATNs were first devised (in Scotland) with psycholinguistic intent, soon adapted (by Woods) for technical purposes, and then picked up (by Kaplan and others) for use in psycholinguistics, memory research, and theoretical syntax. The new formal grammar that eventually resulted was later adopted by a variety of technological NLP researchers—some of whom would hardly recognize a mental process if it punched them on the nose.

A further illustration of this point was provided by the most famous achievement of early NLP: Terry Winograd's program SHRDLU (Winograd 1972; Boden 1977, ch. 6). The name, by the way, wasn't an acronym but a slyly subversive joke. In the publishing technology of the 1960s, one row of the keyboard of a standard linotype typesetting machine consisted of these letters. Typesetters would often signal a mistake by inserting them in a faulty line, so that the proof-readers would easily spot that a mistake had been made. Bad proof-reading might result in this deliberate gibberish being printed in the final text. That fact was made much of in the counter-cultural *MAD* magazine, which was deliberately peppered with multiple occurrences of this nonsense word. As a devotee of *MAD*, Winograd (1946–) decided to use it as the name of his program (T. Winograd, personal communication).

Winograd's program was written, at MIT's AI Laboratory, in the heyday of NewFAI. In 1961 George Ernst (with Minsky and Shannon on his thesis committee) had started to build a robot that would pick up blocks from a table, build them into towers, and put them into and out of a box. And by the late 1960s, the MIT robotics team were trying to integrate computer vision with their robot's problem solving, as well as working on a better robot hand (L. J. Fogel and McCulloch 1970: 256–65). It was in this intellectual environment that Winograd's doctoral research began.

The overall aim was to enable the team to give instructions to the robot in English, and for the robot to respond in real time. But for that to be possible, AI needed to advance in English as well as robotics.

Winograd's thesis title, 'Procedures as a Representation for Data in a Computer Program for Understanding Natural Language', showed that he wasn't interested only in English, or even NLP. To the contrary, he was largely concerned with questions about virtual architectures for AI. Indeed, his concepts of "heterarchy" and "procedural programming" had a huge impact on GOFAI (see Chapter 10.iv.a).

What's more, he wasn't doing psychological AI—except, perhaps, in the most general sense. Although his opening sentence was "When a person sees or hears a sentence, he makes full use of his knowledge and intelligence to understand it," he was careful to stress that SHRDLU wasn't intended as a detailed model of what goes on in people's minds.

Nevertheless, his work was considered so significant by the cognitive science community that an entire issue of the journal *Cognitive Psychology* was devoted to it—and it was simultaneously published in book form by a leading university press. In explaining their unprecedented decision, the psychologist editors said:

Winograd's system is not a "simulation", but it incorporates important ideas about human syntactic, semantic, and problem-solving abilities, and, in particular, about their interactions in understanding natural language. Human intelligence goes far beyond this system, and human language comprehension may turn out to differ in major respects from the means employed here. At the very least, however, Winograd's system should prove an invaluable tool for thinking about what we do when we understand and respond to natural language.

(Winograd 1972, p. vii)

Their reference to "interactions" between syntactic, semantic, and problem-solving abilities, and Winograd's opening sentence (quoted above), indicated the sense in which SHRDLU went far beyond previous NLP work.

Like Woods, Winograd provided a powerful computer parser (based on Halliday's systemic grammar: see Section ix.a). It could even parse long sentences with a syntactic sting in the tail, such as: *How many eggs would you have been going to use in the cake if you hadn't learned your mother's recipe was wrong?* But the most important NLP novelty was that the parser was closely integrated with semantic analysis, problem solving, and world knowledge (see 10.iv.a).

SHRDLU's world knowledge—of a table, a box, and various coloured objects that could be moved by a robot arm—had to be laboriously programmed in. So Bar-Hillel's argument citing *The box is in the pen* still held. As Winograd himself pointed out, many obvious implications of English sentences about boxes, blocks, and tables were invisible to the program.

This was a prime reason why, in the 1980s, he rejected the approach underlying his early work (Winograd 1980*a*; Winograd and Flores 1986). He'd been persuaded, by Fernando Flores and Dreyfus, that various neo-Kantian thinkers were better guides to language than the logicist philosophers whose views imbued GOFAI—and most computational linguistics, too (see Chapter 16.vi–viii). So his 1986 book, co-authored with Flores, highlighted the hermeneutic philosophers Martin Heidegger (sixteen index entries) and Hans-Georg Gadamer (nine), the autopoietic theorist Humberto Maturana (fifteen), and the later Wittgenstein (two). None of these had figured in his earlier work—and many of his erstwhile admirers were aghast. (For more on Winograd's volte-face, see Chapter 11.ii.g.)

The cognitive integration, in SHRDLU, went beyond the fact that syntax, semantics, reasoning, and world knowledge were all included in the one *program*. The crucial point was that they could interact in the processing of a single *sentence*. When interpreting the syntactically ambiguous instruction *Put the blue pyramid on the block in the box*, for instance, the program checked to see whether there was one and only one blue pyramid already sitting on the block, or (if not) whether there was one and only one block already inside the box. (Compare this with Quillian's semantic network, which used *purely lexical* information to disambiguate sentences like *I threw the man in the ring*: 10.iii.a.) And in obeying the instruction, and deciding how to reply to it, SHRDLU might have to do some problem solving, to enable the robot arm (which could pick up only one thing at a time) to get at the blue pyramid concerned.

With the help of predicate logic ("one and only one thing such that..."), also included in the program, Russell's theory of definite descriptions (see 4.iii.c) had been implemented in computational form. However, Winograd defined the program's lexicon, and its syntactic and semantic terms, not as formulae of the predicate calculus but as mini-programs. So his definition of *the* didn't simply assert (as Russell had done) that this word was apt only if there was exactly one thing that fitted the description given. Rather, it instructed the system to go and look for such a thing on encountering the word. As a result, when SHRDLU was asked to "Grasp the pyramid", there being *three* pyramids in the scene, it sensibly replied, "I don't understand which pyramid you mean." However, if a specific pyramid had already been mentioned in the conversation, it assumed that "the pyramid" referred to *that* one. Similarly for *pyramid*, whose definition was a procedure telling SHRDLU to consider—and, if appropriate, to look for—a thing of a certain sort. The syntactic definitions, of *noun*, *adjective*, and the like, were also expressed as procedures rather than assertions.

An ATN too, of course, would look for a 'suitable' word on encountering *the*, or *pyramid*, and would employ some specific procedure to find a noun, or a verb. But these syntactic categories weren't *defined* procedurally. This gave ATNs a generality that SHRDLU lacked, for *some* ATN could be devised for *any* grammar.

SHRDLU thus constituted a pioneering example of "procedural semantics", an approach to meaning and understanding that became widely influential in cognitive science. As we saw in Chapter 7.ii.d and iv.d–e, procedural semantics was taken up in studies of language and problem solving. It also affected work in practical AI (10.iv.a). With respect to technological NLP, the specifics of Winograd's parser were less influential than Woods's ATNs. But Winograd's work had shown the possibility of integrating syntax and semantics (and world knowledge, when available) in understanding language.

This general lesson could be applied by those who came after him, even if they used a different grammar and method of processing. For example, the author of the speech program whose many errors betrayed underlying anxieties (7.ii.c) singled out Winograd, for having enabled him "for the first time . . . to get a handle on how to think about and represent cognitive and linguistic processes" (Clippinger 1977, p. xv). That stuttering program was very different from the grammatically perfect SHRDLU. Nevertheless, SHRDLU was crucial in its ancestry.

Awesome though SHRDLU was, it wasn't quite so impressive as my description, above, has suggested. For one thing, the "conversation" wasn't really a conversation:

> I believe that in the process of debugging I got it to go through the whole dialog in a single run, but I wouldn't want to state that for sure. Most of the work was done in a more fragmentary manner. (T. Winograd, personal communication)

(Each fragment was relatively speedy: the system, when running compiled, was "fast enough to carry on a real-time discourse", with between 5 and 20 seconds needed for the combined interpretation and response; and the display was designed to move at the speed of a real arm: personal communication.)

For another, the system couldn't be used 'off the shelf':

> The program was at times run effectively by others (including Stu Card who was there for a summer and helped get out many of the bugs), but only after they took some trouble to become familiar with it. It wasn't at a level of robustness for walk-up users. (T. Winograd, personal communication)

In this, SHRDLU was far from unique. It was a common failing of early GOFAI programs that they couldn't be run—or developed—by others, because of bugs. Indeed, this was one of the complaints made by McDermott in his famous criticism of AI work in the mid-1970s (11.iii.a).

c. 'What I did on my holiday'

The growing interest in automatic parsing throughout the 1970s and 1980s, which fed and was fed by the theoretical work on grammars outlined in Section ix, wasn't matched in respect of automatic composition. Even a young child can write an essay on 'What I did on my holiday'. But a computer, even if it could go on holiday in the first place, would be very hard put to do so.

The output of many (non-MT) NLP programs, including SHRDLU, was based on pre-assigned templates. Although these could be syntactically varied to some extent, they were essentially similar to the 'canned' responses of very early programs such as ELIZA. Indeed, the problem of how to enable computers to generate syntactically elegant text was rarely addressed. The reasons were part-practical, part-theoretical.

For most practical purposes, the automatic composition of elegant syntax was, and still is, a stylistic luxury. Many NLP programs, such as the human–computer interfaces used by the general public, can get by with very simple syntax. Even MT programs can acceptably simplify the original syntax, provided that intelligibility is retained. Often, the original text (of technical manuals, for instance) can be written in a deliberately simple fashion without loss of meaning. And even programs that generate the text—not just the plot—of stories can 'succeed' by emulating the syntax of stories written for 5-year-olds, not of Marcel Proust.

The theoretical obstacle was a version of Bar-Hillel's problem: subtle use of syntax requires subtle world knowledge. When people use language, they typically generate syntactic structures that are both apt (with respect to the situation being described) and helpful (with respect to intelligibility). In doing so, many questions arise. For instance:

* When is clause subordination appropriate,
* and which event should be referred to by the subordinate clause?
* When are two separate sentences preferable to a single sentence in which two clauses are conjoined?
* When should one employ subjunctives and conditionals?
* How should one do so?
* And how can one choose *appropriate* referring expressions, determiners, modifiers, tense, aspect, and modal verbs?

These intriguing theoretical questions were investigated in the late 1970s by Antony Davey, who related them to various psycholinguistic phenomena. Here, the relevant point is not the psychology, but the fact that he implemented some of his answers in an NLP program (Davey 1978). He showed how syntax could be used, in a principled way, to signal the strategy driving individual games of noughts and crosses (tic-tac-toe).

Thus his program produced the following description of the game depicted in Figure 9.5:

I started the game by taking the middle of an edge, and you took an end of the opposite one. I threatened you by taking the square opposite the one I had just taken, but you blocked my line and threatened me. However, I blocked your diagonal and threatened you. If you had blocked my edge, you would have forked me, but you took the middle of the one opposite the corner I had just taken and adjacent to mine and so I won by completing the edge.

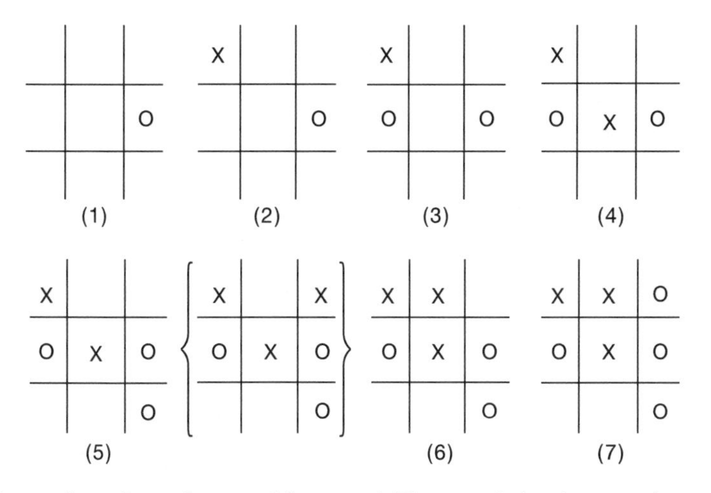

Fig. 9.5. Game of noughts and crosses (tic-tac-toe). The move in brackets wasn't actually made, but was 'imagined' by the program when it described the game depicted in the numbered frames. (Based on an example in A. C. Davey 1978: 18.)

No post-editing was done by Davey: this passage was the actual output of his program. The syntax nicely reflected the structure of attack, defence, and counter-attack informing the game, and helped distinguish strategy from tactics. For instance, it would have been less apt to reverse the last two ideas in the second sentence (as in "... but you threatened me and blocked my line"), or to start it by saying "I took the square opposite the one I had just taken and so threatened you." Many other superficially equivalent alternatives were specifically—and rightly—avoided by the program.

This was the first automatic system capable of generating complex, semantically appropriate, syntactic structures. As such, it was a significant achievement. But, for the reasons outlined above, it had little influence on the development of NLP. Its unprecedented power was unnecessary in most practical contexts, and wasn't achievable in the general case—for composing essays on holidays, for example. Davey had chosen noughts and crosses as his domain precisely because the rules are easily statable, the games are fully describable, and the strategy is readily intelligible. It would be difficult, and often impossible, to adapt his program to other domains.

Different approaches to grammar generation were sometimes considered. For instance, rules governing shifts in the focus of attention (see below) were used in choosing the active or passive voice, pronouns or definite descriptions, and the sequence of sentences within a paragraph (McKeown 1985). But these didn't match the detail of Davey's work. Consequently, today's NLP programs are asymmetrical in their grammatical abilities, much as SHRDLU was. Their expertise in parsing belies their simple-mindedness in sentence construction.

d. Semantic coherence

The maturation of NLP also required a concern for the semantic coherence and rhetorical structure of entire texts. Let's consider the first topic here (the second is discussed in subsection f).

Impressive though it was, Winograd's program had virtually no ability to deal with a text, as opposed to a series of single sentences. Admittedly, anaphora and definite descriptions were interpreted by noting what objects had been referred to in previous sentences. But there was no representation of the theme of the conversation as a whole, nor any prediction as to what the human might say next.

Still less was there any systematic representation of conversational etiquette, or of the distinction between the words uttered and the communicative intention behind them. Eventually, and partly driven by the problems involved in designing usable natural-language interfaces, these issues of semantics and pragmatics came to the fore (see 13.v).

One early attempt to identify the theme of an entire text had been made by Wilks, using CLRU's thesaurus approach (see Section x.d, above). Another, even earlier, was the "General Inquirer", initiated around 1960 by Philip Stone (1936–) at Harvard and colleagues (Stone *et al.* 1962). The method relied on an "affective" classification of the English dictionary, based on tags drawn from Fred Bales's small-group psychology. For example, *union*, *cooperation*, and *love* were coded under the Affiliation tag, whereas *fight* and *war* were classed as Hostility. Counting tag words was straightforward. And hierarchical structures, wherein one semantic tag seemed to be providing the context for another, could be found by list processing—using Yngve's new list-processing language, COMIT.

This approach provided only a very broad analysis—and could be highly misleading, to boot. The computer analysis might show Affiliation governing Hostility, for instance, or Hostility governing Affiliation. However, to decide whether these seeming contradictions reflected actual incoherence, as opposed (for example) to a threat followed by conciliatory remarks, the researcher would have to go back to the text itself.

I myself used a prototype version of the Inquirer in October 1962, as a student in Stone's team analysing the ongoing Khrushchev–Kennedy interchange about the Cuban missile crisis. (The texts were being made immediately available to us in electronic form by an international press association.) The hope was that implicit affective changes in Nikita Khrushchev's attitude might be picked up even before they'd been made explicit by him.

The attempt wasn't successful. For example, Khrushchev's speeches, but not John Kennedy's, scored very high on Affiliation. On checking the original transcripts, the explanation was clear. What Kennedy called "the USA" and "the USSR", Khrushchev (or his translator) called "the *United* States" and "the Soviet *Union*". The problematic words, of course, were then removed. But they might not have been identified in the first place, if the initial scorings hadn't been so highly counter-intuitive. In addition, it was unclear what significance—if any—should be attached to the fact that frequency counts showed one tag hierarchically dominating another.

Both Wilks's system and Stone's suffered from a broad-brush approach. They could identify general themes and overall affective tone, respectively, but couldn't follow the development of a theme from one sentence to the next. And there was no question of their being able to predict the likely content of sentences yet to come.

The first people to model these matters were the social psychologist Robert Abelson and the computational linguist Roger Schank. One reason for their collaboration was that Schank was doing psychological, not technological, AI. He did eventually produce a number of commercially marketed NLP tools, but he saw them—correctly or not—as reflecting how humans process language (see Chapter 7.i.c and ii.d).

By the mid-1960s, Abelson (at Yale) had simulated individual systems of political beliefs, predicting responses to a variety of new inputs. Soon afterwards, he outlined a computational theory of social roles, of motivational–emotional "themes", such as betrayal and cooperation, and of "scripts", such as escape (see Chapter 7.i.c).

By this time, Schank—then at Stanford, though soon to join Abelson at Yale—was doing NLP based on his conceptual dependency (CD) theory. This combined a semantic primitive analysis of verbs with a version of case grammar (Schank 1973; Boden 1977, ch. 7). He aimed not only to represent the meaning of an input sentence, but also to predict the likely topic/s of the following sentences. The verb *hit*, for example, involves an "instrumental" case. If this case slot was left open (as in the sentence *John hit Mary*), the CD representation enabled/predicted the follow-up query "What with?"—where the default inference was that the instrument was the actor's hand.

In their collaborative work at Yale, Schank and Abelson (1977) used the term "scripts" in a new way, to represent the behaviour of various actors in culturally stereotyped situations. Their restaurant script, for example, codified the roles of customer, waitress, and cashier in a hamburger bar. From the early 1970s, Schank and his students combined CD theory with this new theory of scripts to generate questions and answers about (specially composed) texts.

For instance, Wendy Lehnert (1978) wrote a question-answering program that integrated CD representations of verbs with world knowledge about hamburger restaurants. Given a mini-story including the sentence "When the hamburger arrived, it was burnt", the program would infer—what was not made explicit in the input—that the customer who had ordered it neither ate it nor paid for it. These texts provided the story snippets that would soon be used as examples by John Searle (1980) in his discussion of the Chinese room (16.v.c).

Meanwhile, Jim Meehan at Yale, and researchers in other places, had been using Schankian ideas in modelling the generation and summary of simple story plots (Meehan 1975; Rumelhart 1975; Schank and Riesbeck 1981; Lehnert and Ringle 1982). And up to the early 1980s, Schank's student Michael Dyer (at Yale and then UCLA) considered more complex types of story, involving developments of Abelson's ideas about motivational and emotional structures (Dyer 1983).

These Yale-based programs combined linguistics with GOFAI, for they relied heavily on AI ideas about planning in interpreting texts. In this, though in little else, they resembled Winograd's SHRDLU.

For example, a key concept in Dyer's NLP approach was the TAU, or thematic abstraction unit. TAUs were used to organize memory, direct the process of understanding, support analogical reasoning, and enable one story to remind the system of a different one. They were defined as abstract patterns of planning and plan adjustment, involving aspects such as enablement conditions, cost and efficacy, risk, coordination, availability, legitimacy, affect, skill, vulnerability, and (legal) liability. Those dimensions were used by the system to recognize individual episodes in the text as examples of one TAU or another, and different story plots would involve different sets and/or sequences of TAUs. The TAUs included (for instance) *incompetent agent*, *a stitch in time saves nine*, *too many cooks spoil the broth*, *red-handed*, *hidden blessing*, and *hypocrisy*.

As these labels indicate, TAUs were abstractly defined as general planning schemata, applicable to a wide range of texts. As such, they could in principle be used to describe the internal coherence of many different stories. But because of the specific world knowledge needed to consider individual plans in actual texts, Dyer's approach in practice fell foul of the Bar-Hillel problem. (It also fell foul of the McDermott problem: using meaning-rich English phrases, such as 'a stitch in time . . .', as the names for semantically impoverished technical terms: see Chapter 11.iii.a.)

Dyer's program had to be specially designed to anticipate a specific range of stories—which in turn usually had to be specially written. So this approach to NLP, or to knowledge representation in general, was limited in terms of its potential applications. If the text domain wasn't laboriously mapped onto a specific set of planning problems, a Schankian program couldn't generate questions and answers appropriate to it.

In Winograd's (unpublished) invited lecture at the 1973 international AI conference in Los Angeles, when his own name was still on the lips of everyone interested in cognitive science, he pointed out—with characteristic modesty and intellectual generosity—that Schank was addressing important problems which he himself hadn't touched. As for Abelson's work, this was being recommended at much the same time in Minsky's influential, and widely pre-circulated, paper on "frames" (see 10.iii.a). A few years later, Minsky (1977, 1985) would draw on Abelson's ideas again, and on Schank's as well, in his "society" theory of mind (see 12.iii.c).

In short, Schank and Abelson's ideas eventually influenced AI in general (and psycholinguistics too: 7.ii.d), not just NLP.

e. The seductiveness of semantic networks

To say—as Winograd did—that Schank was addressing important problems isn't to say that he was doing so successfully. Quite apart from doubts about semantic primitives in general (see Section viii.c), his own list wasn't drawn up in a principled fashion. It dropped from fourteen to twelve, to eleven. Moreover, his work suffered from an ailment that was all too common in the AI of the time: a logical and linguistic vagueness, grounded in the intuitive use of semantic links.

This charge may seem strange, since computer models can't function by intuition. Any semantic link implemented in a computer program must have some specific definition. But the case is different when one considers the human researcher's interpretation of *diagrams* of semantic networks, such as Figures 9.6 to 9.8. Many such diagrams adorned the NLP literature of the mid-1970s. They flourished in the AI work on knowledge representation, too (see Chapter 10.iii.a). Indeed, they'd been originated by Quillian (1961, 1968) to model the structure of long-term memory.

Quillian's networks represented conceptual associations of *various* kinds, including class membership and similarity (see 10.iii.a). His general approach was taken up by many NLP researchers, including Schank, to represent the meaning of sentences. (It

$$\text{John} \Leftrightarrow \text{hit} \overset{o}{\leftarrow} \text{dog}$$

FIG. 9.6. Conceptual-dependency diagram representing "John hit the dog." Reprinted with permission from Schank and Colby (1973: 193.)

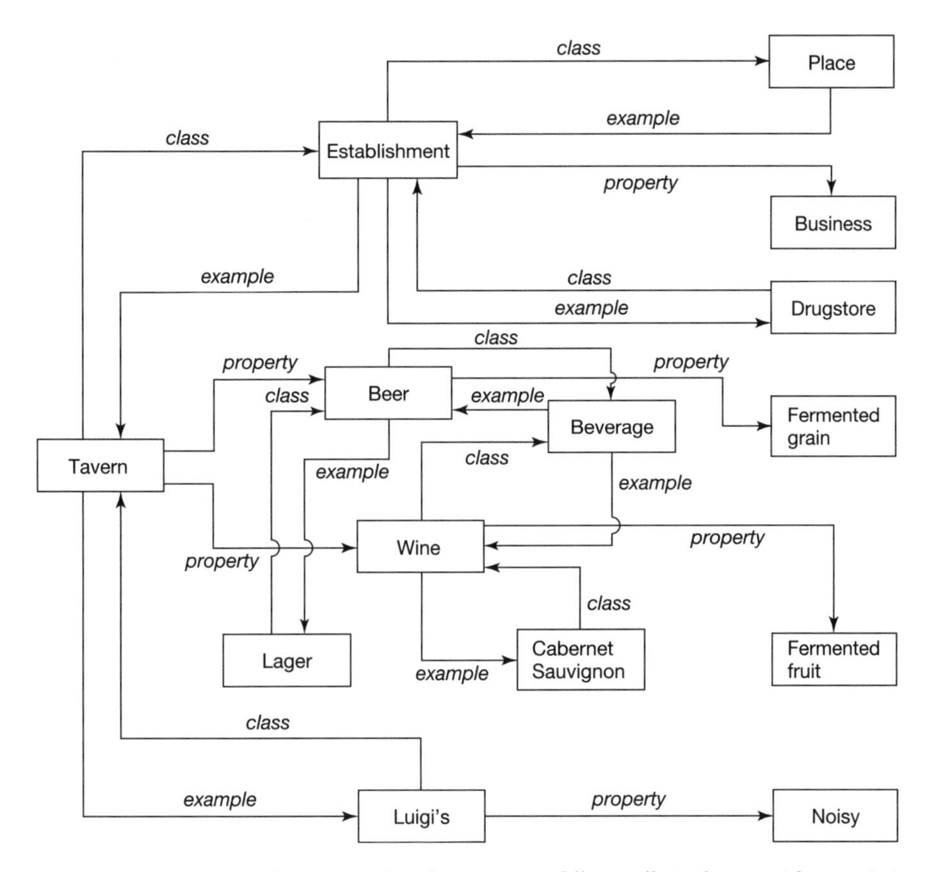

FIG. 9.7. Semantic network representing the concept of "tavern". Redrawn with permission from Lindsay and Norman (1972: 389)

was also applied to vision and to concept learning, by Patrick Winston and John R. Anderson, for instance: see 10.iii.d and 7.iv.c.) But, as semantic networks became increasingly popular, W. A. Woods (1975) rang a warning bell.

Woods pointed out that these representations were usually ambiguous, and—in their naive form—ill-suited to capture many crucial semantic distinctions. In particular, there were many problems concerning relative clauses and quantification:

> [When] one does extract a clear understanding of the semantics of the notation, most of the existing semantic network notations are found wanting in some major respects—notably the representation of propositions without commitment to asserting their truth and in representing various types of intensional descriptions of objects without commitment to their external existence, their external distinctness, or their completeness in covering all of the objects which are presumed to exist in the world. I have also pointed out the logical inadequacies of almost all current network notations for representing quantified information and some of the disadvantages of some logically adequate techniques. (W. A. Woods 1975: 79–80)

Additional problems that needed to be addressed, he said, included (among others) the representation of mass terms, probability, time, tense, and the use of adverbial modification.

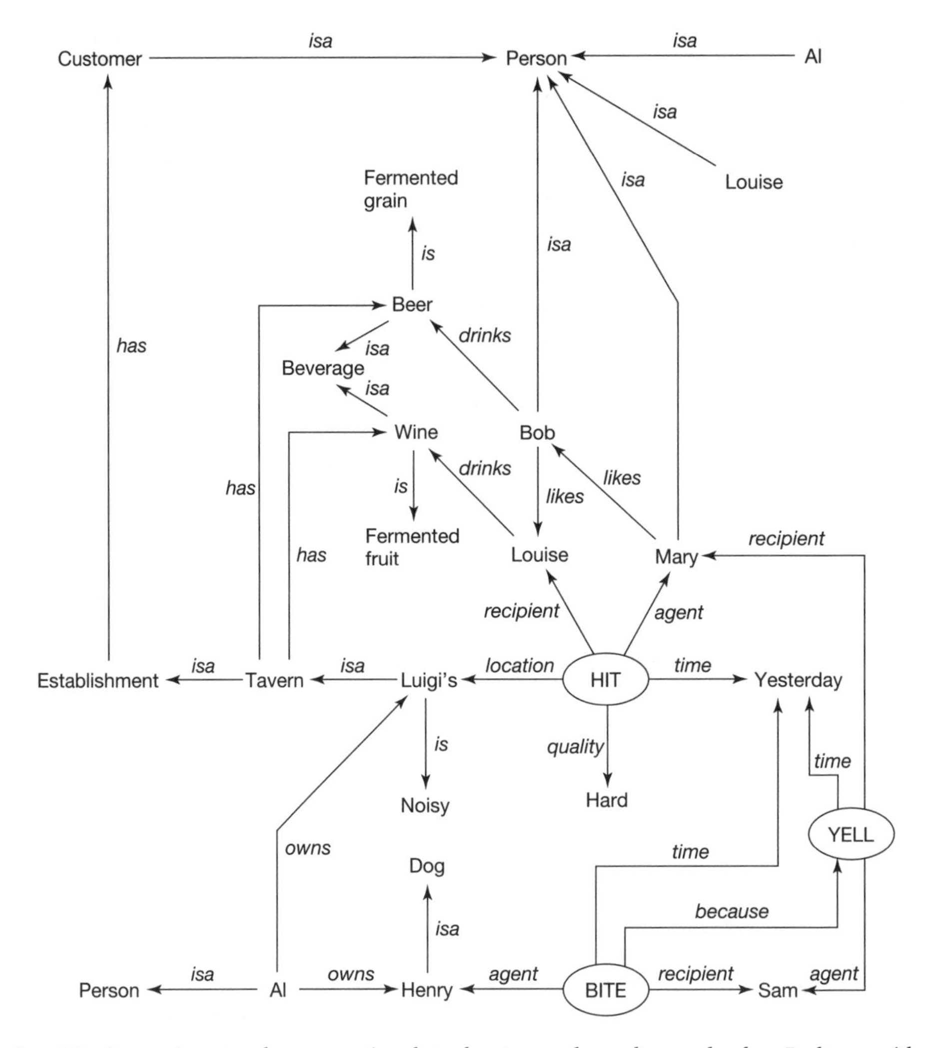

FIG. 9.8. Semantic network representing data about several people—and a dog. Redrawn with permission from Lindsay and Norman (1972: 400)

The justice of Woods's critique can be illustrated by the Schankian CD network shown in Figure 9.6. The link between "hit" and "dog" is labelled as being one of a specific kind, denoting the objective case. And Schank glosses this diagram accordingly, as being equivalent to "John hit the dog". But it could equally well be interpreted as "John hit *a* dog". If Winograd's SHRDLU had relied on this type of representation, it could not have queried the meaning of "Grasp the pyramid" when no specific pyramid had been indicated, nor interpreted that same phrase correctly later in the conversation. Moreover, if even this simple diagram is ambiguous, then one may expect more complex ones (such as Figure 9.7) to be highly problematic.

Where Schank's system was concerned, semantic ambiguity wasn't the only weakness. The system's syntactic power was grossly limited in comparison with that of SHRDLU, or of Woods's ATNs. It couldn't cope with complex auxiliary verbs and nested sentences, as in the query concerning the eggs in the cake (see above). And because it had no representation of logic, it couldn't interpret quantifiers such as *all* or *some*—nor, as we've just seen, definite versus indefinite articles.

Accordingly, those NLP workers who had a healthy respect for logical semantics and grammatical niceties weren't much influenced by Schank's work. Rather, they paid careful attention to syntax, and tried to fill the semantic gaps listed in Woods's critique.

Woods himself, for example, soon improved his LUNAR system in many ways, such as enabling it to interpret the natural-language quantifiers *all*, *every*, *some*, and *most* (W. A. Woods 1978). And others soon improved ATNs by defining a procedure (the "HOLD" function) that enabled them to parse relative clauses in an efficient, and theoretically elegant, way (Wanner and Maratsos 1978). From the early 1980s on, much NLP research implemented highly detailed syntactic theories such as GPSG and its computational cousins (see ix.d–f, above). And some developed theories of logical semantics broadly inspired by Montague, but better fitted to the constraints of actual language use (e.g. Barwise and Perry 1984).

f. Whatever will they say next?

The actual *use* of language involves not only syntax and semantics, but pragmatics too. This became increasingly clear through the 1970s, as computational linguists tried to design human–computer interfaces for interactive tasks (13.v). These included querying databases, planning (of travel itineraries, for instance), and giving/seeking advice on ongoing action (as in repairing a car).

Such applications require a rhetorically coherent conversation, as opposed to a string of isolated sentence pairs. Moreover, the human should be free to use language in everyday ways, and the computer should converse 'naturally' also.

Accordingly, NLP researchers in the mid-1970s began to model how people plan what they are going to say, how they use language to do things other than state facts, how they give or request information in a way appropriate to the interlocutor's state of mind at that moment, and how they keep track of shifts in conversational focus. In doing this, they drew on previous work in the philosophy of language.

Both Wittgenstein (1953) and John Austin (1962*a*) had shown that language can be used to do many different things, and Searle (1969, 1975) had recently formalized some of Austin's ideas. Austin (1911–60) had pointed out, for instance, that some verbs are used not to state facts but to promise, to command, to warn, or to marry. (Consider "I'm warning you...", "I promise", and the bride's "I do".) He believed that there's a finite number of speech acts, perhaps a few hundred, in terms of which language use could be systematically classified.

However, Austin (and Searle) didn't assume a one-to-one match between words and speech acts, because people often use language well suited to one speech act in order to perform another. For example, the verb "I suggest" can be used not to suggest, but to command. An indicative sentence can be used to ask a question ("I don't know the time"), or to make a request ("It's cold in here"). And a Yes–No question can convey a command ("Can you pass the salt?"), or elicit substantive information ("Can you tell me the way to the station?").

It follows that human speakers can—and interactive computers should—not only recognize different speech acts, but also distinguish the literal meaning of the sentence from the communicative intention of the speaker. Consequently, Searle's theory of speech acts was applied by Philip Cohen (a colleague of Woods at BBN) and Raymond Perrault to the computer modelling of conversation (P. R. Cohen and Perrault 1979).

The intellectual benefits flowed in both directions. This NLP work uncovered some flaws in Searle's theory, which was later developed—and axiomatized—as a result (Gazdar 1981*b*; Searle and Vanderveken 1985).

Cohen and Perrault drew on other philosophical work, too. H. Paul Grice (1913–88) had noted that people choose what to say—and what to leave unsaid—by considering not only truth, but also helpfulness, relevance, and brevity (Grice 1957, 1967/1975, 1969). These "conversational postulates" are involved also in understanding someone else's remarks. If they are clearly contravened, the remark will be interpreted differently: as a joke, or sarcasm, for instance. In order to apply them, both speaker and hearer must consider each other's beliefs and interests. What is helpful to one person may be unnecessary or confusing for another, or for the same person at a different time. (Grice's seminal work eventually led to a more elaborate theory of "relevance": see Chapter 7.iii.d.)

An interactive NLP system, then, should be able to adjust its output to suit different human individuals, or one individual at various points in the conversation. It should also include procedures for recognizing and correcting *misunderstandings*—on its own part, or the human's.

In implementing Grice's ideas, Cohen and Perrault used AI planning (Chapter 10.iii.c) to show how the multilevel plans of two conversationalists can be mutually adjusted so that communication takes place as intended. Each of these plans included a (continually updated) model of the other speaker's beliefs and intentions. (One might say that the programmers had sketched a "Theory of Mind" for AI systems: see 7.vi.f and 13.iii.e.) And each plan was used to guide the way in which different topics were 'naturally' introduced, or dropped, as the conversation progressed.

Similarly, Richard Power (1979) modelled the ongoing conversation between two robots cooperating in opening a door, where the relevant information wasn't directly available to both of them (see Chapter 13.iii.e). Only the robot on the bolt-side of the door could see whether the bolt was up or down. Opening the door involved problem-oriented planning concerned with how to get the door open (by moving the bolt). It also required *communicative* activities such as:

* attracting the other robot's attention;
* suggesting,
* and then negotiating, an agreed common goal
* (followed by a series of sub-goals);
* communicating or requesting information accessible to only one robot;
* confirming that this information has been duly noted;
* instructing another agent to do something which one cannot do oneself;
* . . . and so on.

From time to time, a robot would have to find out what the other robot knew and compare that with its own knowledge. This required processes that modelled each robot's changing knowledge and its (continually updated) model of the other's state of mind.

Related work was being done at much the same time by Barbara Grosz (1977; Grosz and Sidner 1979). For instance, she modelled the changing reference of terms such as "it" as the conversational focus shifted. As remarked above, focus shifting was later used in generating appropriate syntax, as well as in planning *when* to talk about *what* (McKeown 1985).

One of the general trends in 1980s NLP research was to integrate the two non-linguistic constraints of *attention* (related to focus of interest) and *intention* (related to plans and goals). An interdisciplinary workshop devoted to this aim was held in Monterey, California, in 1987 (Cohen *et al.* 1990). Many of the papers presented there showed how far NLP had moved on since Power's model of the conversation between the door-opening robots. Others were highly relevant, also, to AI work on planning—and on plan recognition, which involves the identification of the other agent's beliefs and intentions.

Several of the Monterey talks were devoted to one or another aspect of speech act theory. And Searle himself gave a paper in which he argued that "collective intentions" can be formally modelled, but can't be analysed in terms of "individual intentions" (Searle 1990*c*). Indeed, some of his remarks could be paraphrased as asserting the existence of something akin to a "group mind" (see 8.iii, preamble).

In addition, Grosz's (and other) work on the structure of conversations would eventually feed into AI teaching programs (Rickel *et al.* 2002). This enabled (for instance) virtual-reality agents to converse with human technicians about how to operate complex machinery (Chapter 13.vi.b).

None of these examples, of course, could match the human use of language. Never mind fancy syntax. The following interchange (Nicholas Negroponte's ultimate goal for NLP) could well make sense for two intimate friends, but would challenge all but the ideal example of "personalized" computers:

"Okay, where did you hide it?"

"Hide what?"

"You know."

"Where do you think?"

"Oh."

(quoted in Brand 1988: 153)

Evidently, sensible conversation depends on much more than *language*. It often needs knowledge of the situation, and of the speakers, too. Flawless computerized conversation is as unattainable as "perfect" translation and interpretation, at least in the general case.

Nevertheless, the attempt to attain these unattainable goals, and to enable computers to parse sentences of indefinite complexity, led to results of both theoretical and practical importance. Some relevant psychological research was discussed in Chapter 7.ii, and the benefits for philosophy included advances in speech act theory (see above). As for the practical aspects, applications burgeoned.

By the mid-1980s, a fairly superficial survey of "technological" NLP required no fewer than 460 large pages (T. Johnson 1985). Now, it would need many more. For significant progress has been made since then. Today's survey, in addition to updating the older topics, would need to add work on non-verbal pragmatics: not just inflection and tone of voice, but eye-gaze, hand gestures, eyebrow movements and other facial expressions, and general body posture.

All these things, which involve significant cultural variations (think of the Indian way of indicating "Yes": not by nodding, but by inclining the head from side to side), help us to interpret conversations in real life. And all are being mimicked, with more or less success, in applications of virtual reality (VR)—see below, and Chapter 13.vi.

g. A snippet on speech

Some of the technological advances in NLP over the past thirty years concern speech processing, which I've chosen not to discuss. Suffice it to say that a great deal was learnt from the blackboard-based early speech-understanding systems, such as the single-user HEARSAY and its speaker-independent successor HARPY (Reddy *et al.* 1973; Newell *et al.* 1973; Erman and Lesser 1980; Lowerre and Reddy 1980). Indeed, one of today's most successful commercial applications, a voice-recognition program marketed by Dragon Systems, was developed by Janet Baker—whose DRAGON program was a component of HARPY (J. K. Baker 1975).

Many of the difficulties that plagued these pioneering attempts (for an overview, see Barr and Feigenbaum 1981, ch. 5), and which led DARPA to halt their funding in 1976, have now been overcome. This has been done largely by combining rule-based knowledge (of phonological tree structures, for instance) with statistical pattern recognition, and by strengthening the phonetics (e.g. Rowden 1992, chs. 6–8; Stolcke 1997; P. A. Taylor 2000). Sometimes, speech recognizers have relied on self-organized "maps" of quasi-phonemes, rather than built-in phonetic representations (Kohonen 1988; see 14.ix.a).

The results have benefited both speech recognition and speech generation. They've also allowed for these to be combined in speech-to-speech translation.

Today's commercially available speech recognition systems can accept menu-based inputs from virtually any speaker on first hearing, so are widely used in telephone-answering services. HEARSAY's notorious slowness—which caused some disillusionment with AI in official circles at the end of the 1970s (Klatt 1977)—is a thing of the past. (Current speech recognizers rely on adaptive statistical models, not on GOFAI: cf. Chapter 12.) Some inexpensive desktop systems today allow one to dictate at normal speeds, without any unnatural pauses between the words. Their powers of recognition are flexible, to some degree: they can learn extra items to enlarge their vocabulary, and adapt to the individual accent of the user.

Visual speech recognition may be feasible in the near- to mid-future, since only sixteen lip-movements are needed to represent English—all of them used in saying *I thought you really meant it.* Already, "talking heads" can be generated (to 'read' input text) which pronounce words with highly effective lip-synchronization (Lucena *et al.* 2002). The sixteen lip positions may be generic, but can also be based on photographs of a specific person's face (cf. Chapter 13.vi.e).

Speech translation is progressing too. A program is being marketed which can deal with telephoned conference registrations, switching between English, German, and Japanese. Research throughout the 1990s sought to aid face-to-face commercial transactions in English between Germans and Japanese who don't speak the language fluently (W. J. Hutchins 1994, sect. 8). Indeed, the new century saw a detailed report on a speech-to-speech translation system which, unlike its few rivals, works in real time. It's not yet (as I write this page) commercially viable—although it may be, by the time you read this book. But a trustworthy reviewer has described it as highly promising, and refreshingly free of hype (Sampson 2001). This achievement is a long way from HEARSAY, whose linguistic knowledge often outweighed its phonetics—so much so, that it might emit a whole sentence in 'reply' to a cough.

Even virtual reality is taking speech on board (see Chapter 13.vi.d–e). In some computer games, one can ask a dragon, "Can you fly?"—and it does. This VR dragon is no SHRDLU: it's merely recognizing the word *fly*. But its future descendants may be able to use syntax and speech act theory to interpret the input sentence either as a question *or* as a request. Certainly, plausible interfaces for virtual reality will demand no less.

Meanwhile, a system called TESSA has been developed for the British Post Office to enable counter-clerks to communicate with deaf people whose native language is BSL, or British Sign Language (Cox *et al.* 2002). This program translates the clerk's speech into sign language, which is visibly expressed as the finger movements, bodily gestures, and facial expressions of a VR avatar. (The screen-creature even manages to look "rueful" when explaining that there are no more first-class stamps.)

By early 2003, about 400 pre-designed translations were being used by the avatar. The team had discovered early in their research that human clerks can't be trained to speak *only* the precise words defining these 400 phrases. So TESSA listens to what the clerk says and then offers a menu of up to half a dozen 'equivalent' English sentences, from which the person chooses one. It's that one which is displayed in BSL form via the avatar. In short, TESSA combines speech recognition with MT and advanced computer graphics.

Automatic speech synthesis is improving also. The first example, Ray Kurzweil's reading machine for the blind (gratefully used by Stevie Wonder, among others), was announced in January 1975. It was immediately featured on America's TV news: Walter Cronkite even allowed it to read the sign-off message for that evening's newscast. Given that it could read virtually any printed font, it was a superb achievement at the time. But it was most unnatural to listen to. The voice of its Xerox successor (Kurzweil sold the company to Xerox in 1980) is a good deal more pleasant.

In addition, there are now many more types of AI-voice application. For instance, a system recently developed for use in a hospital cardiac unit employs near-naturalistic intonation to reflect the relative importance of many different items of medical data. These are passed automatically from the operating theatre to the nurses busily preparing for the patient's arrival in intensive care (McKeown and Pan 2000). And in March 1999, my daily newspaper reported that the physicist Stephen Hawking (current holder of Isaac Newton's and Charles Babbage's Lucasian Chair) had acquired a new speech synthesizer, much more 'human'—and more British—than the one he'd been using for many years.

By the time this book is published, the choices available will doubtless be better still. I say that not just because technology inevitably moves on, but because of a specific research advance. By the close of the century, NLP researchers at Edinburgh University had produced a "dialect independent" lexicon suitable for synthesizing spoken English in many different local accents (Fitt and Isard 1999; Fitt 2003).

In this lexicon, which is based on the work of the descriptive linguist John Wells (1982), speech sounds are classified in terms of the words in which they occur—as in the "boot" vowel, the "foot" vowel, and so on. The description of a dialect for synthesis purposes specifies which of these vowel classes to merge (for instance, the vowels of "father" and "bother" for many American dialects) and which to keep separate (such as the vowels of "tide" and "tied" in most Scottish dialects). The sounds are 'glued together' to form the sentences required—"sentences", not "words", because the pronunciation of a particular word often varies according to the word that follows it.

The sounds themselves aren't synthesized from scratch, but are retrieved from the recorded speech of someone asked to read a script containing examples of each of the sounds required. In May 2004 the accents that had been synthesized in this way included received pronunciation of British English, several American dialects, and a Scottish, an Irish, and an Australian dialect (S. D. Isard, personal communication). More can, and will, be added. The collection of key phonetic symbols (in 2004, just thirty-four vowels and twenty-seven consonants) may turn out to be adequate. But it can be expanded if necessary.

Whether Hawking would be pleased to hear of this advance—which was licensed for commercial use in 2001, and is already being employed in the 'rVoice' product—is unclear. The newspapers in April 2004 reported that he was distressed about his familiar voice system wearing out, as he felt that its Dalek-like tones had become part of his identity. (As he put it when Intel Chairman Gordon Moore donated £7.5 million for a new science library in Cambridge, "I'm Intel inside myself"—Mialet 2003: 453.) In other words, he seems *not* to want his speech synthesizer to pass the Turing Test (16.ii.c).

The Edinburgh-based synthesizer doesn't quite do that, anyway. For example, dialects also vary in intonation and segment duration, but these features (which are independent of the lexicon) aren't represented in the system. So although it's clear that *this* synthesized voice is Scottish and *that* one Australian, neither sounds quite right to anyone with a good ear for accents. Nevertheless, speech processing has come a long way since HEARSAY.

From artificial voices to artificial intelligence: our next topic is the history of AI. *This* chapter, of course, has already begun that narrative. Language may be less central to intelligence than early AI took it to be (see 13.iii.b and 15.vii.a). Nevertheless, NLP is crucial to the AI project. And Chomsky's wider influence affected AI, as well as psychology and philosophy—as we'll now see.